# Spintronics Handbook: Spin Transport and Magnetism, Second Edition

Semiconductor Spintronics—Volume Two

# Spintronics Handbook: Spin Transport and Magnetism, Second Edition

Semiconductor Spintronics—Volume Two

Edited by
Evgeny Y. Tsymbal and Igor Žutić

**CRC Press**
Taylor & Francis Group
Boca Raton London New York

CRC Press is an imprint of the
Taylor & Francis Group, an **informa** business

CRC Press
Taylor & Francis Group
6000 Broken Sound Parkway NW, Suite 300
Boca Raton, FL 33487-2742

First issued in paperback 2020

© 2019 by Taylor & Francis Group, LLC
CRC Press is an imprint of Taylor & Francis Group, an Informa business

No claim to original U.S. Government works

ISBN-13: 978-1-4987-6960-0 (hbk)
ISBN-13: 978-0-367-77970-2 (pbk)

**Visit the Taylor & Francis web site at**
**http://www.taylorandfrancis.com**

**and the CRC Press web site at**
**http://www.crcpress.com**

Cover illustration courtesy of Markus Lindemann, Nils C. Gerhardt, and Carsten Brenner.

The e-book of this title contains full colour figures and can be purchased here: http://www.crcpress.com/9780429434235. The figures can also be found under the 'Additional Resources' tab.

# Contents

## SECTION IV—Spin Transport and Dynamics in Semiconductors

## SECTION V—Magnetic Semiconductors, Oxides and Topological Insulators

# Foreword

S pintronics is a field of research in which novel properties of materials, especially atomically engineered magnetic multilayers, are the result of the manipulation of currents of spin-polarized electrons. Spintronics, in its most recent incarnation, is a field of research that is almost 30 years old. To date, its most significant technological impact has been in the development of a new generation of ultra-sensitive magnetic recording read heads that have powered magnetic disk drives since late 1997. These magnetoresistive read heads, which use spin-valves based on spin-dependent scattering at magnetic/non-magnetic interfaces and, since 2007, magnetic tunnel junctions (MTJs) based on spin-dependent tunneling across ultra-thin insulating layers, have a common thin film structure. These structures involve "spin engineering" to eliminate the influence of long-range magneto-dipole fields via the use of synthetic or artificial antiferromagnets, which are formed from thin magnetic layers coupled antiferromagnetically via the use of atomically thin layers of ruthenium. These structures involve the discoveries of spin-dependent tunneling in 1975, giant magnetoresistance at low temperatures in Fe/Cr in 1988, oscillatory interlayer coupling in 1989, the synthetic antiferromagnet in 1990, giant magnetoresistance at room temperature in Co/Cu and related multilayers in 1991, and the origin of giant magnetoresistance as being a result of predominant interface scattering in 1991–1993. Together, these discoveries led to the spin-valve recording read head that was introduced by IBM in 1997 and led, within a few years, to a 1,000-fold increase in the storage capacity of magnetic disk drives. This rapid pace of improvement has stalled over the past years as the difficulty of stabilizing tiny magnetic bits against thermal fluctuations whilst at the same time being able to generate large enough magnetic fields to write them, has proved intractable. The possibility of creating novel spintronic magnetic memory-storage devices to rival magnetic disk drives in capacity and to vastly exceed them in performance has emerged in the form of Racetrack Memory. This concept and the physics underlying it are discussed in this book, together with a more conventional spintronic memory, magnetic random access memory (MRAM). MRAM is based on MTJ magnetic memory bits, each one accessed in a two-dimensional cross point array via a transistor. The fundamental concept of MRAM was proposed in 1995 using local fields to write the MTJ elements. This basic concept was proven in 1999 with the subsequent demonstration of large-scale, fully integrated 64 Mbit memory chips in the following decade. Writing these same elements using spin angular momentum from sufficiently large spin-polarized currents passed through the tunnel junction elements emerged in the 1990s and is now key to the development of massive-scale MRAM chips. The second edition of this book discusses these emerging spintronic technologies as well as other breakthroughs and key advances, both fundamental and applied, in the field of spintronics.

Beyond MRAM and Racetrack Memory, this book elucidates other nascent opportunities in spintronics that do not rely directly on magneto-resistive effects, such as fault-tolerant quantum computing, non-Boolean spin-wave logic, and lasers that are enhanced by spin-polarized carriers. It is interesting that spintronics is a field of research that continues to surprise even though the fundamental property of spin was realized nearly a century ago, and the basic concept of spin-dependent scattering in magnetic materials was introduced by Neville Mott just shortly after the notion of "spin" was conceived.

Since the first edition of this book, spintronics has so much evolved that a new name of "spin-orbitronics" has been coined to describe these new discoveries and developments. In the first edition of this book, spin-orbit coupling was regarded rather negatively as a property that leads to mixing between spin-channels and the loss of spin angular momentum from spin currents to the lattice, thereby limiting the persistence of these same spin currents, both temporally and spatially. In this edition, several physical phenomena derived from spin-orbit coupling are shown to be key to the development of several new technologies, such as, in particular, the current-induced motion of a series of magnetic domain walls that underlies Racetrack Memory. This relies especially on the generation of pure spin currents via the spin Hall effect (SHE). The magnitude of the SHE was thought for some time to be very small in conventional metals, but over the past few years, this has rather been shown to be incorrect. Significant and useful SHEs have been discovered in a number of heavy materials where spin-orbit coupling is large. These spin currents can be used to help move domain walls or to help switch the magnetization direction of nanoscale magnets. Whether they can be usefully used for MRAM, however, is still a matter of debate.

Another very interesting development since the first edition of this book is the explosive increase in our understanding and knowledge of topological insulators and their cousins including, most recently, Weyl semi-metals. The number of such materials has increased astronomically and, indeed, it is now understood that a significant fraction of all extant materials are "topological". What this means, in some cases, is that the spin of the carriers is locked to their momentum leading, for example, to the Quantum Spin Hall Effect.

The very concept of these materials is derived from band inversion, which is often due to strong spin-orbit coupling. From a spintronics perspective, the novel properties of these materials can lead to intrinsic spin currents and spin accumulations that are topologically "protected" to a greater or lesser degree. The concept of topological protection is itself evolving.

Distinct from electronic topological effects are topological spin textures such as skyrmions and anti-skyrmions. The latter were only experimentally found 2 years ago. These spin textures are nano-sized magnetic objects that are related to magnetic bubbles, which are also found in magnetic materials with perpendicular magnetic anisotropy but which have boundaries or walls that are innately chiral. The chirality is determined by a vector magnetic exchange – a Dzyaloshinskii–Moriya interaction (DMI) – that is often derived from spin-orbit coupling. The DMI favors orthogonal alignment of neighboring magnetic moments in contrast to conventional ferromagnetic or antiferromagnetic exchange interactions that favor collinear magnetic arrangements. Skyrmion and anti-skyrmion spin textures have very interesting properties that could also be useful for Racetrack Memories. Typically, skyrmions and anti-skyrmions evolve from helical or conical spin textures. The magnetic phase of such systems can have complex dependences on temperature, magnetic field, and strain. Some chiral antiferromagnetic spin textures have interesting properties such as an anomalous Hall effect (AHE), which is derived from their topological chiral spin texture in the absence of any net magnetization. In practice, however, a small unbalanced moment is needed to set the material in a magnetic state with domains of the same chirality in order to evidence the AHE. On the other hand, these same chiral textures can display an intrinsic spin Hall effect whose sign is independent of the chirality of the spin texture.

The DMI interaction can also result from interfaces particularly between heavy metals and magnetic layers. Such interfacial DMIs can give rise to chiral domain walls as well as magnetic bubbles with chiral domain walls – somewhat akin to skyrmions. The tunability of the interfacial DMI via materials engineering makes it of special interest.

Thus, since the first edition of this book, chiral spin phenomena, namely chiral spin textures and domain walls, and the spin Hall effect itself, which is innately chiral, have emerged as some of the most interesting developments in spintronics. The impact of these effects was largely unanticipated. It is not too strong to say that we are now in the age of "chiraltronics"!

Another topic that has considerably advanced since the first edition of this book is the field of what is often now termed spin caloritronics, namely the use of temperature gradients to create spin currents and the use of thermal excitations of magnetic systems, i.e. magnons, for magnonic devices. Indeed, magnons carry spin angular momentum and can propagate over long distances. Perhaps here it is worth mentioning the extraordinarily long propagation distances of spin currents via magnons in antiferromagnetic systems that have recently been realized.

Recently discovered atomically thin ferromagnets reveal how the presence of spin-orbit coupling overcomes the exclusion of two-dimensional ferromagnetism expected from the Mermin-Wagner theorem. These two-dimensional materials, which are similar to graphene in that they can readily be exfoliated from bulk samples, provide a rich platform to study magnetic proximity effects and transform a rapidly growing class of van der Waals materials. Through studies of magnetic materials, it is possible to reveal their peculiar quantum manifestations. Topological insulators can become magnetic by doping with 3d transition metals; the quantum anomalous Hall effect has been discovered in such materials, and heterostructures that consist of magnetic and non-magnetic topological insulators have been used to demonstrate current-induced control of magnetism.

Spintronics remains a vibrant research field that spans many disciplines ranging from materials science and chemistry to physics and engineering. Based on the rich developments and discoveries over the past thirty years, one can anticipate a bountiful future.

**Stuart Parkin**

Director at the Max Planck Institute of Microstructure Physics

Halle (Saale), Germany

and

Alexander von Humboldt Professor, Martin-Luther-Universität

Halle-Wittenberg, Germany

Max Planck Institute of Microstructure Physics

Halle (Saale), Germany

# Preface

The second edition of this book continues the path from the foundations of spin transport and magnetism to their potential device applications, usually referred to as spintronics. Spintronics has already left its mark on several emerging technologies, e.g., in magnetic random access memories (MRAMs), where the fundamental properties of magnetic tunnel junctions are key for device performance. Further, many intricate fundamental phenomena featured in the first edition have since evolved from an academic curiosity into the potential basis for future spintronic devices. Often, as in the case of spin Hall effects, spin-orbit torques, and electrically-controlled magnetism, the research has migrated from the initial low-temperature discovery in semiconductors to technologically more suitable room temperature manifestations in metallic systems. This path from exotic behavior to possible application continues to the present day and is reflected in the modified title of the book, which now explicitly highlights "spintronics," as its overarching scope. Exotic topics of today, for example, pertaining to topological properties, such as skyrmions, topological insulators, or even elusive Majorana fermions, may become suitable platforms for the spintronics of tomorrow. Impressive progress has been seen in the last decade in the field of spin caloritronics, which has evolved from a curious prediction 30 years ago to a vibrant field of research.

Since the first edition, there has been a significant evolution in material systems displaying spin-dependent phenomena, making it difficult to cover even the key developments in a single volume. The initially featured chapter on graphene spintronics is now complemented by a chapter on the spin-dependent properties of a broad range of two-dimensional materials that can form a myriad of heterostructures coupled by weak van der Waals forces and support superconductivity or ferromagnetism even in a single atomic layer. Exciting developments have also been seen in the field of complex oxide heterostructures, where the non-trivial properties are driven by the interplay between the electronic, spin, and structural degrees of freedom. A particular example is the magnetism emerging in two-dimensional electron gases at oxide interfaces composed of otherwise nonmagnetic constituents. The updated structure of a significantly expanded book reflects various materials developments and it is now thematically divided into three volumes, each based on broadly defined metallic and semiconductor systems or their nanoscale and applied aspects.

Spintronics becomes more and more attractive as a viable platform for propelling semiconducting technology beyond its current limits. Various schemes have been proposed to enhance the functionalities of the existing technologies based on the spin degree of freedom. Among them is the voltage control of magnetism, exploiting the nonvolatile performance of ferromagnet-based devices in conjunction with their low-power operation. Another approach is utilizing spin currents carried by magnons to transport and process information. Magnon spintronics involves interesting fundamental physics and offers novel spin wave-based computing technologies and logic circuits. Optical control of magnetism is another approach, which has attracted a lot of attention due to the recent discovery of the all-optical switching of magnetization and its realization at the nanoscale. Chapters on these subjects are included in the new edition of the book.

Nearly nine decades after the discovery of superconducting proximity effects by Ragnar Holm and Walther Meissner, several new chapters now explore how a given material can be transformed through proximity effects whereby it acquires the properties of its neighbors, for example, becoming superconducting, magnetic, topologically non-trivial, or with an enhanced spin-orbit coupling. Such proximity effects not only complement the conventional methods of designing materials by doping or functionalization but can also overcome their various limitations and enable yet more unexplored spintronic applications.

We are grateful both to the authors who set aside their many priorities and contributed new chapters, which have significantly expanded the scope of this book, as well as to those who patiently provided valuable updates to their original chapters and kept this edition even more timely. The completion of the second edition was again greatly facilitated by Verona Skomski, who tirelessly collected authors' contributions and assisted their preparation for the submission to the publisher. We acknowledge the support of NSF-DMR, NSF-MRSEC, NSF-ECCS, SRC, DOE-BES, US ONR which, through the support of our research and involvement in spintronics, has also enabled our editorial work. We are thankful to our families for their support, patience, and understanding during extended periods of time when we remained focused on the completion of this edition.

**Evgeny Y. Tsymbal**
Department of Physics and Astronomy,
Nebraska Center for Materials and Nanoscience, University of Nebraska,
Lincoln, Nebraska 68588, USA

**Igor Žutić**
Department of Physics, University at Buffalo,
State University of New York, Buffalo, New York 14260, USA

# About the Editors

**Evgeny Y. Tsymbal** is a George Holmes University Distinguished Professor at the Department of Physics and Astronomy of the University of Nebraska-Lincoln (UNL), and Director of the UNL's Materials Research Science and Engineering Center (MRSEC). He joined UNL in 2002 as an Associate Professor, was promoted to a Full Professor with Tenure in 2005 and named a Charles Bessey Professor of Physics in 2009 and George Holmes University Distinguished Professor in 2013. Prior to his appointment at UNL, he was a research scientist at University of Oxford, United Kingdom, a research fellow of the Alexander von Humboldt Foundation at the Research Center-Jülich, Germany, and a research scientist at the Russian Research Center "Kurchatov Institute," Moscow. Evgeny Y. Tsymbal's research is focused on computational materials science aiming at the understanding of fundamental properties of advanced ferromagnetic and ferroelectric nanostructures and materials relevant to nanoelectronics and spintronics. He has published over 230 papers, review articles, and book chapters and presented over 180 invited presentations in the areas of spin transport, magnetoresistive phenomena, nanoscale magnetism, complex oxide heterostructures, interface magnetoelectric phenomena, and ferroelectric tunnel junctions. Evgeny Y. Tsymbal is a fellow of the American Physical Society, a fellow of the Institute of Physics, UK, and a recipient of the UNL's College of Arts & Sciences Outstanding Research and Creativity Award (ORCA). His research has been supported by the National Science Foundation, Semiconductor Research Corporation, Office of Naval Research, Department of Energy, Seagate Technology, and the W. M. Keck Foundation.

**Igor Žutić** received his Ph.D. in theoretical physics at the University of Minnesota, after undergraduate studies at the University of Zagreb, Croatia. He was a postdoc at the University of Maryland and the Naval Research Lab. In 2005 he joined the State University of New York at Buffalo as an Assistant

Professor of Physics and got promoted to an Associate Professor in 2009 and to a Full Professor in 2013. He proposed and chaired Spintronics 2001: International Conference on Novel Aspects of Spin-Polarized Transport and Spin Dynamics, at Washington DC. Work with his collaborators spans a range of topics from high-temperature superconductors, Majorana fermions, proximity effects, van der Waals materials, and unconventional magnetism, to the prediction and experimental realization of spin-based devices that are not limited to magnetoresistance. He has published over 100 refereed articles and given over 150 invited presentations on spin transport, magnetism, spintronics, and superconductivity. Igor Žutić is a recipient of the 2006 National Science Foundation CAREER Award, the 2019 State University of New York Chancellor's Award for Excellence in Scholarship and Creative Activities, the 2005 National Research Council/American Society for Engineering Education Postdoctoral Research Award, and the National Research Council Fellowship (2003–2005). His research is supported by the National Science Foundation, the Office of Naval Research, the Department of Energy, Office of Basic Energy Sciences, the Defense Advanced Research Project Agency, and the Airforce Office of Scientific Research. He is a fellow of the American Physical Society.

# Contributors

**Ian Appelbaum**
Department of Physics
University of Maryland
College Park, Maryland

**Ilaria Bergenti**
Institute for Nanostructured
    Materials Studies
Bologna, Italy

**J. M. D. Coey**
Centre for Research on Adaptive
Nanostructures and Nanodevices
Trinity College
Dublin, Ireland

**Scott A. Crooker**
National High Magnetic Field
    Laboratory
Los Alamos National Laboratory
Los Alamos, New Mexico

**Paul A. Crowell**
School of Physics and Astronomy
University of Minnesota
Minneapolis, Minnesota

**Elbio Dagotto**
Department of Physics and
    Astronomy
The University of Tennessee
Knoxville, Tennessee
and
Materials Science and Technology
    Division
Oak Ridge National Laboratory
Oak Ridge, Tennessee

**T. Huong Dang**
Unité Mixte de Physique
Centre National de la Recherche
Scientifique-Thales
Université Paris-Saclay
and
Laboratoire des Solides Irradiés,
Ecole Polytechnique, CNRS and
CEA/DRF/IRAMIS
Institut Polytechnique de Paris
Palaiseau, France

**Valentin Dediu**
Institute for Nanostructured
    Materials Studies
Bologna, Italy

**Shuai Dong**
Department of Physics
Southeast University
Nanjing, China

**H.-J. Drouhin**
Laboratoire des Solides Irradiés,
    Ecole Polytechnique
CNRS and CEA-DRF-IRAMIS
Institut Polytechnique de Paris
Palaiseau, France

**E. Erina**
Laboratoire des Solides Irradiés,
    Ecole Polytechnique
CNRS and CEA-DRF-IRAMIS
Institut Polytechnique de Paris
Palaiseau, France

**Jaroslav Fabian**
Department of Physics
University of Regensburg
Regensburg, Germany

**Sergey D. Ganichev**
Department of Physics
University of Regensburg
Regensburg, Germany

**Jean Marie George**
Unité Mixte de Physique
Centre National de la Recherche
    Scientifique-Thales
Université Paris-Saclay
Palaiseau, France

**Matthew J. Gilbert**
Department of Electrical and
    Computer Engineering
University of Illinois at
    Urbana-Champaign
Urbana, Illinois
and
Department of Electrical
    Engineering Stanford
    University
Stanford, California

**Ewelina M. Hankiewicz**
Institute for Theoretical Physics
    and Astrophysics
Würzburg University
Würzburg, Germany

**Luis E. Hueso**
CIC nanoGUNE Consolider
Basque Foundation for Science
Bilbao, Spain

**Patrick Irvin**
Department of Physics and
    Astronomy
University of Pittsburgh
and
Pittsburgh Quantum Institute
Pittsburgh, Pennsylvania

**Henri Jaffrès**
Unité Mixte de Physique
Centre National de la Recherche
    Scientifique-Thales
Université Paris-Saclay
Palaiseau, France

**Berend T. Jonker**
Magnetoelectronic Materials and
    Devices Branch
Materials Science and Technology
    Division
Naval Research Laboratory
Washington, DC

**Tomás Jungwirth**
Institute of Physics
Academy of Sciences of the Czech
    Republic
Praha, Czech Republic
and
School of Physics and Astronomy
University of Nottingham
Nottingham, United Kingdom

**Abhinav Kandala**
IBM T.J. Watson Research Center
Yorktown Heights, New York

**Jeremy Levy**
Department of Physics and
    Astronomy
University of Pittsburgh
and
Pittsburgh Quantum Institute
Pittsburgh, Pennsylvania

**Fumihiro Matsukura**
Tohoku University
Sendai, Japan

**T. L. Hoai Nguyen**
Institute of Physics, VAST, 10
    Daotan,
Badinh, Hanoi, Vietnam

**Hideo Ohno**
Tohoku University
Sendai, Japan

**Satoshi Okamoto**
Materials Science and Technology
    Division
Oak Ridge National Laboratory
Oak Ridge, Tennessee

**Yun-Yi Pai**
Department of Physics and
    Astronomy
University of Pittsburgh
and
Pittsburgh Quantum Institute
Pittsburgh, Pennsylvania

**Anthony Richardella**
Department of Physics and
    Materials Research Institute
The Pennsylvania State University
University Park, Pennsylvania

**Nitin Samarth**
Department of Physics and
    Materials Research Institute
The Pennsylvania State University
University Park, Pennsylvania

**John Schliemann**
Department of Physics
University of Regensburg
Regensburg, Germany

**Jairo Sinova**
Johannes Gutenberg Universität
    Mainz
Institute of Physics
Mainz, Germany

**Carsten Timm**
Institute of Theoretical Physics
Technische Universität Dresden
Dresden, Germany

**Duy-Quang To**
Laboratoire des Solides Irradiés,
    Ecole Polytechnique
CNRS and CEA-DRF-IRAMIS
Institut Polytechnique de Paris
Palaiseau, France

**Maxim Trushin**
Centre for Advanced 2D Materials
National University of Singapore
Singapore

**Anthony Tylan-Tyler**
Department of Physics and
    Astronomy
University of Pittsburgh
and
Pittsburgh Quantum Institute
Pittsburgh, Pennsylvania

**M. W. Wu**
Hefei National Laboratory for
    Physical Sciences at Microscale
    and Department of Physics
University of Science and
    Technology of China
Hefei, Anhui, China

**Jörg Wunderlich**
Hitachi Cambridge Laboratory
Cambridge, United Kingdom

# Section IV

# Spin Transport and Dynamics in Semiconductors

# 1

# Spin Relaxation and Spin Dynamics in Semiconductors and Graphene

**Jaroslav Fabian and M. W. Wu**

The spin of conduction electrons decays due to the combined effect of spin–orbit coupling and momentum scattering. The spin–orbit coupling couples the spin to the electron momentum that is randomized by momentum scattering off of impurities and phonons. Seen from the perspective of the electron spin, the spin–orbit coupling gives a spin precession, while momentum scattering makes this precession randomly fluctuating, both in magnitude and orientation.

The specific mechanisms for the spin relaxation of conduction electrons were proposed by Elliott [1] and Yafet [2], for conductors with a center of inversion symmetry, and by D'yakonov and Perel' [3], for conductors without an inversion center. In $p$-doped semiconductors, there is in play another spin relaxation mechanism, due to Bir et al. [4]. As this has a rather limited validity we do not describe it here. More details can be found in reviews [5–9].

Before we discuss the two main mechanisms, we introduce a toy model that captures the relevant physics of spin relaxation without resorting explicitly to quantum mechanics: *the electron spin in a randomly* fluctuating *magnetic field*. We will find certain universal qualitative features of the spin relaxation and dephasing in physically important situations.

The next part of this review covers the experimental as well as computational status of the field, discussing the spin relaxation in semiconductors under varying conditions such as temperature and doping density.

## 1.1 TOY MODEL: ELECTRON SPIN IN A FLUCTUATING MAGNETIC FIELD

Consider an electron spin $\mathbf{S}$ (or the corresponding magnetic moment) in the presence of an external time-independent magnetic field $\mathbf{B}_0 = B_0 \mathbf{z}$, giving rise to the Larmor precession frequency $\boldsymbol{\omega}_0 = \omega_0 \mathbf{z}$, and a fluctuating time-dependent field $\mathbf{B}(t)$ giving the Larmor frequency $\boldsymbol{\omega}(t)$ (see Figure 1.1). We assume that the field fluctuates about zero and is correlated on the -timescale of $\tau_c$:

$$\overline{\omega(t)} = 0, \quad \overline{\omega_\alpha(t)\omega_\beta(t')} = \delta_{\alpha\beta}\overline{\omega_\alpha^2}e^{-|t-t'|/\tau_c}. \tag{1.1}$$

Here $\alpha$ and $\beta$ denote the Cartesian coordinates and the overline denotes averaging over different random realizations $\mathbf{B}(t)$. We will see later that such fluctuating fields arise quite naturally in the context of the electron spins in solids.

The following description applies equally to the classical magnetic moment described by the vector $\mathbf{S}$ as well as to the quantum mechanical spin whose expectation value is $\mathbf{S}$. Writing out the torque equation, $\dot{\mathbf{S}} = \boldsymbol{\omega} \times \mathbf{S}$, we get the following equations of motion:

$$\dot{S}_x = -\omega_0 S_y + \omega_y(t)S_z - \omega_z(t)S_y, \tag{1.2}$$

$$\dot{S}_y = \omega_0 S_x - \omega_x(t)S_z + \omega_z(t)S_x, \tag{1.3}$$

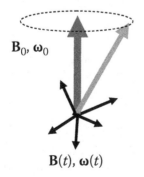

$\mathbf{B}_0, \boldsymbol{\omega}_0$

$\mathbf{B}(t), \boldsymbol{\omega}(t)$

**FIGURE 1.1**   Electron spin precesses about the static $\mathbf{B}_0$ field along $\mathbf{z}$. The randomly fluctuating magnetic field $\mathbf{B}(t)$ causes spin relaxation and spin dephasing.

$$\dot{S}_z = \omega_x(t)S_y - \omega_y(t)S_x. \tag{1.4}$$

These equations are valid for one specific realization of $\boldsymbol{\omega}(t)$. Our goal is to find instead effective equations for the time evolution of the average spin, $\bar{S}(t)$, given the ensemble of Larmor frequencies $\boldsymbol{\omega}(t)$.

It is convenient to introduce the complex "rotating" spins $S_\pm$ and Larmor frequencies $\omega_\pm$ in the $(x,y)$ plane:

$$S_+ = S_x + iS_y, \quad S_- = S_x - iS_y, \tag{1.5}$$

$$\omega_+ = \omega_x + i\omega_y, \quad \omega_- = \omega_x - i\omega_y. \tag{1.6}$$

The inverse relations are

$$S_x = \frac{1}{2}(S_+ + S_-), \quad S_y = \frac{1}{2i}(S_+ - S_-), \tag{1.7}$$

$$\omega_x = \frac{1}{2}(\omega_+ + \omega_-), \quad \omega_y = \frac{1}{2i}(\omega_+ - \omega_-). \tag{1.8}$$

The equations of motion for the spin set $(S_+, S_-, S_z)$ are

$$\dot{S}_+ = i\omega_0 S_+ + i\omega_z S_+ - i\omega_+ S_z, \tag{1.9}$$

$$\dot{S}_- = -i\omega_0 S_- - i\omega_z S_- + i\omega_- S_z, \tag{1.10}$$

$$\dot{S}_z = -\left(\frac{1}{2}i\right)(\omega_+ S_- - \omega_- S_+). \tag{1.11}$$

In the absence of the fluctuating fields, the spin $S_+$ rotates in the complex plane anticlockwise (for $\omega_0 > 0$) while $S_-$ clockwise.

The precession about $B_0$ can be factored out by applying the ansatz*:

$$S_\pm = s_\pm(t)e^{\pm i\omega_0 t}. \tag{1.12}$$

Indeed, it is straightforward to find the time evolution of the set $(s_+, s_-, s_z \equiv S_z)$:

$$\dot{s}_+ = i\omega_z s_+ - i\omega_+ s_z e^{-i\omega_0 t}, \tag{1.13}$$

$$\dot{s}_- = -i\omega_z s_- + i\omega_- s_z e^{i\omega_0 t}, \tag{1.14}$$

$$\dot{s}_z = -\left(\frac{1}{2}i\right)\left(\omega_+ s_- e^{-i\omega_0 t} - \omega_- s_+ e^{i\omega_0 t}\right). \tag{1.15}$$

---

* This is analogous to going to the interaction picture when dealing with a quantum mechanical problem of that type.

The penalty for transforming into this "rotating frame" is the appearance of the phase factors $\exp(\pm i\omega_0 t)$.

The solutions of Equations 1.13 through 1.15 can be written in terms of the integral equations

$$s_+(t) = s_+(0) + i\int_0^t dt'\omega_z(t')s_+(t') - i\int_0^t dt'\omega_+(t')s_z(t')e^{-i\omega_0 t'}, \tag{1.16}$$

$$s_-(t) = s_-(0) - i\int_0^t dt'\omega_z(t')s_-(t') + i\int_0^t dt'\omega_-(t')s_z(t')e^{i\omega_0 t'}, \tag{1.17}$$

$$s_z(t) = s_z(0) - \frac{1}{2i}\int_0^t dt'\left[\omega_+(t')s_-(t')e^{-i\omega_0 t'} - \omega_-(t')s_+(t')e^{i\omega_0 t'}\right]. \tag{1.18}$$

We should now substitute the above solutions back into Equations 1.13 through 1.15. The corresponding expressions become rather lengthy, so we demonstrate the procedure on the $s_+$ component only. We get

$$\dot{s}_+(t) = i\omega_z(t)s_+(0) - \omega_z(t)\int_0^t dt'\omega_z(t')s_+(t')$$

$$+\omega_z(t)\int_0^t dt'\omega_+(t')s_z(t')e^{-i\omega_0 t'} - i\omega_+(t)e^{-i\omega_0 t}s_z(0) \tag{1.19}$$

$$+\frac{1}{2}e^{-i\omega_0 t}\omega_+(t)\int_0^t dt'\left[\omega_+(t')s_-(t')e^{-i\omega_0 t'} - \omega_-(t')s_+(t')e^{i\omega_0 t'}\right].$$

The reader is encouraged to write the analogous equations for $\dot{s}_z$ (that for $\dot{s}_-$ is easy to write since $s_- = s_+^*$).

We now make two approximations. First, we assume that the fluctuating field is rather weak and stay in *the second* order in $\omega$.[*] This allows us to factorize the averaging over the statistical realizations of the field

$$\overline{\omega(t)\omega(t')s(t')} \approx \overline{\omega(t)\omega(t')}\,\overline{s(t')} \tag{1.20}$$

as the spin changes only weakly over the timescale, $\tau_c$, of the changes of the fluctuating fields. This approximation is called the *Born approximation*, alluding to the analogy with the second-order time-dependent perturbation theory in quantum mechanics. Going beyond the Born approximation one would need to execute complicated averaging schemes of the product in Equation 1.20, since $s(t)$ in general depends on $\omega(t' \le t)$.

---

[*] More precisely, we assume that $|\omega(t)|\tau_c \ll 1$, so that the spin does not fully precess about the fluctuating field before the field makes a random change.

As the second assumption, we consider a "coarse-grained" time evolution, meaning that we are effectively averaging $s(t)$ over the timescale of the correlation time $\tau_c$; we are interested in times $t$ much greater than $\tau_c$. That allows us to approximate

$$\int_0^{t \gg \tau_c} dt'\overline{\omega(t)\omega(t')}s(t') \approx \int_0^{t \gg \tau_c} dt'\overline{\omega(t)\omega(t')}\overline{s(t)}, \qquad (1.21)$$

since the correlation function $\overline{\omega(t)\omega(t')}$ is significant in the time interval of $|t-t'| \approx \tau_c$ only. The above approximation makes clear that the spin $s(t)$ is the representative coarse-grained (running-averaged) spin of the time interval $(t-\tau_c, t)$. Equation 1.21 is a realization of the *Markov approximation*. The physical meaning is that the spin $s$ varies only slowly on the timescale of $\tau_c$ over which the correlation of the fluctuating fields is significant. We then need to restrict ourselves to the timescales $t$ larger than the correlation time $\tau_c$. In effect, we will see that in this approximation *the rate of change of the spin at a given time depends on the spin at that time, not on the previous history of the spin.*

Applying the Born–Markov approximation to Equation 1.19, we obtain for the average spin $\overline{s}_+$ the following time evolution equation*:

$$\overline{s}_+ = i\overline{\omega_z(t)}s_+(0) - \int_0^t dt'\overline{\omega_z(t)\omega_z(t')}\overline{s_+(t)}$$

$$+ \int_0^t dt'\overline{\omega_z(t)\omega_+(t')}e^{-i\omega_0 t'}\overline{s_z(t)} - i\overline{\omega_+(t)}e^{-i\omega_0 t}s_z(0) + \frac{1}{2}e^{-i\omega_0 t} \qquad (1.22)$$

$$\times \int_0^t dt'\Big[\overline{\omega_+(t)\omega_+(t')}e^{-i\omega_0 t'}\overline{s_-(t)} - \overline{\omega_+(t)\omega_-(t')}e^{i\omega_0 t'}\overline{s_+(t)}\Big].$$

Using the rules of Equation 1.1, Equation 1.22 simplifies to

$$\overline{s}_+ = -\overline{\omega_z^2}\int_0^t dt'e^{-(t-t')/\tau_c}\overline{s_+(t)} + \frac{1}{2}e^{-i\omega_0 t}$$

$$\qquad (1.23)$$

$$\times \int_0^t dt'\Big[(\overline{\omega_x^2} - \overline{\omega_y^2})e^{-i\omega_0 t'}\overline{s_-(t)} - (\overline{\omega_x^2} + \overline{\omega_y^2})e^{i\omega_0 t'}\overline{s_+(t)}\Big]e^{-(t-t')/\tau_c}.$$

Since we consider the times $t \gg \tau_c$, we can approximate

$$\int_0^t dt'e^{-(t-t')/\tau_c} \approx \int_{-\infty}^t dt'e^{-(t-t')/\tau_c} = \tau_c. \qquad (1.24)$$

---

* The initial values of the spin, $s(0)$, are fixed and not affected by averaging.

Similarly,

$$\int_0^t dt' e^{-(t-t')/\tau_c} e^{-i\omega_0(t\pm t')} \approx \int_{-\infty}^t dt' e^{-(t-t')/\tau_c} e^{-i\omega_0(t\pm t')} = \tau_c \frac{1 \pm i\omega_0\tau_c}{1 + \omega_0^2\tau_c^2}. \quad (1.25)$$

The imaginary parts induce the precession of $s_\pm$, which is equivalent to shifting (renormalizing) the Larmor frequency $\omega_0$. The relative change of the frequency is $(\omega\tau_c)^2$, which is assumed much smaller than one by our Born approximation. We thus keep the real parts only and obtain

$$\dot{\overline{s}}_+ = -\overline{\omega_z^2}\tau_c\overline{s}_+ + \frac{1}{2}\frac{\tau_c}{1+\omega_0^2\tau_c^2}\left[(\overline{\omega_x^2} - \overline{\omega_y^2})\overline{s}_-e^{-2i\omega_0 t} - (\overline{\omega_x^2} + \overline{\omega_y^2})\overline{s}_+\right]. \quad (1.26)$$

Using the same procedure (or simply using $s_- = s_+^*$), we would arrive for the analogous equation for $s_-$:

$$\dot{\overline{s}}_+ = -\overline{\omega_z^2}\tau_c\overline{s}_- + \frac{1}{2}\frac{\tau_c}{1+\omega_0^2\tau_c^2}\left[(\overline{\omega_x^2} - \overline{\omega_y^2})\overline{s}_+e^{2i\omega_0 t} - (\overline{\omega_x^2} + \overline{\omega_y^2})\overline{s}_-\right]. \quad (1.27)$$

Similarly,

$$\dot{\overline{s}}_z = -(\overline{\omega_x^2} + \overline{\omega_y^2})\frac{\tau_c}{1+\omega_0^2\tau_c^2}\overline{s}_z. \quad (1.28)$$

For the rest of the section we omit the overline on the symbols for the spins, so that $S$ will mean the average spin. Returning back to our rest frame of the spins rotating with frequency $\omega_0$, we get

$$\dot{S}_+ = i\omega_0 S_+ - \overline{\omega_z^2}\tau_c\frac{1}{2}\frac{\tau_c}{1+\omega_0^2\tau_c^2}\left[(\overline{\omega_x^2} - \overline{\omega_y^2})S_- - (\overline{\omega_x^2} + \overline{\omega_y^2})S_+\right], \quad (1.29)$$

$$\dot{S}_- = i\omega_0 S_- - \overline{\omega_z^2} - \frac{1}{2}\frac{\tau_c}{1+\omega_0^2\tau_c^2}\left[(\overline{\omega_x^2} - \overline{\omega_y^2})S_+ - (\overline{\omega_x^2} + \overline{\omega_y^2})S_-\right], \quad (1.30)$$

$$\dot{S}_z = -(\overline{\omega_x^2} + \overline{\omega_y^2})\frac{\tau_c}{1+\omega_0^2\tau_c^2}\overline{S}_z. \quad (1.31)$$

Finally, going back to $S_x$ and $S_y$

$$\dot{S}_x = -\omega_0 S_y - \overline{\omega_z^2}\tau_c S_x - \frac{\tau_c}{1+\omega_0^2\tau_c^2}\overline{\omega_y^2}\overline{S}_x, \quad (1.32)$$

$$\dot{S}_y = \omega_0 S_x - \overline{\omega_z^2}\tau_c S_y - \frac{\tau_c}{1+\omega_0^2\tau_c^2}(\overline{\omega_x^2})S_y, \quad (1.33)$$

$$\dot{S}_z = -(\overline{\omega_x^2} + \overline{\omega_y^2})\frac{\tau_c}{1+\omega_0^2\tau_c^2}S_z. \quad (1.34)$$

We can give the above equation a more conventional form by introducing two types of the spin decay times. First, we define the *spin relaxation time* $T_1$ by

$$\frac{1}{T_1} = \left( \overline{\omega_x^2} + \overline{\omega_y^2} \right) \frac{\tau_c}{1 + \omega_0^2 \tau_c^2},$$
(1.35)

and the spin dephasing times $T_2$ by

$$\frac{1}{T_{2x}} = \overline{\omega_z^2} \tau_c + \frac{\overline{\omega_y^2} \tau_c}{1 + \omega_0^2 \tau_c^2},$$
(1.36)

$$\frac{1}{T_{2y}} = \overline{\omega_z^2} \tau_c + \frac{\overline{\omega_x^2} \tau_c}{1 + \omega_0^2 \tau_c^2}.$$
(1.37)

We then write

$$\dot{S}_x = -\omega_0 S_y - \frac{S_x}{T_{2x}},$$
(1.38)

$$\dot{S}_y = \omega_0 S_x - \frac{S_y}{T_{2y}},$$
(1.39)

$$\dot{S}_z = -\frac{S_z}{T_1}.$$
(1.40)

Our fluctuating field is effectively at infinite temperature, at which the average value for the spin in a magnetic field is zero. A more general spin dynamics is

$$\dot{S}_x = -\omega_0 S_y - \frac{S_x}{T_{2x}},$$
(1.41)

$$\dot{S}_y = \omega_0 S_x - \frac{S_y}{T_{2y}},$$
(1.42)

$$\dot{S}_z = -\frac{S_z - S_{0z}}{T_1}.$$
(1.43)

where $S_{0z}$ is the equilibrium value of the spin in the presence of the static magnetic field of the Larmor frequency $\omega_0$ at the temperature at which the environmental fields giving rise to $\omega(t)$ are in equilibrium. The above equations are called the *Bloch equations*.

The spin components $S_x$ and $S_y$, which are perpendicular to the applied static field $\mathbf{B}_0$, decay exponentially on the timescales of $T_{2x}$ and $T_{2y}$, respectively. These times are termed spin *dephasing* times, as they describe the loss of the *phase* of the spin components perpendicular to the static field $\mathbf{B}_0$. They are also often called *transverse* times, for that reason. The time $T_1$ is termed the spin *relaxation* time, as it describes the (thermal) relaxation of the spin to the equilibrium. During the spin relaxation in a static magnetic field, the energy is exchanged with the environment. In the language of statistical physics, the relaxation process establishes the Boltzmann probability distribution for the system. Similarly, dephasing establishes the "random phases" postulate that says that there is no correlation (coherence) among the degenerate states, such as the two transverse spin orientations; in thermal equilibrium such states add incoherently.

For the sake of discussion consider an isotropic system in which

$$\overline{\omega_x^2} = \overline{\omega_y^2} = \overline{\omega_z^2} = \overline{\omega^2}. \tag{1.44}$$

If the static magnetic field is weak, $\omega_0\tau_c \ll 1$, the three times are equal:

$$T_1 = T_{2x} = T_{2y} = \frac{1}{\overline{\omega^2}\tau_c}. \tag{1.45}$$

There is no difference between the spin relaxation and spin dephasing. It is at first sight surprising that the spin relaxation time is inversely proportional to the correlation time. The more random the external field appears, the less the spin decays. We will explain this fact below by the phenomenon of motional narrowing.

In the opposite limit of large Larmor frequency, $\omega_0\tau_c \gg 1$, the spin relaxation rate vanishes

$$\frac{1}{T_1} \approx 2\frac{\overline{\omega^2}}{\omega_0^2}\frac{1}{\tau_c} \to 0, \tag{1.46}$$

while the spin dephasing time is given by what is called *secular broadening*:

$$\frac{1}{T_2} \approx \overline{\omega_z^2}\tau_c. \tag{1.47}$$

If secular broadening is absent, the leading term in the dephasing time will be, as in the relaxation

$$\frac{1}{T_2} \approx \frac{\overline{\omega^2}}{\omega_0^2}\frac{1}{\tau_c}. \tag{1.48}$$

In this limit the spin dephasing rate is proportional to the correlation rate, not to the correlation time.

In many physical situations the fluctuating fields are anisotropic. Perhaps the most case is such that $\omega_z = 0$ while

$$\overline{\omega_x^2} = \overline{\omega_y^2} = \overline{\omega^2}. \tag{1.49}$$

Then, from Equations 1.35 through 1.37, it follows that

$$T_2 = 2T_1. \tag{1.50}$$

This anisotropic spin relaxation is due to the fact that while the transverse spin can be flipped by two modes of the in-plane fluctuating field, the in-plane spins can be flipped by one mode only (the one perpendicular to the spin).

In the cases in which there is no clear distinction between $T_1$ and $T_2$, we often use the symbol

$$\tau_s = T_1 = T_2, \tag{1.51}$$

to describe the *generic spin relaxation.*

### 1.1.1 MOTIONAL NARROWING

The surprising fact that at low magnetic fields the spin relaxation rate is proportional to the correlation time of the fluctuating field (as opposed to its inverse) is explained by *motional narrowing*. Consider the spin transverse to an applied magnetic field and assume that the field has a single magnitude, but can randomly switch directions, between up to down, leading to a random precession of the spin clock- and anticlockwise. In effect, the spin phase executes a random walk. A single step takes the time $\tau_c$, the correlation time of the fluctuating field. After $n$ steps, that is, after the time $t = n\tau$, the standard deviation of the phase will be $\delta\phi = (\omega\tau_c)\sqrt{n}$, the well-known result for a random walk. We call the spin dephasing time the time it takes for $\delta\phi \approx 1$. This happens after the time $\tau_s = \tau_c/(\omega\tau_c)^2$, or $\tau_s = 1/(\omega^2\tau_c)$, which is the result we obtained earlier from the Born–Markov approximation, Equation 1.45.

### 1.1.2 REVERSIBLE DEPHASING, SPIN ENSEMBLE, RANDOM WALK IN INHOMOGENEOUS FIELDS

Our previous calculation was carried out for a single spin in a fluctuating magnetic field. After the decay of the spin components, the information about the original spin is irreversibly lost as we have no information on the actual history of the fluctuating field. Such an irreversible loss of spin is often termed spin *decoherence*. We will see that spin dephasing can be reversible. We will also see that a simple exponential decay, of the type $\exp(-t/\tau_s)$, is not a rule.

#### 1.1.2.1 Reversible Spin Dephasing: Spin Ensemble in Spatially Random Magnetic Field

There are physically relevant cases in which the decay of spin is reversible. A typical example is an *ensemble* of localized spins, each precessing about a static local magnetic field $\mathbf{B}_0 + \mathbf{B}_1$, with varying static components $\mathbf{B}_1$ (of zero average) giving rise to random precession frequencies $\boldsymbol{\omega}_1$ (see Figure 1.2). Another important example is that of the conduction electrons in noncentrosymmetric crystals, such as GaAs, in which the spin–orbit coupling acts as a momentum-dependent magnetic field. The spins of the electrons of different momenta precess with different frequencies. We are interested in the total spin as the sum of the individual spins.

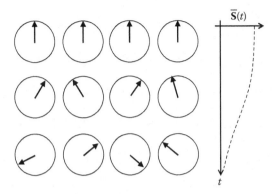

**FIGURE 1.2**   Electron spin precesses about the static $\mathbf{B}_0$ field along $\mathbf{z}$ perpendicular to this page. The spatially fluctuating magnetic field $B_1\mathbf{z}$ causes reversible spin dephasing.

Consider the external field along $z$ direction, and take the fluctuating frequencies from the Gaussian distribution

$$P(\omega_1) = \frac{1}{\sqrt{2\pi\delta\omega^2}} e^{-\omega_1^2/2\delta\omega^2}, \tag{1.52}$$

with zero mean and $\delta\omega_1^2$ variance. Denote the in-plane spin of the electron $a^*$ by $S_x^a$ and $S_y^a$. This spin precesses with the frequency $\omega_0 + \omega_1^a$, leading to the time evolution for the rotating spins:

$$S_\pm^a(t) = S_x^a(t) \pm iS_y^a(t) = S_\pm^a(0)e^{\pm i\omega_0 t}e^{\pm i\omega_1^a t}. \tag{1.53}$$

Suppose at $t = 0$ all the spins are lined up, that is, $S_\pm^a(0) = S_\pm(0)$. The total spin $S_\pm(t)$ is the sum

$$S_\pm(t) = \sum_s S_\pm^a(t) = S_\pm(0)e^{\pm i\omega_0 t}\int_{-\infty}^{\infty} d\omega_1 P(\omega_1)e^{\pm i\omega_1 t}. \tag{1.54}$$

Evaluating the Gaussian integral we get

$$S_\pm(t) = S_\pm(0)e^{\pm i\omega_0 t}e^{-\delta\omega_1^2 t^2/2}. \tag{1.55}$$

The in-plane component vanishes after the time of about $1/\delta\omega_1$, but this dephasing of the spin is reversible, since each individual spin preserves the memory of the initial state. The disappearance of the spin is purely due to the statistical averaging over an ensemble in which the individual spins have, after certain time, random phases. This spin decay is not a simple exponential, but rather Gaussian.

---

$^*$ This discussion applies to nuclear spins as well.

### 1.1.2.2 Spin Echo

We have seen that there are irreversible and reversible effects both present in spin dephasing. It turns out that the reversible effects can be separated out by the phenomenon of the spin echo. Figure 1.3 explains this mechanism in detail. Suppose all the localized spins in our ensemble point in one direction at time $t = 0$. At a later time, $t = T_\pi$, the spins dephase due to the inhomogeneities of the precession frequencies and the total spin is small. We apply a short pulse of an external magnetic field, the so-called $\pi$ pulse, that rotates the spins along the axis parallel to the original spin direction, mapping the spins as $(S_x, S_y) \to (-S_x, S_y)$. At that moment the spins will still be dephased, but the one that is the fastest is now the last, and the one that is the slowest appears as the first. At the time $t = 2T_\pi$ all the spins catch on, producing a large spin signal along the original spin direction. In reality this signal will be weaker than that at $t = 0$ due to the presence of irreversible processes, by $\exp(-2T_\pi/T_2)$. Important, reversible processes are not counted in $T_2$.

### 1.1.2.3 Spin Random Walk in Inhomogeneous Magnetic Field

Another interesting situation appears when we consider the possibility that the spin diffuses through a region of an inhomogeneous precession frequency.* We can model this situation on the system of spins in one dimension, each spin jumping in a time $\tau$ left or right. Suppose the precession frequencies vary in the $x$ direction (see Figure 1.4) as

$$\omega(x) = \omega_0 + \omega'x. \tag{1.56}$$

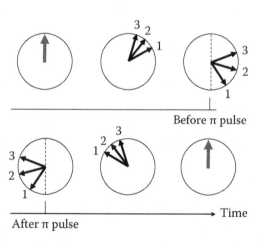

**FIGURE 1.3**  Initially all the spins point up. Due to random precession frequencies the spins soon point in different directions and the average spin vanishes. Applying a $\pi$-pulse rotating the spins along the vertical axis makes the fastest spin the slowest, and vice versa. After the fastest ones catch up again with the slowest, the original value of the average spin is restored. Any reduction from the original value is due to irreversible processes signal.

---

* The inhomogeneity could be due to the magnetic field or to the g-factor.

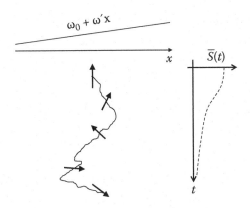

**FIGURE 1.4**   The electron performs a random walk. Its spin precesses by the inhomogeneous magnetic field along the $x$-axis. The average spin dephases to zero with time.

Here $\omega'$ is the gradient of $\omega$. As the electron diffuses, it precesses with the precession frequency $\omega(t) = \omega[x(t)]$, given by the position of the electron $x(t)$ at time $t$. The time evolution for an individual spin is

$$\dot{S}_+ = \omega(t)S_+(t) = \omega(t)S_+(0) + \omega(t)\int_0^\infty dt'\omega(t')S(t').$$  (1.57)

We are interested in the averaged quantities over many realizations of the random walk. We also consider the timescales much greater than the individual random walks steps $\tau$.

The accumulated phase after $N$ steps or $t = N\tau$ time is the sum

$$\phi_N = x_1 + x_2 + \cdots + x_N = N\delta_1 + (N-1)\delta_1 + \cdots + \delta_N.$$  (1.58)

Here $\delta_i = \pm 1$ is a random variable representing the random step left or right. The variance of $\phi_N$ is

$$\sigma_N^2 = \sum_i^{N-1}(N-i)^2 \approx \frac{1}{3}N^3.$$  (1.59)

According to the central limit theorem, $\phi_N$ is distributed normally with the above variance:

$$P(\phi_N) = \frac{1}{\sqrt{2\pi\sigma_N^2}}e^{-\phi_N^2/2\sigma_N^2}.$$  (1.60)

We can now transform the dynamical equation into the ensemble averaging

$$S_+(t) = S_+(0)e^{i\phi(t)} = \int_{-\infty}^\infty d\phi_N e^{i\phi(t)}P(\phi_N) = e^{-\omega_1^2 Dt^3/3},$$  (1.61)

where we denoted the diffusivity by $D = \tau/2$, for the unit length step.

### 1.1.3 QUANTUM MECHANICAL DESCRIPTION

The Born–Markov approximation to the dynamics of a system coupled to an external environment can be cast in the quantum mechanical language. We refer the reader to Ref. [8] for the derivation. The physics of this derivation is the same as what we did in Section 1.1. Only the formalism is different.

We consider the system described by the Hamiltonian

$$H(t) = H_0 + V(t), \tag{1.62}$$

in which $H_0$ is our system *per se* and $V(t)$ is the time-dependent random fluctuating field of zero average and correlation time $\tau_c$:

$$\overline{V(t)} = \overline{V(t)} = 0, \quad \overline{V(t)V(t')} \sim e^{-|t-t'|/\tau_c}. \tag{1.63}$$

The system is fully described by the density matrix $\rho$.

The transformation to the rotating frame is equivalent to going to the interaction picture in quantum mechanics:

$$\rho_I(t) = e^{iH_0t/\hbar} \rho e^{-iH_0t/\hbar}, \tag{1.64}$$

$$V_I(t) = e^{iH_0t/\hbar} V(t) e^{-iH_0t/\hbar}. \tag{1.65}$$

Performing the operations, as outlined in Section 1.1, for the classical model, we arrive at the effective time evolution for the density of state operator:

$$\frac{d\overline{\rho_I(t)}}{dt} = \left(\frac{1}{i\hbar}\right)^2 \int_0^{t \gg \tau_c} dt' \left[\overline{V_I(t),[V_I(t'),\overline{\rho_I(t)}]}\right]. \tag{1.66}$$

The above equation is called the *Master equation* and is the starting equation in many important problems in which a quantum system is in contact with a reservoir (Figure 1.4).

### 1.1.4 SPIN RELAXATION OF CONDUCTION ELECTRONS

We will consider nonmagnetic conductors in zero or weak magnetic fields so that the spin dephasing and spin relaxation times are equal, $T_2 = T_1 = \tau_s$. The formula for the spin relaxation time

$$\frac{1}{\tau_s} = \omega^2 \tau_c, \tag{1.67}$$

that was derived above for the spin in a fluctuating magnetic field applies in a semiquantitative sense (i.e., it gives an order of magnitude estimates and useful trends) to the conduction electron spins as well. We analyze below the D'yakonov–Perel' [3] and the Elliott–Yafet [1, 2] mechanisms.

The D'yakonov–Perel' mechanism is at play in solids lacking a center of spatial inversion symmetry. The most prominent example is the semiconductor GaAs. In such solids the spin–orbit coupling is manifested as some effective magnetic field—*the spin–orbit field*—that depends on the electron momentum. Electrons in different momentum states feel different spin–orbit fields, so that the spin precesses with a given Larmor frequency, until the electron is scattered into another momentum state (see Figure 1.5). As the electron momentum changes on the timescale $\tau$ of the momentum relaxation time, the net effect of the momentum scattering on the spin is to produce random fluctuations of the Larmor frequencies. We have motional narrowing. Since these frequencies are correlated by $\tau_c = \tau$, we arrive at

$$\frac{1}{\tau_s} = \omega_{so}^2 \tau, \tag{1.68}$$

for the spin relaxation time. The magnitude of $\omega_{so}$ is the measure of the strength of the spin–orbit coupling. The spin relaxation rate is directly proportional to the momentum relaxation time—the more the electron scatters, the less its spin dephases.

The Elliott–Yafet mechanism works in systems with and without a center of inversion. It relies on spin-flip momentum scattering. The spin-flip amplitudes are due to spin–orbit coupling, while the momentum scattering is due to the presence of impurities (that also contribute to the spin–orbit coupling), phonons, rough boundaries, or whatever is capable of randomizing the electron momentum. During the scattering events, the spin is preserved (see Figure 1.6). How do we account for such a scenario with our toy model? Consider an electron scattering off of an impurity with a spin flip. This spin flip can be viewed as a precession that occurs during the time of the interaction of the electron with the impurity. Let us say that the scattering takes the time of $\lambda_F/\upsilon_F$, where $\upsilon_F$ is the electron velocity and $\lambda_F$ is the electron wavelength at the Fermi level (considering that it is greater or at most equal to the size $a$ of the impurity—otherwise we could equally put $a/\upsilon_F$). Then, the precession angle $\varphi = \omega_{so}(\lambda_F/\upsilon_F)$, with $\omega_{so}$ denoting the spin–orbit coupling-induced precession frequency. Let us compare this with the

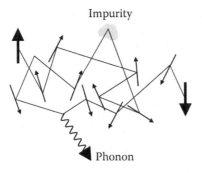

**FIGURE 1.5**   D'yakonov–Perel' mechanism. The electron starts with the spin up. As it moves, its spin precesses about the axis corresponding to the electron velocity. Phonons and impurities change the velocity, making the spin to precess about a different axis (and with different speed). During the scattering event the spin is preserved.

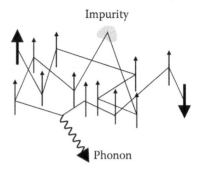

**FIGURE 1.6** Elliott–Yafet mechanism. The electron starts with the spin up. As it scatters off of impurities and phonons, the spin can also flip due to spin–orbit coupling. Between the scattering the spin is preserved. After, say, a million scattering events, the spin will be down.

angle of precession, $\omega\tau$, in the motional narrowing model, Equation 1.67. We see that the spin-flip can be described by the effective precession frequency $\omega = \omega_{so}(\lambda_F/\upsilon_F\tau)$. We then obtain for the spin relaxation rate

$$\frac{1}{\tau_s} = \omega_{so}^2 \frac{\lambda_F^2}{\upsilon_F^2 \tau} \approx \left(\frac{\varepsilon_{so}}{\varepsilon_F}\right)^2 \frac{1}{\tau}. \tag{1.69}$$

Here $\varepsilon_{so} = \hbar\omega_{so}$ and $\varepsilon = (\hbar k_F)\upsilon_F/2$ is the Fermi energy. For the Elliott–Yafet mechanism holds that the more the electron scatters, the more the spin dephases.

Below we discuss the two mechanisms in more detail, providing the formalisms for their investigation.

## 1.2 D'YAKONOV–PEREL' MECHANISM

D'yakonov and Perel' [3] considered solids without a center of inversion symmetry, such as GaAs or InAs. The mechanism also works for electrons at surfaces and interfaces. In such solids, the presence of spin–orbit coupling induces the spin–orbit fields, which give rise to the spin precession. Momentum relaxation then causes the random walk of the spin phases.

### 1.2.1 Spin–Orbit Field

In solids without a center of inversion symmetry, spin–orbit coupling splits the electron energies:

$$\varepsilon_{\mathbf{k},\uparrow} \neq \varepsilon_{\mathbf{k},\downarrow}. \tag{1.70}$$

Only the Kramers degeneracy is left, due to time reversal symmetry:

$$\varepsilon_{\mathbf{k},\uparrow} = \varepsilon_{-\mathbf{k},\downarrow}. \tag{1.71}$$

This energy splitting at a given momentum **k** is conveniently described by the spin–orbit field $\mathbf{\Omega}$, giving a Zeeman-like (but momentum-dependent) energy contribution to the electronic states, described by the additional Hamiltonian (to the usual band structure):

$$H_1 = \frac{\hbar}{2}\mathbf{\Omega}_\mathbf{k} \cdot \boldsymbol{\sigma}. \tag{1.72}$$

The time reversal symmetry requires that the spin–orbit field is an odd function of the momentum:

$$\mathbf{\Omega}_\mathbf{k} = -\mathbf{\Omega}_{-\mathbf{k}}. \tag{1.73}$$

The representative example of a spin–orbit field is the Bychkov–Rashba [10] ($\alpha_{BR}$) and Dresselhaus [11, 12] ($\gamma_D$) couplings in 2d electron gases formed at the zinc-blend heterostructures grown along [001] [8]:

$$H_1 = (\alpha_{BR} + \gamma_D)\sigma_x k_y - (\alpha_{BR} - \gamma_D)\sigma_y k_x. \tag{1.74}$$

The axes are $x = [110]$ and $y = [1\bar{1}0]$. The spin–orbit field is

$$\hbar\mathbf{\Omega}_\mathbf{k} = 2\big[(\alpha_{BR} + \gamma_D)k_y, -(\alpha_{BR} - \gamma_D)k_x\big]. \tag{1.75}$$

This field has the $C_{2v}$ symmetry, reflecting the structural symmetry of the zinc-blend interfaces (such as GaAs/GaAlAs) with the principal axes along [110] and [1$\bar{1}$0].

## 1.2.2 KINETIC EQUATION FOR THE SPIN

Let $\mathbf{s}_\mathbf{k}$ be the electron spin in the momentum state **k**. The time evolution of the spin is then described by

$$\frac{\partial \mathbf{s}_\mathbf{k}}{\partial t} - \mathbf{\Omega}_\mathbf{k} \times \mathbf{s}_\mathbf{k} = -\sum_{\mathbf{k'}} W_{\mathbf{kk'}}\big(\mathbf{s}_\mathbf{k} - \mathbf{s'}_\mathbf{k}\big). \tag{1.76}$$

The left-hand side gives the full time derivative $d\mathbf{s}_\mathbf{k}/dt$, which is due to the explicit change of the spin, its direction, and the change of the spin due to the change of the momentum **k**. The right-hand side is the change of the spin at **k** due to the spin-preserving scattering from and to that state. The scattering rate between the two momentum states **k** and **k'** is $W_{\mathbf{kk'}}$.

There are two timescales in the problem. First, momentum scattering, which occurs on the timescale of the momentum relaxation time $\tau$, makes the spins in different momentum states equal. Second, this uniform spin decays on the timescale of the spin relaxation time $\tau_s$, which we need to find. We assume that $\tau \ll \tau_s$; this assumption is well satisfied in real systems. In principle we could go directly to our model of the electron spin in a fluctuating magnetic field, with the role of the random Larmor precession playing by $\mathbf{\Omega}$, identifying $\tau_c = \tau$. However, it is instructive to see how this two timescale problem is solved directly.

We separate the fast and slow components of the spins as follows:

$$\mathbf{s_k} = \mathbf{s} + \boldsymbol{\xi_k}, \quad \langle \boldsymbol{\xi_k} \rangle = 0. \tag{1.77}$$

The symbol $\langle \cdots \rangle$ denotes averaging over different momenta. Our goal is to find the effective equation for the time evolution of $\mathbf{s}$, which is the actual spin averaged over $\mathbf{k}$. The fast component, $\boldsymbol{\xi_k}$, decays on the timescale of $\tau$ to the value given by the instantaneous value of $\mathbf{s}$. Our goal is to find the effective equation for the time evolution of $\mathbf{s}$, which is the actual spin averaged over $\mathbf{k}$. The fast component, $\boldsymbol{\xi_k}$, decays on the timescale of $\tau$ to the quasistatic value given by the instantaneous value of $\mathbf{s}$.

Upon substituting Equation 1.77 into the kinetic Equation 1.76 and averaging over $\mathbf{k}$, we obtain the equation of motion for the averaged spin

$$\dot{\mathbf{s}} = \langle \boldsymbol{\Omega_k} \times \boldsymbol{\xi_k} \rangle, \tag{1.78}$$

using that $\langle \boldsymbol{\Omega_k} = 0 \rangle$. We need to find $\boldsymbol{\xi_k}$. Since $\mathbf{s}$ is hardly changing on the timescales relevant to $\boldsymbol{\xi_k}$, we write

$$\dot{\boldsymbol{\xi}}_{\mathbf{k}} - \boldsymbol{\Omega_k} \times \mathbf{s} - \boldsymbol{\Omega_k} \times \boldsymbol{\xi_k} = -\sum_{\mathbf{k'}} W_{\mathbf{kk'}} \left( \boldsymbol{\xi_k} - \boldsymbol{\xi}_{\mathbf{k'}} \right). \tag{1.79}$$

We make the following assumption:

$$\Omega \tau \ll 1. \tag{1.80}$$

That is, we assume that the precession is slow on the timescale of the momentum relaxation time (see our note on the Born approximation in the toy model in Section 1.1). For simplicity, we make the relaxation time approximation to model the time evolution of the fast component $\boldsymbol{\xi_k}$:

$$\boldsymbol{\xi_k} - \boldsymbol{\Omega_k} \times \mathbf{s} - \boldsymbol{\Omega_k} \times \boldsymbol{\xi_k} = -\frac{\boldsymbol{\xi_k}}{\tau}. \tag{1.81}$$

From the condition of the quasistatic behavior, $\partial \boldsymbol{\xi_k} / \partial t = 0$, we get up to the first order in $\Omega \tau$ the following solution for the quasistatic $\boldsymbol{\xi_k}$:

$$\boldsymbol{\xi_k} = \tau \left( \boldsymbol{\Omega_k} \times \mathbf{s} \right). \tag{1.82}$$

The above is a realization of coarse graining, in which we effectively average the spin evolution over the timescales of $\tau$.

Substituting the quasistatic value of $\boldsymbol{\xi_k}$ from Equation 1.82 to the time evolution equation for $\mathbf{s}$, Equation 1.78, we find

$$\dot{\mathbf{s}} = \tau \langle \boldsymbol{\Omega_k} \times \left( \boldsymbol{\Omega_k} \times \mathbf{s} \right) \rangle. \tag{1.83}$$

Using the vector product identities, we finally obtain for the individual spin components $\alpha$:

$$\dot{s}_\alpha = \left\langle \Omega_{k\alpha} \Omega_{k\beta} \right\rangle \tau s_\beta - \left\langle \Omega_k^2 \tau \right\rangle s_\alpha. \tag{1.84}$$

These equations describe the effective time evolution of the electron spin in the presence of the momentum-dependent Larmor precession $\mathbf{\Omega_k}$.

For the specific model of the zinc-blend heterostructure with the $C_{2v}$ spin–orbit field, Equation 1.75, we obtain the spin dephasing dynamics

$$\dot{s}_x = \frac{-s_x}{\tau_s}, \quad \dot{s}_y = \frac{-s_y}{\tau_s}, \quad \dot{s}_z = \frac{-s_y}{\tau_z}, \tag{1.85}$$

where:

$$\frac{1}{\tau_{x,y}} = \frac{\left(\alpha_{BR} \pm \gamma_D\right)^2}{\alpha_{BR}^2 + \gamma_D^2} \frac{1}{\tau_s}, \quad \frac{1}{\tau_z} = \frac{2}{\tau_s}, \tag{1.86}$$

and

$$\frac{1}{\tau_s} = \frac{4m}{\hbar^4} \varepsilon_k \left(\alpha_{BR}^2 + \gamma_D^2\right). \tag{1.87}$$

The up (down) sign is for the $s_x$ ($s_y$). The spin relaxation is anisotropic. The maximum anisotropy is for the case of equal magnitudes of the Bychkov–Rashba and Dresselhaus interactions, $\alpha_{BR} = \pm\gamma_D$. In this case one of the spin components does not decay.* The $s_z$ component of the spin relaxes roughly twice faster than the in-plane components.

## 1.2.3 PERSISTENT SPIN HELIX

Let us consider the case of $\alpha_{BR} = \gamma_D = \lambda/2$. According to Equation 1.86, the spin component $s_x$ does not decay, while the decay rates of $s_y$ and $s_z$ are the same:

$$\dot{s}_y = \frac{-2s_y}{\tau_s}, \quad \dot{s}_z = \frac{-2s_z}{\tau_s}. \tag{1.88}$$

It turns out that a particular nonuniform superposition of $s_y$ and $s_z$ can exhibit no decay as well. This superposition has been termed *persistent spin helix* [13].

Let us assume that the spin is no longer uniform, so that the kinetic equation contains the spin gradient as well, due to the quasiclassical change of the electronic positions:

$$\frac{\partial \mathbf{s_k}}{\partial t} - \mathbf{\Omega_k} \times \mathbf{s_k} + \frac{\partial \mathbf{s_k}}{\partial \mathbf{r}} \cdot \mathbf{v_k} = -\sum_{k'} W_{kk'} \left(\mathbf{s_k} - \mathbf{s'_k}\right). \tag{1.89}$$

---

* The decay of that component would be due to higher-order (such as cubic) terms in the spin–orbit fields.

A running spin wave

$$\mathbf{s_k}(\mathbf{r}) = \mathbf{s_k} e^{i\mathbf{q \cdot r}}; \quad \mathbf{s_k} \equiv \mathbf{s_k}(\mathbf{q}), \tag{1.90}$$

then evolves according to

$$\frac{\partial \mathbf{s_k}}{\partial t} - \boldsymbol{\Omega_k} \times \mathbf{s_k} + i(\mathbf{q \cdot v_k}) \mathbf{s_k} = -\sum_{\mathbf{k'}} W_{\mathbf{kk'}} (\mathbf{s_k} - \mathbf{s_k'}). \tag{1.91}$$

We again separate the fast and slow spins as follows:

$$\mathbf{s_k} = \mathbf{s} + \boldsymbol{\xi_k}, \quad \langle \boldsymbol{\xi_k} \rangle = 0. \tag{1.92}$$

For the dynamics of the slow part we get

$$\dot{\mathbf{s}}_\mathbf{k} = \langle \boldsymbol{\Omega_k} \times \boldsymbol{\xi_k} \rangle - i \langle (\mathbf{q \cdot v_k}) \boldsymbol{\xi_k} \rangle. \tag{1.93}$$

Proceeding as in the previous section, assuming that $\mathbf{s}$ is stationary on the timescales relevant to $\boldsymbol{\xi}$, we can write

$$\dot{\boldsymbol{\xi}}_\mathbf{k} - \boldsymbol{\Omega}_\mathbf{k} \times \mathbf{s} - \boldsymbol{\Omega}_\mathbf{k} \times \boldsymbol{\xi_k} + i(\mathbf{q \cdot v_k})\mathbf{s} + i(\mathbf{q \cdot v_k})\boldsymbol{\xi_k}$$
$$= -\sum_{\mathbf{k'}} W_{\mathbf{kk'}} (\boldsymbol{\xi_k} - \boldsymbol{\xi_{k'}}). \tag{1.94}$$

We now work with the following assumptions

$$\Omega\tau \ll 1, \quad q\ell \ll 1, \tag{1.95}$$

where $\ell = g\tau$ is the mean free path. We thus assume that the precession is slow on the timescale of the momentum relaxation time, as well as (this is new here) the electronic motion is diffusive on the scale of the wavelength of the spin wave. In the momentum relaxation approximation, also considering the leading terms according to the conditions in Equation 1.95, we get

$$\dot{\boldsymbol{\xi}}_\mathbf{k} - \boldsymbol{\Omega_k} \times \mathbf{s} + i(\mathbf{q \cdot v_k})\mathbf{s} = -\frac{\boldsymbol{\xi}}{\tau}. \tag{1.96}$$

In the steady state, corresponding to a given $\mathbf{s}(t)$, the solution is

$$\boldsymbol{\xi_k} = \tau(\boldsymbol{\Omega_k} \times \mathbf{s}) - i\tau(\mathbf{q \cdot v_k})\mathbf{s}. \tag{1.97}$$

Substituting to the time evolution equation for $\mathbf{s}$, Equation 1.95, we find

$$\dot{\mathbf{s}} = -2i\tau \langle (\mathbf{q \cdot v_k})(\boldsymbol{\Omega_k} \times \mathbf{s}) \rangle + \tau \langle \boldsymbol{\Omega_k} \times (\boldsymbol{\Omega_k} \times \mathbf{s}) \rangle - \tau \langle (\mathbf{q \cdot v_k})^2 \rangle \mathbf{s}. \tag{1.98}$$

Using the vector product identities and introducing the diffusivity

$$D = \langle \mathbf{v}_{\mathbf{k}\alpha}^2 \rangle \tau, \tag{1.99}$$

where $\alpha$ denote the Cartesian coordinates (we assume an isotropic system), we finally obtain

$$\dot{s}_\alpha = -2i\tau\varepsilon_{\alpha\beta\gamma}q_\delta \langle v_{k\delta}\Omega_{k\beta}\rangle s_\gamma + \langle \Omega_{k\alpha}\Omega_{k\beta}\rangle \tau s_\beta - \langle \Omega_k^2\tau\rangle s_\alpha - Dq^2 s_\alpha. \quad (1.100)$$

For our specific case of

$$\Omega_k = (\lambda k_y, 0, 0), \quad \lambda = \alpha_{BR} + \beta_D, \quad (1.101)$$

we find that $s_x$ decays only via diffusion:

$$\dot{s}_x = -Dq^2 s_x. \quad (1.102)$$

The spin dephasing is ineffective. This is an expected result.

More interesting behavior is found for the two remaining spin components, $s_y$ and $s_z$. These two spin components, transverse to the spin–orbit field, are coupled

$$\dot{s}_y = +2i\frac{m\lambda}{\hbar}q_y D s_z - \left(\tau\Omega^2 + Dq^2\right) s_y, \quad (1.103)$$

$$\dot{s}_z = -2i\frac{m\lambda}{\hbar}q_y D s_y - \left(\tau\Omega^2 + Dq^2\right) s_z, \quad (1.104)$$

and we denoted $\Omega^2 \equiv \langle \Omega_k^2\rangle$. Interestingly, this set of coupled differential equations has only decaying solutions. This is best seen by looking at the rotating spins

$$s_+ = s_y + i s_z, \quad s_- = s_y - i s_z, \quad (1.105)$$

whose time evolutions are uncoupled

$$\dot{s}_+ = -\left(\Omega^2\tau + Dq^2 - \frac{2m\lambda}{\hbar}q_y D\right) s_+, \quad (1.106)$$

$$\dot{s}_- = -\left(\Omega^2\tau + Dq^2 + \frac{2m\lambda}{\hbar}q_y D\right) s_-. \quad (1.107)$$

Considering that

$$\Omega^2\tau = \lambda^2\langle k_y^2\rangle\tau = \lambda^2\left(\frac{m}{\hbar}\right)^2\langle v_y^2\rangle\tau = \lambda^2\left(\frac{m}{\hbar}\right)^2 D, \quad (1.108)$$

we find that the decay of $s_+$ vanishes for the wave-vector

$$q_y^{PSH} = \frac{m\lambda}{\hbar} = \frac{m}{\hbar}(\alpha_{BR} + \beta_D). \quad (1.109)$$

The abbreviation PSH stands for the *persistent spin helix*, which is the spin wave described by $s_+$ at this particular wave-vector (see Figure 1.7). While

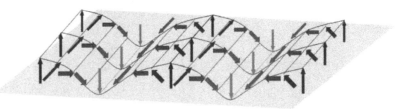

**FIGURE 1.7**   The persistent spin helix is a wave of circularly polarized spin. The sense of polarization, clock or counterclockwise, depends on the relative sign of the Dresselhaus and the Bychkov–Rashba spin–orbit coupling.

individually both $s_y$ and $s_z$ decay at a generic wave-vector (and also in the uniform case, $q = 0$), the spin helix they form does not decay in the approximation of the linear spin–orbit field. The spin wave rotating in the opposite sense, $s_-$, on the other hand, decays. And vice versa for $q_y^{\text{PSH}} = -m\lambda / \hbar$. The persistent spin helix was observed in the spin grating experiment [14].

## 1.3  ELLIOTT–YAFET MECHANISM

Elliott [1] was first to recognize the role of the intrinsic spin–orbit coupling—that coming from the host ions—on spin relaxation. Yafet [2] significantly extended this theory to properly treat electron–phonon spin-flip scattering. The Elliott–Yafet mechanism dominates the spin relaxation of conduction electrons in elemental metals and semiconductors, the systems with space inversion symmetry. In systems lacking this symmetry, the mechanism competes with the D'yakonov–Perel' one; the dominance of one over the other depends on the material in question and specific conditions, such as temperature and doping.

Suppose a nonequilibrium spin accumulates in a nonmagnetic degenerate conductor. The spin accumulation is given as the difference between the chemical potentials for the spin-up and the spin-down electrons. Let us denote the corresponding potentials by $\mu\uparrow$ and $\mu\downarrow$. The nonequilibrium electron occupation function for the spin $\lambda$ is

$$f_{\lambda\mathbf{k}} \approx \frac{1}{e^{\beta(\varepsilon_\mathbf{k} - \mu_\lambda)} + 1} \approx f_\mathbf{k}^0 + \left[-\frac{\partial f_\mathbf{k}^0}{\partial \varepsilon_\mathbf{k}}\right](\mu_\lambda - \varepsilon_F). \tag{1.110}$$

Here

$$f_\mathbf{k}^0 = f^0(\varepsilon_\mathbf{k}) = \frac{1}{e^{\beta(\varepsilon_\mathbf{k} - \varepsilon_F)} + 1}, \tag{1.111}$$

describes the equilibrium degenerate electronic system of the Fermi energy $\varepsilon_F$. The electron density for the spin $\lambda$ is

$$n_\lambda = \int d\varepsilon g_s(\varepsilon) f_\lambda^0(\varepsilon) \approx \frac{n}{2} + \int d\varepsilon g(\varepsilon) \left[-\frac{\partial f^0(\varepsilon)}{\partial \varepsilon}\right](\mu_\lambda - \varepsilon_F). \tag{1.112}$$

Here $g_s(\varepsilon)$ is the electronic density of states, defined per unit volume and per spin, at the Fermi level:

$$g_s = \sum_{\mathbf{k}} \left[ -\frac{\partial f_{\mathbf{k}}^0}{\partial \varepsilon_{\mathbf{k}}} \right]. \tag{1.113}$$

The total electron density $n$ is

$$n = 2 \int d\varepsilon g(\varepsilon) f^0(\varepsilon). \tag{1.114}$$

We assume that the spin accumulation does not charge the system, that is, the charge neutrality is preserved $n_\uparrow + n_\downarrow = n$. This condition is well satisfied in metals and degenerate semiconductors that we consider. We then get

$$\mu_\uparrow + \mu_\downarrow = 2\varepsilon_F. \tag{1.115}$$

The spin density is

$$s = n_\uparrow - n_\downarrow = g_s(\mu_\uparrow - \mu_\downarrow) = g_s \mu_s, \tag{1.116}$$

where $\mu_s$ is the spin quasichemical potential, $\mu_s(\mu\uparrow - \mu\downarrow)$.

The spin relaxation time $T_1$ is defined by the decay law:

$$\frac{ds}{dt} = \frac{dn_\uparrow}{dt} - \frac{dn_\downarrow}{dt} = W_{\uparrow\downarrow} - W_{\downarrow\uparrow} = -\frac{s}{T_1} = -g_s \frac{\mu_s}{T_1}. \tag{1.117}$$

Here $W_{\uparrow\downarrow}$ is the net number of transitions per unit time from the spin $\downarrow$ to $\uparrow$. Similarly, $W_{\downarrow\uparrow}$ expresses the rate of spin flips from $\uparrow$ to $\downarrow$. In the degenerate conductors, the spin decay is directly proportional to the decay of the spin accumulation $\mu_s$:

$$\frac{d\mu_s}{dt} = -\frac{\mu_s}{T_1}. \tag{1.118}$$

### 1.3.1 ELECTRON-IMPURITY SCATTERING

We need to distinguish the cases of the impurity or host-induced spin–orbit coupling. Although the formulas for the calculation of $T_1$ look similar in the two cases, they are nevertheless conceptually different.

#### 1.3.1.1 Spin–Orbit Coupling by the Impurity

If the spin–orbit coupling comes from the impurity potential (in this case the coupling is often termed *extrinsic*), the spin-flip scattering is due to that potential.

The number of transitions per unit time from the spin-up to the spin-down states is

$$W_{\uparrow\downarrow} = \sum_{\mathbf{k}n} \sum_{\mathbf{k}'n'} W_{\mathbf{k}'n'\uparrow, \mathbf{k}n\downarrow} - W_{\mathbf{k}n\downarrow, \mathbf{k}'n'\uparrow}. \tag{1.119}$$

The rate is given by the spin-flip events from down to up minus the ones from up to down. We use the Fermi golden rule to write out the individual scattering rates:

$$W_{\mathbf{k}'n'\uparrow,\mathbf{k}n\downarrow} = \frac{2\pi}{\hbar} f'_{\mathbf{k}n}\left(1 - f'_{\mathbf{k}'n}\right)\left|U_{\mathbf{k}n\uparrow,\mathbf{k}'n'\downarrow}\right|^2 \delta\left(\varepsilon_{\mathbf{k}'n'} - \varepsilon_{\mathbf{k}n}\right), \qquad (1.120)$$

$$W_{\mathbf{k}n\uparrow,\mathbf{k}'n'\downarrow} = \frac{2\pi}{\hbar} f_{\mathbf{k}'n'}\left(1 - f_{\mathbf{k}n}\right)\left|U_{\mathbf{k}'n'\downarrow,\mathbf{k}n\uparrow}\right|^2 \delta\left(\varepsilon_{\mathbf{k}'n'} - \varepsilon_{\mathbf{k}n}\right). \qquad (1.121)$$

Substituting for the occupation numbers the linearized expression, Equation 1.110, and using the definition of the spin relaxation of Equation 1.117, we obtain for the spin relaxation rate the expression

$$\frac{1}{T_1} = \frac{2\pi}{\hbar}\frac{1}{g_s}\sum_{\mathbf{k}\mathbf{k}'}\left|U_{\mathbf{k}n\uparrow,\mathbf{k}'n'\downarrow}\right|^2 \left[-\frac{\partial f^0(\varepsilon_{\mathbf{k}})}{\partial\varepsilon_{\mathbf{k}}}\right]\delta\left(\varepsilon_{\mathbf{k}'n'} - \varepsilon_{\mathbf{k}n}\right). \qquad (1.122)$$

If we define the spin relaxation for the individual momentum state $\mathbf{k}$ as

$$\frac{1}{T_{1\mathbf{k}}} = \frac{2\pi}{\hbar}\sum_{\mathbf{k}'}\left|U_{\mathbf{k}n\uparrow,\mathbf{k}'n'\downarrow}\right|^2 \left[-\frac{\partial f^0(\varepsilon_{\mathbf{k}})}{\partial\varepsilon_{\mathbf{k}}}\right]\delta\left(\varepsilon_{\mathbf{k}'n'} - \varepsilon_{\mathbf{k}n}\right), \qquad (1.123)$$

which has a straightforward interpretation as the spin-flip rate by the elastic impurity scattering to all the possible states $\mathbf{k}'$, we get

$$\frac{1}{T_1} = \left\langle\frac{1}{T_{1\mathbf{k}}}\right\rangle_{\varepsilon_{\mathbf{k}} = \varepsilon_F}, \qquad (1.124)$$

as the average of the individual scattering rates over the electronic Fermi surface.

### 1.3.1.2 Spin–Orbit Coupling by the Host Lattice

If the spin–orbit coupling comes from the host lattice (in this case the coupling is often termed *intrinsic*), the spin-flip scattering is due to the admixture of the Pauli spin-up and spin-down states in the Bloch eigenstates.

How do the Bloch states actually look like in the presence of spin–orbit coupling? Elliott showed that the Bloch states corresponding to a generic lattice wave-vector $\mathbf{k}$ and band $n$ can be written as [1]

$$\Psi_{\mathbf{k},n\Uparrow}(\mathbf{r}) = \left[a_{\mathbf{k}n}(\mathbf{r})|\uparrow\rangle + b_{\mathbf{k}n}(\mathbf{r})|\downarrow\rangle\right]e^{i\mathbf{k}\cdot\mathbf{r}}, \qquad (1.125)$$

$$\Psi_{\mathbf{k},n\Downarrow}(\mathbf{r}) = \left[a^*_{-\mathbf{k}n}(\mathbf{r})|\downarrow\rangle - b^*_{-\mathbf{k}n}(\mathbf{r})|\downarrow\rangle\right]e^{i\mathbf{k}\cdot\mathbf{r}}. \qquad (1.126)$$

The states $|\uparrow\rangle$ and $|\downarrow\rangle$ are the usual Pauli spinors. We can select the two states such that $|a_{\mathbf{k}}n| \approx 1$ while $|b_{\mathbf{k}n}| \ll 1$, due to the weak spin–orbit coupling; this justifies calling the two above states "spin up" ($\Uparrow$) and "spin down" ($\Downarrow$).

In fact, to "prepare" the states for the calculation of the spin relaxation, they need to satisfy

$$\langle \mathbf{k}, n\lambda \mid \sigma_z \mid \mathbf{k}, n\lambda' \rangle = \lambda \delta_{\lambda\lambda'}, \tag{1.127}$$

with $\lambda = \Uparrow, \Downarrow$. That is, the two states should diagonalize the spin matrix $S_z$ (or whatever spin direction one is interested in).

The Bloch states of Equations 1.125 and 1.126 allow for a spin flip even if the impurity does not induce a spin–orbit coupling. Indeed, the matrix element

$$\langle \mathbf{k}, n \Uparrow \mid U \mid \mathbf{k}, n \Downarrow \rangle \sim ab, \tag{1.128}$$

is in general non-zero due to the spin admixture. The spin-flip probability is proportional to $|b|^2$, the spin admixture probability. This quantity is crucial in estimating the spin relaxation in the Elliott–Yafet mechanism. The spin relaxation time $T_1$ in this case can be calculated using the formula Equation 1.122, with

$$U_{\mathbf{k}n\uparrow, \mathbf{k}'n'\downarrow} = U_{\mathbf{k}n\Uparrow, \mathbf{k}'n'\Downarrow}, \tag{1.129}$$

given by Equation 1.128. A useful *rule of thumb* for estimating the spin relaxation time in this case is

$$\frac{1}{T_1} \approx \frac{\langle b_{\mathbf{k}n}^2 \rangle}{\tau_p}, \tag{1.130}$$

where the averaging of the spin admixture probabilities $b^2$ is performed over the Fermi surface (or the relevant energy scales of the problem); $\tau_p$ is the spin-conserving momentum relaxation time. We stress that $b$ is obtained from the states prepared according to Equation 1.127.

## 1.3.2 Electron–Phonon Scattering

The spin flip due to the scattering of electrons off phonons involves the intrinsic spin–orbit potential. The electron Bloch states are the ones given by Equations 1.125 and 1.126, prepared according to Equation 1.127.

The net number of transitions per unit time from the spin-up to the spin-down states is

$$W_{\uparrow\downarrow} = \sum_{\mathbf{k}n} \sum_{\mathbf{k}'n'} \sum_{\mathbf{q}\nu} W_{\mathbf{k}n\uparrow, \mathbf{q}\nu; \mathbf{k}'n'\downarrow} + W_{\mathbf{k}n\uparrow; \mathbf{k}'n'\downarrow, \mathbf{q}\nu}$$
$$- W_{\mathbf{k}'n'\downarrow, \mathbf{q}\nu; \mathbf{k}n\uparrow} - W_{\mathbf{k}'n'\downarrow; \mathbf{k}n\uparrow, \mathbf{q}\nu}. \tag{1.131}$$

We introduced the rates of the spin-flip transitions accompanied by the phonon absorption and emissions as follows. The net transition rate from the single electron state $|\mathbf{k}'n'\downarrow\rangle$ to the electron state $|\mathbf{k}n\uparrow\rangle$, while the phonon of momentum $\mathbf{q}$ and polarization $v$ is emitted, is

$$W_{\mathbf{k}n\uparrow,\mathbf{q}v;\mathbf{k}'n'\downarrow} = \frac{2\pi}{\hbar}\left|M_{\mathbf{k}n\uparrow,\mathbf{q}v;\mathbf{k}'n'\downarrow}\right|^2 f_{\mathbf{k}'n'\downarrow}\left(1 - f_{\mathbf{k}n\uparrow}\right)\delta\left(\varepsilon_{\mathbf{k}n} - \varepsilon_{\mathbf{k}'n'} + \hbar\omega_{\mathbf{q}v}\right). \quad (1.132)$$

Similarly, the net transition rate from the single electron state $|\mathbf{k}'n'\downarrow\rangle$ to the electron state $|\mathbf{k}n\uparrow\rangle$, while the phonon of momentum $\mathbf{q}$ and polarization $v$ is absorbed, is

$$W_{\mathbf{k}n\uparrow;\mathbf{k}'n'\downarrow,\mathbf{q}v} = \frac{2\pi}{\hbar}\left|M_{\mathbf{k}n\uparrow;\mathbf{k}'n'\downarrow\mathbf{q}v}\right|^2 f_{\mathbf{k}'n'\downarrow}\left(1 - f_{\mathbf{k}n\uparrow}\right)\delta\left(\varepsilon_{\mathbf{k}n} - \varepsilon_{\mathbf{k}'n'} - \hbar\omega_{\mathbf{q}v}\right). \quad (1.133)$$

The same way are defined the remaining two rates, $W_{\mathbf{k}'n'\downarrow,\mathbf{q}v;\mathbf{k}n\uparrow}$ for the spin flip from $\mathbf{k}n\uparrow$ to $\mathbf{k}'n'\downarrow$ with the phonon emission, and $W_{\mathbf{k}'n'\downarrow;\mathbf{k}n\uparrow\mathbf{q}v}$, for the phonon absorption:

$$W_{\mathbf{k}'n'\downarrow,\mathbf{q}v;\mathbf{k}n\uparrow} = \frac{2\pi}{\hbar}\left|M_{\mathbf{k}'n'\downarrow,\mathbf{q}v;\mathbf{k}n\uparrow}\right|^2 f_{\mathbf{k}n\uparrow}\left(1 - f_{\mathbf{k}'n'\downarrow}\right)\delta\left(\varepsilon_{\mathbf{k}'n'} - \varepsilon_{\mathbf{k}n} + \hbar\omega_{\mathbf{q}v}\right), \quad (1.134)$$

$$W_{\mathbf{k}'n'\downarrow;\mathbf{k}n\uparrow\mathbf{q}v} = \frac{2\pi}{\hbar}\left|M_{\mathbf{k}'n'\downarrow,\mathbf{q}v;\mathbf{k}n\uparrow,\mathbf{q}v}\right|^2 f_{\mathbf{k}n\uparrow}\left(1 - f_{\mathbf{k}'n'\downarrow}\right)\delta\left(\varepsilon_{\mathbf{k}'n'} - \varepsilon_{\mathbf{k}n} - \hbar\omega_{\mathbf{q}v}\right). \quad (1.135)$$

The calculation of the spin relaxation due to the electron–phonon scattering is rather involved and we cite here only the final result for degenerate conductors:

$$\frac{1}{T_1} = \frac{4\pi}{\hbar}\frac{1}{g_s}\sum_{\mathbf{k}n}\sum_{\mathbf{k}'n'}\sum_{v}\left(-\frac{\partial f_{\mathbf{k}n}^0}{\varepsilon_{\mathbf{k}n}}\right)\frac{\hbar}{2NM}\frac{N^2}{\omega_{\mathbf{k}'-\mathbf{k},v}} \quad (1.136)$$

$$\times\left|\boldsymbol{\varepsilon}_{\mathbf{k}-\mathbf{k}',v}\cdot\left\langle\mathbf{k}n\Uparrow\left|\boldsymbol{\nabla}V\right|\mathbf{k}'n'\Downarrow\right\rangle\right|^2$$

$$\times\left\{\left[n_{\mathbf{q}v} - f_{\mathbf{k}'n'}^0 + 1\right]\delta\left(\varepsilon_{\mathbf{k}n} - \varepsilon_{\mathbf{k}'n'} - \hbar\omega_{\mathbf{q}v}\right)\right.$$

$$\left.+ \left[n_{\mathbf{q}v} + f_{\mathbf{k}'n'}^0\right]\delta\left(\varepsilon_{\mathbf{k}n} - \varepsilon_{\mathbf{k}'n'} + \hbar\omega_{\mathbf{q}v}\right)\right\}. \quad (1.137)$$

The electronic bands are described by the energies $\varepsilon_{\mathbf{k}n}$ of the state with momentum $\mathbf{k}$ and band index $n$. Phonon frequencies are $\omega_{\mathbf{q}v}$, for the phonon of momentum $\mathbf{q}$ and polarization $v$; similarly for the phonon polarization vector $\boldsymbol{\varepsilon}_{\mathbf{q}v}$. We further denoted by $M$ the atomic mass, by $N$ the number of atoms in the lattice, and by $\boldsymbol{\nabla}V$ the gradient of the electron-lattice ion potential. The equilibrium phonon occupation numbers are denoted as $n_{\mathbf{q}v}$, given as

$$n_{\mathbf{q}v} = n(\omega_{\mathbf{q}v}) = \frac{1}{e^{\beta\hbar\omega_{\mathbf{q}v}} - 1}. \quad (1.138)$$

The electronic states $|\mathbf{k}n\sigma\rangle$ are normalized to the whole space.

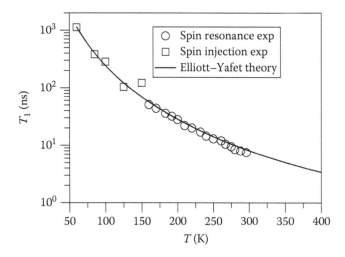

**FIGURE 1.8**  Spin relaxation in silicon. Phonon-induced spin relaxation in silicon results in an approximate $T^3$ power law [15]. Shown are the experimental data from spin resonance [16] and spin injection [17] experiments, and a calculation based on the Elliott–Yafet mechanism [15].

Two types of processes contribute to the phonon-induced spin flips. First, what we call the *Elliott processes* are the Elliott-type of spin flips in which the Bloch states given by Equations 1.125 and 1.126 scatter by the scalar part of the gradient of the electron–ion potential $V$. Second, what we call the *Yafet processes* are the spin flips due to the gradient of the spin–orbit part of the electron–ion potential. These two processes are typically of similar order of magnitude and have to be added coherently in order to obtain $T_1$.

Figure 1.8 shows the experimental data of the spin relaxation in intrinsic (non-degenerate) silicon, obtained by the spin resonance [8, 16] and the spin injection [17] experiments. The calculation based on the Elliott–Yafet mechanism of the phonon-induced spin flips reproduces the experiments [15].

### 1.3.2.1 Yafet Relation

The expected temperature dependence of the phonon-induced spin flips in degenerate conductors, is, according to the Elliott–Yafet mechanism, $1/T_1 \sim T$ at high temperatures (starting roughly at a fraction of the Debye temperature $T_D$) and $1/T_1 \sim T^3$ at low temperatures, in analogy with the conventional spin-conserving electron–phonon scattering. The high-temperature dependence originates from the linear increase of the phonon occupation numbers $n$ with increasing temperature: $n_{q\nu} \sim T$, as $k_B T \gg \omega_{q\nu}$. At $T > T_D$ are all the phonons excited. The low-temperature dependence of the spin-conserving scattering follows from setting the relevant phonon energy scale to $k_B T$. The matrix element

$$\left\langle \mathbf{k}n \Uparrow \middle| \boldsymbol{\nabla} V \middle| \mathbf{k}'n' \Uparrow \right\rangle \sim q, \tag{1.139}$$

which gives the $1/T_1 \sim T^3$ dependence. However, Yafet showed [2] that for the spin-flip matrix element the space inversion symmetry modifies the momentum dependence to

$$\left\langle \mathbf{k}n \Uparrow \left| \boldsymbol{\nabla} V \right| \mathbf{k}'n' \Downarrow \right\rangle \sim q^2, \tag{1.140}$$

so that

$$\frac{1}{T_1} \sim T^5, \tag{1.141}$$

instead of the expected $T^3$. Since the same temperature dependence holds for the phonon-induced electrical resistance $\rho(T)$, we can write

$$\frac{1}{T_1} \sim \rho(T), \tag{1.142}$$

known as the *Yafet relation*.

The relation Equation 1.140 holds if both the Elliott and Yafet processes are taken into account. Individually, they would lead to a linear dependence on $q$. The quantum mechanical interference between these two processes thus significantly reduces the spin-flip probability at low momenta $q$. An example is shown in Figure 1.9. The Elliott and Yafet processes would individually give much stronger spin relaxation than is observed. Their destructive interference can modify $T_1$ by orders of magnitude.

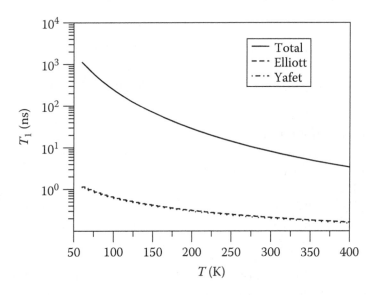

**FIGURE 1.9**    Interference between the Elliott and Yafet processes. Individually the Elliott and Yafet phonon-induced spin relaxation processes give spin relaxation orders of magnitude stronger than the total one. This example is from the calculation of $T_1$ in silicon [15].

# 1.4 RESULTS BASED ON KINETIC-SPIN-BLOCH-EQUATION APPROACH

It was shown by Wu et al. from a full microscopic kinetic-spin-Bloch-equation approach [9] that the single-particle approach is inadequate in accounting for the spin relaxation/dephasing both in the time [18–22] and in the space [23–25] domains. The momentum dependence of the effective magnetic field (the D'yakonov–Perel' term) and the momentum dependence of the spin diffusion rate along the spacial gradient [23] or even the random spin–orbit coupling [26] all serve as inhomogeneous broadenings [19, 20]. It was pointed out that in the presence of inhomogeneous broadening, any scattering, including the carrier–carrier Coulomb scattering, can cause an irreversible spin relaxation/dephasing [19]. Moreover, besides the spin relaxation/dephasing channel the scattering provides, it also gives rise to the counter effect to the inhomogeneous broadening. The scattering tends to drive carriers to a more homogeneous state and therefore suppresses the inhomogeneous broadening. Finally, this approach is valid in both strong and weak scattering limits and also can be used to study systems far away from the equilibrium, thanks to the inclusion of the Coulomb scattering.

In the following subsection, we present the main results based on kinetic-spin-Bloch-equation approach. We first briefly introduce the kinetic spin Bloch equations. Then we review the results of the spin relaxation/dephasing in the time and space domains, respectively.

## 1.4.1 KINETIC SPIN BLOCH EQUATIONS

By using the nonequilibrium Green function method with gradient expression as well as the generalized Kadanoff–Baym ansatz [27], we construct the kinetic spin Bloch equations as follows:

$$\dot{\rho}_{\mathbf{k}}(\mathbf{r},t) = \dot{\rho}_{\mathbf{k}}(\mathbf{r},t)\big|_{\mathrm{dr}} + \dot{\rho}_{\mathbf{k}}(\mathbf{r},t)\big|_{\mathrm{dif}} + \dot{\rho}_{\mathbf{k}}(\mathbf{r},t)\big|_{\mathrm{coh}} + \dot{\rho}_{\mathbf{k}}(\mathbf{r},t)\big|_{\mathrm{scat}}. \tag{1.143}$$

Here $\rho_{\mathbf{k}}(\mathbf{r},t)$ are the density matrices of electrons with momentum $\mathbf{k}$ at position $\mathbf{r}$ and time $t$. The off-diagonal elements of $\rho_{\mathbf{k}}$ represent the correlations between the conduction and valence bands, different subbands (in confined structures), and different spin states. $\dot{\rho}_{\mathbf{k}}(\mathbf{r},t)\big|_{\mathrm{dr}}$ are the driving terms from the external electric field. The coherent terms, in Equation 1.143, $\dot{\rho}_{\mathbf{k}}\big|_{\mathrm{coh}}$ are composed of the energy spectrum, magnetic field, and effective magnetic field from the D'yakonov–Perel' term and the Coulomb Hartree–Fock terms. The diffusion terms $\dot{\rho}_{\mathbf{k}}(\mathbf{r},t)\big|_{\mathrm{dif}}$ come from the spacial gradient. The scattering terms $\dot{\rho}_{\mathbf{k}}(\mathbf{r},t)\big|_{\mathrm{scat}}$ include the spin-flip and spin-conserving electron–electron, electron–phonon, and electron-impurity scatterings. The spin-flip terms correspond to the Elliot–Yafet and/or Bir–Aronov–Pikus mechanisms. Detailed expressions of these terms in the kinetic spin Bloch equations depend on the band structures, doping situations, and dimensionalities [9] and can be found in the literature for different cases, such as intrinsic quantum wells [18], $n$-type quantum wells with [21, 28–31] and without [22, 32–34] electric field, $p$-type quantum wells [35–37], quantum wires [38, 39], quantum dots [40], and bulk materials [41] in the spacial uniform case and quantum wells in spacial nonuniform case [23, 24, 42, 43]. By numerically solving the kinetic spin Bloch equations with all the scattering explicitly included, one is able to obtain the time evolution and/or spacial distribution of the density matrices, and hence all the measurable quantities, such as mobility, diffusion constant, optical relaxation/dephasing time, spin relaxation/dephasing time, spin diffusion length, as well as hot-electron temperature, can be determined from the theory without any fitting parameters.

## 1.4.2 SPIN RELAXATION/DEPHASING

In this subsection we present the main understandings of the spin relaxation/dephasing added to the literature from the kinetic-spin-Bloch-equation approach. We focus on three related issues: (i) the importance of the Coulomb interaction to the spin relaxation/dephasing; (ii) spin dynamics far away from the equilibrium; and (iii) qualitatively different behaviors from those wildly adopted in the literature.

First we address the effect of the Coulomb interaction. Based on the single-particle approach, it has been long believed that the Coulomb scattering is irrelevant to the spin relaxation/dephasing [44]. It was first pointed out by Wu and Ning [19] that in the presence of inhomogeneous broadening in spin precession, that is, the spin precession frequencies are **k**-dependent, any scattering, including the spin-conserving scattering, can cause irreversible spin dephasing. This inhomogeneous broadening can come from the energy-dependent $g$-factor [19], the D'yakonov–Perel' term [20], the random spin–orbit coupling [26], and even the momentum dependence of the spin diffusion rate along the spacial gradient [23]. Wu and Ning first showed that with the energy-dependent $g$-factor as an inhomogeneous broadening, the Coulomb scattering can lead to irreversible spin dephasing [19]. In [001]-grown $n$-doped quantum wells, the importance of the Coulomb scattering for spin relaxation/dephasing was proved by Glazov and Ivchenko [45] by using perturbation theory and by Weng and Wu [21] from the kinetic-spin-Bloch-equation approach. In a temperature-dependent study of the spin dephasing in [001]-oriented $n$-doped quantum wells, Leyland et al. experimentally verified the effects of the electron–electron Coulomb scattering by closely measuring the momentum scattering rate from the mobility [46]. By showing the momentum relaxation rate obtained from the mobility cannot give the correct spin relaxation rate, they showed the difference comes from the Coulomb scattering. Later Zhou et al. even predicted a peak from the Coulomb scattering in the temperature dependence of the spin relaxation time in a high-mobility low-density $n$-doped (001) quantum well [29]. This was later demonstrated by Ruan et al. experimentally [47].

Figure 1.10 shows the temperature dependence of the spin relaxation time of a 7.5 nm $GaAs/Al_{0.4}Ga_{0.6}As$ quantum well at different electron and impurity densities [29]. For this small well width, only the lowest subband is needed in the calculation. It is shown in the figure that when the

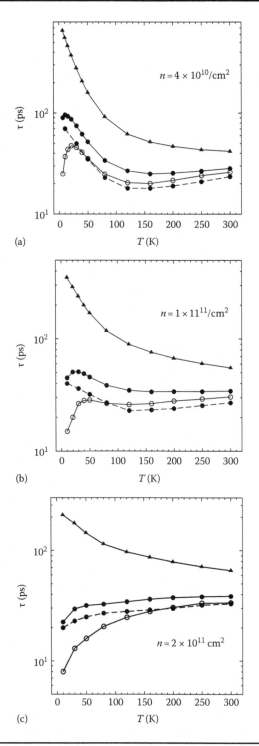

**FIGURE 1.10** Spin relaxation time $\tau$ versus the temperature $T$ with well width $a = 7.5$ nm and electron density $n$ being (a) $4 \times 10^{10}$ cm$^{-2}$, (b) $1 \times 10^{11}$ cm$^{-2}$, and (c) $2 \times 10^{11}$ cm$^{-2}$, respectively. Solid curves with triangles: impurity density $n_i = n$; solid curves with dots: $n_i = 0.1n$; solid curves with circles: $n_i = 0$; dashed curves with dots: $n_i = 0.1n$ and no Coulomb scattering. (After Zhou, J. et al., *Phys. Rev. B* 75, 045305, 2007. With permission.)

electron-impurity scattering is dominant, the spin relaxation time decreases with increasing temperature monotonically. This is in good agreement with the experimental findings [48] and a nice agreement of the theory and the experimental data from 20 to 300 K is given in Ref. [29]. However, it is shown that for sample with high mobility, that is, low-impurity density, when the electron density is low enough, there is a peak at low temperature. This peak, located around the Fermi temperature of electrons $T_F^e = E_F / k_B$, is identified to be solely due to the Coulomb scattering [29, 49]. It disappears when the Coulomb scattering is switched off, as shown by the dashed curves in the figure. This peak also disappears at high impurity densities. It is also noted in Figure 1.10c that for electrons of high density so that $T_F^e$ is high enough and the contribution from the electron–longitudinal optical-phonon scattering becomes marked, the peak disappears even for sample with no impurity and the spin relaxation time increases with temperature monotonically. The physics leading to the peak is due to the crossover of the Coulomb scattering from the degenerate to the non-degenerate limit. At $T < T_F^e$, electrons are in the degenerate limit and the electron–electron scattering rate $1/\tau_{ee} \propto T^2$. At $T > T_F^e$, $1/\tau_{ee} \propto T^{-1}$ [45, 50]. Therefore, at low electron density so that $T_F^e$ is low enough and the electron-acoustic phonon scattering is very weak compared with the electron–electron Coulomb scattering, the Coulomb scattering is dominant for high mobility sample. Hence, the different temperature dependence of the Coulomb scattering leads to the peak. It is noted that the peak is just a feature of the crossover from the degenerate to the non-degenerate limit. The location of the peak also depends on the strength of the inhomogeneous broadening. When the inhomogeneous broadening depends on momentum linearly, the peak tends to appear at the Fermi temperature. A similar peak was predicted in the electron spin relaxation in $p$-type GaAs quantum well and the hole spin relaxation in (001) strained asymmetric Si/SiGe quantum well, where the electron and hole spin relaxation times both show a peak at the hole Fermi temperature $T_F^h$ [36, 37]. When the inhomogeneous broadening depends on momentum cubically, the peak tends to shift to a lower temperature. It was predicted that a peak in the temperature dependence of the electron spin relaxation time appears at a temperature in the range of $(T_F^e / 4, T_F^e / 2)$ in the intrinsic bulk GaAs [41] and a peak in the temperature dependence of the hole spin relaxation time at $T_F^h / 2$ in $p$-type Ge/SiGe quantum well [37]. Ruan et al. demonstrated the peak experimentally in a high-mobility low-density GaAs/Al$_{0.35}$Ga$_{0.65}$As heterostructure and showed a peak appears at $T_F^e / 2$ in the spin relaxation time versus temperature curve [47].

For larger well width, the situation may become different in the non-degenerate limit. Weng and Wu calculated the spin relaxation/dephasing for (001) GaAs quantum wells with larger well width and high mobility by including the multi-subband effect [28]. It is shown that for small/large well width so that the linear/cubic Dresselhaus term is dominant, the spin relaxation/dephasing time increases/decreases with the temperature. This is because with the increase of temperature, both the inhomogeneous broadening and

the scattering get enhanced. The relative importance of these two competing effects is different when the linear/cubic term is dominant [28]. Jiang
and Wu further introduced strain to change the relative importance of the
linear and cubic D'yakonov–Perel' terms and showed the different temperature dependences of the spin relaxation time [51]. This prediction has been
realized experimentally by Holleitner et al. where they showed that in $n$-type
two-dimensional InGaAs channels, when the linear D'yakonov–Perel' term
is suppressed, the spin relaxation time decreases with temperature monotonically [52]. Another interesting prediction related to the multi-subband
effect is related to the effect of the inter-subband Coulomb scattering. From
the calculation Weng and Wu found out that although the inhomogeneous
broadening from the higher subband of the (001) quantum well is much
larger, due to the strong inter-subband Coulomb scattering, the spin relaxation times of the lowest two subbands are identical [28]. This prediction
has later been verified experimentally by Zhang et al., who studied the spin
dynamics in a single-barrier heterostructure by time-resolved Kerr rotation
[53]. By applying a gate voltage, they effectively manipulated the confinement of the second subband, and the measured spin relaxation times of the
first and second subbands are almost identical at large gate voltage. Lü et al.
showed that due to the Coulomb scattering, $T_2 = T_2^*$ in (001) GaAs quantum
wells for a wide temperature and density regime [54]. It was also pointed
out by Lü et al. that in the strong (weak) scattering limit, introducing the
Coulomb scattering will always lead to a faster (slower) spin relaxation/
dephasing [35].

Another important effect from the Coulomb interaction to the spin
relaxation/dephasing comes from the Coulomb Hartree–Fock contribution
in the coherent terms of the kinetic spin Bloch equations. Weng and Wu
[21] first pointed out that at a high spin polarization, the Hartree–Fock term
serves as an effective magnetic field along the $z$-axis, which blocks the spin
precession. As a result, the spin relaxation/dephasing time increases dramatically with the spin polarization. They further pointed out that the spin relax-
ation/dephasing time decreases with temperature at high spin polarization
in quantum well with small well width, which is in contrast to the situation
with small spin polarizations. These predictions have been verified experimentally by Stich et al. in an $n$-type (001) GaAs quantum well with high
mobility [55, 56]. By changing the intensity of the circularly polarized lasers,
Stich et al. measured the spin dephasing time in a high mobility $n$-type GaAs
quantum well as a function of initial spin polarization. Indeed they observed
an increase of the spin dephasing time with the increased spin polarization,
and the theoretical calculation based on the kinetic spin Bloch equations
nicely reproduced the experimental findings when the Hartree–Fock term
was included [55]. It was also shown that when the Hartree–Fock term is
removed, one does not see any increase of the spin dephasing time. Later,
they further improved the experiment by replacing the circular-polarized
laser pumping with the elliptic polarized laser pumping. By doing so, they
were able to vary the spin polarization without changing the carrier density.
Figure 1.11 shows the measured spin dephasing times as function of initial

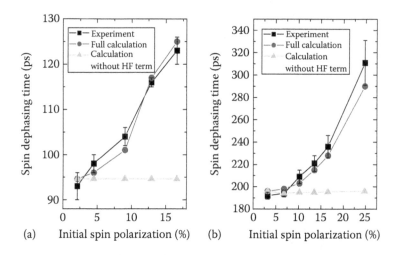

**FIGURE 1.11**    (a) The spin dephasing times as a function of initial spin polarization for constant, *low* excitation density and variable polarization degree of the pump beam. The measured spin dephasing times are compared to calculations with and without the Hartree–Fock (HF) term, showing its importance. (b) The spin dephasing times measured and calculated for constant, *high* excitation density and variable polarization degree. (After Stich, J. et al., *Phys. Rev. B 76*, 205301, 2007. With permission.)

spin polarization under two fixed pumping intensities, together with the theoretical calculations with and without the Coulomb Hartree–Fock term. Again the spin dephasing time increases with the initial spin polarization as predicted and the theoretical calculations with the Hartree–Fock term are in good agreement with the experimental data [56]. Moreover, Stich et al. also confirmed the prediction of the temperature dependences of the spin dephasing time at low and high spin polarizations [56]. Figure 1.12a shows

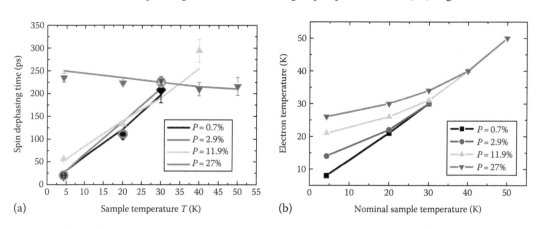

**FIGURE 1.12**    (a) Spin dephasing time as a function of sample temperature, for different initial spin polarizations. The measured data points are represented by solid points, while the calculated data are represented by lines of the same grayscale. (b) Electron temperature determined from intensity-dependent photoilluminance measurements as a function of the nominal sample temperature, for different pump beam fluence and initial spin polarization, under experimental conditions corresponding to the measurements shown in (a). The measured data points are represented by solid points, while the curves serve as guide to the eye. (After Stich, J. et al., *Phys. Rev. B 76*, 205301, 2007. With permission.)

the measured temperature dependences of the spin dephasing time at different initial spin polarizations. As predicted, the spin dephasing time increases with increasing temperature at small spin polarization but decreases at large spin polarization. The theoretical calculations also nicely reproduced the experimental data. The hot-electron temperatures in the calculation were taken from the experiment (Figure 1.12b). The effective magnetic field from the Hartree–Fock term has been measured by Zhang et al. from the sign switch of the Kerr signal and the phase reversal of Larmor precessions with a bias voltage in a GaAs heterostructure [57]. Korn et al. [58] also estimated the average effect by applying an external magnetic field in the Faraday configuration, as shown in Figure 1.13a, for the same sample reported earlier [55, 56]. They compared the spin dephasing times of both large and small spin polarizations as function of external magnetic field. Due to the effective magnetic field from the Hartree–Fock term, the spin relaxation times are different under small external magnetic field but become identical when the magnetic field becomes large enough. From the merging point, they estimated that the mean value of the effective magnetic field is below 0.4 T. They further showed that this effective magnetic field from the Hartree–Fock term cannot be compensated by the external magnetic field, because it does not break the time-reversal symmetry and is therefore not a genuine magnetic field, as said earlier. This can be seen from Figure 1.13b that the spin relaxation time at large spin polarization shows identical external magnetic field dependences when the magnetic field is parallel or antiparallel to the growth direction.

We now turn to discuss the spin relaxation/dephasing far away from the equilibrium. In fact, the spin relaxation/dephasing of high spin polarization addressed earlier is one of the cases far away from the equilibrium. Another case is the spin dynamics in the presence of a high in-plane electric field. The spin dynamics in the presence of a high in-plane electric field was first studied by Weng et al. [22] in GaAs quantum well with only the lowest subband by solving the kinetic spin Bloch equations. To avoid the "runaway" effect

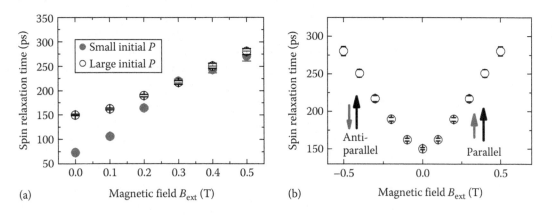

**FIGURE 1.13** (a) Spin dephasing times as a function of an external magnetic field perpendicular to the quantum well plane for small and large initial spin polarization. (b) Same as (a) for large initial spin polarization and both polarities of the external magnetic field. (After Korn, T. et al., *Adv. Solid State Phys.* 48, 143, 2009. With permission.)

[59], the electric field was calculated up to 1 kV/cm. Then, Weng and Wu further introduced the second subband into the model and the in-plane electric field was increased up to 3 kV/cm [28]. Zhang et al. included $L$ valley and the electric field was further increased up to 7 kV/cm [32]. The effect of in-plane electric field to the spin relaxation in system with strain was investigated by Jiang and Wu [51]. Zhou et al. also investigated the electric-field effect at low lattice temperatures [29].

The in-plane electric field leads to two effects: (a) It shifts the center of mass of electrons to $\mathbf{k}_d = m^* \mathbf{v}_d = m^* \mu \mathbf{E}$ with $\mu$ representing the mobility, which further induces an effective magnetic field via the D'yakonov–Perel' term [22]; (b) the in-plane electric field also leads to the hot-electron effect [60]. The first effect induces a spin precession even in the absence of any external magnetic field and the spin precession frequency changes with the direction of the electric field in the presence of an external magnetic field [22, 32]. The second effect enhances both the inhomogeneous broadening and the scattering, two competing effects leading to rich electric-field dependence of the spin relaxation/dephasing and thus spin manipulation [22, 28, 29, 32, 33, 41, 51].

Finally we address some issues of which the kinetic-spin-Bloch-equation approach gives qualitatively different predictions from those widely used in the literature. These issues include the Bir–Aronov–Pikus mechanism, the Elliot–Yafet mechanism, and some density/temperature dependences of the spin relaxation/dephasing time.

It has long been believed in the literature that for electron relaxation/dephasing, the Bir–Aronov–Pikus mechanism is dominant at low temperature in $p$-type samples and has important contribution to intrinsic sample with high photo-excitation [61–67]. This conclusion was made based on the single-particle Fermi golden rule. Zhou and Wu reexamined the problem using the kinetic-spin-Bloch-equation approach [30]. They pointed out that the Pauli blocking was overlooked in the Fermi golden rule approach. When electrons are in the non-degenerate limit, the results calculated from the Fermi golden rule approach are valid. However, at low temperature, electrons can be degenerate and the Pauli blocking becomes very important. As a result, the previous approaches always overestimated the importance of the Bir–Aronov–Pikus mechanism at low temperature. Moreover, the previous single-particle theories underestimated the contribution of the D'yakonov–Perel' mechanism by neglecting the Coulomb scattering. Both made the Bir–Aronov–Pikus mechanism dominate the spin relaxation/dephasing at low temperature. Later, Zhou et al. performed a thorough investigation of electron spin relaxation in $p$-type (001) GaAs quantum wells by varying impurity, hole and photoexcited electron densities over a wide range of values [36], under the idea that very high impurity density and very low photoexcited electron density may effectively suppress the importance of the D'yakonov–Perel' mechanism and the Pauli blocking. Then the relative importance of the Bir–Aronov–Pikus and D'yakonov–Perel' mechanisms may be reversed. This indeed happens as shown in the phase-diagram-like picture in Figure 1.14 where the relative importance

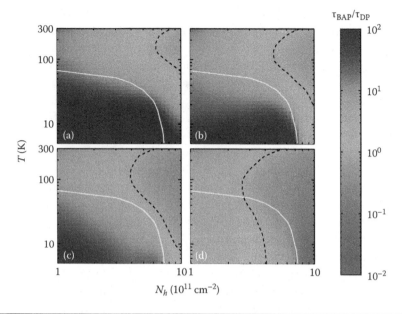

**FIGURE 1.14** Ratio of the spin relaxation time due to the Bir–Aronov–Pikus mechanism to that due to the D'yakonov–Perel' mechanism, $\tau_{BAP}/\tau_{DP}$, as function of temperature and hole density with (a) $N_i = 0$, $N_{ex} = 10^{11}$ cm$^{-2}$; (b) $N_i = 0$, $N_{ex} = 10^9$ cm$^{-2}$; (c) $N_i = N_h$, $N_{ex} = 10^{11}$ cm$^{-2}$; (d) $N_i = N_h$, $N_{ex} = 10^9$ cm$^{-2}$. The black dashed curves indicate the cases satisfying $\tau_{BAP}/\tau_{DP} = 1$. Note the smaller the ratio $\tau_{BAP}/\tau_{DP}$ is, the more important the Bir–Aronov–Pikus mechanism becomes. The yellow solid curves indicate the cases satisfying $\partial_{\mu_h}[N_{LH^{(1)}} + N_{HH^{(2)}}]/\partial_{\mu_h}N_h = 0.1$. In the regime above the yellow curve the multi-hole-subband effect becomes significant. (After Zhou, Y. et al., *New J. Phys.* 11, 113039, 2009. With permission.)

of the Bir–Aronov–Pikus and D'yakonov–Perel' mechanisms is plotted as function of hole density and temperature at low and high impurity densities and photo-excitation densities. For the situation of high hole density, they even included multi-hole subbands as well as the light hole band. It is interesting to see from the figures that at relatively high photo-excitations, the Bir–Aronov–Pikus mechanism becomes more important than the D'yakonov–Perel' mechanism only at high hole densities and high temperatures (around hole Fermi temperature) when the impurity is very low (zero in Figure 1.14a). Impurities can suppress the D'yakonov–Perel' mechanism and hence enhance the relative importance of the Bir–Aronov–Pikus mechanism. As a result, the temperature regime is extended, ranging from the hole Fermi temperature to the electron Fermi temperature for high hole density. When the photo-excitation is weak so that the Pauli blocking is less important, the temperature regime where the Bir–Aronov–Pikus mechanism is important becomes wider compared to the high excitation case. In particular, if the impurity density is high enough and the photo-excitation is so low that the electron Fermi temperature is below the lowest temperature of the investigation, the Bir–Aronov–Pikus mechanism can dominate the whole temperature regime of the investigation at sufficiently high hole density, as shown in Figure 1.14d. The corresponding spin relaxation times of each mechanism under high or low-impurity and photo-excitation densities are demonstrated in Figure 1.14. They also discussed the density dependences of spin relaxation with some intriguing properties related to the high hole subbands [36]. The predicted Pauli-blocking effect in the Bir–Aronov–Pikus mechanism has been partially demonstrated experimentally by Yang et al. [68]. They showed by increasing the pumping density that the temperature dependence of the spin dephasing time deviates from the one from the Bir–Aronov–Pikus mechanism and the peaks at high excitations agree well with those predicted by Zhou and Wu [30].

Another widely accepted but incorrect conclusion is related to the Elliot–Yafet mechanism. It is widely accepted in the literature that the Elliot–Yafet mechanism dominates spin relaxation in $n$-type bulk III–V semiconductor at low temperature, while the D'yakonov–Perel' mechanism is important at high temperature [5, 7, 8, 69, 70]. Jiang and Wu pointed out that the previous understanding is based on the formula that can only be used in the non-degenerate limit. Moreover, the momentum relaxation rates are calculated via the approximated formula for mobility [69]. By performing an accurate calculation via the kinetic-spin-Bloch-equation approach, they showed that the Elliot–Yafet mechanism is *not* important in III–V semiconductors, including even the narrow-band InAs and InSb [41]. Therefore, the D'yakonov–Perel' mechanism is the only spin relaxation mechanism for $n$-type III–V semiconductors in metallic regime.

Jiang and Wu have further predicted a peak in the density dependence of the spin relaxation/dephasing time in $n$-type III–V semiconductors where the spin relaxation/dephasing is limited by the D'yakonov–Perel' mechanism [41]. Previously, the non-monotonic density dependence of spin lifetime was observed in low-temperature ($T \lesssim 5$ K) experiments, where the localized electrons play a crucial role and the electron system is in the insulating regime or around the metal-insulator transition point [71]. Jiang and Wu found, for the first time, that the spin lifetime in *metallic* regime is also *non-monotonic*. Moreover, they pointed out that it is a *universal* behavior for *all* bulk III–V semiconductors at *all* temperatures where the peak is located at $T_F \sim T$ with $T_F$ being the electron Fermi temperature. The underlying physics for the non-monotonic density dependence in metallic regime can be understood as following: In the non-degenerate regime, as the distribution is the Boltzmann one, the density dependence of the inhomogeneous broadening is marginal. However, the scattering increases with the density. Consequently, the spin relaxation/dephasing time increases with the density. However, in the degenerate regime, due to the Fermi distribution, the inhomogeneous broadening increases with the density much faster than the scattering does. As a result, the spin relaxation/dephasing time decreases with the density. Similar behavior was also found in two-dimensional system [31, 37] where the underlying physics is similar. The predicted peak was later observed by Krauß et al. [72], as shown in Figure 1.15 where theoretical calculation based on the kinetic spin Bloch equations nicely reproduced the experimental data by Shen [73].

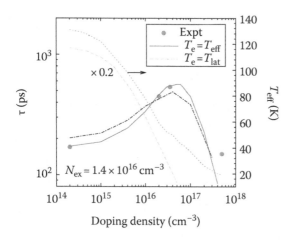

**FIGURE 1.15** Electron spin relaxation times from the calculation via the kinetic-spin-Bloch-equation approach (red solid curve) and from the experiment [72] (blue ●) as function of the doping density. The green dashed curve shows the results without hot-electron effect. The hot-electron temperature used in the computation is plotted as the blue dotted curve. (Note the scale is on the right-hand side of the frame.) The chain curve is the calculated spin relaxation time with a fixed hot-electron temperature 80 K, and $N_{ex} = 6 \times 10^{15}$ cm$^{-3}$. (After Shen, K., *Chin. Phys. Lett.* 26, 067201, 2009. With permission.)

### 1.4.3 SPIN DIFFUSION/TRANSPORT

By solving the kinetic spin Bloch equations together with the Poisson equation self-consistently, one is able to obtain all the transport properties such as the mobility, charge diffusion length, and spin diffusion/injection length without any fitting parameter. It was first pointed out by Weng and Wu [23] that the drift-diffusion equation approach is inadequate in accounting for the spin diffusion/transport. It is important to include the off-diagonal term between opposite spin bands $\rho_{k\uparrow\downarrow}$ in studying the spin diffusion/transport. With this term, electron spin precesses along the diffusion and therefore $\mathbf{k}\cdot\nabla_{\mathbf{r}}\rho_{\mathbf{k}}(\mathbf{r},t)$ in the diffusion terms $\dot{\rho}_{\mathbf{k}}(\mathbf{r},t)|_{\text{dif}}$ provides an additional inhomogeneous broadening. With this additional inhomogeneous broadening, any scattering, including the Coulomb scattering, can cause an irreversible spin relaxation/dephasing [23]. Unlike the spin precession in the time domain where the inhomogeneous broadening is determined by the effective magnetic field from the D'yakonov–Perel' term, $\mathbf{h}(\mathbf{k})$, in spin diffusion and transport it is determined by

$$\mathbf{\Omega_k} = \frac{|\,g\mu_B\mathbf{B} + \mathbf{h}(\mathbf{k})\,|}{k_x}, \qquad (1.144)$$

provided the diffusion is along the $x$-axis [42]. Here, the magnetic field is in the Voigt configuration. Therefore, even in the absence of the D'yakonov–Perel' term $\mathbf{h}(\mathbf{k})$, the magnetic field *alone* can provide an inhomogeneous broadening and leads to the spin relaxation/dephasing in spin diffusion and transport. This was first pointed out by Weng and Wu back to 2002 [23] and has been realized experimentally by Appelbaum et al. in bulk silicon [74, 75], where there is no D'yakonov–Perel' spin–orbit coupling due to the center inversion symmetry. Zhang and Wu further investigated the spin diffusion and transport in symmetric Si/SiGe quantum wells [43].

When $B = 0$ but the D'yakonov–Perel' term is present, then the inhomogeneous broadening for spin diffusion and transport is determined by $\mathbf{\Omega_k} = \mathbf{h(k)}/k_x$. In (001) GaAs quantum well where the D'yakonov–Perel' term is determined by the Dresselhaus term [76], the average of $\mathbf{\Omega_k}$ reads $\langle\mathbf{\Omega_k}\rangle = C\left(\langle k_y^2\rangle - \langle k_z^2\rangle, 0, 0\right)$ with $C$ being a constant. For electrons in quantum well, this value is not zero. Therefore, the spacial spin oscillation due to the Dresselhaus effective magnetic field survives even at high temperature when the scattering is strong. This effect was first predicted by Weng and Wu by showing that a spin pulse can oscillate along the diffusion in the absence of the magnetic field at very high temperature [24]. Detailed studies were carried out later on this effect [25, 42, 77]. The spin oscillation without any applied magnetic field in the transient spin transport was later observed experimentally by Crooker and Smith in strained bulk system [78]. Differing from the two-dimensional case, in bulk, the average of $\mathbf{\Omega_k}$ from the Dresselhaus term is zero, since $\langle\mathbf{\Omega_k}\rangle = C\left(\langle k_y^2\rangle - \langle k_z^2\rangle, 0, 0\right) = 0$ due to the symmetry in the $y$- and $z$-directions. This is consistent with the experimental result that there is no spin oscillation for the system without stress. However, when the stress is applied, an additional spin–orbit coupling, namely, the coupling of electron spins to the strain tensor, appears, which is linear in momentum [5]. This additional spin–orbit coupling also acts as an effective magnetic field. Therefore, once the stress is applied, one can observe spacial spin oscillation even when there is no applied magnetic field [78].

Cheng and Wu further developed a new numerical scheme to calculate the spin diffusion/transport in GaAs quantum wells with very high accuracy and speed [42]. It was discovered that due to the scattering, especially the Coulomb scattering, $T_2 = T_2^*$ is valid even in the spacial domain. This prediction remains yet to be verified experimentally. Moreover, as the inhomogeneous broadening in spin diffusion is determined by $|\mathbf{h(k)}|/k_x$ in the absence of magnetic field, the period of the spin oscillations along the $x$-axis is independent of the electric field perpendicular to the growth direction of the quantum well [42], which is different from the spin precession rate in the time domain [22]. This is consistent with the experimental findings by Beck et al. [79].

Cheng et al. applied the kinetic-spin-Bloch-equation approach to study the spin transport in the presence of competing Dresselhaus and Rashba fields [80]. When the Dresselhaus and Bychkov–Rashba [10] terms are both important in semiconductor quantum well, the total effective magnetic field can be highly anisotropic and spin dynamics is also highly anisotropic in regard to the spin polarization [81]. For some special polarization direction, the spin relaxation time is extremely large [81–84]. For example, if the coefficients of the linear Dresselhaus and Bychkov–Rashba terms are equal to each other in (001) quantum well of small well width and the cubic Dresselhaus term is not important, the effective magnetic field is along the

[110] direction for all electrons. For the spin components perpendicular to the [110] direction, this effective magnetic field flips the spin and leads to a finite spin relaxation/dephasing time. For spin along the [110] direction, this effective magnetic field cannot flip it. Therefore, when the spin polarization is along the [110] direction, the Dresselhaus and Bychkov–Rashba terms cannot cause any spin relaxation/dephasing. When the cubic Dresselhaus term is taken into account, the spin dephasing time for spin polarization along the [110] direction is finite but still much larger than other directions [85]. The anisotropy in the spin direction is also expected in spin diffusion and transport. When the Dresselhaus and Bychkov–Rashba terms are comparable, the spin injection length $L_d$ for the spin polarization perpendicular to [110] direction is usually much shorter than that for the spin polarization along [110] direction. In the ideal case when there are only the linear Dresselhaus and Bychkov–Rashba terms with identical strengths, spin injection length for spin polarization parallel to the [110] direction becomes infinity [82, 83]. This effect has promoted Schliemann et al. to propose the nonballistic spin-field-effect transistor [83]. In such a transistor, a gate voltage is used to tune the strength of the Bychkov–Rashba term and therefore control the spin injection length. However, Cheng et al. pointed out that spin diffusion and transport actually involve both the spin polarization and spin transport directions [80]. The latter has long been overlooked in the literature. In the kinetic-spin-Bloch-equation approach, this direction corresponds to the spacial gradient in the diffusion term $[\dot{\rho}_k(\mathbf{r},t)|_{\text{dif}}]$ and the electric field in the drifting term $[\dot{\rho}_k(\mathbf{r},t)|_{\text{dr}}]$. The importance of the spin transport direction has not been realized until Cheng et al. pointed out that the spin transport is highly anisotropic not only in the sense of the spin polarization direction but also in the spin transport direction when the Dresselhaus and Bychkov–Rashba effective magnetic fields are comparable [80]. They even predicted that in (001) GaAs quantum well with identical linear Dresselhaus and Bychkov–Rashba coupling strengths, the spin injection along [$\bar{1}$10] or [110]* can be infinite *regardless of* the direction of the spin polarization. This can be easily seen from the inhomogeneous broadening Equation 1.144, which well defines the spin diffusion/transport properties. For the spin diffusion/transport in a (001) GaAs quantum well with identical Dresselhaus and Bychkov–Rashba strengths (the schematic is shown in Figure 1.16 with the transport direction chosen along the $x$-axis), the inhomogeneous broadening is given by [80]

$$\mathbf{\Omega}_k = \left\{ 2\beta \left( \sin\left(\theta - \frac{\pi}{4}\right) + \cos\left(\theta - \frac{\pi}{4}\right)\frac{k_y}{k_x} \right)\hat{\mathbf{n}}_0 \right.$$
$$\left. + \gamma \left( \frac{k_x^2 - k_y^2}{2}\sin 2\theta + k_x k_y \cos 2\theta \right)\left( \frac{k_y}{k_x}, -1, 0 \right) \right\}, \tag{1.145}$$

---

* [10] or [110] depends on the relative signs of the Dresselhaus and Rashba coupling strengths.

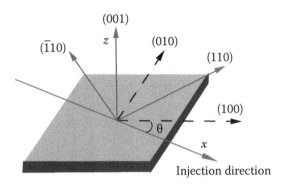

**FIGURE 1.16** Schematic of the different directions considered for the spin polarizations [(110), ($\bar{1}$10), and (001)-axes] and spin -diffusion/injection ($x$-axis). (After Cheng, J.L. et al., *Phys. Rev. B* 75, 205328, 2007. With permission.)

with $\theta$ being the angle between the spin transport direction ($x$-axis) and [001] crystal direction. It can be split into two parts: the zeroth-order term (on $k$), which is always along the same direction of $\hat{\mathbf{n}}_0$, and the second-order term, which comes from the cubic Dresselhaus term. If the cubic Dresselhaus term is omitted, the effective magnetic fields for all $\mathbf{k}$ states align along $\hat{\mathbf{n}}_0$ (crystal [110]) direction. Therefore, if the spin polarization is along $\hat{\mathbf{n}}_0$, there is no spin relaxation even in the presence of scattering since there is no spin precession. Nevertheless, it is interesting to see from Equation 1.145 that when $\theta = 3\pi/4$, that is, the spin transport is along the [$\bar{1}$10] direction, $\mathbf{\Omega_k} = 2m^*\beta\hat{\mathbf{n}}_0$ is independent of $\mathbf{k}$ if the cubic Dresselhaus term is neglected. Therefore, in this special spin transport direction, there is no inhomogeneous broadening in the spin transport for *any* spin polarization. The spin injection length is therefore infinite regardless of the direction of spin polarization. This result is highly counterintuitive, considering that the spin relaxation times for the spin components perpendicular to the effective magnetic field are finite in the spacial uniform system. The surprisingly contradictory results, that is, the finite spin relaxation/dephasing time versus the infinite spin injection length, are due to the difference in the inhomogeneous broadening in spacial uniform and nonuniform systems. For genuine situation, due to the presence of the cubic term, the spin injection length is still finite and the maximum spin injection length does not happen at the identical Dresselhaus and Bychkov–Rashba coupling strengths, but shifted by a small amount due to the cubic term [80]. However, there is strong anisotropy in regard to the spin polarization and spin injection direction, as shown in Figure 1.17. This predication has not yet been realized experimentally. However, very recent experimental findings on spin helix [13, 14] have provided strong evidence to support this predication [86].

Now we turn to the problem of spin grating. Transient spin grating, whose spin polarization varies periodically in real space, is excited optically by two non-collinear coherent light beams with orthogonal linear polarization [14, 87–89]. Transient spin grating technique can be used to study the spin transport since it can directly probe the decay rate of nonuniform spin

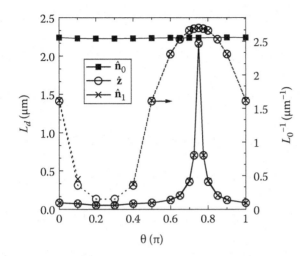

**FIGURE 1.17** Spin diffusion length $L_d$ (solid curves) and the inverse of the spin oscillation period $L_0^{-1}$ (dashed curves) for identical Dresselhaus and Rashba coupling strengths as functions of the injection direction for different spin polarization directions $\hat{n}_0$, $\hat{z}$ and $\hat{n}_1$ ($\hat{n}_1 = \hat{z} \times \hat{n}_0$, i.e., crystal direction $[\bar{1}10]$) at $T = 200$ K. It is noted that the scale of the spin oscillation period is on the right-hand side of the frame. (After Cheng, J.L. et al., *Phys. Rev. B* 75, 205328, 2007. With permission.)

distributions. Spin diffusion coefficient $D_s$ can be obtained from the transient spin grating experiments [87–89]. In the literature, the drift-diffusion model was employed to extract $D_s$ from the experimental data. With the drift-diffusion model, the transient spin grating was predicted to decay exponentially with time with a decay rate of $\Gamma_q = D_s q^2 + 1/\tau_s$, where $q$ is the wave-vector of the spin grating and $\tau_s$ is the spin relaxation time [87, 88]. However, this result is not accurate since it neglects the spin precession, which plays an important role in spin transport, as first pointed out by Weng and Wu [22]. Indeed, experimental results show that the decay of transient spin grating takes a double-exponential form instead of single exponential one [14, 87, 89]. Also the relation that relates the spin injection length with the spin diffusion coefficient $D_s$ and the spin relaxation time $\tau_s$, i.e., $L_s = 2\sqrt{D_s \tau_s}$ from the drift-diffusion model should be checked. In fact, if this relation is correct, the above prediction of infinite spin injection length at certain spin injection direction for any spin polarization in the presence of identical Dresselhaus and Bychkov–Rashba coupling strengths cannot be correct.

Weng et al. studied this problem from the kinetic-spin-Bloch-equation approach [90]. By first solving the kinetic spin Bloch equations analytically by including only the elastic scattering, that is, the electron-impurity scattering, they showed that the transient spin grating should decay double exponentially with two decay rates $\Gamma_\pm$. In fact, none of the rates is quadratic in $q$. However, the average of them reads [90]

$$\Gamma = \frac{(\Gamma_+ + \Gamma_-)}{2} = \frac{Dq^2 + 1}{\tau'_s} \qquad (1.146)$$

with $1/\tau'_s = (1/\tau_s + 1/\tau_{s1})/2$, which differs from the current widely used formula by replacing the spin decay rate by the average of the in- and out-of-plane relaxation rates. The difference of these two decay rates is a linear function of the wave-vector $q$ when $q$ is relatively large:

$$\Delta\Gamma = cq + d, \tag{1.147}$$

with $c$ and $d$ being two constants. The steady-state spin injection length $L_s$ and spin precession period $L_0$ are then [90]

$$L_s = \frac{2D_s}{\sqrt{|c^2 - 4D_s(1/\tau'_s - d)|}}, \tag{1.148}$$

$$L_0 = \frac{2D_s}{c}. \tag{1.149}$$

They further showed that the above relation Equations 1.146 and 1.147 are valid even including the inelastic electron–electron and electron-phonon scatterings by solving the full kinetic spin Bloch equations, as shown in Figure 1.18. A good agreement with the experimental data [89] of double exponential decays $\tau_\pm = \Gamma_\pm^{-1}$ is shown in Figure 1.19. Finally it was shown that the infinite spin injection length predicted by Cheng et al. in the presence of identical Dresselhaus and Bychkov–Rashba coupling strengths [80] addressed earlier can exactly be obtained from Equation 1.148 as in that special case $\tau'_s = \tau_s$, $d = 0$, and $c = 2\sqrt{D_s/\tau_s}$ [90]. However, $\sqrt{D_s\tau_s}$ always remains

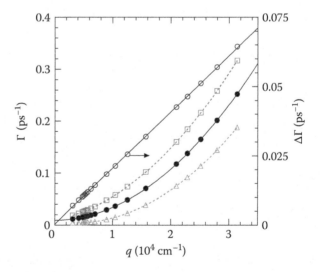

**FIGURE 1.18** $\Gamma = (\Gamma_+ + \Gamma_-)/2$ and $\Delta\Gamma = (\Gamma_+ + \Gamma_-)/2$ versus $q$ at $T = 295$ K. Open boxes/triangles are the relaxation rates ($\Gamma_{+/-}$ calculated from the full kinetic spin Bloch equations. Filled/open circles represent $\Gamma$ and $\Delta\Gamma$ respectively. Noted that the scale for $\Delta\Gamma$ is on the right-hand side of the frame. The solid curves are the fitting to $\Gamma$ and $\Delta\Gamma$, respectively. The dashed curves are guide to eyes. (After Weng, M.Q. et al., *J. Appl. Phys.* 103, 063714, 2008. With permission.)

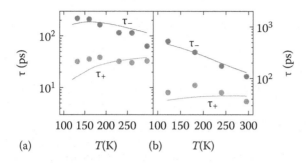

**FIGURE 1.19** Spin relaxation times $\tau_\pm$ versus temperature for (a) high-mobility sample with $q = 0.58 \times 10^4 \, cm^{-1}$ and (b) low-mobility sample with $q = 0.69 \times 10^4 \, cm^{-1}$. The dots are the experiment data from Ref. [90]. (After Weng, M.Q. et al., *J. Appl. Phys.* 103, 063714, 2008. With permission.)

finite unless $\tau_s = \infty$. Therefore, Equations 1.146 through 1.149 give the correct way to extract the spin injection length from the spin grating measurement.

## 1.5 SPIN RELAXATION IN GRAPHENE

Spin relaxation in graphene has a convoluted history with many ends still open [91]. Based on the mechanisms of spin relaxation described earlier in this chapter, our first guess would be that electrons in graphene lose their spins at scattering events with impurities and phonons. The spin-flip would be facilitated by spin-orbit coupling, native in graphene. This is the Elliott–Yafet mechanism. The other potential mechanism, Dyakonov–Perel, is nominally absent here, since graphene alone has space inversion symmetry and thus possesses no spin-orbit fields. Consequently, spin precession—which is needed for the Dyakonov–Perel mechanism to work—is absent. Since graphene is made of light carbon atoms, spin-orbit coupling in graphene is rather weak, about 10 μeV [92]. This intrinsic mechanism is expected to give spin relaxation times in the range of 10 ns–1 μs, depending on the specifics of electron-phonon coupling. The presence of phonons in the spin relaxation mechanism would also lead to a distinct temperature dependence of the spin relaxation rate. As is usual with the Elliott–Yafet mechanism, the spin relaxation rate would be also proportional to the momentum relaxation rate.

What about graphene on a substrate? Wouldn't the substrate break space inversion symmetry to induce large enough spin-orbit fields to relax electron spins efficiently according to the Dyakonov–Perel mechanism? In most experiments on spin relaxation, graphene is on substrates, typically $SiO_2$. Such a substrate could directly induce spin-orbit fields in graphene due to weak van der Waals hybridization, but also due to unintended charges on the substrate; the charges generate electric fields which couple with spin-orbit coupling to induce Rashba fields. The former effect has not been investigated in detail yet, but it is expected that spin-orbit fields due to hybridization of the substrate orbitals with graphene $p_z$ orbitals are no greater than 10 s of μeV. The effect of charges on a substrate was investigated in [93]. The outcome of this theoretical work is that spin relaxation due to substrates is relatively weak, giving the spin lifetimes of some microseconds. Also, since the substrate causes spin-orbit fields that lie in the plane (graphene sheet), the spin relaxation rate would be strongly anisotropic: out-of-plane spin would decay twice as fast as in-plane spins. This is a very distinct feature which can be probed experimentally. There have been many theoretical modelings of spin relaxation in graphene based on spin-orbit coupling [94–105], most giving spin relaxation times for realistic conditions in the 10 ns–μs range.

The outcomes of spin-injection experiments came as a surprise. In the first spin-injection experiment [106], the spin relaxation time in graphene was extracted to be about 100 ps [107], see also Chapter 4, Volume 3. The momentum relaxation time was also some 100 fs. Subsequent experiments [107–111] have confirmed this order of magnitude for $\tau_s$. The current status is that the spin relaxation times in graphene on common substrates and without special treatments, are 0.1–1 ns, essentially independent of temperature. This is orders of magnitude less than expected theoretically, based on spin-orbit coupling mechanisms, as discussed above. Recent experiments have also shown that the spin relaxation time is highly isotropic [112]. There is then little doubt that neither phonons (absent temperature dependence) nor spin-orbit fields due to substrates (absent anisotropy) cause spin relaxation of the experimental samples. This excludes both Elliott–Yafet and Dyakonov–Perel mechanisms. What, then, causes the observed ultrafast spin relaxation of electrons in graphene?

Thus far, the only theory that consistently explains experimental data on spin relaxation in graphene is resonant scattering of electrons off magnetic moments. Magnetic moments in graphene appear due to defects, such as vacancies or covalently bonded adatoms and admolecules [113]. There is no need for transition metal ions. Thus, it is expected that, say, polymer treatment of graphene devices will result in residual bonded organic molecules which will give a magnetic moment. It has been shown theoretically [114] that less than 1 ppm of hydrogen adatoms, for example, would lead to spin relaxation time of the order of 100 ps. The energy dependence of the spin relaxation rate agrees very well with experiment. The same theory also explains the rather puzzling behavior of the spin relaxation in bilayer graphene [115]. The comparison between the theory and experiments for single and bilayer graphene is shown in Figure 1.20. There have been additional theoretical works confirming that resonant scattering off magnetic moments is sufficient to give experimentally observed spin lifetimes [116–119].

However, there are graphene samples in which spin relaxation is greatly inhibited. Namely, graphene on hBN, or graphene encapsulated by hBN, exhibit spin relaxation times on the order of 1–10 ns [122–124]. There is not yet much investigation of these samples but it appears that the increase of spin relaxation times is not accompanied by an increase in the momentum relaxation time. This indicates that the scatterers causing the momentum relaxation and spin relaxation are decoupled, which is the hallmark of the resonant scattering by magnetic moments. But the spin lifetimes of 10 ns

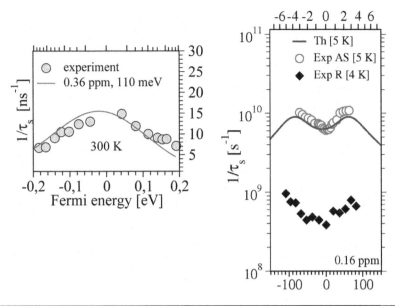

**FIGURE 1.20** Spin relaxation rate as calculated using the mechanism of resonant scattering off local magnetic moments. Left: spin relaxation rate as a function of the Fermi energy for single-layer graphene. The solid line is the calculation for 0.36 ppm of magnetic moments, at 300 K, and with electron-hole puddles smearing of 110 meV. The experimental data are filled circles [120]. The figure adapted from [114]. Right: the same but for bilayer graphene. Theoretical calculations are performed for 0.16 ppm of hydrogen adatoms. The experimental data are from [110] (empty circles) and from [121] (filled squares), with indicated temperature. The Fermi-level smearing is 23 meV (the smearing is lower in bilayer graphene due to the greater electronic density of states). The disagreement between the two experimental results for bilayer graphene underlines the extrinsic character of spin relaxation in graphene. (After Kochan, D. et al., *Phys. Rev. Lett.* 115, 196601, 2015. With permission.)

can already be marginally large enough to be limited by other mechanisms, possibly Dyakonov-Perel and Elliott-Yafet. Further experimental investigations, especially of the temperature dependence and anisotropy, would help to settle the open questions.

While graphene is an exciting spintronics material, mainly because the spin diffusion length is typically greater than a micron (despite the ultrashort spin relaxation time), there has been considerable excitement recently about graphene on two-dimensional transition-metal dichalcogenides (TMDC). Namely, it is predicted by the *ab initio* theory that spin-orbit coupling of graphene on TMDCs such as $MoS_2$ or $WSe_2$, is about 1–10 meV [125, 126], several orders of magnitude more than in pristine graphene. This opens great possibilities for spintronics and some of the opportunities in TMDCs and other 2D materials are discussed in Volume 3, Chapter 5. There have already been experiments measuring the spin relaxation time of electrons in graphene on TMDCs, resulting in some controversy. On the one hand, weak localization measurements seem to yield spin relaxation times on the order of picoseconds [127, 128]. On the other hand, conventional Hanle effect measurements in spin-transport circuits give spin relaxation times of tens of picoseconds [129, 130].

One peculiar feature of the induced spin-orbit coupling is the emergence of the spin-valley "locking", inherited by graphene from TMDCs. Namely, the electron spin in K and K' valleys is opposite. One can imagine a fictitious magnetic field, out-of-plane, producing Zeeman splitting, which is opposite at the two K valleys. In addition, there is the usual Rashba field due to the TMDC substrate, again in meV range. The presence of the out-of-plane valley fields substantially alters the spin relaxation anisotropy. Indeed, if there is only the Rashba interaction, which makes in-plane spin-orbit fields, the out-of-plane spins relax faster. But in the presence of valley Zeeman fields, which are out-of-plane, it is the in-plane spins that decay faster: see Figure 1.21. for an illustration. It was theoretically predicted [131] that the spin anisotropy, expressed as the ratio of the out-of-plane to in-plane spin relaxation times, $\frac{\tau_{s,\perp}}{\tau_{s,\parallel}}$, is 10–100 in graphene on TMDCs. This prediction has been swiftly confirmed experimentally [118, 132].

What about other 2D materials? At the moment, we have perhaps the best understanding of the spin relaxation of conduction electrons in phosphorene. Phosphorene, a single (or multiple) layer of black phosphorous is a direct band gap semiconductor which has space inversion symmetry. Like

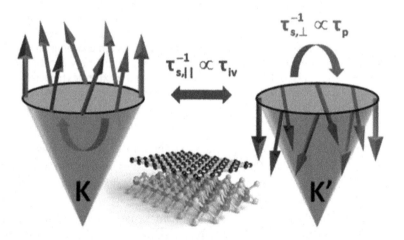

**FIGURE 1.21**    Why is spin relaxation anisotropic in graphene on TMDCs. There are two valleys, K and K', where conduction electrons in graphene live. The Rashba spin-orbit fields in combination with somewhat greater out-of-plane valley Zeeman fields give a specific spin texture for electron spins at the two valleys, as shown. Intravalley scattering, with the characteristic momentum relaxation time $\tau_p$ mainly relaxes out-of-plane spins, since the effective fluctuating spin-orbit field lies in the plane (the out-of-plane component at a given valley is fixed, up or down). The Dyakonov–Perel mechanism of spin relaxation then gives the spin relaxation time, $\tau_{s,\perp}$, proportional to the momentum relaxation time, $\tau_p$, as indicated. On the other hand, intervalley scattering is governed by the relecation time $\tau_{iv}$. The in-plane spins will be mainly relaxed by the fluctuating Zeeman valley fields in intervalley scattering, and the Dyakonov–Perel mechanism yields $\tau_{s,\parallel}$, proportional to $\tau_{iv}$. (After Cummings, A.W. et al., *Phys. Rev. Lett.* 119, 206601, 2017. With permission.)

graphene, it is expected that the spin relaxation is due to spin-orbit coupling and momentum scattering by phonons and impurities. In other words, the mechanism of spin relaxation should be Elliott–Yafet. This is indeed what is observed experimentally in the first electrical spin injection experiment into an ultrathin black phosphorous [134]. The experiment finds a distinct increase of the spin relaxation rate with increasing temperature, which is a signature of phonons-induced spin relaxation. The observed spin relaxation times are nanoseconds, which is rather long, making phosphorene a promising candidate for spintronics applications. The observed spin relaxation time agrees also very well with microscopic theory [135]. The theory further predicts a large spin relaxation anisotropy, due to the anisotropic crystalline structure of phosphorene. This is yet to be observed in experiment.

## ACKNOWLEDGMENTS

This work was supported by DFG SFB 689 and SPP1286, Natural Science Foundation of China under Grant No. 10725417, the National Basic Research Program of China under Grant No. 2006CB922005, and the Knowledge Innovation Project of Chinese Academy of Sciences.

## REFERENCES

1. R. J. Elliott, Theory of the effect of spin–orbit coupling on magnetic resonance in some semiconductors, *Phys. Rev.* **96**, 266 (1954).
2. Y. Yafet, F. Seitz, and D. Turnbull (Eds.), G-factors and spin-lattice relaxation of conduction electronics, *Solid State Physics*, vol. 14, Academic, New York, p. 2, 1963.
3. M. I. D'yakonov and V. I. Perel', Spin relaxation of conduction electrons in noncentrosymmetric semiconductors, *Sov. Phys. Solid State* **13**, 3023 (1971).
4. G. L. Bir, A. G. Aronov, and G. E. Pikus, Spin relaxation of electrons due to scattering by holes, *Sov. Phys. JETP* **42**, 705 (1976).
5. F. Meier and B. P. Zakharchenya (Eds.), *Optical Orientation*, North-Holland, New York, 1984.
6. J. Fabian and S. Das Sarma, Spin relaxation of conduction electrons, *J. Vac. Sci. Technol. B* **17**, 1708 (1999).
7. I. Žutić, J. Fabian, and S. Das Sarma, Spintronics: Fundamentals and applications, *Rev. Mod. Phys.* **76**, 323 (2004).
8. J. Fabian, A. Matos-Abiague, C. Ertler, P. Stano, and I. Žutić, Semiconductor spintronics, *Acta Phys. Slov.* **57**, 565 (2007).
9. M. W. Wu, J. H. Jiang, and M. Q. Weng, Spin dynamics in semiconductors, *Phys. Rep.* **493**, 61 (2010).
10. Y. A. Bychkov and E. I. Rashba, Properties of a 2D electron gas with lifted spectral degeneracy, *JETP Lett.* **39**, 78 (1984).
11. G. Dresselhaus, Spin–orbit coupling effects in zinc blende structures, *Phys. Rev.* **100**, 580 (1955).
12. M. I. Dyakonov and V. Y. Kachorovskii, Spin relaxation of two-dimensional electrons in noncentrosymmetric semiconductors, *Sov. Phys. Semicond.* **20**, 110 (1986).
13. B. A. Bernevig, J. Orenstein, and S. C. Zhang, Exact SU(2) symmetry and persistent spin helix in a spin–orbit coupled system, *Phys. Rev. Lett.* **97**, 236601 (2007).
14. J. D. Koralek, C. P. Weber, J. Orenstein et al., Emergence of the persistent spin helix in semiconductor quantum wells, *Nature* **458**, 610 (2009).

15. J. L. Cheng, M. W. Wu, and J. Fabian, The spin relaxation of conduction electrons in silicon, *Phys. Rev. Lett.* **104**, 016601 (2010).

16. D. J. Lepine, Spin resonance of localized and delocalized electrons in phosphorus-doped silicon between 20 and 30 K, *Phys. Rev. B* **2**, 2429 (1970).

17. B. Huang and I. Appelbaum, Coherent spin transport through a 350 micron thick silicon wafer, *Phys. Rev. Lett.* **99**, 177209 (2007).

18. M. W. Wu and H. Metiu, Kinetics of spin coherence of electrons in an undoped semiconductor quantum well, *Phys. Rev. B* **61**, 2945 (2000).

19. M. W. Wu and C. Z. Ning, A novel mechanism for spin dephasing due to spin-conserving scatterings, *Eur. Phys. J. B* **18**, 373 (2000).

20. M. W. Wu, Spin dephasing induced by inhomogeneous broadening in D'yakonov–Perel' effect in a n-doped GaAs quantum well, *J. Phys. Soc. Jpn.* **70**, 2195 (2001).

21. M. Q. Weng and M. W. Wu, Spin dephasing in n-type GaAs quantum wells, *Phys. Rev. B* **68**, 075312 (2003).

22. M. Q. Weng, M. W. Wu, and L. Jiang, Hot-electron effect in spin dephasing in n-type GaAs quantum wells, *Phys. Rev. B* **69**, 245320 (2004).

23. M. Q. Weng and M. W. Wu, Longitudinal spin decoherence in spin diffusion in semiconductors, *Phys. Rev. B* **66**, 235109 (2002).

24. M. Q. Weng and M. W. Wu, Kinetic theory of spin transport in n-type semiconductor quantum wells, *J. Appl. Phys.* **93**, 410 (2003).

25. M. Q. Weng, M. W. Wu, and Q. W. Shi, Spin oscillations in transient diffusion of a spin pulse in n-type semiconductor quantum wells, *Phys. Rev. B* **69**, 125310 (2004).

26. E. Ya. Sherman, Random spin–orbit coupling and spin relaxation in symmetric quantum wells, *Appl. Phys. Lett.* **82**, 209 (2003).

27. H. Haug and A. P. Jauho, *Quantum Kinetics in Transport and Optics of Semiconductors*, Springer, Berlin, 1996.

28. M. Q. Weng and M. W. Wu, Multisubband effect in spin dephasing in semiconductor quantum wells, *Phys. Rev. B* **70**, 195318 (2004).

29. J. Zhou, J. L. Cheng, and M. W. Wu, Spin relaxation in n-type GaAs quantum wells from a fully microscopic approach, *Phys. Rev. B* **75**, 045305 (2007).

30. J. Zhou and M. W. Wu, Spin relaxation due to the Bir–Aronov–Pikus mechanism in intrinsic and p-type GaAs quantum wells from a fully microscopic approach, *Phys. Rev. B* **77**, 075318 (2008).

31. J. H. Jiang, Y. Zhou, T. Korn, C. Schüller, and M. W. Wu, Electron spin relaxation in paramagnetic Ga(Mn)As quantum wells, *Phys. Rev. B* **79**, 155201 (2009).

32. P. Zhang, J. Zhou, and M. W. Wu, Multivalley spin relaxation in the presence of high in-plane electric fields in n-type GaAs quantum wells, *Phys. Rev. B* **77**, 235323 (2008).

33. J. H. Jiang, M. W. Wu, and Y. Zhou, Kinetics of spin coherence of electrons in n-type InAs quantum wells under intense terahertz laser fields, *Phys. Rev. B* **78**, 125309 (2008).

34. P. Zhang and M. W. Wu, Effect of nonequilibrium phonons on hot-electron spin relaxation in n-type GaAs quantum wells, *Europhys. Lett.* **92**, 47009 (2010).

35. C. Lü, J. L. Cheng, and M. W. Wu, Hole spin dephasing in P-type semiconductor quantum wells, *Phys. Rev. B* **73**, 125314 (2006).

36. Y. Zhou, J. H. Jiang, and M. W. Wu, Electron spin relaxation in P-type GaAs quantum wells, *New J. Phys.* **11**, 113039 (2009).

37. P. Zhang and M. W. Wu, Hole spin relaxation in [001] strained asymmetric Si/SiGe and Ge/SiGe quantum wells, *Phys. Rev. B* **80**, 155311 (2009).

38. C. Lü, U. Zülicke, and M. W. Wu, Hole spin relaxation in P-type GaAs quantum wires investigated by numerically solving fully microscopic kinetic spin Bloch equations, *Phys. Rev. B* **78**, 165321 (2008).

39. C. Lü, H. C. Schneider, and M. W. Wu, Electron spin relaxation in n-type InAs quantum wires, *J. Appl. Phys.* **106**, 073703 (2009).

40. J. H. Jiang, Y. Y. Wang, and M. W. Wu, Reexamination of spin decoherence in semiconductor quantum dots from the equation-of-motion approach, *Phys. Rev. B* **77**, 035323 (2008).

41. J. H. Jiang and M. W. Wu, Electron-spin relaxation in bulk III–V semiconductors from a fully microscopic kinetic spin Bloch equation approach, *Phys. Rev. B* **79**, 125206 (2009).

42. J. L. Cheng and M. W. Wu, Spin diffusion/transport in n-type GaAs quantum wells, *J. Appl. Phys.* **101**, 073702 (2007).

43. P. Zhang and M. W. Wu, Spin diffusion in Si/SiGe quantum wells: Spin relaxation in the absence of D'yakonov–Perel' relaxation mechanism, *Phys. Rev. B* **79**, 075303 (2009).

44. M. E. Flatté, J. M. Bayers, and W. H. Lau, Spin dynamics in semiconductors, in *Spin Dynamics in Semiconductors*, D. D. Awsehalom, D. Loss, and N. Samarth (Eds.), Springer, Berlin, 2002.

45. M. M. Glazov and E. L. Ivchenko, Precession spin relaxation mechanism caused by frequent electron–electron collisions, *JETP Lett.* **75**, 403 (2002).

46. M. A. Brand, A. Malinowski, O. Z. Karimov et al., Precession and motional slowing of spin evolution in a high mobility two-dimensional electron gas, *Phys. Rev. Lett.* **89**, 236601 (2002); W. J. H. Leyland, R. T. Harley, M. Henini, A. J. Shields, I. Farrer, and D. A. Ritchie, Energy-dependent electron–electron scattering and spin dynamics in a two-dimensional electron gas, *Phys. Rev. B* **77**, 205321 (2008).

47. X. Z. Ruan, H. H. Luo, Y. Ji, Z. Y. Xu, and V. Umansky, Effect of electron–electron scattering on spin dephasing in a high-mobility low-density two-dimensional electron gas, *Phys. Rev. B* **77**, 193307 (2008).

48. Y. Ohno, R. Terauchi, T. Adachi, F. Matsukura, and H. Ohno, Electron spin relaxation beyond D'yakonov–Perel' interaction in GaAs/AlGaAs quantum wells, *Physica E* **6**, 817 (2000).

49. F. X. Bronold, A. Saxena, and D. L. Smith, Semiclassical kinetic theory of electron spin relaxation in semiconductors, *Phys. Rev. B* **70**, 245210 (2004).

50. G. F. Giulianni and G. Vignale, *Quantum Theory of the Electron Liquid*, Cambridge University Press, Cambridge, UK, 2005.

51. L. Jiang and M. W. Wu, Control of spin coherence in n-type GaAs quantum wells using strain, *Phys. Rev. B* **72**, 033311 (2005).

52. A. W. Holleitner, V. Sih, R. C. Myers, A. C. Gossard, and D. D. Awschalom, Dimensionally constrained D'yakonov–Perel' spin relaxation in n-InGaAs channels: Transition from 2D to 1D, *New J. Phys.* **9**, 342 (2007).

53. F. Zhang, H. Z. Zheng, Y. Ji, J. Liu, and G. R. Li, Spin dynamics in the second subband of a quasi–two-dimensional system studied in a single-barrier heterostructure by time-resolved Kerr rotation, *Europhys. Lett.* **83**, 47007 (2008).

54. C. Lü, J. L. Cheng, M. W. Wu, and I. C. da Cunha Lima, Spin relaxation time, spin dephasing time and ensemble spin dephasing time in n-type GaAs quantum wells, *Phys. Lett. A* **365**, 501 (2007).

55. D. Stich, J. Zhou, T. Korn et al., Effect of initial spin polarization on spin dephasing and the electron g factor in a high-mobility two-dimensional electron system, *Phys. Rev. Lett.* **98**, 176401 (2007).

56. D. Stich, J. Zhou, T. Korn et al., Dependence of spin dephasing on initial spin polarization in a high-mobility two-dimensional electron system, *Phys. Rev. B* **76**, 205301 (2007).

57. F. Zhang, H. Z. Zheng, Y. Ji, J. Liu, and G. R. Li, Electrical control of dynamic spin splitting induced by exchange interaction as revealed by time-resolved Kerr rotation in a degenerate spin-polarized electron gas, *Europhys. Lett.* **83**, 47006 (2008).

58. T. Korn, D. Stich, R. Schulz, D. Schuh, W. Wegscheider, and C. Schüller, Spin dynamics in high-mobility two-dimensional electron system, *Adv. Solid State Phys.* **48**, 143 (2009).

59. A. P. Dmitriev, V. Y. Kachorovskii, and M. S. Shur, High-field transport in a dense two-dimensional electron gas in elementary semiconductors, *J. Appl. Phys.* **89**, 3793 (2001).

60. E. M. Conwell, *High Field Transport in Semiconductors*, Pergamon, Oxford, 1972.

61. T. C. Damen, L. Vina, J. E. Cunningham, J. Shah, and L. J. Sham, Subpicosecond spin relaxation dynamics of excitons and free carriers in GaAs quantum wells, *Phys. Rev. Lett.* **67**, 3432 (1991).

62. J. Wagner, H. Schneider, D. Richards, A. Fischer, and K. Ploog, Observation of extremely long electron-spin-relaxation times in p-type δ-doped GaAs/$Al_xGa_{1-x}As$ double -heterostructures, *Phys. Rev. B* **47**, 4786 (1993).

63. H. Gotoh, H. Ando, T. Sogawa, H. Kamada, T. Kagawa, and H. Iwamura, Effect of electron–hole interaction on electron spin relaxation in GaAs/AlGaAs quantum wells at room temperature, *J. Appl. Phys.* **87**, 3394 (2000).

64. T. F. Boggess, J. T. Olesberg, C. Yu, M. E. Flatté, and W. H. Lau, Room-temperature electron spin relaxation in bulk InAs, *Appl. Phys. Lett.* **77**, 1333 (2000).

65. S. Hallstein, J. D. Berger, M. Hilpert et al., Manifestation of coherent spin precession in stimulated semiconductor emission dynamics, *Phys. Rev. B* **56**, R7076 (1997).

66. P. Nemec, Y. Kerachian, H. M. van Driel, and A. L. Smirl, Spin-dependent electron many-body effects in GaAs, *Phys. Rev. B* **72**, 245202 (2005).

67. H. C. Schneider, J.-P. Wüstenberg, O. Andreyev et al., Energy-resolved electron spin dynamics at surfaces of p-doped GaAs, *Phys. Rev. B* **73**, 081302 (2006).

68. C. Yang, X. Cui, S.-Q. Shen, Z. Xu, and W. Ge, Spin relaxation in submonolayer and monolayer InAs structures grown in a GaAs matrix, *Phys. Rev. B* **80**, 035313 (2009).

69. P. H. Song and K. W. Kim, Spin relaxation of conduction electrons in bulk III–V semiconductors, *Phys. Rev. B* **66**, 035207 (2002).

70. D. D. Awschalom, D. Loss, and N. Samarth (Eds.), *Semiconductor Spintronics and Quantum Computation*, Springer-Verlag, Berlin, 2002; M. I. D'yakonov (Ed.), *Spin Physics in Semiconductors*, Springer, Berlin, 2008, and references therein.

71. R. I. Dzhioev, K. V. Kavokin, V. L. Korenev et al., Low-temperature spin relaxation in n-type GaAs, *Phys. Rev. B* **66**, 245204 (2002).

72. M. Krauß, R. Bratschitsch, Z. Chen, S. T. Cundiff, and H. C. Schneider, Ultrafast spin dynamics in optically excited bulk GaAs at low temperatures, *Phys. Rev. B* **81**, 035213 (2010).

73. K. Shen, A peak in density dependence of electron spin relaxation time in n-type bulk GaAs in the metallic regime, *Chin. Phys. Lett.* **26**, 067201 (2009).

74. I. Appelbaum, B. Huang, and D. J. Monsma, Electronic measurement and control of spin transport in silicon, *Nature* **447**, 295 (2007).

75. B. Huang, L. Zhao, D. J. Monsma, and I. Appelbaum, 35% magnetocurrent with spin transport through Si, *Appl. Phys. Lett.* **91**, 052501 (2007).

76. G. Dresselhaus, Spin–orbit coupling effects in zinc blende structures, *Phys. Rev.* **100**, 580 (1955).

77. L. Jiang, M. Q. Weng, M. W. Wu, and J. L. Cheng, Diffusion and transport of spin pulses in an n-type semiconductor quantum well, *J. Appl. Phys.* **98**, 113702 (2005).

78. S. A. Crooker and D. L. Smith, Imaging spin flows in semiconductors subject to electric, magnetic, and strain fields, *Phys. Rev. Lett.* **94**, 236601 (2005).

79. M. Beck, C. Metzner, S. Malzer, and G. H. Döhler, Spin lifetimes and strain-controlled spin precession of drifting electrons in GaAs, *Europhys. Lett.* **75**, 597 (2006).

80. J. L. Cheng, M. W. Wu, and I. C. da Cunha Lima, Anisotropic spin transport in GaAs quantum wells in the presence of competing Dresselhaus and Rashba spin–orbit coupling, *Phys. Rev. B* **75**, 205328 (2007).

81. N. S. Averkiev and L. E. Golub, Giant spin relaxation anisotropy in zinc-blende heterostructures, *Phys. Rev. B* **60**, 15582 (1999).

82. N. S. Averkiev, L. E. Golub, and M. Willander, Spin relaxation anisotropy in two-dimensional semiconductor systems, *J. Phys. Condens. Matter* **14**, R271 (2002).

83. J. Schliemann, J. C. Egues, and D. Loss, Nonballistic spin-field-effect transistor, *Phys. Rev. Lett.* **90**, 146801 (2003).

84. R. Winkler, Spin orientation and spin precession in inversion-asymmetric quasi-two-dimensional electron systems, *Phys. Rev. B* **69**, 045317 (2004).

85. J. L. Cheng and M. W. Wu, Spin relaxation under identical Dresselhaus and Rashba coupling strengths in GaAs quantum wells, *J. Appl. Phys.* **99**, 083704 (2006).

86. K. Shen and M. W. Wu, Infinite spin diffusion length of any spin polarization along direction perpendicular to effective magnetic field from Dresselhaus and Rashba spin–orbit couplings with identical strengths in (001) GaAs quantum wells, *J. Supercond. Nov. Magn.* **22**, 715 (2009).

87. C. P. Weber, N. Gedik, J. E. Moore, J. Orenstein, J. Stephens, and D. D. Awschalom, Observation of spin Coulomb drag in a two-dimensional electron gas, *Nature* **437**, 1330 (2005).

88. A. R. Cameron, P. Rickel, and A. Miller, Spin gratings and the measurement of electron drift mobility in multiple quantum well semiconductors, *Phys. Rev. Lett.* **76**, 4793 (1996).

89. C. P. Weber, J. Orenstein, B. A. Bernevig, S.-C. Zhang, J. Stephens, and D. D. Awschalom, Nondiffusive spin dynamics in a two-dimensional electron gas, *Phys. Rev. Lett.* **98**, 076604 (2007).

90. M. Q. Weng, M. W. Wu, and H. L. Cui, Spin relaxation in n-type GaAs quantum wells with transient spin grating, *J. Appl. Phys.* **103**, 063714 (2008).

91. W. Han, R. K. Kawakami, M. Gmitra, and J. Fabian, Graphene spintronics, *Nat. Nanotechnol.* **9**, 794 (2014).

92. M. Gmitra, S. Konschuh, and C. Ertler, C. Ambrosch-Draxl, and J. Fabian, Band-structure topologies of graphene: Spin-orbit coupling effects from first principles, *Phys. Rev. B* **80**, 235431 (2009).

93. C. Ertler, S. Konschuh, M. Gmitra, and J. Fabian, Electron spin relaxation in graphene: The role of the substrate, *Phys. Rev. B* **80**, 041405 (2009).

94. D. Van Tuan, F. Ortmann, D. Soriano, S. O. Valenzuela, and S. Roche, Pseudospin-driven spin relaxation mechanism in graphene, *Nat. Phys.* **10**, 857 (2014).

95. P. Zhang and M. W. Wu, Electron spin diffusion and transport in graphene, *Phys. Rev. B* **84**, 045304 (2011).

96. S. Fratini, D. Gosálbez-Martínez, P. Merodio Cámara, and J. Fernández-Rossier, Anisotropic intrinsic spin relaxation in graphene due to flexural distortions, *Phys. Rev. B* **88**, 115426 (2013).

97. D. Huertas-Hernando, F. Guinea, and A. Brataas, Spin-orbit-mediated spin relaxation in graphene, *Phys. Rev. Lett.* **103**, 146801 (2009).

98. P. R. Struck and G. Burkard, Effective time-reversal symmetry breaking in the spin relaxation in a graphene quantum dot, *Phys. Rev. B* **82**, 125401 (2010).

99. H. Ochoa, A. H. Castro Neto, and F. Guinea, Elliot-Yafet mechanism in graphene, *Phys. Rev. Lett.* **108**, 206808 (2012).

100. V. K. Dugaev, E. Sherman, and J. Barnas, Spin dephasing and pumping in graphene due to random spin-orbit interaction, *Phys. Rev. B* **83**, 085306 (2011).

101. S. Jo, D.-K. Ki, D. Jeong, H.-J. Lee, and S. Kettemann, Spin relaxation properties in graphene due to its linear dispersion, *Phys. Rev. B* **84**, 075453 (2011).

102. H. Dery, P. Dalal, L. Cywinski, and L. J. Sham, Spin-based logic in semiconductors for reconfigurable large-scale circuits, *Nature* **447**, 573 (2007).

103. B. Dóra, F. Murányi, and F. Simon, Electron spin dynamics and electron spin resonance in graphene, *Eur. Phys. Lett.* **92**, 17002 (2010).

104. L. Wang and M. W. Wu, Electron spin relaxation in bilayer graphene, *Phys. Rev. B* **87**, 205416 (2013).

105. D. V. Fedorov M. Gradhand, S. Ostanin et al., Impact of electron-impurity scattering on the spin relaxation time in graphene: A first-principles study, *Phys. Rev. Lett.* **110**, 156602 (2013).

106. N. Tombros, C. Jozsa, M. Popinciuc, H. T. Jonkman, and B. J. van Wees, Electronic spin transport and spin precession in single graphene layers at room temperature, *Nature* **448**, 571 (2007).

107. A. Avsar, T.-Y. Yang, S. Bae et al., Toward wafer scale fabrication of graphene based spin valve devices, *Nano Lett.* **11**, 2363 (2011).

108. N. Tombros, S. Tanabe, A. Veligura, et al., Anisotropic spin relaxation in graphene, *Phys. Rev. Lett.* **101**, 046601 (2008).

109. K. Pi, W. Han, K. M. McCreary, A. G. Swartz, Y. Li, and R. K. Kawakami, Manipulation of spin transport in graphene by surface chemical doping, *Phys. Rev. Lett.* **104**, 187201 (2010).

110. T.-Y. Yang, J. Balakrishnan, F. Volmer et al., Observation of long spin-relaxation times in bilayer graphene at room temperature, *Phys. Rev. Lett.* **107**, 047206 (2011).

111. W. Han and R. K. Kawakami, Spin relaxation in single-layer and bilayer graphene, *Phys. Rev. Lett.* **107**, 047207 (2011).

112. B. Raes, J. E. Scheerder, M. V. Costache et al., Determination of the spin-lifetime anisotropy in graphene using oblique spin precession, *Nat. Commun.* **7**, 11444 (2016).

113. O. V. Yazyev, Emergence of magnetism in graphene materials and nanostructures, *Rep. Prog. Phys.* **73**, 056501 (2010).

114. D. Kochan, M. Gmitra, and J. Fabian, Spin relaxation mechanism in graphene: Resonant scattering by magnetic impurities, *Phys. Rev. Lett.* **112**, 116602 (2014).

115. D. Kochan, S. Irmer, M. Gmitra, and J. Fabian, Resonant scattering by magnetic impurities as a model for spin relaxation in bilayer graphene, *Phys. Rev. Lett.* **115**, 196601 (2015).

116. J. Bundesmann, D. Kochan, F. Tkatschenko, J. Fabian, and K. Richter, Theory of spin-orbit-induced spin relaxation in functionalized graphene, *Phys. Rev. B* **92**, 081403 (2015).

117. J. Wilhelm, M. Walz, and F. Evers, Ab initio pin-flip conductance of hydrogenated graphene nanoribbons: Spin-orbit interaction and scattering with local impurity spins, *Phys. Rev. B* **92**, 014405 (2015).

118. V. G. Miranda, E. R. Mucciolo, and C. H. Lewenkopf, Spin relaxation in disordered graphene: Interplay between puddles and defect-induced magnetism, *J. Phys. Chem. Solids* (2017). doi:https://doi.org/10.1016/j.jpcs.2017.10.022

119. D. Soriano, D. V. Tuan, S. M.-M. Dubois et al., Spin transport in hydrogenated graphene, *2D Mater.* **2**, 022002 (2015).

120. M. Wojtaszek, I. J. Vera-Marun, T. Maassen, and B. J. van Wees, Enhancement of spin relaxation time in hydrogenated graphene spin-valve devices, *Phys. Rev. B* **87**, 081402 (2013).

121. W. Han and R. K. Kawakami, Spin relaxation in single-layer and bilayer graphene, *Phys. Rev. Lett.* **107**, 047207 (2011).

122. M. Drögeler, F. Volmer, M. Wolter et al., Nanosecond spin lifetimes in single- and few-layer graphene–hBN heterostructures at room temperature, *Nano Lett.* **14**, 6050 (2014).

123. M. H. D. Guimarães, P. J. Zomer, J. Ingla-Aynés, J. C. Brant, N. Tombros, and B. J. van Wees, Controlling spin relaxation in hexagonal BN-encapsulated graphene with a transverse electric field, *Phys. Rev. Lett.* **113**, 086602 (2014).

124. M. Drögeler, C. Franzen, F. Volmer et al., Spin lifetimes exceeding 12 ns in graphene nonlocal spin valve devices, *Nano Lett.* **16**, 3533 (2016).

125. M. Gmitra and J. Fabian, Graphene on transition-metal dichalcogenides: A platform for proximity spin-orbit physics and optospintronics, *Phys. Rev. B* **92**, 155403 (2015).

126. M. Gmitra, D. Kochan, P. Högl, and J. Fabian, Trivial and inverted Dirac bands and the emergence of quantum spin Hall states in graphene on transition-metal dichalcogenides, *Phys. Rev. B* **93**, 155104 (2016).

127. Z. Wang, D.-K. Ki, H. Chen, H. Berger, A. H. MacDonald, and A. F. Morpurgo, Strong interface-induced spin-orbit interaction in graphene on $WS_2$, *Nat. Commun.* **6**, 8339 (2015).

128. B. Yang, M.-F. Tu, J. Kim et al., Tunable spin–orbit coupling and symmetry-protected edge states in graphene/$WS_2$, *2D Mater.* **3**, 031012 (2016).

129. S. Omar and B. J. van Wees, Graphene-$WS_2$ heterostructures for tunable spin injection and spin transport, *Phys. Rev. B* **95**, 081404 (2017).

130. A. Dankert and S. P. Dash, Electrical gate control of spin current in van der Waals heterostructures at room temperature, *Nat. Commun.* **8**, 16093 (2017).

131. A. W. Cummings, J. H. Garcia, J. Fabian, and S. Roche, Giant spin lifetime anisotropy in graphene induced by proximity effects, *Phys. Rev. Lett.* **119**, 206601 (2017).

132. T. S. Ghiasi, J. Ingla-Aynés, A. A. Kaverzin, and B. J. van Wees, Large proximity-induced spin lifetime anisotropy in transition-metal dichalcogenide/graphene heterostructures, *Nano Lett.* **17**, 7528 (2017).

133. A. Avsar, J. Y. Tan, M. Kurpas et al., Gate-tunable black phosphorus spin valve with nanosecond spin lifetimes, *Nat. Phys.* **13**, 888 (2017).

134. M. Kurpas, M. Gmitra, and J. Fabian, Spin-orbit coupling and spin relaxation in phosphorene: Intrinsic versus extrinsic effects, *Phys. Rev. B* **94**, 1155423 (2016).

# Electrical Spin Injection and Transport in Semiconductors

**Berend T. Jonker**

# 2.1 INTRODUCTION

The field of semiconductor electronics has been based exclusively on the manipulation of charge. The remarkable advances in performance have been due in large part to size scaling, i.e. systematic reduction in device dimensions, enabling significant increases in circuit density. This trend, known as Moore's Law [1], is likely to be curtailed in the near future by practical and fundamental limits. Consequently, there is keen interest in exploring new avenues and paradigms for future technologies. *Spintronics*, or the use of carrier spin as a new degree of freedom in an electronic device, represents one of the most promising candidates for this paradigm shift [2, 3]. Like charge, spin is an intrinsic fundamental property of an electron, and is one of the alternative state variables under consideration on the International Technology Roadmap for Semiconductors (ITRS) for processing information in the new ways that will be required beyond the ultimate scaling limits of the existing silicon-based complementary metal-oxide-semiconductor (CMOS) technology [4].

Semiconductor-based spintronics offers many new avenues and opportunities which are inaccessible to metal-based spintronic structures. This is due to the characteristics for which semiconductors are so well-known: the existence of a band gap that can often be tuned over a significant range in ternary compounds; the accompanying optical properties on which a vast optoelectronic industry is based; and the ability to readily control carrier concentrations and transport characteristics via doping, gate voltages, and band offsets. Coupling the new degree of freedom of carrier spin with the traditional band gap engineering of modern electronics offers opportunities for new functionality and performance for semiconductor devices (examples can be found in Chapters 16 and 17, Volume 3).

There are four essential requirements for implementing a semiconductor-based spintronics technology:

1. Efficient *electrical injection* of spin-polarized carriers from an appropriate contact into the semiconductor heterostructure.
2. Adequate *spin diffusion lengths and spin lifetimes* within the semiconductor host medium.
3. Effective *control and manipulation* of the spin system to provide the desired functionality.
4. Efficient *detection* of the spin system to provide the output.

Although the focus of this chapter is nominally on the first factor, each of these factors will necessarily be addressed to some extent, as each plays a role in any experimental research effort.

The generation of spin-polarized carrier populations in semiconductors has historically been accomplished by optical excitation, an approach referred to as "optical pumping" or "optical orientation" [5]. This has been the basis for decades of research exploring spin-dependent phenomena in semiconductor hosts, beginning with the pioneering work of Lampel [6], and enabling the development of new spectroscopic techniques which provide extraordinary insight into spin properties [7, 8]. Injection of spin-polarized electrons from a scanning tunneling microscope tip in ultra-high vacuum was first reported in 1992 [9], and provides a highly localized probe, suitable for basic research. However, a broad technology based on spin manipulation in a semiconductor host requires an efficient and practical means of spin injection and detection. This is best accomplished *via a discrete electrical contact*. Such a contact defines the extent of the spin source (or detector), provides an interface between the spin system and the canonical electronic input/output parameters of voltage and current, and offers a very simple and direct means of implementing spin injection/detection compatible with existing device design and fabrication technology [10–12].

In this chapter, we will discuss the factors relevant to spin injection and transport in semiconductors, review the necessary concepts, and illustrate these by way of examples, to provide both a practical guide and an overview of the current state of the art. We will conclude with a brief section (Section 2.6) on electrical injection and *detection* of spin accumulation using multiterminal lateral transport geometries in silicon, where an operation well above room temperature has recently been demonstrated.

## 2.2 SPIN POLARIZATION BY OPTICAL PUMPING

The earliest method of generating a spin-polarized population of carriers within a semiconductor was by optical excitation using circularly polarized light, often referred to as "optical pumping" [5]. A photon carries a unit of angular momentum either parallel or antiparallel to its propagation direction, and the corresponding polarization is referred to as positive ($\sigma+$) or negative ($\sigma-$) helicity, respectively. Absorption of a photon whose energy is equal to or greater than the band gap promotes an electron from the valence band (VB) to the conduction band (CB). If the light is circularly polarized, the electron and hole must share the angular momentum transferred in a manner that satisfies conservation laws [13].

In a direct gap zinc-blende semiconductor such as GaAs, the CB is $s$-like in character and twofold spin degenerate, and the electron can occupy states with values of spin angular momentum $m_j = \pm1/2$. The VB is $p$-like in character and fourfold spin degenerate in bulk material, so that the hole can occupy states with values of angular momentum $m_j = \pm1/2, \pm3/2$, corresponding to "light hole" (LH) and "heavy hole" (HH) states, respectively (Figure 2.1a). For photons incident along the surface normal, interband transitions at the zone center ($k_{//} = 0$) satisfy the selection rule $\Delta m_j = \pm1$, reflecting absorption of the photon's original angular momentum. The probability of a transition involving a LH or HH state is weighted by the square of the corresponding matrix element connecting it to the appropriate electron state, so that HH transitions are three times more likely than LH transitions. Thus, absorption of photons with angular momentum +1 produces three "spin-down" ($m_j = -1/2$) electrons for every one "spin-up" ($m_j = +1/2$) electron, resulting in an electron population with a spin polarization of 50% in a bulk material, where the HH and LH

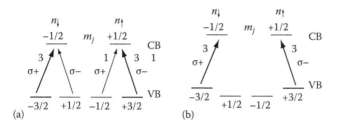

**FIGURE 2.1** Optical transitions allowed due to absorption (or emission—reverse the arrows) of circularly polarized light for (a) bulk material and (b) a structure in which the light/heavy hole band degeneracy is lifted, as in a quantum well or strained material. The HH transitions are three times more probable than the LH.

states are degenerate. In a quantum well (QW), either quantum confinement or strain can lift this degeneracy, as illustrated in Figure 2.1b, where the HH states are at lower energy (nearer the center of the band gap). In this case, circularly polarized light whose energy is just sufficient to excite these states (and not the LH states) can in principle produce an electron population, which is 100% spin-polarized. The electron spin lifetime is much longer than that of the hole in most materials, and the holes are generally assumed to depolarize instantaneously. Spin relaxation and lifetimes are discussed in Chapter 1, Volume 2.

Note that the resulting electron spin is oriented parallel or antiparallel to the propagation direction of the incident photon (typically the surface normal), consistent with the angular momentum of the absorbed photon. In addition, the orientation of carrier spin is typically locked along the surface normal in a quantum well or can be modified by strain. These are important factors to bear in mind when performing and interpreting an experiment.

## 2.3 SPIN POLARIZATION BY ELECTRICAL INJECTION: GENERAL CONSIDERATIONS

A discrete electrical contact to create, manipulate, and detect spin currents and populations in a semiconductor host enables the development of a much broader range of technologies than that allowed by optical excitation alone. A discrete contact clearly defines the source/detector dimensions and is readily scalable, an indispensable attribute, in contrast with optical approaches which are diffraction limited. Existing electronics is based upon a voltage or current for input and output. A spin contact must, therefore, provide a suitable interface between these conventional parameters and the spin system to be utilized for information processing or sensing in the semiconductor. A magnetic material has a spin-polarized band structure and density of states (DOS), and is well suited for such a contact—the electron population at the Fermi energy, $E_F$, naturally exhibits a net spin polarization, which serves as a source for electrical injection under appropriate bias into an unpolarized host.

The idea that a spin current could in principle accompany a charge current was first hypothesized by Aronov and Pikus in 1976 [14], when they noted that the flow of electric current $I_e$ from a ferromagnetic (FM) metal should exhibit a net spin polarization (magnetization) due to its spin-polarized DOS. If this charge current flows into a nonmagnetic material, it simultaneously corresponds to a flow of magnetization, since each electron also bears the fundamental unit of magnetization, $\mu_B$, the Bohr magneton. This magnetization current $I_m$ is simply written as

$$I_m = \eta \mu_B \left( \frac{I_e}{e} \right), \tag{2.1}$$

where:

$e$ is the electron's charge

$0 \leq \eta \leq 1$ is a phenomenological factor to account for the fact that the Fermi surface of the ferromagnet may contain both spin bands (or to account for possible spin scattering at the interface)

This remarkably simple and intuitive relation provided the seed from which the burgeoning field of spin injection and transport has sprung.

A contact for spin injection/detection should exhibit some specific characteristics. In general, one desires high spin injection efficiency at low bias power extending well above room temperature for practical device applications. While it is common to think of these properties as attributes of the contact material alone, it is essential to note that the interface between the contact and the semiconductor is equally important and is often more challenging to understand. Thus, the following should be considered as desirable attributes for the combination of spin contact material and the corresponding interface:

✦ A Curie temperature $T_c > 400\,\mathrm{K}$, although certain specialized applications (e.g. cooled infrared detectors) may permit $T_c$'s as low as $100\,\mathrm{K}$.

✦ High remanent magnetization, i.e. the magnetization retained at zero field. The nonvolatile behavior essential for many applications (memory, optical isolators, reprogrammable logic) is predicated upon a remanence that is a substantial fraction of the saturation magnetization.

✦ The magnetization should be readily switched at modest power expense, important for applications which require fast and routine switching (memory, logic). Historically, this has meant a low coercive field ($H_c < 200\,\mathrm{Oe}$), but more recent interest is focused on reducing the critical current required to switch the free layer in a spin torque transfer element (see Chapters 7–9 and 15 in Volume 1 for a discussion of spin torque effects), or utilizing an electric field to assist in magnetization reversal [15].

✦ A highly polarized DOS at $E_F$. This is often viewed to first order as the source term when discussing spin injection from any contact, and higher polarizations produce larger signal levels. Half metals such as select Heusler alloys ideally exhibit 100% spin polarization at $E_F$, and are of keen interest, although this polarization is rarely, if ever, achieved in practice.

✦ The contact/interface should have a reasonably low specific resistance to allow the usable current densities. A 100% spin-polarized current of $10^{-3}$ A cm$^{-2}$ may not be as useful as a 30% spin-polarized current of 10 A cm$^{-2}$, particularly in generating spin accumulation in semiconductor structures (with the exception of single-electron devices). High resistances also degrade signal-to-noise ratios, compromising device characteristics such as operating frequency. The contact resistance will be a compromise involving consideration of many factors.

✦ The contact/interface should offer spin injection which is robust against defects. Defects at complex heterointerfaces are to be expected—their role in spin scattering is poorly understood and poses an immediate challenge for the development of spin transport devices.

✦ The contact/interface should be thermodynamically stable, and thermally stable at temperatures typically encountered for growth, processing, and device operation.

The above list is neither exhaustive nor intended to be an absolute yardstick by which to gauge the merits of a particular contact/interface—performance and intended application define the latter. It is useful, however, to keep these items in mind as we continue to develop the spin contacts, corresponding interface, and the new functionality that semiconductor spintronic devices offer.

Two classes of magnetic materials have been utilized as contacts for electrical spin injection into semiconductors: *semiconductors* and *metals* [10, 12]. FM metals are familiar to all and offer some significant advantages, but also present some challenges for efficient utilization as spin contacts—they will be discussed in later paragraphs. Magnetic semiconductors are less familiar, but have, in fact, been studied for decades. They have enjoyed a resurgence of research interest since 1990, and were successfully used before FM metals to electrically inject spin-polarized carriers into another semiconductor. An overview of recent developments in III–V and oxide magnetic semiconductors appears in Chapters 9 and 12, Volume 2.

## 2.4 SEMICONDUCTOR/SEMICONDUCTOR ELECTRICAL SPIN INJECTION

### 2.4.1 Magnetic Semiconductors: Material Properties

Magnetic semiconductors simultaneously exhibit semiconducting properties and typically either paramagnetic or FM order. The coexistence of these properties in a single material provides fertile ground for fundamental studies, and offers exciting possibilities for a broad range of applications. The use of a magnetic semiconductor as a spin contact enables the design of a spin-injecting semiconductor/semiconductor interface guided by known principles of bandgap engineering (CB and VB offsets, doping, and carrier transport) and epitaxial growth (lattice match, interface structure, and materials compatibility). Therefore, they seem ideal candidates for incorporation in semiconductor spintronic devices. However, their magnetic properties to date fall well short of those required for practical devices—e.g. the Curie temperature is typically well below room temperature, despite much effort to increase it.

*Ferromagnetic semiconductors* (FMS) were intensively studied from 1950 to 1970—classic examples include the europium chalcogenides (e.g. EuO, S, and Se) and the chalcogenide spinels, in which the magnetic element forms a significant fraction of the atomic constituents (see Figure 2.2a) [16, 17]. Simple transport experiments have demonstrated that the exchange splitting of the band edges below the Curie temperature could be used as a spin-dependent potential barrier, which selectively passes one spin component while blocking the other, leading to current with a net spin polarization. This "spin filter effect" was demonstrated in EuS- and EuSe-based single barrier heterostructures in 1967 by Esaki et al. [18], and has been recently revisited [19, 20] (spin filtering is reviewed in Chapter 14, Volume 1). However, device applications languished due to low Curie temperatures and the inability to incorporate these materials in thin-film form with mainstream semiconductor device materials.

Recent work successfully incorporated epilayers of $n$-type $CdCr_2Se_4$, another classic magnetic semiconductor, with GaAs-based heterostructures by molecular beam epitaxy (MBE) [21] and demonstrated electrical spin injection [22]. $CdCr_2Se_4$ is a direct gap ($E_g = 1.3\,eV$) chalcogenide spinel, which is reasonably lattice matched to technologically important materials such as Si and GaP (~1.7% tensile mismatch), and to GaAs (5.2% tensile mismatch), assuming one uses an effective lattice constant of $a_o/2$ for the $CdCr_2Se_4$. Single crystal epilayers grown on both GaP(001) and GaAs(001) substrates are $n$-type ($n \sim 10^{18}\,cm^{-3}$), FM with the easy magnetization axis in plane (along the GaAs(110)), and have a Curie temperature of 132 K [21]. Subsequent work successfully demonstrated electrical injection of spin-polarized electrons from $CdCr_2Se_4$ contacts into GaAs [22] using the spin-polarized light-emitting diode (spin-LED) approach, described in detail below.

Oxide MBE has been used to successfully grow epitaxial films of EuO on Si(001) [23] and GaN(0001) substrates [24]. EuO has a highly spin-polarized band structure and a large specific Faraday rotation. Andreev reflection

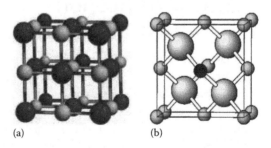

(a)                    (b)

**FIGURE 2.2** Lattice models of a classic magnetic semiconductor EuO (a), and a diluted magnetic semiconductor $Ga_{1-x}Mn_xAs$ (b).

measurements show that the electron spin polarization in these films exceeds 90%. However, no spin injection from EuO into either substrate has been reported to date. It should be noted that EuO has a Curie temperature, $T_c$, of only 69 K, a common shortcoming of most FMS compounds, which essentially precludes their use in practical device applications.

Pronounced magnetic behavior can be introduced into some semiconductors by incorporating a small amount (~1%–10%) of certain magnetic impurities into the lattice. These materials are referred to as "*diluted magnetic semiconductors*" (DMS), because the material is formed by diluting a host semiconductor lattice with a substitutional magnetic impurity, most typically Mn, to form an alloy (see Figure 2.2b). Strong exchange interactions between the magnetic impurity and the host carriers lead to magnetic behavior. Some of these materials are truly FM, while others are paramagnetic yet still exhibit very useful and highly flexible magnetic properties.

The paramagnetic compounds are commonly based on the II–VI or IV–VI semiconductors, where there is a high degree of solubility for the magnetic element. Classic examples include $Zn_{1-x}Mn_xSe$, $Cd_{1-x}Mn_xTe$, and $Pb_{1-x}Mn_xSe$. Comprehensive reviews may be found in Refs. [25, 26]. These materials are often called "*semimagnetic semiconductors*" (SMS) because their paramagnetic behavior and Zeeman splitting of the CB and VB are dramatically enhanced over what might be expected from the magnetic ions alone. This enhancement originates from very large exchange interactions between the $s$- and $p$-like carriers of the CB and VB of the host, and the $d$ electrons of the substitutional magnetic impurity. This $sp$-$d$ exchange leads to a tremendous amplification of the Zeeman splitting of the band edges in an applied magnetic field, and produces amplified magneto-optical properties such as giant Faraday rotation, as well as other field dependent effects. If this Zeeman splitting is parametrized by $\Delta E = g^* u_B H$, then the effective value of the host $g$-factor is increased from 2, typical of most hosts to a value of $g^* \sim 100$ or greater [25]. For modest fields, the spin splitting significantly exceeds $k_B T$ at low temperature. For example, the splitting of the $m_j = +1/2$ and $-1/2$ electron states in $Zn_{0.94}Mn_{0.06}Se$ is ~10 meV at 3 T and 4.2 K, so that the CB effectively forms a completely polarized source of $m_j = -1/2$ electrons, and this spin polarization can be tuned by the external magnetic field. These properties are certainly attractive for a spin-injecting contact.

Oestreich et al. initially proposed the use of an SMS as a spin-injecting contact [27]. They used time-resolved photoluminescence (PL) to demonstrate that *optically* excited carriers became spin aligned in a $Cd_{1-x}Mn_xTe$ layer on a picosecond time scale, and that the spin-polarized electrons were transferred into an adjacent CdTe layer with little loss in spin polarization. However, this pronounced behavior is limited to low temperatures ($T < 25$ K) and relatively highly applied magnetic fields ($H > 0.5$ T), making these paramagnetic semiconductors less attractive for device fabrication. Nevertheless, bulk crystals have found application as magnetically tunable optical polarizers operating at room temperature [28], where the long optical path length compensates for the severe reduction in $g$-factor and magnetization at higher temperatures.

A concerted effort was made to introduce magnetic order into semiconductor compounds already recognized for device applications by alloying small amounts of Mn with GaAs and InAs to form new DMS materials. Nonequilibrium growth at relatively low substrate temperatures by MBE permits incorporation of Mn at levels well above the solubility limit of the host lattice. Pioneering work led to the discovery of spontaneous FM order in DMS alloys such as $In_{1-x}Mn_xAs$ in 1989 [29] and $Ga_{1-x}Mn_xAs$ in 1996 [30, 31]. Since Mn acts as both the magnetic element and an acceptor, they are $p$-type. After much research [32–34], these new FMS materials now exhibit Curie temperatures up to 82 K [35] and 200 K [36], respectively. Although their optical and electronic properties are not nearly as clean and controllable as their nonmagnetic hosts, these materials have been vigorously studied for their potential in future spin-dependent semiconductor device technologies. For example, $Ga_{1-x}Mn_xAs$ has been used as a source of spin-polarized holes in resonant tunneling diodes (RTD) [37, 38], LED [39, 40], and for current-induced magnetization switching [41]. Electric field control of FM order has been demonstrated in $In_{1-x}Mn_xAs$ [42], $Mn_xGe_{1-x}$ [43], and $Ga_{1-x}Mn_xAs$ [44] heterostructures, demonstrating one of the highly unusual multiferroic properties of these materials and portending a host of new applications.

The theory based on a mean field Zener model has predicted that FM order should be stabilized in a wide variety of semiconductor hosts when diluted or alloyed with Mn at concentrations of ~5% and sufficiently high hole densities [45, 46]. Subsequent work has indicated that other magnetic atoms should produce similar effects. This has stimulated a groundswell of research activity to synthesize new diluted FMS compounds, with the goal of achieving technologically attractive materials [47]. Unfortunately, progress has been hampered on both the theoretical and experimental fronts. Although the mean field model works well for $Ga_{1-x}Mn_xAs$ for which it was developed, it is clear that it is not appropriate for many other semiconductor hosts—the relevant mechanisms leading to potential magnetic order and the proper choice of theoretical framework depend strongly upon the position of the electronic states introduced by the magnetic atom relative to the band edges of the host semiconductor. New theoretical approaches are now being developed and appear promising [48].

The origins of ferromagnetic order in even the canonical ferromagnetic semiconductor, $Ga_{1-x}Mn_xAs$, are still debated and remain a topic of current research effort [49]. Further evidence supporting the mean field Zener model in which the Fermi energy lies within the valence band and magnetic order is mediated by extended hole states [50] is countered by evidence that the Fermi energy is instead located within a separate Mn-derived impurity band of localized states which stabilize ferromagnetic order [51, 52]. The picture is complicated by the structural disorder introduced by the presence of Mn and the low growth temperatures required to incorporate Mn into the lattice, and the impact of such disorder is very difficult to model. The basic physics and potential device applications are covered in several recent reviews [53, 54].

The principal obstacle in experimental efforts to synthesize new diluted FMS materials has been the formation of unwanted phases due to the relatively low solubility of the magnetic atoms (Mn, Cr, Fe, Co) in most of the hosts considered (e.g. the group III-As, III-Sb, and III-N families). In many cases, nanoscale and microscale precipitates of known FM bulk phases form, which are exceedingly difficult to detect by the usual structural probes of x-ray diffraction and electron microscopy, but are all too readily detected with standard magnetometry measurements. A major challenge facing the experimentalist is to utilize characterization techniques which discriminate against the potential presence of such precipitates, clearly distinguish the contribution of the FMS material from that of the precipitates, and directly probe the key characteristics expected for the FMS itself. Electrical spin injection from the FMS provides one such litmus test, since it is enabled by the net carrier spin polarization intrinsic to a true FMS.

### 2.4.2 MAGNETIC SEMICONDUCTORS AS SPIN-INJECTING CONTACTS: INITIAL TRANSPORT STUDIES

As noted earlier, the use of a magnetic semiconductor as a contact enables design of a spin injection semiconductor device guided by known principles of bandgap engineering and epitaxial growth. This has greatly facilitated demonstration of electrical spin injection from such contacts.

The first report of electrical injection of spin-polarized carriers from a diluted FMS, $p$-type GaMnAs, was obtained for RTD structures [37, 38]. A $p$-$Ga_{0.965}Mn_{0.035}As$ emitter layer was grown by MBE on a GaAs/AlAs/ GaAs-QW/AlAs/p-GaAs(001) RTD structure—the corresponding VB diagram is shown as the inset of Figure 2.3. The structure was biased to inject holes from the GaMnAs emitter, and resonant tunneling occurs at bias voltages where the emitter VB edge is aligned with one of the confined LH or HH states in the GaAs QW (labeled LH1, HH2, etc.). As the temperature is lowered below the GaMnAs Curie temperature (~70 K), the resonant peak labeled HH2 splits, and the temperature dependence of the splitting mirrors that of the GaMnAs magnetization. The authors attribute this to the spontaneous exchange splitting of the GaMnAs VB edge below $T_c$ and resonant tunneling from the spin-polarized hole states, which accompanies the onset of FM order in the emitter. Thus, by selecting the bias voltage, one can use the RTD to preferentially transmit one hole spin state or the other. The fact that other resonant peaks did *not* split was tentatively attributed to some selection rule process, and is likely due to the orbital character/spin orientation of the GaAs QW hole states relative to the in-plane magnetization of the GaMnAs emitter contact. Although the RTD provides some qualitative indication of electrical spin injection, it is difficult to obtain more specific and very basic information such as the net spin polarization achieved.

### 2.4.3 SPIN-POLARIZED LIGHT-EMITTING DIODE: SPIN-LED

In order to study and develop electrical spin injection from a discrete contact, one must first have a reliable means of both *detecting* the presence of spin-polarized carriers and of *quantifying* the spin polarization achieved.

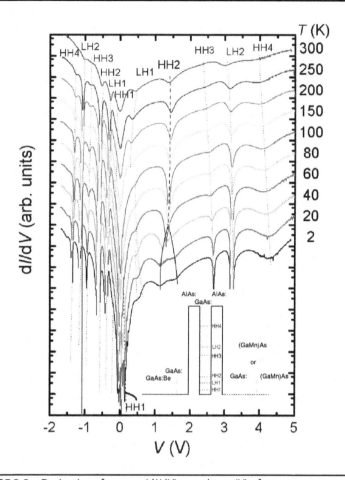

**FIGURE 2.3**   Derivative of current (d$I$/d$V$) vs. voltage ($V$) of a resonant tunneling diode with FM (Ga,Mn)As emitter. The labeling indicates the relevant resonant state in the GaAs well. When holes are injected from the (Ga,Mn)As side (positive bias), a spontaneous splitting of resonant peaks HH2 is observed below 80 K. The transition temperature of (Ga,Mn)As is expected to be 70 K. The splitting is attributed to spin splitting of (Ga,Mn)As VB states. (After Ohno, H. *Science* 281, 951, 1998. With permission.)

A simple LED structure takes advantage of one of the distinguishing characteristics of semiconductors—radiative recombination of carriers—and provides a powerful platform for this purpose. In a normal LED, electrons and holes recombine in the vicinity of a *p–n* junction or QW to produce light when a forward bias current flows. This light is unpolarized, because all carrier spin states are equally populated. However, if electrical injection produces a spin-polarized carrier population within the semiconductor, the same selection rules discussed above for optical pumping also describe the radiative recombination pathways allowed. Inspection of Figure 2.1 reveals that if injected carriers retain their spin polarization, radiative recombination results in the emission of circularly polarized light. A simple analysis based on these selection rules provides *a quantitative and model independent measure* of the spin polarization of the carriers participating.

A schematic of such a spin-LED [55] is shown in Figure 2.4 for surface and edge-emitting geometries. As an example, we consider injection of spin-polarized electrons from a magnetic contact layer, which recombine with holes supplied from the substrate. The net circular polarization $P_{circ}$ of the light emitted is readily determined from the measured intensities of the positive and negative helicity components of the electroluminescence (EL), $I(\sigma+)$ and $I(\sigma-)$, respectively Equation 2.2. These in turn are directly related to the occupation of the carrier states. Assuming a spin-polarized electron population and an unpolarized hole population in a bulk-like sample—i.e. all of the hole states are at the same energy and thus have the same probability of being occupied—a general expression for the degree of circular polarization in the Faraday geometry (Figure 2.4a) follows directly from Figure 2.1a. $P_{circ}$ can be written in terms of the relative populations of the electron spin states $n_\uparrow$ ($m_j = +1/2$) and $n_\downarrow$ ($m_j = -1/2$), where $0 \leq n \leq 1$, and $n_\uparrow + n_\downarrow = 1$.

$$
\begin{aligned}
P_{circ} &= \frac{[I(\sigma+) - I(\sigma-)]}{[I(\sigma+) + I(\sigma-)]} \\
&= 0.5 \frac{(n_\downarrow - n_-)}{(n_\downarrow + n_-)} \\
&= 0.5 P_{spin}.
\end{aligned}
\tag{2.2}
$$

The optical polarization is directly related to the electron spin polarization $P_{spin} = (n_\downarrow - n_\uparrow)/(n_\downarrow + n_\uparrow)$ at the moment of radiative recombination, and has a maximum value of 0.5 due to the bulk degeneracy of the HH and LH bands.

In a QW, the HH and LH bands are separated in energy by quantum confinement, which modifies Equation 2.2 and significantly impacts the analysis. The HH/LH band splitting is typically several meV even in shallow QWs, and is much larger than the thermal energy at low temperature (~0.36 meV at 4.2 K), so that the LH states are at higher energy and are not occupied (Figure 2.1b). For typical $Al_xGa_{1-x}As/GaAs$ QW structures with Al concentration $0.03 \leq x \leq 0.3$ and a 15 nm QW, a simple calculation yields a value of 3–10 meV for the HH/LH splitting [56]. Thus, only the HH levels

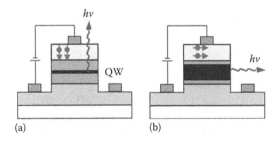

(a)                    (b)

**FIGURE 2.4**    Schematic of spin-LEDs showing relative orientation of electron spins and light propagation direction appropriate for deducing the spin polarization of the electrons participating in the radiative recombination process for (a) surface-emitting and (b) edge-emitting geometries. The holes are assumed to be unpolarized in these examples.

participate in the radiative recombination process at low temperature, as shown in Figure 2.1b, and $P_{circ}$ is calculated as before.

$$P_{circ} = \frac{(n_\downarrow - n_-)}{(n_\downarrow + n_-)} = P_{spin}. \tag{2.3}$$

In this case, $P_{circ}$ is equal to the electron spin polarization in the well, and can be as high as 1.

The use of a QW offers several distinct advantages over a $p$–$n$ junction in this approach. The QW provides a specific spatial location within the structure where the spin polarization is measured, and hence depth resolution. Varying the distance of the QW from the injecting interface may then provide a measure of spin transport lengths. This feature was utilized by Hagele et al. to obtain a lower bound of 4 μm for spin diffusion lengths in optically pumped GaAs at 10 K [57]. In addition, the light emitted from the QW has an energy characteristic of the QW structure, and may therefore be easily distinguished from spectroscopic features arising from other areas of the structure or impurity-related emission.

Note that the carrier spin polarization, $P_{spin}$, determined by this procedure is the spin polarization *at the instant and location* at which radiative recombination occurs in the structure (Figure 2.5). After injection at the interface, the spin-polarized carriers (a) must first transport some distance to the point of radiative recombination (e.g. to the QW), and then (b) wait some time $\tau_r$ characteristic of the structure before radiative recombination occurs, where $\tau_r$ is the radiative lifetime. Spin relaxation is likely to occur during both of these processes—scattering events during transport lead to loss of polarization, and the spin polarization decays exponentially with time as $\exp(-t/\tau_s)$, where $\tau_s$ is the spin lifetime (see Chapter 1, Volume 1 for a detailed discussion of spin relaxation). It is difficult to correct for (a) to obtain the spin polarization at the point of injection, $P_{inj}$, without very detailed knowledge of the interface structure and relevant scattering mechanisms. However, a straightforward procedure can be used to correct for (b).

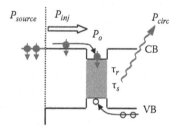

**FIGURE 2.5** Band diagram illustrating lifetimes and polarizations in a spin-LED. The electron spin polarization in the electrical contact, $P_{source}$, results in some net injected spin polarization $P_{inj}$ just inside the semiconductor. Through drift and diffusion, this electron spin reaches the QW and produces a spin polarization $P_o$. After radiative recombination, this is manifested as a circular polarization in the EL, $P_{circ}$. If the VB degeneracy is lifted, $P_{circ} = P_{spin}$. The initial QW spin polarization can be obtained from a rate equation analysis, $P_o = P_{spin} (1 + \tau_r/\tau_s)$.

Essentially, one takes a snapshot of the spin system with $\tau_r$ as the shutter speed. If $\tau_r$ was much shorter than $\tau_s$, this would provide an accurate measure of $P_{spin}$ in much the same way that a short exposure time or shutter speed is used to capture action photographs with a camera. This is rarely the case, however, and the spin system decays over the time $\tau_r$. Therefore, the result measured represents a lower bound for the carrier spin polarization achieved by electrical injection. A simple rate equation analysis may be applied to obtain a more accurate measure of the initial carrier spin polarization, $P_o$, that exists when the electrically injected carriers enter the region of radiative recombination (e.g. the QW). $P_o$ is given by [58]

$$P_o = P_{\text{spin}}\left(1+\frac{\tau_r}{\tau_s}\right), \tag{2.4}$$

where $P_{spin}$ is the value determined experimentally as described above. This provides a first-order correction for what is effectively the instrument response function of the LED to serve as a spin detector and the efficiency of the radiative recombination process in the particular structure and material utilized.

The value of $\tau_r/\tau_s$ can be determined by a simple PL measurement using near band-edge circularly polarized excitation (optical pumping) [5, 58]. Assuming that both LH and HH states are excited, the initial spin polarization produced by optical pumping is $P_o = 0.5$ (see Figure 2.1a). One measures the circular polarization of the PL, $P_{PL}$, and uses Equation 2.4 to solve for $\tau_r/\tau_s$: $0.5 = P_{PL}(1+\tau_r/\tau_s)$. Typical values are $1 \leq \tau_r/\tau_s \leq 10$, so that this correction can be significant.

Thus, the spin-LED serves as a *polarization transducer*, effectively converting carrier spin polarization, which is difficult to measure by any other method, to an optical polarization, which can be easily and accurately measured using standard optical spectroscopic techniques. The existence of circularly polarized EL demonstrates successful electrical spin injection (subject to appropriate control experiments), and an analysis of the circular polarization using these fundamental selection rules provides a quantitative assessment of carrier spin polarization in the QW without resorting to a specific model. Note that this approach measures the spin polarization of the carrier *population* in the semiconductor (electron or hole) achieved by electrical injection, and not the spin polarization of the injected *current*. Detailed knowledge of the transport mechanism and spin scattering is necessary to connect the two.

A number of conditions facilitate application of the spin-LED approach:

1. The analysis of the measurements is considerably more reliable if the experimental geometry is appropriate for the optical selection rules [5, 13]. The hole spin, electron spin, and the optical emission/analysis axes (or projections thereof) must be colinear to extract information on carrier spin polarization from the circular-polarized EL. The hole states in a QW have preferred orientations due to

quantum confinement and reduced symmetry [59, 60]: the $k = 0$ HH orbital angular momentum is oriented entirely along the growth direction ($z$-axis), whereas the $k = 0$ LH orbital angular momentum has non-zero projections in all three directions. Analysis of the HH QW exciton requires that the injected electron spin must be along the $z$-axis *and* the optical measurement must also be performed along the same axis, meaning that surface-emitted rather than edge-emitted light should be analyzed (Figure 2.4a) [61]. Thus, the Faraday rather than the Voigt geometry must be used. If the injected spins are oriented in-plane, then an edge-emission geometry (Figure 2.4b) may be utilized to analyze the EL *if* the radiative recombination region is bulk-like, or if one measures the LH exciton in a QW. An example of spin analysis for edge emission from a spin-LED is provided in Ref. [62].

2. The radiative recombination region should not be highly strained. Strain modifies the selection rules, compromising the quantitative relationship between $P_{circ}$ and $P_{spin}$. Strain may produce optical polarization in the absence of carrier spin polarization, or reduce $P_{circ}$ below that expected from the corresponding $P_{spin}$. While no system is perfect due to lattice mismatch, thermal expansion coefficients, and contact issues, the $Al_xGa_{1-x}As$/GaAs QW system comes very close due to the very small variation of lattice constant with Al concentration. In contrast, the lattice constant varies rapidly with In concentration, and strain is almost unavoidable in the GaAs/$In_xGa_{1-x}As$ QW system. Other factors complicate the use of InGaAs in spin-LEDs— the introduction of In, with its stronger spin–orbit interaction, may reduce spin lifetimes [63], and the relatively large $g$-factor introduces significant magnetic field-dependent effects, which complicate interpretation of the EL polarization [64].

3. The origin of the EL must be correctly identified to determine its spatial origin within the structure, and to confirm that it derives from recombination processes for which the selection rules are valid [65, 66], i.e. spin-conserving processes such as free exciton or free electron recombination. In practice, this means that the EL must be spectroscopically resolved and standard analyses applied to assist in the identification of the emission peaks. Note that bound exciton and impurity-related emission typically involve non-spin-conserving processes, and therefore cannot be used.

4. The VB degeneracy must be correctly identified, as described above. In a QW, the LH/HH splitting can be readily calculated for many materials [56], and should be compared to the measurement temperature to determine their relative contributions.

To briefly summarize, a quantitative and model independent determination of the initial spin polarization $P_o$ can be obtained from a spin-LED experiment if a few straightforward conditions are met: one employs the proper experimental geometry, identifies the radiative transition producing the EL, and determines a value for $\tau_r/\tau_s$ for the samples under consideration. This is all relatively easy to do in the $Al_xGa_{1-x}As/GaAs$ QW system, and, with some additional care, in the $GaAs/In_xGa_{1-x}As$ QW system.

### 2.4.4 MAGNETIC SEMICONDUCTORS AS SPIN-INJECTING CONTACTS: SPIN-LED STUDIES

#### 2.4.4.1 $Zn_{1-x}Mn_xSe$

As an example illustrating the concepts and procedures above, we consider electrical spin injection from a diluted magnetic semiconductor contact, $n$-type $Zn_{0.94}Mn_{0.06}Se$, into an AlGaAs/GaAs QW/AlGaAs LED structure, as first described by Fiederling et al. [67] and Jonker et al. [68]. $Zn_{1-x}Mn_xSe$ was chosen as the contact material for a number of reasons. It can readily be doped $n$-type, allowing one to focus on *electron* transport, and the giant Zeeman splitting described above provides an essentially 100% spin-polarized electron population at modest applied magnetic fields, albeit at low temperature. $Zn_{1-x}Mn_xSe$ forms high-quality epitaxial films on GaAs due to a close lattice match, and the CB offset can be tailored to facilitate electron flow from the $Zn_{1-x}Mn_xSe$ into the $Al_yGa_{1-y}As/GaAs$ structure by suitable choice of the Mn and Al concentrations [68].

A flat band diagram of the structure is shown in Figure 2.6 illustrating the band alignments and the spin splitting of the CB and VB edges of the $Zn_{1-x}Mn_xSe$ with applied magnetic field. The ZnMnSe spin polarization may be varied simply by varying the applied field, a very useful handle for experimental studies. At sufficiently high fields, the CB spin splitting $(\Delta E = g^*u_B H \sim 10\,\text{meV})$ is much larger than the measurement temperature ($\sim 0.4\,\text{meV}$), so that the $Zn_{1-x}Mn_xSe$ contact is essentially 100% spin-polarized. The samples were grown by MBE and processed into surface-emitting LED mesas 200–400 μm in diameter using standard photolithographic techniques. Figure 2.7 shows a schematic cross-section and photograph of the final devices. The top metallization to the ZnMnSe consists of concentric Au rings to help insure uniform current distribution, leaving most of the mesa

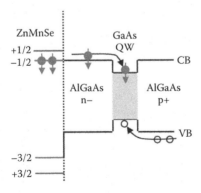

**FIGURE 2.6**   ZnMnSe/AlGaAs spin-LED band diagram, illustrating the giant Zeeman splitting of the ZnMnSe band edges and electrical injection of spin-polarized electrons into the GaAs QW.

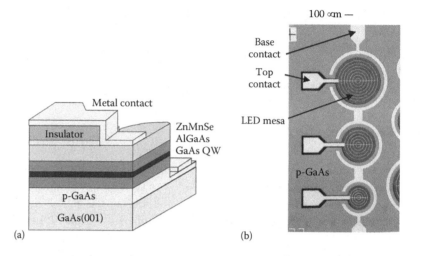

**FIGURE 2.7**   (a) Schematic cross-section of spin-LED illustrating the structure and processing steps. (b) Photograph of completed surface-emitting devices. The active mesa areas (dark circular regions) are 200, 300, and 400 μm in diameter. (After Ohno, H. *Science* 281, 951, 1998. With permission.)

surface optically transparent. Details of the MBE growth and device fabrication may be found in reference [68].

The EL was measured by electrically biasing the LEDs to inject electrons from the *n*-ZnMnSe into the GaAs QW at current densities of ~0.01–1.0 A cm$^{-2}$. Representative EL spectra from such a structure— (in this case with a multiple QW LED ($3 \times (20$ nm $Al_{0.08}Ga_{0.92}As/10$ nm GaAs))—are shown in Figure 2.8 for selected values of the applied field. The light emitted along the surface normal and magnetic field direction (Faraday geometry) was analyzed for σ+ and σ− circular polarization and spectroscopically resolved. The energy of the emission confirms that the radiative recombination occurs in the GaAs QW via the HH ground state exciton (note that other tests are applied to confirm this identification— see Ref. [65]). As noted in the previous section (condition (1)), quantum confinement locks the HH spins along the *z*-axis (surface normal), and therefore, both the injected electron spins and the axis for optical analysis must be oriented along the *z*-axis as well to interpret any circular polarization observed in terms of spin polarization. Edge emission will give a spurious or null result [61].

At zero field, the σ+ and σ− components are identical, as expected, since no spin polarization yet exists in the ZnMnSe. As the magnetic field increases, the ZnMnSe bands split into spin states due to the giant Zeeman effect, and spin-polarized electrons are injected into the AlGaAs/GaAs LED structure. The corresponding spectra exhibit a large difference in intensity between the σ+ and σ− components, demonstrating that the spin-polarized electrons successfully reach the GaAs QW. The circular polarization, $P_{circ} = [I(\sigma+) - I(\sigma-)]/[I(\sigma+) + I(\sigma-)]$, increases with applied field as the ZnMnSe

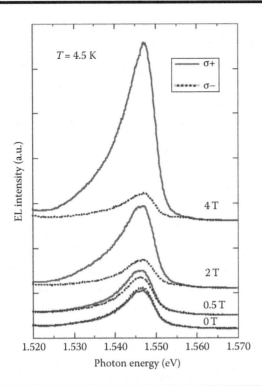

**FIGURE 2.8** EL spectra from a surface-emitting spin-LED with a $Zn_{0.94}Mn_{0.06}Se$ contact for selected values of applied magnetic field, analyzed for positive ($\sigma+$) and negative ($\sigma-$) circular polarization. The magnetic field is applied along the surface normal (Faraday geometry). The spectra are dominated by the HH exciton. Typical operating parameters are 100 μA and 2.5 V.

polarization increases, and saturates at a value of ~80% at 4 T. From the discussion leading to Equation 2.3, the spin polarization of the electron population in the GaAs QW, $P_{spin}$, is therefore 80%.

This value is the QW spin polarization at the time of radiative recombination, i.e. after a time $\sim\tau_r$ has elapsed. A value for the initial QW spin polarization, $P_o$, can be obtained using Equation 2.4 to correct for the effect that the relative values of the radiative and spin lifetimes have on $I(\sigma+)$ and $I(\sigma-)$. Independent PL measurements using circularly polarized excitation provide a value of $\tau_r/\tau_s$, as described previously. The PL is analyzed for $\sigma+$ and $\sigma-$ helicity, and exhibits a polarization $P_{PL} = 40\%$. Applying $0.5 = P_{PL}(1 + \tau_r/\tau_s)$ gives $\tau_r/\tau_s = 0.25 \pm 0.05$. Equation 2.4 then gives $P_o \sim 100\%$, demonstrating that a well-ordered ZnMnSe/AlGaAs interface is essentially transparent to spin transport. Thus, an all-electrical process using a discrete contact can produce a carrier spin polarization within the semiconductor that equals or exceeds that achieved via optical pumping.

Several control experiments rule out contributions to the circular polarization measured from spurious sources, and are an essential part of any such experiment. LEDs with nonmagnetic $n$-ZnSe contact layers show no circular polarization with magnetic field, as expected. The circular dichroism resulting from transmission through the ZnMnSe is negligible because the GaAs QW emission wavelength is very far from that corresponding to the band gap of $Zn_{0.94}Mn_{0.06}Se$. PL data from the GaAs QW excited with linearly polarized light from the same LED mesa structures used for the EL studies show little polarization, providing a very effective built-in reference for each mesa LED. Such dichroism effects could be much larger for emission energies very near the ZnMnSe band gap, where strong absorption occurs, and must be considered when designing the LED structure.

### 2.4.4.2 $Ga_{1-x}Mn_xAs$

Spin-LED structures have also been fabricated with $p$-type GaMnAs as the spin contact. The first report [39] utilized a 300 nm $Ga_{0.955}Mn_{0.045}As/$ GaAs/10 nm $In_{0.13}Ga_{0.87}As$ QW/$n$-GaAs(001) heterostructure in which spin-polarized holes injected from the GaMnAs radiatively recombined in the strained InGaAs QW with unpolarized electrons from the substrate. Because the magnetization of the GaMnAs (and nominal spin orientation of the injected holes) was in-plane, they used an edge-emission geometry with the optical axis parallel to the magnetization. They measured the EL at the QW ground state transition, and observed a 1% polarization, which exhibited the same temperature dependence as the GaMnAs magnetization. A reliable interpretation in terms of hole spin injection and polarization, however, is compromised by the use of a strained QW as the radiative recombination region—as discussed previously, (1) the orbital angular momentum of the QW HH ground state is oriented along the growth direction ($z$-axis), orthogonal to the nominal in-plane orientation of the spins injected from the GaMnAs, and (2) strain in the QW leads to admixture of states and modifies the selection rules. Inducing an out-of-plane magnetization in the GaMnAs (e.g. by applying a magnetic field) or utilizing a structure with a bulk-like recombination region (Figure 2.4b) may have alleviated these issues.

More definitive measurements were performed on identical sample structures in 2002 using the Faraday geometry (surface emission) to address the first issue noted above [40, 69]. The EL spectra at $T = 5$ K and the field dependence of the circular polarization for selected temperatures are shown in Figure 2.9. The field dependence tracks the out-of-plane (hard axis) magnetization of the GaMnAs (Figure 2.9b), and the saturation value depends upon the thickness, $d$, of the undoped GaAs spacer layer between the GaMnAs injector and the InGaAs QW. The circular polarization is 4% for $d = 70$ nm, and increases to 7% for $d = 20$ nm, attributed to the hole spin diffusion length effects in the GaAs spacer. These data indicate that the spin polarization of the QW hole population is at least 7% at the measurement temperature. The polarization decreases with increasing temperature, and disappears by 62 K—the Curie temperature of the GaMnAs injector. In Ref. [69], the use of small mesas to force the easy axis of the GaMnAs contact out of plane permitted the demonstration of hole spin injection without an applied magnetic field in the surface emission geometry.

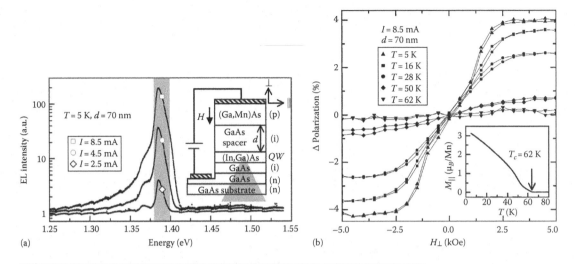

**FIGURE 2.9** (a) Spectrally resolved EL intensity along the growth direction for several bias currents, $I$. Inset shows device schematic and EL collection geometries. (b) Temperature dependence of the relative changes in the energy-integrated [gray shaded area in (a)] polarization $\Delta P$ as a function of out-of-plane magnetic field. When $T < 62$ K, polarization saturates at $H_\perp \sim 2.5$ kOe. Inset shows $M(T)$, indicating that the polarization is proportional to the magnetic moment. (After Young, D.K. et al., *Appl. Phys. Lett.* 80, 1598, 2002. With permission.)

### 2.4.4.3 Spin Scattering by Interface Defects

Spin scattering by interface defects is an important issue in heteroepitaxial systems, and can rapidly suppress the spin polarization achieved by electrical injection. A comprehensive understanding of this process remains a major challenge for future research. Initial work addressed the effect of interface nanostructure on the spin injection efficiency at the ZnMnSe/AlGaAs interface [70]. High resolution transmission electron microscopy (TEM) was used to assess the interface morphology in spin-LED structures, and found that the most prevalent defects were stacking faults (SF) in <111> directions nucleating at or near the ZnMnSe/AlGaAs interface, where they formed line defects (Figure 2.10a). The electron spin polarization in the GaAs QW determined from the circular polarization $P_{circ}$ of the EL exhibited an approximately linear dependence on the density of interface defects, as shown in Figure 2.10b. This correlation was explained by a model that incorporated spin–orbit (Elliot–Yafet) scattering—the model showed that the asymmetric potential of the interface defect results in strong spin-flip scattering in the forward direction. A simple expression, $P_{circ} \sim 1 - 4r_o n$, gave excellent agreement with the experimental data with no adjustable parameters, where $r_o \sim 10$ nm is the Thomas–Fermi screening length and $n$ is the measured defect density. These results provided the first experimental demonstration that interface defect structure limits spin injection efficiency in the diffusive transport regime. It is interesting to note that strong spin injection persists even when the misfit dislocation density greatly exceeds values which would be fatal for conventional devices such as III–V-based diode lasers and field effect transistors. This is especially reassuring, since

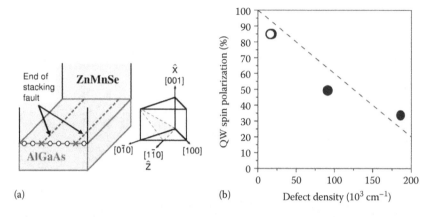

(a)                                      (b)

**FIGURE 2.10**    (a) Diagram illustrating the linear interface defects resulting from the intercept of (111)-type SF planes and the interface plane. Only one of the four possible (111)-type SF planes is shown for clarity. (b) Correlation of GaAs QW spin polarization with stacking fault density. Two data points nearly overlap at 85% polarization. The error bars are comparable to the symbol size. The dashed line is the calculated result with no adjustable parameters.

interface misfit dislocations are a generic defect routinely encountered in heteroepitaxial device structures.

## 2.5 FERROMAGNETIC METAL/SEMICONDUCTOR ELECTRICAL SPIN INJECTION: OPTICAL DETECTION WITH SPIN-LED STRUCTURES

FM metals offer most of the properties desired for a practical spin-injecting contact material: reasonable spin polarizations at $E_F$ (~40%–50%), high Curie temperatures, low coercive fields, fast switching times, and a well-developed material technology due to decades of research and development driven in large part by the magnetic storage industry. An FM contact introduces *nonvolatile* and *reprogrammable* operation in a very natural way, due to both the material's intrinsic magnetic anisotropy and the ability to tailor the magnetic characteristics in numerous ways. Spin-torque transfer switching is rapidly emerging as the mechanism of choice for manipulating the contact magnetization, and offers many advantages for spintronic applications, such as embedded memory or field programmable gate arrays (see Chapters 7, 8, Volume 1, and Chapter 4, Volume 2 for an overview). This emerging technology can readily be incorporated in semiconductor spintronic devices. In addition, metallization is a standard process in any semiconductor device fabrication line, so that an FM metallization could easily be incorporated into existing processing schedules.

Initial efforts to inject spin-polarized carriers from an FM metal contact into a semiconductor were not very encouraging. Several groups reported a change in voltage or resistance on the order of 0.1%–1%, which they attributed to spin accumulation or transport in the semiconductor [71–73]. Such small effects, however, make it difficult to either unambiguously confirm spin injection or successfully implement new device concepts. In addition, some have argued that these measurements were compromised by contributions from anisotropic magnetoresistance, or a local Hall effect, which can easily contribute a signal of 1%–2% [74–76]. Thus, great care must be taken in the experimental design to recognize and eliminate such spurious contributions.

Zhu et al. [77] utilized optical rather than electrical detection in a spin-LED structure (the spin-LED is described in Section 2.4.3) consisting of an Fe Schottky contact to a GaAs/In$_{0.2}$Ga$_{0.8}$As QW LED detector. They were unable to observe any clear difference in intensity when they analyzed the EL for positive (σ+) or negative (σ–) helicity polarization, in contrast with earlier results for a ZnMnSe magnetic semiconductor contact, as described in Section 2.4.4.1. However, by examining the high and low energy tails of the EL peak (which they attributed to LH and HH contributions, respectively) using pulsed current injection and lock-in detection techniques, with some background subtraction, they identified a signal which they attributed to electrical spin injection from the Fe contact. They concluded that a spin polarization of ~2% had been achieved in the GaAs, and found that this signal was independent of temperature from 25 to 300 K. The lack of temperature dependence raises some question as to their interpretation of this signal, since the spin lifetime is known to decrease rapidly with temperature for GaAs(001) structures [78].

Two fundamental issues are key to understanding, successfully demonstrating, and optimizing electrical spin injection from an FM metal into a semiconductor—the role of band symmetries in facilitating spin transmission across a heterointerface, and the impact of the large difference in conductivity between the metal and semiconductor.

## 2.5.1 Band Symmetries and Spin Transmission

The first issue can be stated rather simply: the symmetries and orbital composition of the bands participating in the transport process in the metal must be compatible with those of the semiconductor for optimum spin transmission to occur. Theoretical work has explored the electronic structure at the FM metal/semiconductor interface in an effort to elucidate the role of band structure in spin injection [79–82]. This work has emphasized the importance of matching the *symmetries* as well as the energies of the bands between metal and semiconductor to optimize spin injection efficiency.

Some initial insight can be gained by examining the bulk band structures. Figure 2.11 shows selected portions of the band structure along (001) for Fe, GaAs [82, 83], and Si [84]. For typical semiconductors of interest, including GaAs, InAs, GaP, ZnSe, and Si, the CB for the (001) surface exhibits $\Delta_1$ symmetry near the zone center (dashed curves), with significant $s$ and $p_z$-orbital contributions. The Fe *majority* spin band (dashed curve), which crosses $E_F$ midway across the zone, is also of $\Delta_1$ symmetry, with significant $s$-, $p_z$-, and $d_{z^2}$-orbital contributions. The $s$ and $p_z$ components have large spatial extent, and the $p_z$ and $d_{z^2}$ orbitals point directly into the semiconductor, leading to strong overlap with the corresponding states comprising the

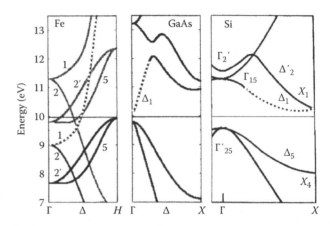

**FIGURE 2.11** Band structures along (001) for Fe, GaAs, and Si. The majority spin band of Fe crossing the Fermi level is of $\Delta_1$ symmetry (labeled "1" in the plot, dashed curve), and the CB edge of both GaAs and Si (dashed curve) is also of $\Delta_1$ symmetry. The Fe minority spin bands are of different symmetry ($\Delta_2$, $\Delta_5$, labeled "2" and "5"). This is expected to enhance transmission of majority spin electrons from the Fe, and suppress minority spin injection, leading to highly polarized spin currents in the semiconductor. The plots for Fe and GaAs are taken from Ref. [76]—the underestimate of the GaAs band gap is characteristic of the computational approach used. The plot for Si is derived from Ref. [78], and the energy scale is expanded relative to that used for Fe and GaAs. (After Wunnicke, O. et al., *Phys. Rev. B* 65, 241306(R), 2002; Chelikowsky, J.R. et al., *Phys. Rev. B* 14, 556, 1976. With permission.)

semiconductor CB. The corresponding Fe $\Delta_1$ *minority* spin band is 1.3 eV above $E_F$ and therefore will not contribute to transport. In contrast, the Fe *minority* spin bands which cross $E_F$, exhibit quite different symmetries ($\Delta_2$, $\Delta_5$), with orbital components that do not couple strongly to the semiconductor. Thus, the CB of the semiconductor is well-matched in orbital composition/symmetry and energy to the majority spin bands of the Fe, facilitating propagation of majority spin electrons, while a poor match exists to the Fe minority spin bands.

Theoretical treatments have addressed several FM metal/semiconductor systems, and generally assume a well-ordered interface so that the electron momentum parallel to the interface plane, $k_{//}$, is conserved to simplify the calculation. For the case of spin injection from Fe into GaAs(001), calculations show that these band symmetries play a critical role, leading to a significant enhancement of transmission from the $\Delta_1$ Fe majority spin band at $E_F$, and a suppression of transmission from the Fe minority spin bands ($\Delta_2$, $\Delta_{2'}$, $\Delta_5$) [82, 85]. Consequently, majority spin electrons are preferentially transmitted from Fe into the GaAs, while minority spins are blocked, producing a significant enhancement of spin injection efficiency. In such cases, the metal/semiconductor interface essentially serves as a *band structure spin filter* due to basic issues of band symmetry and orbital composition.

Similar arguments can be made for the case of Fe and Si(001). As seen in Figure 2.11, the bottom of the Si(001) CB is also of $\Delta_1$ symmetry, enhancing transmission of Fe $\Delta_1$-band majority spin electrons, and suppressing minority spin current. Calculations assuming an ideal Fe/Si(001) interface indicated that the current injected from the Fe contact should be strongly spin-polarized [86]. Similar conclusions were obtained for Fe/InAs(001) [87], and earlier for epitaxial Fe/MgO tunnel barriers by Butler et al. [88].

The band bending which occurs at the metal/semiconductor interface complicates the picture, but can be included in the calculation to some extent. Wunnicke et al. [82] and Mavropoulos [86] simulated the effect of a Schottky barrier by introducing a potential step between the Fe and the semiconductor, and found that the current remained strongly polarized.

Thus, the knowledge of the interface band structure is important for understanding and optimizing spin injection efficiency. The *physical structure* of the interface is a key ingredient here. The perfectly abrupt interface typically assumed for calculation is unlikely to exist in practice. Intermixing leading to compound formation or disorder will alter the band symmetries and strongly impact the arguments above. Symmetry breaking due to disorder at the Fe/InAs(001) interface in the form of Fe atoms occupying In or As interface sites was shown to rapidly suppress the spin filtering effect and resultant high spin polarization in the InAs by opening more channels for minority spin transport [81].

It should be noted that compound formation per se is not necessarily detrimental, provided that bands with appropriate symmetries and/or orbital composition are preserved or created. It is indeed possible that the spin injection efficiency at certain interfaces with poorly matched band symmetries will improve with compound formation which produces bands of more favorable orbital composition. However, disorder due to random intermixing, poorly ordered compound formation, or defects will, in general, lead to mixing of states, compromising any state-specific transmission, and reducing the spin injection efficiency.

## 2.5.2 CONDUCTIVITY MISMATCH: DESCRIPTION OF THE PROBLEM

The second fundamental issue presenting an unanticipated challenge to utilizing an FM metal as a spin-injecting contact on a semiconductor is the very large difference in conductivity between the two materials. Simply stated, the ability of the semiconductor to accept carriers is independent of spin, and much smaller (lower conductivity) than that of the metal to deliver them. Consequently, equal numbers of spin-up and spin-down electrons are injected, regardless of the FM metal's initial polarization, resulting in essentially zero spin polarization in the semiconductor.

This issue, commonly referred to in the literature as the problem of "conductivity mismatch," can readily be understood in the context of the canonical two-channel model typically used to describe spin-polarized current flow in FM metals [89]. In this model, spin-up or "majority spin" electrons flow in one channel, while spin-down or "minority spin" electrons flow in the other.* Using the analogy of water flow through a hose for electrical

---

* For a discussion of this terminology, see Ref. [90].

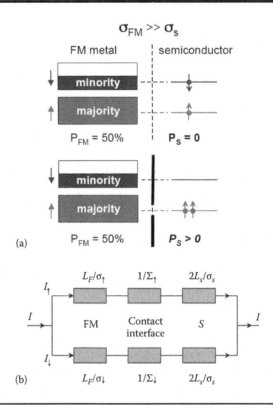

**FIGURE 2.12**    (a) Schematic of the two-channel model of spin transport illustrating the conductivity mismatch issue, which must be considered for spin injection from an FM metal into a semiconductor. A spin-selective interface resistance provides a solution, as shown in the lower panel. (b) Equivalent resistor model (due to A.G. Petukhov). (After Jonker, B.T. et al., *MRS Bull.* 28, 740, 2003. With permission.)

resistance, the relative conductivities of these channels in each material can be represented by the diameter of the hose, as shown in Figure 2.12a. In the metal, the diameter of both spin-up and spin-down hoses is very large (high conductivity), while in the semiconductor, the diameter is very small (low conductivity). The problem of conductivity mismatch and its impact on spin injection is thus reduced to the intuitive physical picture of attempting to effectively transfer water from a sewer pipe to a drinking straw.

In the FM metal, the carriers are partially spin-polarized (by definition), so that one of the pipes is nearly full while the other may be nearly empty. In the nonpolarized, low carrier density semiconductor, the two spin channels are of equal diameter and partially full. If one now considers the flow of water (spin-polarized current) from the sewer pipes (metal) to the drinking straws (semiconductor), it is immediately apparent that the comparatively small conductivity of the semiconductor limits current flow. While one drop of water may be transferred from the majority spin sewer pipe of the metal to the corresponding majority spin drinking straw of the semiconductor, an identical amount will flow in the minority spin channel, even though the metal pipe may be nearly empty, due to the limited conductivity of the semiconductor. This results in *zero* spin polarization in the semiconductor. It is apparent that this will be the case regardless of how nearly empty the metal minority spin band may be, due to the very limited ability of the semiconductor bands to accept current flow. This is the essence of the conductivity mismatch issue.

This simple picture provides rather surprising insight into efforts to transfer spin-polarized carriers between these two materials. In the diffusive transport regime, significant spin injection can occur for only two conditions: (1) the FM metal must be 100% spin-polarized (minority channel completely empty), or (2) the conductivity of the FM metal and semiconductor must be closely matched. No FM metal meets either of these criteria. While half-metallic materials offer 100% spin polarization in principle [91, 92], defects such as antisites or interface structure rapidly suppress this value [93].

Model calculations by several groups [76, 94–98] have addressed this issue and provide a quantitative treatment. These authors extended the work of van Son et al. [99] and Valet and Fert [89] to calculate the spin polarization achieved in a semiconductor due to transport of spin-polarized carriers from the ferromagnet. An equivalent resistor circuit explicitly incorporates the conductivities of the FM metal, interface, and semiconductor [11, 94], as shown in Figure 2.12b, and permits discussion of spin injection efficiency in terms of these physical parameters in the classical diffusive transport regime. The resistances representing the FM metal and semiconductor are given by $L_F/\sigma_F$ and $L_s/\sigma_s$, where $L$ and $\sigma$ are the spin diffusion lengths and conductivities, respectively, with $\sigma_F = \sigma_\uparrow + \sigma_\downarrow$ summing the two FM spin channels. The interface conductivity $\Sigma = \Sigma_\uparrow + \Sigma_\downarrow$ is assumed to be spin dependent. With some algebra and effort, the spin injection coefficient $\gamma$ can be shown to be

$$\gamma = \left( \frac{I_- - I_\downarrow}{I} \right) = \frac{\left( r_F \dfrac{\Delta\sigma}{\sigma_F} + r_c \dfrac{\Delta\Sigma}{\Sigma} \right)}{\left( r_F + r_S + r_c \right)}, \tag{2.5a}$$

where:

$$r_F = L_F \frac{\sigma_F}{4\sigma_- \sigma_\downarrow}, \tag{2.5b}$$

$$r_c = \frac{\Sigma}{4\Sigma_- \Sigma_\downarrow}, \tag{2.5c}$$

$$r_s = \frac{L_s}{\sigma_s}, \tag{2.5d}$$

are the effective resistances of the ferromagnet, contact interface, and semiconductor, respectively, $I_\uparrow$ and $I_\downarrow$ are the majority and minority spin current, $\Delta\sigma = \sigma_\uparrow - \sigma_\downarrow$ and $\Delta\Sigma = \Sigma_\uparrow - \Sigma_\downarrow$.

For an FM metal, $r_F/(r_F + r_s + r_c) \ll 1$, and the contribution of the first term in Equation 2.5a is negligible. The second term is significant only if $\Delta\Sigma \neq 0$ and $r_c \gg r_F, r_s$. Thus, two criteria must be satisfied for significant spin injection to occur across the interface between a typical FM metal and a semiconductor: the interface resistance $r_c$ must (1) be spin selective, and (2) dominate the series resistance in the near-interface region.

### 2.5.3 CONDUCTIVITY MISMATCH: PRACTICAL SOLUTION

A tunnel barrier between the FM metal and semiconductor satisfies both these criteria, and was suggested as a potential solution to the conductivity mismatch problem by several groups [76, 94–97]. The spin selectivity ($\Delta\Sigma \neq 0$) comes naturally from the spin-polarized DOS of the FM metal at $E_F$, which serves as the source term in a mathematical treatment of the tunneling process. The resistance of the tunnel barrier is readily controlled by its thickness, and can easily be the largest in the series resistance such that $r_c \gg r_F, r_s$. A detailed calculation was provided by Rashba [95], and a more comprehensive overview was provided recently by Fert et al. [100].

A tunnel barrier can be introduced at a metal/semiconductor interface in at least two ways: tailoring the band bending in the semiconductor, which typically leads to Schottky barrier formation, or physically inserting a discrete insulating layer such as an oxide. Examples of both are presented below for GaAs and Si.

#### 2.5.3.1 Tailored Schottky Barrier as a Tunnel Barrier

One avenue is to take advantage of the band bending which occurs at the metal/semiconductor interface. This approach exploits a natural characteristic of the interface, and avoids the use of a discrete barrier layer and the accompanying problems with pinholes and thickness uniformity. Schottky contacts are also routine ingredients in semiconductor technology. In the case of an $n$-type semiconductor, electrons are transferred into the metal, depleting the semiconductor interfacial region and causing the CB to bend upward, forming a pseudo-triangular-shaped barrier with a quadratic falloff with distance into the semiconductor [101]. The depletion width associated with the Schottky barrier depends upon the doping level of the semiconductor, and is generally far too large to allow tunneling to occur. For example, in $n$-GaAs, the depletion width is on the order of 100 nm for $n \sim 10^{17}$ cm$^{-3}$, and 40 nm for $n \sim 10^{18}$ cm$^{-3}$ [102]. However, this width can be tailored by the doping profile used at the semiconductor surface [103]. Heavily doping the surface region can reduce the depletion width to a few nanometers, so that electron tunneling from the metal to the semiconductor becomes a highly probable process under reverse bias.

This approach was first utilized to achieve large electrical spin injection from Fe epilayers into AlGaAs/GaAs QW LED structures [10, 104, 105]. The $n$-type doping profile of the surface AlGaAs was designed by solving Poisson's equation with several criteria in mind: (a) minimize the Schottky barrier depletion width to facilitate tunneling; (b) use a minimum amount of dopant and heavily doped regions to accomplish this, since high $n$-doping is associated with stronger spin scattering and short spin lifetimes [106, 107]; and (c) avoid formation of an electron "puddle" or accumulation region (i.e. pushing $E_F$ above the CB edge), which may either dilute the polarization of the electrons injected from the Fe contact or contribute to spin scattering. A schematic of the doping profile and resultant barrier is shown in Figure 2.13a, and the band diagram resulting from the Poisson equation solution for the full LED structure is illustrated in Figure 2.13b. The doping

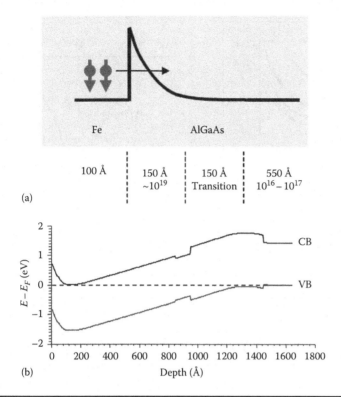

**FIGURE 2.13** Design of an Fe Schottky barrier and spin-LED using a doping profile to facilitate tunneling of spin-polarized electrons from the Fe through the Schottky tunnel barrier. (a) Doping profile and resultant reduction of depletion width at the Fe/AlGaAs interface, and (b) Poisson equation solution corresponding to doping profile and AlGaAs/GaAs QW/AlGaAs structure described in the text.

of the top 150 Å of $n$-type $Al_{0.1}Ga_{0.9}As$ was chosen to be $n = 1 \times 10^{19}\,cm^{-3}$ to minimize the depletion width, followed by a 150 Å transition region, while the rest was $n = 1 \times 10^{16}\,cm^{-3}$ with a 100 Å dopant setback at the QW. The LED structure consisted of 850 Å $n$-$Al_{0.1}Ga_{0.9}As$/100 Å undoped GaAs/500 Å $p$-$Al_{0.3}Ga_{0.7}As$/$p$-GaAs buffer layer on a $p$-GaAs(001) substrate. The width of the GaAs QW was chosen to be 100 Å to ensure separation of the LH and HH levels and corresponding excitonic spectral features, an important consideration for quantitative interpretation of the data, as discussed earlier. Details of the growth may be found elsewhere [104, 105].

### 2.5.3.1.1 Confirmation of Tunneling: Analysis of the Transport Mechanism

To demonstrate that the tailored doping profile reduces the depletion width sufficiently to produce a tunnel barrier, it is necessary to analyze the current–voltage ($I$–$V$) characteristics of the Fe/AlGaAs Schottky contact and apply well-known criteria to identify the dominant transport process. Such criteria should also be applied when a discrete oxide barrier is used. The mere addition of a layer intended to serve as a nominal tunnel barrier does not ensure that transport occurs by tunneling—indeed, pinholes are a chronic problem, and extensive work in the metal/insulator/metal tunnel

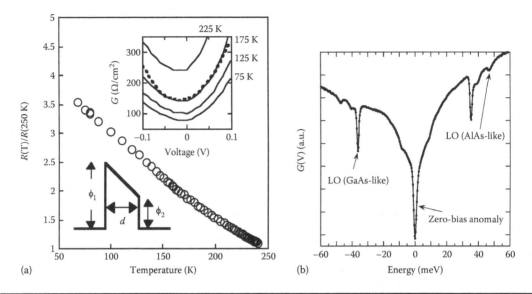

**FIGURE 2.14** (a) Inset: series of conductance curves taken at different temperatures. The dotted lines are representative fits to the data. Parameters for the fitting are defined in the schematic of the tunnel barrier. Normalized ZBR as a function of temperature for an Fe/Al$_{0.1}$Ga$_{0.9}$As Schottky barrier contact, showing minimal temperature dependence consistent with tunneling. (b) Conductance vs. applied voltage at 2.7 K. A zero bias anomaly, as well as phonon peaks attributed to GaAs-like and AlAs-like LO phonons, is clearly visible.

junction community has shown that their presence cannot be ruled out easily [108–110]. Pinholes form low-resistance areas, which essentially short out the high resistance tunnel barrier layer. Application of the "Rowell criteria" for tunneling [108, 111], and observation of phonon signatures and a zero bias anomaly in the low temperature conductance spectra provide clear, unambiguous tests to determine whether tunneling is the dominant transport mechanism.

There are three Rowell criteria. The first—the conductance ($G = dI/dV$) should have an exponential dependence on the thickness of the barrier—cannot be readily applied in this case due to the nonrectangular shape of the barrier and variations of the barrier width with bias. The second criterion states that the conductance should have a parabolic dependence on the voltage and can be fit with known models, e.g. a Simmons (symmetric barrier) [112] or Brinkman, Dynes, and Rowell (BDR) model (asymmetric barrier) [113]. The inset to Figure 2.14a shows $G$–$V$ data at a variety of temperatures, and a representative fit (dashed line) using the BDR model. Parameters of this asymmetric barrier model are defined in the diagram. Fits to the data between ±100 meV at several different temperatures yield an average barrier thickness of $d = 29$ Å, and barrier heights of $\phi_1 = 0.46$ eV and $\phi_2 = 0.06$ eV.* The large potential difference between the two sides of the barrier is physically

---

* Notice that the voltage range here is much smaller than the 1~2 V applied to the LED in the spin injection experiment. In the transport measurement, the voltage drop is primarily across the tunnel barrier interface, while in the LED, there are a variety of series resistances and contact resistances, as well as the band gap of GaAs that need to be considered.

consistent with a triangular Schottky barrier tunnel junction. Although $\phi_1$ is lower than might be expected for an Fe/AlGaAs interface, image force lowering of the barrier due to the highly degenerate nature of the AlGaAs can lead to reduction of the barrier by >0.3 eV [103]. The "goodness" of the fits, energy range considered, and deviation of the fit parameters from "known" physical characteristics of the barrier are typical of similar treatments in the literature [108, 114]. Therefore, one may conclude that the second Rowell criterion is satisfied.

While the first two criteria are routinely invoked as proof of tunneling, it has been argued that *neither* can reliably distinguish tunneling from contributions due to spurious effects such as pinholes [108–110]. It has been shown that the $G$–$V$ data can be fit with reasonable parameters even when tunneling was not the dominant transport path [108]. Jönsson-Åkerman et al. have presented convincing evidence that the *third* Rowell criterion is indeed a definitive confirmation of tunneling [108]. This criterion states that the zero bias resistance (ZBR), i.e. the slope of the $I$–$V$ curve at zero bias, should exhibit a weak, insulating-like temperature dependence. ZBR data are shown in Figure 2.14a as a function of temperature for the tailored Fe/AlGaAs Schottky contact, and clearly exhibit such a weak dependence over a wide range of temperature. Thus, the third Rowell criterion is also satisfied, confirming that single-step tunneling is the dominant conduction mechanism.

Further evidence for tunneling may be provided by the observation of phonon signatures and a pronounced zero bias anomaly in the conductance spectra.* For transport across an ohmic contact, electron energies can never reach a value higher than a few kT above $E_F$, regardless of the applied bias, and phonon modes are not observed because the injected electron energy is too low to excite them. A tunnel barrier enables injection of electrons at higher energies that are sufficient to excite phonons, thereby enabling a corresponding spectroscopy [115]—the observation of such features then provides further proof for tunneling.

The conductance spectrum for an Fe/Al$_{0.1}$Ga$_{0.9}$As sample at 2.7 K is shown in Figure 2.14b, and exhibits two distinct features between 30 and 50 meV. In AlGaAs, two sets of longitudinal-optical (LO) phonon modes are present, one GaAs-like and the other AlAs-like, with energies of 36 and 45 meV, respectively. The observed features agree very well with these nominal values, and are labeled accordingly. In addition, the relative intensities of the GaAs and AlAs phonon interactions are positively correlated with the relative Ga:Al content [116]—the observed features exhibit an intensity ratio of ~10:1, further confirming their identity.

At the lowest temperatures, the $G$–$V$ data have a pronounced feature at zero bias. Zero bias anomalies are generally observed in semiconductor tunneling devices [115]. Although poorly understood, they have been attributed to inelastic scattering effects arising from acoustic phonons and barrier defects, and are closely associated with tunneling. The observation of such a feature in these data provides further evidence for tunneling.

---

* See, e.g., Ref. [115].

### *2.5.3.1.2 Examples: Spin Injection via Tailored Schottky Tunnel Barrier*

The Fe/AlGaAs/GaAs/AlGaAs LED structures described above (Figure 2.13) provide conclusive evidence for efficient spin injection from an FM metal into a semiconductor, using the tunnel barrier to circumvent the conductivity mismatch [104, 105]. These samples were processed to form surface emitting LEDs (Figures 2.4a and 2.7b), annealed at 200°C to improve the interface structure, and biased to inject spin-polarized electrons from the Fe through the tailored Schottky tunnel barrier and into the semiconductor, where they radiatively recombine with unpolarized holes. The EL spectra are shown in Figure 2.15a, where the light emitted along the surface normal (Faraday geometry) is analyzed for σ+ and σ− circular polarization for selected values of applied magnetic field. The spectra are dominated by the GaAs QW HH exciton, with a linewidth of 5 meV, clearly identifying the location where radiative recombination occurs.

With no applied magnetic field, the Fe magnetization (easy axis) and corresponding electron spin orientation are entirely in the plane of the thin film. Although spin injection may indeed occur, it cannot be detected via surface emission because the average electron spin along the surface normal ($z$-axis) is zero, and the σ+ and σ− components are nearly coincident, as expected. The magnetic field is applied to rotate the Fe magnetization (and electron spin orientation in the Fe) out of plane, so that any net electron spin polarization can be manifested as circular polarization in the EL via the quantum selection rules. As the magnetic field increases, the component of Fe magnetization and electron spin polarization along the $z$-axis

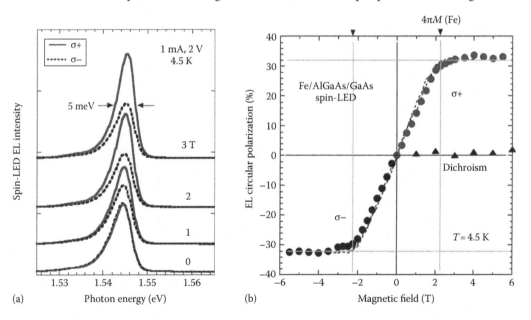

(a)   (b)

**FIGURE 2.15**   EL data from an Fe Schottky tunnel barrier spin-LED. (a) EL spectra for selected values of applied magnetic field, analyzed for positive and negative helicity circular polarization. (b) Magnetic field dependence of $P_{circ} = P_{spin}$. The dashed line shows the out-of-plane Fe magnetization obtained with SQUID magnetometry and scaled to fit the EL data. The triangles indicate the measured background contribution, including dichroism, using PL from an undoped reference sample.

continuously increase, and the corresponding spectra exhibit a substantial difference in intensity of the σ+ and σ– components—this difference rapidly increases with field, signaling successful electrical spin injection.

The field dependence of the circular polarization $P_{circ} = [I(\sigma+) - I(\sigma-)]/[I(\sigma+) + I(\sigma-)]$, where $I(\sigma+)$ and $I(\sigma-)$ are the EL component peak intensities when analyzed as σ+ and σ–, respectively, is summarized in Figure 2.15b. $P_{circ}$ rapidly increases with field and directly tracks the out-of-plane magnetization of the Fe film obtained by independent magnetometry measurements (dashed line). $P_{circ}$ saturates at a value of 32% at a magnetic field value characteristic of the Fe contact, $B = 2.2$, $T = 4\pi M_{Fe}$, where the Fe magnetization is saturated out-of-plane. Thus, the electron spin orientation is preserved during injection from the Fe contact, with a net spin polarization $P_{spin} = P_{circ} = 32\%$ at the moment of radiative recombination in the GaAs QW.

Significant polarization is observed to nearly room temperature. Preliminary analysis of the temperature dependence of $P_{circ}$ shows that it is dominated by that of the QW spin lifetimes, indicating that the injection process itself is independent of temperature [104], as expected for tunneling. Previous work has shown that the electron spin relaxation in a GaAs QW generally occurs more rapidly with increasing temperature [5, 78], suppressing the measured circular polarization—the GaAs(001) QW is simply an imperfect spin detector, with a strong temperature dependence of its own. As noted earlier, the optical polarization measured depends upon the values of the spin and radiative carrier lifetimes. Both have been extensively studied, vary with temperature, and depend upon the physical parameters of the structure and the "quality" of the sample material.

One can determine the initial spin polarization $P_o$ that exists at the moment the electrically injected electrons enter the GaAs QW by independently determining the value of the ratio $\tau_r/\tau_s$ by PL measurements and applying Equation 2.4, as described previously for the case of spin injection from ZnMnSe films. These measurements yield $\tau_r/\tau_s = 0.78 \pm 0.05$, resulting in a value $P_o = P_{spin} (1 + \tau_r/\tau_s) = 57\%$. It is interesting to note that this value exceeds the nominal spin polarization of bulk Fe (~45%), clearly demonstrating that the bulk spin polarization of FM metals does not represent a limit to the spin polarization that can be achieved via electrical injection. Effects such as spin filtering at the Fe/AlGaAs interface and spin accumulation in the GaAs QW enable the generation of highly polarized carrier populations, and can be exploited in the design and operation of semiconductor spintronic devices.

A number of control experiments must be performed to rule out spurious effects. For example, LED structures fabricated with a nonmagnetic metal contact showed little circular polarization and very weak field dependence, eliminating contributions from Zeeman splitting in the semiconductor itself. Possible contributions to the measured $P_{circ}$ arising from magnetic dichroism as the light emitted from the QW passes through the Fe film may be determined both analytically and directly measured. This contribution was calculated to be 0.9% using well-established models at the appropriate wavelength for the thickness of the Fe film [117]. This contribution was also directly measured by independent PL measurements on undoped

Fe/AlGaAs/GaAs QW test structures. Linearly polarized laser excitation creates unpolarized electrons and holes in the AlGaAs/GaAs, which emit unpolarized light when they recombine in the GaAs QW. This light passes through the Fe film, and any circular polarization measured is therefore due to a combination of dichroism in the Fe, field-induced splittings in the QW, or other background effects. The solid triangles in Figure 2.15b summarize these measurements, and show that the sum of any such contributions is <1%. Note that the effect measured due to electron spin injection is over 30 times larger.

The sign of the polarization demonstrates that $m_j = -1/2$ electrons injected from the Fe Schottky tunnel contact dominate the radiative recombination process in the QW when the magnetic field is applied along the surface normal (Figure 2.1b). This is confirmed by the results from the ZnMnSe-based spin-LEDs, where the CB spin-splitting is known and the σ+ component also dominates for a similar field orientation. In the nomenclature of the magnetic metals community [90, 118], such electrons are referred to as "majority spin" or "spin-up," even though the *spin* is antiparallel to the net magnetization. In this community, the term "spin-up" is used to describe an electron whose *moment* (rather than spin) is parallel to the magnetization. The corresponding state is at lower energy and more populated, and is therefore synonymously referred to as the "majority spin" state. Such carriers are designated here as "$n_m\uparrow$" where the subscript "$m$" is used to unambiguously indicate the convention used in this community [90]. Since the electron's spin and moment are antiparallel [119, 120], the *spin* of a "spin-up" electron is actually antiparallel to the magnetization. Similarly, the terms "minority spin" and "spin-down" (designated $n_m\downarrow$) are used synonymously to refer to electrons whose *moment* is antiparallel to the magnetization, and therefore have a higher energy than the "majority spin" electrons.

The experimental observation that the polarization of the current injected from the Fe contact through the Schottky tunnel barrier corresponds to majority spin in Fe is consistent with the pioneering work of Meservey and Tedrow [118], and with the model proposed by Stearns [121]. While one might reasonably expect majority spin injection to dominate, recent work has shown that a number of factors are likely to contribute to the spin polarization of the tunneling current from an FM contact, including barrier thickness, band bending, and details of bonding at the interface, which affect the interface electronic structure [122–124].

The interface atomic structure and its correlation with spin injection efficiency was addressed for the Fe/AlGaAs/GaAs system by combining the spin-LED results described above with high resolution TEM analysis of spin-LED samples and density functional theory (DFT) calculations of the Fe/AlGaAs interface [125]. The TEM analysis showed that the as-grown Fe/AlGaAs interface exhibited some degree of disorder, as shown in Figure 2.16a. The chemical order and coherence of the interface could be significantly improved by a low temperature anneal (200°C for 10 min), resulting in a highly ordered interface (Figure 2.16b) and a 44% increase in spin injection efficiency. The annealing temperature was well below the

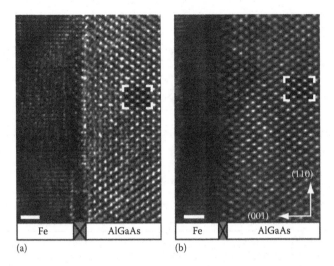

**FIGURE 2.16**    HRTEM images ([–110] cross section) of an Fe/AlGaAs spin-LED sample. (a) As-grown, exhibiting 18% spin polarization, and (b) following a mild post growth anneal, exhibiting a 26% spin polarization. Image simulations are inset (white brackets). The rectangles on the bottom of the images indicate Fe, AlGaAs, and the interfacial region (gray box with cross). Scale bars equal 1.0 nm. The contrast variation across the Fe regions is a result of changes in thickness.

~600°C growth temperature of the AlGaAs/GaAs QW structure, and therefore did not affect the characteristics of the semiconductor away from the Fe interface.

Phase and $Z$-contrast images were compared with those simulated from four interface models determined by DFT to be likely low energy candidates. Three of these models, shown in Figure 2.17 (from left to right: abrupt, partially intermixed, and fully intermixed), were previously proposed and studied theoretically [126]. The fourth model contained several monolayers of $Fe_3GaAs$ (a known stable alloy in the Fe-Ga-As phase diagram) sandwiched as an interlayer between the GaAs and Fe. For each model, $Z$-contrast images were simulated with software [127] using the relaxed atomic coordinates determined by the DFT calculations. Based upon a mathematical comparison of line scans of the atomic intensity profiles, both parallel and perpendicular to the interface between the experimental and simulated images (Figure 2.18), the authors concluded that the interface of the annealed sample forms by intermixing of the Fe and AlGaAs occurring on a single atomic plane, resulting in an interface with alternating Fe and As atoms (see Figure 2.18b, and middle model structure in Figure 2.17).

The 44% increase in spin injection efficiency observed experimentally was attributed to the greater tunneling efficiency for spin-polarized electrons across the chemically and structurally coherent annealed interface. As discussed previously (Section 2.5.1), band symmetries play an important role in spin injection from a metal to a semiconductor. Strong spin filtering is expected to occur at the ideal Fe/GaAs(001) interface (abrupt with no intermixing), because the $\Delta_1$ symmetry of the bulk Fe majority-spin state near $E_F$ matches that of the bulk GaAs CB states, enhancing transmission

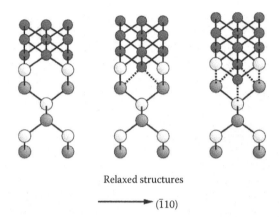

Relaxed structures

$\longrightarrow$ ($\bar{1}$10)

**FIGURE 2.17**    Low energy structures calculated by DFT after layer relaxation for the As-terminated Fe/GaAs(001) interface. Left—abrupt interface; middle—partially intermixed, with one plane consisting of alternating Fe and As atoms; and right—fully intermixed. The gray and light gray spheres represent Ga and As atoms, and the dark gray spheres represent Fe atoms. Highly strained bonds ~15%–20% longer than ideal are shown as dotted lines. (After Erwin, S.C. et al., *Phys. Rev. B* 65, 205422, 2002. With permission.)

of majority spin electrons [82]. The symmetries of the Fe minority-spin bands do not match the GaAs, so that these states decay very quickly, suppressing minority spin transmission. The same analysis was applied to the partially intermixed and fully intermixed interface models reported in Ref. [126] and considered in the TEM analysis above. The results show no significant change in the $\Delta_1$ decay rate between the abrupt and partially intermixed models [125], indicating that both should enable highly polarized spin injection. However, a significantly faster decay of the majority spin $\Delta_1$ state into the GaAs was found for the fully intermixed model, indicating lower spin polarization of the injected carriers, and consistent with the measured polarizations from the spin-LED samples.

This picture of the Fe/AlGaAs interface structure is consistent with earlier detailed studies of the initial nucleation, interface formation, and magnetic properties of Fe films grown on the As-dimer terminated GaAs(001) $2\times4$ and $c(4\times4)$ surfaces reported by Kneedler et al. [128–130] and Thibado et al. [131]. These authors concluded that the interface that was ultimately formed after a few monolayers of Fe deposition was planar with little intermixing, and characterized by Fe–As bonding. Excess As diffuses through the Fe film and segregates to the surface. Complementary studies further indicated that the electrical character of the Fe/GaAs interface was not dominated by As antisite defects [132, 133], consistent with surface segregation of As. An overview of the growth and magnetic properties may be found in Ref. [134].

This tailored doping profile and Schottky tunnel barrier has been widely used in subsequent studies to elucidate spin injection, transport, and detection in a variety of FM metal/AlGaAs/GaAs heterostructures [135–140]. Generation of *minority spin* electrons has been reported for a limited range

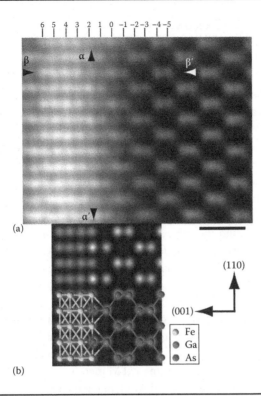

(a)

(b)

**FIGURE 2.18** *Z*-contrast TEM images of the Fe/AlGaAs(001) [−110] interface from spin-LED samples. (a) Experimental image from a sample with 26% spin polariza-tion. The arrowheads, α–α′ and β–β′, indicate the location and direction of the line profiles parallel and perpendicular to the interface, respectively, as used in the analysis. (b) Simulated image of a partially intermixed interface and inset ball-and-stick model from which it was calculated (Fe atoms appear in yellow, As in blue, and Ga in red). The scale bar equals 0.5 nm.

of bias conditions in Fe/AlGaAs/GaAs structures with interface doping pro-files and structures nominally identical to those described above [135, 137]. Two possible theoretical explanations have been offered. The first invokes formation of interface states which mediate minority spin transmission when the structure is biased so that electrons flow from the GaAs into the Fe (spin extraction) [124]. Note that this calculation was based on the perfectly abrupt Fe/GaAs interface model (Figure 2.17 left image), rather than the partially intermixed model (Figure 2.17 middle image), as concluded from the TEM analysis above. The specific structure of the interface will have a significant impact on the formation and character of interface states, and such calculations can provide further insight by addressing other interface structures. The second explanation invokes formation of bound states which form in the CB QW produced by the heavy doping near the semiconductor interface [123]. The relative contribution of these bound states to the spin transport process can also lead to the generation of minority spin accumula-tion in the GaAs. These results underscore the fact that several factors need to be considered in developing a more comprehensive understanding of spin transmission across a heterointerface. The tailored Schottky tunnel barrier approach has also been used very recently for spin injection into ZnSe [141] and Si [142].

### 2.5.3.2 Spin Injection via a Discrete Layer as a Tunnel Barrier

Discrete insulating layers may also be used to form a tunnel barrier between the FM metal and the semiconductor. These provide the more familiar and well-studied rectangular potential barrier, but also introduce an additional interface into the structure and raise issues of pinholes and uniformity of layer thickness. FM metal/oxide/metal structures have been extensively studied, since they form the basis for a tunneling spectroscopy used to determine the metal spin polarization [143], and for tunnel magnetoresistance devices being developed for nonvolatile memory [144–147]. Reviews of this area may be found in Chapters 10 and 11, Volume 1. Various oxides have been used as tunnel barriers for spin injection into a semiconductor, including $Al_2O_3$, $SiO_2$, MgO, and $Ga_2O_3$. Examples of each are presented below.

#### 2.5.3.2.1 Al₂O₃/AlGaAs-GaAs

Spin-polarized tunneling has been well-studied in metal/$Al_2O_3$/metal heterostructures. Careful measurements using a superconductor such as Al for one contact have provided quantitative information on the spin polarization of the tunneling current from many FM metals [143, 148, 149].

Several groups have utilized $Al_2O_3$ barriers in FM metal/$Al_2O_3$/AlGaAs-GaAs spin-LED structures with great success [150–153]. Motsnyi et al. [150] used a CoFe contact and the oblique Hanle effect to measure and analyze the EL, and reported values for the GaAs electron spin polarization $P_{spin}$ of 14%–21% at 80 K, and 6% at 300 K. In addition, they were able to experimentally determine the radiative and spin lifetimes of the GaAs itself. When they corrected their measured values for this "instrument response function" of the GaAs detector as discussed previously, they obtained a value $P_o = 16\%$ at 300 K (this quantity is denoted by "Π" in their notation). Manago and Akinaga obtained spin polarizations of ~1% using either Fe, Co, or NiFe contacts [152]. van't Erve et al. obtained a value of $P_{spin} = 40\%$ at 5 K for samples, which utilized a lightly $n$-doped ($10^{16}$ cm$^{-3}$) $Al_{0.1}Ga_{0.9}As$ layer adjacent to the $Al_2O_3$ [153]. However, they reported lower operating efficiency (higher bias voltages and currents) than for the Schottky barrier-based spin-LEDs of Refs. [104, 105]. This was subsequently remedied by including the same semiconductor doping profile shown in Figure 2.13 before the growth of the $Al_2O_3$ layer, so that high $P_{spin} = 40\%$ was achieved at $T = 5$ K and bias conditions comparable to those utilized for the Schottky tunnel barrier devices. They determined the efficiency of their GaAs spin detector by optical pumping to obtain the value $\tau_r/\tau_s = 0.75$, resulting in a value $P_o = P_{spin}(1 + \tau_r/\tau_s) = 70\%$—a value that again significantly exceeds the spin polarization of bulk Fe.

#### 2.5.3.2.2 Al₂O₃/Si

One of the first demonstrations of spin injection into Si utilized an $Al_2O_3$ tunnel barrier in an Fe/$Al_2O_3$/Si(001) $n$–$i$–$p$ spin-LED heterostructure [154]. While Si is clearly an attractive candidate for spintronic devices, its indirect band gap makes it much more difficult to probe optically, and led many to believe that the optical spectroscopic techniques, which had proven so productive when applied to the III–Vs, would be unable to provide much insight into spin-dependent behavior in Si. Although the indirect gap complicates application of polarized optical techniques used routinely in GaAs, it is worth noting that the first successful optical pumping experiments to induce a net electron/nuclear spin polarization were performed in Si rather than in a direct gap material by Lampel [6].

Several fundamental properties of Si make it an ideal host for spin-based functionality. Spin–orbit effects producing spin relaxation are much smaller in Si than in GaAs due to the lower atomic mass and the inversion symmetry of the crystal structure itself. The dominant naturally occurring isotope, Si [28], has no nuclear spin, suppressing hyperfine interactions. Consequently, spin lifetimes are relatively long in Si, as demonstrated by an extensive literature on both donor-bound [155] and free electrons [156, 157]. In addition, silicon's mature technology base and overwhelming dominance of the semiconductor industry make it an obvious choice for implementing spin-based functionality. Several spin-based Si devices have indeed been proposed, including transistor structures [158, 159] and elements for application in quantum computation/information technology [160].

Jonker et al. electrically injected spin-polarized electrons from a thin FM Fe film through an $Al_2O_3$ tunnel barrier into an Si(001) $n–i–p$ doped heterostructure, and observed circular polarization of the EL [154]. This signals that the electrons retain a net spin polarization at the time of radiative recombination, which they estimated to be ~30%, based on simple arguments of momentum conservation. This interpretation was confirmed by similar measurements on Fe/$Al_2O_3$/Si/AlGaAs/GaAs QW structures in which the spin-polarized electrons injected from the Fe drift under applied field from the Si across an air-exposed interface into the AlGaAs/GaAs structure and recombined in the GaAs QW. In this case, the polarized EL can be quantitatively analyzed using the standard selection rules, yielding an electron spin polarization of 10% in the GaAs. More recently, the theory provided a quantitative link between the circular polarization of the Si EL and the electron spin polarization. These results are summarized in the following paragraphs.

The Si $n–i–p$ LED structures were grown by MBE, transferred in air, and introduced to a second chamber for deposition of the Fe/$Al_2O_3$ tunnel contact. An $n$-doping level ~$2 \times 10^{18}$ cm$^{-3}$ at 300 K was chosen to prevent carrier freeze-out at lower temperatures. This is well below the metal–insulator transition of $5.6 \times 10^{18}$ cm$^{-3}$. Details of the growth and processing may be found in Ref. [154]. A schematic of the sample structure, corresponding band diagram of the spin-injecting interface, and a photograph of the processed surface-emitting LED devices are shown in Figure 2.19a. A high resolution cross-sectional TEM image of the Fe/$Al_2O_3$/Si contact interface region is shown in Figure 2.19b, and reveals a uniform and continuous oxide tunnel barrier with relatively smooth interfaces. The Fe film is polycrystalline on the nominally amorphous $Al_2O_3$ layer.

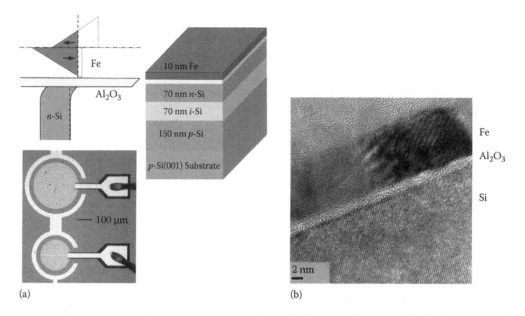

(a)                                    (b)

**FIGURE 2.19**    Fe/Al$_2$O$_3$/Si spin-LED structures. (a) Band diagram of the spin-injecting interface and schematic of the sample structure, with an optical photograph of the processed spin-LED devices. The light circular mesas are the active LED regions. (b) Cross-sectional TEM image of the spin-injecting interface region.

Typical EL spectra from a surface-emitting Fe/Al$_2$O$_3$/Si $n-i-p$ spin-LED structure are shown in Figure 2.20a and b for $T = 5-80$ K, analyzed for σ+ and σ− circular polarization. At 5 K, the spectra are dominated by features arising from electron-hole recombination accompanied by transverse acoustic (TA) or transverse optical (TO) phonon emission. Subsequent analysis to determine the origin of the emission features has resulted in a revised identification of the EL peaks observed. This analysis revealed that the TO/TA peaks occur in correlated pairs, with the peaks at 1105 meV ($TA_s$) and 1065–1070 meV ($TO_s$) due to recombination in the $p$-Si substrate ($p \sim 10^{19}$ cm$^{-3}$), and the peaks at 1090 meV ($TA_1$) and 1050 meV ($TO_1$) likely arising from recombination in the interface region. At 80 K, the $TO_s$ feature dominates. At zero field, no circular polarization is observed because the Fe magnetization and corresponding electron spin orientation lie in-plane and orthogonal to the light propagation direction. As discussed previously, although spin injection may occur, it cannot be detected with this orthogonal alignment. Therefore, a magnetic field is applied to rotate the Fe spin orientation out-of-plane, and the main spectral features each exhibit circular polarization, as shown by the difference between the red (σ+) and blue (σ−) curves in the 3 T spectra at $T = 5$, 50, and 80 K.

The magnetic field dependence of $P_{circ}$ for each feature is summarized in Figure 2.20c. As the Fe magnetization (and majority electron spin orientation) rotates out of plane with increasing field, $P_{circ}$ increases and saturates above ~2.5 T with average values of 3.7%, 3.5%, and 1.9% for the $TA_s$, $TA_1$, and $TO_1$ features, respectively, at 5 K, and 2.1% and 2% for the $TO_s$ feature at 50 and 80 K. Note that, for each feature, $P_{circ}$ tracks the magnetization of the Fe contact, shown as a solid line scaled to the data, indicating that the spin orientation of the electrons that radiatively recombine in the Si directly reflects that of the electron spin orientation in the Fe contact. The field at which the Fe magnetization saturates out of plane is characteristic of Fe ($H_{sat} = 4\pi M_{Fe} = 2.2\,T$), and is unaffected by lateral patterning, which might alter the in-plane coercive field. This also argues against possible compound formation (e.g. FeSi, etc.) at the interface due to pinholes, which would lead to a different field dependence. The monotonic decrease in $P_{circ}$ from the higher energy ($TA_s$) to the lowest energy feature ($TO_1$) as seen in Figure 2.20c is consistent with spin relaxation expected to accompany electron energy relaxation. The sign of $P_{circ}$ indicates that Fe majority spin electrons dominate, as was the case for the tailored Schottky Fe/AlGaAs and Fe/Al$_2$O$_3$/AlGaAs contacts. Data from the requisite reference samples used to determine background effects such as dichroism show only a weak paramagnetic field dependence $\sim$0.1%/T, ruling out spurious contributions. These data together unambiguously demonstrate that spin-polarized electrons are electrically injected from the Fe contact into the Si heterostructure.

An analysis of the EL to extract the electron spin polarization $P_{spin}$ in the Si is complicated by the indirect gap character and the participation of phonons in the radiative recombination process. Since the electron spin angular momentum must be shared by the photons and any phonons involved in radiative recombination, the photons will carry away only some fraction of the spin angular momentum of the initial spin-polarized electron population. Thus, our experimentally measured value of $P_{circ}$ will result in a significant underestimate of the corresponding Si electron polarization. A second significant factor affecting $P_{circ}$ is the very long radiative lifetime typical of indirect gap materials—as noted previously, $P_{spin}$ decays exponentially with a characteristic time $\tau_s$ before radiative recombination occurs. Despite these issues, the broader rule of conservation of momentum can be used to estimate a *lower bound* for the initial electron spin polarization, and the same rate equation analysis leading to Equation 2.4 applies to correct for these lifetime effects: $P_o = P_{spin}\,(1 + \tau_r/\tau_s) > P_{circ}\,(1 + \tau_r/\tau_s)$. Typical values for $\tau_r$ in doped Si at low temperature are 0.1–1 ms [161, 162]. The radiative lifetime of emission features associated specifically with acoustic phonon emission was measured to be 480 μs for $1\,K < T < 5\,K$ [161]. The spin lifetime for free electrons well above the metal–insulator transition has been determined to be $\sim$1 μs in both bulk [156] and modulation doped samples [157], and is expected to increase at lower electron densities [156]. Therefore, using $P_{circ} \sim 0.03$ (Figure 2.20c), $\tau_r = 100\,\mu s$ and $\tau_s = 10\,\mu s$, a conservative estimate is given by $P_o > P_{circ}(1 + \tau_r/\tau_s) \sim 0.3$ or 30%, comparable to that achieved in GaAs.

This value is supported by similar experiments on Fe/Al$_2$O$_3$/80 nm $n$-Si/80 nm $n$-Al$_{0.1}$Ga$_{0.9}$As/10 nm GaAs QW/200 nm $p$-Al$_{0.3}$Ga$_{0.7}$As heterostructures [154], in which electrons injected from the Fe into the Si drift across the Si/Al$_{0.1}$Ga$_{0.9}$As interface with applied bias and radiatively recombine in the GaAs QW. In this case, the standard quantum selection rules can be rigorously applied to quantify the electron spin polarization, which

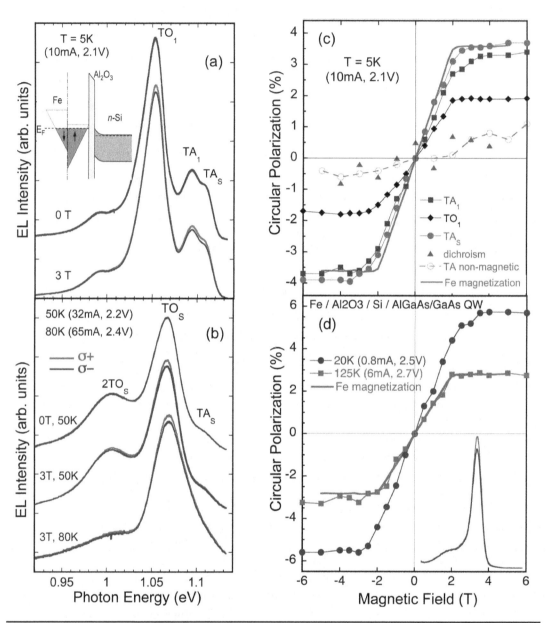

**FIGURE 2.20**    See EL spectra from surface-emitting Fe/Al$_2$O$_3$/Si $n$–$i$–$p$ spin-LED structures, analyzed for positive ($\sigma+$) and negative ($\sigma-$) helicity circular polarization at temperatures of (a) 5 K, and (b) 50 and 80 K. The spectra are dominated by features arising from electron-hole recombination accompanied by transverse acoustic (*TA*) and transverse optical (*TO*) phonon emission. (c) The magnetic field dependence of the circular polarization for each feature in the Si spin-LED spectrum at 5 K. (d) Inset shows the EL spectrum from Fe/Al$_2$O$_3$/80 nm $n$-Si/$n$-Al$_{0.1}$Ga$_{0.9}$As/GaAs QW/$p$-Al$_{0.3}$Ga$_{0.7}$As spin-LEDs at $T$ = 20 K and $H$ = 3 T. The dominant feature is the GaAs QW free exciton, which exhibits strong polarization. The magnetic field dependence of this circular polarization is plotted for $T$ = 20 and 125 K. The out-of-plane Fe magnetization curve appears in panel (c) and (d) as a solid line for reference.

eventually reaches the GaAs [163]. The EL spectrum at 20 K and 3 T is shown as insert in Figure 2.20d, and is dominated by the free exciton emission at 1.54 eV from the GaAs QW. The field dependence of the polarization of this feature, $P_{circ}$(GaAs), is shown in Figure 2.20d. $P_{circ}$(GaAs) again tracks the magnetization of the Fe contact, and saturates at values of 5.6% at 20 K and 2.8% at 125 K. Separate optical pumping measurements on the GaAs QW provide a direct measure of $\tau_r/\tau_s$ (GaAs QW) = 0.8. Thus, the spin polarization of the electrons in the GaAs is $P_{GaAs} = P_{circ}(1 + \tau_r/\tau_s) = 0.1$ or 10% at 20 K. This is consistent with our estimate above of $P_{Si} \sim 30\%$, and establishes a firm lower bound for $P_{Si} > 10\%$. It is indeed remarkable that the electrons injected from the Fe contact drift through the Si, cross the Si/AlGaAs interface and still retain a significant spin polarization, given (1) the relatively poor crystalline quality of Si epilayers on GaAs (lattice mismatch 3.9%); (2) the heterovalent interface structure; (3) the 0.3 eV CB offset (Si band lower than $Al_{0.1}Ga_{0.9}As$) [164]; and (4) the fact that the sample surface was exposed to air before growth of the Si, and then again before growth of the Fe/$Al_2O_3$ contact.

Recent theory provides a fundamental and more rigorous interpretation of the circular polarization of the TO and TA features in the Si spin-LED EL spectra, and thereby an independent determination of the electron spin polarization achieved in the Si by electrical injection. Li and Dery [165] have shown that the circular polarization of these features is directly related to the electron spin polarization for a given doping level in the region where radiative recombination occurs. For $p = 10^{19}$ cm$^{-3}$, they conclude that the TA feature should have a maximum circular polarization of 13% for a 100% spin-polarized electron population. Thus, the measured value $P_{circ} \sim 3.5\%$ for the $TA_s$ feature (Figure 2.20c) indicates an electron spin polarization of 27%. The estimates of the electron spin polarization summarized here are in remarkably good agreement.

The EL intensity decreases rapidly with increasing temperature. It is too low to reliably analyze at temperatures higher than those shown in Figure 2.20 for emission from either the Si (80 K) or GaAs (125 K) even though clear polarization remains—a further limitation of the LED as a spin detector. Subsequent work using electrical rather than optical spin detection confirms that spin injection in the FM metal/$Al_2O_3$/Si system persists to room temperature, as described in the next section.

The $Al_2O_3$ layer serves as both a tunnel barrier and a diffusion barrier, preventing interdiffusion between the Fe and Si which is likely to occur. Spin injection is also observed in Fe/Si $n$–$i$–$p$ samples when the Fe is deposited directly on the Si, where the Schottky barrier serves as a tunnel barrier [166]. However, the net spin polarization achieved is lower due to the inhomogeneous character of the interface that is formed.

Grenet et al. have also used the spin-LED approach to demonstrate spin injection into Si [167]. They employed a (Co/Pt)/$Al_2O_3$ tunnel barrier contact, where the Co/Pt exhibits strong perpendicular magnetic anisotropy with the remanent magnetization along the surface normal. This obviates the need for an applied magnetic field to saturate the out-of-plane magnetization, as required for the Fe/Al$_2$O$_3$ contacts described above. The entire structure consisted of (3 nm Pt/1.8 nm Co)/2 nm Al$_2$O$_3$/100 nm $n$-Si/10 nm Si$_{0.7}$Ge$_{0.3}$ QW/550 nm $p$-Si/$p$-Si(001), with 50 nm dopant setbacks at the QW. The strained Si$_{0.7}$Ge$_{0.3}$ QW served as the radiative recombination region, which produces a Type II band alignment with strong hole localization in the Si$_{0.7}$Ge$_{0.3}$. They measure a maximum (typical) circular polarization $P_{circ}$ = 3% (1.2%) in the EL signal from the QW, which is nearly constant at 200 K. The magnetic field dependence of $P_{circ}$ tracks the Co/Pt magnetization, confirming that the injected electrons retain their spin orientation from the Co/Pt contact.

### 2.5.3.2.3  SiO$_2$/Si

SiO$_2$ has been the gate dielectric of choice for generations of Si metal-oxide-semiconductor devices, because it is easy to form and provides the low interface state density necessary for device operation. It has been the cornerstone of the vast Si electronics industry, because no other semiconductor has a robust native oxide which forms such an electrically stable self-interface. As such, it is an obvious choice to use as the tunnel barrier in a Si-based spin transport device. However, there are few reports of its use as a spin tunnel barrier in any structure. Smith et al. reported a smaller-than-expected magnetoresistance of 4% at 300 K in Co/SiO$_2$/CoFe tunnel junctions, which they attributed to overoxidation at the metal interfaces [168]. A composite NiFe/SiO$_2$/Al$_2$O$_3$ tunnel barrier was used as the emitter in a magnetic tunnel transistor by Park et al. [169]. They reported a tunnel spin polarization of 27% at 100 K, which was lower than the value of 34% obtained for an NiFe/Al$_2$O$_3$ emitter in the same structure. No data were reported for a tunnel barrier consisting of SiO$_2$ alone.

Li et al. successfully used SiO$_2$ as a spin tunnel barrier on Si, and recently reported spin injection from Fe through SiO$_2$ into a Si $n$–$i$–$p$ heterostructure, producing an electron spin polarization in the Si, $P_{Si}$ > 30% at 5 K [170]. The samples were similar to those described above and consist of 10 nm Fe/2 nm SiO$_2$/70 nm $n$-Si/70 nm undoped Si/150 nm $p$-Si/$p$-Si(001) substrate, forming a spin-LED structure identical to that described above. The SiO$_2$ layer was formed by natural oxidation of the $n$-Si surface assisted by illumination from an ultraviolet lamp. A TEM image of the Fe/SiO$_2$/Si interface region appears in Figure 2.21a, and shows a reasonably uniform SiO$_2$ layer with well-defined interfaces. The SiO$_2$ appears amorphous, as expected, while the Fe is polycrystalline.

Transport measurements based on the Rowell criteria confirmed that conduction from the Fe through the SiO$_2$ occurred by tunneling. The conductance exhibits a parabolic dependence upon the bias voltage, as shown in Figure 2.21b for several temperatures. Good fits (solid lines) to the conductance data are obtained within an energy range of ±100 meV using the BDR model [113], which yield an effective barrier height of ∼1.7 eV and a barrier thickness of ∼21 Å at 10 K. The temperature dependence of the ZBR, expressed as $R_0$(T)/$R_0$(300 K), where $R_0$ is the slope of the $I$–$V$ curve at zero bias, is summarized in Figure 2.21c. The ZBR exhibits a weak temperature dependence indicative of single-step tunneling rather than the exponential

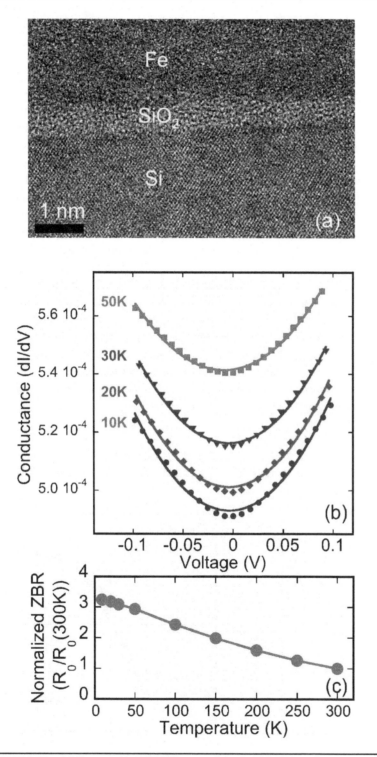

**FIGURE 2.21**    (a) Cross-sectional TEM image of the Fe/SiO$_2$/Si interface region. (b) Conductance curves for transport through an Fe/SiO$_2$/$_n$-Si(001) contact exhibiting parabolic behavior, and fits using the BDR model. (c) Zero bias resistance of the contact as a function of temperature. The weak temperature dependence indicates single-step tunneling through a pinhole-free SiO$_2$ tunnel barrier.

dependence of thermionic emission. A weak temperature dependence of the ZBR has been shown to be a reliable indicator for a pinhole-free tunnel barrier [108], as discussed earlier in this chapter.

EL spectra at $T = 5$ K and $B = 0$ and 3 T are shown in Figure 2.22a, analyzed for σ+ and σ– circular polarization. The spectra are very similar to those of Figure 2.20a (Fe/Al$_2$O$_3$/Si $n–i–p$ spin-LEDs), and are dominated by TA and TO phonon-mediated recombination, with specific peak identifications as discussed in Section 2.5.3.2.2. Although emission is initially detected at a bias of 1.6 V, higher biases are typically used to reduce data acquisition time, as the polarization measured is relatively independent of the bias. A magnetic field along the surface normal rotates the Fe magnetization out of plane, enabling the injected electron spin polarization to be manifested as circular polarization in the EL. At $B = 3$ T, where the Fe magnetization is saturated out of plane, the spectra exhibit polarization for each of the spectral features, with $P_{circ} \sim 3\%$. The magnetic field dependence of $P_{circ}$ for the three main peaks in the spectra is summarized in Figure 2.22b, and the behavior and discussion parallel that of Figure 2.20c—the electrons that tunnel from the Fe through the SiO$_2$ to eventually radiatively recombine in the Si clearly retain a net spin polarization from the Fe. Applying the same rate equation analysis, discussed above, yields a value of 30% as a lower bound for the electron spin polarization achieved in the silicon for temperatures up to 50 K. A similar value is obtained by applying the phonon-mediated recombination theory of Li and Dery [165] to the $TA_s$ feature.

### 2.5.3.2.4 MgO/AlGaAs-GaAs

MgO has been widely used in metal/oxide/metal tunnel structures, and more recently in metal spin torque transfer devices (see Chapters 7, 8, and 15,

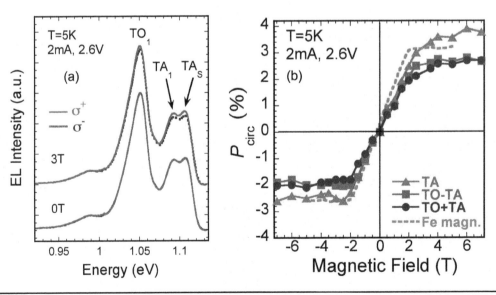

**FIGURE 2.22**    (a) EL spectra at 5 K from a Si $n–i–p$ structure with an Fe/SiO$_2$ contact at zero field and 3 T, analyzed for σ+ and σ– circular polarization. (b) Magnetic field dependence of $P_{circ}$ for the $TA_s$, $TA_1$, and $TO_1$ features at 5 K. $P_{circ}$ consistently tracks the out-of-plane magnetization of the Fe (dashed line).

Volume 1 for a comprehensive overview). Its rock-salt structure is closely lattice-matched to bcc-Fe, and high-quality epitaxial growth has been demonstrated. The band symmetry arguments originally applied to the Fe/MgO/Fe system by MacLaren, Butler et al. [80, 81, 88] show that the Fe $\Delta_1$ majority spin band couples well to the propagating state in the MgO barrier, while the minority spin bands do not, resulting in highly spin-polarized current. Thus, Fe/MgO is an attractive candidate as a spin-injecting contact for a semiconductor such as GaAs or Si (see Figure 2.11).

Jiang et al. demonstrated robust spin injection up to room temperature with CoFe/MgO tunnel contacts on AlGaAs/GaAs QW/AlGaAs(001) spin-LED structures [171]. A cross-sectional TEM image of the tunnel contact region is shown in Figure 2.23a. The experimental procedure and analysis are identical to that described previously in Section 2.5.3.1.2 (Figure 2.15) for Fe Schottky tunnel/AlGaAs/GaAs QW/AlGaAs spin-LED structures [104, 105]. The EL spectra analyzed for σ+ and σ− polarization for temperatures of 100 and 290 K are shown in Figure 2.23b and c, and exhibit significant circular polarization of 52% and 32%, respectively, after background subtraction when the Fe magnetization is saturated along the surface normal.

The authors argue that these high values for $P_{circ}$ are due to improved spin injection efficiency at higher temperatures than either the Fe/Al$_2$O$_3$ or Fe Schottky tunnel contacts provide. Because spin-dependent tunneling is expected to have a weak temperature dependence, a more likely explanation lies in the temperature dependence of the spin and radiative lifetimes of the GaAs QW—the *detector* efficiency rather than the spin injection efficiency was much higher due to the particular characteristics of the GaAs (which can vary significantly from one MBE machine to another). This was explicitly investigated by Salis et al., who used time-resolved optical techniques to measure the lifetimes for these same Fe/MgO spin-LED samples and found that they were strongly temperature dependent, as shown in Figure 2.24 [172]. Using the values at 10 K of $P_{circ}$ ($P_{EL}$ in Figure 2.24b) = 0.47, $\tau_r = 300\,ps$, and $\tau_s = 560\,ps$ (Figure 2.24a), from Equation 2.4, we find $P_o = P_{circ}(1 + \tau_r/\tau_s) = 0.72$. For the second sample they studied (see Figure 3 of Ref. [172]), at 10 K, the corresponding values are $P_{circ}$ ($P_{EL}$) = 0.32, $\tau_r = 500\,ps$, $\tau_s = 700\,ps$, and $P_o = 0.55$. These values for the initial spin polarization achieved in the GaAs QW are very similar to those obtained using either the Fe/GaAs Schottky tunnel barrier ($P_o = 0.57$) or Fe/Al$_2$O$_3$ spin-injecting contacts ($P_o = 0.70$), described in Sections 2.5.3.1.2 and 2.5.3.2.1, indicating that all three tunnel barriers are of comparable efficiency for spin injection. These data underscore the importance of quantifying the detector response function. Nevertheless, these results conclusively show that efficient electrical spin injection persists to room temperature and presents no obstacle for future development of semiconductor spintronic devices.

Lu et al. investigated the effect of MgO barrier thickness and crystallinity on the spin injection efficiency in CoFeB/MgO/AlGaAs/GaAs spin-LED structures [173]. They found that the injection efficiency as determined from $P_{circ}$ increased as the MgO thickness was increased from 1.4 to 4.3 nm. They attributed this to a suppression of band bending due to hole accumulation in

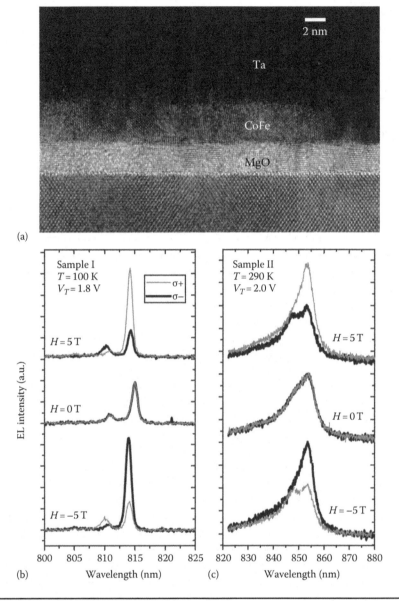

(a)

(b)

(c)

**FIGURE 2.23**    (a) TEM image of the CoFe/MgO/AlGaAs/GaAs spin-LED structure. Electroluminescence spectra from the spin-LEDs are shown at temperatures of (b) 100 K and (c) 290 K. (After Jiang, X. et al., *Phys. Rev. Lett.* 94, 056601, 2005. With permission.)

the AlGaAs at the MgO interface for the thicker barriers, which prevented the holes from tunneling into the FM injector. They also found that the spin injection efficiency was higher for MgO grown at 300°C where it exhibited some degree of crystalline texture vs. amorphous films grown at room temperature, which they attributed to contributions from the band symmetry matching arguments discussed earlier in this chapter. Truong et al. combined pulsed electrical input and time-resolved optical spectroscopy to analyze the temporal response of the spin polarization, following injection with

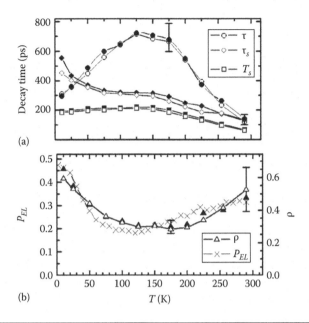

**FIGURE 2.24** (a) Measured radiative ($\tau$), spin ($\tau_s$) and combined lifetimes as a function of temperature for the spin-LEDs of Figure 2.23. (b) Measured circular polarization $P_{EL}$ and spin detector efficiency $\rho = T_s/\tau_r$ as a function of temperature for $B = 0$ (open symbols) and 0.8 T (closed symbols). (After Salis, G. et al., *Appl. Phys. Lett.* 87, 262503, 2005. With permission.)

these same spin-LED structures [174]. They concluded that the buildup time of the electron spin polarization in the GaAs QW was much faster than the rise time of the EL signal, which was limited to ~700 ps due to parasitics in the device circuit.

### 2.5.3.2.5 GaO$_x$/AlGaAs-GaAs

Unlike Si, the native oxide of GaAs is difficult to grow with proper stoichiometry and is less stable than $SiO_2$. Hence a metal-oxide-semiconductor technology, which allows inversion mode operation in a GaAs device, has never blossomed. Previous efforts reported a very low interface state density ($10^{10}$–$10^{11}$ cm$^{-2}$ eV$^{-1}$) at the GaO$_x$/GaAs interface [175], comparable to that of $SiO_2$/Si, making GaO$_x$ an attractive candidate as a gate insulator. Saito et al. utilized GaO$_x$ as the spin tunnel barrier in Fe/GaO$_x$/AlGaAs/GaAs QW spin-LED structures [176]. They fabricated surface emitting devices and measured the circular polarization of the EL as a function of applied magnetic field (Faraday geometry) and temperature. The experimental procedure and analysis are again identical to that described previously in Section 2.5.3.1.2 (Figure 2.15) for Fe Schottky tunnel/AlGaAs/GaAs QW structures. They find that $P_{circ}$ tracks the out-of-plane magnetization of the Fe contact with a maximum value $P_{circ} = P_{spin} = 20\%$ at $T = 2$ K. Using a typical value for $\tau_r/\tau_s \sim 1$ from the literature, they calculate the initial spin polarization in the GaAs: $P_o = P_{spin}(1 + \tau_r/\tau_s) = 40\%$. $P_{circ}$ decreased rapidly with temperature to a value of 2% at 300 K, as found previously for GaAs QW based spin-LEDs with both Fe Schottky and Fe/$Al_2O_3$ tunnel barrier contacts, and this is likely due to the temperature dependence of $\tau_r$ and $\tau_s$ [10]. Thus, the temperature dependence observed is that of the GaAs QW detector rather than that of the spin injection process through $GaO_x$.

In summary, the use of a tunnel barrier allows one to circumvent the conductivity mismatch issue, enabling the use of FM metals as spin-injecting contacts into a semiconductor. FM metals introduce many desirable attributes to a device technology, including intrinsic nonvolatile and reprogrammable operating characteristics. They exhibit reasonable spin polarizations at $E_F$ ($\sim$40%–50%), high Curie temperatures, low coercive fields, fast switching times, and an advanced materials technology, which permits one to tailor the magnetic characteristics for a particular application. An FM metal/tunnel barrier contact injects spin-polarized electrons near the semiconductor CB edge with high efficiency, where the electron energy is determined by the bias applied across the tunnel barrier ($\sim$10–100 meV). The realization of efficient electrical injection and significant spin polarization using a simple tunnel barrier compatible with "back-end" semiconductor processing should greatly facilitate progress in the development of semiconductor spintronic devices.

### 2.5.4 HOT ELECTRON SPIN INJECTION

We conclude this section by describing a completely different avenue toward electrical injection of spin-polarized carriers from an FM metal into a semiconductor, which also circumvents the conductivity mismatch issue. This is the process of hot electron injection, which builds on the development of the spin-valve transistor and magnetic tunnel transistor [175–179]. This topic is covered in detail in Chapter 3, Volume 2, and we include a brief discussion here for the sake of providing a more complete overview on electrical spin injection.

Electrical spin injection into a semiconductor by hot electron injection was first demonstrated by Jiang et al. using a $GaAs/In_{0.2}Ga_{0.8}As$ multiple QW structure as the spin detector, incorporating the spin-LED approach [180]. A schematic of the heterostructure is shown in Figure 2.25, and consists of a metal emitter/$Al_2O_3$ tunnel barrier/FM metal base deposited on GaAs, which forms the top layer of the $GaAs/In_{0.2}Ga_{0.8}As$ structure on a $p$-GaAs(001) substrate. The FM metal base layer forms a Schottky barrier contact with the GaAs. Electrons from the metal emitter (which need not be FM) tunnel through the $Al_2O_3$ tunnel barrier and enter the FM metal base at an energy well above $E_F$ determined by the emitter-base bias voltage. Electrons that are minority spin relative to the base magnetization are rapidly scattered due to their short mean free path and lose energy in the FM base, relaxing to $E_F$ where they are blocked from entering the semiconductor by the Schottky barrier. Although majority spin electrons have a longer mean free path, most are also scattered in the base, relax to $E_F$ and are blocked by the Schottky barrier. But a small fraction of the majority spin electrons do not scatter, and retain sufficient energy to surmount the Schottky barrier and enter the semiconductor, forming a spin-polarized current. These electrons

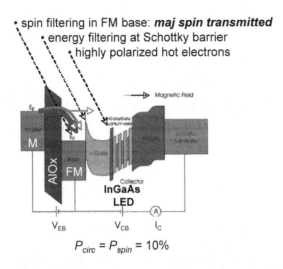

$$P_{circ} = P_{spin} = 10\%$$

**FIGURE 2.25** Schematic energy band diagram of a magnetic tunnel transistor merged with a QW LED collector. The emitter/base bias $V_{EB}$ controls the energy of the injected hot electrons, while the collector/base bias $V_{CB}$ can be used to adjust the band bending of the LED. The emitter consists of 5 nm $Co_{84}Fe_{16}$. The 2.2 nm thick $Al_2O_3$ tunnel barrier is formed by reactive sputtering of Al in the presence of oxygen. The base layer consists of 3.5 nm $Ni_{81}Fe_{19}$ and 1.5 nm $Co_{84}Fe_{16}$ with the NiFe layer adjacent to the GaAs. (After Jiang, X. et al., *Phys. Rev. Lett.* 90, 256603, 2003. With permission.)

thermalize to the CB and recombine with unpolarized holes in the InGaAs QWs. The circular polarization of the resultant EL provides an assessment of the spin injection.

Using the procedures described above, Jiang et al. measure a value $P_{circ} = 10\%$ at $T = 1.4\,$K when the FM base magnetization is rotated out of plane by an applied field (Faraday geometry), signaling successful spin injection using this hot electron approach. The value of $P_{circ}$ is somewhat lower than measured in other spin-LED experiments, and the authors suggest two likely reasons: (1) the injected hot electrons lose a significant amount of polarization during the process of thermalization to the bottom of the CB, and (2) after the hot electrons enter the QW region, further spin relaxation can occur in the QWs before recombination. The measured result, therefore, sets a lower bound for the injected electron spin polarization.

This hot electron injection approach was later adopted by Appelbaum et al. to demonstrate electrical spin injection into Si [181]. The structure is similar to that of Figure 2.25 (substituting Si for GaAs/AlGaAs), but they replace the InGaAs QW detector section with a second FM metal layer, which serves as a spin analyzer, functioning in the same way as described above for the FM emitter. This enables them to detect spin currents by measuring changes in the current through the device, as one changes the orientation of the magnetization of emitter and analyzer layers from parallel to antiparallel. Rather than measuring an optical polarization or a voltage produced by spin accumulation [136], this approach measures the change in current. A detailed treatment may be found in Chapter 3, Volume 2.

Thus, the combination of hot electron injection, spin filtering in the FM base or analyzer layer, and energy filtering by a Schottky barrier results in preferential transmission of majority spin current and a blocking of the minority spin current, enabling spin injection and analysis. A practical concern with this approach is that it is very inefficient—most of the current flows in the emitter-base circuit (or FM analyzer circuit), while a very small fraction $\sim10^{-4}$ enters the semiconductor as spin-polarized current. This is a recognized problem with the magnetic tunnel transistor [179].

## 2.6 FERROMAGNETIC METAL/ SEMICONDUCTOR ELECTRICAL SPIN INJECTION: ELECTRICAL DETECTION

### 2.6.1 GENERAL CONSIDERATIONS AND NON-LOCAL DETECTION

The preceding sections summarize rapid progress in developing a practical and technologically attractive methodology for electrical spin injection, enabling one to produce electron populations in a semiconductor with spin polarizations exceeding those of typical FM metals. It was repeatedly noted that the temperature dependence observed experimentally was dominated by the spin *detector*, e.g. the temperature dependence of the radiative and spin lifetimes in the spin-LED structures. Indeed, spin injection by tunneling is expected to depend only weakly upon temperature. Optical detection of spin polarization in a semiconductor has proven to be invaluable as a research tool, and will enable a subset of applications such as spin lasers (see Chapter 16, Volume 3). However, it is not practical for the broader range of electronic applications, where a voltage input/output with minimal temperature dependence is required to interface to the rest of the circuit. Thus, utilizing *an FM metal as a spin detector* has great technological appeal.

The issue of conductivity mismatch has significant implications for electrical spin detection, and places constraints on the range of contact resistances at the FM metal/semiconductor interface, which permit electrical detection of spin accumulation. This is reviewed in detail in Refs. [97, 100], and in Chapter 1, Volume 1 by Fert. The specific cases of FM metal contacts as spin detectors on GaAs and on Si were discussed by Dery et al. [182] and Min et al. [183], respectively. The criteria for the conventional two-terminal magnetoresistance device are particularly demanding, and have yet to be experimentally realized in a conventional semiconductor such as Si or GaAs.

The concept of *non-local* spin detection was introduced by Johnson and Silsbee in 1985 in their study of spin accumulation in metals [184] (see Chapter 5, Volume 1), and developed extensively by others for spin detection in metals [185, 186], GaAs [136, 137], Si [187, 188], and graphene [189]. It is based upon the fact that spin diffuses away from the point of generation independently of the direction of charge flow, i.e. *a pure spin current flows outside of the charge current path*. This process is similar to that of heat diffusing away from a hot contact on a surface produced, e.g. by the tip of a soldering gun. This is illustrated in Figure 2.26, where an FM metal tunnel

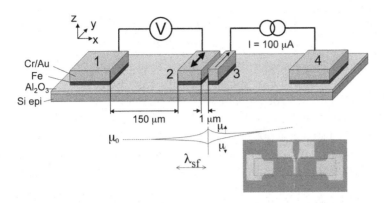

**FIGURE 2.26**  Schematic layout of four terminal non-local spin valve (NLSV) device and depiction of the spin-dependent electrochemical potential. A spin-polarized current is injected at contact 3 and changes in the electrochemical potential are measured non-locally at contact 2, placed 1 μm from contact 3. The difference in shape (100 × 24 μm) and (100 × 6 μm) for contact 2 and 3, respectively, allows for independent control of the magnetizations of the two contacts. The large reference contacts (100 × 150 μm) are spaced 150 μm (several spin flip lengths) from the injector and detector contact. The inset shows an image of the actual device. (After Jansen, R., *J. Phys. D* 36, R289, 2003. With permission.)

barrier contact (labeled 3 in Figure 2.26) is used to inject a spin-polarized charge current into the semiconductor. A second FM contact (labeled 2) is placed outside of the spin-polarized charge current path defined by the bias applied between contacts 3 and 4. Contact 2 senses the spin-splitting of the chemical potential, $\Delta\mu = \mu_{up} - \mu_{down}$, produced by the spin accumulation resulting from spin diffusing away from the FM injector, contact 3. The spin accumulation is the greatest directly underneath the injector contact, and decays with increasing distance from the contact with a certain spin diffusion length, $\lambda_{sf}$. The voltage measured at the detector contact 2 is proportional to $\Delta\mu$ and the projection of the semiconductor spin polarization onto the contact magnetization. Therefore, one can use some mechanism to manipulate the semiconductor spin polarization (magnitude or orientation) to encode or process information with the pure spin system, and directly read out the result as a voltage at this second contact. There are two key concepts to be recognized: (1) *the non-local geometry enables one to separate and distinguish a pure spin current from a spin-polarized charge current*, and (2) *the spin accumulation produces a real voltage, which can be sensed by an FM contact*. The following section illustrates these processes, and summarizes recent work on spin injection and non-local electrical detection in Si. Non-local spin injection/detection in GaAs is covered in Chapter 6, Volume 2, and in graphene in Chapters 4 and 5, Volume 3.

Spin accumulation can also be generated and detected using a single FM metal tunnel barrier contact, which serves as both the spin injector and detector in a simpler three terminal geometry with two nonmagnetic reference contacts [136, 190]. This approach is illustrated in Sections 2.6.3 and 2.6.4, using FM metal/oxide tunnel barrier contacts on both Si and GaAs.

## 2.6.2 Non-Local Spin Injection and Detection in Silicon: Pure Spin Currents

The non-local spin valve (NLSV) geometry of Figure 2.26 has been used to generate and detect pure spin current in silicon in a lateral transport device, and determine the corresponding spin lifetimes via the Hanle effect [187, 188]. The silicon samples were 250 nm thick epilayers grown by MBE on insulating Si(001) substrates at 500°C. The films were $n$-doped to achieve an electron concentration of $2-3 \times 10^{18}\,\text{cm}^{-3}$ at room temperature, below the metal–insulator transition ($n_{MIT} \sim 5 \times 10^{18}\,\text{cm}^{-3}$), but high enough to avoid carrier freeze-out at low temperature. The samples were exposed to atmosphere upon removal from the MBE system. After a 10% HF acid etch and deionized water rinse to produce a hydrogen passivated surface, the samples were loaded into a second MBE system, heated to desorb the hydrogen, and a 1.5 nm $Al_2O_3$ tunnel barrier and 10 nm polycrystalline Fe film were deposited to form the spin-injecting tunnel barrier contacts, as described in Ref. [154].

Conventional photolithography and wet chemical etching were used to define multiple NLSV structures, each consisting of four $Fe/Al_2O_3$ contacts spaced as shown in Figure 2.26. Contacts 2 ($24 \times 100\,\mu\text{m}$) and 3 ($6 \times 100\,\mu\text{m}$) are separated by 1 μm and serve as the spin detector and injector, respectively. The different aspect ratios provide different coercive fields with the easy axis along the long axis so that the contact magnetizations can be aligned parallel or antiparallel. Contacts 1 and 4 are reference contacts ($100 \times 150\,\mu\text{m}$) located several spin diffusion lengths away from the transport channel to ensure that the spin polarization is zero at these spin grounds.

When a negative bias is applied to contact 3, a spin-polarized electron current is injected from the Fe into the Si (Figure 2.27a) to produce (a) a spin-polarized charge current, which flows to contact 4 due to the applied bias, and (b) a pure spin current, which diffuses isotropically. The electron spin is majority spin oriented in the plane of the surface due to the in-plane magnetization of the Fe injector contact (and opposite the magnetization of the injector by convention). These spin currents produce spin accumulation in the Si described by the splitting of the spin-dependent electrochemical potential, $\Delta\mu = \mu_{up} - \mu_{down}$, which decreases with distance from the point of injection, as discussed above and shown explicitly in Figure 2.28a. The

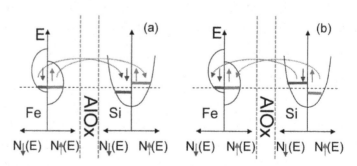

**FIGURE 2.27** Schematic illustration of (a) injection and (b) extraction of spins from silicon by means of spin-dependent tunneling.

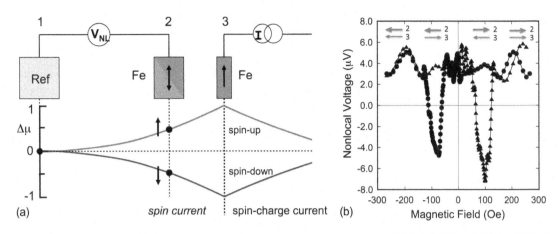

**FIGURE 2.28**    (a) Schematic of the spin splitting of the electrochemical potential $\Delta\mu$ corresponding to spin accumulation in the silicon produced by spin injection from contact 3 in the four terminal non-local spin valve geometry (reference contact 4 is not shown). Depending upon the direction of its magnetization, the non-local voltage $V_{NL}$ measured at contact 2 measures either the spin-up or spin-down branch of the chemical potential relative to the reference contact. (b) Non-local voltage vs. in-plane magnetic field, for an injection current of $-100\,\mu A$ at $10\,K$ in an $Fe/Al_2O_3/n$-Si non-local spin valve structure. Two levels corresponding to the parallel and antiparallel remanent states are clearly visible. The triangles (circles) correspond to increasing (decreasing) the in-plane magnetic field. The arrows indicate the relative orientation of the detector (2) and injector (3) contact magnetizations. (After Jiang, X. et al., *Phys. Rev. Lett.* 90, 256603, 2003. With permission.)

pure spin diffusion current is analyzed at contact 2, which is outside of the charge current path ("non-local"), where $\Delta\mu$ is manifested as a voltage. No charge current flows in the detection circuit defined by reference contact 1 and magnetic detector contact 2, thus excluding spurious contributions from anisotropic magnetoresistance and local Hall effects.

The non-local voltage $V_{NL}$ measured at the detector contact 2 is sensitive to the relative orientation of the contact magnetization and that of the spin polarization in the Si beneath it. In Figure 2.28b, $V_{NL}$ is plotted as a function of an applied in-plane magnetic field $B_y$ used to switch the magnetization of the injector and detector. At large negative field, the magnetizations are parallel, and the majority spins injected into the Si are parallel to the electron spin orientation of the detector, resulting in a positive voltage at contact 2 corresponding to the difference between the point indicated on the "spin-up" curve under contact 2 in Figure 2.28a and the spin ground. As the field is increased (triangles), contact 2 with the lower coercive field switches so that the injector/detector magnetizations are antiparallel. The orientation of the spin current diffusing to contact 2 is now antiparallel to that of the detector, producing an abrupt change at $B_y \sim 50\,Oe$ and a plateau in the non-local voltage, with the value corresponding to the difference between the point indicated on the "spin-down" curve under contact 2 (Figure 2.28a) and the spin ground. As $B_y$ is increased further, contact 3 also switches so that injector and detector contact magnetizations are again parallel, and the non-local voltage abruptly changes to its original value. The process is repeated as the field is swept in the opposite direction (circles, Figure 2.28b), producing an output characteristic similar to that seen in metal pseudo spin-valve structures [191], and confirming remanent (nonvolatile) parallel and antiparallel contact orientations.

When a positive bias is applied at contact 3, electrons flow from the Si into the Fe contact (Figure 2.27b). The higher conductivity of the Fe majority spin channel results in efficient majority spin current, leaving an accumulation of *minority* spin electrons in the Si and a reversal of the spin-splitting of the chemical potential. This bias configuration is often referred to as "spin extraction [192]", since majority spins are *extracted* from the semiconductor. A minority rather than majority pure spin current now diffuses from the injector to the detector contact 2, resulting in an inversion of the non-local voltage peaks for antiparallel alignment of injector and detector contact magnetizations, as seen in Figure 2.29. The magnitude of the non-local voltage is roughly linear with the magnitude of the bias current.

Thus, the orientation of the spin in the Si, and the corresponding splitting of the chemical potential, can be reversed either by switching the contact magnetization or by changing the polarity of the bias voltage on the injecting contact.

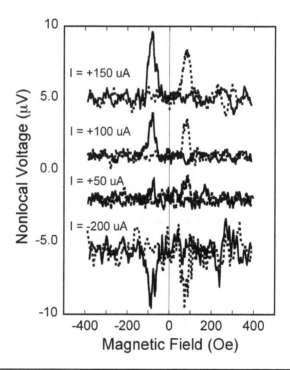

**FIGURE 2.29** Non-local voltage vs. in-plane magnetic field at 10 K for several values of the injection current. Graphs are offset for clarity. For negative bias, electrons are injected from Fe into the Si channel, and the change in non-local voltage is consistent with majority spin injection. For positive bias, electrons are extracted from the silicon into the Fe contact. The majority spins are more readily extracted, resulting in the accumulation of minority spins in the silicon. A change in sign is seen for non-local voltage peaks for the antiparallel state, consistent with minority spin accumulation. (After Jiang, X. et al., *Phys. Rev. Lett.* 90, 256603, 2003. With permission.)

Further confirmation of spin transport and another avenue enabling manipulation of the spin in the silicon is provided by the Hanle effect [193], in which a magnetic field perpendicular to the surface (electron spin) induces precession and dephasing of the spins in the silicon, resulting in a modulation and eventual suppression of the detected NL MR signal. The Hanle effect is widely regarded as the gold standard of spin injection and transport, providing definitive proof of the existence of a spin-polarized electron population [136, 137, 181, 186, 187]. The Hanle effect curve at $T = 10$ K and a bias current of $-100 \mu$A (spin injection) is shown in Figure 2.30. The magnetizations of the injector and detector contacts are placed in the parallel remanent state, and a magnetic field (Hanle field, $B_z$) is applied perpendicular to the surface of the device. As the Hanle field increases from zero, the injected spins in the silicon precess around $B_z$ during transit to the detector contact, resulting in both dephasing and an increasing degree of antiparallel alignment relative to the detector spin orientation, producing a corresponding decrease in the magnitude of the non-local signal. Variations in transit times in the transport channel due to the width of the injector and detector contacts truncate

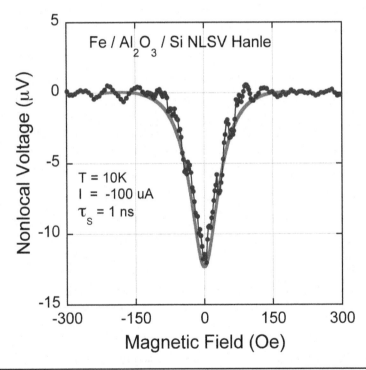

**FIGURE 2.30**   The Hanle effect curve at $T = 10$ K and a bias current of $-100 \mu$A (spin injection) obtained for the four terminal non-local spin valve contact geometry with Fe/Al$_2$O$_3$/Si contacts. The spin in the $n$-Si transport channel ($n \sim 2 - 3 \times 10^{18}$ cm$^{-3}$) precesses during transit between injector and detector contacts due to an applied perpendicular magnetic field. The magnetizations of the injector and detector contacts are in the parallel remanent state. The solid line is a fit to the data using the model described in the text, and yields a spin lifetime of 1 ns and diffusion constant of 10 cm$^2$/s.

the full Hanle curve and suppress further precessional oscillations in the non-local voltage. If the injector bias is reversed to operate in the spin extraction mode, or the injector and detector contacts are placed in the antiparallel remanent state, the sign of the Hanle curve reverses [187, 188]. Hanle curves are visible to measurement temperatures of ~150 K for the NLSV geometry and doping levels $n \sim 2 \times 10^{18}$ cm$^{-3}$, used in these samples.

The Hanle lineshape can be fit analytically to obtain the spin lifetime in this pure spin current. A fit to the data using an approach similar to that of Ref. [137] (and discussed in Chapter 6, Volume 2) is shown by the smooth curve in Figure 2.30, and yields a spin lifetime of ~1 ns at $T = 10$ K. The lifetimes for majority and minority spins (referenced to the injector contact 3) in the Si channel are comparable, as expected for a nonmagnetic semiconductor. This value is shorter than in the bulk at a comparable electron density and temperature. The relatively short spin lifetimes are attributed to the fact that this lateral transport geometry probes spin diffusion near the Si/Al$_2$O$_3$ interface, where surface scattering and interface states are likely to produce more rapid spin relaxation than in the bulk.

NLSV measurements using the geometry of Figure 2.26 were also reported recently in degenerately doped $n$-Si (doped well above the metal–insulator transition) by Ando et al. using Fe$_3$Si/Si Schottky tunnel barrier contacts [194], and by Sasaki et al. using Fe/MgO tunnel barrier contacts [195]. Both groups observed the non-local magnetoresistance behavior (Figure 2.28b) due to reversal of the contact magnetization. However, neither reported a Hanle signal, and thus provide no information on spin lifetimes. By measuring the dependence of the NLSV signal on injector/detector contact spacing, Sasaki et al. obtained an estimate for the spin diffusion length of 2.25 μm at $T = 8$ K and a carrier density $n \sim 1 \times 10^{20}$ cm$^{-3}$. They also report that a measurable NL MR signal persists to $T = 120$ K.

Thus, an FM metal/Al$_2$O$_3$ tunnel barrier contact can be effectively used as both a spin injector and a spin detector for Si, providing a direct voltage readout of the Si spin polarization in a lateral transport device. With the non-local geometry, one can generate and detect a pure spin current that flows outside of the charge current path. The spin orientation of this pure spin current is controlled in one of three ways: (a) by switching the magnetization of the FM metal injector contact; (b) by changing the polarity of the bias on the "injector" contact, which enables the generation of either majority or minority spin populations in the Si, and provides a way to electrically manipulate the injected spin orientation without changing the magnetization of the contact itself; and (c) by inducing spin precession through application of a small perpendicular magnetic field. The contact bias and magnetization, together with spin precession, allow full control over the orientation of the spin in the silicon channel and subsequent detection as a voltage, demonstrating that information can be transmitted and processed with pure spin currents in silicon. The FM contacts provide nonvolatile functionality as well as the potential for reprogrammability. This is accomplished in a lateral transport geometry using lithographic techniques compatible with existing device geometries and fabrication methods.

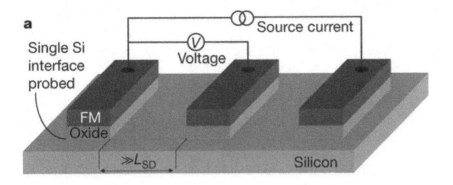

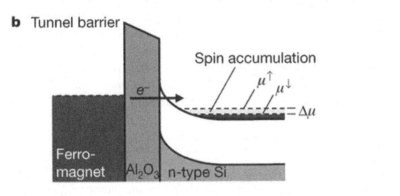

**FIGURE 2.31**    (a) Three terminal measurement geometry for injection and detection of spin accumulation under left contact. (b) Energy band diagram of the junction, depicting the ferromagnet, tunnel barrier, and *n*-type Si conduction and VBs. (After Dash, S.P. et al., *Nature* 462, 491, 2009. With permission.)

### 2.6.3 Three Terminal Spin Injection and Detection in Si at Room Temperature: Spin-Polarized Charge Current

A simpler three terminal geometry can also be used to generate and measure spin accumulation in a semiconductor [136, 190], as illustrated in Figure 2.31 for Si. In this case a single FM tunnel barrier contact serves as both the spin injector and detector, and the spin accumulation $\Delta\mu$ measured is that which develops directly under the contact in the presence of a spin-polarized charge current. As discussed in the next section, *this spin accumulation does not necessarily develop in the semiconductor itself*—it may, in fact, develop in the interface states or trap states within the oxide tunnel barrier. Thus, spin injection into the semiconductor is not required to generate spin accumulation and corresponding voltage signal in this measurement geometry. The density of such states depends upon the procedures used to form the interface and oxide layer, and may vary widely. Spin accumulation in such states may indeed prove to be just as interesting and technologically relevant as in the semiconductor, but may complicate and compromise spin injection/transport/detection in the semiconductor channel.

This three terminal approach was used by Dash et al. [196] to demonstrate spin accumulation and electrical detection in heavily $n$- and $p$-doped Si at room temperature. They used FM metal/$Al_2O_3$/Si tunnel barrier contacts for which spin injection into Si had been successfully demonstrated [154]. In contrast with previous work, their Si was degenerately doped, with $n$-($p$-) doping of $1.8 \times 10^{19}\,cm^{-3}$ ($4.8 \times 10^{18}\,cm^{-3}$), well above the metal–insulator transition. The Hanle effect was used to observe the spin accumulation and determine the corresponding spin lifetimes directly under the contact as a function of bias and temperature.

A cross-sectional diagram of the contact interface region illustrating the spin injection, accumulation, and subsequent precession induced by the perpendicular magnetic field $B_z$ is shown in Figure 2.32a. At $B_z = 0$, the spin accumulation $\Delta\mu$ builds and reaches a static equilibrium value. For $B_z \neq 0$, the spins precess at the Larmor frequency $\omega_L = g\mu_B B_z/\hbar$, resulting in a reduction of the net spin accumulation due to precessional dephasing (the Hanle effect). Here, $g$ is the Lande $g$-factor, $\mu_B$ the Bohr magneton, and $\hbar$ is the reduced Planck's constant. Thus, the voltage at the detector contact decreases as the magnitude of $B_z$ increases with an approximately Lorentzian lineshape given by $\Delta\mu(B) = \Delta\mu(0)/[1 + (\omega_L \tau_s)^2)]$, and the spin lifetime $\tau_s$ is obtained from fits to this lineshape.

The Hanle data showing spin accumulation at the $Ni_{80}Fe_{20}$/$Al_2O_3$/Si contact are shown in Figure 2.32b and c as a function of temperature. A clear signal due to spin accumulation is observed at room temperature, and increases at lower temperatures. The magnitude of the spin accumulation $\Delta\mu$ at the tunnel interface is obtained from the magnitude of the Hanle signal $\Delta V = TSP \times \Delta\mu/2e$, where $TSP = 0.3$ is the known tunneling spin polarization [183] for $Al_2O_3$/$Ni_{80}Fe_{20}$ at 300 K. Thus, $\Delta\mu = 1.2\,meV$ at 300 K, and increases to a value of 3 meV at 5 K. Dash et al. [196] note that their measured voltage and these $\Delta\mu$ values are about two orders of magnitude larger than that expected from the model of Fert and Jaffrès for spin injection and accumulation in the semiconductor itself [97, 100]. They suggest that lateral inhomogeneity of the tunnel current leads to higher localized spin injection current densities, but an alternate explanation involving accumulation in localized interface states [190] must also be considered, as discussed below. $Al_2O_3$/Si interfaces formed by sputter- or electron-beam deposition of the oxide are known to exhibit large interface state densities [197].

The data are well fit by a simple Lorentzian of the form described above, as shown by the solid line in Figure 2.32b. The full width at half maximum corresponds to $\omega_L = 1/\tau_s$, yielding a spin lifetime $\tau_s = 142\,ps$ at 300 K for the heavily doped $n$-Si with a measured electron density of $1.8 \times 10^{19}\,cm^{-3}$. The lifetime does not change significantly with temperature to 5 K. Similar results were obtained for $p$-doped Si, with the surprising result that the hole spin lifetimes deduced were a factor of two longer (~270 ps at 300 K) than the electron spin lifetimes, in marked contrast to what is typically reported for III–V semiconductors such as GaAs. This is inconsistent with the known physics of spin relaxation in simple cubic semiconductors like Si and GaAs.

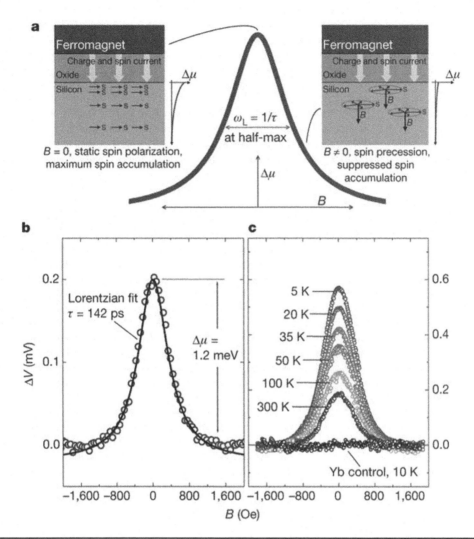

**FIGURE 2.32**    (a) Illustration of the Hanle effect, in which spin precession produced by a perpendicular magnetic field leads to a decay of the spin accumulation as measured by the voltage at the injection/detector terminal in the three terminal geometry. A plot of the voltage vs. magnetic field gives a pseudo-Lorentzian lineshape from which the spin lifetime can be determined. (b) Hanle curves shown for measurement temperatures of 5–300 K. A spin lifetime of 142 ps is obtained from the fit (solid line), and is independent of temperature. (After Dash, S.P. et al., *Nature* 462, 491, 2009. With permission.)

The degeneracy of the light hole and heavy hole states in the valence band results in very rapid hole spin depolarization on the sub-picosecond time scale, so that hole spin lifetimes are much shorter than those of the electron. The long hole spin lifetimes reported, and the fact that they are longer than the nominal electron spin lifetimes measured using similar samples and methodology, raise concern as to whether the spin signal measured derives from carriers in the silicon or from localized states at the oxide interface. The latter will be discussed in Section 2.6.4.

More recent work [198], has utilized $Co_{0.9}Fe_{0.1}/SiO_2$ and $Ni_{0.8}Fe_{0.2}/SiO_2$ tunnel barrier contacts on Si(001) [170] in a similar three terminal geometry, and observed spin accumulation and the Hanle effect to 500 K. This is a significant temperature milestone, since commercial and military electronic systems must operate at temperatures well above room temperature. The temperature 500 K easily exceeds the maximum temperature specification required to meet commercial (360 K), industrial (375 K), and even military (400 K) ratings. $SiO_2$ is highly desirable as a tunnel barrier for at least two reasons: (1) the $SiO_2$/Si interface is well-developed and known to have a low level of interface trap states, and (2) the existing semiconductor processing infrastructure is very familiar and comfortable with this system, facilitating future transition.

The measurements were performed on silicon-on-insulator (SOI) samples with a doping levels of $n(As) = 3 \times 10^{18}$ cm$^{-3}$ to $6 \times 10^{19}$ cm$^{-3}$, above and below the metal–insulator transition. The $SiO_2$ tunnel barrier was formed in situ by plasma oxidation, followed immediately by sputter-beam deposition of a 10 nm $Co_{0.9}Fe_{0.1}$ film, which was subsequently patterned to define the magnetic contacts. Nonmagnetic reference contacts were deposited by lift-off. The Hanle data obtained are shown in Figure 2.33a. The signal amplitude is nearly 0.2 mV at low temperature, and signal is clearly visible at 500 K. The signal amplitude is in good agreement with that predicted by theory—the measured spin resistance-area product $A\Delta V_{3T}(B_z = 0)/I$ is within a factor of 2–3 of the simplest model, which assumes that the observed spin accumulation occurs in the bulk [190] ($A$ is the contact area, $I$ is the bias current, and $\Delta V_{3T}(B_z = 0)$ is the measured Hanle amplitude).

The Hanle curves exhibit the classic pseudo-Lorentzian lineshape expected, and can be fit very well with the Lorentzian model, described above. The spin lifetimes thus obtained are ~100 ps and essentially

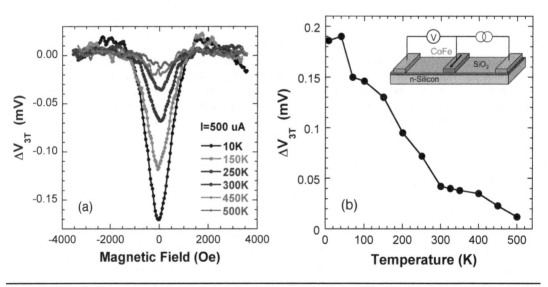

**FIGURE 2.33**    (a) Hanle curves obtained at various temperatures to 500 K for $Co_{0.9}Fe_{0.1}$/$SiO_2$ tunnel barrier contacts on Si(001). The amplitude decreases with increasing temperature. The linewidth and corresponding spin lifetime show little variation with temperature. (b) Temperature dependence of the amplitude of the Hanle signal. The inset shows the three terminal measurement geometry.

independent of temperature. This is similar to spin lifetimes measured in bulk Si at similar carrier densities of $n = 3.4 \times 10^{19}$ cm$^{-3}$ (As), and attributed to the metallic character of the sample, where impurity dominated spin–orbit interaction dominates spin relaxation for a wide temperature range [199]. The amplitude of the signal $\Delta V_{3T}(B_z = 0)$ decreases monotonically with temperature, as shown in Figure 2.33b. This behavior is inconsistent with that of the spin lifetime, indicating that other factors, such as spin injection/detection efficiency, may be responsible. Thus, spin precessional effects in this technologically dominant material are readily observed at the elevated temperatures necessary for utilization in commercial electronic systems.

## 2.6.4 THREE TERMINAL SPIN INJECTION AND DETECTION IN GaAs: INTERPRETATION AND THE ROLE OF LOCALIZED STATES

A clear interpretation of such three terminal Hanle data is still being developed. In contrast with the NLSV (four terminal) geometry, where the spin current *must flow through the semiconductor* before producing the spin accumulation enabling detection at a second contact, in the three terminal geometry, it is not always clear that spin transport/accumulation occurs in the semiconductor itself.

Significant insight was provided by recent experiments of Tran et al. [190] who used the three terminal geometry and Hanle measurements with Co/Al$_2$O$_3$/$n$-GaAs(001) samples. The GaAs 50 nm thick channel was heavily $n$-doped (Si) at $5 \times 10^{18}$ cm$^{-3}$ (well above the metal–insulator transition of $2 \times 10^{17}$ cm$^{-3}$) on a semi-insulating substrate, with a 15 nm thick surface layer at $2 \times 10^{19}$ cm$^{-3}$ to minimize the depletion region and facilitate spin injection. They observed a Hanle signal $\Delta V \sim 1.2$ meV at $T = 10$ K (Figure 2.34a), corresponding to a spin accumulation $\Delta \mu = 2e \, (\Delta V / \gamma) \sim 6$ meV (where $\gamma = 0.4$ is the tunnel spin polarization for Co/Al$_2$O$_3$), much larger than that expected from the predictions of the standard theory of spin injection in the diffusive regime for a single ferromagnet/semiconductor interface [97, 100]. They concluded that their measured signal did *not* originate from spin accumulation and precession in the GaAs channel. Rather, states localized near the Al$_2$O$_3$/GaAs interface served as a spin accumulation layer with long spin lifetimes leading to exceptionally large signal levels, as illustrated in Figure 2.34b. These localized states participate in a sequential tunneling process, and act as a confining layer which enhances the spin accumulation. The thickness of the layer containing these states was estimated to be ~1 nm. The authors speculate that the origin of these states may be either ionized Si donors within the depletion layer or interface states at the Al$_2$O$_3$/GaAs interface. The oxide/GaAs interface is well-known to have a high density of interface states, which has thwarted the development of a metal-oxide-semiconductor transistor technology [200]. Thus, the voltage measured in such a three-terminal geometry will include contributions from spin accumulation in localized states associated with the tunnel barrier and corresponding interfaces, and may

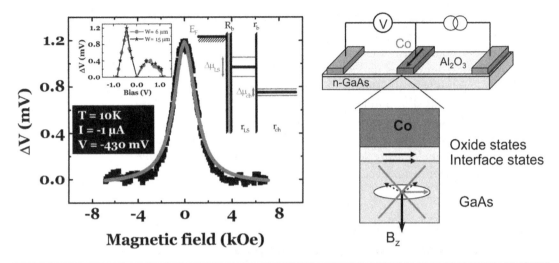

**FIGURE 2.34** (a) Hanle signal obtained in a three terminal geometry for a $15 \times 196\,\mu m^2$ Co/Al$_2$O$_3$ tunnel barrier contact to GaAs. A small quadratic background has been subtracted. The solid line is a fit to the data using the model described assuming spin accumulation in interface states. The left inset shows the bias dependence of the Hanle signal, and the right inset illustrates the spin splitting of the electrochemical potential in the interfacial region. (After Tran, M. et al., *Phys. Rev. Lett.* 102, 036601, 2009. With permission.) (b) Schematic of the measurement geometry and cross-section of the region under the magnetic tunnel barrier contact, illustrating the location of interface states which the authors concluded produced the spin accumulation and strong Hanle signal measured.

not demonstrate spin injection or accurately reflect spin accumulation in the semiconductor itself. Four terminal (NLSV) measurements from the same samples, in which the spin current must flow through the semiconductor to a second detector contact, should provide valuable insight and clarification.

## 2.7 GRAPHENE AS A TUNNEL BARRIER FOR SPIN INJECTION/DETECTION IN SILICON

As discussed in the preceding sections, a tunnel barrier between a ferromagnetic metal and the semiconductor circumvents the large conductivity mismatch and enables electrical spin injection and detection. Extensive effort has been directed towards developing appropriate tunnel barriers for spin contacts, with most work focusing on a reverse-biased ferromagnetic Schottky barrier or an insulating oxide layer such as Al$_2$O$_3$ or MgO. However, metal Schottky barriers and oxide layers are susceptible to interdiffusion, interface defects and trapped charge, which have been shown to compromise spin injection, transport and detection. Ferromagnetic metals readily form silicides even at room temperature, and diffusion of the ferromagnetic species into the silicon creates magnetic scattering sites, limiting spin diffusion lengths and spin lifetimes in the silicon. Even a well-developed and widely utilized oxide such as SiO$_2$ is known to have defects and trapped or mobile charge, which limit both charge and spin-based performance. Such approaches also result in contacts with high resistance–area (RA) products, exacerbating practical issues of local heating and power consumption. More importantly, previous work has shown that smaller RA products within a select window of values are essential for efficient spin injection and detection [97,100].

An ideal tunnel barrier should exhibit several key material characteristics:

- a uniform and planar habit with well-controlled thickness,
- minimal defect and trapped charge density,
- a low RA product for minimal power consumption,
- chemical stability,
- compatibility with both the ferromagnetic metal and the semiconductor of choice,
- thermal stability to ensure minimal diffusion to/from the surrounding materials at the temperatures required for device processing.

Graphene, an atomically thin honeycomb lattice of carbon, offers a compelling option. Although it is very conductive in plane, it exhibits low conductivity perpendicular to the plane [201]. Its $sp^2$ bonding results in a highly uniform, defect-free layer that is chemically inert, thermally robust, and essentially impervious to diffusion [202]. Recent work has in fact demonstrated that single-layer graphene can be used as a spin tunnel barrier between two metals in a magnetic tunnel junction [203, 204].

In this section, we show that a ferromagnetic metal/monolayer graphene contact serves as a spin-polarized tunnel barrier contact that provides efficient electrical spin injection and detection in planar silicon [205, 206] and silicon nanowires [207] using both the three terminal and four terminal nonlocal spin valve transport geometries described previously. The graphene barrier also enables one to achieve contact RA products that fall within the critical window of values required for practical devices. We demonstrate electrical injection and detection of spin accumulation in silicon above room temperature, and show that the corresponding spin lifetimes correlate with the silicon donor concentration, confirming that the spin accumulation measured occurs in the silicon and not in the graphene or interface trap states. The RA products are three orders of magnitude lower than those achieved with oxide tunnel barrier contacts on silicon substrates with identical doping levels.

## 2.7.1 THREE-TERMINAL DEVICE GEOMETRY

Graphene was grown by low-pressure chemical vapor deposition within copper foil 'enclosures' [208], and transferred onto hydrogen-passivated $n$-type silicon (001) substrates with electron density $n = 1 \times 10^{19}$, $3 \times 10^{19}$ and $6 \times 10^{19}$ $cm^{-3}$. $Ni_{80}Fe_{20}$ was deposited on top of the graphene via standard liftoff techniques to form the device structures illustrated in Figure 2.35. Details of the fabrication process may be found in Ref. [205].

Current-voltage ($I$-$V$) curves for several devices at room temperature are shown in Figure 2.36a, including reference contacts consisting of NiFe deposited directly onto the hydrogen-passivated Si substrate, and onto $Al_2O_3$

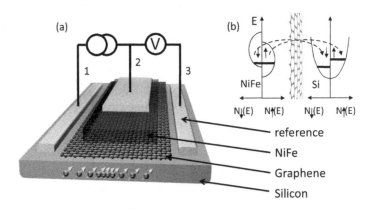

**FIGURE 2.35** Schematic of the samples. (a) Monolayer graphene serves as a tunnel barrier between the FM metal contact and the Si substrate. Contacts 1 and 3 are ohmic Ti/Au contacts. (b) Schematic illustrating spin injection and spin accumulation.

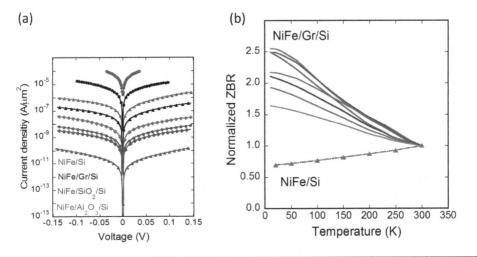

**FIGURE 2.36** Electrical characteristics of the NiFe/tunnel barrier/Si contacts. (a) The current-voltage curves for NiFe/Si (red), NiFe/graphene/Si (black), NiFe/ 2 nm SiO2/Si (green), and NiFe/$Al_2O_3$/Si (blue) contacts at RT. The Si doping concentration is $1 \times 10^{19}$ cm$^{-3}$ (triangles), $3 \times 10^{19}$ cm$^{-3}$ (diamonds), and $6 \times 10^{19}$ cm$^{-3}$ (circles). (b) The normalized zero bias resistance (ZBR) shows a weak insulator-like temperature dependence for the NiFe/graphene contacts, confirming tunnel transport. Each solid color line is from a different contact, illustrating the reproducibility of the data. The ohmic NiFe/Si contacts exhibit metallic behavior; results for three contacts are shown by triangles, red-dashed and green-dashed lines.

and $SiO_2$ tunnel barriers. The highest resistance is obtained for the NiFe/$SiO_2$/Si samples and the lowest for the NiFe/Si reference samples. The corresponding RA products on the Si ($6 \times 10^{19}$ cm$^{-3}$) substrate are 0.4 k$\Omega$ um$^2$ for direct contact (which is ohmic), 6 k$\Omega$ um$^2$ for the graphene barrier, and 15 M$\Omega$ um$^2$ for the $SiO_2$ barrier, demonstrating that the graphene barrier produces an RA product orders of magnitude smaller than the oxides for thicknesses necessary to produce a pinhole free tunnel barrier.

Figure 2.36b shows the temperature dependence of the zero bias resistance (ZBR) for NiFe/monolayer graphene/Si ($n = 6 \times 10^{19}$ cm$^{-3}$) contacts. The weak temperature dependence confirms that transport occurs by tunneling through a pin-hole-free tunnel barrier, as discussed in Section 2.5.3.3.1. The $Al_2O_3$ and $SiO_2$ contacts also show a weak, insulating-like dependence on temperature, confirming that they form pinhole free tunnel barriers. In contrast, the direct NiFe/Si contact exhibits metallic behavior.

Spin accumulation and precession directly under the magnetic tunnel barrier contact were measured using the Hanle effect in the three terminal geometry as described in Section 2.6.3, and the results are shown in Figure 2.37. The top three curves in Figure 2.37a are the Hanle measurements of the graphene, $SiO_2$ and $Al_2O_3$ tunnel barrier samples at room temperature. All three samples show a clear Hanle effect, evidenced by a Lorentzian line shape caused by field-induced precession of the spin population, confirming successful spin accumulation. Note that the graphene barrier produces the highest spin signal, yet has a significantly lower RA product than the oxide barrier contacts. The bottom two curves in Figure 2.37a show results for additional reference samples consisting of the NiFe/Si and a non-magnetic/graphene/Si sample, measured at 10 K. No Hanle effect is observed even at these low temperatures, as expected. The temperature dependence of the Hanle signal for the NiFe/graphene/Si ($1 \times 10^{19}$ cm$^{-3}$) samples is shown in Figure 2.37b. Much larger signals are observed at low temperature, and the amplitude decreases monotonically with temperature.

No evidence for a Hanle-like signal is observed when the magnetic field is applied in-plane along the long axis of the magnetic contact. The appearance

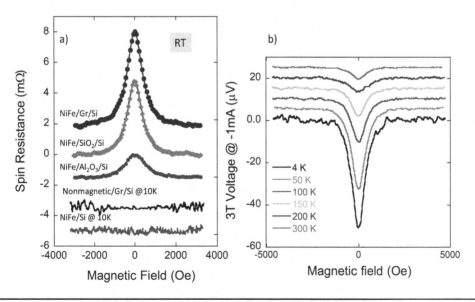

**FIGURE 2.37** Hanle spin precession measurements. (a) Room temperature Hanle data for spin injection NiFe/Al$_2$O$_3$/Si ($3 \times 10^{19}$) (red), NiFe/SiO$_2$/Si ($3 \times 10^{19}$) (blue), and NiFe/Graphene/Si ($1 \times 10^{19}$) (green). Also shown are the control samples; nonmagnetic/ graphene/Si ($1 \times 10^{19}$) and NiFe/Si ($1 \times 10^{19}$) measured at 10 K. (b) Temperature dependent Hanle data for NiFe/Graphene/Si ($1 \times 10^{19}$).

of such an "inverted" Hanle effect has been attributed to precessional dephasing produced by random fringe fields arising from surface/interface roughness of the magnetic contact [209], although other factors may contribute [210]. The absence of an inverted Hanle signal implies that the NiFe/graphene tunnel barrier produces a more uniform magnetic film and interface which suppress such effects relative to that provided by the NiFe/oxide interface.

Values for the spin lifetime are obtained from fits to the Hanle curves [205] using the Lorentzian described above. The spin lifetime depends strongly on contact bias and silicon donor density, and weakly on temperature due to the metallic character of the silicon [199]. The spin lifetime decreases with increasing donor density, as expected from electron spin resonance (ESR) measurements on bulk silicon [211, 212]. This is shown explicitly in Figure 2.38, where we plot the spin lifetime obtained from three-terminal Hanle data as a function of electron density for four different tunnel barrier materials (graphene, $Al_2O_3$, MgO and $SiO_2$) and three different magnetic metal contacts (Fe, CoFe and NiFe). At $T = 4$ K, fits to the Hanle data yield spin lifetimes of 140 ps and 105 ps for the NiFe/graphene/Si($1 \times 10^{19}$) and Si($6 \times 10^{19}$) samples, respectively. These values agree well with those reported for NiFe/$SiO_2$ tunnel barrier contacts [198], and show a clear correlation with the character of the silicon.

The spin lifetime measured with the three-terminal Hanle geometry shows a clear dependence only on electron density, and the dependence is consistent with literature ESR data on bulk silicon [211, 212]. The spin lifetime is completely independent of the tunnel barrier material or magnetic

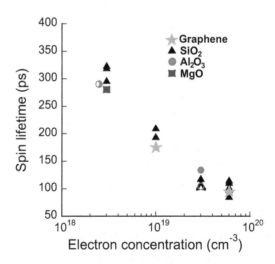

**FIGURE 2.38**   Spin lifetimes obtained from three-terminal Hanle measurements at 10 K as a function of the Si electron density for the tunnel barrier materials indicated and different ferromagnetic metal contacts (Fe, CoFe, NiFe). The symbol shape distinguishes the tunnel barrier material: triangles – $SiO_2$, circles – $Al_2O_3$, and stars – graphene. Solid symbols correspond to devices with $Ni_{0.8}Fe_{0.2}$ contacts, half-solid symbols to Fe contacts, and open symbols to $Co_{0.9}Fe_{0.1}$ contacts. The spin lifetimes show a pronounced dependence on the Si doping level, and little dependence on the choice of tunnel barrier or magnetic metal.

metal used for the contact. The values for the graphene tunnel barriers fall directly on the curve. These data confirm that the spin accumulation occurs in the silicon, and not in the graphene or possible interface trap states. The measured spin lifetimes are shorter than those in bulk silicon because they reflect the environment directly beneath the contact, where the reduced symmetry and increased scattering from the interface are likely to produce additional spin scattering, as discussed in Ref. [198].

The magnetic contact's conventional RA product is an important parameter in determining the practical application of a spin-based semiconductor device (note that this is the standard resistance–area product and not the spin resistance–area product). Calculations have shown that significant local (two-terminal) magnetoresistance (MR) can be achieved only if the contact RA product falls within a specific range which depends upon the Si channel conductivity, the spin lifetime, and the contact spacing and width [97, 100, 213]. The RA products of tunnel barrier contacts to date have been much larger than required, making such devices unattainable. Previous work to lower the RA product utilized a low-work function metal such as gadolinium at the tunnel barrier interface, but no spin accumulation in the semiconductor was demonstrated [214]. In contrast, the low RA products provided by the graphene tunnel barriers fall well within this window, and enable realization of advanced spintronic devices.

We calculate the range of optimum RA products and the corresponding local MR as a function of the Si electron density using the methodology of Ref. [97] and the contact geometry shown as the inset to Figure 2.39. Further details can be found in Ref. [205]. The geometric parameters are chosen to be consistent with the node anticipated for Si device technology within the next 5 years. The color code in Figure 2.39 identifies the range of useful MR and the corresponding window of contact RA products required. Tunnel barrier contacts of $FM/Al_2O_3$ and $FM/SiO_2$ fabricated in our lab on identical substrates have been shown to produce significant spin accumulation in Si, but have RA products that are too high to generate usable local MR for oxide thicknesses sufficient to produce a pinhole free barrier. In contrast, using monolayer graphene as the tunnel barrier lowers the RA product by orders of magnitude, and values for the NiFe/graphene contacts on bulk wafers fall well within the range required to generate high local MR. Reducing the RA product also has a positive effect on the electrical properties of the spin device, as lowering the resistance reduces noise and increases the speed of an electrical circuit [215].

## 2.7.2 Four-Terminal Non-Local Spin Valve Geometry

Magnetic metal/graphene tunnel barrier contacts have also been employed in the four-terminal non-local spin valve (NLSV) geometry (described in Sections 2.6.1 and 2.6.2) to generate pure spin currents in silicon and determine spin lifetimes from the corresponding Hanle effect. The spin lifetimes are the same as those obtained from the three-terminal geometry devices described above, providing further confirmation that the spin accumulation

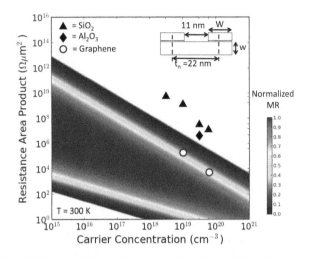

**FIGURE 2.39** Resistance–area product window for local magnetoresistance. Calculation of the local (two-terminal) magnetoresistance (MR) as a function of the conventional resistance–area product of the contact and the Si electron density for the device geometry shown in the inset, using the theory of Ref. [11]. Data points are the resistance–area products measured for our ferromagnetic metal/ tunnel barrier/Si contacts using 2 nm $SiO_2$ (triangles), 1.5 nm $Al_2O_3$ (diamond) and monolayer graphene (circles) tunnel barriers prepared from identical Si wafers in our laboratory. The ferromagnetic metal/graphene resistance–area products fall within the window of useful magnetoresistance values. W = w = 11 nm.

indeed occurs in the silicon channel rather than in trap states localized at the interface or in some oxide. The existence of such localized trap states and their potential contribution to spin accumulation has been reported for other spin tunnel barrier structures [190], particularly those purposefully fabricated to maximize the density of such states [216, 217].

Non-local spin valve devices were fabricated using $Ni_{80}Fe_{20}$/monolayer graphene tunnel barriers on the same silicon substrates described above and in Refs. [205] and [206]. Monolayer graphene grown by chemical vapor deposition was transferred onto hydrogen passivated Si(001) wafers, and NiFe injector and detector contacts defined by electron-beam lithography and liftoff techniques. The contacts were either 200 nm and 1 μm wide, or 200 nm and 200 nm wide, with edge-to-edge separations of 200 nm or 400 nm. These contacts were then used as a hard mask to etch away the exposed graphene by a light oxygen plasma, confirmed by a resistance check. A schematic of the device and a scanning electron microscope (SEM) image of the injector/detector area are shown in Figure 2.40.

Spin-polarized charge current injected at the NiFe/graphene contact 2 generates a spin accumulation in the silicon beneath the contact. Spin diffusion causes a pure spin current to flow in the silicon channel away from the point of injection, and a corresponding spin splitting of the electrochemical potential is detected as a voltage on the NiFe/graphene detector contact 3. The results of several devices on a silicon wafer with $n = 1 \times 10^{19}$ cm$^{-3}$ are shown in Figure 2.41 at T = 10 K. The non-local resistance is defined to be

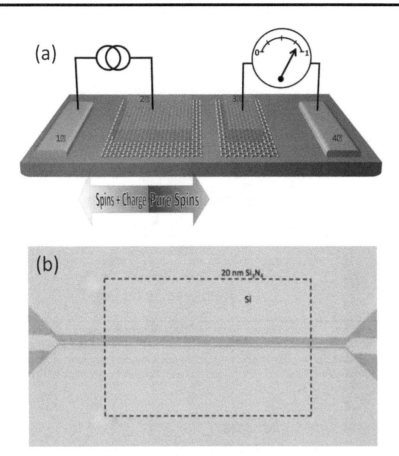

**FIGURE 2.40** (a) Schematic of non-local spin valve device. Contacts 1 and 4 are ohmic contacts applied directly to the Si substrate, and contacts 2 (1 μm wide) and 3 (200 nm wide) are Au capped $Ni_{80}Fe_{20}$/graphene tunnel barriers for electrical spin injection and detection. (b) SEM image of the active area of the NLSV device, dashed line indicates 20 μm × 30 μm window, contacts are 200 nm and 1 μm wide separated edge-to-edge by 200 nm.

the voltage measured at the non-local detector contact 3 divided by the bias current applied between contacts 1 and 2. An in plane magnetic field is used to switch the magnetizations of the two electrodes independently. When the magnetizations of injector and detector contacts are parallel, a low non-local resistance is measured, and when the magnetizations are antiparallel a high non-local resistance is measured – this is the classic signature of a NLSV device. A net change in the non-local resistance of 1–15 milli-ohm was measured at 10K, with the larger values exhibited by devices in which both injector and detector contacts were 200 nm.

Application of an out-of-plane magnetic field causes the spins to precess and dephase as they diffuse towards the detector contact 3, and the non-local voltage decreases with increasing magnetic field due to this Hanle effect. The non-local Hanle data are shown in Figure 2.42 after background subtraction, and can be modeled using the integral spin diffusion approach described in Section 2.6.2. The model fits the experimental data well, as shown by the red

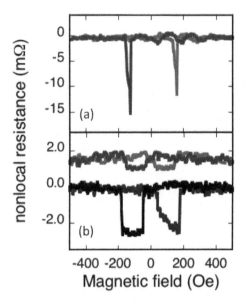

**FIGURE 2.41** (a) NLSV measurement at 10 K where both injector and detector are 200 nm wide with an edge-to-edge separation of 200 nm. (b) The same for devices with a 1 μm wide injector and a 200 nm wide detector, with an edge-to-edge separation of 200 nm. The silicon electron concentration is $1 \times 10^{19}$ cm$^{-3}$.

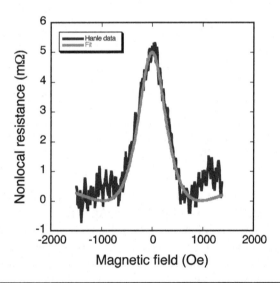

**FIGURE 2.42** Non-local Hanle data after background subtraction at $T = 10$ K. The solid line is a fit to the data and yields a spin lifetime of 140 ps. The silicon electron density is $n = 1 \times 10^{19}$ cm$^{-3}$.

curve in Figure 2.42, and yields values for the diffusion constant $D = 2.9$ cm$^2$/sec and the silicon spin lifetime of $140 \pm 10$ ps. This spin lifetime is the same as that obtained from the three-terminal data described above for the same electron density, providing further confirmation that the three-terminal data are not corrupted by parasitic effects such as localized trap states.

These three-terminal and NLSV results demonstrate that a ferromagnetic metal/monolayer graphene contact serves as a spin-polarized tunnel barrier contact with a low RA product, a crucial requirement enabling future semiconductor spintronic devices. Utilizing multilayer rather than monolayer graphene in such structures may provide much higher values of the tunnel spin polarization due to the band structure derived spin filtering effects that have been predicted for selected ferromagnetic metal/multilayer graphene structures [218, 219]. Such an increase will improve the performance of semiconductor spintronic devices by providing higher signal-to-noise ratios and corresponding operating speeds.

### 2.7.3 SPIN TRANSPORT IN SILICON NANOWIRES

Semiconductor nanowires (NWs) provide an avenue to further reduce the ever-shrinking dimensions of transistors. Realization of spin-based Si NW devices requires efficient electrical spin injection and detection, which depend critically on the interface resistance between a ferromagnetic metal contact and the NW. This is especially problematic with semiconducting NWs because of the exceedingly small contact area, which can be of order 100 nm$^2$. Research has shown that standard oxide tunnel barriers can provide good spin injection into planar Si structures, but such contacts grown on NWs are often far too resistive to yield reliable and consistent results. The graphene tunnel barriers described above offer a solution—they form low RA product tunnel contacts on bulk Si, while providing spin-injecting properties superior to those of oxide tunnel barriers.

In this section we describe fabrication of nanoscale spintronic devices comprised of a Si nanowire channel with low-resistance graphene tunnel barriers for electrical spin injection and detection of pure spin currents. Both four-terminal NLSV and three-terminal (3T) geometries are used to probe spin accumulation and transport. In contrast with previous work on semiconductor NWs which reported no Hanle effect [220–224], Hanle spin precession is observed here in both measurements [207]. Because spurious effects can produce spin-valve-like behavior even in the non-local geometry [225], these Hanle data provide confirmation of spin accumulation and pure spin transport, enabling direct measurements of spin lifetimes and diffusion lengths in these challenging NW structures.

Si nanowires were grown using gold seeded vapor-liquid-solid techniques with a <111> growth axis and phosphorous doped to an electron concentration $n \sim 3 \times 10^{19}$ cm$^{-3}$ [226]. They were removed from the growth substrate by sonication and dispersed on a Si$_3$N$_4$/Si substrate. Appropriate nanowires $\sim$ 100–150 nm in diameter were chosen optically, and two outer ohmic contacts were made using electron-beam lithography. NiFe/graphene tunnel barriers were fabricated as described above to form the devices illustrated in Figure 2.43 on which both 3T and NLSV measurements could be made. These magnetic contacts, labeled 2 and 3, are 1 μm and 200 nm wide respectively, separated edge-to-edge by 200 nm. These contacts are then used as a hard mask to remove the excess graphene by

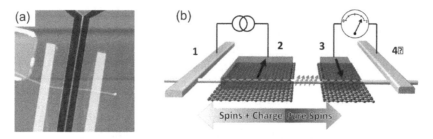

**FIGURE 2.43** (a) False color atomic force microscopy image of a silicon nanowire with the four contacts used in the spin measurements. The ferromagnetic metal/graphene tunnel barrier contacts used to inject and detect spin appear as blue, the gold ohmic reference contacts appear as yellow, and the light green line is the silicon nanowire transport channel. The bright dot on the end of the nanowire is the gold nanoparticle used to seed the nanowire growth. (b) Schematic of the four terminal nanowire device in the non-local spin valve geometry. A spin-polarized charge current is injected at the NiFe/graphene contact 2, generating a pure spin current that flows to the right within the silicon nanowire. This spin current generates a voltage that is detected on NiFe/graphene contact 3.

a light $O_2$ plasma etch. Measurements of the *I-V* curves and temperature dependence of the zero bias resistance were used to confirm the tunneling character of the NiFe/graphene contacts. Further fabrication details may be found in Ref. [207].

The spin injecting properties and spin-RA products of the NiFe/graphene tunnel barrier contacts were initially assessed using the 3T geometry in which a single FM contact is used to both generate and detect spin accumulation in the area of the NW immediately beneath the contact. A current is applied between contacts 1 and 3 (see Figure 2.44a), and injection of spin-polarized electrons from the NiFe/graphene contact 3 produces a net spin accumulation in the Si nanowire. This results in a spin splitting of the electrochemical potential which is detected as a voltage measured across contacts 3 and 4. When a magnetic field $B_z$ is applied along the surface normal and perpendicular to the electron spin direction, the spins precess and dephase, reducing the net spin accumulation and the corresponding voltage at contact 3, as illustrated in Figure 2.44b. A plot of the detector voltage versus the applied field reveals a clear Hanle signal with the expected Lorentzian lineshape, as shown in Figure 2.44c. Fits to the data as described in section 2.6.3 yield a spin lifetime of $260 \pm 10$ ps, a factor of 2 longer than that obtained in planar Si devices of similar doping level with either oxide [198] or graphene tunnel barrier contacts [205]. No evidence for a Hanle-like signal is observed when the magnetic field is applied in-plane.

These spin lifetimes are an order of magnitude shorter than the 2.5 ns lifetime obtained from electron spin resonance measurements on bulk samples of similar carrier density [199]. We attribute this to the proximity of the FM contact interface where magnetic fringe fields due to contact edges or closure domains may lead to strong local spin precession and dephasing [227, 228]. In addition, spins can scatter at the FM contact interface, or readily diffuse between the Si and FM metal contact through the relatively transparent tunnel barrier.

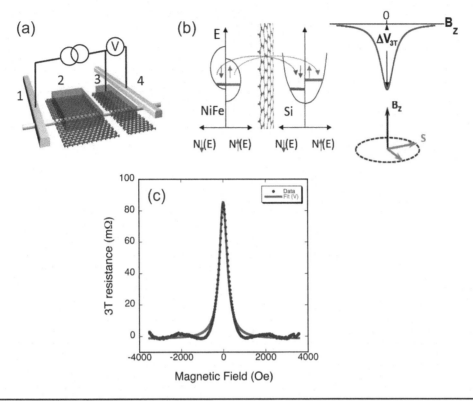

**FIGURE 2.44**   Three terminal Hanle measurements. (a) Schematic of the 3T measurement; (b) Simplified diagram illustrating spin injection from the NiFe across the graphene tunnel barrier producing spin accumulation in the Si NW, and the Hanle spin precession measurement. $N\uparrow(E)$ and $N\downarrow$ (E) are the majority and minority spin density of states, respectively. (c) 3T Hanle measurements for a 150 nm diameter NW at 10 K and $-200\ \mu A$ bias current (spin injection) after a quadratic background subtraction and plotted as resistance. The solid line is a fit of the data.

One expects a much larger enhancement of the spin lifetime in a true one-dimensional (1D) structure relative to bulk material due to the discontinuous density of states that reduces available spin relaxation channels [229]. However, Si nanowires are expected to exhibit true 1D behavior only for diameters less than ~ 10 nm [230, 231]. The diameter of the nanowires used here and in measurements by other groups [220–224] are all much larger due to limitations in fabrication, and therefore too wide to exhibit an ideal 1D density of states and the corresponding full enhancement in spin lifetime. The fact that the Hanle-derived spin lifetime is not *reduced* relative to values obtained from similar measurements on bulk planar samples indicates that surface scattering due to the much larger surface-to-volume ratio for the NW does not play a more significant role at these relatively high doping levels, or that another relaxation mechanism dominates.

The same samples are used in the NLSV geometry of Figure 2.43b to demonstrate the generation and detection of pure spin current in the NW channel produced by the NiFe/graphene contacts. As described in Sections 2.6.1 and 2.6.2, the pure spin current flowing from the injector to detector

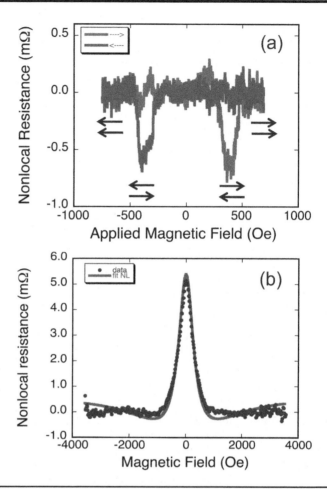

**FIGURE 2.45** Non-local spin-valve and Hanle measurements. (a) NLSV data on 150 nm wide nanowire, contacts are 200 nm and 1 μm wide with a 200 nm edge-to-edge spacing. The arrows indicate the relative orientation of the magnetizations of the injector and detector contacts, and the blue and red traces indicate increasing and decreasing magnetic field sweeps. (b) NLSV Hanle data for the Si NW for parallel alignment of contacts 2 and 3 after background subtraction. All data are for a bias current of –200 μA at $T = 10$ K

contact creates a spin accumulation and spin-splitting of the electrochemical potential which appears as a voltage on the detector contact. The results are shown in Figure 2.45a. The non-local resistance is defined to be the voltage measured at the non-local detector contact divided by the injector contact bias current. An in plane magnetic field is used to switch the magnetizations of the two electrodes independently due to their different coercive fields, producing parallel and antiparallel alignments. When the magnetizations of injector and detector contacts are parallel, a low non-local resistance is measured, and when the magnetizations are antiparallel a high non-local resistance is measured—this is the classic signature of a NLSV device. A net change in the non-local resistance of ~ 0.7 mΩ is measured at 10K.

The magnetizations of the two contacts reverse cleanly at well-defined coercive fields consistent with shape anisotropy, so that both parallel and antiparallel alignment of the injector/detector magnetizations are realized. This is attributed to the smooth bridging of the NW by the graphene layer, producing a more gradual variation in the contact topography to which the NiFe film conforms. In contrast, magnetic metal/oxide tunnel contacts applied directly to the NW exhibit poorly defined switching characteristics attributed to local domain patterns arising from the more complex morphology created as the contact attempts to conform to the NW sidewalls [221].

Application of an out-of-plane magnetic field causes the spins to precess and dephase as they diffuse towards the detector contact, and the non-local voltage decreases with increasing magnetic field due to this Hanle effect. The NLSV Hanle data are shown in Figure 2.45b, and can be modeled using the integral spin diffusion approach described in Section 2.6.2. The model fits the experimental data well, as shown by the red curve in Figure 2.45b, and yields values for the diffusion constant $D = 0.00047$ m$^2$/sec and the NLSV spin lifetime of $190 \pm 10$ ps. This effective spin lifetime is shorter than the value of $260 \pm 10$ ps obtained in the 3T geometry, although the reasons are unclear at present. The spin diffusion length is calculated as $L_{SD} = [D\,\tau_s]^{0.5} \sim$ 300 nm, which is significantly shorter than results reported previously [220, 221]. However, these were based only on the amplitude of the NLSV data rather than direct determination of the spin lifetime provided by the Hanle measurements.

Achieving good contacts remains one of the largest technological barriers to exploiting geometrical nanowires for both charge and spin-based applications. The single layer graphene barriers described here provide a high quality tunnel barrier with a low RA product free of pinholes that smoothly bridges the nanowire. This results in clean magnetic switching characteristics for the magnetic contacts, and enables the first observation of Hanle spin precession of both spin accumulation in the nanowire in the 3T geometry, and of the pure spin current generated in the nanowire transport channel using the NLSV geometry. The Hanle data provide direct measurements of the spin lifetime and spin diffusion lengths in these nanoscale spintronic devices.

## 2.8 OUTLOOK AND CRITICAL RESEARCH ISSUES

A great deal of progress has been made in spin injection and transport in semiconductors in the last 15 years. The contributions from many groups summarized here and elsewhere have rapidly advanced the state-of-the-art in electrical injection, detection, and spin manipulation in semiconductors of immediate technological interest such as GaAs and Si. There are several important areas that require further in-depth research.

Interfaces continue to play a critical role in at least two ways. First, they impact spin injection efficiency from a given contact, which has been the primary topic of this chapter. Second, they impact spin lifetimes in lateral transport structures or in any thin film, nanoscale device. While spin lifetimes have been reasonably well-studied in bulk Si, there is very little information available on these lifetimes near interfaces. In Si, the reduced symmetry will enable spin relaxation via mechanisms which are symmetry-forbidden in the bulk, such as D'yakonov–Perel'. For all semiconductors, the specific interface structure, defects and impurities, and band bending are expected to play a role, but the relative significance of these contributions is yet to be determined.

Contact resistance plays a major role in the relative efficiency of a spin-injecting contact, as illustrated by several model calculations, and described in Sections 2.5.2 and 2.5.3. While a large value seems, at first glance, to be desirable for spin injection, this will severely limit the spin-polarized current flow through the contact into the semiconductor, and thereby limit the spin accumulation achieved. Therefore, a compromise must be reached. If one considers electrical *detection*, further constraints apply as discussed by Min et al. [183]—for the case of moderately doped Si ($n \sim 10^{16}\,\mathrm{cm}^{-3}$), contact resistances of order $10^{-2}$–$10^{-5}\,\Omega\,\mathrm{cm}^2$ will be required. Similar considerations apply to other materials such as GaAs, although the specific values vary. These values are achievable even with tunnel barriers, but will be strongly case-specific and application-dependent.

Utilizing intrinsic 2D layers such as graphene or hexagonal boron nitride as tunnel contacts offers many advantages over conventional materials deposited by vapor deposition (e.g. $Al_2O_3$ or MgO), enabling a path to highly scaled electronic and spintronic devices. The use of multilayer rather than single layer graphene in such structures may provide much higher values of the tunnel spin polarization because of band structure derived spin filtering effects predicted for selected ferromagnetic metal/multilayer graphene structures [218, 219]. This increase would further improve the performance of nanowire spintronic devices by providing higher signal to noise ratios and corresponding operating speeds, advancing the techological applications of nanowire devices.

## ACKNOWLEDGMENTS

It is a pleasure to acknowledge the many colleagues and collaborators who have made significant contributions through either experimental work, theory, or discussion: Brian Bennett, Steve Erwin, Aubrey Hanbicki, George Kioseoglou, Jim Krebs, Connie Li, Athos Petrou, Andre Petukhov, Philip Thompson, Olaf van't Erve, and many others. I wish to especially acknowledge Gary Prinz, who provided me with the opportunity to enter and contribute to the field of magnetoelectronics. Various aspects of the work presented here were supported by the Office of Naval Research, DARPA and core program at the Naval Research Laboratory. This effort would not have been possible without the long-term basic research support of both the Office of Naval Research and NRL.

# REFERENCES

1. G. Moore, Cramming more components onto integrated circuits, *Electronics* **38** (April 19, 1965).
2. G. A. Prinz, Magnetoelectronics, *Science* **282**, 1660 (1998).
3. S. A. Wolf, D. D. Awschalom, R. A. Buhrman et al., Spintronics: A spin-based electronics vision for the future, *Science* **294**, 1448 (2001).
4. International Technology Roadmap for Semiconductors, *Executive Summary, Emerging Research Devices*, and *Emerging Research Materials*, 2009. http://www.itrs.net.
5. F. Meier and B. P. Zakharchenya (Eds.), *Optical Orientation*, vol. 8, North-Holland, New York, 1984.
6. G. Lampel, Nuclear dynamic polarization by optical electronic saturation and optical pumping in semiconductors, *Phys. Rev. Lett.* **20**, 491 (1968).
7. S. A. Crooker, J. J. Baumberg, F. Flack, N. Samarth, and D. D. Awschalom, Terahertz spin precession and coherent transfer of angular momenta in magnetic quantum wells, *Phys. Rev. Lett.* **77**, 2814 (1996).
8. M. J. Stevens, A. L. Smirl, R. D. R. Bhat, A. Najmaie, J. E. Sipe, and H. M. van Driel, Quantum interference control of ballistic pure spin currents in semiconductors, *Phys. Rev. Lett.* **90**, 136603 (2003).
9. S. F. Alvarado and P. Renaud, Observation of spin-polarized-electron tunneling from a ferromagnet into GaAs, *Phys. Rev. Lett.* **68**, 1387 (1992).
10. B. T. Jonker, Progress toward electrical injection of spin-polarized electrons into semiconductors, *Proc. IEEE* **91**, 727 (2003).
11. B. T. Jonker, S. C. Erwin, A. Petrou, and A. G. Petukhov, Electrical spin injection and transport in semiconductor spintronic devices, *MRS Bull.* **28**, 740 (2003).
12. B. T. Jonker and M. E. Flatte, Electrical spin injection and transport in semiconductors, in *Nanomagnetism, Volume 1: Ultrathin Films, Multilayers and Nanostructures*, D. L. Mills and J. A. C. Bland (Eds.); in *Contemporary Concepts of Condensed Matter Science*, E. Burstein, M. L. Cohen, D. L. Mills, and P. J. Stiles (Eds.), Elsevier, Amsterdam, the Netherlands, p. 227, 2006.
13. C. Weisbuch and B. Vinter, *Quantum Semiconductor Structures*, Academic Press, New York, p. 65, 1991.
14. A. G. Aronov and G. E. Pikus, Spin injection into semiconductors, *Fiz. Tekh. Poluprovodn.* **10**, 1177 (1976) [*Sov. Phys. Semicond.* **10**, 698 (1976)]; A. G. Aronov, *Pis'ma Zh. Eksp. Teor. Fiz.* **24**, 37 (1976) [*JETP Lett.* **24**, 32 (1976)].
15. D. Chiba, M. Sawicki, Y. Nishitani, Y. Nakatani, F. Matsukura, and H. Ohno, Magnetization vector manipulation by electric fields, *Nature* **455**, 515 (2008).
16. C. Haas, Magnetic semiconductors, *CRC Crit. Rev. Solid State Sci.* **1**, 47 (1970).
17. E. L. Nagaev, Photoinduced magnetism and conduction electrons in magnetic semiconductors, *Phys. Stat. Sol. B* **145**, 11 (1988).
18. L. Esaki, P. J. Stiles, and S. von Molnar, Magnetointernal field emission in junctions of magnetic insulators, *Phys. Rev. Lett.* **19**, 852 (1967).
19. J. S. Moodera, X. Hao, G. A. Gibson, and R. Meservey, Electron–spin polarization in tunnel junctions in zero applied field with ferromagnetic EuS barriers, *Phys. Rev. Lett.* **61**, 637 (1988).
20. J. S. Moodera, R. Meservey, and X. Hao, Variation of the electron spin polarization in EuSe tunnel junctions from zero to near 100% in a magnetic field, *Phys. Rev. Lett.* **70**, 853 (1993).
21. Y. D. Park, A. T. Hanbicki, J. E. Mattson, and B. T. Jonker, Epitaxial growth of an n-type ferromagnetic semiconductor $CdCr_2Se_4$ on GaAs (001) and GaP (001), *Appl. Phys. Lett.* **81**, 1471 (2002).
22. G. Kioseoglou, A. T. Hanbicki, J. M. Sullivan et al., Electrical spin injection from an n-type ferromagnetic semiconductor into a III–V device heterostructure, *Nat. Mater.* **3**, 799 (2004).

23. J. Lettieri, V. Vaithyanathan, S. K. Eah et al., Epitaxial growth and magnetic properties of EuO on (001)Si by molecular beam epitaxy, *Appl. Phys. Lett.* **83**, 975 (2003).

24. A. Schmehl, V. Vaithyanathan, A. Herrnberger et al., Epitaxial integration of the highly spin-polarized ferromagnetic semiconductor EuO with silicon and GaN, *Nat. Mater.* **6**, 882 (2007).

25. J. K. Furdyna and J. Kossut, Diluted magnetic semiconductors, in *Semiconductors and Semimetals*, vol. 25, R. K. Willardson and A. C. Beer (series Eds.), Academic Press, New York, 1988.

26. N. Samarth and J. K. Furdyna, Diluted magnetic semiconductors, *Proc. IEEE* **78**, 990 (1990).

27. M. Oestreich, J. Hubner, D. Hagele et al., Spin injection into semiconductors, *Appl. Phys. Lett.* **74**, 1251 (1999).

28. K. Onodera, T. Masumoto, and M. Kimura, 980 nm compact optical isolators using $Cd_{1-x-y}Mn_xHg_yTe$ single crystals for high power pumping laser diodes, *Electron. Lett.* **30**, 1954 (1994).

29. H. Munekata, H. Ohno, S. von Molnar, A. Segmüller, L. L. Chang, and L. Esaki, Diluted magnetic III–V semiconductors, *Phys. Rev. Lett.* **63**, 1849 (1989).

30. J. De Boeck, R. Oesterholt, A. Van Esch et al., Nanometer-scale magnetic MnAs particles in GaAs grown by molecular beam epitaxy, *Appl. Phys. Lett.* **68**, 2744 (1996).

31. H. Ohno, A. Shen, F. Matsukura et al., (Ga, Mn)As: A new diluted magnetic semiconductor based on GaAs, *Appl. Phys. Lett.* **69**, 363 (1996).

32. S. J. Potashnik, K. C. Ku, S. H. Chun, J. J. Berry, N. Samarth, and P. Schiffer, Effects of annealing time on defect-controlled ferromagnetism in $Ga_{1-x}Mn_xAs$, *Appl. Phys. Lett.* **79**, 1495 (2001).

33. K. W. Edmonds, K. Y. Wang, R. P. Campion et al., High-Curie-temperature $Ga_{1-x}Mn_xAs$ obtained by resistance-monitored annealing, *Appl. Phys. Lett.* **81**, 4991 (2002).

34. T. Hayashi, Y. Hashimoto, S. Katsumota, and Y. Iye, Effect of low-temperature annealing on transport and magnetism of diluted magnetic semiconductor (Ga, Mn)As, *Appl. Phys. Lett.* **78**, 1691 (2001).

35. Y. Yuan, Y. Wang, K. Gao et al., High Curie temperature and perpendicular magnetic anisotropy in homoepitaxial InMnAs films, *J. Phys. D Appl. Phys.* **48**, 235002 (2015).

36. L. Chen, X. Yang, F. Yang et al., Enhancing the Curie temperature of ferromagnetic semiconductor (Ga,Mn)As to 200 K via nanostructure engineering, *Nano Lett.* **11**, 2548 (2011).

37. H. Ohno, Making nonmagnetic semiconductors ferromagnetic, *Science* **281**, 951 (1998).

38. H. Ohno, N. Akiba, F. Matsukura, A. Shen, K. Ohtani, and Y. Ohno, Spontaneous splitting of ferromagnetic (Ga,Mn)As valence band observed by resonant tunneling spectroscopy, *Appl. Phys. Lett.* **73**, 363 (1998).

39. Y. Ohno, D. K. Young, B. Beschoten, F. Matsukura, H. Ohno, and D. D. Awschalom, Electrical spin injection in a ferromagnetic semiconductor heterostructure, *Nature* **402**, 790 (1999).

40. D. K. Young, E. Johnston-Halperin, D. D. Awschalom, Y. Ohno, and H. Ohno, Anisotropic electrical spin injection in ferromagnetic semiconductor heterostructures, *Appl. Phys. Lett.* **80**, 1598 (2002).

41. M. Yamanouchi, D. Chiba, F. Matsukura, and H. Ohno, Current-induced domain-wall switching in a ferromagnetic semiconductor structure, *Nature* **428**, 539 (2004).

42. H. Ohno, D. Chiba, F. Matsukura et al., Electric field control of ferromagnetism, *Nature* **408**, 944 (2000).

43. Y. D. Park, A. T. Hanbicki, S. C. Erwin et al., A group-IV ferromagnetic semiconductor: $Mn_xGe_{1-x}$, *Science* **295**, 651 (2002).

44. A. M. Nazmul, S. Kobayashi, S. Sugahara, and M. Tanaka, Control of ferromagnetism in Mn delta doped GaAs based semiconductor heterostructures, *Physica E* **21**, 937–943 (2004).

45. T. Dietl, H. Ohno, F. Matsukura, J. Cibert, and D. Ferrand, Zener model description of ferromagnetism in zinc-blende magnetic semiconductors, *Science* **287**, 1019 (2000).

46. T. Dietl, H. Ohno, and F. Matsukura, Hole-mediated ferromagnetism in tetrahedrally coordinated semiconductors, *Phys. Rev. B* **63**, 195205 (2001).

47. A. H. MacDonald, P. Schiffer, and N. Samarth, Ferromagnetic semiconductors: Moving beyond (Ga,Mn)As, *Nat. Mater.* **4**, 195 (2005).

48. H. Raebiger, S. Lany, and Z. Zunger, Control of ferromagnetism via electron doping in $In_2O_3$:Cr, *Phys. Rev. Lett.* **101**, 027203 (2008).

49. N. Samarth, Ferromagnetic semiconductors: Battle of the bands, *Nat. Mater.* **11**, 360 (2012).

50. S. Souma, L. Chen, R. Oszwałdowski et al., Fermi level position, Coulomb gap, and Dresselhaus splitting in (Ga,Mn)As. *Sci. Rep.* **6**, 27266 (2016).

51. M. Dobrowolska K. Tivakornsasithorn, X. Liu et al., Controlling the Curie temperature in (Ga,Mn)As through location of the Fermi level within the impurity band, *Nat. Mater.* **11**, 444 (2012).

52. M. Kobayashi I. Muneta, Y. Takeda et al., Unveiling the impurity band induced ferromagnetism in the magnetic semiconductor (Ga,Mn)As, *Phys. Rev. B* **89**, 205204 (2014).

53. T. Jungwirth J. Wunderlich, V. Novák et al., Spin-dependent phenomena and device concepts explored in (Ga,Mn)As, *Rev. Mod. Phys.* **86**, 855 (2014).

54. T. Dietl and H. Ohno, Dilute ferromagnetic semiconductors: Physics and spintronic structures, *Rev. Mod. Phys.* **86**, 187 (2014).

55. B. T. Jonker, Polarized optical emission due to decay or recombination of spin-polarized injected carriers, U.S. patent 5,874,749 (filed June 23, 1993, awarded February 23, 1999 to U.S. Navy).

56. G. Bastard, *Wave Mechanics Applied to Semiconductor Heterostructures*, Halsted, Halsted, New York, p. 109, 1988.

57. D. Hagele, M. Oestreich, W. W. Ruhle, N. Nestle, and K. Eberl, Spin transport in GaAs, *Appl. Phys. Lett.* **73**, 1580 (1998).

58. M. I. D'yakonov and V. I. Perel, Chapter 2: Theory of optical spin orientation of electrons and nuclei in semiconductors, in *Optical Orientation*, F. Meier and B. P. Zakharchenya (Eds.), North-Holland, Amsterdam, the Netherlands, p. 274, 1984.

59. R. W. Martin, R. J. Nicholas, G. J. Rees, S. K. Haywood, N. J. Mason, and P. J. Walker, Two-dimensional spin confinement in strained-layer quantum wells, *Phys. Rev. B* **42**, 9237 (1990).

60. S. A. Crooker, D. D. Awschalom, J. J. Baumberg, F. Flack, and N. Samarth, Optical spin resonance and transverse spin relaxation in magnetic semiconductor quantum wells, *Phys. Rev. B* **56**, 7574 (1997).

61. R. Fiederling, P. Grabs, W. Ossau, G. Schmidt, and L. W. Molenkamp, Detection of electrical spin injection by light-emitting diodes in top- and side-emission configurations, *Appl. Phys. Lett.* **82**, 2160 (2003).

62. O. M. J. van't Erve, G. Kioseoglou, A. T. Hanbicki, C. H. Li, and B. T. Jonker, Remanent electrical spin injection from Fe into AlGaAs/GaAs light emitting diodes, *Appl. Phys. Lett.* **89**, 072505 (2006).

63. A. Tackeuchi, O. Wada, and Y. Nishikawa, Electron spin relaxation in InGaAs/InP multiple quantum wells, *Appl. Phys. Lett.* **70**, 1131 (1997).

64. A. F. Isakovic, D. M. Carr, J. Strand, B. D. Schultz, C. J. Palmstrom, and P. A. Crowell, Optical pumping in ferromagnet-semiconductor heterostructures: Magneto-optics and spin transport, *Phys. Rev. B* **64**, 161304(R) (2001).

65. B. T. Jonker, A. T. Hanbicki, Y. D. Park et al., Quantifying electrical spin injection: Component-resolved electroluminescence from spin-polarized light-emitting diodes, *Appl. Phys. Lett.* **79**, 3098 (2001).

66. G. Kioseoglou, A. T. Hanbicki, B. T. Jonker, and A. Petrou, Bias-controlled hole degeneracy and implications for quantifying spin polarization, *Appl. Phys. Lett.* **87**, 122503 (2005).

67. R. Fiederling, M. Kelm, G. Reuscher et al., Injection and detection of a spin-polarized current in a light-emitting diode, *Nature* **402**, 787 (1999).

68. B. T. Jonker, Y. D. Park, B. R. Bennett, H. D. Cheong, G. Kioseoglou, and A. Petrou, Robust electrical spin injection into a semiconductor heterostructure, *Phys. Rev. B* **62**, 8180 (2000).

69. D. K. Young, J. A. Gupta, E. Johnston-Halperin, R. Epstein, Y. Kato, and D. D. Awschalom, Optical, electrical and magnetic manipulation of spins in semiconductors, *Semicond. Sci. Technol.* **17**, 275 (2002).

70. R. M. Stroud, A. T. Hanbicki, Y. D. Park et al., Reduction of spin injection efficiency by interface defect spin scattering in ZnMnSe/AlGaAs-GaAs spin-polarized light emitting diodes, *Phys. Rev. Lett.* **89**, 166602 (2002).

71. P. R. Hammar, B. R. Bennett, M. J. Yang, and M. Johnson, Observation of spin injection at a ferromagnet–semiconductor interface, *Phys. Rev. Lett.* **83**, 203 (1999).

72. C.-M. Hu, J. Nitta, A. Jensen, J. B. Hansen, and H. Takayanagi, Spin-polarized transport in a two dimensional electron gas with interdigital-ferromagnetic contacts, *Phys. Rev. B* **63**, 125333 (2001).

73. S. Gardelis, C. G. Smith, C. H. W. Barnes, E. H. Linfield, and D. A. Ritchie, Spin-valve effects in a semiconductor field-effect transistor: A spintronic device, *Phys. Rev. B* **60**, 7764 (1999).

74. F. G. Monzon, H. X. Tang, and M. L. Roukes, Magnetoelectronic phenomena at a ferromagnet–semiconductor interface, *Phys. Rev. Lett.* **84**, 5022 (2000).

75. B. J. van Wees, Comment on observation of spin injection at a ferromagnet–semiconductor interface, *Phys. Rev. Lett.* **84**, 5023 (2000).

76. A. T. Filip, B. H. Hoving, F. J. Jedema, B. J. van Wees, B. Dutta, and S. Borghs, Experimental search for the electrical spin injection in a semiconductor, *Phys. Rev. B* **62**, 9996 (2000).

77. H. J. Zhu, M. Ramsteiner, H. Kostial, M. Wassermeier, H.-P. Schönherr, and K. H. Ploog, Room temperature spin injection from Fe into GaAs, *Phys. Rev. Lett.* **87**, 016601 (2001).

78. R. C. Miller, D. A. Kleinman, W. A. Nordland Jr., and A. C. Gossard, Luminescence studies of optically pumped quantum wells in GaAs-Al$_x$Ga$_{1-x}$ As multilayer structures, *Phys. Rev. B* **22**, 863 (1980).

79. W. H. Butler, X.-G. Zhang, X. Wang, J. van Ek, and J. M. MacLaren, Electronic structure of FM/semiconductor/FM spin tunneling structures, *J. Appl. Phys.* **81**, 5518 (1997).

80. J. M. MacLaren, W. H. Butler, and X.-G. Zhang, Spin-dependent tunneling in epitaxial systems: Band dependence of conductance, *J. Appl. Phys.* **83**, 6521 (1998).

81. J. M. MacLaren, X.-G. Zhang, W. H. Butler, and X. Wang, Layer KKR approach to Bloch-wave transmission and reflection: Application to spin-dependent tunneling, *Phys. Rev. B* **59**, 5470 (1999).

82. O. Wunnicke, Ph. Mavropoulos, R. Zeller, P. H. Dederichs, and D. Grundler, Ballistic spin injection from Fe(001) into ZnSe and GaAs, *Phys. Rev. B* **65**, 241306(R) (2002).

83. P. Mavropoulos, O. Wunnicke, and P. H. Dederichs, Ballistic spin injection and detection in Fe/semiconductor/Fe junctions, *Phys. Rev. B* **66**, 024416 (2002).

84. J. R. Chelikowsky and M. L. Cohen, Nonlocal pseudopotential calculations for the electronic structure of eleven diamond and zinc-blende semiconductors, *Phys. Rev. B* **14**, 556 (1976).

85. S. Vutukuri, M. Chshiev, and W. H. Butler, Spin-dependent tunneling in FM/semiconductor/FM structures, *J. Appl. Phys.* **99**, 08K302 (2006).

86. P. Mavropoulos, Spin injection from Fe into Si(001): Ab initio calculations and role of the Si complex band structure, *Phys. Rev. B* **78**, 054446 (2008).

87. M. Zwierzycki, K. Xia, P. J. Kelly, G. E. W. Bauer, and I. Turek, Spin injection through an Fe/InAs interface, *Phys. Rev. B* **67**, 092401 (2003).

88. W. H. Butler, X.-G. Zhang, T. C. Schulthess, and J. M. MacLaren, Spin-dependent tunneling conductance of Fe/MgO/Fe sandwiches, *Phys. Rev. B* **63**, 054416 (2001).

89. T. Valet and A. Fert, Theory of the perpendicular magnetoresistance in magnetic multilayers, *Phys. Rev. B* **48**, 7099 (1993).

90. B. T. Jonker, A. T. Hanbicki, D. T. Pierce, and M. D. Stiles, Spin nomenclature for semiconductors and magnetic metals, *J. Magn. Magn. Mater.* **277**, 24 (2004).

91. R. A. de Groot, New class of materials: Half-metallic ferromagnets, *Phys. Rev. Lett.* **50**, 2024 (1983).

92. W. E. Pickett and J. S. Moodera, Half metallic magnets, *Phys. Today* **54**(5), 39 (May 2001).

93. D. Orgassa, H. Fujiwara, T. C. Schulthess, and W. H. Butler, First-principles calculation of the effect of atomic disorder on the electronic structure of the half-metallic ferromagnet NiMnSb, *Phys. Rev. B* **60**, 13237 (1999).

94. G. Schmidt, D. Ferrand, L. W. Molenkamp, A. T. Filip, and B. J. van Wees, Fundamental obstacle for electrical spin injection from a ferromagnetic metal into a diffusive semiconductor, *Phys. Rev. B* **62**, R4790 (2000).

95. E. I. Rashba, Theory of electrical spin injection: Tunnel contacts as a solution of the conductivity mismatch problem, *Phys. Rev. B* **62**, R16267 (2000).

96. D. L. Smith and R. N. Silver, Electrical spin injection into semiconductors, *Phys. Rev. B* **64**, 045323 (2001).

97. A. Fert and H. Jaffrès, Conditions for efficient spin injection from a ferromagnetic metal into a semiconductor, *Phys. Rev. B* **64**, 184420 (2001).

98. Z. G. Yu and M. Flatte, Electric-field dependent spin diffusion and spin injection into semiconductors, *Phys. Rev. B* **66**, 201202(R) (2002).

99. P. C. van Son, H. van Kempen, and P. Wyder, Boundary resistance of the ferromagnetic-nonferromagnetic metal interface, *Phys. Rev. Lett.* **58**, 2271 (1987).

100. A. Fert, J.-M. George, H. Jaffrès, and R. Mattana, Semiconductors between spin-polarized source and drain, *IEEE Trans. Electron. Dev.* **54**, 921 (2007).

101. E. H. Rhoderick and R. H. Williams, *Metal-Semiconductor Contacts*, 2nd edn., Clarendon. Oxford, UK, 1988.

102. M. Ilegems, Properties of III–V layers, in *The Technology and Physics of Molecular Beam Epitaxy*, E. H. C. Parker (Ed.), Plenum, New York, p. 119, 1985.

103. S. M. Sze, *Physics of Semiconductor Devices*, 2nd edn., Wiley, New York, p. 294, 1981.

104. A. T. Hanbicki, B. T. Jonker, G. Itskos, G. Kioseoglou, and A. Petrou, Efficient electrical spin injection from a magnetic metal/tunnel barrier contact into a semiconductor, *Appl. Phys. Lett.* **80**, 1240 (2002).

105. A. T. Hanbicki, O. M. J. van't Erve, R. Magno, G. Kioseoglou, C. H. Li, and B. T. Jonker, Analysis of the transport process providing spin injection through an Fe/AlGaAs Schottky barrier, *Appl. Phys. Lett.* **82**, 4092 (2003).

106. J. M. Kikkawa and D. D. Awschalom, Resonant spin-amplification in n-type GaAs, *Phys. Rev. Lett.* **80**, 4313 (1998).

107. R. I. Dzhioev, K. V. Kavokin, V. L. Korenev et al., Low-temperature spin relaxation in n-type GaAs, *Phys. Rev. B* **66**, 245204 (2002).

108. B. J. Jönsson-Åkerman, R. Escudero, C. Leighton, S. Kim, I. K. Schuller, and D. A. Rabson, Reliability of normal-state current-voltage characteristics as an indicator of tunnel-junction barrier quality, *Appl. Phys. Lett.* **77**, 1870 (2000).

109. D. A. Rabson, B. J. Jönsson-Åkerman, A. H. Romero et al., Pinholes may mimic tunneling, *J. Appl. Phys.* **89**, 2786 (2001).

110. U. Rüdiger, R. Calarco, U. May et al., Temperature dependent resistance of magnetic tunnel junctions as a quality proof of the barrier, *J. Appl. Phys.* **89**, 7573 (2001).

111. E. Burstein and S. Lundqvist (Eds.), *Tunneling Phenomena in Solids*, Plenum, New York 1969.

112. J. G. Simmons, Generalized formula for the electric tunnel effect between similar electrodes separated by a thin insulating film, *J. Appl. Phys.* **34**, 1793 (1963).

113. W. F. Brinkman, R. C. Dynes, and J. M. Rowell, Tunneling conductance of asymmetrical barriers, *J. Appl. Phys.* **41**, 1915 (1970).

114. S. H. Chun, S. J. Potashnik, K. C. Ku, P. Schiffer, and N. Samarth, Spin-polarized tunneling in hybrid metal-semiconductor magnetic tunnel junctions, *Phys. Rev. B* **66**, 100408(R) (2002).

115. E. L. Wolf, *Principles of Electron Tunneling Spectroscopy*, Oxford University Press, New York, 1989.

116. A. M. Andrews, H. W. Korb, N. Holonyak, C. B. Duke, and G. G. Kleiman, *Phys. Rev. B* **5**, 2273 (1972).

117. J. J. Krebs and G. A. Prinz, *NRL Memo Rep. 3870*, Naval Research Lab, Washington, DC, 1978.

118. R. Meservey and P. M. Tedrow, Spin-polarized electron tunneling, *Phys. Rep.* **238**, 173–234 (1999). See 200–202.

119. C. Kittel, *Introduction to Solid State Physics*, 5th edn., Wiley, New York, pp. 438–439, 1976.

120. N. W. Ashcroft and N. D. Mermin, *Solid State Physics*, Holt, Rinehart, Winston, New York, pp. 654, 661, 1976.

121. M. B. Stearns, Simple explanation of tunneling spin polarization of Fe, Co, Ni and its alloys, *J. Magn. Magn. Mater.* **5**, 167 (1977).

122. J. M. De Teresa, A. Barthelemy, A. Fert, J. P. Contour, F. Montaigne, and P. Seneor, Role of metal-oxide interface in determining the spin polarization of magnetic tunnel junctions, *Science* **286**, 507 (1999).

123. H. Dery and L. J. Sham, Spin extraction theory and its -relevance to spintronics, *Phys. Rev. Lett.* **98**, 046602 (2007).

124. A. N. Chantis, K. D. Belashchenko, D. L. Smith, E. Y. Tsymbal, M. van Schilfgaarde, and R. C. Albers, Reversal of spin polarization in Fe/GaAs(001) driven by resonant surface states: First-principles calculations, *Phys. Rev. Lett.* **99**, 196603 (2007).

125. T. J. Zega, A. T. Hanbicki, S. C. Erwin et al., Determination of interface atomic structure and its impact on spin transport using Z-contrast microscopy and density-functional theory, *Phys. Rev. Lett.* **96**, 196101 (2006).

126. S. C. Erwin, S.-H. Lee, and M. Scheffler, First-principles study of nucleation, growth, and interface structure of Fe/GaAs, *Phys. Rev. B* **65**, 205422 (2002).

127. E. J. Kirkland, *Advanced Computing in Electron Microscopy*, Plenum, New York, 1998.

128. E. Kneedler, P. M. Thibado, B. T. Jonker et al., Epitaxial growth, structure, and composition of Fe films on GaAs(001)-2×4, *J. Vac. Sci. Technol. B* **14**, 3193 (1996).

129. E. M. Kneedler and B. T. Jonker, Kerr effect study of the onset of magnetization in Fe films on GaAs(001)-2×4, *J. Appl. Phys.* **81**, 4463 (1997).

130. E. M. Kneedler, B. T. Jonker, P. M. Thibado, R. J. Wagner, B. V. Shanabrook, and L. J. Whitman, Influence of substrate surface reconstruction on the growth and magnetic properties of Fe on GaAs(001), *Phys. Rev. B* **56**, 8163 (1997).

131. P. M. Thibado, E. Kneedler, B. T. Jonker, B. R. Bennett, B. V. Shanabrook, and L. J. Whitman, Nucleation and growth of Fe on GaAs(001)-2×4 studied by scanning tunneling microscopy, *Phys. Rev. B* **53**, 10481 (1996).

132. B. T. Jonker, O. J. Glembocki, R. T. Holm, and R. J. Wagner, Enhanced carrier lifetimes and suppression of midgap states in GaAs at a magnetic metal interface, *Phys. Rev. Lett.* **79**, 4886 (1997).

133. C. S. Gworek, P. Phatak, B. T. Jonker, E. R. Weber, and N. Newman, Pressure dependence of Cu, Ag, and Fe n-GaAs Schottky barrier heights, *Phys. Rev. B* **64**, 045322 (2001).

134. B. T. Jonker, Electrical spin injection into semiconductors, in *Ultrathin Magnetic Structures IV: Applications of Nanomagnetism*, B. Heinrich and J. A. C. Bland (Eds.), Springer, Berlin, Germany, 2005.

135. S. A. Crooker, M. Furis, X. Lou et al., Imaging spin transport in lateral ferromagnet/semiconductor structures, *Science* **309**, 2191 (2005).

136. X. Lou, C. Adelmann, M. Furis, S. A. Crooker, C. J. Palmstrøm, and P. A. Crowell, Electrical detection of spin accumulation at a ferromagnet-semiconductor interface, *Phys. Rev. Lett.* **96**, 176603 (2006).

137. X. Lou, C. Adelmann, S. A. Crooker et al., Electrical detection of spin transport in lateral ferromagnet-semiconductor devices, *Nat. Phys.* **3**, 197 (2007).

138. M. C. Hickey, C. D. Damsgaard, S. N. Holmes et al., Spin injection from $Co_2MnGa$ into an InGaAs quantum well, *Appl. Phys. Lett.* **92**, 232101 (2008).

139. M. Holub, J. Shin, D. Saha, and P. Battacharya, Electrical spin injection and threshold reduction in a semiconductor laser, *Phys. Rev. Lett.* **98**, 146603 (2007).

140. P. Kotissek, M. Bailleul, M. Sperl et al., Cross-sectional imaging of spin injection into a semiconductor, *Nat. Phys.* **3**, 872 (2007).

141. A. T. Hanbicki, G. Kioseoglou, M. A. Holub, O. M. J. van't Erve, and B. T. Jonker, Electrical spin injection from Fe into ZnSe(001), *Appl. Phys. Lett.* **94**, 082507 (2009).

142. G. Kioseoglou, A. T. Hanbicki, R. Goswami et al., Electrical spin injection into Si: A comparison between Fe/Si Schottky and $Fe/Al_2O_3$ tunnel contacts, *Appl. Phys. Lett.* **94**, 122106 (2009).

143. R. Meservey and P. M. Tedrow, Spin polarized electron tunneling, *Phys. Rep.* **238**, 173–234 (1994).

144. J. S. Moodera, L. R. Kinder, T. M. Wong, and R. Meservey, Large magnetoresistance at room temperature in ferromagnetic thin film tunnel junctions, *Phys. Rev. Lett.* **74**, 3273 (1995).

145. J. Daughton, Magnetic tunneling applied to memory, *J. Appl. Phys.* **81**, 3758 (1997).

146. S. S. P. Parkin, C. Kaiser, A. Panchula et al., Giant tunneling magnetoresistance at room temperature with MgO(100) tunnel barriers, *Nat. Mater.* **3**, 862 (2004).

147. S. Yuasa, T. Nagahama, A. Fukushima, Y. Suzuki, and K. Ando, Giant room temperature magnetoresistance in single crystal Fe/MgO/Fe magnetic tunnel junctions, *Nat. Mater.* **3**, 868 (2004).

148. J. S. Moodera, J. Nowak, L. R. Kinder et al., Quantum well states in spin-dependent tunnel structures, *Phys. Rev. Lett.* **83**, 3029 (1999).

149. D. J. Monsma and S. S. P. Parkin, Spin polarization of tunneling current from ferromagnet/$Al_2O_3$ interfaces using copper-doped aluminum superconducting films, *Appl. Phys. Lett.* **77**, 720 (2000).

150. V. F. Motsnyi, J. De Boeck, J. Das et al., Electrical spin injection in a ferromagnet/tunnel barrier/semiconductor heterostructure, *Appl. Phys. Lett.* **81**, 265 (2002).

151. V. F. Motsnyi, P. Van Dorpe, W. Van Roy et al., Optical investigation of electrical spin injection into semiconductors, *Phys. Rev. B* **68**, 245319 (2003).

152. T. Manago and H. Akinaga, Spin-polarized light-emitting diode using metal/insulator/semiconductor structures, *Appl. Phys. Lett.* **81**, 694 (2002).

153. O. M. J. van't Erve, G. Kioseoglou, A. T. Hanbicki et al., Comparison of Fe/Schottky and $Fe/Al_2O_3$ tunnel barrier contacts for electrical spin injection into GaAs, *Appl. Phys. Lett.* **84**, 4334 (2004).

154. B. T. Jonker, G. Kioseoglou, A. T. Hanbicki, C. H. Li, and P. E. Thompson, Electrical spin injection into silicon from a ferromagnetic metal/tunnel barrier contact, *Nat. Phys.* **3**, 542 (2007).

155. G. Feher and E. A. Gere, Electron spin resonance experiments on donors in silicon: II. Electron spin relaxation effects, *Phys. Rev.* **114**, 1245 (1959).

156. V. Zarifis and T. G. Castner, ESR linewidth behavior for barely metallic n-type silicon, *Phys. Rev. B* **36**, 6198 (1987).

157. A. M. Tyryshkin, S. A. Lyon, W. Jantsch, and F. A. Schäffler, Spin manipulation of free two-dimensional electrons in Si/SiGe quantum wells, *Phys. Rev. Lett.* **94**, 126802 (2005).

158. S. Sugahara and M. Tanaka, A spin metal-oxide-semiconductor field effect transistor using half-metallic contacts for the source and drain, *Appl. Phys. Lett.* **84**, 2307 (2004).

159. C. L. Dennis, C. V. Tiusan, J. F. Gregg, G. J. Ensell, and S. M. Thompson, Silicon spin diffusion transistor: Materials, physics and device characteristics, *IEEE Proc. Circuits Dev. Syst.* **152**, 340 (2005).

160. E. Yablonovitch, H. W. Jiang, H. Kosaka, H. D. Robinson, D. S. Rao, and T. Szkopek, Quantum repeaters based on Si/SiGe heterostructures, *Proc. IEEE* **91**, 761 (2003).

161. M. L. W. Thewalt, S. P. Watkins, U. O. Ziemelis, E. C. Lightowlers, and M. O. Henry, Photoluminescence lifetime, absorption and excitation spectroscopy measurements on isoelectronic bound excitons in beryllium-doped silicon, *Solid State Commun.* **44**, 573 (1982).

162. P. Yu and M. Cardona, *Fundamentals of Semiconductors*, Springer-Verlag, Berlin, Germany, 1996.

163. I. Žutić, J. Fabian, and S. C. Erwin, Spin injection and detection in silicon, *Phys. Rev. Lett.* **97**, 026602 (2006).

164. J. C. Costa, F. Williamson, T. J. Miller et al., Barrier height variation in Al/GaAs Schottky diodes with a thin silicon interfacial layer, *Appl. Phys. Lett.* **58**, 382 (1991).

165. P. Li and H. Dery, Theory of spin-dependent phonon-assisted optical transitions in silicon, *Phys. Rev. Lett.* **105**, 037204 (2010).

166. G. Kioseoglou, A. T. Hanbicki, R. Goswami et al., Electrical spin injection into Si: A comparison between Fe/Si Schottky and Fe/$Al_2O_3$ tunnel contacts, *Appl. Phys. Lett.* **94**, 122106 (2009).

167. L. Grenet, M. Jamet, P. Noé et al., Spin injection in silicon at zero magnetic field, *Appl. Phys. Lett.* **94**, 032502 (2009).

168. D. J. Smith, M. R. McCartney, C. L. Platt, and A. E. Berkowitz, Structural characterization of thin film ferromagnetic tunnel junctions, *J. Appl. Phys.* **83**, 5154 (1998).

169. B. G. Park, T. Banerjee, B. C. Min, J. C. Lodder, and R. Jansen, Tunnel spin polarization of $Ni_{80}Fe_{20}/SiO_2$ probed with a magnetic tunnel transistor, *Phys. Rev. B* **73**, 172402 (2006).

170. C. Li, G. Kioseoglou, O. M. J. van't Erve, P. E. Thompson, and B. T. Jonker, Electrical spin injection into Si(001) through an $SiO_2$ tunnel barrier, *Appl. Phys. Lett.* **95**, 172102 (2009).

171. X. Jiang, R. Wang, R. M. Shelby et al., Highly spin polarized room temperature tunnel injector for semiconductor spintronics using MgO(100), *Phys. Rev. Lett.* **94**, 056601 (2005).

172. G. Salis, R. Wang, X. Jiang et al., Temperature independence of the spin-injection efficiency of a MgO-based tunnel spin injector, *Appl. Phys. Lett.* **87**, 262503 (2005).

173. Y. Lu, V. G. Truong, P. Renucci et al., MgO thickness dependence of spin injection efficiency in spin-light emitting diodes, *Appl. Phys. Lett.* **93**, 152102 (2008).

174. V. G. Truong, P.-H. Binh, P. Renucci et al., High speed pulsed electrical spin injection in spin-light emitting diode, *Appl. Phys. Lett.* **94**, 141109 (2009).

175. M. Passlack, M. Hong, J. P. Mannaerts, J. R. Kwo, and L. W. Tu, Recombination velocity at oxide–GaAs interfaces fabricated by in situ molecular beam epitaxy, *Appl. Phys. Lett.* **68**, 3605 (1996).

176. H. Saito, J. C. Le Breton, V. Zayets, Y. Mineno, S. Yuasa, and K. Ando, Efficient spin injection into semiconductor from an Fe/GaO$_x$ tunnel injector, *Appl. Phys. Lett.* **96**, 012501 (2010).

177. D. J. Monsma, J. C. Lodder, Th. J. A. Popma, and B. Dieny, Perpendicular hot electron spin-valve effect in a new magnetic field sensor: The spin-valve transistor, *Phys. Rev. Lett.* **74**, 5260 (1995).

178. K. Mizushima, T. Kinno, T. Yamauchi, and K. Tanaka, Energy-dependent hot electron transport across a spin-valve, *IEEE Trans. Magn.* **33**, 3500 (1997).

179. R. Jansen, The spin-valve transistor: A review and outlook, *J. Phys. D* **36**, R289 (2003).

180. X. Jiang, R. Wang, S. van Dijken et al., Optical detection of hot-electron spin injection into GaAs from a magnetic tunnel transistor source, *Phys. Rev. Lett.* **90**, 256603 (2003).

181. I. Appelbaum, B. Huang, and D. J. Monsma, Electronic measurement and control of spin transport in silicon, *Nature* **447**, 295 (2007).

182. H. Dery, Ł. Cywiński, and L. J. Sham, Lateral diffusive spin transport in layered structures, *Phys. Rev. B* **73**, 041306(R) (2006).

183. B. C. Min, K. Motohashi, J. C. Lodder, and R. Jansen, Tunable spin-tunnel contacts to silicon using low work function ferromagnets, *Nat. Mater.* **5**, 817 (2006).

184. M. Johnson and R. H. Silsbee, Interfacial charge-spin coupling: Injection and detection of spin magnetization in metals, *Phys. Rev. Lett.* **55**, 1790 (1985).

185. F. J. Jedema, A. T. Filip, and B. J. van Wees, Electrical spin injection and accumulation at room temperature in an all-metal mesoscopic spin valve, *Nature* **410**, 345 (2001).

186. F. J. Jedema, H. B. Heersche, A. T. Filip, J. J. A. Baselmans, and B. J. vanWees, Electrical detection of spin precession in a metallic mesoscopic spin valve, *Nature* **416**, 713 (2002).

187. O. M. J. van't Erve, A. T. Hanbicki, M. Holub et al., Electrical injection and detection of spin-polarized carriers in silicon in a lateral transport geometry, *Appl. Phys. Lett.* **91**, 212109 (2007).

188. O. M. J. van't Erve, C. Awo-Affouda, A. T. Hanbicki, C. H. Li, P. E. Thompson, and B. T. Jonker, Information processing with pure spin currents in silicon: Spin injection, extraction, modulation and detection, *IEEE Trans. Electron. Dev.* **56**, 2343 (2009).

189. N. Tombros, C. Jozsa, M. Popinciuc, H. T. Jonkman, and B. J. van Wees, Electronic spin transport and spin precession in single graphene layers at room temperature, *Nature* **448**, 571 (2007).

190. M. Tran, H. Jaffrès, C. Deranlot et al., Enhancement of the spin accumulation at the interface between a spin-polarized tunnel junction and a semiconductor, *Phys. Rev. Lett.* **102**, 036601 (2009).

191. A. V. Pohm, B. A. Everitt, R. S. Beech, and J. M. Daughton, Bias field and end effects on the switching thresholds of "pseudo spin valve" memory cells, *IEEE Trans. Magn.* **33**, 3280 (1997).

192. A. M. Bratkovsky and V. V. Osipov, Efficient spin extraction from nonmagnetic semiconductors near forward-biased ferromagneticsemiconductor modified junctions at low spin polarization of current, *J. Appl. Phys.* **96**, 4525 (2004).

193. M. I. D'yakonov and V. I. Perel, Chapter 2: Theory of optical spin orientation of electrons and nuclei, in *Optical Orientation, Modern Problems in Condensed Matter Science*, vol. 8, F. Meier and B. P. Zakharchenya (Eds.), North-Holland, Amsterdam, the Netherlands, p. 39, 1984.

194. Y. Ando, K. Hamaya, K. Kasahara et al., Electrical injection and detection of spin-polarized electrons in silicon through an Fe$_3$Si/Si Schottky tunnel barrier, *Appl. Phys. Lett.* **94**, 182105 (2009).

195. T. Sasaki, T. Oikawa, T. Suzuki, M. Shiraishi, Y. Suzuki, and K. Tagami, Electrical spin injection into silicon using MgO tunnel barrier, *Appl. Phys. Express* **2**, 053003 (2009).

196. S. P. Dash, S. Sharma, R. S. Patel, M. P. de Jong, and R. Jansen, Electrical creation of spin polarization in silicon at room temperature, *Nature* **462**, 491 (2009).

197. L. Manchanda, M. D. Morris, M. L. Green et al., Multi-component high-K gate dielectrics for the silicon industry, *Microelectron. Eng.* **59**, 351 (2001), and references therein.

198. C. Li, O. M. J. van't Erve, and B. T. Jonker, Electrical injection and detection of spin accumulation in silicon at 500K with magnetic metal/silicon dioxide contacts, *Nat. Commun.* **2**, 245 (2011), doi:10.1038/ncomms1256.

199. Y. Ochiai and E. Matsuura, ESR in heavily doped n-type silicon near a metal-nonmetal transition, *Phys. Stat. Sol. (a)* **38**, 243 (1976).

200. M. Passlack, R. Droopad, Z. Yu et al., Screening of oxide/GaAs interfaces for MOSFET applications, *IEEE Trans. Electron. Dev. Lett.* **29**, 1181 (2008), and references therein.

201. K. S. Krishnan and N. Ganguli, Large anisotropy of the electrical conductivity of graphite, *Nature* **144**, 667 (1939).

202. S. Chen L. Brown, M. Levendorf et al., Oxidation resistance of graphene-coated Cu and Cu/Ni alloy, *ACS Nano* **5**, 1321 (2011).

203. E. Cobas, A. L. Friedman, O. M. J. van't Erve, J. T. Robinson, and B. T. Jonker, Graphene as a tunnel barrier: Graphene-based magnetic tunnel junctions, *Nano Lett.* **12**, 3000 (2012).

204. E. D. Cobas O. M. van't Erve, S. F. Cheng et al., Room temperature spin filtering in metallic ferromagnet/multilayer graphene/ferromagnet junctions, *ACS Nano* **10**, 10357 (2016).

205. O. M. J. van't Erve, A. L. Friedman, E. Cobas, C. H. Li, J. T. Robinson, and B. T. Jonker, Low-resistance spin injection into silicon using graphene tunnel barriers, *Nat. Nanotechnol.* **7**, 737 (2012).

206. O. M. J. van't Erve A. L. Friedman, E. Cobas et al., A graphene solution to conductivity mismatch: Spin injection from ferromagnetic metal / graphene tunnel contacts into silicon, *J. Appl. Phys.* **113**, 17C502 (2013).

207. O. M. J. van't Erve A. L. Friedman, C. H. Li et al. Spin transport and Hanle effect in silicon nanowire using graphene tunnel barriers, *Nat. Commun.* **6**, 7541 (2015).

208. X. Li C. W. Magnuson, A. Venugopal et al., Large-area graphene single crystals grown by low-pressure chemical vapor deposition of methane on copper, *J. Am. Chem. Soc.* **133**, 2816 (2011).

209. S. P. Dash S. Sharma, J. C. Le Breton et al., Spin precession and inverted Hanle effect in a semiconductor near a finite-roughness ferromagnetic interface, *Phys. Rev. B* **84**, 054410 (2011).

210. Y. Aoki, M. Kameno, Y. Ando et al., Investigation of the inverted Hanle effect in highly doped Si, *Phys. Rev. B* **86**, 081201(R) (2012).

211. J. H. Pifer, Microwave conductivity and conduction-electron spin-resonance linewidth of heavily doped Si:P and Si:As, *Phys. Rev. B* **12**, 4391 (1975).

212. V. Zarifis and T. G. Castner, ESR linewidth behavior for barely metallic n-type silicon, *Phys. Rev. B* **36**, 6198 (1987).

213. H. Dery, L. Cywinski, and L. J. Sham, Lateral diffusive spin transport in layered structures, *Phys. Rev. B* **73**, 041306(R) (2006).

214. B.-C. Min, K. Motohashi, C. Lodder, and R. Jansen, Tunable spin-tunnel contacts to silicon using low-work-function ferromagnets, *Nat. Mater.* **5**, 817 (2006).

215. L. Cywinski, H. Dery, and L. J. Sham, Electric readout of magnetization dynamics in a ferromagnet–semiconductor system, *Appl. Phys. Lett.* **89**, 042105 (2006).

216. O. Txoperena, Y. Song, L. Qing et al., Impurity-assisted tunneling magnetoresistance under a weak magnetic field, *Phys. Rev. Lett.* **113** (2014).
217. O. Txoperena, M. Gobbi, A. Bedoya-Pinto et al., How reliable are Hanle measurements in metals in a three-terminal geometry? *Appl. Phys. Lett.* **102**, 192406 (2013).
218. V. M. Karpan G. Giovannetti, P. A. Khomyakov et al., Graphite and graphene as perfect spin filters, *Phys. Rev. Lett.* **99**, 176602 (2007); ibid., Theoretical prediction of perfect spin filtering at interfaces between close-packed surfaces of Ni or Co and graphite or grapheme, *Phys. Rev. B* **78**, 195419 (2008).
219. O. V. Yazyev and A. Pasquarello, Magnetoresistive junctions based on epitaxial graphene and hexagonal boron nitride, *Phys. Rev. B* **80**, 035408 (2009).
220. Y.-C. Lin, Y. Chen, A. Shailos and Y. Huang, Detection of spin polarized carrier in silicon nanowire with single crystal MnSi as magnetic contacts, *Nano Lett.* **10**, 2281 (2010).
221. S. Zhang, S. A. Dayeh, Y. Li, S. A. Crooker, D. L. Smith, and S. T. Picraux, Electrical spin injection and detection in Silicon nanowires through oxide tunnel barriers, *Nano Lett.* **13**, 430 (2013).
222. J. Tarun S. Huang, Y. Fukuma et al., Demonstration of spin valve effects in silicon nanowires, *J. App. Phys.* **109**, 07C508 (2011).
223. J. Tarun S. Huang, Y. Fukuma et al., Temperature evolution of spin-polarized electron tunneling in silicon nanowire–permalloy lateral spin valve system, *Appl. Phys. Express* **5**, 045001 (2012).
224. E.-S. Liu, J. Nah, K. M. Varahramyan, and E. Tutuc, Lateral spin injection in germanium nanowires, *Nano Lett.* **10**, 3297 (2010).
225. R. Nakane, S. Sato, S. Kokutani, and M. Tanaka, Appearance of anisotropic magnetoresistance and electric potential distribution in Si-base multi-terminal devices with Fe electrodes, *IEEE Magn. Lett.* **3**, 3000404 (2012).
226. Y. Cui, L. J. Lauhon, M. S. Gudiksen, J. F. Wang, and C. M. Lieber, Diameter-controlled synthesis of single-crystal silicon nanowires, *Appl. Phys. Lett.* **78**, 2214 (2001).
227. C. Awo-Affouda O. M. van't Erve, G. Kioseoglou et al., Contributions to Hanle lineshapes in Fe/GaAs non-local spin valve transport, *Appl. Phys. Lett.* **94**, 102511 (2009).
228. Y. Ando S. Yamada, K. Kasahara et al., Effect of the magnetic domain structure in the ferromagnetic contact on spin accumulation in silicon, *Appl. Phys. Lett.* **101**, 232404 (2012).
229. H. Song and H. Dery, Analysis of phonon-induced spin relaxation processes in silicon, *Phys. Rev. B* **86**, 085201 (2012).
230. A. Tilke, F. Simmel, H. Lorenz, R. Blick, and J. Kotthaus, Quantum interference in a one-dimensional silicon nanowire, *Phys. Rev. B* **68**, 075311 (2003).
231. A. Lherbier, M. P. Persson, Y.-M. Niquet, F. Triozon, and S. Roche, Quantum transport length scales in silicon-based semiconducting nanowires: Surface roughness effects, *Phys. Rev. B* **77**, 085301 (2008).

# 3

# Spin Transport
# in Si and Ge

*Hot Electron Injection and*
*Detection Experiments*

**Ian Appelbaum**

## 3.1 INTRODUCTION

The spin injection and detection limitations posed by conductivity and lifetime mismatch between semiconductors and ferromagnetic (FM) metals [1–4] has been variously solved by lifting the constraints of ohmic transport with the insertion of a tunnel barrier [5, 6] or the use of ferromagnetic semiconductors [7, 8]. However, for many semiconductors, there are additional materials-dependent obstacles to the observation of spin transport. The materials properties of elemental group-IV semiconductors silicon (Si) and germanium (Ge), pose problems such as alloy formation at metallic contacts and difficulties with the use of optical methods for spin detection due to indirect bandgap [9, 10].

In this chapter, we discuss how all these issues can be circumvented by using a mechanism for spin injection and detection that differs fundamentally from ohmic transport in that (i) the inelastic mean-free-path (mfp) is not the smallest length scale in the device, and (ii) electrochemical potentials cannot be uniquely defined in thermal equilibrium. With these techniques, the physical length scale of the metallic electron injection contacts is shorter than the mfp, and conduction occurs through states far above the Fermi level (as compared to the thermal energy $k_B T$), far out of thermal equilibrium. Because this transport mode utilizes electrons with high kinetic energy that do not suffer appreciable inelastic scattering, it is known as "Ballistic Hot Electron Transport." Both spin injection and detection are performed all-electronically, and interfacial structure [11] plays only a minor role, enabling observation and study of spin transport in materials such as Si and Ge which had been previously excluded from the field. Related background material on spin injection and transport can be found in Chapter 5, Volume 1, Chapters 2 and 6, Volume 2, and on spin relaxation in Chapter 1, Volume 2, in this book.

## 3.2 HOT ELECTRON GENERATION AND COLLECTION

There are several techniques for generating hot electrons for injection from a metal into a semiconductor. Since metals have a very high density of electrons at and below the Fermi energy $E_F$, all must rely on an electrically rectifying barrier to eliminate transport of these thermalized electrons across the metal-semiconductor interface, which would dilute the injected hot electron current. This barrier is ideally created by the difference in work functions of the metal and the electron affinity of the semiconductor [12–15], but in reality its energetic height $q\phi$, where $q$ is the elementary charge, is determined more by the details of surface states which lie deep in the bandgap of the semiconductor that pin the Fermi level [16]. There is, of course, always a leakage current due to thermionic emission over this "Schottky" barrier at non-zero temperature $T$ given by the Richardson–Dushman expression $\propto T^2 \exp(-q\phi / k_B T)$; because typical barrier heights are in the range 0.6–0.8 eV [17] for Si and Ge, hot electron collection with Schottky barriers is often performed at temperatures below ambient conditions to reduce current leakage to negligible levels.

Some of the earliest hot electron work in metal-semiconductor devices involved the use of subbandgap photon absorption in the metal, which scattered electrons from a state with energy $E < E_F$ to another state with energy $E + \hbar\omega > E_F$, where $\hbar\omega$ is the photon energy. If the Schottky barrier height is $q\phi$, then electrons with initial energy $E > E_F + q\phi - \hbar\omega$ can cross the interface, assuming they are injected or generated within a mfp (typically in the range 1–100 nm) of it. Because this is similar to the photoelectric effect (except that the electrons cross an internal interface) the method is known as internal photoemission (IPE). By varying the photon energy, Schottky barrier heights [18] (and even buried heterojunction band offsets [19]) can be determined in a model-independent way by measuring the turn-on threshold of the collected hot electron current without biasing the device, and the hot electron current can be independently tuned by changing the illumination intensity.

In a later experiment, hot electrons were generated electrically using a forward-biased Schottky diode, and again collected by another (lower height) Schottky interface in a semiconductor-metal-semiconductor (SMS) device. Because this provides a narrow distribution of energies with width $\approx k_B T$, the hot electron mean free path of metal thin films could reliably be measured [20]. However, in contrast to Internal Photoemission, the energy itself cannot be changed since this is fixed by the emitter Schottky barrier height [21, 22].

The use of tunnel junctions (TJs, which consist of two metallic conductors separated by a thin insulator) [23–28] remedies this problem. The electrostatic potential energy $qV$ provided by voltage bias $V$ tunes the energy of hot electrons emitted from the cathode, and the exponential energy dependence of quantum-mechanical tunneling assures a narrow distribution. Hot electrons thus created can be collected by the Schottky barrier if $qV > q\phi$. Because of its robust insulating native oxide, aluminum (Al) is often employed during tunnel junction fabrication; although in principle any insulator can be used, it has been empirically found that the best are $Al_xO$, MgO, and AlN [29, 30]. The first proposal for a tunnel-junction hot electron injector was by Mead, who suggested another tunnel junction or the vacuum level as a collector [31–33]. Very soon thereafter, Spratt, Schwartz, and Kane showed that a semiconductor collector could be used to realize this device with a $Au/Al_xO/Al$ tunnel junction [34]. Heiblum's THETA device is an all-semiconductor version of this design [35, 36].

All three techniques (IPE, SMS, and TJ), along with illustrations of their corresponding hot electron energetic distributions, are shown in Figure 3.1.

## 3.3  SPIN-POLARIZED HOT ELECTRON TRANSPORT

The inelastic mfp of hot electrons in ferromagnetic metal films is spin-dependent: majority ("spin up") electrons have a longer mfp than minority ("spin down") electrons. Therefore, an initially unpolarized hot electron current will become spin-polarized by spin-selective scattering during ballistic

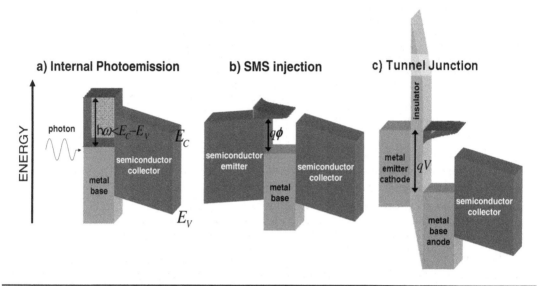

**a) Internal Photoemission**   **b) SMS injection**   **c) Tunnel Junction**

**FIGURE 3.1** Illustration of band diagrams for three different hot electron generation techniques: internal photoemission (IPE), semiconductor-metal-semiconductor (SMS) metal-base transistor, and tunnel junction (TJ) emission. The energetic distributions of the hot electrons generated (shown in red) are roughly homogeneous up to $\hbar\omega$ above the collector Fermi energy for IPE, peaked at the fixed Schottky barrier height of the emitter ($q\phi$) for SMS, and peaked near the emitter cathode Fermi energy $qV$ (where $V$ is the applied voltage and $q$ is the elementary charge) for TJ.

transport through a ferromagnetic metal film, where the polarization is given by

$$P = \frac{e^{-\frac{l}{\lambda_{\text{maj}}}} - e^{-\frac{l}{\lambda_{\text{maj}}}}}{e^{-\frac{l}{\lambda_{\text{maj}}}} + e^{-\frac{l}{\lambda_{\text{maj}}}}}, \tag{3.1}$$

where:

$l$     is the FM film thickness
$\lambda_{\text{maj}}$   is the majority spin mfp
$\lambda_{\text{min}}$   is the minority spin mfp

This "ballistic spin-filtering" effect can be used not only for spin polarization at injection, but also for spin analysis of a hot electron current at detection, much as an optical polarizing filter can be used both for electric field polarization and analysis of photons by changing the relative orientation of the optical axis.

A device making direct use of this effect is called the spin-valve transistor (SVT) [37–40]. It is a SMS metal-base transistor device [20] with a multilayer base consisting of at least two ferromagnetic layers (spin polarizer and analyzer) whose magnetizations can be set in a parallel or antiparallel configuration; switching between these two relative orientations induces a large change in the collector current of several hundred percent because of the ballistic spin-filtering effect.

Other non-SMS SVT-related designs have been demonstrated as well. To generate hot electrons, Monsma et al. also suggested the use of a magnetic tunnel junction together with a Schottky hot electron collector in their initial work introducing the SVT [37]; this was then later demonstrated using solid-state devices [41, 42], and with metal-vacuum-metal tunnel junctions using STM/BEEM [43]. In 2002, Van Dijken et al. suggested that moving the ferromagnetic polarizer to the emitter cathode of the tunnel junction comprises a new device, which they named the Magnetic Tunnel Transistor (MTT) [44]. (This configuration was, however, first proposed by Monsma in 1998, without describing it as a new device distinct from the SVT) [45]. Optical methods such as IPE with subbandgap illumination [46, 47] and photogeneration via interband transition with super-bandgap illumination [48] have also been used to demonstrate hot electron spin-valve effects.

Although we can speculate that the hot electrons presumably maintain some residual spin polarization after collection by the Schottky barrier [49], no spin detection after transport through the semiconductor was ever demonstrated in the above works, so spin injection success is unknown. The first successful use of spin-polarized ballistic hot electron injection was not reported until 2003, by Jiang et al. [50], where a MTT was used for injection into GaAs, and circular polarization analysis of band edge light emitted from InGaAs quantum wells was used for detection (a "spinLED"). However, because of the optical selection rules demanding a polar "Faraday" geometry, a large perpendicular magnetic field was required to induce a perpendicular magnetization of the magnetic layers in the MTT injector, so spin precession measurements (necessary to unambiguously confirm spin transport [51, 52]) could not be done.

After injection across the interface, the hot electrons must relax to the conduction band edge before transport and recombination. Because GaAs is a non-centrosymmetric zinc blende lattice that does not preserve inversion symmetry, Dresselhaus spin-orbit terms in the Hamiltonian break the spin degeneracy of the conduction band away from $\bar{k} = 0$ and the Dyakonov–Perel effect [53] can induce strong spin depolarization [54]. For this reason, injection near the conduction band edge in states close to the Brillouin zone center is strongly preferred, or application of hot electron spin injection to a material without this scattering mechanism is desired.

The group-IV elemental semiconductors silicon and germanium form in the diamond lattice, which *does* preserve inversion symmetry, and hence eliminates the deleterious Dyakonov–Perel mechanism. All subsequent work presented in this chapter will therefore focus on our experience developing hot electron techniques for spin injection, transport, and detection in these two materials, and will highlight the resulting knowledge gained pertaining to their spin-dependent electronic structure.

## 3.4  SPINS IN SILICON

Spin-polarized electrons in Si were first studied decades ago using resonance between Zeeman-split levels in a large magnetic field. Of interest in this field of electron spin resonance (ESR) was not only the conduction electrons but also electrons bound to donors [55–60]. The microwave absorption line positions (or electrically detected magnetic resonance (EDMR) [61, 62] conductance features) give the gyromagnetic ratio $\gamma = g\mu_B / \hbar$, where $g$ is the g-factor and $\mu_B$ is the Bohr magneton, and their linewidths give a measure of the inverse of the spin lifetime [59, 63].

The gyromagnetic ratio in Si corresponds to a g-factor very close to that of a bare electron in vacuum ($g \approx 2$) [64], and observed lifetimes were relatively long (several ns even at ambient), with very narrow linewidths. This apparent superiority of Si over other semiconductors is due to several fortuitous intrinsic properties which make it a relatively simple spin system [65–67].

It is often pointed out that since SOC scales as $Z^4$ in atomic systems, silicon's low atomic number ($Z_{Si} = 14$) leads to a reduced spin-orbit coupling. This is only partly true. The geometric symmetry of the diamond lattice and the proximity of the CBM to the $X$-point is arguably more important. For example, the small SOC correction to g-factor can be explained by the fact that double orbital degeneracy imposed by non-symmorphic elements of the symmetry group (those containing partial translations) at the $X$-point is preserved even after inclusion of SOI perturbation. Diamagnetic corrections to g-factor, which can be calculated via evaluation of operator matrix elements with remote bands [68]

$$\langle n|g_{\text{orbit}}|n''\rangle = \frac{4i}{m} \sum_{n' \neq n} \frac{\langle \psi_n \hat{\pi}_x \psi_{n'} \rangle \langle \psi_{n'} \hat{\pi}_y \psi_{n''} \rangle}{E_n - E_{n'}}, \tag{3.2}$$

are therefore zero *for every band* at the $X$-point, because the energy denominators are the same, whereas the numerators change sign between orbitally degenerate states. This is why a large valence band split-off is required for conduction band g-factor renormalization in zinc blende.

At the Si conduction band valley minimum only 15% along the $\Delta$-axis away from $X$ toward $\Gamma$, the double orbital degeneracy of $\Delta_5$ valence band states (which transform like axial vector components $\{zx, zy\}$ [69]) $\approx 4\text{eV}$ below the conduction band is indeed broken by SOI. However, because this state is so close to $X$ where it vanishes, the splitting is only $\approx 2\text{ meV}$. Thus, even though the numerators are non-zero for momentum $\hat{\pi}_{x,y}$ transverse to the valleys (conduction band is $\Delta_1$ that transforms like $z$), the total contribution from Equation 3.2 is tiny and only several parts in $10^3$.

The long spin lifetime in Si (especially for electrons bound in neutral donors) was exploited to achieve spontaneous emission of tunable microwaves in the late 1950s [70], and suggested much later as a fundamental ingredient in solid-state quantum computation [71]. In fact, Si was the first material with which optical orientation (generation of spin-polarized electrons in the conduction band through interband transitions induced by circularly polarized super-bandgap photon illumination [72]) was demonstrated [73, 74], yet the importance of achieving true spin transport in Si was often overlooked in reviews of the field [75]. This milestone was accomplished convincingly for the first time only in 2007, using the ballistic hot electron techniques described in Section 3.5 [76].

Due to its apparent advantages over other semiconductors, many groups tried to demonstrate phenomena attributed to spin transport in Si [77–85], but this was typically done with ohmic FM-Si contacts and two-terminal magnetoresistance measurements or in transistor-type devices [86, 87] which are bound to fail due to the "fundamental obstacle" for ohmic spin injection mentioned in Section 3.1 [1, 3, 4, 6]. Although weak spin-valve effects can be found in the literature, no evidence of spin precession is available, so the signals measured are ambiguous at best [51, 52]. Indeed, although magnetic exchange coupling across ultrathin tunneling layers of Si was seen, no spin-valve magnetoresistance was observed [88].

These failures were addressed by pointing out that only in a narrow window of FM-Si Resistance-Area (RA) product was a large spin polarization and hence large magnetoresistance expected in all-electrical two-terminal devices [89]. Subsequently, several efforts to tune the FM-Si interface resistance were made [90–94]. However, despite the ability to tune the RA product by over 8 orders of magnitude and even into the anticipated high-MR window (for instance, by using a low-work-function Gd layer), no evidence that this approach has been fruitful for Si can be found, and the theory has been confirmed only for the case of low-temperature-grown 5 nm-thick GaAs [95, 96].

Subsequent to the first demonstration of spin transport in Si [76], optical detection (circular electroluminescence analysis [97]) was shown to indicate spin injection from a FM and transport through only several tens of nm of Si, using first an $Al_xO$ tunnel barrier [98] and then tunneling through the Schottky barrier [99], despite the indirect bandgap. These methods required a large perpendicular magnetic field to overcome the large in-plane shape anisotropy of the FM contact, but later a perpendicular anisotropic magnetic multilayer was shown to allow spin injection into Si at zero external magnetic field [100]. While control samples with nonmagnetic injectors do show negligible spin polarization, again no evidence of spin precession was presented.

More recently, four-terminal non-local measurements on Si devices have been made, for instance with $Al_xO$ [101] or MgO [102, 103] tunnel barriers, or Schottky contacts using ferromagnetic silicide injector and detector [104]. Since these initial demonstrations, substantial progress has been made by Shiraishi's Osaka/Kyoto group [105–107] using this method.

Demonstrations of electrically injected spin accumulation in nonmagnetic materials are considered reliable when measured in a non-local four-terminal geometry. When performed correctly, the experiment electrically measures the decay of spin polarization in the nonmagnetic bulk material between the injection and detection regions. Therefore, it is advantageous to have a submicron separation between the ferromagnetic injector and detector electrodes if the spin diffusion length of the nonmagnetic bulk material is not much longer than one micron. To mitigate this requirement, some researchers have recently resorted to a local three-terminal measurement wherein one ferromagnetic electrode is used for both injection and detection of the spin signal in elemental semiconductors and complex oxides [108–114]. The results of these experiments, however, are often inconsistent with our understanding of spin transport phenomena in these materials [115–119]. And indeed, recent mounting evidence show that Pauli blocking during transport through localized defect/impurity states successfully accounts for the observed phenomena, without any need to invoke actual spin accumulation in the semiconductor [120–124]. It should be noted, however, that with clean MBE-grown interfaces, signals consistent with actual spin injection (much smaller in magnitude and with linewidths reflecting the expected bulk spin lifetime in heavily-doped Si) have been observed [125], as in the original work applying a local three-terminal method to GaAs [126].

As mentioned in Section 3.1, another way to overcome the conductivity/lifetime mismatch for spin injection is to use a carrier-mediated ferromagnetic semiconductor heterointerface. The interfacial quality may have a strong effect on the injection efficiency, so epitaxial growth will be necessary. Materials such as dilute magnetic semiconductors Mn-doped Si [127], Mn-doped chalcopyrites [10, 128–131], or the "pure" ferromagnetic semiconductor EuO [132] have all been suggested, but none as yet have been demonstrated as spin injectors for Si. It should be noted that while their intrinsic compounds are indeed semiconductors, due to the carrier-mediated nature of the ferromagnetism, it is seen only in highly (i.e. degenerately) doped and essentially metallic samples in all instances.

Another alternative injection method, spin pumping via ferromagnetic resonance with microwave absorption, was reported not only for electrons in n-Si [133], but also for holes in p-Si [134, 135]. We note that from a theoretical perspective, holes in the valence band of any bulk cubic semiconductor have spin lifetimes on the order of the momentum lifetime (ps or less), since spin-orbit interaction spin-mixes the light hole states with up/down probabilities of 1/3 and 2/3 so that Elliott–Yafet scattering among these states and the degenerate heavy holes causes highly efficient spin flips.

Several experimental [136, 137] and theoretical [138] works also addressed spin-polarized electrons and holes [139, 140] in Si/SiGe 2DEGs, but this is outside the scope of the present introduction; we deal here only with spin transport in 3D bulk Si, but do note that SiGe alloys could very well become useful as a way of tailoring spin transport devices to make use of local spin-orbit effects.

## 3.5 BALLISTIC HOT ELECTRON INJECTION AND DETECTION DEVICES

Two types of tunnel junctions have been employed for injection of spin-polarized electrons into non-degenerate semiconductors. The first used ballistic spin filtering of initially unpolarized electrons from a nonmagnetic Al cathode by a ferromagnetic anode base layer in direct contact to undoped, 10-micron-thick single-crystal Si(100) [76]. Despite SVT measurements suggesting the possibility of 90% spin polarization in the metal [40, 141], only 1%

polarization was found after injection into and transport through the Si. It was discovered later that a nonmagnetic Cu interlayer spacer could be used to increase the polarization to approximately 37% [142], likely due to Si's tendency to readily form spin-scattering "magnetically-dead" alloys (silicides) at interfaces with ferromagnetic metals.

Despite the possibilities for high spin polarization with these ballistic spin filtering injector designs, the short hot electron mfps in FM thin film anodes causes a very small injected charge current on the order of 100nA, with an emitter electrostatic potential energy approximately 1 eV above the Schottky barrier and a contact area approximately $100 \times 100\, \mu\mathrm{m}^2$. Because the injected spin density and spin current are dependent on the product of spin polarization and charge current, this technique is not ideal for transport measurements. Therefore, an alternative injector utilizing a FM tunnel-junction cathode and nonmagnetic anode (which has a larger mfp) was used for approximately ten times greater charge injection and hence larger spin signals, despite somewhat smaller potential spin polarization of approximately 15% [143, 144]. These injectors can be thought of as one-half of a magnetic tunnel junction [145], with a spin polarization proportional to the Fermi-level density of states spin asymmetry, rather than exponentially dependent on the spin-asymmetric mfp as is the case with ballistic spin filtering described above.

Although the injection is due to ballistic transport in the metallic contact, the conduction band mfp is typically only on the order of 10 nm [146], *so the vast majority of the subsequent transport to the detector over a length scale of tens [76, 143, 142], hundreds [144, 147], or thousands [148] of microns occurs at the conduction band edge following momentum relaxation* [149]. Typically, relatively large accelerating voltages are used so that the dominant transport mode is carrier drift; the presence of rectifying Schottky barriers on either side of the transport region assures that the resulting electric field does nothing other than determine the drift velocity of spin polarized electrons, hence the transit time [150, 151]—there are no spurious (unpolarized) currents induced to flow. Furthermore, undoped transit layers are primarily used; otherwise band bending would create a confining potential and increase the transit time, potentially leading to excessive depolarization [152].

The ballistic hot electron spin detector is comprised of a semiconductor-FM-semiconductor structure (both Schottky interfaces), fabricated using UHV metal-film wafer bonding (a spontaneous cohesion of ultra-clean metal film surfaces which occurs at room-temperature and nominal force in ultra-high vacuum) [38, 153]. After injection and transport through the semiconductor, spin-polarized electrons are ejected from the conduction band over the Schottky barrier and into hot electron states far above the Fermi energy. Again, because the mfp in FMs is larger for majority-spin (i.e. parallel to magnetization) hot electrons, the number of electrons coupling with conduction band states in a n-Si collector on the other side (which has a smaller Schottky barrier height due to contact with Cu) is dependent on the final spin polarization and the angle between spin and detector magnetization.

Quantitatively, we expect a contribution to our signal from each electron having spin orientation θ with respect to the detector magnetization

$$\propto \cos^2 \frac{\theta}{2} e^{-\frac{l}{\lambda_{maj}}} + \sin^2 \frac{\theta}{2} e^{-\frac{l}{\lambda_{min}}}$$

$$= \frac{1}{2}\left[\left(e^{-\frac{l}{\lambda_{maj}}} - e^{-\frac{l}{\lambda_{min}}}\right)\cos\theta + \left(e^{-\frac{l}{\lambda_{maj}}} + e^{-\frac{l}{\lambda_{min}}}\right)\right]. \tag{3.3}$$

Because the exponential terms are constants, this has the simple form ∝ cosθ + *const.*; in the following, we disregard the constant term, as it is spin-independent. The spin transport signal is thus the (reverse) current flowing across the n-Si collector Schottky interface. In essence, this device (whose band diagram is schematically illustrated in Figure 3.2) can be thought of as a split-base tunnel-emitter SVT with several hundred to thousands of microns of Si between the FM layers.

### 3.5.1 EXPERIMENTS ON SILICON

Two types of measurements are typically made: "spin valve" in a magnetic field parallel to the plane of magnetization, and spin precession in a magnetic field perpendicular to the plane of magnetization. The former allows the measurement of the difference in signals between parallel (P) and antiparallel (AP) injector/detector magnetization and hence is a straightforward way of determining the conduction electron spin polarization,

$$P = \frac{I_P - I_{AP}}{I_P + I_{AP}}. \tag{3.4}$$

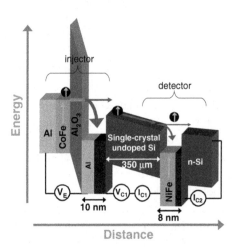

**FIGURE 3.2** Schematic band diagram of a four-terminal (two for TJ injection and two for FM SMS detection) ballistic hot electron injection and detection device with a 350 μm-thick Si transport layer.

Typical spin-valve measurement data, indicating ≈ 8% spin polarization after transport through 350 μm undoped Si, is shown in Figure 3.3.

Measurements in perpendicular magnetic fields reveal the average spin orientation after transit time $t$ through the Si, due to precession at frequency $\omega = g\mu_B B / \hbar$, where $B$ is magnetic field. If the transit time is determined only by drift, i.e. $t = \dfrac{L}{\mu E}$, we expect our spin transport signal to behave $\propto \cos g\mu_B Bt / \hbar$. However, due to transit time uncertainty $\Delta t$ caused by random diffusion, there is likewise an uncertainty in spin precession angle $\Delta\theta = \omega\Delta t$ which increases as the magnetic field (and hence $\omega$) increases. When this uncertainty approaches $2\pi$ rad, the spin signal is fully suppressed by a cancellation of contributions from antiparallel spins, a phenomenon called spin "dephasing", or the Hanle effect [154].

On a historical note, our device is essentially a solid-state analog of experiments performed in the 1950s that were used to determine the g-factor of the free electron in vacuum using Mott scattering as spin polarizer and analyzer and spin precession during time-of-flight in a solenoid [64]. In our case, we already know the g-factor (from e.g. ESR lines), so our experiments in strong drift electric fields where spin dephasing is weak can be used to measure transit time with $t = h / g\mu_B B_{2\pi}$, where $B_{2\pi}$ is the magnetic field period of the observed precession oscillations, despite the fact that we make DC measurements, not time-of-flight [150, 151]. Typical spin-precession data, indicating transit time of approximately 12ns to cross 350 μm undoped Si in an electric field of ≈ 580 V/cm, is shown in Figure 3.4a.

One important application of this transit time information from spin precession is to correlate it to the spin polarization determined from spin-valve measurements and Eq. 3.4 to extract spin lifetime. By varying the internal electric field, we change the drift velocity and hence average transit time. A reduction of polarization is seen with an increase in average transit time (as in Figure 3.5a) that we can fit well to first order using an exponential-decay model $P \propto e^{-t/\tau}$, and extract the timescale $\tau$ [144]. In this way, we have observed spin lifetimes of approximately 1 μs at 60 K

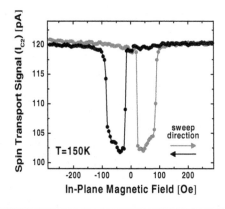

**FIGURE 3.3**    In-plane magnetic field measurements show the "spin-valve" effect and can be used to calculate the spin polarization after transport in Si.

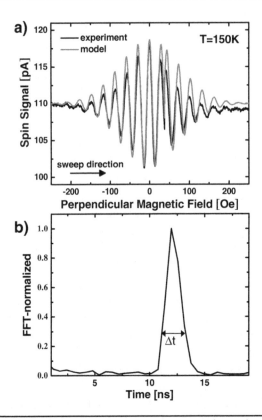

**FIGURE 3.4** (a) A typical spin precession measurement shows the coherent oscillations due to drift and the suppression of signal amplitude ("dephasing") as the precession frequency rises. Our model simulates this behavior well. (b) The real part of the Fourier transform of the precession data in (a) reveals the spin current arrival distribution.

in 350 μm-thick transport devices [155].* In Figure 3.5b, the temperature dependence of spin lifetime is compared to the $T^{-5/2}$ power law predicted by Yafet [60]. This power law arose from considering that the intravalley spin-flip matrix element is quadratic in phonon momentum $q$; Fermi's golden rule gives the lowest-order spin-flip transition rate as this matrix element squared ($q^4 \propto E^2 \propto T^2$) times the familiar $\sqrt{E} \propto \sqrt{T}$ density of states in 3-dimensions.

Yafet's approximation was improved by the numerical calculations of Cheng et al. which were found to be closer to $T^{-3}$ [156]. As Li and Dery [157] and Song and Dery [158] point out, however, the spin lifetime dependence on temperature is fundamentally not well-captured by a simple power law because it is primarily due to intervalley scattering with large-momentum $f$-process [159] phonons at high temperature and intervalley scattering with small-momentum acoustic phonons at low temperature.

---

* In Ref. [144], a more conservative estimate of the spin lifetime (e.g. 520 ns at 60 K) was obtained by fitting to the transit time dependence of an alternative quantity expected to be proportional to the spin polarization, rather than using Eq. 3.4 directly.

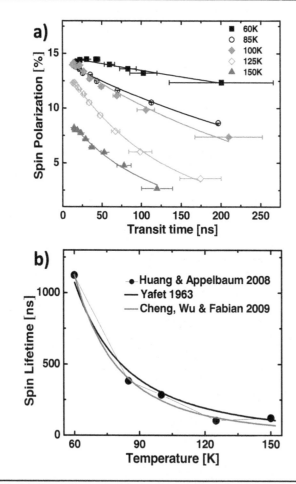

**FIGURE 3.5**    (a) Fitting the normalized spin signal from in-plane spin-valve measurements to an exponential decay model using transit times derived from spin precession measurements at variable internal electric field yields measurement of spin lifetimes in undoped bulk Si. (b) The experimental spin lifetime values obtained as a function of temperature are compared to Yafet's $T^{-5/2}$ power law for indirect-bandgap semiconductors [60] and Cheng et al.'s $T^{-3}$ derived from a full band structure theory [156].

## 3.6 SPINS IN GERMANIUM

Germanium shares the same diamond lattice with silicon and is also an indirect-gap multivalley semiconductor, but spin-orbit coupling in the conduction band of this material is very different. While it is true that the atomic number is much greater (32 as opposed to 14 for Si), the primary reason for stronger spin-orbit coupling is that the conduction band minimum lies at the $L$-point, rather than close to the $X$-point where the symmetry group retains non-symmorphic lattice symmetry elements that preserve orbital degeneracy.

Once again, we can illustrate the effects of spin-orbit-coupling by analyzing the dominant diamagnetic contributions to the Landé g-factor. Germanium's $L_1$ conduction band is primarily affected by the SOI-induced

$\approx 0.2$ eV splitting of the $L_3'$ valence band, $\approx 2.2$ eV below it [160, 161]. The energy denominators in Equation 3.2 are thus substantially different, so that a quite incomplete cancellation occurs in the sum. Furthermore, the wave function symmetries ($L_3'$ states are odd with respect to transverse directions, whereas $L_1$ is even, and both are even with respect to the valley axis) lead to non-zero numerators only for magnetic moments pointed along the valley axis. The transverse diamagnetic correction therefore vanishes to lowest order and $g_\perp \approx 2$, but the longitudinal contribution is substantial so that $g_\parallel \sim 1$ [162].

In a system like this with conduction band valley degeneracy and anisotropic Landé g-factor, an unusual effect can occur [163]: for an electron in a valley whose axis is oriented along $\hat{z}$ at an angle $\theta$ with an external magnetic field $\vec{B}$, we can choose $\hat{x}$ in the plane of $\hat{z}$ and $\vec{B}$, such that the Zeeman Hamiltonian governing spin state evolution is

$$\mathcal{H} = \mu_B \left( g_\parallel B\cos\theta\sigma_z + g_\perp B\sin\theta\sigma_x \right), \tag{3.5}$$

where $\mu_B$ is the Bohr magneton, and $\sigma_{x,z}$ are the $2\times2$ Pauli spin-1/2 matrices. This seemingly trivial Hamiltonian can be algebraically transformed into an equivalent picture for a free electron with g-factor $g_0 \approx 2$:

$$\mathcal{H} = \mu_B g_0 \frac{B}{g_0} \left( \left( g_\parallel - g_\perp \right)\cos\theta\sigma_z + g_\perp\sin\theta\sigma_x + g_\perp\cos\theta\sigma_z \right)$$
$$= g_0\mu_B \left( B\frac{g_\parallel - g_\perp}{g_0}\cos\theta\sigma_z + \frac{g_\perp}{g_0}\vec{B}\cdot\vec{\sigma} \right). \tag{3.6}$$

Note that this transformed Hamiltonian indicates that the electron spin acts as if it were a free electron in a renormalized magnetic field $\frac{g_\perp}{g_0}\vec{B}$, *plus another magnetic field, oriented along the valley axis*, with magnitude $|B|\frac{g_\parallel - g_\perp}{g_0}\cos\theta$. This additional field is randomized during the fast intervalley scattering process; time-dependent perturbation theory shows that it opens a new channel of spin relaxation, even if the external magnetic field is perfectly aligned with the initial spin orientation.

This extraordinary mechanism is reminiscent of the Dyakonov–Perel spin relaxation process which dominates in non-centrosymmetric semiconductor crystal lattices [53, 72], for example the III-V compound semiconductors like GaAs. In that case, broken *spatial inversion* symmetry allows spin-orbit interaction to cause a momentum-dependent spin splitting (Dresselhaus spin-orbit coupling [164]); *intra*valley scattering during spin precession about this random effective magnetic field leads to depolarization. In the anisotropic g-factor mechanism described above, the origin of the additional random field is rooted instead in the broken *time reversal* symmetry induced by the real external magnetic field, and *inter*valley scattering allows g-factor anisotropy to drive its fluctuation between four different orientations. This subtle phenomenon can be experimentally verified in appropriate spin transport devices by the suppression of spin polarization with longitudinal magnetic field in the spin-valve effect.

The long-distance germanium spin transport devices we fabricated to observe this phenomenon [165] are nominally identical in operation to their silicon counterparts that use the same ballistic hot electron transport methods [144, 166]. However, substantial changes to the fabrication procedure were required due to the details of mechanical, chemical, and metallurgical properties of Ge. The transport layer of >40 $\Omega$cm 325 ± 25 μm-thick Ge(001) was bonded to a CoFe(2 nm)/NiFe(5 nm)/Cu(3 nm)/n-Si spin detector structure, whose layer sequence was chosen to maximize the Schottky barrier height on the Ge side [167], and minimize barrier height on the n-Si side to facilitate the ballistic transport of electrons from Ge, through the metal, and into the Si conduction band.

Figure 3.6a shows experimental results from these devices in an in-plane magnetic field quasi-statically swept through both injector and detector magnetic thin film coercive fields at a temperature of 41 K. A linear background has been subtracted for clarity to show only the spin-dependent current $\Delta I_{C2}$ for magnetic field orientation along the in-plane <110> and <100> directions. Both data show prominent magnetic field-dependent spin depolarization with a profile very different from the ordinary spin-valve effect, for example in Figure 3.3.

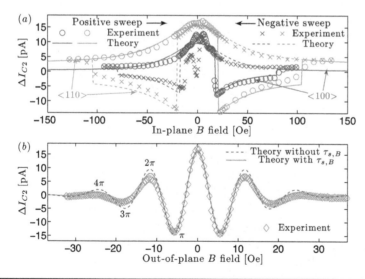

**FIGURE 3.6** Experimental and simulated spin signal $I_{C2}$ vs applied magnetic field B, in an accelerating electric field caused by a voltage of $V_{C1} = 0.6$ V over the 325 μm transport distance in undoped Ge at a temperature of 41 K. Panel (a) shows the spin-valve effect for the $\vec{B}$ field along both $\langle 110 \rangle$ (in blue) and $\langle 100 \rangle$ (in red) due to switching injector or detector magnetizations in an in-plane B field. The round (cross) markers are experimental results when B is swept in the positive (negative) direction, with the solid (dashed) curve corresponding to theoretical simulation. Panel (b) shows coherent spin precession in an out-of-plane B field. The diamond markers are experiment results. The solid (dashed) curve is the theoretical simulation with (without) the g-factor anisotropy-induced depolarization. (After Li, P. et al., *Phys. Rev. Lett.* 111, 257204, 2013. With permission.)

With a model for this $B$-dependent total spin relaxation rate, we are able to simulate the spin-valve experiment data, using a drift-diffusion model [168, 169] that takes into account the transit time uncertainty, and hence spin orientation distribution at the detector. The Elliott–Yafet spin lifetime $\tau_{s,ph}$ is a free fitting parameter in this theory, allowing us to determine relaxation rates from a single spin-valve measurement—this is not possible with ordinary spin-valve measurements in materials where this mechanism is absent. As can be seen by the direct comparison in Figure 3.6a, the theory matches the experimental result for both <110> and <100> in-plane magnetic field orientations very well when $\tau_{s,ph} = 258$ ns. This long spin lifetime at 41 K is the consequence of vanishing intravalley spin flips due to time reversal and spatial inversion symmetries at the $L$-point up to cubic order in phonon wave-vector [170].

Figure 3.6b shows data and the corresponding drift-diffusion simulation results for a measurement in out-of-plane magnetic field, perpendicular to the magnetic and spin axis, causing coherent precession and an oscillating spin detector signal. In this geometry, we must include both spin-lattice (depolarization) and spin-spin (dephasing) relaxation. Clearly, the theoretical simulation matches experimental data very well. For comparison, we also show the simulated result excluding the $g$-factor anisotropy-induced contribution to the spin relaxation. Its discrepancy with experimental data is not apparent at low precession angles in small fields, but becomes prominent at subsequent extrema corresponding to $2\pi$, $3\pi$ and $4\pi$ rad rotations when $B$ increases.

Because the electron temperature is easily decoupled from the lattice temperature in transport conditions at finite electric field in this material [166], the spin lifetimes extracted from fitting the spin-valve depolarization features are typically lower than those obtained by correlating zero-magnetic-field polarization with mean transit time from spin precession data [144, 166] except at the lowest accelerating voltages and temperatures. Figure 3.7 compares the temperature dependence of spin lifetimes at several internal electric fields to the theoretical Elliott–Yafet prediction [170], which applies to the germanium electron-phonon system at thermal equilibrium. The spin-valve-obtained lifetimes systematically drop with increasing electric field, and are noticeably temperature-independent in high electric fields, unlike the "Larmor-clock"-derived values [171] from fitting precession and minor-loop spin polarization data at $V_{C1} < 0.6$ V. The origin of low-temperature spin lifetime suppression seen in these data is likely due to extrinsic effects, as has been observed in electron spin resonance studies of Si [59], and our own observation of inelastic exchange scattering with neutral donors [172].

## 3.7 OUTLOOK

There has been significant progress in using ballistic hot electron spin injection and detection techniques for spin transport studies in Si and Ge. However, there are limitations of these methods. For example, the small ballistic transport transfer ratio is typically no better than $10^{-3} - 10^{-2}$; the low injection

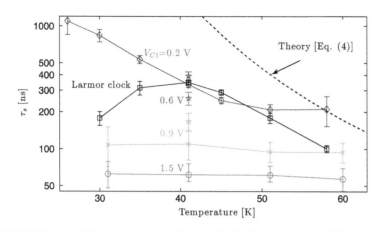

**FIGURE 3.7** Temperature dependence of spin lifetime in undoped Ge. Low-field measurements at $V_{C1} = 0.2$ V yield a monotonically increasing spin lifetime with decreasing temperature, similar to the Elliott–Yafet theoretical prediction [170]. Electric-field-induced intervalley scattering causes enhanced suppression of lifetimes. Lifetimes obtained by correlating transit time (Larmor clock from spin precession measurements) and zero-magnetic-field polarization (from minor-loop spin-valve measurements) show low-temperature depolarization. (After Li, P. et al., *Phys. Rev. Lett.* 111, 257204, 2013. With permission.)

currents and detection signals obtained will result in sub-unity gain and limit direct applications of these devices. In addition, our reliance on the ability of Schottky barriers to serve as hot electron filters presently limits device operation temperatures to approximately 200 K—although materials with higher Schottky barrier heights could extend this closer to room temperature—and also limits application to only non-degenerately doped semiconductors. Carrier freeze-out in the n-Si spin detection collector at approximately 20 K introduces a fundamental low-temperature limit as well [169].

Despite these shortcomings, there are also unique capabilities afforded by this method, such as independent control over internal electric field and injection current, and spectroscopic control over the injection energy level [173]. Unlike, for instance, optical techniques, other semiconductor materials should be equally well suited to study with these methods. The purpose is to use these devices as tools to understand spin transport properties for the design of spintronic devices, just as the Haynes–Shockley experiment [174] enabled the design of electronic minority-carrier devices such as the bipolar junction transistor.

There is still much physics to be done with ballistic hot electron spin injection and detection. For instance, spin control via time-dependent resonant fields is apparently feasible [175], and application of strain along the valley axes can suppress intervalley scattering to greatly enhance the spin lifetime [176, 177]. Similar fabrication techniques as used for vertical devices can be used to assemble lateral spin transport devices, where in particular very long transit lengths [148] and the effects of an electrostatic gate to control the proximity to a Si/SiO$_2$ interface can be investigated [178–180]. The massive spin lifetime suppression induced by the interface that was observed in this work suggests that such devices can be used to explore the effects of broken inversion symmetry [53] and paramagnetic [62] or charged [181, 182] defects on conduction electron spins at the semiconductor/insulator interface. In addition, this opportunity for sensitive characterization may be of importance to the electronics community as they continue to push MOSFET scaling toward its ultimate limits.

Hopefully, more experimental groups will develop the technology necessary to compete in this wide-open field. Theorists, too, are eagerly invited to address topics such as whether this injection technique fully circumvents the "fundamental obstacle" because of a remaining Sharvin-like effective resistance [183], or whether it introduces anomalous spin dephasing [169].

## ACKNOWLEDGMENTS

I.A. would especially like to thank Douwe Monsma for introducing him to the field, and Chagaan Baatar and Henryk Temkin for crucial early encouragement and continued support.

Since beginning work on this subject in 2006, many students and postdocs have made essential contributions. Biqin Huang, Hyuk-Jae Jang, and Jing Li deserve special thanks for fabrication and measurement efforts which have made much of this work possible. Pengke Li's combined experimental and theoretical strengths have been invaluable in developing a deeper understanding of the physics. Fruitful collaboration with Prof. Hanan Dery has similarly been an essential ingredient in the success of this research.

This work has been financially supported by the Office of Naval Research, DARPA/MTO, Defense Threat Reduction Agency, and the National Science Foundation.

## REFERENCES

1. M. Johnson and R. Silsbee, Thermodynamic analysis of interfacial transport and of the thermomagnetoelectric system, *Phys. Rev. B* **35**, 4959 (1987).
2. M. Johnson and R. Silsbee, Spin-injection experiment, *Phys. Rev. B* **37**, 5326 (1988).
3. G. Schmidt, D. Ferrand, L. Molenkamp, A. Filip, and B. van Wees, Fundamental obstacle for electrical spin injection from a ferromagnetic metal into a diffusive semiconductor, *Phys. Rev. B* **62**, R4790 (2000).
4. G. Schmidt, Concepts for spin injection into semiconductors—A review, *J. Phys. D* **38**, R107 (2005).
5. J. C. Slonczewski, Conductance and exchange coupling of two ferromagnets separated by a tunneling barrier, *Phys. Rev. B* **39**, 6995 (1989).
6. E. Rashba, Theory of electrical spin injection: Tunnel contacts as a solution of the conductivity mismatch problem, *Phys. Rev. B* **62**, R16267 (2000).
7. R. Fiederling, M. Keim, G. Reuscher et al., Injection and detection of a spin polarized current in a n-i-p light emitting diode, *Nature* **402**, 787 (1999).
8. Y. Ohno, D. K. Young, B. Beschoten et al., Electrical spin injection in a ferromagnetic semiconductor heterostructure, *Nature* **402**, 790 (1999).
9. I. Žutić, J. Fabian, and S. Das Sarma, Spintronics: Fundamentals and applications. *Rev. Mod. Phys.* **76**, 323 (2004).

10. I. Žutić, J. Fabian, and S. C. Erwin, Spin injection and detection in silicon, *Phys. Rev. Lett.* **97**, 026602 (2006).

11. T. J. Zega, A. T. Hanbicki, S. C. Erwin et al., Determination of interface atomic structure and its impact on spin transport using Z-contrast microscopy and density-functional theory., *Phys. Rev. Lett.* **96**, 196101 (2006).

12. W. Schottky, Semiconductor theory of the barrier film, *Naturwissenschaften* 26 (1938).

13. W. Schottky, Deviations from Ohm's law in semiconductors, *Phys. Z* 41 (1940).

14. N. F. Mott, Note on the contact between a metal and an insulator or semiconductor, *Proc. Cambridge Philos. Soc.* **34**, 568 (1938).

15. N. F. Mott, The theory of crystal rectifiers, *Proc. R. Soc. Lond. A* **171**, 27 (1939).

16. J. Tersoff, Schottky barrier heights and the continuum of gap states, *Phys. Rev. Lett.* **52**, 465 (1984).

17. S. Sze, *Physics of Semiconductor Devices*, 2nd edn., Wiley-Interscience, New York, 1981.

18. C. Crowell, L. Howarth, W. Spitzer, and E. Labate, Attenuation length measurements of hot electrons in metal films, *Phys. Rev.* **127**, 2006 (1962).

19. M. Heiblum, M. I. Nathan, and M. Eizenberg, Energy band discontinuities in heterojunctions measured by internal photoemission, *Appl. Phys. Lett.* **47**, 503 (1985).

20. C. Crowell and S. Sze, Hot electron transport and electron tunneling in thin film structures , in *Physics of Thin Films*, vol. 4, G. Hass and R. Thun (Eds.), Academic press, New York, 1967.

21. C. Crowell and S. Sze, Quantum-mechanical reflection of electrons at metalsemiconductor barriers: Electron transport in semiconductor-metal-semiconductor structures, *J. Appl. Phys.* **37**, 2683 (1966).

22. S. Sze and H. Gummel, Appraisal of semiconductor-metal-semiconductor transistor, *Solid-State Electron.* **9**, 751 (1966).

23. A. Sommerfeld and H. Bethe, Electronentheorie der metalle, in *Handbuch der Physik*, vol. 24, S. Flugge (Ed.), Springer, Berlin, 1933.

24. C. Duke, *Tunneling in Solids*, Academic, New York, 1969.

25. J. Fisher and I. Giaever, Tunneling through thin insulating layers, *J. Appl. Phys.* **32**, 172 (1961).

26. I. Giaever, Energy gap in superconductors measured by electron tunneling, *Phys. Rev. Lett.* **5**, 147 (1960).

27. J. G. Simmons, Generalized thermal j-v characteristic for electric tunnel effect, *J. Appl. Phys.* **35**, 2655 (1964).

28. W. F. Brinkman, R. C. Dynes, and J. M. Rowell, Tunneling conductance of asymmetrical barriers, *J. Appl. Phys.* **41**, 1915 (1970).

29. J. S. Moodera, J. Nassar, and G. Mathon, Spin-tunneling in ferromagnetic junctions, *Annu. Rev. Mater. Sci.* **29**, 381 (1999).

30. Y. Yang, P. F. Ladwig, Y. A. Chang et al., Thermodynamic evaluation of the interface stability between selected metal oxides and Co, *J. Mater. Res.* **19**, 1181 (2004).

31. C. Mead, The tunnel-emission amplifier, *Proc. IRE* **48**, 359 (1960).

32. C. Mead, Operation of tunnel-emission devices, *J. Appl. Phys.* **32**, 646 (1961).

33. C. A. Mead, Transport of hot electrons in thin gold films, *Phys. Rev. Lett.* **8**, 56 (1962).

34. J. Spratt, R. Schwartz, and W. Kane, Hot electrons in metal films: Injection and collection, *Phys. Rev. Lett.* **6**, 341 (1961).

35. M. Heiblum, Tunneling hot-electron transfer amplifier: A hot-electron GaAs device with current gain, *Appl. Phys. Lett.* **47**, 1105 (1985).

36. M. Heiblum, Direct observation of ballistic transport in GaAs, *Phys. Rev. Lett.* **55**, 2200 (1985).

37. D. Monsma, J. Lodder, T. Popma, and B. Dieny, Perpendicular hot electron spin-valve effect in a new magnetic field sensor: The spin-valve transistor, *Phys. Rev. Lett.* **74**, 5260 (1995).

38. D. Monsma, R. Vlutters, and J. Lodder, Room temperature-operating spin-valve transistors formed by vacuum bonding, *Science* **281**, 407 (1998).

39. I. Appelbaum, K. Russell, D. Monsma et al., Luminescent spin-valve transistor, *Appl. Phys. Lett.* **83**, 4571 (2003).

40. R. Jansen, The spin-valve transistor: A review and outlook, *J. Phys. D* **36**, R289 (2003).

41. K. Mizushima, T. Kinno, T. Yamauchi, and K. Tanak, Energy-dependent hot electron transport across a spin-valve, *IEEE Trans. Magn.* **33**, 3500 (1997).

42. G. Tae, J. Eom, J. Song, and K. Kim, Spin-polarized hot electron injection into two-dimensional electron gas by magnetic tunnel transistor, *Jpn. J. Appl. Phys.* **46**, 7717 (2007).

43. W. Rippard and R. Buhrman, Spin-dependent hot electron transport in Co/Cu thin films, *Phys. Rev. Lett.* **84**, 971 (2000).

44. S. van Dijken, X. Jiang, and S. S. P. Parkin, Room temperature operation of a high output current magnetic tunnel transistor, *Appl. Phys. Lett.* **80**, 3364 (2002).

45. D. Monsma, The spin-valve transistor, PhD thesis, U. Twente, 1998. (ISBN:903651049X).

46. I. Appelbaum, D. Monsma, K. Russell, V. Narayanamurti, and C. Marcus, Spin-valve photo-diode, *Appl. Phys. Lett.* **83**, 3737 (2003).

47. B. Huang and I. Appelbaum, Perpendicular hot-electron transport in the spin-valve photodiode, *J. Appl. Phys.* **100**, 034501 (2006).

48. B. Huang, I. Altfeder, and I. Appelbaum, Spin-valve phototransistor, *Appl. Phys. Lett.* **90**, 052503 (2007).

49. S. Wolf, D. Awschalom, R. Buhrman et al., Spintronics: A spin-based electronics vision for the future, *Science* **294**, 1488 (2001).

50. X. Jiang, R. Wang, S. van Dijken et al., Optical detection of hot-electron spin injection into GaAs from a magnetic tunnel transistor, *Phys. Rev. Lett.* **90**, 256603 (2003).

51. F. Monzon, H. Tang, and M. Roukes, Magnetoelectronic phenomena at a ferromagnet-semiconductor interface, *Phys. Rev. Lett.* **84**, 5022 (2000).

52. M. Johnson and R. Silsbee, Interfacial charge-spin coupling: Injection and detection of spin magnetization in metals, *Phys. Rev. Lett.* **55**, 1790 (1985).

53. M. Dyakonov and V. Perel, Spin relaxation of conduction electrons in noncentrosymmetric semiconductors, *Sov. Phys. Solid State* **13**, 3023 (1972).

54. X. Jiang, R. Wang, S. van Dijken et al., Optical detection of hot-electron spin injection into GaAs from a magnetic tunnel transistor, *IBM Res. Dev.* **50**, 111 (2006).

55. A. M. Portis, A. F. Kip, C. Kittel, and W. H. Brattain, Electron spin resonance in a silicon semiconductor, *Phys. Rev.* **90**, 988 (1953).

56. G. Feher, Electron spin resonance experiments on donors in silicon 1. Electronic structure of donors by the electron nuclear double resonance technique, *Phys. Rev.* **114**, 1219 (1959).

57. G. Feher and E. Gere, Electron spin resonance experiments on donors in silicon 2. electron spin relaxation effects, *Phys. Rev.* **114**, 1245 (1959).

58. D. Wilson and G. Feher, Electron spin resonance experiments on donors in silicon 3. investigation of excited states by application of uniaxial stress and their importance in relaxation processes, *Phys. Rev.* **124**, 1068 (1959).

59. D. Lepine, Spin resonance of localized and delocalized electrons in phosphorus-doped silicon between 20 and 300 degrees K, *Phys. Rev. B* **2**, 2429 (1970).

60. Y. Yafet, G-factors and spin-lattice relaxation of conduction electrons in *Solid State Physics-advances in Research and Applications*, vol. 14, F. Seitz and D. Turnbull (Eds.), Academic Press, New York, 1963.

61. R. Ghosh and R. Silsbee, Spin-spin scattering in a silicon two-dimensional electron gas, *Phys. Rev. B* **46**, 12508 (1992).

62. M. Xiao, I. Martin, E. Yablonovitch, and H. Jiang, Electrical detection of the spin resonance of a single electron in a silicon field-effect transistor, *Nature* **430**, 435 (2004).

63. J. Fabian, A. Matos-Abiague, C. Ertler, P. Stano, and I. Žutić, Semiconductor spintronics, *Acta Phys. Slovaca* **57**, 565 (2007).

64. W. Louisell, R. Pidd, and H. Crane, An experimental measurement of the gyromagnetic ratio of the free electron, *Phys. Rev.* **94**, 7 (1954).

65. V. Sverdlov and S. Selberherr, Silicon spintronics: Progress and challenges, *Phys. Rep.* **585**, 1 (2015). ISSN 0370-1573. Silicon spintronics: Progress and challenges.

66. R. Jansen, Silicon spintronics, *Nat. Mater.* **11**, 400 (2012).

67. I. Žutić and J. Fabian, Spintronics: Silicon twists, *Nature* **447**, 269 (2007).

68. P. Li and I. Appelbaum, Symmetry, distorted band structure, and spin-orbit coupling of group-III metal-monochalcogenide monolayers, *Phys. Rev. B* **92**, 195129 (2015).

69. M. Dresselhaus, G. Dresselhaus, and A. Jorio, *Group Theory: Application to the Physics of Condensed Matter*, Springer, Berlin, 2008.

70. G. Feher, J. P. Gordon, E. Buehler, E. A. Gere, and C. D. Thurmond, Spontaneous emission of radiation from an electron spin system, *Phys. Rev.* **109**, 221 (1958).

71. B. Kane, A silicon-based nuclear spin quantum computer, *Nature* **393**, 1331 (1998).

72. M. Dyakonov and V. Perel, Spin orientation of electrons associated with the interband absorption of light in semiconductors, *Sov. Phys. JETP* **33**, 1053 (1971).

73. G. Lampel, Nuclear dynamic polarization by optical electronic saturation and optical pumping in semiconductors, *Phys. Rev. Lett.* **20**, 491 (1968).

74. N. Bagraev, L. Vlasenko, and R. Zhitnikov, Optical orientation of $Si^{29}$ nuclei in n-type silicon and its dependence on the pumping light intensity, *Sov. Phys. JETP* **44**, 500 (1976).

75. D. Awschalom and M. E. Flatté, Challenges for semiconductor spintronics, *Nat. Phys.* **3**, 153 (2007).

76. I. Appelbaum, B. Huang, and D. J. Monsma, Electronic measurement and control of spin transport in silicon, *Nature* **447**, 295 (2007).

77. Y. Q. Jia, R. C. Shi, and S. Y. Chou, Spin-valve effects in nickel/silicon/nickel junctions, *IEEE Trans. Magn.* **32**, 4707 (1996).

78. K. I. Lee, H. J. Lee, J. Y. Chang et al., Spin-valve effect in an FM/Si/FM junction, *J. Mater. Sci.* **16**, 131 (2005).

79. W. J. Hwang, H. J. Lee, K. I. Lee et al., Spin transport in a lateral spin-injection device with an FM/Si/FM junction, *J. Magn. Magn. Mater.* **272–276**, 1915 (2004).

80. S. Hacia, T. Last, S. F. Fischer, and U. Kunze, Study of spin-valve operation in permalloy-$SiO_2$-Silicon nanostructures, *J. Supercond.* **16**, 187 (2003).

81. S. Hacia, T. Last, S. F. Fischer, and U. Kunze, Magnetotransport study of nanoscale permalloy-si tunnelling structures in lateral spin-valve geometry, *J. Phys. D* **37**, 1310 (2004).

82. C. Dennis, C. Sirisathitkul, G. Ensell, J. Gregg, and S. M. Thompson, High current gain silicon-based spin transistor, *J. Phys. D* **36**, 81 (2003).

83. C. Dennis, J. Gregg, G. Ensell, and S. Thompson, Evidence for electrical spin tunnel injection into silicon, *J. Appl. Phys.* **100**, 043717 (2006).

84. C. Dennis, C. Tiusan, R. Ferreira, et al., Tunnel barrier fabrication on Si and its impact on a spin transistor, *J. Magn. Magn. Mater.* **290**, 1383 (2005).

85. J. Gregg, I. Petej, E. Jouguelet, and C. Dennis, Spin electronics—a review, *J. Phys. D* **35**, R121 (2002).

86. S. Sugahara and M. Tanaka, A spin metal-oxide-semiconductor field-effect transistor using half-metallic-ferromagnet contacts for the source and drain, *Appl. Phys. Lett.* **84**, 2307 (2004).

87. M. Cahay and S. Bandyopadhyay, Room temperature silicon spin-based transistors, in *Device Applications of Silicon Nanocrystals and Nanostructures*, N. Koshida (Ed.), Springer, New York, 2009.

88. R. Gareev, L. Pohlmann, S. Stein, et al., Tunneling in epitaxial Fe/Si/Fe structures with strong antiferromagnetic interlayer coupling, *J. Appl. Phys.* **93**, 8038 (2003).

89. A. Fert and H. Jaffrès, Conditions for efficient spin injection from a ferromagnetic metal into a semiconductor, *Phys. Rev. B* **64**, 184420 (2001).

90. B. Min, J. Lodder, R. Jansen, and K. Motohashi, Cobalt-Al$_2$O$_3$-silicon tunnel contacts for electrical spin injection into silicon, *J. Appl. Phys.* **99**, 08S701 (2006).

91. B. Min, K. Motohashi, C. Lodder, and R. Jansen, Tunable spin-tunnel contacts to silicon using low-work-function ferromagnets, *Nat. Mater.* **5**, 817 (2006).

92. K. Wang, J. Stehlik, and J.-Q. Wang, Tunneling characteristics across nanoscale metal ferric junction lines into doped Si, *Appl. Phys. Lett.* **92**, 152118 (2008).

93. T. Uhrmann, T. Dimopoulos, H. Brückl et al., Characterization of embedded MgO/ferromagnet contacts for spin injection in silicon, *J. Appl. Phys.* **103**, 063709 (2008).

94. T. Dimopoulos, D. Schwarz, T. Uhrmann et al., Magnetic properties of embedded ferromagnetic contacts to silicon for spin injection, *J. Phys. D* **42**, 085004 (2009).

95. R. Mattana, J.-M. George, H. Jaffrès, et al., Electrical detection of spin accumulation in a *p*-type GaAs quantum well, *Phys. Rev. Lett.* **90**, 166601 (2003).

96. A. Fert, J.-M. George, H. Jaffres, and R. Mattana, Semiconductors between spin-polarized sources and drains, *IEEE Trans. Elec. Dev.* **54**, 921 (2007).

97. P. Li and H. Dery, Theory of spin-dependent phonon-assisted optical transitions in silicon, *Phys. Rev. Lett.* **105**, 037204 (2010).

98. B. Jonker, G. Kioseoglou, A. Hanbicki, C. Li, and P. Thompson, Electrical spin-injection into silicon from a ferromagnetic metal/tunnel barrier contact. *Nat. Phys.* **3**, 542 (2007).

99. G. Kioseoglou, A. T. Hanbicki, R. Goswami, et al., Electrical spin injection into Si: A comparison between Fe/Si Schottky and Fe/Al$_2$O$_3$tunnel contacts, *Appl. Phys. Lett.* **94**, 122106 (2009).

100. L. Grenet, M. Jamet, P. Noé et al., Spin injection in silicon at zero magnetic field, *Appl. Phys. Lett.* **94**, 032502 (2009).

101. O. van't Erve, A. Hanbicki, M. Holub, C. Li, C. Awo-Affouda et al. Electrical injection and detection of spin-polarized carriers in silicon in a lateral transport geometry, *Appl. Phys. Lett.* **91**, 212109 (2007).

102. T. Sasaki, T. Oikawa, T. Suzuki et al., Electrical spin injection into silicon using MgO tunnel barrier, *Appl. Phys. Express* **2**, 53003 (2009).

103. T. Sasaki, T. Oikawa, T. Suzuki, et al. Temperature dependence of spin diffusion length in silicon by Hanle-type spin precession *Appl. Phys. Lett.* **96**, 122101 (2010).

104. Y. Ando, K. Hamaya, K. Kasahara et al., Electrical injection and detection of spin-polarized electrons in silicon through an Fe$_3$Si/Si Schottky tunnel barrier, *Appl. Phys. Lett.* **94**, 182105 (2009).

105. M. Shiraishi, Y. Honda, E. Shikoh et al., Spin transport properties in silicon in a nonlocal geometry, *Phys. Rev. B* **83**, 241204 (2011).

106. M. Kameno, Y. Ando, T. Shinjo et al., Spin drift in highly doped n-type Si, *Appl. Phys. Lett.* **104**, 092409 (2014).

107. T. Sasaki, Y. Ando, M. Kameno et al., Spin transport in nondegenerate Si with a spin MOSFET structure at room temperature, *Phys. Rev. Appl* **2**, 034005 (2014).

108. S. Dash, S. Sharma, R. Patel, M. de Jong, and R. Jansen, Electrical creation of spin polarization in silicon at room temperature, *Nature* **462**, 491 (2009).

109. A. Dankert, R. S. Dulal, and S. P. Dash, Efficient spin injection into silicon and the role of the Schottky barrier, *Sci. Rep.* **3**, 3196 (2013).

110. C. Li, O. van't Erve, and B. Jonker, Electrical injection and detection of spin accumulation in silicon at 500K with magnetic metal/silicon dioxide contacts, *Nat. Commun.* **2**, 245 (2011).

111. N. W. Gray and A. Tiwari, Room temperature electrical injection and detection of spin polarized carriers in silicon using MgO tunnel barrier, *Appl. Phys. Lett.* **98**, 102112 (2011).

112. A. Jain, J.-C. Rojas-Sanchez, M. Cubukcu et al., Crossover from spin accumulation into interface states to spin injection in the germanium conduction band, *Phys. Rev. Lett.* **109**, 106603 (2012).

113. W. Han, X. Jiang, A. Kajdos et al., Spin injection and detection in lanthanum- and niobium-doped $SrTiO_3$ using the Hanle technique, *Nat. Commun.* **4**, 2134 (2013).

114. K.-R. Jeon, B.-C. Min, Y.-H. Park, S.-Y. Park, and S.-C. Shin, Electrical investigation of the oblique Hanle effect in ferromagnet/oxide/semiconductor contacts, *Phys. Rev. B* **87**, 195311 (2013).

115. O. Txoperena, M. Gobbi, A. Bedoya-Pinto et al., How reliable are Hanle measurements in metals in a three-terminal geometry? *Appl. Phys. Lett.* **102**, 192406 (2013).

116. O. Txoperena and F. Casanova, Spin injection and local magnetoresistance effects in three-terminal devices, *J. Phys. D* **49**, 133001 (2016).

117. H. N. Tinkey, P. Li, and I. Appelbaum, Inelastic electron tunneling spectroscopy of local "spin accumulation" devices, *Appl. Phys. Lett.* **104**, 232410 (2014).

118. I. Appelbaum, H. N. Tinkey, and P. Li, Self-consistent model of spin accumulation magnetoresistance in ferromagnet/insulator/semiconductor tunnel junctions, *Phys. Rev. B* **90**, 220402 (2014).

119. H. N. Tinkey, H. Dery, and I. Appelbaum, Defect passivation by proton irradiation in ferromagnet-oxide-silicon junctions, *Appl. Phys. Lett.* **109**, 142407 (2016).

120. Y. Song and H. Dery, Magnetic-field-modulated resonant tunneling in ferromagnetic-insulator-nonmagnetic junctions, *Phys. Rev. Lett.* **113**, 6 (2014).

121. O. Txoperena, Y. Song, L. Qing et al., Impurity-assisted tunneling magnetoresistance under a weak magnetic field, *Phys. Rev. Lett.* **113**, 146601 (2014).

122. H. Inoue, A. Swartz, N. Harmon et al., Origin of the magnetoresistance in oxide tunnel junctions determined through electric polarization control of the interface, *Phys. Rev. X* **5**, 7 (2015).

123. Z. Yue, M. C. Prestgard, A. Tiwari, and M. E. Raikh, Resonant magnetotunneling between normal and ferromagnetic electrodes in relation to the three-terminal spin transport, *Phys. Rev. B* **91**, 195316 (2015).

124. M. Tran, H. Jaffrès, C. Deranlot et al., Enhancement of the spin accumulation at the Interface between a spin-polarized tunnel junction and a semiconductor, *Phys. Rev. Lett.* **102**, 036601 (2009).

125. Y. Ando, Y. Maeda, K. Kasahara et al., Electric-field control of spin accumulation signals in silicon at room temperature, *Appl. Phys. Lett.* **99**, 132511 (2011).

126. X. Lou, C. Adelmann, M. Furis et al., Electrical detection of spin accumulation at a ferromagnet-semiconductor interface, *Phys. Rev. Lett.* **96**, 176603 (2006).

127. M. Bolduc, C. Awo-Affouda, A. Stollenwerk et al., Above room temperature ferromagnetism in Mn-ion implanted Si, *Phys. Rev. B* **71**, 033302 (2005).

128. S. C. Erwin and I. Žutić, Tailoring ferromagnetic chalcopyrites, *Nat. Mater.* **3**, 410 (2004).

129. S. Cho, S. Choi, G. Cha et al., Synthesis of new pure ferromagnetic semiconductors: $MnGeP_2$ and $MnGeAs_2$, *Solid State Commun.* **129**, 609 (2004).

130. S. Cho, S. Choi, G.-B. Cha et al., Room-temperature ferromagnetism in $Zn_{1-x}Mn_xGeP_2$ semiconductors, *Phys. Rev. Lett.* **88**, 257203 (2002).

131. Y. Ishida, D. D. Sarma, K. Okazaki et al., In situ photoemission study of the room temperature ferromagnet $ZnGeP_2$:Mn, *Phys. Rev. Lett.* **91**, 107202 (2003).

132. A. Schmehl, V. Vaithyanathan, A. Herrnberger et al., Epitaxial integration of the highly spin-polarized ferromagnetic semiconductor EuO with silicon and GaN, *Nat. Mater.* **6**, 882 (2007).

133. Y. Pu, P. M. Odenthal, R. Adur, et al., Ferromagnetic resonance spin pumping and electrical spin injection in silicon-based metal-oxide-semiconductor heterostructures, *Phys. Rev. Lett.* **115**, 246602 (2015).

134. E. Shikoh, K. Ando, K. Kubo et al., Spin-pump-induced spin transport in *p*-type Si at room temperature, *Phys. Rev. Lett.* **110**, 127201 (2013).

135. K. Ando and E. Saitoh, Observation of the inverse spin Hall effect in silicon, *Nat. Commun.* **3**, 629 (2012).

136. S. Ganichev, S. Danilov, V. Bel'kov et al., Pure spin currents induced by spin-dependent scattering processes in SiGe quantum well structures, *Phys. Rev. B* **75**, 155317 (2007).

137. M. Friesen, P. Rugheimer, D. Savage et al., Practical design and simulation of silicon-based quantum-dot qubits, *Phys. Rev. B* **67**, 121301 (2003).

138. P. Zhang and M. Wu, Spin diffusion in Si/SiGe quantum wells: Spin relaxation in the absence of Dyakonov-Perel relaxation mechanism, *Phys. Rev. B* **79**, 075303 (2009).

139. N. Bagraev, N. Galkin, W. Gehlhoff et al., Spin interference in silicon three-terminal one-dimensional rings, *J. Phys. D* **18**, L567 (2006).

140. P. Zhang and M. Wu, Hole spin relaxation in strained asymmetric Si/SiGe and Ge/SiGe quantum wells, *Phys. Rev. B* **80**, 155311 (2009).

141. S. van Dijken, X. Jiang, and S. S. P. Parkin, Giant magnetocurrent exceeding 3400% in magnetic tunnel transistors with spin-valve base layers, *Appl. Phys. Lett.* **83**, 951 (2003).

142. B. Huang, D. J. Monsma, and I. Appelbaum, Experimental realization of a silicon spin field-effect transistor, *Appl. Phys. Lett.* **91**, 072501 (2007).

143. B. Huang, L. Zhao, D. J. Monsma, and I. Appelbaum, 35% magnetocurrent with spin transport through Si, *Appl. Phys. Lett.* **91**, 052501 (2007).

144. B. Huang, D. J. Monsma, and I. Appelbaum, Coherent spin transport through a 350 micron thick silicon wafer, *Phys. Rev. Lett.* **99**, 177209 (2007).

145. J. S. Moodera, L. R. Kinder, T. M. Wong, and R. Meservey, Large magnetoresistance at room temperature in ferromagnetic thin film tunnel junctions, *Phys. Rev. Lett.* **74**, 3273 (1995).

146. L. D. Bell, S. J. Manion, M. H. Hecht et al., Characterizing hot-carrier transport in silicon heterostructures with the use of ballistic-electron-emission microscopy, *Phys. Rev. B* **48**, 5712 (1993).

147. J. Li, B. Huang, and I. Appelbaum, Oblique Hanle effect in semiconductor spin transport devices, *Appl. Phys. Lett.* **92**, 142507 (2008).

148. B. Huang, H.-J. Jang, and I. Appelbaum, Geometric dephasing-limited Hanle effect in long-distance lateral silicon spin transport devices, *Appl. Phys. Lett.* **93**, 162508 (2008).

149. J. Li and I. Appelbaum, Inelastic spin depolarization spectroscopy in silicon, *J. Appl. Phys.* **114**, 033705 (2013).

150. C. Canali, C. Jacoboni, F. Nava, G. Ottaviani, and A. Quaranta, Electron drift velocity in silicon, *Phys. Rev. B* **12**, 2265 (1975).

151. C. Jacoboni, C. Canali, G. Ottaviani, and A. Quaranta, A review of some charge transport properties of silicon, *Solid State Electron.* **20**, 77 (1977).

152. H.-J. Jang, J. Xu, J. Li, B. Huang, and I. Appelbaum, Non-ohmic spin transport in n-type doped silicon, *Phys. Rev. B* **78**, 165329 (2008).

153. I. Altfeder, B. Huang, I. Appelbaum, and B. C. Walker, Self-assembly of epitaxial monolayers for vacuum wafer bonding, *Appl. Phys. Lett.* **89**, 223127 (2006).

154. W. Hanle, The magnetic influence on the polarization of resonance fluorescence, *Z. Physik* **30**, 93 (1924).

155. B. Huang, Vertical transport silicon spintronic devices, PhD thesis, U. Delaware, 2008.

156. J. L. Cheng, M. W. Wu, and J. Fabian, Theory of the spin relaxation of conduction electrons in silicon, *Phys. Rev. Lett.* **104**, 016601 (2010).

157. P. Li and H. Dery, Spin-orbit symmetries of conduction electrons in silicon, *Phys. Rev. Lett.* **107**, 107203 (2011).

158. Y. Song and H. Dery, Analysis of phonon-induced spin relaxation processes in silicon, *Phys. Rev. B* **86**, 085201 (2012).

159. F. J. Morin, T. H. Geballe, and C. Herring, Temperature dependence of the piezoresistance of high-purity silicon and germanium, *Phys. Rev.* **105**, 525 (1957).

160. J. Tauc and E. Antončík, Optical observation of spin-orbit interaction in germanium, *Phys. Rev. Lett.* **5**, 253 (1960).

161. M. Cardona and H. S. Sommers, Effect of temperature and doping on the reflectivity of germanium in the fundamental absorption region, *Phys. Rev.* **122**, 1382 (1961).

162. L. M. Roth and B. Lax, *g* factor of electrons in germanium, *Phys. Rev. Lett.* **3**, 217 (1959).

163. J.-N. Chazalviel, Spin relaxation of conduction electrons in highly-doped *n*-type germanium at low temperature, *J. Phys. Chem. Solids* **36**, 387 (1975). ISSN 0022-3697.

164. G. Dresselhaus, Spin-orbit coupling effects in zinc blende structures, *Phys. Rev.* **100**, 580 (1955).

165. P. Li, J. Li, L. Qing, H. Dery, and I. Appelbaum, Anisotropy-driven spin relaxation in germanium, *Phys. Rev. Lett.* **111**, 257204 (2013).

166. J. Li, L. Qing, H. Dery, and I. Appelbaum, Field-induced negative differential spin lifetime in silicon, *Phys. Rev. Lett.* **108**, 157201 (2012).

167. Y. Zhou, M. Ogawa, X. Han, and K. L. Wang, Alleviation of Fermi-level pinning effect on metal/germanium interface by insertion of an ultrathin aluminum oxide, *Appl. Phys. Lett.* **93**, 202105 (2008).

168. S. Crooker, M. Furis, X. Lou et al., Imaging spin transport in lateral ferromagnet/semiconductor structures, *Science* **309**, 2191 (2005).

169. B. Huang and I. Appelbaum, Spin dephasing in drift-dominated semiconductor spintronics devices, *Phys. Rev. B* **77**, 165331 (2008).

170. P. Li, Y. Song, and H. Dery, Intrinsic spin lifetime of conduction electrons in germanium, *Phys. Rev. B* **86**, 085202 (2012).

171. B. Huang and I. Appelbaum, Time-of-flight spectroscopy via spin precession: The Larmor clock and anomalous spin dephasing in silicon, *Phys. Rev. B* **82**, 241202 (2010).

172. L. Qing, J. Li, I. Appelbaum, and H. Dery, Spin relaxation via exchange with donor impurity-bound electrons, *Phys. Rev. B* **91**, 241405 (2015).

173. B. Huang, D. J. Monsma, and I. Appelbaum, Spin lifetime in silicon in the presence of parasitic electronic effects, *J. Appl. Phys.* **102**, 013901 (2007).

174. J. Haynes and W. Shockley, The mobility and life of injected holes and electrons in germanium, *Phys. Rev.* **81**, 835 (1951).

175. C. C. Lo, J. Li, I. Appelbaum, and J. J. L. Morton, Microwave manipulation of electrically injected spin-polarized electrons in silicon, *Phys. Rev. Appl.* **1**, 014006 (2014).

176. H. Dery, Y. S. Song, P. Li and I. Zutic, Silicon spin communication., *Appl. Phys. Lett.* **99**, 082502 (2011).

177. J.-M. Tang, B. T. Collins, and M. E. Flatté, Electron spin-phonon interaction symmetries and tunable spin relaxation in silicon and germanium, *Phys. Rev. B* **85**, 045202 (2012).

178. H.-J. Jang and I. Appelbaum, Spin polarized electron transport near the Si/SiO$_2$ Interface, *Phys. Rev. Lett.* **103**, 117202 (2009).

179. J. Li and I. Appelbaum, Modeling spin transport in electrostatically-gated lateral-channel silicon devices: Role of interfacial spin relaxation, *Phys. Rev. B* **84**, 165318 (2011).

180. J. Li and I. Appelbaum, Lateral spin transport through bulk silicon, *Appl. Phys. Lett.* **100**, 162408 (2012).

181. E. Y. Sherman, Random spin–orbit coupling and spin relaxation in symmetric quantum wells, *Appl. Phys. Lett.* **82**, 209 (2003).

182. J. Shim, K. Raman, Y. Park et al., Large spin diffusion length in an amorphous organic semiconductor, *Phys. Rev. Lett.* **100**, 226603 (2008).

183. E. Rashba, Inelastic scattering approach to the theory of a magnetic tunnel transistor source, *Phys. Rev. B* **68**, 241310 (2003).

# 4

# Tunneling Magnetoresistance, Spin-Transfer and Spinorbitronics with (Ga,Mn)As

**J.-M. George, D. Quang To, T. Huong Dang, E. Erina, T. L. Hoai Nguyen, H.-J. Drouhin, and H. Jaffrès**

This chapter reviews some fundamental properties of tunneling spin-current, spin-transport of holes and spin-transfer of angular momenta in the valence band of magnetic III–V tunneling heterojunctions involving the ferromagnetic semiconductor (Ga,Mn)As. This material is further discussed in Chapter 9, Volume 2 and has recently been the focus of some extended reports where readers could find additional details of the electronic structure [1, 2]. Discussion complementing this chapter can be found throughout this book, for example, on electric-field control of magnetism in Chapter 13, Volume 2, spin torque and spin-orbit torque in Chapters 7–9, Volume 2, Hall effects in Chapter 8, Volume 2, topological insulators in Chapters 14 and 15, Volume 2, and magnetic quantum dots in Chapter 6, Volume 3.

The present review includes some more up-to-date developments since 2012 with (Ga,Mn)As involving principles of current-in-plane spin-orbit torques [3], anatomy of spin transport at interface, as well as spin-Hall magnetoresistance (SMR) and its unidirectional character(U-SMR) [4]. These phenomena together with their experimental evidence are associated with new physical properties, and these are made possible via the recent strong interest in the novel *spinorbitronics* area. Moreover, (Ga,Mn)As like (Ge,Mn) for group IV semiconductors, remains a unique prototype group III–V semiconductor which demonstrates non-zero exchange interactions and carrier-mediated ferromagnetism. This makes possible the development of spinorbitronics devices with their particular interest and properties appearing as soon as both exchange strengths and spin-orbit interactions (SOI) come into play. However, the full metallic character of (Ga,Mn)As described by exchange-split holes in the valence bands (VB) of the III–V host is still under debate, owing to some latest results and experiments. The picture of exchange-split delocalized host of holes interacting with localized Mn *via* the Zener-like picture of ferromagnetism is strongly debated, favoring an impurity band description which seems to be demonstrated by recent spectroscopic tunneling experiments. Although this issue is still unsolved, (Ga,Mn)As remains an important prototype material to evidence new types of spinorbitronics phenomena supported by III–V hosts, at least up to its Curie temperature, $T_C$ of about 200 K [1, 2, 5–8]. Very recently, it has been reported that the (In,Fe)As compound could support ferromagnetism in its conduction band (CB) which may be very attractive [9] for spintronic applications in the CB of semiconducting heterostructures.

Among the strong interest in the field of spintronics and today's spinor-bitronics with semiconductors [10, 11], this current research has its roots in the work of Tanaka et al. from the early 2000s [12]. This research is generally motivated by (1) the wealth of the physics involving transport of spin-orbit coupled states carrying quantized angular momenta in epitaxial hetero-structures; (2) its application to spin-transfer torque phenomena and its deep fundamental understanding; (3) the need to use materials and tunneling devices providing an efficient spin injection into group III–V semiconduc-tors [13–16] and able to circumvent the conductivity mismatch issue [17]; and possibly (4) to use the high coherence of hole spin states in a semicon-ductor quantum dot [18, 19].

Generating a current of angular momenta (both *spin* and *orbital*) in a semiconductor and converting the latter information into an electrical signal via tunnel injection [20] or tunnel filtering is still challenging. This appears as two important prerequisites before any spin manipulation. Despite a rela-tive low $T_C$, currently reaching 200 K nowadays, the $p$-type ferromagnetic semiconductor (Ga,Mn)As, largely described in Chapter 9, Volume 2 [21] and already well documented [1, 2, 21–27], offers the possibility of a good inte-gration with conventional III–V species and related heterostructures. The (Ga,Mn)As ferromagnetism properties may, in the simple case, be semi-quan-titatively understood by the so-called $p$-$d$ Zener approach [28]. This describes a mean-field approach for the average exchange interactions between delocal-ized $p$-type carriers and $3d$ Mn local magnetization in the host III–V semi-conducting host [29, 30]; although a recent alternative approach focuses on an impurity-band mostly uncoupled with the host [31–34]. In addition, (Ga,Mn)As exhibits a specific anisotropic character of its different valence bands, location of exchange interactions, SOI, and of an anisotropic strain-field [29] which makes it very attractive for further functionalities. In that sense, III–V heterostructures including (Ga,Mn)As as a ferromagnetic reservoir often represent model systems to study some new properties of spin-injection.

On the other hand, spin-currents, SOI, and their interplay are the fun-damentals of intriguing physical phenomena like the spin-Hall effect (SHE) and the inverse spin-Hall effect (ISHE) [35–41], and inverse Edelstein effects (IEE) [42–44] as recently observed with Rashba-states or topological insula-tors (TI). These phenomena manifest themselves by a left/right asymmetry in the scattering process of spin-polarized carriers along the transverse direc-tion of their flow, giving rise to spin-to-charge conversion and vice versa. These are currently the basis of new functionalities like spin-orbit torques (SOT) at magnetic/SOI interfaces [3, 45–47] in mind of magnetic switching using in-plane currents (CIP). In that context, investigations of both SOI and exchange strengths in solids and at interfaces [48, 49] is of a prime impor-tance e.g. for the determination of the generalized spin-mixing conductances for spin-transfer. Concomitantly with the numerous literature devoted to spin-Hall effects in metals and conductors, a mechanism of tunnel-Hall effects in magnetic tunnel junctions (MTJs) [50, 51] and a mechanism of tun-neling planar Hall effect (TPHE) emerging at ferromagnet FM/TI junctions [52] have recently been proposed. Those qualitatively differ from the SHE in terms of the relevant geometry, the *forward/backward scattering* in the present case, and/or the magnetization configuration. In particular, the latter effect is maximized for a planar magnetization parallel to the applied bias, in the longitudinal geometry, where these other Hall effects vanish. This is also the case for the U-SMR effects arising at heavy metal (SOI)/ferromagnetic interfaces [53], semiconductor/ferromagnetic semiconductors [4] and topological insulators/ferromagnetic interfaces [54]. In the latter case, the non-linear resistance contribution vs. the direction of the transverse magnetization or in-plane current direction, as evidenced experimentally (U-SMR), have been ascribed to carrier asymmetry-scattering (of the same symmetry property) and to a second order conductivity component derived from it.

In that spirit, we have recently investigated theoretically the particular properties of skew-tunneling of carriers at semiconductor interfaces or magnetic tunnel junction involving exchange and SOI in the host semiconductor material [50, 51] like possibly played by (Ga,Mn)As or (Ge,Mn) ferromagnetic injectors [55]. The prototype system is a simple interface or a tunnel junctions made of the same ferromagnetic layers in an antiparallel (*AP*) magnetic configuration. It may admit a difference in the carrier transmission (or diffusion) upon their incidence, that is upon their parallel wave-vector $\pm k_\parallel$, vs. the reflection plane defined by the magnetization and the surface normal via SOI. This new symmetry term in the electronic transport has an impact, even small, on the conductivity once the Boltzmann diffusive equation is developed at the second order of the electric field, and as recently experimentally observed [54]. Moreover, such transport asymmetry may also be associated with a pure interfacial (tunnel) *topological* transverse Hall current after summation of the electron channels over the Fermi surface [50, 51]. This properties of carrier transport occurs in the CB involving SOI-Dresselhaus interactions or in the VB getting benefit of the atomic SOI. In all cases, calculations go in favor of a robust forward tunnel scattering asymmetry linked to a certain transport chirality in the tunneling region. This specificity of spin-transport with spin-orbit split electronic states may be of a significant importance to explain the specific non-linear U-SMR effects, as revealed recently at (Ga,Mn)As/GaAs:Mn interfaces with current in-plane geometry [4].

This chapter is divided into five sections: (1) In the first section, we will present results of tunneling magnetoresistance (TMR) obtained on (Ga,Mn)As/III–V/(Ga,Mn)As junctions [12, 56–59] studied hereafter for spin injection problem. TMR results on hybrid (Ga,Mn)As/Insulator(I)/transition metal junctions will be also be briefly discussed [60–62]. In this section, quantum size coherent tunneling phenomena and tunneling spectroscopy features observed on resonant (Ga,Mn)As [63] and with III–V (In,Ga)As quantum wells (QWs) resonant structures [64] will be addressed, to question the coherent nature, or not, of transport in heterojunctions and related systems. This is actually an important topic focused on finding the adequate resonant position levels of (Ga,Mn)As QWs in order to determine the correct band structure [31, 65]. We then tackle the role of different material properties, mainly hole filling and exchange energy, on the TMR properties and will establish phase diagrams that correlate these parameters to the TMR. (2) The second section will be devoted to tunneling anisotropic magnetoresistance (TAMR) effects [66–71] including at least one (Ga,Mn)As electrode. TAMR generally manifests itself by a significant variation of the tunneling current with respect to the magnetization direction of the ferromagnetic electrode (for a review on TAMR, the readers can refer to Matos-Abiague and Fabian [72, 73]). In the present case, this physical phenomena originates from the strong anisotropy of the hole fermi surfaces of the different (Ga,Mn)As subbands involved in the tunneling transport of holes. (3) The third section will be devoted to the description of the skew tunneling effect and related anomalous tunnel hall effect involving ferromagnetic semiconductors [50, 51, 55] and investigated theoretically using analytical developments together with advanced 30-band $\mathbf{k} \cdot \mathbf{p}$ tunneling methods. (4) In the fourth part, we will describe spin transfer torque experiments [74–77] which have been demonstrated in these systems [3, 59, 79–82] accounting for the conservation of the angular momenta of holes carried by the current for perpendicular-to-plane and in-plane current injection. (5) In the last part, we will describe the skew tunneling and derive in detail the anomalous tunneling Hall effect. The investigations will be first made via the relevant perturbation calculations of the scattering process based on the Green's function techniques developed in the first order of the SOI parameter. We also demonstrate results of transport asymmetry implemented with advanced 30-band $\mathbf{k} \cdot \mathbf{p}$ methods adapted to the tunneling transport and using relevant wave function and wave current matching conditions that one compares to more standard $\mathbf{k} \cdot \mathbf{p}$ codes with excellent agreement.

In the appendix, we will develop the modeling via advanced $\mathbf{k} \cdot \mathbf{p}$ kinetic exchange approach, the angular momentum transport in the valence band of semiconductors heterostructures. Concerning (Ga,Mn)As this includes, in a mean-field approach [29, 83], the average $p$-$d$ interaction between spin up and spin down holes in the VB of the semiconductor parametrized by the spin-exchange parameter and added to the Kinetic or Kohn–Luttinger Hamiltonian [84]. Within the envelope function approximation, the tunneling transport of holes in the VB of semiconductors will be described in detail according to the procedure given by Petukhov et al. [85, 86] for matching hole wave functions and derivatives at each interface. This approach constitutes the generalization to magnetic systems of the boundary conditions for hole envelope wave functions and tunneling current examined about 20 years ago by Wessel and Altarelli for semiconducting heterojunctions [87, 88].

## 4.1 TUNNELING MAGNETORESISTANCE IN THE VALENCE-BAND OF *P*-TYPE SEMICONDUCTORS

### 4.1.1 INTRODUCTION

The quantum mechanical tunnel effect is one of the oldest quantum phenomena and still continues to enrich our understanding of many fields in physics. The generality of the tunneling phenomenon is such that every dedicated textbook on quantum mechanics discusses tunneling through a potential barrier, and the possibility for an electron to tunnel through such a barrier with a characteristic imaginary velocity and imaginary wave-vector. Despite this long history and many experimental investigations, the tunneling process of electrons (or holes) continues to amaze physicists in the solid-state community. For spin-dependent tunneling between two ferromagnetic materials, first proposed and reported [89] in 1975, and subsequently observed, it has taken two decades to be reliably demonstrated at room temperature [90]. The large TMR effects possible in ferromagnet tunnel junctions and MTJs led to the development of nonvolatile magnetic random access memories (MRAMs) and a new generation of magnetic read-heads in hard disk-drives (HDD), for example, see Chapters 12–14, Volume 3. For sufficiently thin barriers, typically a few nanometers of insulating layer, carriers possess a finite probability to be detected on the other side of the potential barrier, due to the coupling between the evanescent wave functions on each side of the barrier. With the condition that the density of states (DOS) of the electrodes is spin polarized in the case of ferromagnets, and in a simple picture of a tunneling current proportional to the DOS within each spin channel, the transmission probability is spin dependent, that is, depends on the relative orientation of the two magnetizations. This is the so-called TMR effect. Additionally, when dealing with tunneling of holes in the valence band of semiconductors, one can expect an influence of the large spin-orbit coupling on the transmission probability through the barrier as well as a possible deflection of the characteristic carrier wave-vector leading to the so-called TAMR effects. Today, and as largely discussed in this chapter, one of the implementation deals with the properties of tunneling and transmission of carriers involving spin-orbit coupled states beyond only the spin degree of freedom.

## 4.1.2 Overview of TMR Phenomena Including (Ga,Mn)As

The first generation of MTJs has integrated amorphous oxide as tunnel barrier, and spin-dependent tunneling effects were mainly described in terms of the phenomenological Jullière model [89]. This has begun to change with the synthesis of high-quality epitaxial Fe/MgO/Fe MTJs [91, 92] and related junctions showing up huge TMR ratio explained thanks to spin-symmetry filtering of the electron wave function during their tunneling transfer [93, 94], see Chapters 11–14, Volume 2. From a fundamental point of view, the ferromagnetic semiconductor (Ga,Mn)As provided, at about the same time, new opportunities to study spin-polarized transport phenomena in semiconductors heterojunctions [12]. One of the interests in (Ga,Mn)As lies in the wealth and varieties of its electronic valence band structure, location of exchange interactions, and strong spin-orbit coupling. In this vein, advantages of semiconductor tunnel junctions are fourfold: (1) III–V heterostructures can be epitaxially grown in a wide variety of tunnel devices with abrupt interfaces and with atomically controlled layer thicknesses; (2) junctions can be easily integrated with other III–V structures and devices; (3) many structural and band parameters are controllable, like barrier thickness and barrier height, allowing the engineering of any band profile; and (4) one can introduce quantum heterostructures much more easily than in any other material system.

The first large spin-valve TMR ratio (>70% low temperature) observed on III–V-based (Ga,Mn)As/AlAs/(Ga,Mn)As MTJs [12] were obtained by Tanaka and Higo in 2001 (Figure 4.1). One of the first interesting properties of these epitaxially grown MTJs was the evidence of a rapid drop of TMR with increasing the AlAs barrier thickness demonstrating a change of the effective tunneling spin-polarization of holes with barrier thickness. This was ascribed, in the frame of spin-orbit nearest neighbor $sp^3s^*$ tight-binding model of tunneling, to the conservation of the in-plane wave-vector parallel to the interface during the tunneling transfer process due to translation invariance. Some theoretical investigations performed within the 6×6 multiband $\mathbf{k} \cdot \mathbf{p}$ theory emphasize the specific role of the spin-orbit coupling within the barrier to possibly explain such a loss of TMR with the III–V barrier thickness [95, 96]. However, other calculations performed by Krstajic and Peeters gave opposite conclusions [97].

Beyond single tunnel junctions, TMR obtained on (Ga,Mn)As-based III–V double barrier structures, (Ga,Mn)As/AlAs/GaAs/AlAs/(Ga,Mn)As [98]

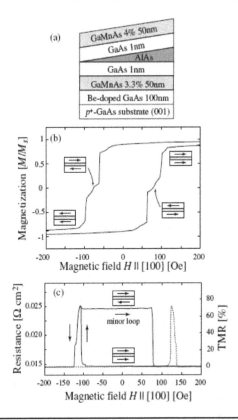

**FIGURE 4.1** (a) Schematic illustration of a wedge-type ferromagnetic semiconductor trilayer heterostructure sample grown by LT-MBE. (b) Magnetization of a $Ga_{0.96}Mn_{0.04}As$/AlAs 3-nm/$Ga_{0.967}Mn_{0.033}As$ trilayer measured by SQUID at 8 K. (c) TMR curves at 8 K of the trilayer 200 μm in diameter. Bold solid and dashed curves were obtained by sweeping the magnetic field from positive to negative, and from negative to positive, respectively. A minor loop is shown by a thin solid curve. In both (b) and (c), the magnetic field was applied along the [100] axis in the plane. (After Tanaka, M. et al., *Phys. Rev. Lett.* 87, 026602, 2001. With permission.)

and (Ga,Mn)As/AlAs/InGaAs/AlAs/(Ga,Mn)As [64] were integrating (Ga,Mn)As/AlAs junctions as spin injector/detector devices. These devices allowed to demonstrate spin injection phenomena into III–V materials in the current perpendicular to plane (CPP) configuration. However, such spin-valve effects observed on double barrier structures with in-plane magnetization still raise questions about the symmetry-state of the hole wave function injected in the quantum QW whose natural quantization axis is the growth direction. In the case of experiments of Mattana et al. [98], the formation of hydrogenic impurity states inside the QW, Ga vacancies, for instance, of quasi-spherical or cubic symmetry was suggested. These symmetry hole states of extended envelope wave function have been described theoretically in the $\mathbf{k} \cdot \mathbf{p}$ framework by Baldereschi and Lipari in the early seventies [99–101] and could give rise to resonant tunneling effects in the quantum well [102]. Such resonant tunneling effects was already observed in our group through As antisites deep levels in MnAs/GaAs/MnAs tunnel junctions synthesized at low temperature [103, 104]. Nevertheless, the oscillatory character of the TMR highlighted by Ohya et al. [64] on structures of higher epitaxial quality is more impressive with regard to the hole symmetry-state. A part of the answer lies in the possible existence of two orthogonal mixed-spin hole states on resonant levels in the QW. This conclusion should be supported by future theoretical investigations. Several kinds of III–V (Ga,Mn)As-based tunnel junctions were developed hereafter, including thin GaAs layer [56, 57, 81, 82] or (In,Ga)As barriers [58, 80] specifically grown for their low resistance area product with a view to proceeding to spin transfer experiments. More surprising are the results of Xiang et al. [105] about the observation of small in-plane magnetoresistance effects on (Ga,Mn)As/highly Be-doped GaAs/(Ga,Mn)As trilayers [105] where GaAs should play the role of barrier for holes.

In parallel to tunneling transport experiments, (Ga,Mn)As/GaAs Esaki diodes were used to demonstrate spin injection by electrical means on lateral structures with a non-local detection [106], and by optical techniques [107, 108] giving more than 80% for the degree of helicity of the light emitted from these spin-light emitting diodes (spin-LEDs). Calculations performed in the frame of a spin-dependent tight-binding Hamiltonian, including the exchange interactions within (Ga,Mn)As, were then employed to understand such high spin injection efficiency [109]. On the other hand, Chun et al. [60] demonstrated spin-polarized tunneling effects on hybrid MnAs/AlAs/(Ga,Mn)As structures, as well as Saito et al. with hybrid Fe/ZnSe/(Ga,Mn)As and Fe/GaO/(Ga,Mn)As junctions [61, 62]. This shows the possibility to observe large TMR ratio with bipolar *n/i/p* structures and consequently to convert spin-polarized electrons into spin-polarized holes and vice versa.

From a more fundamental point of view, could tunneling of holes through a thin (Ga,Mn)As layer and through the whole III–V heterojunction be realistically described by coherent processes over quite long layer thickness, as was recently observed with tunneling spectroscopic methods [31, 63, 65]? Specifically, clear resonant tunneling effect were observed through a relatively thin AlAs/(Ga,Mn)As/(Al,Ga)As resonant tunneling diode with varying QW thickness ranging from 3.8 to 20 nm (Figure 4.2). These experiments seems to support the picture of the coherent nature of spin-polarized holes tunneling processes through relatively thin (Ga,Mn)As QWs, despite the low-temperature growth procedure (250°C) often described to be detrimental for the observation of such phenomena. Moreover, these

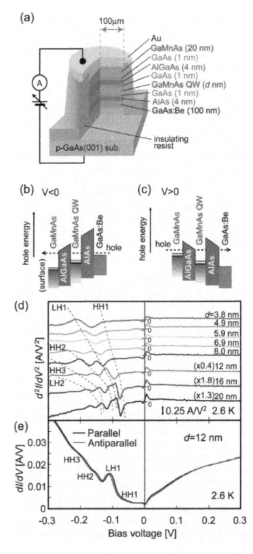

**FIGURE 4.2**    (a) Schematic device structure of the $Ga_{0.95}Mn_{0.05}As$ 20-nm/GaAs 1-nm/$Al_{0.5}Ga_{0.5}As$ 4-nm/GaAs 1-nm/$Ga_{0.95}Mn_{0.05}As$ $d$-nm/GaAs 1-nm/AlAs 4-nm/ GaAs:Be 100-nm resonant tunneling diode (RTD) junction. (b) and (c) Schematic band diagrams of the resonant tunneling diode (RTD) junction when the bias polarity is negative and positive, respectively. (d) $d^2I/dV^2$-$V$ characteristics of these RTD junctions with various QW thicknesses $d$ in parallel magnetization at 2.6 K. Numbers in the parentheses express the magnification ratio for the vertical axis. (e) $dI/dV$-$V$ characteristics of the junction with $d = 12$ nm at 2.6 K in parallel (blue curve) and antiparallel (red curve) magnetization. (After Tanaka, M. et al., *Phys. Rev. Lett.* 87, 026602, 2001. With permission.).

experiments are well explained by 4×4 valence-band $\mathbf{k} \cdot \mathbf{p}$ [84] model including *p-d* exchange interactions, thus demonstrating the validity of the coherent approach to describe tunneling transport of holes in a wide family of (Ga,Mn)As/III–V heterostructures.

### 4.1.3 MATERIAL AND CHEMICAL TRENDS

#### 4.1.3.1 Atomic Picture of Mn-Doped GaAs Host (See Appendix A)

According to optical studies, Mn in GaAs is known to form a shallow acceptor center characterized by a moderate binding energy [111] of 110 meV, and a small magnitude of the energy difference of the order of 8 (±3) meV between the states corresponding to the spin of the bound hole parallel and antiparallel to the Mn spin [111]. This relative small atomic *p-d* exchange energy originates from the long extend of the bound hole wave function where the Mn *d* states are mostly localized. A direct consequence is that the average exchange integral $\Delta_{\text{exc}}$ is expected to be enhanced with increasing the Mn content *x*. The average exchange interaction in (Ga,Mn)As reads $\Delta_{\text{exc}} = -\dfrac{5}{2} x N_0 \beta \, \text{eV}$, where $N_0$ is the concentration of cations and $N_0 \beta = -\dfrac{16}{S}\left(\dfrac{1}{-\Delta+U} + \dfrac{1}{\Delta}\right) \times \left(\dfrac{1}{3}(pd\sigma) - \dfrac{2\sqrt{3}}{9}(pd\pi)^2\right) < 0$ is the exchange integral found by treating the *p-d* hybridization as a perturbation in the configuration interaction picture [112], giving rise to antiferromagnetic interactions between *p* and *d* shells (Figure 4.3). Here, *S* is the localized *d* spin, U is the characteristic Coulomb interaction in the *d* shell, $\Delta$ is the characteristic energy difference between *p* and majority spin *d* orbitals, whereas $(pd\sigma)$ and $(pd\pi)$ are characteristic Slater-Koster hopping integrals. The value of $N_0\beta = -1.2 \, \text{eV}$ ($\beta = -54 \, \text{meV nm}^3$) is generally accepted from core level photoemission measurements for (Ga,Mn)As with a $T_C$ close to 60 K [113].

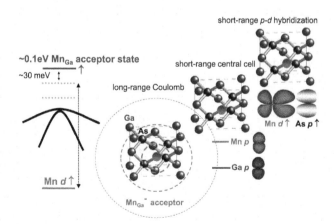

**FIGURE 4.3** Schematic illustration of the long-range Coulomb and the two short-range potentials each contributing 30 meV to the binding energy of the (Ga,Mn) acceptor. (After Larsson, B.E. et al., *Phys. Rev. B*, 37, 4137, 1988. With permission.)

### 4.1.3.2 Electronic and Ferromagnetic Properties

(Ga,Mn)As that we focus on in this chapter is the most extensively studied III–V dilute magnetic semiconductor (DMS). Ferromagnetism in (Ga,Mn) As requires Mn content of $x \gtrsim 1\%$ [114–116], whereas the metal-to-insulator transition occurs at slightly higher content ($x > 1\%$) [115–117]. Efforts to achieve the growth of Mn-doped GaAs at high concentration ($x > 1\%$) by means of nonequilibrium molecular beam epitaxy (MBE) at low substrate temperatures has led to the development of metallicity and ferromagnetism on (Ga,Mn)As with respective $T_C$ of 7.5 K and 60 K. More recently, the $T_C$ have been pushed up towards 190–200 K [5–8] by improvement in the MBE technique deposition, allowing Mn concentrations up to nominally $x \simeq 16\%$ per cation site [1, 5–8, 118]. Using delta-doped doping technique with Mn, i.e. mainly containing two-dimensional Mn-rich layer, $T_C \simeq 250$ K has been reached [119].

Owing to a relatively large distance between magnetic ions in DMS, no direct exchange coupling between $d$-like orbitals localized on Mn ions is expected. Thus, rather indirect coupling involving band states accounts for spin-spin interactions. In this section we describe the effects of short-range superexchange that dominates in the absence of carriers and competes with long-range carrier-mediated interactions if the concentration of band carriers is sufficiently high. The double-exchange mechanism contributes when relevant magnetic ions are in two different charge states [28, 120–122], and if the system of transition metal electrons is on the metallic side or in the vicinity of the Anderson–Mott transition. Double exchange may indeed be considered as a strong coupling limit of the $p$-$d$ Zener model, but does not necessarily lead to the formation of an impurity band [123]. Within this model, $T_C$ attains a maximum if the impurity band is half-filled, so that (in the case relevant here) the concentrations of Mn-$2p$ and Mn-$3p$ ions are approximately equal. Ferromagnetic coupling mediated by bound holes in the strongly localized regime was also predicted within a tight-binding approximation for a neutral or singly ionized pair of substitutional Mn acceptors in GaAs, neglecting on-site Coulomb repulsion U and intrinsic antiferromagnetic interaction between Mn-$2p$ spins [124].

Zener noted the role of band carriers in promoting ferromagnetic ordering in the 1950s. This ordering can be viewed as driven by the lowering of the carriers' energy associated with their redistribution between spin subbands, split by the $sp$-$d$ exchange coupling to the localized spins in DMS. The sign of this interaction between localized spins oscillates with the spin-spin distance like the Ruderman–Kittel–Kasuya–Yosida (RKKY) interaction. However, the Zener and RKKY models were found to be equivalent within the mean-field approximation. These approximations are valid as long as the period of RKKY oscillations $R = \pi / k_F$ is large compared to an average distance between localized spins. Hence, the technically simpler mean-field Zener approach is meaningful in the regime usually relevant to DMS for a lower hole concentration. Owing to higher DOS and larger exchange coupling to Mn spins, holes are considerably more efficient in mediating spin-dependent interactions between localized spins in DMS. This hole-mediated Zener–RKKY ferromagnetism is enhanced by exchange interactions within the carrier liquid. Such interactions account for ferromagnetism of metals (the Stoner mechanism) and contribute to the magnitude of the $T_C$ in DMS. From the aforementioned arguments and mean-field theory, the Curie Temperature of DMS, $T_C = T_F - T_{AF}$ may be written as:

$$T_C = x_{eff} N_0\, S(S+1) \mathcal{A}_F \tilde{\chi}_c \beta^2\, / (3 k_B) - T_{AF} \tag{4.1}$$

where:

$T_{AF}$    is the antiferromagnetic temperature due to superexchange interactions

$x_{eff}$    is the effective Mn impurity concentration

$N_0$    in Equation 4.1 as defined before is the cation concentration

$S$    is the spin quantum number of the local magnetization

$\mathcal{A}_F$    is an enhancement factor

$\beta$    is the average exchange integral

$\tilde{\chi}_c = m^*_{DOS} k_F\, / (\pi^2 \hbar^2)$ is the DOS for intraband spin excitations at the Fermi level

For a detailed review, refer to Ref. [1]. Figure 4.4 displays the Curie temperature $T_C$ vs hole concentration $p$ normalized by the effective Mn density ($x_{eff}$ and hole concentration $p = N_0 x_{eff}$, respectively) for annealed (Ga,Mn) As thin films, where squares and stars represent samples with hole density $p$ determined from a high field Hall effect and ion channeling measurements [125], respectively. The solid curve shows the tight-binding theory [126].

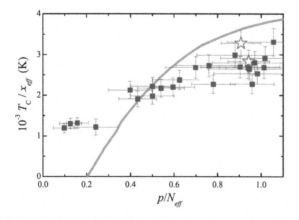

**FIGURE 4.4**  Curie temperature $T_C$ vs hole concentration p normalized by the effective Mn density ($T_C = x_{eff}$ and hole concentration $p = N_0 x_{eff}$, respectively) for annealed (Ga,Mn)As thin films, where squares and stars represent samples with hole density p determined from a high field Hall effect and ion channeling measurements [125], respectively. The solid curve shows the tight-binding theory [127]. (After Wang, M. et al., *Phys. Rev. B* 87, 121301, 2013. With permission.)

### 4.1.3.3 Impact on Tunneling Transport Properties

From a point of view of material properties, questions remain on the general trends of TMR vs. exchange interactions, $\Delta_{exc} = -\frac{5}{2} x N_0 \beta$, where $x N_0$ is the concentration of Mn atoms as well as hole band filling within (Ga,Mn) As. Also, what are the possible effects of the low-temperature (L-T) growth procedure used for the synthesis of these MTJs on the band lineup, VB offset and related barrier heights of heterostructures integrating (Ga,Mn) As? Quite astonishingly and as shown by non-linear I–V characteristics recorded on junctions, both (Ga,As) and (In,Ga)As materials play the role of a tunnel barrier for holes injected from/into (Ga,Mn)As [56–58, 128]. The same qualitative features have been demonstrated through optical measurement of the hole chemical potential in ferromagnetic (Ga,Mn)As/GaAs heterostructures by photoexcited resonant tunneling [129]. These results go in favor of a band-edge discontinuity due to the smaller band gap of (Ga,Mn) As compared to GaAs [130] and indicates a pinning of the Fermi level deep inside the band gap of the (Ga,As) host. A part of the answer lies in the incorporation of $n$-type double-donor As antisites during the low-temperature growth procedure [27] that governs, by part, the pinning of the Fermi level at a higher energy position than expected neighboring the midgap of GaAs. The second reason is due to the positive coulombic-exchange potential experienced by holes and introduced by Mn species playing the role of hydrogenoid centers for holes orbiting around it [26]. This is at the origin of an impurity-band formation at smaller or intermediate doping level in the host bandgap [131], and there is currently an intense debate about the position of the Fermi level relatively to the impurity band [132]. While infrared measurements [133], dichroism (MCD) [134], reflectivity [33], and tunneling spectroscopy [31, 34, 65] experimental and theoretical [135] investigations seem to support the scenario of a detached impurity band, recent low-temperature conductivity [136] and angle-resolved photoemission spectroscopy (ARPES) measurements [30], as well as $\mathbf{k} \cdot \mathbf{p}$ and tight-binding calculations [112] validate the approach of a valence-band picture for (Ga,Mn)As more compatible with a $\mathbf{k} \cdot \mathbf{p}$ treatment of its electronic properties. In the following, we will then emphasize general phase diagrams established from our $\mathbf{k} \cdot \mathbf{p}$ calculations and giving the TMR vs. the hole concentration or (Ga,Mn) As fermi level the exchange energy $\Delta_{exc}$ (or equivalently, the spin splitting parameter $B_G = \Delta_{exc} / 6$ introduced by Dietl et al. [29]) and also will compare some experimental data to our model.

We are going to focus on our own experimental results. Figure 4.5a, and b display TMR results acquired at low-bias (1 mV) and low-temperature (3 K) on $Ga_{0.93}Mn_{0.07}As$(80 nm)/$In_{0.25}Ga_{0.75}As$ (6 nm)/$Ga_{0.93}Mn_{0.07}As$ (15 nm) tunnel junctions. These are patterned in micronic [58] (Figure 4.5b) (standard UV lithography) and submicronic pillars [80] (Figure 4.5a) fabricated by e-beam lithography for spin transfer experiments. Concerning the e-beam lithography process, the fabrication procedure is discussed in section III, which is devoted to spin transfer experiments. These samples exhibit 120% (micronic) and 150% (submicronic) TMR ratios respectively after annealing,

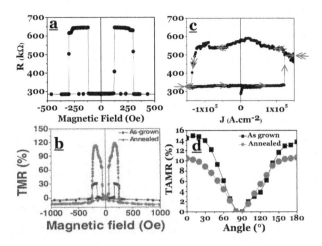

**FIGURE 4.5**    (a) Tunneling magnetoresistance (TMR) measured on 700 nm × 200 nm nanopillar $Ga_{0.93}Mn_{0.07}As$(80 nm)/$In_{0.25}Ga_{0.75}As$ (6 nm)/$Ga_{0.93}Mn_{0.07}As$ (15 nm) at 1 mV and at 3 K. (b) TMR obtained on 128 μm² junction patterned by standard UV lithography on as-grown and annealed samples. (c) Current induced magnetization reversal (CIMS) experiments on submicronic tunnel junction (a) at 3 K. (d) Tunneling anisotropic magnetoresistance (TAMR) experiments measured at 3 K on annealed sample (b). The reference for angle 0° is the direction along the growth axis ($\hat{m} \parallel [001]$).

which are among the highest values found at this temperature with (Ga,Mn) As-based MTJs [12, 56, 57]. The TMR is assumed to be larger for submicronic junctions owing to a better homogeneity of the magnetization configuration. One can notice that the TMR ratio is only 30% for as-grown micronic junctions before any annealing treatment. This highlights a particular change of the (Ga,Mn)As properties. Although the magnetic properties derive from the whole magnetic layers (volume effect), whereas the tunneling process is more sensitive to interfaces, it seems possible, however, to draw some qualitative conclusions. A comparative study of the transport and magnetic properties of these stacks have clearly demonstrated that the rise of TMR observed after annealing conjugated to the decrease of the specific resistance × area (RA) product from $5 \times 10^{-2}$ to $3 \times 10^{-3}$ Ω cm² was correlated to the change of the top (Ga,Mn)As layer properties [58]. In addition, the drop of the RA product conjugated to a reduction of the coercive field (softening of the magnetic film) observed on the top (Ga,Mn)As layer [58, 137–139] unambiguously stem from an increase of the hole concentration in (Ga,Mn) As due to the annealing. The consequences are: (1) a change of the Fermi energy within (Ga,Mn)As; (2) a reduction of the (In,Ga)As barrier height; and (3) an increase of the exchange energy $\Delta_{exc}$ within the top (Ga,Mn)As layer. With a typical exchange integral of $-\frac{5}{2} x N_0 \beta = -1.2$ eV [113], the average spin-splitting in (Ga,Mn)As can be estimated in the mean field approach from the saturation magnetization according to $\Delta_{exc} = 6B_G = \dfrac{A_F \beta M_S}{g \mu_B}$ [29] where $A_F$ is the hole Fermi liquid parameter. Considering that only the top

(Ga,Mn)As layer is affected by thermal treatment, it is then possible to evaluate the increase of the spin splitting $B_G$ parameter from 17 meV to 24 meV [58] and then $\Delta_{exc}$ from 100 meV up to 140 meV for a characteristic $T_C$ of 60 K [58]. We will see in the following section that such range of values for $\Delta_{exc}$ is also compatible with TAMR experiments performed on (Ga,Mn)As-based resonant tunneling devices.

### 4.1.4 EXPERIMENTS VS. 6-BAND K · P THEORY OF TUNNELING SPIN-TRANSPORT

We present in (Figure 4.6) our results of $\mathbf{k} \cdot \mathbf{p}$ calculations (details are given in Appendix B) concerning specific (RA) products of Ga,Mn)As/In$_{0.25}$Ga$_{0.75}$As(6 nm)/(Ga,Mn)As junction vs. the valence band offset (VBO), $d_B$, between (Ga,Mn)As and GaAs or (In,Ga)As. The VBO determines the effective barrier height $\phi$ according to $d_B = -\varepsilon_F + \phi$ (left inset of (Figure 4.6). The top of paramagnetic (Ga,Mn)As valence band ($\Delta_{exc} = 0$) is taken here as the reference for energy, as well as for the Fermi energy. While tunneling experiments estimate the barrier height between metallic (Ga,Mn)As and high-temperature (H-T) grown GaAs in the range between 100 meV [71] and 90 meV [56], photoemission spectra gave a barrier height $\varphi$ of 0.45 eV for L-T grown GaAs [140]. This is in agreement with our 6-band $\mathbf{k} \cdot \mathbf{p}$ model giving a RA product $\simeq 10^{-3}\,\Omega\,\text{cm}^2$ for annealed sampled for $d_B = -0.55$ eV eV and with experimental results of Chiba et al. [56]. Concerning (In,Ga)As, the band offset between In$_{0.25}$Ga$_{0.75}$As and GaAs grown at high temperature is known to be less than 50 meV for samples grown at optimized temperatures [141]. Nevertheless, this VBO has to be taken larger of the order of $d_B = -0.7$ eV

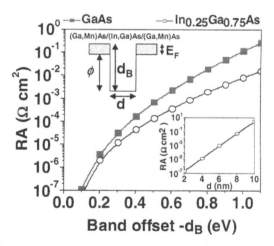

**FIGURE 4.6**    Calculated Resistance × Area (RA) product of trilayer structures vs. the band offset $d_B$ between the ferromagnetic semiconductor and a 6 nm barrier of GaAs or (In,Ga)As. The physical parameters used for (Ga,Mn)As are $\Delta_{exc} = 6B_G = 120$ meV and $n = 1.5 \times 10^{20}$ cm$^{-3}$. Insets: (Bottom) RA product as a function of the (In,Ga)As barrier width $d$. This gives a quasi-exponential variation of the RA product vs. the barrier thickness. (Top) Valence band profile of the heterojunctions.

to match with our calculations. This determines the effective barrier height at about 0.6 eV. The reason is that the real value of VBO (and barrier height) is known to depend on (1) the growth temperature that affects the nature and density of the dangling bonds at interfaces between two semiconductors [142]; and (2) the local density of ionized defects in the barrier (such as As antisites). These generally lead to VB bending susceptible to modify the average barrier height. The bottom inset of (Figure 4.6) gives the exponential dependence of the (RA) product of (Ga,Mn)As/(In,Ga)As/(Ga,Mn) As junctions $\propto \exp(2\kappa d)$ vs. the (In,Ga)As barrier width corresponding to an effective imaginary wave-vector $\kappa_{\text{InGaAs}} \simeq 1.15\,\text{nm}^{-1}$. Experimental results on AlAs barriers with larger effective hole mass (section IV) give an attenuation transmission factor of $\kappa_{\text{AlAs}} \simeq 2.9\,\text{nm}^{-1}$ ($\kappa$ is the evanescent wave in the barrier corresponding to light holes (LH) in the barrier) [12].

Figure 4.7a displays in color code the calculated TMR of Ga,Mn)As/ $\text{In}_{0.25}\text{Ga}_{0.75}\text{As}$(6 nm)/(Ga,Mn)As junctions vs. the spin splitting parameter $B_G$ and Fermi energy $\varepsilon_F$ of (Ga,Mn)As. $B_G$ represents the exchange splitting between two consecutive hole bands at the $\Gamma_8$ symmetry point of the VB. Also plotted on these phase diagrams are the energy of the four first bands at the $\Gamma_8$ point, as well as three different iso hole concentration lines corresponding to $1 \times 10^{20}\,\text{cm}^{-3}$, $3.5 \times 10^{20}\,\text{cm}^{-3}$ and $5 \times 10^{20}\,\text{cm}^{-3}$ as a guide for the readers. High TMR values, up to several hundred percents, can be expected either for spin splitting parameter $B_G$ larger than several tens of meV or for low carrier concentration, that is, when only the first subband is involved in the tunneling transport. This corresponds to a quasi half-metallic character for (Ga,Mn)As. Starting from the first subband ($n = 1$) and increasing the carrier concentration to fill the consecutive lower subbands ($n = 2,3,4$), up and down spin populations start to mix up, leading to a significant drop of TMR. For high carrier concentration ($n = 4$), small TMR is expected which may anticipate difficulties to realize higher $T_C$ and higher TMR. We specify that for low values of spin splitting and Fermi energy, ferromagnetic phase induced by carrier delocalization may not exist (top right corner of the diagram) which is, of course, not taken into account in our 6-band $\mathbf{k} \cdot \mathbf{p}$ model using a basis of propagative envelope wave function. Comparing with our experiments, our estimated value of $B_G$ gives a good qualitative agreement for TMR as illustrated by the trajectory represented in Figure 4.7a between point 1 (as-grown sample) and point 2 (after annealing). TMR ratio obtained in as-grown sample (Figure 4.5b) are well reproduced for a hole concentration in $\text{Ga}_{0.93}\text{Mn}_{0.07}\text{As}$ approaching $10^{20}\,\text{cm}^{-3}$, in rather good agreement with the one measured for single (Ga,Mn)As layer and already reported [143]. From the literature, annealing procedures are known to remove Mn interstitial atoms and increase hole concentration [27]. This should lead to a reduction of the effective barrier height, even if the position of the topmost (Ga,Mn)As valence bands is expected to move upper in the GaAs band gap due to an increase of the exchange term $\Delta_{\text{exc}}$. The strong reduction of the RA product in parallel to a rise of TMR is consistent with such assumption. Nevertheless, the hole concentration extracted here after annealing, $\sim 1.7 \times 10^{20}\,\text{cm}^{-3}$, appears to be slightly smaller compared to the value derived

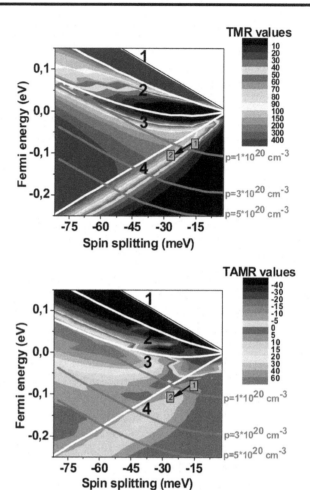

**FIGURE 4.7** Tunnel magnetoresistance TMR (a) and TAMR (b) vs. Fermi energy and spin splitting for a 6 nm (In,Ga)As barrier and a band offset of $d_B = 700$ meV. White lines represent the four bands at the $\Gamma_8$ point. Gray lines indicate the Fermi energy for different hole concentrations.

from Hall effect measurements. The existence of a possible concentration gradient can be at the origin of such discrepancy [144–145]. A reduction of the hole concentration at the interfaces with the barrier due to a significant charge transfer between $p$-type (Ga,Mn)As and $n$-type (In,Ga)As (excess of As antisites) can also be invoked [146, 147].

## 4.2 TUNNELING ANISOTROPIC MAGNETORESISTANCE IN THE VALENCE BAND OF *P*-TYPE SEMICONDUCTORS

### 4.2.1 Overview of TAMR Phenomena

In the absence of spin-orbit coupling, spin and orbital quantum numbers are fully uncoupled. Consequently, no change of the tunnel conductance is expected when the magnetization of the ferromagnet is rotated in space. This assertion starts to change when the spin-orbit coupling is introduced. The combination of spin-orbit coupling and exchange interactions in magnetic tunnel junctions leads to TAMR effects. This can be understood as a variation of the tunneling current vs. the magnetization direction of the ferromagnetic electrode(s). TAMR phenomena can be divided into two different classes according to whether (i) the magnetization always remains perpendicular to the tunneling current leading to in-plane TAMR effects [66, 67]; or (ii) the magnetization rotates from in-plane to out-of-plane directions e.g. parallel to the tunneling current (out-of-plane TAMR) [58, 68]. The physical origin is similar to that of anisotropic agnetoresistance (AMR) effects observed on a single ferromagnetic layer except that in magnetic junctions, the transport of spin-polarized carriers has a tunneling character. In that sense, TAMR can be viewed as a natural extension of AMR in the tunneling regime of transport. In principle, the latter geometry is more favorable to the observation of a larger *intrinsic* TAMR signal, due to a high selectivity in *k*-space by tunneling filtering. First discovered on (Ga,Mn)As-based III–V MTJs, TAMR has since been demonstrated in transition metal-based nanodevices [148–151], magnetic tunnel contacts [152–153], and MTJs [154–156]. TAMR is now the basis of theoretical proposals on resonant tunneling device [157]. For a review on TAMR phenomena, readers can refer to Matos Abiague and Fabian [72, 73].

The strong spin-orbit coupling $\lambda_{so}$ acting in the valence-band of III–V compounds is very favorable to the observation of TAMR phenomena. Significant TAMR effects have been observed on (Ga,Mn)As-based tunnel junctions in the respective in-plane [66, 158] and out-of-plane geometry [58, 68]. This also includes huge TAMR effects obtained in GaAs samples doped with low Mn content [67] at the vicinity of the metal-to-insulator (MIT) transition (Figure 4.8). This effect originates from the occurence of an Efros– Shklovskii gap in (Ga,Mn)As at low bias and low temperature [30] controlled by the extent of the hole wave function on the Mn acceptor atoms within a certain depletion region [159]. Using $\mathbf{k} \cdot \mathbf{p}$ calculations, it has been shown that this extent could depend crucially on the direction of magnetization in (Ga,Mn)As. This was explained by the strong anisotropy of hole envelope function around a single Mn in (Ga,As) in a low Mn doping regime where anisotropic exchange interactions mediate a significant difference in the size of hole wave function between $\mathbf{r} \parallel \hat{m}$ and $\mathbf{r} \perp \hat{m}$. This argument holds also in the reciprocal space ($\hat{k}$) as emphasized in the following section. In the present case, this leads to a magnetization-dependent hole integral transfer tuning the system from an insulating phase into a metallic state by changing the magnetization direction. This anisotropy of the hole envelope wave function around a single Mn center calculated by an exchange-including tight-binding method leading to huge TAMR effects near the MIT transition is sketched in Figure 4.9 [160, 161].

An unconventional TAMR bias dependence, with sign reversal, has been evidenced on (Ga,Mn)As/GaAs($n^+$) Esaki diodes [69]. This exhibits a crossover between positive and negative TAMR signals. This particular feature was also observed in our group (Figure 4.9a) on (Ga,Mn)As/AlAs junctions with opposite convention for the definition of TAMR giving rise to an

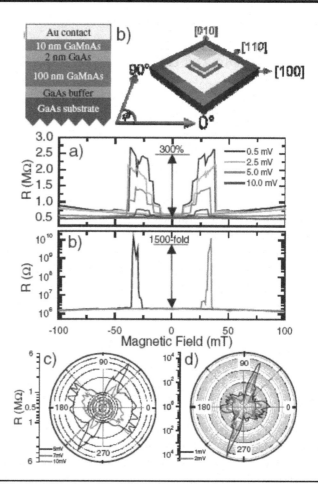

**FIGURE 4.8** Amplification of the TAMR effect at low bias voltage and lox tempera-tures. (a) TAMR along $\phi = 65°$ at 4.2 K for various bias voltages. (b) Very large TAMR along $\phi = 95°$ at 1.7 K and 1 mV bias. (c), (c) $\phi$ at various bias at 1.7 K. (Top) Layer stack under study. (After Ruster, C. et al., *Phys. Rev. Lett.* 94, 027203, 2005. With permission.)

opposite sign. The TAMR changes from positive values at positive bias to negative values for negative bias, that is, when holes are injected from the GaAs:Be electrode towards lowest hole energy subbands corresponding to $n = 4$ (inset of (Figure 4.10)).

## 4.2.2 TAMR WITH HOLES

Physical phenomena explaining TAMR with holes can be understood with the following arguments:

1. The thick tunnel barriers (AlAs, GaAs, InGaAs) transmit easily any LH-projection along the zone axis (or growth direction labeled by $\hat{z}$) and switch off any HH-components because of their well larger effec-tive mass.
2. As emphasized below, the exchange interactions mediate a strong anisotropic LH$\leftrightarrow$HH mixing. Since holes are mainly spin-polarized

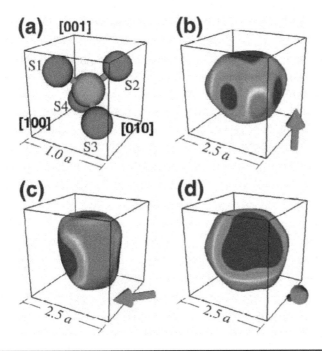

**FIGURE 4.9** Atomic structure close to a single Mn atom in substitional site in GaAs host. As atoms are indicated by S1, S2, S3 and S4. Local spin density surfaces for Mn magnetization along [001] (b), [1̄10] (c) and [111] (d) showing the anisotropy of the hole envelope wave function. (After Tang, J.M. et al., *Phys. Rev. Lett.* 92, 047201, 2004. With permission.)

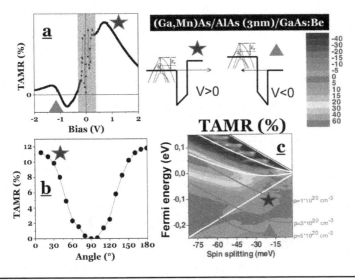

**FIGURE 4.10** (a) Out-of-plane TAMR vs. bias acquired on a (Ga,Mn)As/AlAs(3 nm)/GaAs:Be tunnel junction. Positive biases correspond to hole injected from (Ga,Mn)As towards GaAs:Be. (b) Angular variation of the resistance acquired at a positive bias of +1 V (star) displaying a positive TAMR. The angle 0° corresponds to $\hat{m} \parallel [001]$ (growth axis). (c) Calculation of the corresponding TAMR with the same parameters for (Ga,Mn)As as the one used for Figure 4.6. The injection in the negative bias regime implies an of holes towards at high energy towards the lowest heavy hole band (triangle) thus giving rise to a negative TAMR.

along their propagation direction (at least in the limit of a large spin-orbit coupling $\lambda_{so} \gg \varepsilon_F$) the LH $\leftrightarrow$ HH conversion is efficient for wave-vectors $\hat{k}$ perpendicular to the magnetization $\hat{m}$. This derives from the introduction of non-diagonal exchange terms in the overall Hamiltonian $H_h$ [29].

3. Such arguments explain the large positive TAMR associated to the LH subbands near the $\Gamma$ point ($k \parallel \hat{z}$): LH-to-HH conversion along $\hat{z}$ is forbidden for out-of-plane magnetization ($\hat{m} \parallel \hat{z}$) while it is possible for in-plane magnetization ($\hat{m} \parallel \hat{x}$). Inversely, the same argument yields a large negative TAMR for subbands of HH symmetry near the $\Gamma$ point.

Our calculations performed on (Ga,Mn)As/AlAs(3-nm)/(Ga,Mn)As tunnel junctions seem to corroborate such conclusions. In Figure 4.10c, one can observe a change of sign for TAMR on crossing the third subband, that is, when the LH subbands become dominant in the tunneling transport (star in Figure 4.10c). The first subband clearly gives a negative contribution to TAMR, as does the fourth one, which have the effect of reversing the overall TAMR sign at higher energy (triangle in Figure 4.10c). This originates from the predominant heavy hole character of these subbands, an in-plane magnetization allowing, through off diagonal components, a possible heavy to light hole conversion, and then a larger transmission through the barrier [71]. This argument is reversed for the second and third subbands, with the result that TAMR becomes positive when the $n = 2$ and $n = 3$ subbands are dominant in the tunneling transport. This was theoretically established through tight-binding treatment of TAMR [162]. Reducing the hole concentration through hydrogenation technics should give the possibility of probing this possible crossover from positive to negative TAMR [163]. Coming back to experiments on (Ga,Mn)As/(In,Ga)As/(Ga,Mn)As junctions, taking into account conjugate TMR and TAMR values obtained before and after annealing, one can roughly evaluate the projection of the corresponding signal trajectories in the $[\varepsilon_F, B_G]$ plane followed during annealing (Figure 4.7b). A good qualitative agreement can be found even though symmetrical junctions were simulated in order to restrict the number of parameters.

### 4.2.3 TUNNELING SPECTROSCOPY IN THE VALENCE BANDS OF (GA,MN)AS: RESONANT TAMR IN DOUBLE TUNNEL JUNCTIONS

In order to go a little further in understanding TAMR phenomena with holes, we will now discuss results of resonant TAMR effects observed on AlAs/GaAs/AlAs RTDs spin-filtered by a (Ga,Mn)As electrode [71]. In this work, large TAMR oscillations correlated to the *on/out-of* resonance of QW's quantized states have been observed up to 60 K (inset of Figure 4.11a) corresponding to the $T_C$ of (Ga,Mn)As-based RTDs consisting of 50-nm (Ga$_{0.94}$, Mn$_{0.06}$) As/with a 5-nm undoped GaAs spacer/5-nm AlAs barrier/6-nm GaAs QW/5-nm AlAs barrier/5-nm undoped GaAs spacer/50-nm Be

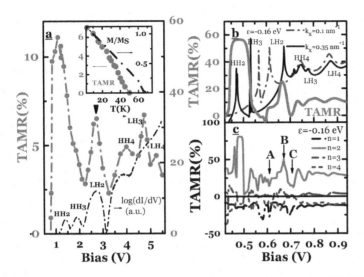

**FIGURE 4.11**    (a) TAMR–V signal (dash-dot red line) and dI/dV–V (dashed line) acquired on the 6-nm GaAs QW. The left axis displays the measured TAMR; the right one takes into account the up-renormalization (see text). Inset of (a) TAMR-T acquired at 2.75 V (LH2 peak) and M-T. (b) TAMR–V calculated (straight red line) and transmission coefficient (unit non shown) calculated at an average energy $\varepsilon_h = -160$ meV and for respective small $k_{\parallel}$ (dotted blue line) and intermediate $k_{\parallel}$ (dashed black line). (c) energy & band-resolved TAMR–V calculated for $\varepsilon_h = -160$ meV. (After Elsen, M. et al., *Phys. Rev. Lett.* 99, 127203, 2007. With permission.)

doped ($p = 1 \times 10^{17}$ cm$^{-3}$) GaAs/50-nm Be doped ($p = 1 \times 10^{17}$ cm$^{-3} \rightarrow 2 \times 10^{19}$ cm$^{-3}$) GaAs/$p^+$ GaAs substrates, from top to bottom. The whole structure was grown at an optimal temperature of 600°C, except the top (Ga,Mn) As layer which was synthesized at 250°C to avoid the formation of MnAs precipitates. Figure 4.10a displays resonant tunnel conductance (dI/dV–V curves in logarithmic scale) and TAMR vs. bias acquired on the 6-nm QW 16 μm² mesa under a saturation field of 6 kOe. Positive bias still corresponds to holes injected from the (Ga,Mn)As layer into GaAs:Be. Note, however that, (i) the differential conductivity is broadened near the resonances due to the large hole density within the (Ga,Mn)As injector; and that (ii) the two first HH1 and LH1 resonances are hardly visible. Comparing with previous results of Ohno et al. [164], a series of negative differential conductance peaks is clearly evidenced with successive contributions of heavy (HH$n$) and light hole (LH$n$) hole states. TAMR data have been obtained by plotting the relative difference of the tunneling current, $j_T$, measured respectively for out-of plane ($j_T^{[001]}$), and in-plane (Ga,Mn)As magnetization ($j_T^{[100]}$) according to $TAMR = \left[ j_T^{[001]} - j_T^{[100]} \right] / j_T^{[100]}$ ($\hat{z} = [001]$ is the growth direction).

For positive bias $V > 0$, we very interestingly observe positive TAMR oscillations characterized by successive peaks and dips whose positions are strongly correlated to the successive resonances. Note, however, that the resonance appears at bias much larger than twice the quantized energy of the discrete levels in the quantum wells. This discrepancy originates from the introduction of a resistance $R_{DZ}$ in series with the double barrier structure in the GaAs:Be electrode because of a gradient in the doping [71]. While the measured TAMR amplitude appears rather moderate (10% or less), the overall *intrinsic* TAMR signal is much larger, by a factor of $\approx 5$, (right scale of Figure 4.11a) when one takes into account the presence of such nonmagnetic serial resistance $R_{DZ}$. Inset of Figure 4.11a displays the temperature dependence of the TAMR peak acquired at 2.75 V, corresponding to resonant tunneling into the LH2 state. The signal amounts to 7% at 4 K and decreases gradually with temperature to vanish at about 50 K close to the (Ga,Mn)As Curie temperature (M-T curve). This demonstrates that the oscillatory behavior of TAMR is induced by the exchange interaction within (Ga,Mn)As layer. One can thus discard (1) any magneto-tunneling effects deriving from the quantization of Landau levels on QW states, as well as (2) deflection of the hole parallel wave-vector by Lorentz force in the plane of the junction [165]. Moreover, both mechanisms would hardly appear at such intermediate field (6 kOe). Our conclusion is that such an oscillatory TAMR behavior reflects the spin-dependent transmission coefficients of the polarized holes injected from (Ga,Mn)As, thus emphasizing the VB anisotropy of the (Ga,Mn)As. Note, however, that this can include an extension of mechanism (2) i.e a deviation of the hole wave-vector along the in-plane $\hat{y} = [010]$ direction normal to the in-plane $\hat{x} = [100]$ magnetization driven by spin-orbit interactions. We will come back to this very important point at the end of this section.

On the basis on the aforementioned arguments on differential transmission of HH and LH-components, one can suggest that the oscillation character of TAMR on/out-of resonance may result from the 4th heavy hole (Ga,Mn)As subband contribution. To this end, we have plotted in Figure 4.11b the transmission amplitude of holes of average energy $\varepsilon_h = -100$ meV ($-160$ meV from the top of the valence band) corresponding to $p = 2 \times 10^{20}$ cm$^{-3}$ for two different parallel wave-vectors $k_{\parallel}$, respectively small $k_x = 0.1$ nm$^{-1}$ (dot curve) and intermediate $k_x = 0.35$ nm$^{-1}$ (straight curve) across the double barrier structure. $\varepsilon_h = -160$ meV is then assumed to be the elastic energy transmitted during the tunneling process. The resulting TAMR shape, displayed in Figure 4.11c, is in excellent agreement with the experimental data exhibiting the successive peaks and dips correlated with the successive positions of resonances. In addition, the magnitude of TAMR is also well-reproduced after renormalization by a factor of 5 (right scale of Figure 4.11a). We observe that most of the TAMR peaks correspond to a specific resonance at intermediate $k_{\parallel}$ (what we call *out-of*-resonance) whereas the TAMR dips match generally with a specific resonance at small $k_{\parallel}$ (what we call *on*-resonance). We then considered the respective contributions of the different holes subbands to the TAMR. The four band resolved TAMR signals extracted for $\varepsilon_h = -160$ meV (still normalized by the *total* tunneling current for an in-plane magnetization) are plotted in Figure 4.10c. The band indexes $n = 1, 4$ *resp.* $n = 2, 3$ stand for the HH *resp.* LH symmetries near the $\Gamma$ point. Quite remarkably, the main contribution to the TAMR oscillations derives from the first LH subband ($n = 2$) exhibiting the same trends as the summed signal

(Figure 4.11b) as well as the experimental data (Figure 4.11a). Let us focus on the specific LH2 resonance (Figure 4.11c) where we hereafter designate $V$ as the effective bias applied on the double barrier structure. The second subband contribution to TAMR is small, about 10%, *on*-resonance ($V = 0.59$ eV: point A in Figure 4.11c, increases up to about 60% *out-of*-resonance for intermediate $k_{\parallel}$ ($V = 0.68$ eV: point B) before vanishing at even larger bias and larger $k_{\parallel}$ ($V = 0.71$ eV: point C).

The following demonstration is based on the calculations of the band-selected tunnel transmissions mapped onto the different (Ga,Mn)As Fermi surfaces as sketched in Figures 4.12 and 4.13. These figures display the different band-selected transmission for respective in-plane (Figure 4.12: $\hat{m} \parallel \hat{x}$) and out-of-plane magnetization (Figure 4.13: $\hat{m} \parallel \hat{z}$). The transmission coefficients $\sum_{n'} \mathbf{T}_{n'}^{k_{\parallel}, \hat{m}}(\varepsilon_h)$ mapped on the two-dimensional $k_{\parallel}$ Fermi surfaces for the two magnetization configurations are plotted in Figures 4.12 and 4.13a–l for the four subbands. In both figures, the indices a–d correspond to the mapping of transmission calculated for a bias $V = 0.59$ eV, indices e–h for a bias $V = 0.68$ eV and indices i–l for a bias $V = 0.71$ eV.

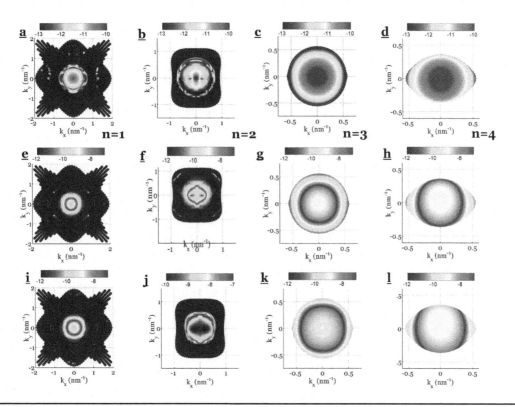

**FIGURE 4.12** Band-selected ($n = 1 - 4$) tunnel transmissions (in color logarithmic scale) mapped onto the different (Ga,Mn)As Fermi surfaces calculated for in-plane magnetization ($\hat{m} \parallel \hat{x}$). The different columns corresponds to a specific selected band. Calculations have been performed for a bias applied to the double barrier structure equal to $V = 0.59$ eV (a-d: first line), $V = 0.68$ eV ((e)–(h): second line) and $V = 0.71$ eV ((i)–(l): third line) around the LH2 resonance.

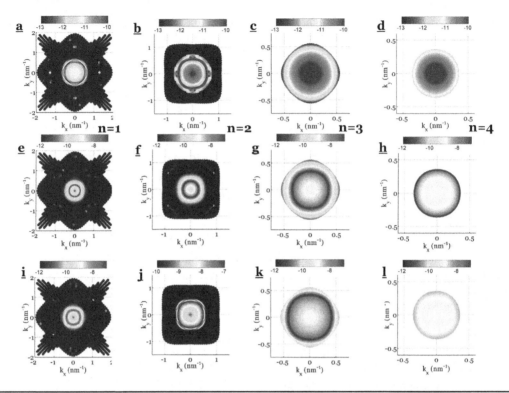

**FIGURE 4.13** Band-selected ($n = 1–4$) tunnel transmissions (in color logarithmic scale) mapped onto the different (Ga,Mn)As Fermi surfaces calculated for in-plane magnetization ($\hat{m} \parallel \hat{x}$). The different columns corresponds to a specific selected band. Calculations have been performed for a bias applied to the double barrier structure equal to $V = 0.59$ eV ((a)–(d): first line), $V = 0.68$ eV ((e-h): second line) and $V = 0.71$ eV ((i-l): third line) around the LH2 resonance.

1. *On LH2 resonance* ($V = 0.59$ eV): *point A)* On Figures 4.11 and 4.12a–d, the bright spot with a relative small radius at the center indicates that the tunnel transmission is mainly driven by small $k_{\parallel}$. The band-resolved TAMRs obtained by $k_{\parallel}$-integration approaches +125% (*resp.* −38%) for $n = 2$ (*resp. n = 4*), whereas it gives −38% (resp. +3%) for $n = 1$, (*resp. n = 3*). Correspondingly, the weighted tunneling currents calculated for an in-plane magnetization are respectively 18% and 22% for $n = 2$ and $n = 4$ (30% for both $n = 1$ and $n = 3$). The overall TAMR is less than 10%, since *the negative HH contribution (n = 1, 4) is partially compensated by the positive LH one (n = 2)*.

2. *Out-of-resonance at $V = 0.68$ eV: point B)* The tunneling transmission in the reciprocal space is now mainly confined within a ring of an intermediate radius from the zone axis e.g. driven by intermediate $k_{\parallel} \approx 0.35$ nm$^{-1}$ (Figures 4.12 and 4.13e–h. Due to the finite extension of the minority spin HH hole Fermi surface ($n = 4$), the resonance condition can be only partially achieved for this particular band: the tunnel conductance falls off. Each weighted tunnel conductivity is then 33%, 30%, 33% and 4% for $n = 1–4$ whereas the corresponding TAMRs equals now −30%, +180%, 20% and −90%. The consequence

is an overall increase of the TAMR up to 60%, due to the deficit of negative TAMR from the minority spin HH hole subband. The overall results are gathered in Table 4.1

3. *At higher bias $V = 0.71$ eV: point C)* The tunnel transmission is driven by holes of even larger $k_{\parallel}$ ($\approx 0.5$ nm$^{-1}$) away from the zone axis $\hat{z}$ (Figures 4.11 and 4.12i–l). It results in a strong LH↔HH mixing along $\hat{z}$ which, in turn, strongly reduces selected TAMR down to –22%, +40%, 9% and –55% for $n = 1 - 4$ (Figure 4.11c). The weighted tunnel conductivities are respectively 60%, 30%, 7% and 3%. As a result, the overall TAMR is low due to the strong reduction of the second subband LH contribution (from +180% to +40%), as shown in Figures 4.11j and 4.12j. The conclusion is that the oscillatory behavior of TAMR originates from the two merging mechanisms established at the beginning of the section and based on the HH and LH character of each subbands. One can then argue that acquiring a large $k_{\parallel}$ *out-of*-resonance naturally mixes the LH and HH characters resulting in a significant TAMR drop (*from 180% to 40%*). Conversely, the argument (*i*) yields a large negative TAMR for subbands of HH symmetry at intermediate or small $k_{\parallel}$ before decreasing at larger $k_{\parallel}$. The ensemble of band-selected *intrinsic* TAMR and their corresponding contribution to the resonant tunneling current are reported in Table 4.2 for respective *on* and *off*-resonant transmission on LH2 peak. The overall results are gathered in Table 4.2.

In addition, the high-resolution of resonance peaks obtained at negative bias $V < 0$ (not shown), that is, when holes are injected from GaAs:Be into (Ga,Mn)As, allows the investigation of other superimposed effects related

**TABLE 4.1**

**Band-Restricted TAMR and Contribution to the LH2 on-Resonance Tunneling Current ($V = 0.59$ eV). The Kinetic Energy of Holes in (Ga,Mn)As is Equal to 160 meV. (HH) and (LH) Indicates the Heavy or Light Character for Holes Near the $\Gamma_8$ Point**

|  | $n=1$ (HH) | $n=2$ (LH) | $n=3$ (LH) | $n=4$ (HH) |
|---|---|---|---|---|
| TAMR | –38% | +125% | +3% | –38% |
| Current | 30% | 18% | 30% | 22% |

**TABLE 4.2**

**Band-Restricted TAMR and Contribution to the LH2 *off*-Resonance Tunneling Current ($V = 0.68$ eV). The Kinetic Energy of Holes in (Ga,Mn)As is Equal to 160 meV. (HH) and (LH) Indicates the Heavy or Light Character for Holes Near the $\Gamma_8$ Point**

|  | $n=1$ (HH) | $n=2$ (LH) | $n=3$ (LH) | $n=4$ (HH) |
|---|---|---|---|---|
| TAMR | –30% | +180% | +20% | –90% |
| Current | 33% | 30% | 33% | 4% |

to new TAMR phenomena. A significant resonant-TAMR signal has been highlighted in asymmetric (Ga,Mn)As-based double barrier structure at negative bias [166]. Such experiments exhibiting a magnetization-dependent shift in bias of the consecutive HH$n$ and LH$n$ resonances seem to originate from a significant modulation of the Fermi level of (Ga,Mn)As imposed by the anisotropic strain field. In that sense, this could be closely related to Coulomb-blockade magnetoresistive (CB-TAMR) phenomena [167] in the in-plane geometry, although in the latter case, a full understanding of CB-TAMR is missing.

## 4.3 SKEW-TUNNELING AND ANOMALOUS TUNNEL HALL EFFECT–UNIDIRECTIONAL SPIN-HALL MAGNETORESISTANCE WITH (GA,Mn)AS

In order to introduce the notion of skew-tunneling phenomena, note that (Figure 4.12a–l) display in general a clear non-cubic asymmetry of the tunneling coefficients for in-plane magnetization ($\hat{m} \parallel \hat{x}$). This results from the action of the spin-orbit coupling on hole motion during their tunneling transfer. This can be viewed as a generalization of the Lorentz force to magnetic systems by which a deviation of the hole wave-vector along the in-plane $\hat{y} = [010]$ direction normal to the $\hat{x} = [100]$ magnetization vector is expected [64]. From the general expression of the spin-orbit coupling $\mathcal{H}_{so} = -\lambda_{so}(\vec{\nabla} V(\vec{r}) \times \vec{p}) \cdot \vec{S}$, it results in a potential discontinuity due to band offset ($\vec{\nabla} V(\vec{r}) \propto \vec{z}$) which introduces a supplementary energy term $\mathcal{H}_{so} \propto -\lambda_{so}(\vec{z} \times \vec{p}) \cdot \vec{S} = -\lambda_{so}(\vec{S} \times \vec{z}) \cdot \vec{p}$ breaking the in-plane symmetry as soon as $\vec{S}$ admits a non-vanishing in-plane component. This term is at the origin of the transmission asymmetry between $\pm k_y$ in-plane wave-vectors. In the envelope function approximation, this effect is implicitly included in the boundary conditions for the wave functions and their derivative (current operator). Such an asymmetry in the transmission coefficient is at the origin of what should be designated as *tunneling (Spin) Hall effect* (Figure 4.14).

On the hand, recent spin-Hall magnetoresistance measurements with an unidirectional character (U-SMR) have been evidenced in two types of bilayer systems either metallic bilayers Co/Pt involving Pt [53] and the (Ga,Mn)As in (Ga,Mn)As/GaAs:Be systems [4] (see Figure 4.15). The same kind of effect, i.e. a current-direction dependent contribution to the resistance, was very recently reported on magnetic or nonmagnetic topological insulator (TI) heterostructures, $Cr_x(Bi_{1-y}Sb_y)_{2-x}Te_3 / (Bi_{1-y}Sb_y)_2Te_3$, effect of several orders of magnitude larger than in other systems (Ref. [54] and Figure 4.15). From the magnetic field and temperature dependence, the UMR is identified to originate from the asymmetric scattering of electrons by magnons [15(a)]. In particular, the large magnitude of UMR is an outcome of spin-momentum locking and a small Fermi wave number at the surface of TI. In particular, this unidirectional character of the magnetoresistance manifests itself in a change of the resistance by switching the magnetization

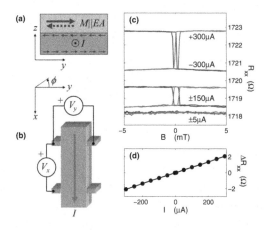

**FIGURE 4.14**    (a) Schematic of the U-SMR phenomenon. Thin arrows represent the SHE-induced spin polarization by non Ferromagnetic Mn-doped GaAs; thick arrows represent the easy-axis (EA) magnetization of the (Ga,Mn)As ferromagnet. (b) Schematic of the device and measurement geometry. (c) Longitudinal resistance measurements at 130 K and different amplitudes and signs of the applied current as a function of the external magnetic field. Steps correspond to the 180° magnetization reversal. (d) Difference between non-linear resistance states vs. current by switching to opposite transverse magnetizations, set by sweeping the magnetic field from negative or positive values to the zero field, as a function of the applied current. (After Olejník, K. et al., *Phys. Rev. B* 91, 180402(R), 2015. With permission.)

transverse to the current flow from one direction to the other. Equivalently, this magnetoresistance contribution changes linearly with the injected in-plane current and its sign follows the sign of current. It clearly reveals a novel symmetry in the field of magnetoresistance.

This is explained by some chirality arguments incorporated in the equation of transport (Boltzmann). If there exists a non-symmetric probability of magnon-diffusion for electrons or holes on the Fermi surface of the TI depending on their incident wave-vectors ($\pm k_{\parallel}$) along the current direction perpendicular to the (transverse) magnetization, the resistance acquires a supplementary non-linear contribution proportional to the current.

Indeed, the Boltzmann equation, $\dfrac{df(k,t)}{dt} = 0$, for an in-plane current (CIP geometry) derived at the second order of the electric field writes for the respective fermi distribution $f$ and its out-of-equilibrium part $g$ (Fermi surface displacement) as:

$$\delta(\varepsilon - \varepsilon_F)ev_F E\cos\theta + \frac{\partial\delta(\varepsilon - \varepsilon_F)}{2\partial\varepsilon}\tau(\theta)(ev_F E)^2\cos^2\theta = \frac{g(\theta)}{\tau(\theta)} \qquad (4.2)$$

with $v_F$, the fermi velocity, and $E_x$ the longitudinal electric field (along the $x$ direction). This leads to an additional $E^2$-dependence contribution to the current-distribution $g$ once the momentum relaxation time $\tau$ depends on the scattering angle along the respective forward and backward directions ($\theta$ angle parameter). This angle dependence on $\theta$ is

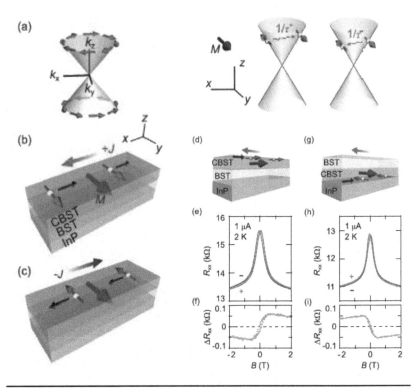

**FIGURE 4.15**   (a) Schematic diagram of spin-momentum locking of the surface Dirac state in TI and principle of magnon-asymmetry scattering process. (b), (c) Schematic illustration of the concept for UMR in TI heterostructures $Cr_x(Bi_{1-y}Sb_y)_{2-x}Te_3 / (Bi_{1-y}Sb_y)_2Te_3$ (CBST=BST) on InP substrate under +J (b) and -J (c) dc current. Here, magnetic field, magnetization, and dc current are along the in-plane direction, where dc current is applied perpendicular to the magnetization direction. (d) Schematic illustration of a "normal" CBST=BST heterostructure. (e) Magnetic field dependence of resistance $R_{xx}$ for the sample depicted in (d), measured under $J=1$ µA (red) and $J=-1$ µA (blue) at 2 K. (f) Difference of the resistance $\Delta R_{xx}$ of plus and minus current shown in (e), (g)–(i). The same as (d)–(f) for the "inverted" BST/CBST heterostructure.

responsible for a supplementary longitudinal conductivity term on the form $\Delta\sigma_{xx}^{(2)} = 1/\rho_{xx}^{(2)} = \dfrac{3}{4}\sigma^{(1)}\dfrac{\mu E\delta\tau_{an.}}{v_F\bar{\tau}}$ where $\delta\tau_{an.}$ is the anisotropic part of the scattering time between forward and backward electronic scattering and where $\sigma^{(1)}$ is the linear conductivity. We will give some particular insights on the properties of the anisotropic part of the scattering time $\delta\tau_{an}$ further on in the next chapter on the focus of the (Ga,Mn)As compound.

Such a carrier diffusion asymmetry on the Fermi surface may then lead to U-SMR effects when one considers that SOI are present in one of the bilayers, at least [50, 51]. We have recently investigated, by theory, this kind of asymmetry phenomena in tunneling transport in the CB of III–V semiconductors contacted by a ferromagnetic (Ga,Mn)As layer [51]. The result is quite spectacular, and in particular, in some selected points of the Brillouin zone. In the case of the VB, one expects even larger asymmetry of transmission/reflection/diffusion in tunneling owing to the larger strength of the SOI.

This effect should also manifest itself by a surface anomalous tunneling Hall effect when summed over the Fermi surface as evidenced in ferromagnetic GeMn nanocolumn embedded in Ge [55]. The next part is devoted to the description of tunneling scattering asymmetry (Figure 4.16) with holes in the VB of (Ga,Mn)As-based III–V heterostructures constituting a new type of spinorbitronics phenomenon. We will consider the most favorable case of bilayer or tunnel junctions, that is, a given interface between the two materials, with same ferromagnetic contacts in the AP magnetic configuration. We work in the current perpendicular-to-plane geometry for the transport.

### 4.3.1 Skew Transmission and Skew Tunneling in the VB from Atomic SOI: A Simplified Perturbative Approach

We will now prove, by analytical and numerical methods, that the asymmetry of tunneling transmission and related anomalous tunnel Hall effect (or skew tunneling) also occurs in the VB of semiconductor junctions involving core atomic SOI, $H_{SO} = \dfrac{\hbar}{4m_0^2 c^2}(\nabla \mathfrak{U} \times \mathbf{p}) \cdot \boldsymbol{\sigma}$. Quite astonishingly, the tunneling scattering asymmetry may then observed in $p$-type orbitals without invoking the assist of any odd-potential spin-orbit like played a *Rashba* or *Dresselhaus* field. We demonstrate that such a chiral property in the transport using (Ga,Mn)As [23, 29, 83] may be understood via a perturbative scattering approach. This type of chirality phenomena in transport may explain the giant Hall-effect observed in GeMn with a nanocolumn growth [55], and

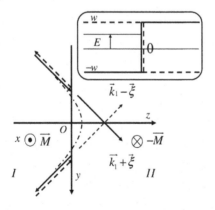

**FIGURE 4.16** Scheme of scattering (or transmission) asymmetry process at an exchange-SOI interface of a bilayer with AP magnetizations **M** and − **M** along x. The propagation direction of carriers (straight arrow) is along z with propagative wave-vector $k_1$ whereas the in-plane incident component $k_∥ = +xi$ (heavy line) or $k_∥ = −\xi$ (dashed line) is along y; *xyz* forms a direct frame. The dash-dotted curve denotes the evanescent waves, either reflected or transmitted. Carriers with a $k_∥ = +\xi$ in-plane wave-vector component are more easily transmitted than those carrying $−\xi$. Top right inset: Energy profile of the exchange step; E is the longitudinal kinetic energy along z and 2w is the exchange splitting in the magnetic materials. (After Dang, T.H. et al., *Phys. Rev., B* 92, 060403(R), 2015. With permission.)

the recent U-SMR observed with (Ga,Mn)As [4] or with topological insulators [54] via magnon-assisted asymmetric diffusion process. We define the asymmetry of tunneling transmission, $\mathcal{A}$, for different in-plane component of the incident wave-vector as:

$$\mathcal{A} = \frac{T_{k_\parallel} - T_{-k_\parallel}}{T_{k_\parallel} + T_{-k_\parallel}} \tag{4.3}$$

In the case of holes:

1. $\mathcal{A}$ appears without specific odd-symmetry potentials but combined effects of *pure atomic spin-orbit and exchange interactions* through the matching rules.
2. The VB brings a supplementary complexity in terms of LH-to-HH band-mixing, each of them being doubly degenerated due to the spin degree of freedom. LH and HH-subbands generally mix together at oblique incidence upon reflection/transmission at interfaces.
3. Under some assumptions, a simple perturbative method may be implemented in the VB along the guideline developed for the CB. We consider a $6 \times 6$ Kane $\{X, Y, Z\} \otimes \{\uparrow, \downarrow\}$ basis functions. By rotation, this basis can be made equivalent to the $6 \times 6$ $\mathbf{k} \cdot \mathbf{p}$ Luttinger basis [29]. Here, *one considers atomic SOI only in the barrier (perturbation) and neglects SOI in the contacts.* The complete investigation including SOI in electrodes will reveal the perfect agreement between the effect and size to expect with the analytical approach.

The expression for the correction to the scattering (or transmission coefficient) and its asymmetry after perturbation by the SOI $V_{SO}$ in a multiband picture (case of VB) is:

$$\delta t_{(nm)}^{\sigma'\sigma} = \sum_l \int_0^d \left[ \Psi_{out}^{(n)0\sigma'}(z') \right]^* \frac{V_{SO,nm}^{\sigma'\sigma}(z')}{2E} \Psi_{in}^{(m)0\sigma}(z')\, d\left(k_{(n)}z'\right), \tag{4.4}$$

where $(m)$, and $(n)$ are the subscript corresponding to the multiband structure of the respective *ingoing* and *outgoing* waves and where $\left[ \Psi_{out}^{(n)0\sigma'}(z') \right]$ generally admits a complex form through LH-HH band-mixing. However, for relatively thick barriers, on can disregard the transmission of any HH character. We will restrict our calculations on the LH-bands only. E is the kinetic incoming kinetic energy.

Once the atomic SOI is introduced, $V_{SO}^{\sigma'\sigma} = \lambda_{SO}\mathbf{L} \cdot \mathbf{S} = \lambda_{SO}(\hat{L}_z.\hat{S}_z + \hat{L}_+\hat{S}_- + \hat{L}_-\hat{S}_+)$ where $\hat{L}$ and $\hat{S}$ are the respective orbital and spin operator, the above expression becomes:

$$\delta t_{(nm)}^{\downarrow\uparrow} = \frac{\lambda_{SO}}{2E} \sum_l \int_0^d \left[ \Psi_{out}^{(n)0\downarrow}(z') \right]^* [\hat{L}^+]_{lm} \Psi_{in}^{(m)0\uparrow}(z')\, d\left(k_{(n)}z'\right), \tag{4.5}$$

for an ingoing spin ↑ carrier ($\Psi_{in}^{(m)0\uparrow}$). The result is that the tunneling transmission is connected to the coupling between the (*in*) and (*out*) orbital moments ($L_+$ operator) in the tunnel barrier (chirality).

## 4.3.2 Transmission and Chirality within the 6-Band k · p Kane Model

From Equation 4.5, we have calculated the tunneling transitions from the *ingoing* LH state into the *outgoing* Z state. The result is [168]:

$$\delta T_{LH\to Z}^{\downarrow\uparrow} = \frac{8K_{LH}}{\lambda_{HH}^2 K_{HH}(1+\lambda_{LH}^2)(1+\lambda_{HH}^2)}\left(\frac{\Delta_{SO}}{3E}\right)^2 \exp\left(-2\lambda K_{LH}d\right)$$
$$\times \left(1+\frac{\mathcal{M}k_\parallel}{\mathcal{L}K_{LH}}\right)^2 \tag{4.6}$$

$$\delta T_{Z\to LH}^{\downarrow\uparrow} = \delta T_{LH\to Z}^{\downarrow\uparrow} \tag{4.7}$$

$d$ is the barrier thickness, $K_{LH,HH}$ are the propagative wave-vectors in the contacts for LH and HH and $\lambda K_{LH,HH}$ are the corresponding evanescent wave-vectors in the barrier for the incoming waves (in unit of $K_{LH,HH}$). $\lambda_{LH,HH}$ are then unitless. The transmission is zero at the limit of an infinite HH-mass (large $K_{HH}$) [168]. The limit of the zero-transmission has been observed numerically in the case of 3 nm thick barrier in Figure 4.3 of Ref. [51] of our previous paper in the limit of a small kinetic energy (and then small HH-to-LH mixing). Nonetheless, the asymmetry $\mathcal{A}$, derived from Eqs (4.7) still exists and is given by:

$$\mathcal{A}_{Z\to LH} = \frac{2\lambda_{LH}K_{LH}\mathcal{M}\mathcal{L}k_\parallel}{\left(\mathcal{M}k_\parallel\right)^2 + \left(\mathcal{L}\lambda_{LH}K_{LH}\right)^2} \tag{4.8}$$

SOI-assisted tunneling transitions involve the coupling of the orbital moment, as described by Equation 4.8, with a different probability and efficiency for respective left and right incidence. (Figure 4.17) displays the asymmetry $\mathcal{A}$ Equation 4.8 vs. the hole energy $\varepsilon$ calculated analytically for two different $k_\parallel$ wave-vectors, $\xi = k_\parallel = 10^{-2}, 5\times10^{-3}$ nm$^{-1}$ (black straight line). It is compared with the results of numerical calculations (brown circle symbols) implemented in a $6\times6$ **k·p** model in the limit of infinite HH-mass ($\mathcal{M} - \mathcal{L} = 2$). Those are in perfect agreement which proves again the power of the perturbative scattering method used here. From a vanishing value at zero kinetic energy ($K_{LH} = 0$), $\mathcal{A}$ gradually increases vs. energy when $K_{LH}$ increases. The maximum is reached at a certain energy, just below the barrier step, corresponding to $\lambda_{LH}K_{LH} = \frac{\mathcal{M}}{\mathcal{L}}k_\parallel$ and then $\mathcal{A}$ decreases down to

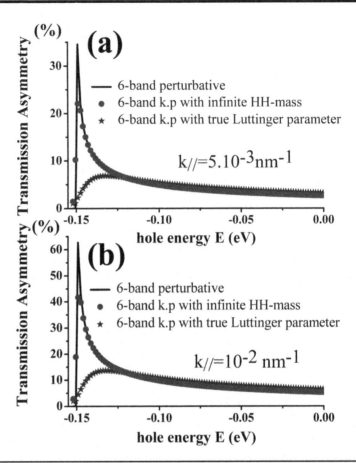

**FIGURE 4.17**    Asymmetry of transmission $\mathcal{A}$ for holes vs. hole energy obtained by perturbative scattering methods (black straight line) and numerical computations (brown circle symbols) with a 6-band tunneling code for the case of a $p$-type magnetic ⇑ GaAs/3 nm GaAs barrier/$p$-type magnetic ⇓ GaAs. The barrier height is equal to the exchange step of 0.3 eV. The calculation is performed for two different hole incident $\xi = k_{\parallel} = 10^{-2}, 5 \times 10^{-3}$ nm$^{-1}$ in the limit of infinite HH-mass ($\mathcal{M} - \mathcal{L} = 2$). Results corresponding to real Luttinger parameters for GaAs are also represented (purple star symbols).

zero when $\lambda K_{LH}$ becomes small (no evanescent states). The comparison with the situation of real 6-band Luttinger parameters for GaAs is also displayed in (Figure 4.17) (purple star symbol) with a good agreement as far as the kinetic energy for holes is sufficiently small to sustain an evanescent wave in the barrier. This also demonstrates that numerical computations are mandatory to correctly describe the amplitude of $\mathcal{A}$ in some real physical situations, like e.g. the case of (Ga,Mn)As/GaAs/(Ga,Mn)As MTJs like described throughout this review.

### 4.3.3 TRANSMISSION ASYMMETRY IN THE VB FROM A 30-BAND K · P POINT OF VIEW

Figure 4.18 display the results of the two-dimensional (2-D) map calculations, in the ($k_{\parallel} = k_x, k_y$) reciprocal space of the 1st Brillouin zone, of the hole transmission coefficient calculated for a $p$-type (Ga,Mn)As⇑/GaAs 3 nm/ (Ga,Mn)As⇓ magnetic tunnel junction of barrier height $\Phi = 0.3$ eV. The hole total energy is fixed at $\varepsilon = -0.25$ eV counted from the top of the VB. These calculations are based on the multiband scattering-matrix technique developed in Ref. [168], known to be more robust than the transfer-matrix method developed for the 6-band model [58, 85]. The results of the 30-band calculations on the skew-tunneling effects are displayed in Figure 4.18d. We have checked that the 14-band model provides similar data with $P' = 0$ and $\Delta' = 0$ with Figure 4.18c or without Figure 4.18b ghost-band treatment (see Appendix B) as well as the 6-band Luttinger approach Figure 4.18a. The map of the transmission coefficients are then very comparable in each case showing an equivalent transmission coefficient mapping and corresponding transport asymmetry. Note that the maximum scale of the hole transmission for the 14 band, 14 ghost-band, and 30 ghost-band (see Appendix B) are rather similar ($15 \times 10^{-3}$) whereas the 6-band Luttinger model gives are slightly different ($45 \times 10^{-3}$) because of the limit of validity of the effective Hamiltonian model.

Figure 4.19 displays the hole transmission asymmetry $\mathcal{A}$ vs. hole energy $\varepsilon$ in the case of a 3-nm-thick tunnel barrier for $k_{\parallel} = 0.05$ nm$^{-1}$. The agreement is almost perfect between the different multiband $\mathbf{k} \cdot \mathbf{p}$ codes proving again the relevance of our 30-band $\mathbf{k} \cdot \mathbf{p}$ tunneling treatment. The energy range covers the valence spin subbands, namely, starting from the highest

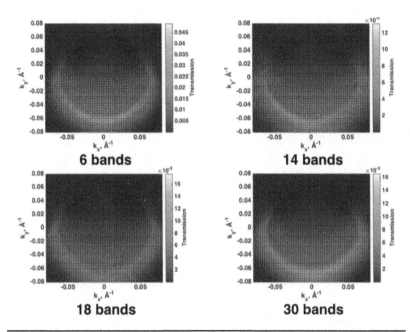

**FIGURE 4.18** Two-dimensional maps of the transmission coefficient $T$ calculated in the VB for a $p$-type magnetic tunnel junction in the $AP$ state by respective $6 \times 6$, $14 \times 14$, $18 \times 18$ (14-ghost-band) and $30 \times 30$ ghost-band $\mathbf{k} \cdot \mathbf{p}$ methods. The parameters are: exchange energy $6w = 0.3$ eV, *total* kinetic energy $\varepsilon = 0.25$ eV counted from the top of the VB; band parameters of the 14-band $\mathbf{k} \cdot \mathbf{p}$ model taken from Ref. [169], and the one for the 30-band from Ref. [170]. The barrier thickness is 3 nm. (After Dang, T.H. et al., *J. Magn. Magn. Mater.* 459, 37, 2018. With permission.)

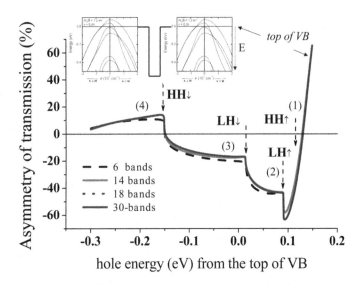

**FIGURE 4.19**    *(Bottom)*: Transmission asymmetry $\mathcal{A}$ vs. *total* energy $\varepsilon$ calculated in the VB for a *p*-type MTJ in the *AP* state by respective $6 \times 6$, $14 \times 14$, $18 \times 18$ (14-ghost-band) and $30 \times 30$ ghost-band $\mathbf{k} \cdot \mathbf{p}$ methods. The parameters are: $6w = 0.3$ eV, parallel wave-vector $\xi = k_{\parallel} = 0.05$ nm$^{-1}$, barrier thickness $d = 3$ nm, and barrier height $0.3$ eV. The energy zero corresponds to the nonmagnetic upper-valence-band maximum.

energy, the up-spin heavy-hole band ($HH \uparrow$), the up-spin light-hole band ($LH \uparrow$), the down-spin light-hole band ($LH \downarrow$), the down-spin heavy-hole band ($HH \downarrow$). The up-spin split-off band ($SO \uparrow$) and the down-spin split-off band ($SO \downarrow$) was considered in a previous contribution [51]. We refer to points (1) to (4) marked by vertical arrows for discussing the contribution from holes emitted from the different spin subbands in Region I to the current injected in Region II. For instance, with these parameters, the energy of the $HH \uparrow$ [$HH \downarrow$] maximum, corresponds to 0.15 eV [−0.15 eV], the energy origin being taken at the top of the valence band of the nonmagnetic material, and is indicated by point (1) [(4)].

Correspondingly, one observes an almost fully negative transmission asymmetry in this energy range for predominant majority spin-up injection, that is, as far as $HH \downarrow$ does not contribute to the current. At more negative energy [$\mathcal{E} < - 0.15$ eV: point (4)], a sign change of $\mathcal{A}$ occurs at the onset of $HH \downarrow$ (in the upper left inset, see the step in the transmission coefficient, which reaches almost +50%). Moreover, calculations give that the asymmetry $\mathcal{A}$ remains positive after crossing $SO \uparrow$ before turning negative again once crossing $SO \downarrow$ [51]. $\mathcal{A}$ then changes sign twice at characteristic energy points corresponding to a sign change of the injected particle spin. We have performed the same kind of calculation for a simple contact [51]. It is remarkable that $\mathcal{A}$, although smaller, keeps the same trends as for the 3-nm tunnel junction, except for a change of sign, showing a subtle dependence of the exchange coupling on the barrier thickness. Without tunnel junction, $\mathcal{A}$

abruptly disappears as soon as $SO \downarrow$ contributes to tunneling [circle region] i.e., when evanescent states disappear. In the case of tunnel junction, $\mathcal{A}$, although small, persists in this energy range and this should be related to the evanescent character of the tunneling wave function in the barrier.

## 4.4 SPIN-TRANSFER IN THE VALENCE BAND OF *P*-TYPE SEMICONDUCTORS

### 4.4.1 SHORT OVERVIEW OF SPIN TRANSFER PHENOMENA WITH OUT-OF-PLANE AND IN-PLANE CURRENT INJECTION

The magnetic moment of a ferromagnetic body can be reversed or re-oriented by transfer of the spin angular momentum carried by a spin-polarized current. This concept of spin transfer has been introduced by Slonczewski [74] and Berger [75], before being confirmed by extensive experiments on pillar-shaped magnetic trilayers; for a review, the reader can refer to Chapters 7 and 8, Volume 1 [77]. Most current-induced magnetization switching (CIMS) experiments have been performed on purely metallic trilayers [171–176], such as Co/Cu/Co, in lithographically patterned nanopillars with detection of the magnetic switching by giant magnetoresistance effects. Typical critical current density $J_C$ required for magnetization reversal in these systems has been of the order of $10^7$ A.cm$^{-2}$ or higher. There have been several reports on CIMS in transition metal MTJs with low junction resistance at critical current density below $10^6$ A.cm$^{-2}$ [177–179]. CIMS experiments on tunnel junctions bring new physical problems [180], and are also of particular interest for their promising application to the switching of the MTJ-MRAM (Magnetic Random Access Memory). For a review of spin transfer in MgO-based MTJs, the reader can refer to the paper of Katine and Fullerton [181] and references therein. (Ga,Mn)As is known to have small magnetization of less than 0.05 T and a high spin-polarization of holes which must result in the reduction of critical current according to the Slonczewski spin-transfer torque model. The results show that a low current density, of the order of $10^5$ A.cm$^{-2}$, is needed for CIMS with (Ga,Mn)As [79, 80] showing the interest of magnetic semiconductors for spin transfer. As emphasized by Chiba et al. [182], the specific VB structure and SOI has to be taken into account and should have the effect of mixing the spin states of carriers. In the following, we present results of CIMS experiments obtained on our own (Ga,Mn)As MTJs whose structure and TMR properties are described in section II, comparable to the one obtained in Ohno's group [79]. We also discuss open questions related to the VB electronic structure compared to the basis applicable to metallic CPP-GMR structures.

On the other hand, considering the spin-orbit torques (SOT), the combination of exchange and SOI in (Ga,Mn)As makes this materials very important. Indeed, more generally, magnetization switching at the interface between ferromagnets and SHE nonmagnetic materials controlled by a current and related *current-induced* torques are of a particular interest. In that sense, the size and symmetry of the SOI at the interface with relevant materials with surface broken symmetry, like using (Ga,Mn)As (III,V), deserves some clear investigations for potential future applications. With that in mind, SOI current induced torques have been already demonstrated with (Ga,Mn)As [3, 183] and more recently, the specific role of the antidamping Slonczewski-like torque in the ferromagnetic resonance regime of spin-transfer. This particularly emphasizes the role of the two types, *Rashba* and *Dresselhaus*, symmetry-like terms originating from the unidirectional character of the interface, together with the symmetry breaking from $T_d$ to $C_{2v}$ symmetry group.

### 4.4.2 Spin Transfer with Holes in (Ga,Mn)As Tunnel Devices

Concerning the fabrication of our nanoscale MTJs, a circular resist mask (height = 200 nm, diameter = 700 nm) is first defined by e-beam lithography and oxygen plasma etching. A pillar is then etched down to the conducting GaAs buffer layer by ion beam etching and a $Si_3N_4$ layer is sputtered to cover the bottom electrode. The next step is the planarization of the surface by spin-coating. The nitride layer on top of the pillar is then removed by ion etching. Finally, the top of the pillar is cleaned with an oxygen plasma, and top electrodes are fabricated by a liftoff process. For CIMS experiments, the application of a significant bias of about 800 mV in our devices is needed to observe spin transfer phenomena, so this switching cannot be detected directly by a clear change of resistance. Instead we used the following procedure. Starting, for example, from a parallel (P) configuration at $V = 0$, the voltage is increased step by step, and, after each step, brought back to 20 mV to compare with that found at 20 mV before the step. We can check in this way whether the magnetic configuration has been irreversibly switched. Of course, only irreversible switchings can be detected and the reversible changes of the "steady precession regime" [77, 184] cannot be detected. Examples of results obtained with this procedure are shown in Figure 4.4c. By comparing with the MR curve at 20 mV of Figure 4.5a, one sees that the magnetic configuration is switched irreversibly from an almost parallel (P) to an almost antiparallel (AP) configuration by a positive current density (current flowing from the thin magnetic layer to the thick one), $j_{c+} = 1.23 \times 10^5$ A cm$^{-2}$ ($V_{c+} = 810$ mV) at 3 K and $j_{c+} = 0.939 \times 10^{-5}$ A cm$^{-2}$ ($V_{c+} = 680$ mV) at 30 K. The configuration is then switched back to parallel by a negative current above a threshold current density of the same order ($j_{c-} = -1.37 \times 10^5$ A cm$^{-2}$ at 3 K and $j_{c-} = -0.986 \times 10^5$ A cm$^{-2}$ at 30 K). Opposite current directions for the P to AP and AP to P transitions is the characteristic behavior of switching by spin transfer [77, 171–179]. Reversing the initial orientation of the magnetizations does not reverse the sign of the switching current, which confirms that Oersted field effects can be ruled out. By more complex cycling procedures that would take too long to describe, we have checked that, as in standard CIMS experiments, the transition is due to the magnetic switching of the thin layer. The fact that the bottom thin layer is the free layer is also consistent with the spin-transfer torque model as the total magnetic moment of the bottom layer is less than that of the thicker top layer in (Ga,Mn)As as already reported. Further experiments [80] have demonstrated that CIMS on (Ga,Mn)As-based MTJs could be obtained only in a small field window of 2–3 Oe around 13 Oe at 3 K, which may be explained by the combined effects of the Joule heating and temperature dependence of the magnetic properties. Indeed, from SQUID and TMR measurements, we know that the anisotropy and coercive field of the thin layer drops to a couple of Oe above 30 K, or, equivalently, by heating the sample with a bias voltage value of about 700 mV. This study has established a particular phase diagram (Figure 4.20) for CIMS vs. the amplitude of the external magnetic field applied. This study has shown that the heating of the sample and the manifestation of a reversible switching regime [77, 172, 174, 176] (which cannot be detected by the experimental protocol used) can prevent any switching of (Ga,Mn)As out-of a certain narrow field window.

### 4.4.3 PHASE DIAGRAM FOR CIMS WITH (GA,MN)AS

A significant difference with respect to the behavior with magnetic metals is that CIMS curves similar to those of Figure 4.4c can be obtained only in a small field window of 2–3 Oe around 13 Oe at 3 K. Let us first say that, from the analysis of the shifts of minor MR cycles, we found that the dipolar field generated by the thick (Ga,Mn)As layer and acting on the thin layer is close to 13 Oe at 3 K, so that $H_{\text{app}} = 13$ Oe corresponds to approximately an effective field $H_{\text{eff}} = 0$ Oe on the thin layer when the magnetic moment of the thick one is in its former direction (case of Figure 4.20c). The behavior of Figure 4.20c in a field range of a couple of Oe around $H_{\text{eff}} = 0$ Oe can be explained by the combined effects of the Joule heating and temperature dependence of the magnetic properties. From SQUID and

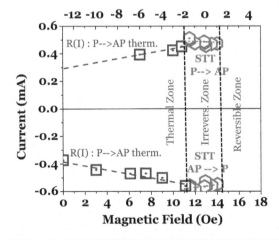

**FIGURE 4.20**    Phase diagram displaying the different regime of magnetization reversal vs. applied field that are thermal switching, irreversible switching (CIMS), and reversible switching. The top abscise corresponds to the effective field equal to the applied field added to the stray field equaling −13 Oe.

TMR measurements, we know that the anisotropy and coercive field of the thin layer drops to a couple of Oe above 30 K or, equivalently, by heating the sample with a bias voltage value of about 700 mV [58]. Suppose, for example, a P state with both magnetizations in the direction of positive $H$ with $H_{app} = 9$ Oe or, equivalently, $H_{eff} \cong -4$ Oe. If, at a current I* smaller than the critical current for CIMS, the heating of the sample reduces the magnitude of the coercive field of the thin layer to 4 Oe or below, the effective field of −4 Oe switches the configuration to AP when the current reaches I* or −I*. This behavior, with switching to AP by positive as well as by negative currents, is what we observe for $H_{app}$ below the window centered on $H_{app} = 13$ Oe. Assume now a same parallel (P) state with $H_{app} = 17$ Oe or, equivalently $H_{eff} = +4$ Oe. If the heating of the sample reduces the anisotropy field below 4 Oe, we enter the regime of reversible switching [77, 78, 172, 174, 176] that we cannot detect. Consistently, we cannot detect any switching for $H_{app}$ above the window. We conclude that, with respect to CIMS with magnetic metals, the dramatic variation of the properties of (Ga,Mn)As with temperature, combined with the heating of the sample, introduces significant complications in CIMS experiments with (Ga,Mn)As and limits the observation of CIMS by spin transfer to a narrow field range. These arguments give rise to the phase diagram shown in Figure 4.20 and highlight the three different regions that are thermal switching, irreversible switching (CIMS), and reversible switching. We conclude that, with respect to CIMS with magnetic metals, the dramatic variation of the properties of (Ga,Mn)As with temperature, combined with the heating of the sample, introduces significant complications in CIMS experiments with (Ga,Mn)As and limits the observation of CIMS by spin transfer to a narrow field range.

The spin-transfer and spin-torques in MTJ can be described by the sum of two different components: the so-called Slonczewski-like, or in-plane torque, torque $T_{IP}$, and the field-like, or out-of-plane torque $T_{OOP}$. Contrary to spin-transfer phenomena in metallic nanopillars, whereby the field-like torque is generally smaller than the Slonczewski or antidamping-like torque [185], the two components are of the same order of magnitude in tunnel barriers because of the coherent nature of the spin-currents within the tunnel barriers [186–189]. However, these two types of torques differs by their specific symmetry with the voltage variations according to:

$$T_{IP} = T_{IP,0}\mathbf{m} \times (\mathbf{s} \times \mathbf{m}) \simeq A_{IP,0}V\mathbf{m} \times (\mathbf{s} \times \mathbf{m}) \tag{4.9}$$

$$T_{OOP} = T_{OOP,0}\mathbf{m} \times \mathbf{s} \simeq A_{OOP,0}V^2(\mathbf{s} \times \mathbf{m}) \tag{4.10}$$

The difference in the symmetry between the two components makes it possible to distinguish them experimentally in spin-diode experiments [190, 191] or via spin-transfer-torque spin-Hall effects (STT-SHE) experiments [46, 47]. The result is that, under the action of these two torques, the general magnetization dynamics of the local magnetization $\mathbf{m}$ follows the generalized Landau-Lifshitz equation according to [74, 78]:

$$\frac{\partial \mathbf{m}}{\partial t} = -\gamma \mathbf{m} \times \mathbf{H} + \mathcal{T}_{OOP,0}\mathbf{m} \times \mathbf{s} + \mathcal{T}_{IP,0}\mathbf{m} \times (\mathbf{s} \times \mathbf{m}) + \alpha \mathbf{m} \times \frac{\partial \mathbf{m}}{\partial t}, \quad (4.11)$$

where:

$\gamma$  is the gyromagnetic factor

$\alpha$  is the phenomenological damping constant

Figure 4.21 displays our latest measurements on STT acquired on $400 \times 200$ nm$^2$ (Ga,Mn)As (10 nm)/GaAs(6 nm)/(Ga,Mn)As(50 nm) tunnel nanojunctions at 10 K prepared for STT experiments in current-perpendicular-to-plane geometry. In these experiments, we have acquired the $R(V)$ switching cycles for different protocols (after positive and negative saturation) and at fixed magnetic field. For example, let us comment first on the first set of sub-figures at the top. The upper left figures corresponds to the characteristic $R(H)$ cycles after preparing the samples to positive saturation ($|H_{sat}| = 7$ KOe) (negative saturation), then decreasing to zero field and applying a positive (negative) magnetic field, before applying the bias for describing the switching cycles. The $R(H)$ curves means that a positive moderate bias (less than 2 V) has for effect to apply a reversible in-plane torque, via $\mathcal{T}_{IP}$, linear in V switching the soft (Ga,Mn)As magnetization in its opposite direction $AP$ state. A larger positive bias (larger than 2 V) will have the effect of switching back this magnetization in its formal direction though $\mathcal{T}_{OOP}$ (thanks to the field-like torque symmetry). A perfect symmetry exists when the sample is prepared starting from the negative saturation (negative field range of the corresponding sub-figures). The upper right figures corresponds

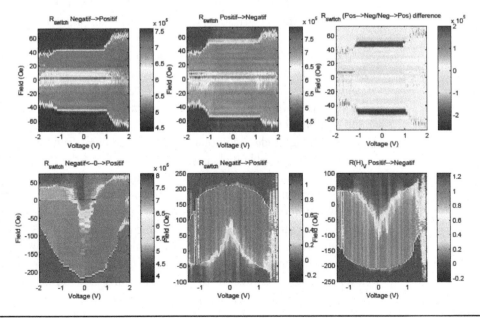

**FIGURE 4.21** Phase Diagram for spin-transfer-torque (STT) experiments acquired on $400 \times 200$ nm$^2$ (Ga,Mn) As(10 nm)/GaAs(6 nm)/(Ga,Mn)As(50 nm) tunnel nanojunctions. The current is injected in the perpendicular direction for these STT-experiments. The $T_C$ of the trilayer structure is 150 K and the measurements are performed at 10 K.

to the same protocol experiments, except that now the sample is prepared in its relative AP state (by operating a minor loop) by the following preparation field-cycle ($\pm H_{sat} \rightarrow \mp 100 Oe \rightarrow H$ where $H$ is the fixed magnetic field while swiping in bias) before applying the bias. The critical value of the field $\pm 40$ Oe from the main difference between the two types of cycles (critical point) corresponds to the value of the antiparallel dipolar stray field from the thick hard (Ga,Mn)As layer to the thinner soft (Ga,Mn)As layer.

More interestingly, the bottom panels of Figure 4.21 emphasize the role of $\mathcal{T}_{OOP}$ at larger bias in the reorientation of the hard (Ga,Mn)As layer along the softer magnetization (at higher bias) to make the two magnets parallel in any case. In other words, the ensemble of these figures demonstrates that the external-field necessary to switch the *hard* (Ga,Mn)As layer along the softer magnetization (that has already switched) in the $P$ (parallel) state is reduced due to the $V^2$-dependence of the $\mathcal{T}_{OOP}$ torques. The $V^2$ shape of the phase diagram in the bottom part of the corresponding figures seems to reveal the $V^2$ dependence of the field-like torque like discussed above.

### 4.4.4 DISCUSSION

The order of magnitude of the switching current densities for MTJs ($\sim 10^5$ A.cm$^{-2}$) is consistent with the model of Slonczewski [74], when one takes into account the magnitude of the (Ga,Mn)As magnetization, well reduced compared to transition metals. This argument was already put forward by Chiba et al. [79, 182]. In the simplest approximation, the switching currents at zero (or low) field are expected to be given by [173, 174]: $I^{P(AP)} = G^{P(AP)} \alpha t M [2\pi M + H_K)$, where $G^{P(AP)}$ is a coefficient depending on the current spin polarization, $\alpha$ is the Gilbert damping coefficient, $t$ is the layer thickness, and $H_K$ is the anisotropy field. The main difference with transition metals originates from the factor $t M [2\pi M + H_K]$. With $t = 15$ nm, $M = 0.035$ T, and a negligible $H_K$ after heating, for the junctions of this paper, and $t = 2.5$ nm, $M = 1.78$ T, $H_k = 0.02$ T for a typical Co/Cu/Co pillar [171–173] we find that the factor $t M [2\pi M + H_K]$ is larger by a factor of 500 for the metallic structure. This mainly explains the difference by two orders of magnitude. Additional factors should come from the current spin polarization (probably higher with (Ga,Mn)As) and the Gilbert coefficient (probably larger by almost an order of magnitude for (Ga,Mn)As grown at low temperature) [192], but a quantitative prediction is still a challenging task. What about the sign of the switching currents in (Ga,Mn)As devices which is the same as for standard metallic pillars [77, 171, 173, 174]? This is consistent with what we expect from the valence bands structure of (Ga,Mn) As, whereby the majority of electron spins (and no hole $p$ state) at the Fermi level are aligned parallel to the local Mn spins ($d$ state) due to the negative $p$-$d$ exchange interaction parametrized by the exchange integral $\beta$ [29, 81]. Complementary experiments [81, 82] performed on double (Ga,Mn) As-based MTJs in ms-pulsed current injection regime have demonstrated the possibility of reducing the critical current densities to less than $3 \times 10^4$ A.cm$^{-2}$, that is, by more than a factor of three compared to single (Ga,Mn)As

MTJs. This is due, in part, to the interfacial torque applied at both side of the middle (Ga,Mn)As sensitive layer and imposed by the symmetric structure.

A question lies in the observation of switching phenomena by spin transfer at bias ($\approx$ 800 mV) corresponding to a quasi-vanishing TMR, and then to a loss of the spin-polarization of holes with bias. Although in standard MTJs with metallic electrodes, the decrease of the MR with the bias is generally ascribed to electron-magnon scattering (emission and annihilation of magnons) [193] and to other types of spin-dependent inelastic scatterings [194], calculations performed by Sankowski et al. [162] put forward a natural decrease of TMR with bias due to the band structure of (Ga,Mn)As. In the former case, even if the spin current is strongly affected by spin-flip effects, the transverse component of the spin-polarized tunneling current cannot be lost and should finally be transferred to the magnetic moment of the layer, as proposed by Levy and Fert [195]. On the other hand, concerning the band structure itself, recent theoretical investigations on torques and TMR in MTJs performed in the frame of Keldysh formalism have demonstrated the possibility of conciliating increase of spin transfer efficiency and decrease of TMR in a same device [187, 188]. The same kind of calculation remains to be investigated in a multiband formalism adapted to VB of semiconducting tunnel junctions, and, in particular, the role of respective Slonczewski and field contributions of the torque in (Ga,Mn)As-based MTJs. The efficiency of the current induced torque exerted by general angular momenta ($J$) and no more spin ($S$), including the role of spin-orbit interactions as detailed in the following Appendix D, also remains to be studied theoretically. However, a deeper description of spin-transfer phenomena and spin-orbit torque with (Ga,Mn)As in both in-plane and current perpendicular-to-plane geometry will require the extension of the generalized spin-mixing conductance involving spin-orbit currents.

## 4.5 SUMMARY

To conclude, we have reviewed different properties of spin-polarized tunneling transport involving the (Ga,Mn)As DMS. It has been shown that a relatively good agreement can be found between experimental TMR and TAMR properties and Dietl theory for DMS in a molecular field approach for (Ga,Mn)As ferromagnetism. The general description of spin-polarized tunnel transport, including (Ga,Mn)As is fully described using standard $\mathbf{k} \cdot \mathbf{p}$ theory and Laudauer Buttiker formalism. In particular, it has been possible to understand the large modification of TMR properties as a function of annealing procedure. Indeed, it is now commonly accepted that an optimized annealing procedure, by the reduction of As interstitial at the free interface, leads in parallel to an increase of the carrier density and an increase of the spin exchange interaction. Further experiments exploring a wider region of the phase diagram established theoretically are required to refine our understanding of the (Ga,Mn)As band structure. A more advanced technological control of the material will be needed to extract precisely (Ga,Mn)As characteristic parameters. More generally, our $\mathbf{k} \cdot \mathbf{p}$ framework seems to catch the essential trends of the spin dependent tunneling phenomena involving holes emitted from (Ga,Mn)As. As an example, the sizeable TAMR effects originating from the anisotropy of the electronic band structure have been finely reproduced. Exploring the TAMR signal on a resonant quantum well has revealed the intricate competition of the different (Ga,Mn)As subbands. The results indicate that a respective positive and negative contribution can be expected from LH and HH bands. More generally, these different contributions emphasize the band curvature of (Ga,Mn)As with HH and LH characters near the $\Gamma$ point, assuming that the LH character is favorably transmitted by the tunnel barrier.

In the last part we have demonstrated CIMS phenomena in nanoscale MTJs integrating (Ga,Mn)As. Thanks to the particular magnetic properties of (Ga,Mn)As, a smaller current density compared to transition metal is required to reverse the magnetization. In the future, a deep understanding of spin transfer phenomena involving holes will imply the development of more sophisticated models, like the one developed in the standard $\mathbf{k} \cdot \mathbf{p}$ approach. In parallel, time-resolved, or frequency-resolved, experiments performed in the reversible regime of spin transfer will be certainly helpful for a complete description of the inner mechanisms for spin transfer. Moreover, we believe that the specificity of carrier-mediated ferromagnetism of (Ga,Mn) As will give new opportunities to explore spin-dependent phenomena, taking advantage of semiconductor functionalities. Despite the difficulties in the route for room temperature operations for this class of material, further spin-related studies with ferromagnetic semiconducting heterojunctions seem necessary to pursue fundamental physics, as well as to establish promising concepts in other systems.

Moreover, we have reported on theoretical investigations of a new type of chiral-tunnel effects in semiconductors junctions and heterostructures mediated by spin-orbit and exchange interactions. Our developments are primary based on Green's function formalism involved in the tunneling transport at the first order of the perturbation theory. The analytical calculations, performed for electrons in the CB with Dresselhaus interactions and holes in the VB with core atomic spin-orbit interactions are favorably compared with 6-band and advanced 30-band $\mathbf{k} \cdot \mathbf{p}$ tunneling models. This study represents an extension of our previous contribution [51] dealing with the role, on the electronic forward and backward transmission-reflection asymmetry. Such forward scattering asymmetry also should lead to skew-tunneling effects involving the branching of evanescent states in the barrier. Recent experiments involving non-linear resistance variations vs. the transverse magnetization direction or current direction in the in-plane current geometry may be invoked by the phenomenon we discuss.

## ACKNOWLEDGMENTS

J. Jaffrès gives special thanks to the students of the École Polytechnique taking part to the research laboratory project (P.R.L.), J. Boust, M. Dujany and M. Le Dantec. T.H.D. acknowledges Idex Paris-Saclay and Triangle de la Physique for funding.

# APPENDIX A: EXCHANGE INTERACTIONS IN (GA,Mn)AS: FROM THE Mn ISOLATED ATOM TO THE METALLIC PHASE

From general group-theory arguments, in the effective mass approximation [99, 101], nonmagnetic shallow acceptors can be described by hydrogenic states of fundamental symmetry term $1S_{3/2}$ of binding energy equal to 28 meV for GaAs. These are characterized by a total angular momentum $\mathbf{F} = \mathbf{L} + \mathbf{J} = 3/2$, which is a constant of motion where $\mathbf{L}$ is the angular momentum of the envelope wave function. The result is that the fundamental $1S_{3/2}$ wave function is $\Phi(S_{3/2}) = f_0(r)|L = 0, J = 3/2, F = 3/2, F_z\rangle + g_0(r)|L = 2, J = 3/2, F = 3/2, F_z\rangle$. According to optical studies, Mn is known to form a shallow acceptor center in GaAs of $A^0 = d^5 + h$ electronic configuration characterized by a binding energy [110] of 110 meV, and an energy difference of the order of 10 ($\pm 3$) meV between the $J = 1$ and $J = 2$ $h$–$d^5$ states [110]. In this picture, the $J = S + j$ quantum number constant of motion is the sum of the $d^5$ Mn spin angular momentum $S = 5/2$ and the $j = 3/2$ hole angular momentum [196, 197]. In the $S - j$ exchange coupling scheme where the exchange interaction is $J_{exc}\,\mathbf{S} \cdot \mathbf{j}$, the energy difference between the extrema $J = 1$ and $J = 4$ states is equal to $9J_{exc}$, whereas it gives $2J_{exc}$ between the two successive $J = 1$ and $J = 2$ states. It follows that the energy difference between the states corresponding to the spin of the bound hole parallel and antiparallel to the Mn spin can be estimated to 45 meV. This relative small $p$-$d$ exchange energy originates from the relatively long extent of the bound hole wave function where the Mn $d$ states are mostly localized within an effective bohr radius $a_0^* \propto \varepsilon / m^* \simeq 0.8$ nm and corresponding to an effective volume of 3 nm³, as well as an effective Mn concentration approaching $x_{loc} = 1.35\%$. A direct consequence is that the average exchange integral $\Delta_{exc}$ is expected to be enhanced with increasing the Mn content $x$ above this threshold Mn concentration $x_{loc} = 1.35\%$.

In the metallic regime and in the **S-s** exchange coupling scheme, the average exchange interaction in (Ga,Mn)As reads $\Delta_{exc} = -5/2xN_0\beta$, $N_0$ is the concentration of cations and $N_0\beta = -(16/S)$ $\left( \dfrac{1}{-\Delta_{eff} + U_{eff}} + \dfrac{1}{\Delta_{eff}} \right) \times \left( \dfrac{1}{3}(pd\sigma) - \dfrac{2\sqrt{3}}{9}(pd\pi) \right)^2 < 0$ is the exchange integral found by treating the $p$-$d$ hybridization as a perturbation in the configuration interaction picture [111] giving rise to antiferromagnetic interactions between $p$ and $d$ shells. Here, $S$ is the localized $d$ spin, $U_{eff} = E(d^{n-1}) + E(d^{n+1}) - 2E(d^n)$ is the characteristic 3$d$-3$d$ Coulomb interaction, $\Delta_{eff} = E(\underline{L}d^n) - E(d^{n-1})$ is the ligand-to-3$d$ charge transfer energy. On the other hand, $(pd\sigma)$ and $(pd\pi)$ are characteristic Slater–Koster hopping integrals. The value of $N_0\beta = -1.2$ eV ($\beta = -54$ meV nm³) is generally admitted from core level photoemission measurements for (Ga,Mn)As with a $T_C$ close to 60 K [113] corresponding to $x_{eff} \simeq 4\%$. (Figure 4.13) display the 4-different exchange-splitted (Ga,Mn)As subbands calculated for a hole density $p = 1.7 \times 10^{20}$ cm$^{-3}$ and an average exchange energy of $\Delta_{exc} = 120$ meV.

From a point of view of material properties, questions remain on the general trends of TMR vs. exchange interactions $\Delta_{exc} = -5/2xN_0\beta$ where $xN_0$ is the concentration of Mn atoms as well as hole band filling within (Ga,Mn)As. Also, what are the possible effects of the low-temperature (L-T) growth procedure used for the synthesis of these MTJs on the band lineup, valence band offset, and related barrier heights of heterostructures integrating (Ga,Mn)As? Remarkably, and as shown by non-linear I–V characteristics recorded on junctions, both (Ga,As) and (In,Ga)As materials play the role of a tunnel barrier for holes injected from/into (Ga,Mn)As [56–58, 128]. The same qualitative features have been demonstrated through optical measurement of the hole chemical potential in ferromagnetic (Ga,Mn)As/GaAs heterostructures by photoexcited resonant tunneling [129]. These results favor a band-edge discontinuity due to the smaller band gap of (Ga,Mn)As compared to GaAs [130], and indicate a pinning of the Fermi level deep inside the band gap of the (Ga,As) host. A part of the answer lies in the incorporation of $n$-type double-donor As antisites during the L-T growth procedure [27] that partly governs the pinning of the Fermi level at a higher energy position than expected neighboring the midgap of GaAs. The second reason is due to the positive coulombic-exchange potential experienced by holes, and introduced by Mn species playing the role of hydrogen centers for holes orbiting around it [26]. This is at the origin of an impurity-band formation at smaller or intermediate doping level in the host bandgap [131], and an intense debate remains about the position of the Fermi level relative to the impurity band [132]. While infrared measurements [133] and dichroism (MCD) [134] experimental and theoretical studies [135] seem to support the scenario of a detached impurity band, recent low-temperature conductivity measurements [136] validate the approach of a VB picture more compatible with a $\mathbf{k} \cdot \mathbf{p}$ treatment of its electronic properties.

# APPENDIX B: DESCRIPTION OF THE 6-BAND k · p THEORY

*The $\mathbf{k} \cdot \mathbf{p}$ methods: generalities.*

We first give some general information about the $\mathbf{k} \cdot \mathbf{p}$ method we used throughout this work. The $\mathbf{k} \cdot \mathbf{p}$ method is known to be very efficient at accurately describing the properties of the electronic structure of a semiconductor near the $\Gamma$ point [198–202]. It may be implemented with a $2 \times 2$-band model for the CB, a $6 \times 6$-band Luttinger model for the VB in an effective Hamiltonian description via the Luttinger parameters $\gamma_i$ [203]. However, an 8-band model is needed to describe the coupling between the CB and the VB, whereas a 14-band model is mandatory to properly account for the absence of inversion symmetry in the $T_d$-symmetry group [204–206].

Beyond that, an extended 30-multiband treatment becomes necessary [170, 207–210] to correctly describe the full BZ of indirect band gap semiconductors like AlAs, Si, Ge, and related compounds. Its treatment requires the inclusion of remote bands in the Hamiltonian basis, whereas the description of the tunneling transport in a 14- or a 30-multiband approach requires resolution of the issue of the spurious bands inherent to band truncation [210–212].

The general $\mathbf{k} \cdot \mathbf{p}$ Hamiltonian in the crystal writes:

$$H_{SC} = \frac{p^2}{2m_0} + \mathfrak{U}(r) + \frac{\hbar}{4m_0^2 c^2} (\nabla \mathfrak{U} \times \mathbf{p}) \cdot \sigma, \qquad (4.12)$$

where:

$H_{SO} = \dfrac{\hbar}{4m_0^2 c^2} (\nabla \mathbf{U} \times \mathbf{p}) . \sigma$ is the atomic-SO term

$\mathfrak{U}$     is the lattice periodic potential

$m_0$    is the free electron mass

$c$      is speed of light

The wave function is the solution of the Schrödinger equation $H_{SC}\Psi = E\Psi$, with the Bloch form $\Psi_{n,\mathbf{k}}(\mathbf{r}) = e^{i\mathbf{k}.\mathbf{r}} \varphi_{n\mathbf{k}}(\mathbf{r})$ satisfying

$$\left[ H_{SC} + \breve{k}^2 + \frac{\hbar}{m_0} \mathbf{k} \cdot \mathbf{p} + \frac{\hbar^2}{4m_0^2 c^2} (\nabla \mathfrak{U} \times \mathbf{k}) \cdot \sigma \right] \varphi_{n\mathbf{k}} = E_{n\mathbf{k}} \, \varphi_{n\mathbf{k}}, \qquad (4.13)$$

where $\breve{k}^2 = (\hbar^2 / 2m_0)k^2$ is the free-electron energy. The atomic-like wave function at $\mathbf{k} = 0$, $\varphi_{n\mathbf{k}}(\mathbf{r})$, are assumed to be known as well as their symmetry. We denote $\varphi_n = \varphi_{n(\mathbf{k}=0)}(\mathbf{r})$ and $E_n = E_{n(\mathbf{k}=0)}$ with $H_{SC} \varphi_n = E_n \varphi_n$.

The wave functions at $\mathbf{k} \neq 0$ can be expanded as a series of $\varphi_n$

$$\varphi_{n\mathbf{k}}(\mathbf{r}) = \sum_{\mathbf{k}} C_{n\mathbf{k}} \varphi_n. \qquad (4.14)$$

where $\{\varphi_n\}$ is a set of basis functions, $\langle \varphi_m | A | \varphi_n \rangle = (1 / L) \int_\Omega \varphi_m^*(\mathbf{r}) A \varphi_n(\mathbf{r}) d\mathbf{r}$, with $\Omega$ the crystal volume.

## K · P Approach for Describing Metallic (Ga,Mn)As

The purpose of this section is to give some insights about the procedure to calculate transport properties of spin-polarized holes in the valence band of heterojunctions integrating the (Ga,Mn)As ferromagnetic semiconductor in the frame of the $\mathbf{k} \cdot \mathbf{p}$ theory. This includes the contribution of the $p$-$d$ exchange interaction in the molecular field approximation. For zinc-blende semiconductors, we take into account explicitly the four $\Gamma_8$ and the two $\Gamma_7$ valence subbands, for which we choose the basis functions in the form according to Dietl et al. [29]:

$$u_1 = \left| J = \frac{3}{2}, j_z = \frac{3}{2} \right\rangle = \frac{1}{\sqrt{2}} (X + iY) \uparrow$$

$$u_2 = \left| J = \frac{3}{2}, j_z = \frac{1}{2} \right\rangle = i\frac{1}{\sqrt{6}} \left[ (X + iY) \downarrow - 2Z \uparrow \right]$$

$$u_3 = \left| J = \frac{3}{2}, j_z = -\frac{1}{2} \right\rangle = i\frac{1}{\sqrt{6}} \left[ (X - iY)\uparrow + 2Z\downarrow \right]$$

$$u_4 = \left| J = \frac{3}{2}, j_z = -\frac{3}{2} \right\rangle = i\frac{1}{\sqrt{2}} (X - iY)\downarrow$$

$$u_5 = \left| J = \frac{1}{2}, j_z = \frac{1}{2} \right\rangle = \frac{1}{\sqrt{3}} \left[ (X + iY)\downarrow + Z\uparrow \right]$$

$$u_6 = \left| J = \frac{1}{2}, j_z = -\frac{1}{2} \right\rangle = i\frac{1}{\sqrt{3}} \left[ -(X - iY)\uparrow + Z\downarrow \right] \tag{4.15}$$

where X, Y and Z denote Kohn–Luttinger amplitudes, which for the symmetry operations of the crystal point group transform as $p_x$, $p_y$ and $p_z$ wave functions of the hydrogen atom.

Added to the Kohn–Luttinger kinetic Hamiltonian [84], the hole Hamiltonian $H_h$ includes a $p$-$d$ exchange term introduced by the interaction between the localized Mn magnetization and the holes derived in the mean-field approximation thus giving [29, 83]:

$$H_h = -(\gamma_1 + 4\gamma_2)k^2 + 6\gamma_2 \sum_{\alpha} L_\alpha^2 k_\alpha^2 +$$

$$+6\gamma_3 \sum_{\alpha \neq \beta} (L_\alpha L_\beta + L_\beta L_\alpha)k_\alpha k_\beta + \lambda_{so}\vec{L}\vec{S} + \Delta_{exc}\hat{m}\vec{S} \tag{4.16}$$

$\alpha = \{x, y, z\}$, $L_\alpha$ are $l = 1$ angular momentum operators, $\vec{S}$ is the vectorial spin operator, $\hat{m}$ the unit magnetization vector, and $\gamma_i$ are Luttinger parameters of the host semiconductor. $\Delta_{exc}$ represents the spin-splitting between the heavy holes at the $\Gamma_8$ point. The detailed form of the $6\times6$ Hamiltonian is given in Ref. [29]. We chose $\Delta_{exc} = 120$ meV in agreement with the value extracted from the magnetization saturation of (Ga,Mn)As [58]. Note that the standard $4\times4$ **k** · **p** Hamiltonian projected on the $\Gamma_8$ ($J = \frac{3}{2}$) basis can be written in a cubic form as:

$$H_h = \left(\gamma_1 + 5/2\gamma_2\right)\kappa^2 I_4 - 2\gamma_2 \left(\kappa \cdot J\right)^2$$

$$+2(\gamma_3 - \gamma_2) \sum_{i<j} \left[ \kappa_i \kappa_j \left( J_i J_j + J_j J_i \right) \right] + \Delta_{exc}\hat{m}\vec{S} \tag{4.17}$$

in unit of $\hbar^2/2m_0$ where $J_i$ is the $4\times4$ angular momentum operator in the subspace $J = 3/2$ along the direction $i$ (Table 4.3).

Figure 4.22 depicts the four different Fermi surfaces of (Ga,Mn)As corresponding to spin-polarized HH and LH-bands at the $\Gamma$ point, and calculated by the 6-band **k** · **p** method. These have been calculated for an exchange parameter $B_G$ of 40 meV and a Fermi energy equaling −0.22 eV counted from the top of the (Ga,Mn)As valence band. The magnetization is along the [001] direction. The less-filled bands display a strong anisotropic character due to the interplay between the spin-orbit and exchange interactions; i.e. when the kinetic energy is small compared to the latter interactions.

**TABLE 4.3**
**Luttinger Parameters $\gamma$, Spin-Orbit Splitting Δ0 (eV) and Valence-Band Offset VBO (eV) Used**

|  | $\gamma_1$ | $\gamma_2$ | $\gamma_3$ | $\Delta_0$ | *VBO* |
|---|---|---|---|---|---|
| GaAs | 6.85 | 2.1 | 2.9 | 0.34 | 0 |
| AlAs | 3.25 | 0.64 | 1.21 | 0.275 | −0.64 |
| InAs | 20.4 | 8.3 | 9.1 | 0.38 |  |
| In0.25Ga0.75As | 10.24 | 8.3 | 4.45 | 0.35 | −0.7 (Low-T) |
| (Ga,Mn)As | 6.85 | 2.1 | 2.9 | 0.34 | +0.22 |

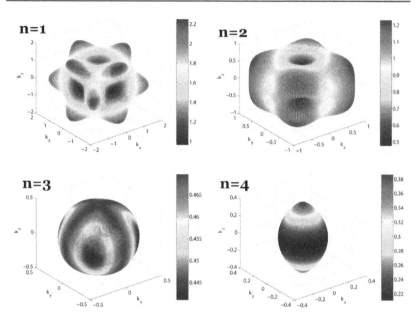

**FIGURE 4.22** The four different Fermi surfaces of (Ga,Mn)As calculated in the **k · p** formalism for a Fermi energy equal −135 meV counted from the top of the valence band ($p = 1 \times 10^{20}$ cm$^{-3}$) and an exchange interaction $\Delta_{exc} = 120$ MeV (calculated without strain). The magnetization is along the [001] growth crystalline axis. The color code scales the Fermi wavevector (in nm$^{-1}$) along the corresponding crystalline axis. The corresponding band structure of (Ga,Mn)As in the reciprocal space can be found in Refs. [12, 25] (After Tanaka, M. and Higo, Y., *Phys. Rev. Lett.* **87**, 026602, 2001; Matsukura, F. at al., III–V ferromagnetic semiconductors, in *Handbook of Magnetic Materials*, vol. 14, K. H. J. Buschow (Ed.), North Holland, Amsterdam, pp. 1–87, 2002. With permission.).

## The 6-Band k · p Kane Model (Issued from the 14-Band)

In a 6-band model like that we use for the scattering perturbative treatment, $\{X,Y,Z\} \otimes \{\uparrow, \downarrow\}$ or linear combinations of them, represent the basis functions in Equation 4.14. This effective 6-band model derives from a larger 14-band basis set $\{X_C, Y_C, Z_C, S, X, Y, Z\} \otimes \{\uparrow, \downarrow\}$ from Lowdin downfolding. The **k · p** term is diagonal in spin and allows (i) the coupling terms between the $\Gamma_6$ and the $\{\Gamma_7, \Gamma_8\}$ wave function as $\langle S | p_x | iX \rangle = \varpi$; (ii) the coupling terms between the $\{\Gamma_7, \Gamma_8\}$ and the higher *CBs* of *p*-type symmetry $\{\Gamma_{7C}, \Gamma_{8C}\}$ as

$\langle X | p_y | i Z_C \rangle = \varpi_X$ as well as (iii) the coupling terms, in the $T_d$ symmetry group, between the $\Gamma_6$ and the $\{\Gamma_{7C}, \Gamma_{8C}\}$ wave function as $\langle S | p_x | i X_C \rangle = \varpi'$, and where $\varpi, \varpi'$, and $\varpi_X$ are real. $\varpi' = 0$ in $O_h$ as considered here. The $\mathbf{k} \cdot \mathbf{p}$ parameters are introduced according to $P = \dfrac{\hbar}{m_0} \varpi, P' = \dfrac{\hbar}{m_0} \varpi', P_X = \dfrac{\hbar}{m_0} \varpi_X$ with respective characteristic energies $E_P = \dfrac{2m_0}{\hbar^2} P^2, E_{P'} = \dfrac{2m_0}{\hbar^2} P'^2, E_{P_X} = \dfrac{2m_0}{\hbar^2} P_X^2$.

The SOI is introduced according to the work of Koster [201]. We resume here the couplings which may differ from zero: (i) the core-SOI in the higher CB, $\Delta^C = \left( \dfrac{3\hbar^2}{4m_0^2 c^2} \right) \left\langle X_C \left| \dfrac{\partial \mathfrak{U}}{\partial x} p_y - \dfrac{\partial \mathfrak{U}}{\partial y} p_x \right| i Y_C \right\rangle$, (ii) the SOI in the VB,

$\Delta = \left( \dfrac{3\hbar^2}{4m_0^2 c^2} \right) \left\langle X \left| \dfrac{\partial \mathfrak{U}}{\partial x} p_y - \dfrac{\partial \mathfrak{U}}{\partial y} p_x \right| i Y \right\rangle$, and (iii) the SOI caused by the lack of inversion center in the $T_d$ symmetry-group, $\Delta' = \left( \dfrac{3\hbar^2}{4m_0^2 c^2} \right) \left\langle X \left| \dfrac{\partial \mathfrak{U}}{\partial x} p_y - \dfrac{\partial \mathfrak{U}}{\partial y} p_x \right| i Y_C \right\rangle$ (in $O_h$, $\Delta' = 0$).

The $p$-$d$ exchange interactions in the VB one needs is introduced in a mean-field approximation from Refs. [29, 83], like $H_{\mathrm{exc}} = 3B_G\, \hat{\mathbf{s}} \cdot \hat{\mathbf{M}}$, where $3B_G$ represents the average energy interaction among holes, $\mathbf{s}$ the hole spin and $\mathbf{M}$ the 3-$d$ localized spin.

## THE 6-BAND K · P KANE MODEL: EIGENVECTORS AND EFFECTIVE MASS IN III–V HETEROSTRUCTURES

One considers a magnetic tunnel barrier of thickness $d$ and barrier height ($\Phi = 2w$) equaling the exchange potential in order to avoid scattering of evanescent waves in the barrier. What are the tunneling properties and related asymmetry of transmission? One assumes the continuity of the effective mass over the whole heterostructure. This can mimic the case of GaSb barrier sandwiched between two (Ga,Mn)As ferromagnetic electrodes with a larger SOI strength in the barrier. The true physical situation of SOI in the III–V, electrodes (e.g. (Ga,Mn)As or (Ge,Mn)) and barrier, will be treated numerically in a second step for comparison (see Ref. [168]).

The $6 \times 6$ Hamiltonian in the three different regions (left electrode, barrier, and right electrode) then writes:

$$\hat{H}_0 = \begin{cases} \hat{H}_{kp} + \hat{H}_{\mathrm{exc}} & \text{for } x < 0 \text{ or } x > a, \\ \hat{H}_{kp} - V_0 & \text{for } 0 < x < a, \end{cases} \tag{4.18}$$

where $\hat{H}_{kp}$ represents the kinetic energy, $\hat{H}_{\mathrm{exc}} = w\hat{\sigma}_x$ the exchange potential with $w$ the exchange strength, and where $V_0$ is the barrier height (band discontinuity).

$$\hat{H}_{k.p} = \begin{pmatrix} \hat{H}_{k.p}^{\uparrow\uparrow} & 0 \\ 0 & \hat{H}_{k.p}^{\downarrow\downarrow} \end{pmatrix}, \tag{4.19}$$

with $\hat{H}_{k.p}^{\downarrow\downarrow} = \hat{H}_{k.p}^{\uparrow\uparrow}$, and

$$H_{k.p}^{\uparrow\downarrow} =$$

$$
\begin{pmatrix}
 & |x\uparrow\rangle & |y\uparrow\rangle & |z\uparrow\rangle \\
\langle x\uparrow| & \begin{aligned}&\frac{\hbar^2}{2m_0}(k_x^2+k_y^2)\\&-\frac{E_{PX}}{E_{5C}-E_8}k_y^2-\frac{E_P}{E_6-E_8}k_x^2\end{aligned} & -k_xk_y\left(\frac{E_{PX}}{E_{5C}-E_8}+\frac{E_P}{E_6-E_8}\right) & 0 \\
\langle y\uparrow| & -k_xk_y\left(\frac{E_{PX}}{E_{5C}-E_8}+\frac{E_P}{E_6-E_8}\right) & \begin{aligned}&\frac{\hbar^2}{2m_0}(k_x^2+k_y^2)\\&-\frac{E_{PX}}{E_{5C}-E_8}k_x^2-\frac{E_P}{E_6-E_8}k_y^2\end{aligned} & 0 \\
\langle z\uparrow| & 0 & 0 & \begin{aligned}&\frac{\hbar^2}{2m_0}(k_x^2+k_y^2)\\&-\frac{E_{PX}}{E_{5C}-E_8}(k_x^2+k_y^2)\end{aligned}
\end{pmatrix}
$$

$$(4.20)$$

We then introduce the $\mathcal{M}$ and $\mathcal{L}$ parameters according to the notation of Ref. [208]

$$M = \left(\frac{E_P}{E_6-E_8}+\frac{E_{PX}}{E_{5C-8}}\right), \tag{4.21}$$

$$L = \left(\frac{E_P}{E_6-E_8}-\frac{E_{PX}}{E_{5C-8}}\right), \tag{4.22}$$

and then choose the parameters to make *HH dispersion nearly flat*, $\frac{\hbar^2}{2m_0}\approx(M-L)/2$. Within this approach, under oblique incidence, LH and HH bands weakly mix together on reflection/transmission process. Consequently, without SOI, the HH and LH states can be almost transmitted like free carriers with respective effective masses $m_{HH}^*$, $m_{LH}^*$.

The core atomic SOI $\hat{H}_{SO}=\frac{\hbar}{4m_0^2c^2}(\nabla\mathfrak{U}\times\hat{\mathbf{p}})\cdot\hat{\boldsymbol{\sigma}}$, acting afterwards as a perturbation, writes in the present basis:

$$\hat{H}_{SO} =$$

$$
\begin{pmatrix}
 & X\uparrow & Y\uparrow & Z\uparrow & X\downarrow & Y\downarrow & Z\downarrow \\
X\uparrow & 0 & -i\Delta/3 & 0 & 0 & 0 & \Delta/3 \\
Y\uparrow & i\Delta/3 & 0 & 0 & 0 & 0 & -i\Delta/3 \\
Z\uparrow & 0 & 0 & 0 & -\Delta/3 & i\Delta/3 & 0 \\
X\downarrow & 0 & 0 & -\Delta/3 & 0 & i\Delta/3 & 0 \\
Y\downarrow & 0 & 0 & -i\Delta/3 & -i\Delta/3 & 0 & 0 \\
Z\downarrow & \Delta/3 & i\Delta/3 & 0 & 0 & 0 & 0
\end{pmatrix},
$$

$$(4.23)$$

In each layer, the bare Hamiltonian is diagonal within the spin index $\sigma$. The condition for an infinite heavy-hole mass is $(\mathcal{M}-\mathcal{L})/2\approx 1$ where

$$\mathcal{M} = \left( \frac{E_P}{E_6 - E_8} + \frac{E_{PX}}{E_{5C-8}} \right) \quad \text{and} \quad \mathcal{L} = \left( \frac{E_P}{E_6 - E_8} - \frac{E_{PX}}{E_{5C-8}} \right) \quad \text{is, in that case, the}$$

inverse of the LH effective mass [208]. One can show that the three eigenvectors corresponding to the same eigenvalue $E$ are of respective orbital-symmetry $|Z\rangle_{L,R}$, $|Y\rangle_{L,R}$ and $\left| X + \dfrac{\mathcal{M}\xi}{\mathcal{L}K}Y \right\rangle_{L,R}$ in the left (L) and right (R) magnetic

contacts and $|Z\rangle_{TB}$, $|Y\rangle_{TB}$ and $\left| X - i\dfrac{\mathcal{M}\xi}{\mathcal{L}\lambda K}Y \right\rangle_{TB}$ in the tunnel barrier region ($\hat{x}$ is the direction of the magnetization, $\hat{y}$ the direction of $k_{\parallel}$, and $\hat{z}$ the direction of tunneling).

With our chosen parameter of infinite HH-mass, $|HH_1\rangle = |Z\rangle$ and $|HH_2\rangle = |Y\rangle$ are the two different HH-orbital states of same (infinite) mass, carrying no current flux. On the other hand, the third eigenstate, $|LH\rangle = \left| X + \dfrac{\mathcal{M}\xi}{\mathcal{L}K}Y \right\rangle$ and $|LH\rangle = \left| X - i\dfrac{\mathcal{M}\xi}{\mathcal{L}\lambda K}Y \right\rangle$ are of a LH-symmetry with a non-zero mass $m_{lh} = \mathcal{L}^{-1}$. LH-to-HH is the only possible mixing after reflection-transmission upon barrier interface.

Under our assumption, it results in two main non-zero tunnel transitions mediated by the SOI in the barrier, *resp.* $\langle Z | \hat{L}_+ | LH \rangle$ and $\langle LH | \hat{L}_+ | Z \rangle$ for respective $|Z\rangle$ and $|LH\rangle$ ingoing waves, the latter process admitting both LH → LH and LH → HH tunneling branching inside the barrier through the non-perfect orthogonality between LH and HH states under oblique incidence.

Note that, among all states (eigenvectors), only $\left| X - i\dfrac{\mathcal{M}\xi}{\mathcal{L}\lambda K}Y \right\rangle_{TB}$ carries a

non-zero orbital moment equaling $\langle LH_{TB} | \hat{L}_z | LH_{TB} \rangle = \dfrac{\lambda K\xi\mathcal{L}\mathcal{M}}{(\lambda K\mathcal{L})^2 + (\xi\mathcal{M})^2} \mu_B$ ($\mu_B$ is the Bohr magneton).

# APPENDIX C: THE 30-BAND FULL-BZ k · p TREATMENT OF THE SKEW TUNNELING

We have implemented a robust tunneling 30-multiband $\mathbf{k} \cdot \mathbf{p}$ calculation method, beyond the 14-band treatment, to calculate the properties of chiral transmission in tunnel junction over the whole BZ if needed. We describe here the method we employed and compare the results to standard tunneling codes.

The requirement to implement a 30-band full-BZ $\mathbf{k} \cdot \mathbf{p}$ treatment originates from the need to describe the tunneling transport in the valleys close at the edge of the first BZ ($1 - BZ$). In that context, Cardona and Pollak [213] used a 15-function basis, without SOI, to describe the band dispersion throughout the whole BZ without supplementary Luttinger parameters. Cavassilas [214] used a 20-function basis, including the spin, and introduced two bands named $s*$ and pseudo-Luttinger parameters to mimic the $d$ levels along the ideas of Vogl et al. for LCAO. This 20-band model contains ten adjustable parameters to describe the $s^*$ bands, nine coupling parameters for the $T_d$ group (only six for $O_h$) and six pseudo-Luttinger parameters, i.e. 25 adjustable parameters. However, this 20-band Hamiltonian do not give access to the $L$ valley of the second CB. Richard et al. [170] then proposed a 30-band model to describe $T_d$ or $O_h$ groups involving SOI. The 15 states of the real crystal correspond to the [000], $(2\pi/a)$[111], and $(2\pi/a)$[200] plane-wave states of free electrons in the "empty" germanium lattice. The large gap between the $(2\pi/a)$[200], and $(2\pi/a)$[200] plane waves (more than 15 eV) suggests that these 15 states are sufficient to obtain a correct energy band diagram. Details of the 30-band method can be found in Refs. [170, 208, 209] for III–V and Ref. [207] for group IV semiconductors. The typical energy of the bands one considers at the $\Gamma$ symmetry point are given in Table 4.4 for GaAs.

The $O_h$ symmetry-group (Si and Ge), admits ten $\mathbf{k} \cdot \mathbf{p}$ matrix elements of interest and seven more compared to the 8-band description among: $P_{Xd} = \langle X_C | p_y | iZ_d \rangle$, $P_3 = \langle D_1 | p_x | iX \rangle$, $P_{3d} = \langle D_1 | p_x | iX_d \rangle$, $P_2 = \langle S_2 | p_x | iX \rangle$, $P_{2d} = \langle S_2 | p_x | iX_d \rangle$, $P_S = \langle S_v | p_x | iX_C \rangle$, $P_U = \langle S_U | p_x | iX_C \rangle$ whereas, for the $T_d$ symmetry group: $P'$, $P'_d = \langle X_d | p_x | iZ \rangle$, $P'_3 = \langle D_1 | p_x | iX_C \rangle$, $P'_2 = \langle S_2 | p_x | iX_C \rangle$, $P'_S = \langle S_v | p_x | iX \rangle$, $P'_{Sd} = \langle S_v | p_x | iX_d \rangle$, $P'_U = \langle S_U | p_x | iX \rangle$, $P'_{Ud} = \langle S_U | p_x | iX_d \rangle$, $P_S = \langle S_v | p_x | iX_C \rangle$, $P_U = \langle S_U | p_x | iX_C \rangle$ are also non-zero. The parameters for GaAs and AlAs compounds we used are gathered in Table 4.5.

The supplementary atomic SOI is introduced via the following couplings: (i) $\Delta_d = \dfrac{3\hbar}{4m_0^2 c^2} \langle X_d | \dfrac{\partial V}{\partial x} p_y - \dfrac{\partial V}{\partial y} p_x | iY_d \rangle$, (ii) the coupling between the two different multiplets $\left( \Gamma_7, \Gamma_8 \right)$ and $\left( \Gamma_{7d}, \Gamma_{8d} \right)$, $\Delta_{dso} = \dfrac{3\hbar}{4m_0^2 c^2} \langle X_d | \dfrac{\partial V}{\partial x} p_y - \dfrac{\partial V}{\partial y} p_x | iY \rangle$, (iii) the coupling between the $\left( \Gamma_{7C}, \Gamma_{8C} \right)$ multiplet and the $\left( \Gamma_{7d}, \Gamma_{8d} \right)$ which stems from $\Gamma_{5C}$ levels and $\Gamma_8$ level which stem for $\Gamma_3$, $\Delta_{3C} = \dfrac{3\hbar}{4m_0^2 c^2} \langle D_1 | \dfrac{\partial V}{\partial x} p_z - \dfrac{\partial V}{\partial y} p_y | iX_C \rangle$. For the $T_d$ group, there are some additional SOI terms beyond $\Delta'$: (i) the coupling inside $\left( \Gamma_7, \Gamma_8 \right)$ multiplets $\Delta'_{Cd} = \dfrac{3\hbar}{4m_0^2 c^2} \langle X_d | \dfrac{\partial V}{\partial x} p_y - \dfrac{\partial V}{\partial y} p_x | iY_C \rangle$, (ii) the coupling inside $\left( \Gamma_7, \Gamma_8 \right)$

## TABLE 4.4
### $k = 0$ Energy Level (eV) used in the 30-Band k·p Model

|  | $\Gamma_{8C}$ | $\Gamma_{7C}$ | $\Gamma_6$ | $\Gamma_8$ | $\Gamma_7$ | $\Gamma_{6v}$ | $\Gamma_{6q}$ | $\Gamma_{8d}$ | $\Gamma_{7d}$ | $\Gamma_{8-3}$ | $\Gamma_{6u}$ |
|---|---|---|---|---|---|---|---|---|---|---|---|
| GaAs | 4.569 | 4.488 | 1.519 | 0 | −0.341 | −12.55 | 13.64 | 11.89 | 11.89 | 10.17 | 8.56 |
| AlAs | 4.69 | 4.54 | 3.13 | 0 | −0.30 | −12.02 | 13.35 | 12.57 | 12.34 | 9.12 | 7.65 |

## TABLE 4.5
### Dipole Matrix Elements of GaAs and AlAs in a 30-Band k·p Model given in Unit of Energy in Definition of Ref. [170]. Energies E and Matrix Components P are Linked by $E = 2m_0/\hbar^2 P^2$

|  | $E_P$ | $E_{PX}$ | $E_{Pd}$ | $E_{PXd}$ | $E_{P3}$ | $E_{P3d}$ | $E_{P2}$ | $E_{P2d}$ | $E_{PS}$ | $E_{PU}$ | $E'_P$ |
|---|---|---|---|---|---|---|---|---|---|---|---|
| GaAs | 22.37 | 17.65 | 0.01 | 4.344 | 4.916 | 8.88 | 0.032 | 23.15 | 2.434 | 19.63 | 0.0656 |
| AlAs | 19.14 | 14.29 | 0.01 | 8.49 | 3.99 | 9.29 | 0.032 | 15.01 | 1.79 | 16.00 | 0.14 |

which stems from the $\Gamma_{5C}$ levels and $\Gamma_8$ level and from the $\Gamma_3$,

$$\Delta_3' = \frac{3\hbar}{4m_0^2 c^2} \langle D_1 | \frac{\partial V}{\partial x} p_y - \frac{\partial V}{\partial y} p_x | iX \rangle, \Delta_{3d}' = \frac{3\hbar}{4m_0^2 c^2} \langle D_1 | \frac{\partial V}{\partial x} p_z - \frac{\partial V}{\partial y} p_y | iX_d \rangle, \text{as}$$

well as (iii) possible supplementary Luttinger parameters in order to account for the effects of more remote bands.

*Spurious states in the 14- and 30-bands* $\mathbf{k} \cdot \mathbf{p}$ *approach. A Novel 'ghost-band' method.*

The appearance of spurious states at large $k$, within (14-band) or outward (30-band) the $1 - BZ$, arises from the truncation procedure inherent to the $\mathbf{k} \cdot \mathbf{p}$ method. Those generally $\hat{y}$ crossing the gap, are detrimental for the calculation of the elastic tunneling transmissions occurring at constant energy $E$. As a result, they have to be removed from the gap without any possibility to suppress them because of the *matching rules*. The method we have adopted is the generalization of the procedure given by Ref. [212] to the 14- and 30-band problems. In order to extend the procedure to a wider region within the BZ away from the $\Gamma$ point, we propose the so-called *ghost-band method*. The method consists of adding supplementary off-diagonal coupling terms ($P_{\text{off}} = \alpha^{ij} k_z^2$) via a set of $\alpha^{ij}$ parameter conveniently chosen, and where $k_z$ is directed along the current flow ($z$). The idea is to utilize the *off-diagonal* formalism [212], but operate at the edge of the $1 - BZ$. We report here the main issues:

1. New fictitious bands (the ghost-bands) of adequate symmetry are introduced in order to branch-on $P_{\text{off}}$. These couplings are calculated in order to leave unchanged the properties of the physical bands in the vicinity of the valleys. These ghost-bands mimic, on average, the other physical truncated remote bands. Their mean position in energy, higher than the $S$-type CB, have to be set by error (on the wave function components and energy difference) minimization procedures. $P_{\text{off}}$ allows inversion of the concavity of the spurious bands at large $k$, thus removing the gap-crossing.

2. The $\mathbf{k} \cdot \mathbf{p}$ Hamiltonian $\mathcal{H}_{k.p}$ involving the spurious state is changed into a novel Hamiltonian $\tilde{\mathcal{H}}_{k.p}$ with $\tilde{\mathcal{H}}_{k.p} + S^{-1} V_{\text{off}} S \, k_z^2$ with $\alpha^{ij} = \left[ S^{-1} V_{\text{off}} S \right]^{ij}$ and where $S$ is the matrix of eigenstates at the point where the off-diagonal coupling with ghost-bands operates (near the edge of the $1 - BZ$).

3. The $\alpha^{ij}$ parameters chosen are calculated to leave totally unchanged the bare Hamiltonian near the $\Gamma$ point. This is performed via effective Hamiltonian analyses. The main argument is that $\alpha^{ij} k_z^2 \ll P k_z$ at small $k_z$.

4. The $\alpha^{ij}$ parameters chosen are calculated to leave totally unchanged the current-operator $(1/\hbar)(\partial \hat{H}/\partial k_z)$ near the $\Gamma$ point from $2\alpha^{ij} k_z \ll P$ at small $k_z$.

5.  In order to optimize the transport in the $X$ or $L$–Valleys of the CB in the vicinity of the $1 - BZ$ edge, one needs to apply $V_{off}$ between spurious and ghost bands in the local basis where the Hamiltonian is diagonal (matrix of eigenstates $\mathcal{S}^{\alpha\beta}$) in order to leave unchanged the remaining physical bands. It has the result that:

    a.  The electronic and transport properties are not affected at the vicinity of the $\Gamma$ point for all the CB, VB, HH, LH, and SO bands by $V_{off}$.

    b.  The electronic and transport properties of the CB are not affected at the vicinity of the point where $V_{off}$ is introduced (close to the first $BZ$-edge). The tunneling current mediated by a finite evanescent wave-vectors will be only poorly affected by the method. Those *ghost-band evanescent states* correspond to very large evanescent wave-vectors.

In order to check the validity of our 30 ghost-band approach, we have calculated the in-plane energy dispersion of holes in an 5.1 nm AlAs/6.2 nm GaAs/5.1 nm AlAs QW along the $X$ direction with our 30-band k · p tunneling code with bulk 30-band parameters of III–V, GaAs and AlAs, extracted from Richard et al. [170] (Figure 4.23 left). One compares the numerical results obtained that way with effective 6-band k · p calculations displayed, at right, for the two $X$ and $K$ directions. Results are in excellent agreement which proves the relevance of our 30-ghost band approach for the tunneling issues in heterostructures. We focus on the particular point that the AlAs barriers admit an indirect gap along the $X$-valley which is perfectly taken into account in our 30-band tunneling model.

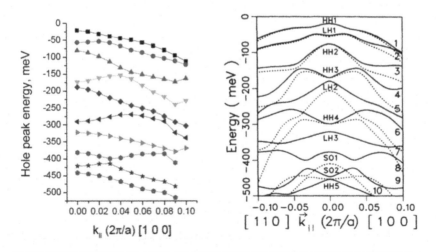

**FIGURE 4.23** Comparison between the in-plane hole energy dispersion for a AlAs/6.21 nm GaAs/AlAs quantum well (QW) derived from our 30-band tunnel k · p code (left) along the $X$ direction of the BZ, and the one derived from Ref. [215] obtained with a 6-band model along both $X$ and $K$ directions.

## APPENDIX D: THEORY OF SPIN-TRANSPORT AND SPIN-TRANSFER WITHIN THE k · p FRAMEWORK

Our calculations of the transmission coefficients are based on the multiband transfer matrix technique developed in Ref. [85, 86] and applied to the hole $6 \times 6$ valence band $\mathbf{k} \cdot \mathbf{p}$ Hamiltonian $H_h$. We use matching conditions for the wave functions and wavecurrent given by Smith and Mailhiot [88]. To calculate the transmission coefficient, we will use the transfer-matrix technique [85, 86] or the layer-by-layer S-scattering matrix formalism [216], and represent the device as a stack of two-dimensional flat-band interior layers of *constant* average electrostatic potential. The 0th (L) and $(N+1)$th (R) layers are semi-infinite emitter and collector. We will take into account elastic processes only, i.e. assume that the hole energy $\varepsilon_h$ and lateral momentum $k_\parallel$ are conserved during tunneling. Substituting $k_z \rightarrow -i\hbar \partial / \partial z$ into the Kane Hamiltonian matrix, we can find wave function eigenvectors $\psi_{n,\alpha}$ depending on $k_\parallel$ in the $\alpha th$ flat-band region ($\alpha$) according to:

$$\psi_n(z) = \sum_{\alpha=1 \rightarrow 6} [A_{n,\alpha}^+ \upsilon(k_{n,\alpha}^+) \exp(i k_{n,\alpha}^+ z) + A_{n,\alpha}^- \upsilon(k_{n,\alpha}^-) \exp(i k_{n,\alpha}^- z)] \quad (4.24)$$

where $\psi_{n,\alpha}$ and $\upsilon_{n,\alpha}$ are six-component column vectors and we separated all our solutions into two subsets with $k_{n,\alpha}^+$ corresponding to either traveling waves carrying the probability current from left to right or to evanescent waves decaying to the right and $k_{\alpha,n}^-$ corresponding to their left counterparts. Since the Kramers symmetry is broken in a magnetic region, $k_n^- \neq -k_n^+$ for real $k$ (traveling waves), however, complex $k$ always occur in complex conjugated pairs, i.e., $k_n^- = k_n^{+*}$. The latter condition is a consequence of the Hermitian character of the hole Hamiltonian. This condition is necessary to ensure that the current across the device is in the steady state. It results also that, in the $\mathbf{k} \cdot \mathbf{p}$ approach, the current operator in the $z$ direction writes $\hat{j} = \frac{1}{\hbar} \nabla_k \hat{H}_h$ [85, 86].

The boundary conditions to match at a single interface between two different materials are:

1. the continuity of the six components of the envelope function according to $\psi_n^+ + \sum_{\bar{n}} r_{n,\bar{n}} \psi_{\bar{n}}^- = \sum_{n'} t_{n,n'} \psi_{n'}^+$ where the subscript $t_{n,n'}$ ($r_{n,\bar{n}}$) refers to the respective transmission (reflection) amplitude from *incident* (*n*), *reflected* ($\bar{n}$) and *transmitted* (*n'*) waves together with

2. the continuity of the six components of the current wave-vector according to $\hat{j} \psi_n^+ + \sum_{\bar{n}} r_{n,\bar{n}} \hat{j} \psi_{\bar{n}}^- = \sum_{n'} t_{n,n'} \hat{j} \psi_{n'}^+$.

The boundary conditions at each interface $z_\alpha$ ($\alpha = 0...N$), corresponding to the continuity of the wave function and the current can be expressed in matrix form as follows:

$$\chi_\alpha \pi^\alpha(z_\alpha) A_\alpha = \chi_{\alpha+1} \pi^{\alpha+1}(z_{\alpha+1}) A_{\alpha+1} \quad (4.25)$$

where $\chi_\alpha$ is the $12 \times 12$ matrix containing the 12 eigenvectors $\upsilon(k_{n,\alpha})$ as well as the 12 current vectors $\hat{j}_z(k_{n,\alpha})\upsilon(k_{n,\alpha})$

$$\chi_\alpha = \begin{bmatrix} \cdot & \upsilon(k_{n,\alpha}^+) & \cdot & \upsilon(k_{n,\alpha}^-) & \cdot \\ \cdot & \hat{j}_z(k_{n,\alpha}^+)\upsilon(k_{n,\alpha}^+) & \cdot & \hat{j}_z(k_{n,\alpha}^-)\upsilon(k_{n,\alpha}^-) & \cdot \end{bmatrix} \qquad (4.26)$$

and where $\pi_{\nu,\mu}^\alpha = \exp(ik_{n,\alpha}z)\delta_{\nu\mu}$ is the propagation matrix within the layer $\alpha$.

We adopt the following convention for enumeration of the lead layers and all related quantities: $0 \equiv L$ (emitter) and $N + 1 \equiv R$ (collector). The system of linear Eqs. above can be solved as

$$\begin{bmatrix} A_L^+ \\ A_L^- \end{bmatrix} = \mathcal{M} = \begin{bmatrix} A_R^+ \\ A_R^- \end{bmatrix} = \begin{bmatrix} \mathcal{M}_{++} & \mathcal{M}_{+-} \\ \mathcal{M}_{-+} & \mathcal{M}_{--} \end{bmatrix}\begin{bmatrix} A_R^+ \\ A_R^- \end{bmatrix} \qquad (4.27)$$

where the $12 \times 12$ transfer matrix

$$\mathcal{M} = \chi_L^{-1}\left[ \Pi_{n=1..N}\chi_\alpha\pi^\alpha(z_{n+1} - z_n)\chi_\alpha^{-1} \right]\chi_R\pi^{N+1}(z_N) \qquad (4.28)$$

is partitioned according to the hole propagation direction in the emitter (L) and collector (R) and where the boundary condition is given by $A_R^- = 0$. It follows that the respective $6 \times 6$ amplitude transmission and reflection matrices are written respectively $t_{n,n'} = [\mathcal{M}_{++}]^{-1}$ and $r_{n,n'} = [\mathcal{M}_{+-}] \times [\mathcal{M}_{++}]^{-1}$.

Nevertheless, one must be aware that the **k · p** theory of bulk semiconductors is based on the consideration of the symmetry of the zone-center eigenstates and a small number of interband matrix elements of the momentum: it generally ignores the problem of the intracell shape and localization of the Bloch functions. However, when the crystal potential is modified on the scale of a unit cell, as, for example, in the case of an abrupt interface, it is intuitively evident that the matrix elements of the perturbation will be strongly dependent on the local properties of the Bloch functions [205, 217, 218]. In general, the reduced point symmetry $C_{2v}$ of a zinc-blende-based (001) interface allows mixing between heavy-hole and light-hole states even under normal incidence. However, this HH-LH non-diagonal mixing for the wave function matching at one interface are calculated to be smaller than 10% for the elastic hole energy considered [205]. More refined calculations of spin-polarized hole transmission trough zinc-blende (001) heterojunction would require tight-binding treatment, as proposed by Sankowski et al. [162].

Assuming that the in-plane wave-vector $k_\parallel$ is conserved, the spin-polarized transmission $t_{n,n'}$ and reflection $r_{n,n'}$ amplitudes through a specific heterojunction can be determined by a matrix numerical procedure detailed in Refs. [85, 86]. The transmission coefficients $\mathbf{T}_{n,n'}^{k_\parallel,\hat{m}}$ and elastic tunneling current $j_T^{\hat{m}}$ follows on the shell $(\varepsilon, k_\parallel)$ according to Landauer–Buttiker formula:

$$\mathbf{T}_{n,n'}^{k_\parallel,\hat{m}} = [t_{n,n'}^{k_\parallel,\hat{m}}]^* t_{n,n'}^{k_\parallel,\hat{m}} \frac{< \psi_{n'}^{k_\parallel,\hat{m}} \mid \hat{j} \mid \psi_{n'}^{k_\parallel,\hat{m}} >}{< \psi_n^{k_\parallel,\hat{m}} \mid \hat{j} \mid \psi_n^{k_\parallel,\hat{m}} >} \qquad (4.29)$$

$$j_T^{\hat{m}} = \frac{e}{h} \sum_{k_\parallel, n, n'} \int_{\varepsilon_F}^{\varepsilon_m} \mathbf{T}_{n,n'}^{k_\parallel, \hat{m}}(\varepsilon, eV) d\varepsilon \tag{4.30}$$

where $\varepsilon$ is the hole energy considered whereas $\varepsilon_m$ refers to the lowest absolute hole energy involved in the transport.

The TAMR $= [j_T^{[001]} - j_T^{[100]}) / [j_T^{[100]})$ is obtained numerically after summation of the tunneling current Equation 4.3 over $n$, $n'$ and $k_\parallel$ and integration on the energy shell $\varepsilon$ taking $\varepsilon_F = -180$ meV and $\varepsilon_m = -140$ meV (considering a typical energy broadening for (Ga,Mn)As of 40 meV). In the calculations performed throughout the article, when not explicitly written, the Fermi level $\varepsilon_F$ of (Ga,Mn)As was fixed at $-180$ meV from the top of the (Ga,Mn)As valence band, thus giving a reasonable hole concentration of about $2 \times 10^{20}$ cm$^{-3}$. Concerning the heterostructure itself, the average VBO between (Ga,Mn)As and AlAs (GaAs), varying linearly with bias, was fixed at 0.7 eV (0.22 eV) at zero bias, corresponding to an effective barrier height of 0.58 eV (0.1 eV). Note that the biaxial strain Hamiltonian $H_{BS}$ can be also added in the Hamiltonian form $H_h$ when strain and stress effect need to be considered.

## GENERAL ARGUMENTS ON SPIN TRANSFER PHENOMENA

Concerning the calculation of spin torque in III–V heterojunctions, one can start from the above equation giving the time derivative of the angular momenta density operator $\hat{\mu}_i = \hat{\rho}\hat{J}_i$ at a certain position inside the heterostructure where $\hat{J}_i$ is the angular momentum operator along direction $i$ and $\vec{J}$ the velocity operator:

$$\frac{\partial \hat{\mu}_i}{\partial t} = -\nabla\left(\vec{j}\,\hat{J}_i\right) + \hat{\rho}\frac{\partial \hat{J}_i}{\partial t}$$

$$\frac{\partial \hat{\mu}}{\partial t} = -\nabla\left(\vec{j}\,\hat{J}_i\right) + \frac{\Delta_{exc}}{\hbar}\hat{m} \times \hat{\mu} \tag{4.31}$$

Summing the latter equation over the different hole states included in the transport gives out (after having considered by the time derivative of the angular moment operator is null in the limit of coherent elastic transport):

$$\left\langle \nabla\left(\vec{j}\,\hat{J}_i\right) \right\rangle = \frac{\Delta_{exc}}{\hbar}\hat{m} \times \langle \hat{\mu} \rangle \tag{4.32}$$

This gives *in fine* the torque at a given interface (int.) after integration over the semi half-infinite structure:

$$\left\langle \vec{j}\,\hat{J}_i \right\rangle_{int.} = \vec{\mathcal{H}}_{exc} \times \langle \vec{\mu}_J \rangle \tag{4.33}$$

where:

$\vec{\mathcal{H}}_{exc}$   is the average exchange field experienced by holes

$\mu_J$    is the magnetic moment vector (per unit surface) injected in the thin layer subject to the torque by the spin-polarized current

We recover here the general results established by Kalitsov et al. [186] in the case of a transport in the CB using the Keldysh framework.

This demonstrates the presence of two intertwined contributions $\langle \vec{j} \, \hat{j}_i \rangle_x$ (parallel to the layer) and $\langle \vec{j} \, \hat{j}_i \rangle_z$ (perpendicular to the layer) for the spin torque ($\hat{y}$ is the magnetization direction) as derived for symmetric [187, 188] and asymmetric barriers [189] and demonstrated experimentally in epitaxial MgO barrier MTJs [190, 191]. These two components of the transverse spin currents are also responsible for the particular angular dependence of the tunneling transmission, as in the case of the conduction band [219, 220]. These two components of the torque are linked to the average value of the two components of the angular momenta carried by the hole current [186]. Within the general circuit-theory formalism developed by Braatas et al. [221] and adapted to metallic trilayers, one can establish that these two contributions in magnetic tunnel junctions to the torque can be linked to the magnetic moment in the barrier (generalization of the spin accumulation) through the real and imaginary part of the tunnel mixing conductance $G_{\uparrow\downarrow}$. These should be calculated in a future work.

## Spin Transfer Phenomena with SOI

We decompose now the total angular momentum **J** as a sum of spin and orbital momentum to give $\mathbf{J} = \mathbf{l} + \mathbf{s}$. In order to determine the efficiency of spin-transfer and spin-orbit-torque, one needs to compute the dynamics of the spin-polarized carriers with spin **s** and local $3d$ transition metal magnetization **M** according to their dynamics. Starting with the overall carrier-Hamiltonian:

$$\mathcal{H} = \mathcal{H}_0 + \mathcal{H}_{\text{exc}} + \mathcal{H}_{SO}, \tag{4.34}$$

where:

$\mathcal{H}_0$    is the spin-independent part (e.g. kinetic part)

$\mathcal{H}_{\text{exc}} = -J_{\text{exc}}\mathbf{s} \cdot \mathbf{M}$ is the exchange Hamiltonian between the carrier spin **s** and the local magnetization **M**

$J_{\text{exc}}$    is the exchange constant

$\mathcal{H}_{SO}$    is the spin-orbit term

From a quantum mechanical picture, the dynamics of the spin **s** inside the ferromagnetic layers is written:

$$\frac{\partial \mathbf{s}}{\partial t} = \frac{1}{i\hbar}[\mathbf{s}, \mathcal{H}_{\text{exc}}] + \mathcal{P}_s \mathbf{n} - \frac{\mathbf{s}}{\tau_s}$$

$$\frac{\partial \mathbf{s}}{\partial t} = \frac{J_{\text{exc}}}{\hbar}\mathbf{s} \times \mathbf{M} + \mathcal{P}_s\mathbf{n} - \frac{\mathbf{s}}{\tau_s}, \tag{4.35}$$

$$\frac{\partial \mathbf{M}}{\partial t} = -\frac{\partial \mathbf{s}}{\partial t} = \frac{J_{\text{exc}}}{\hbar}\mathbf{M} \times \mathbf{s} \tag{4.36}$$

if $\mathcal{H}_{SO}$ is neglected, and where $\mathcal{P}_s$ is the rate of the spin injected from an external source (current, SHE spin-torque) at the interface with the ferromagnet, $\mathbf{n}$ is the vectorial direction of the spin injected, and $\tau_s$ is a characteristic longitudinal spin-flip time inside the ferromagnet. The latter equation, derived from general quantum mechanics arguments, gives the rule for the conservation of the total angular momentum shared between $\mathbf{s}$ and $\mathbf{M}$. Here, we have used:

$$\frac{\partial <\sigma>}{\partial t} = \frac{1}{i\hbar}<[\sigma,\mathcal{H}]>, \tag{4.37}$$

where $<...>$ represents a certain quantum-mechanical averaging over the nonequilibrium carrier states and $<\sigma>= \mathbf{s}$ and where $\sigma$ is the corresponding Pauli matrix.

In the case of a long spin-lifetime $\tau_s$ compared to the precession length $\tau_{\text{exc}} = \frac{\hbar}{J_{\text{exc}}}$, one has:

$$\mathbf{s} = \mathcal{P}_s\frac{\hbar}{J_{\text{exc}}}(\mathbf{n} \times \mathbf{M}), \tag{4.38}$$

giving in the end:

$$\frac{\partial \mathbf{M}}{\partial t} = \mathcal{P}_s\mathbf{M} \times (\mathbf{n} \times \mathbf{M}), \tag{4.39}$$

One integrating over the total thickness $t$ of the ferromagnet, one can express that:

$$\frac{\partial \mathbf{M}}{\partial t} = \mathcal{J}_s\mathbf{M} \times (\mathbf{n} \times \mathbf{M}), \tag{4.40}$$

where $\mathcal{J}_s$ is the spin-current injected from the external source. One recovers then the standard expression of the Slonczewski-type like torque or anti-damping torque.

In the case of a very short spin-lifetime $\tau_s$ compared to the precession length $\tau_{\text{exc}} = \frac{\hbar}{J_{\text{exc}}}$, one has:

$$\mathbf{s} = \mathcal{P}_s\,\tau_s\mathbf{n}, \tag{4.41}$$

giving in the end:

$$\frac{\partial \mathbf{M}t}{\partial t} = \mathcal{J}_s\frac{J_{\text{exc}}\tau_s}{\hbar}\mathbf{M} \times \mathbf{n}, \tag{4.42}$$

and one recovers the field-like torque component.

$H_{SO}$ introduces an additional precession term in the bulk or at interfaces (e.g. Rashba) which can lead to local spin-memory loss [222] of the longitudinal component (that is parallel to the magnetization) and to spin-decoherence of its transverse component responsible for spin-transfer. According to this, the SOI fields may have the effect, via local (interfacial) spin-precession of incoming spin-polarized carriers, to decrease the efficiency of the spin-torque (STT or SOT) reduced from the expected maximum amplitude, e.g.

$\dfrac{\partial \mathbf{M}t}{\partial t} = \mathcal{J}_s \mathbf{M} \times (\mathbf{n} \times \mathbf{M})$. This total transverse spin-current is generally divided

into a spin-torque current in the volume ($v$) of the ferromagnet and at its

interface ($S$), and expressed as $\displaystyle\int_{v+S} \dfrac{J_{exc}}{\hbar} \mathbf{M} \times <\sigma> dz$, and a spin-current dis-

sipated in the lattice (and then lost) because now $\mathcal{J}_f = \displaystyle\int_{v+S} \dfrac{J_{exc}}{\hbar} \mathbf{M} \times <\sigma> dz$

is no longer fulfilled. The presence of the SOI at surface and in bulk requires a re-examination of the generalized spin-mixing conductance as proposed recently in the case of metallic multilayers [223].

# REFERENCES

1. T. Dietl and H. Ohno, Dilute ferromagnetic semiconductors: Physics and spintronic structures, *Rev. Mod. Phys.* **86**, 187–251 (2014).
2. T. Jungwirth, J. Wunderlich, V. Novak et al., Spin-dependent phenomena and device concepts explored in (Ga,Mn)As, *Rev. Mod. Phys.* **86**, 855–896 (2014).
3. H. Kurebayashi, J. Sinova, D. Fang et al., An antidamping spin-orbit torque originating from the Berry curvature, *Nat. Nanotechnol.* **9**, 211–217 (2014).
4. K. Olejnik, V. Novak, J. Wunderlich, and T. Jungwirth, Electrical detection of magnetization reversal without auxiliary magnets, *Phys. Rev. B* **91**, 180402(R) (2015).
5. V. Novak, K. Oleijnik, J. Wunderlich et al., Curie point singularity in the temperature derivative of resistivity in (Ga,Mn)As, *Phys. Rev. Lett.* **101**, 077201 (2008).
6. M. Wang, R. P. Campion, A. W. Rushforth, K. W. Edmonds, C. T. Foxon, and B. L. Gallagher, Achieving high curie temperature in (Ga,Mn)As, *Appl. Phys. Lett.* **93**, 132103 (2008).
7. L. Chen, S. Yan, P. F. Xu et al., Low-temperature magnetotransport behaviors of heavily Mn-doped (Ga,Mn)As films with high ferromagnetic transition temperature, *Appl. Phys. Lett.* **95**, 182505 (2015).
8. P. Nemec, V. Novak, N. Tesarova et al., The essential role of carefully optimized synthesis for elucidating intrinsic material properties of (Ga,Mn)As, *Nat. Commun.* **4**, 1422 (2013).
9. L. D. Anh, N. Hai, and M. Tanaka, Observation of spontaneous spin-splitting in the band structure of an n-type zinc-blende ferromagnetic semiconductor, *Nat. Commun.* **7**, 13810, (2016).
10. I. Žutić, J. Fabian, and S. Das Sarma, Spintronics: Fundamentals and applications, *Rev. Mod. Phys.* **76**, 323–410 (2004).
11. C. Chappert, A. Fert, and F. Nguyen Van Dau, The emergence of spin electronics in data storage, *Nat. Mater.* **6**, 813–823 (2007).
12. M. Tanaka and Y. Higo, Large tunneling magnetoresistance in GaMnAs/AlAs/GaMnAs ferromagnetic semiconductor tunnel junctions, *Phys. Rev. Lett.* **87**, 026602 (2001).

13. E. Rashba, Theory of electrical spin injection: Tunnel junction as a solution of the conductivity mismatch problem, *Phys. Rev. B* **62**, R16267–R16270 (2000).

14. A. Fert, H. Jaffrès, Conditions for efficient spin injection from a ferromagnetic metal into a semiconductor, *Phys. Rev. B* **64**, 184420 (2001).

15. H. Jaffrès and A. Fert, Spin injection from a ferromagnetic metal into a semiconductor, *J. Appl. Phys.* **91**, 8111–8113 (2002).

16. A. Fert, J.-M. George, H. Jaffrès, and R. Mattana, Semiconductors between spin-polarized sources and drains, *IEEE Trans. Elec. Dev.* **54**, 921–932 (2007).

17. G. Schmidt, D. Ferrand, L. W. Molenkamp et al., Fundamental obstacle for electrical spin injection from a ferromagnetic metal into a diffusive semiconductor, *Phys. Rev. B* **62**, R4790–R4793 (2000).

18. B. Eble, C. Testelin, P. Desfonds et al., Hole nuclear spin interaction in quantum dots, *Phys. Rev. Lett.* **102**, 146601 (2009).

19. D. Brunner, B. D. Gerardot, A. Dalgarno et al., A coherent single-hole spin in a semiconductor, *Science* **325**, 70–72 (2009).

20. M. Oltscher, M. Ciorga, M. Utz, D. Schuh, D. Bougeard, and D. Weiss, Electrical spin injection into high mobility 2D systems, *Phys. Rev. Lett.* **113**, 236602 (2014).

21. M. Tanaka, S. Ohya, and P. N. Hai, Recent progress in III-V based ferromagnetic semiconductors: Band structure, Fermi level, and tunneling transport, *Appl. Phys. Rev.* **1**, 011101 (2014).

22. H. Ohno, Properties of ferromagnetic III–V semiconductors, *J. Magn. Magn. Mater.* **200**, 110–129 (1999).

23. T. Dietl, H. Ohno, F. Matsukura, J. Cibert, and D. Ferrand, Zener model description of ferromagnetism in zinc-blende magnetic semiconductors, *Science* **287**, 1019–1022 (2000).

24. H. Ohno and F. Matsukura, A ferromagnetic III–V semiconductor: (Ga,Mn)As, *Solid State Commun.* **117**, 179–186 (2001).

25. F. Matsukura, H. Ohno, and T. Dietl, III–V ferromagnetic semiconductors, in *Handbook of Magnetic Materials*, vol. 14, K. H. J. Buschow (Ed.), North Holland, Amsterdam, pp. 1–87, 2002.

26. J. Konig, J. Schliemann, T. Jungwirth, and A. H. MacDonald, Ferromagnetism in (III,V) semiconductors, in *Electronic Structure and Magnetism of Complex Material*, D. J. Singh and D. A. Papaconstantopoulos (Eds.), Material Science, Springer, Berlin, pp. 163–211, 2002.

27. T. Jungwirth, J. Sinova, J. Masek, J. Kucera, and A. H. M. MacDonald, Theory of ferromagnetic (III,Mn)V semiconductors, *Rev. Mod. Phys.* **78**, 809–864 (2006).

28. C. Zener, Interaction between the *d* shells in the transition metals, *Phys. Rev.* **81**, 440–444 (1951).

29. T. Dietl, H. Ohno, and F. Matsukura, Hole-mediated ferromagnetism in tetrahedrally coordinated semiconductors, *Phys. Rev.* **B63**, 195205 (2001).

30. S. Souma, L. Chen, R. Oszwaldowski et al., Fermi level position, Coulomb gap, and Dresselhaus splitting in (Ga,Mn)As, *Sci. Rep.* (2016), doi:10.1038, srep27266.

31. S. Ohya, I. Muneta, Y. Xin, K. Tanaka, and M. Tanaka, Valence-band structure of ferromagnetic semiconductor (In,Ga,Mn)As, *Phys. Rev. B* **86**, 094418 (2012).

32. M. Kobayashi, I. Muneta, Y. Takeda et al., Unveiling the impurity band induced ferromagnetism in the magnetic semiconductor (Ga,Mn)As, *Phys. Rev. B* **89**, 205204 (2014).

33. T. Ishii, T. Kawazoe, Y. Hashimoto et al., Electronic structure near the Fermi level in the ferromagnetic semiconductor (Ga,Mn)As studied by ultrafast time-resolved light-induced reflectivity measurements, *Phys. Rev. B* **93**, 241303 (2016).

34. I. Muneta, S. Ohya, H. Terada, and M. Tanaka, Sudden restoration of the band ordering associated with the ferromagnetic phase transition in a semiconductor, *Nat. Commun.* **7**, 12013 (2016).

35. Y. K. Kato, R. C. Myers, A. C. Gossard, and D. D. Awschalom, Observation of the spin Hall effect in semiconductors, *Science* **306**, 1910–1913 (2004).

36. J. Sinova, D. Culcer, Q. Niu, N. A. Sinitsyn, T. Jungwirth, and A. H. MacDonald, Universal intrinsic spin Hall effect, *Phys. Rev. Lett.* **92**, 126603 (2004).

37. J. Wunderlich, B. Kaestner, J. Sinova, and T. Jungwirth, Experimental observation of the spin-Hall effect in a two-dimensional spin-orbit coupled semiconductor system, *Phys. Rev. Lett.* **94**, 047204 (2005).

38. S. Murakami, N. Nagaosa, S. C. Zhang, Dissipationless quantum spin current at room temperature, *Science* **301**, 1348–1351 (2003).

39. H.-A. Engel, B. I. Halperin, and E. I. Rashba, Theory of spin Hall conductivity in n-doped GaAs, *Phys. Rev. Lett.* **95**, 166605 (2005).

40. N. Nagaosa, J. Sinova, S. Onoda, A. H. MacDonald, and N. P. Ong, Anomalous Hall effect, *Rev. Mod. Phys.* **82**, 1539–1592 (2010).

41. J. Sinova, S. O. Valenzuela, J. Wunderlich, C. H. Back, and T. Jungwirth, Spin Hall effects, *Rev. Mod. Phys.* **87**, 1213–1259 (2015).

42. J. C. Rojas Sanchez, L. Vila, G. Desfonds et al., Spin-to-charge conversion using Rashba coupling at the interface between non-magnetic materials, *Nat. Commun.* **4**, 2944 (2013).

43. E. Lesne, Y. Fu, S. Oyarzun et al., Highly efficient and tunable spin-to-charge conversion through Rashba coupling at oxide interfaces, *Nat. Mater.* **15**, 1261–1266 (2016).

44. S. Oyarzun, A. K. Nandy, F. Rortais et al., Evidence for spin-to-charge conversion by Rashba coupling in metallic states at the Fe/Ge(111) interface, *Nat. Commun.* **7**, 873 (2016).

45. I. M. Miron, G. Gaudin, S. Auffret et al., Current-driven spin torque induced by the Rashba effect in a ferromagnetic metal layer, *Nat. Mater.* **9**, 230–234 (2010).

46. L. Liu, O. J. Lee, T. J. Gudmundsen, D. C. Ralph, and R. A. Buhrman, Current-induced switching of perpendicularly magnetized magnetic layers using spin torque from the spin Hall effect, *Phys. Rev. Lett.* **109**, 096602 (2012).

47. K. Garello, I. M. Miron, C. O. Avci et al., Symmetry and magnitude of spin-orbit torques in ferromagnetic heterostructures, *Nat. Nanotechnol.* **8**, 587–593 (2013).

48. O. Krupin, G. Bihlmayer, K. Starke et al., Rashba effect at magnetic metal surfaces, *Phys. Rev. B* **71**, 201403 (2005).

49. G. Bihlmayer, Y. Koroteev, Y. M. Echenique et al., The Rashba-effect at metallic surfaces, *Surf. Sci.* **600**, 3888–3891 (2006).

50. A. Matos-Abiague and J. Fabian, Tunneling anomalous and spin Hall effects, *Phys. Rev. Lett.* **115**, 056602 (2015).

51. T. H. Dang, H. Jaffrès, T. L. Hoai Nguyen, and H.-J. Drouhin, Giant forward-scattering asymmetry and anomalous tunnel Hall effect at spin-orbit-split and exchange-split interfaces, *Phys. Rev. B* **92**, 060403(R) (2015); T. Huong Dang, D. Quang To, E. Erina, H. Jaffrès, V. I. Safarov, H. Jaffres, and H.-J. Drouhin, Theory of the anomalous tunnel Hall Effect at ferromagnet-semiconductor junctions, *J. Magn. Magn. Mater.* **459**, 37–42 (2018).

52. B. Scharf, A. Matos-Abiague, J.-E. Han, E. M. Hankiewicz, and I. Žutić, Tunneling planar hall effect in topological insulators: Spin valves and amplifiers, *Phys. Rev. Lett.* **117**, 166806 (2016).

53. C. O. Avci, K. Garello, A. Ghosh, M. Gabureac, S. F. Alvarado, and P. Gambardella, Unidirectional spin Hall magnetoresistance in ferromagnet/normal metal bilayers, *Nat. Phys.* **11**, 570–575 (2015).

54. K. Yasuda, A. Tsukazaki, R. Yoshimi, K. S. Takahashi, M. Kawasaki, and Y. Tokura, Large unidirectional magnetoresistance in a magnetic topological insulator, *Phys. Rev. Lett.* **117**, 127202 (2016).

55. M. Jamet, A. Barski, T. Devillers, and V. Poydenot, High-Curie-temperature ferromagnetism in self-organized $Ge_{1-x}Mn_x$ nanocolumns, *Nat. Mater.* **5**, 653–659 (2006).

56. D. Chiba, F. Matsukura, and H. Ohno, Tunneling magnetoresistance in (Ga,Mn)As-based heterostructures with a GaAs barrier, *Physica E* **21**, 966–969 (2004).

57. R. Mattana, M. Elsen, J.-M. George et al., Chemical profile and magnetoresistance of $Ga_{1-x}Mn_xAs/GaAs/AlAs/GaAs/Ga_{1-x}Mn_xAs$ tunnel junctions, *Phys. Rev. B* **71**, 075206 (2005).

58. M. Elsen, H. Jaffrès, R. Mattana et al., Spin-polarized tunneling as a probe of the electronic properties of $Ga_{1-x}Mn_xAs$ heterostructures, *Phys. Rev. B* **76**, 144415 (2007).

59. J.-M. George and H. Jaffrès, Spin-transfer with GaMnAs/InGaAs/GaMnAs tunnel junction, (unpublished).

60. S. H. Chun, S. J. Potashnik, K. C. Ku, P. Schiffer, and N. Samarth, Spin-polarized tunneling in hybrid metal-semiconductor magnetic tunnel junctions, *Phys. Rev. B* **66**, 100408 (2002).

61. H. Saito, S. Yuasa, K. Ando, Y. Hamada, and Y. Suzuki, Spin-polarized tunneling in metal-insulator-semiconductor $Fe/ZnSe/Ga_{1-x}Mn_xAs$ magnetic tunnel diodes, *Appl. Phys. Lett.* **89**, 232502 (2006).

62. H. Saito, A. Yamamoto, S. Yuasa, and K. Ando, High tunneling magnetoresistance in $Fe/GaO_x/Ga_{1-x}Mn_xAs$ with metal/insulator/semiconductor structure, *Appl. Phys. Lett.* **93**, 172515 (2008).

63. S. Ohya, N. Hai, Y. Mizuno, and M. Tanaka, Quantum-size effect and Tunneling Magnetoresistance in ferromagnetic-semiconductor quantum heterostructure, *Phys. Rev. B* **75**, 155328 (2007).

64. S. Ohya, N. Hai, and M. Tanaka, Tunneling magnetoresistance in GaMnAs/AlAs/InGaAs/AlAs/GaMnAs double-barrier magnetic tunnel junctions, *Appl. Phys. Lett.* **87**, 012105 (2005).

65. S. Ohya, I. Muneta, N. Hai, and M. Tanaka, valence-band structure of ferromagnetic semiconductor GaMnAs studied by spin-dependent resonant tunneling spectroscopy, *Phys. Rev. Lett.* **104**, 167204 (2010).

66. C. Gould, C. Ruster, T. Jungwirth et al., Tunneling anisotropic magnetoresistance: A spin-valve-like tunnel magnetoresistance using a single magnetic layer, *Phys. Rev. Lett.* **93**, 117203 (2004).

67. C. Ruster, C. Gould, T. Jungwirth et al., Very Large Tunneling anisotropic magnetoresistance of a (Ga,Mn)As/GaAs/(Ga,Mn)As stack, *Phys. Rev. Lett.* **94**, 027203 (2005).

68. H. Saito, S. Yuasa, and K. Ando, Origin of the tunnel anisotropic magnetoresistance in $Ga_{1-x}Mn_xAs/ZnSe/Ga_{1-x}Mn_xAs$ magnetic tunnel junctions of II–VI/III–V heterostructures, *Phys. Rev. Lett.* **95**, 086604 (2005).

69. R. Giraud, M. Gryglas, L. Thevenard, A. Lemaitre, and G. Faini, Voltage-controlled tunneling anisotropic magnetoresistance of a ferromagnetic $p^{++}$-(Ga,Mn)As/$n^+$-GaAs Zener-Esaki diode, *Appl. Phys. Lett.* **87**, 242505 (2005).

70. A. Giddings, M. Khalid, T. Jungwirth et al., Large tunneling anisotropic magnetoresistance in (Ga,Mn)As nanoconstrictions, *Phys. Rev. Lett.* **94**, 127202 (2005).

71. M. Elsen, H. Jaffrès, R. Mattana et al., Exchange-mediated anisotropy of (Ga,Mn)As valence-band probed by resonant tunneling spectroscopy, *Phys. Rev. Lett.* **99**, 127203 (2007).

72. A. Matos Abiague and J. Fabian, Anisotropic tunneling magnetoresistance and tunneling anisotropic magnetoresistance: Spin-orbit coupling in magnetic tunnel junctions, *Phys. Rev. B* **79**, 155303 (2009).

73. A. Matos Abiague, M. Gmitra, and J. Fabian, Angular dependence of the tunneling anisotropic magnetoresistance in magnetic tunnel junctions, *Phys. Rev. B* **80**, 045312 (2009).

74. J. Slonczewski, Current-driven excitations of magnetic multilayers, *J. Magn. Magn. Mater.* **159**, L1–L7 (1996).

75. L. Berger, Emission of spin waves by a magnetic multilayer traversed by a current, *Phys. Rev. B* **54**, 9353–9358 (1996).

76. A. Fert, A. Barthelemy, J. Ben Youssef et al., Review of recent results on spin polarized tunneling and magnetic switching by spin injection, *Mater. Sci. Eng. B* **84**, 1–9 (2001).

77. M. Stiles and J. Miltat, *Spin Dynamics in Confined Magnetic Structures III*, B. Hillebrands and A. Thiaville (Eds.), Springer, vol. 101, pp 225–308 (2006).

78. D. C. Ralph and M. D. Stiles, Spin transfer torques, *J. Magn. Magn. Mater.* **30**, 1190–1216 (2008).

79. D. Chiba, Y. Sato, T. Kita, F. Matsukura, and H. Ohno, Current-driven magnetization reversal in a ferromagnetic semiconductor (Ga,Mn)As/GaAs/(Ga,Mn) As tunnel junction, *Phys. Rev. Lett.* **93**, 216602 (2004).

80. M. Elsen, O. Boulle, J.-M. George et al., Spin transfer experiments on (Ga,Mn) As/(In,Ga)As/ (Ga,Mn)As tunnel junctions, *Phys. Rev. B* **73**, 035303 (2006).

81. J. Okabayashi, M. Watanabe, H. Toyao, T. Yamagushi, and J. Yoshino, Pulse-width dependence in current-driven magnetization reversal using GaMnAs-based double-barrier magnetic tunnel junction, *J. Supercond. Nov. Magn.* **20**, 443–446 (2007).

82. M. Watanabe, J. Okabayashi, H. Toyao, T. Yamagushi, and J. Yoshino, Current-driven magnetization reversal at extremely low threshold current density in (Ga,Mn)As-based double-barrier magnetic tunnel junctions, *Appl. Phys. Lett.* **92**, 082506 (2008).

83. M. Abolfath, T. Jungwirth, J. Brum, and A. H. MacDonald, Theory of magnetic anisotropy in $III_{1-x}Mn_xV$ ferromagnets, *Phys. Rev. B* **63**, 054418 (2001).

84. J.-M. Luttinger and W. Kohn, Motion of electrons and holes in perturbed periodic fields, *Phys. Rev.* **97**, 869–883 (1955).

85. A. G. Petukhov, A. N. Chantis, and D. O. Demchenko, Resonant enhancement of tunneling magnetoresistance in double-barrier magnetic heterostructures, *Phys. Rev. Lett.* **89**, 107205 (2002).

86. A. G. Petukhov, D. O. Demchenko, and A. N. Chantis, Electron spin polarization in resonant interband tunneling devices, *Phys. Rev. B* **68**, 125332 (2003).

87. R. Wessel and M. Altarelli, Resonant tunneling of holes in double-barrier heterostructures in the envelope-function approximation, *Phys. Rev. B* **39**, 12802–12807 (1989).

88. D. L. Smith and C. Mailhiot, Theory of semiconductor superlattice electronic structure, *Rev. Mod. Phys.* **62**, 173–234 (1990).

89. M. Jullière, Tunneling between ferromagnetic films, *Phys. Lett. A* **54**, 225–226 (1975).

90. J. S. Moodera, L. R. Kinder, T. M. Wong, and R. Meservey, Large magnetoresistance at room temperature in ferromagnetic thin film tunnel junctions, *Phys. Rev. Lett.* **74**, 3273–3277 (1995).

91. S. S. P. Parkin, C. Kaiser, A. Panchula et al., Giant tunneling magnetoresistance at room temperature with MgO(100) tunnel barriers, *Nat. Mater.* **3**, 862–867 (2004).

92. S. Yuasa, T. Nagahama, A. Fukushima, Y. Suzuki, and K. Ando, Giant room-temperature magnetoresistance in single-crystal Fe/MgO/Fe magnetic tunnel junctions, *Nat. Mater.* **3**, 868–871 (2004).

93. W. H. Butler, X.-G. Zhang, T. C. Schulthess, and J. M. MacLaren, Spin-dependent tunneling conductance of Fe/MgO/Fe sandwiches, *Phys. Rev. B* **63**, 054416 (2001).

94. J. Mathon and A. Umerski, Theory of tunneling magnetoresistance of an epitaxial Fe/MgO/Fe(001) junction, *Phys. Rev. B* **63**, 220403 (2001).

95. L. Brey, C. Tejedor, and J. Fernandez-Rossier, Tunnel magnetoresistance in GaMnAs: Going beyond Julliere formula, *Appl. Phys. Lett.* **85**, 1996 (2004).

96. L. Brey, J. Fernandez-Rossier, and C. Tejedor, Spin depolarization in the transport of holes across $Ga_{1-x}Mn_xAs/Ga_yAl_{1-y}$ As/$p$-GaAs, *Phys. Rev. B* **70**, 235334 (2004).

97. P. Krstajic and F. M. Peeters, Spin-dependent tunneling in diluted magnetic semiconductor trilayer structures, *Phys. Rev. B* **72**, 125350 (2005).

98. R. Mattana, J.-M. George, H. Jaffrès et al., Electrical detection of spin accumulation in a $p$-Type GaAs quantum well, *Phys. Rev. Lett.* **90**, 166601 (2003).

99. N. O. Lipari and A. Baldereschi, Angular momentum theory and localized states in solids. Investigation of shallow acceptor states in semiconductors, *Phys. Rev. Lett.* **25**, 1660–1664 (1970).

100. A. Baldereschi and N. O. Lipari, Spherical model of shallow acceptor states in semiconductors, *Phys. Rev. B* **8**, 2697–2709 (1973).

101. A. Baldereschi and N. O. Lipari, Cubic contributions to the spherical model of shallow acceptor states, *Phys. Rev. B* **9**, 1525–1539 (1973).

102. J. M. George, H. Jaffrès, R. Mattana et al., Electrical spin detection in GaMnAs-based tunnel junctions: Theory and experiments, *Mol. Phys. Rep.* **40**, 23–33 (2004).

103. V. Garcia, H. Jaffrès, M. Eddrief, M. Marangolo, V. H. Etgens, and J.-M. George, Resonant tunneling magnetoresistance in MnAs/III–V/MnAs junctions, *Phys. Rev. B* **72**, 081303(R) (2005).

104. V. Garcia, H. Jaffrès, J.-M. George, M. Marangolo, M. Eddrief, and V. H. Etgens, Spectroscopic measurement of spin-dependent resonant tunneling through a 3D disorder: The case of MnAs/GaAs/MnAs junctions, *Phys. Rev. Lett.* **97**, 246802 (2006).

105. G. Xiang, B. L. Sheu, M. Zhu, P. Schiffer, and N. Samarth, Noncollinear spin valve effect in ferromagnetic semiconductor trilayers, *Phys. Rev. B* **76**, 035324 (2007).

106. M. Ciorga, A. Einwanger, U. Wurstbauer, D. Schuh, W. Wegscheider, and D. Weiss, Electrical spin injection and detection in lateral all-semiconductor devices, *Phys. Rev. B* **79**, 165321 (2009).

107. P. Van Dorpe, Z. Liu, W. Van Roy et al., Very high spin polarization in GaAs by injection from a (Ga,Mn)As Zener diode, *Appl. Phys. Lett.* **84**, 3495–3497 (2004).

108. M. Kohda, T. Kita, Y. Ohno, F. Matsukura, and H. Ohno, Bias voltage dependence of the electron spin injection studied in a three-terminal device based on a (Ga,Mn)As/$n^+$-GaAs Esaki diode, *Appl. Phys. Lett.* **89**, 012103 (2006).

109. P. Van Dorpe, W. Van Roy, J. De Boeck et al., Voltage-controlled spin injection in a (Ga,Mn)As/(Al,Ga)As Zener diode, *Phys. Rev. B* **72**, 205322 (2005).

110. J. Masek, F. Maca, J. Kudrnovsky et al., Microscopic analysis of the valence band and impurity band theories of (Ga,Mn)As, *Phys. Rev. Lett.* **105**, 227202 (2010).

111. M. Linnarsson, E. Janzén, B. Monemar, M. Kleverman, and A. Thilderkvist, Electronic structure of the GaAs:$Mn_{Ga}$ center, *Phys. Rev. B* **55**, 6938–6944 (1997).

112. B. E. Larsson, K. C. Hass, H. Ehrenreich, and A. E. Carlsson, Theory of exchange interactions and chemical trends in diluted magnetic semiconductors, *Phys. Rev. B* **37**, 4137–4154 (1988).

113. J. Okabayashi, A. Kimura, O. Rader et al., Core-level photoemission study of $Ga_{1-x}Mn_xAs$, *Phys. Rev. B* **58**, R4211–R4214 (1998).

114. F. Matsukura, H. Ohno, A. Shen, and Y. Sugarawa, Transport properties and origin of ferromagnetism in (Ga,Mn)As, *Phys. Rev. B* **57**, R2037–R2040 (1998).

115. S. J. Potashnik, K. C. Ku, R. Mahendiran et al., Saturated ferromagnetism and magnetization deficit in optimally annealed $Ga_{1-x}Mn_xAs$, *Phys. Rev. B* **66**, 012408 (2002).

116. R. P. Campion, K. W. Edmonds, L. X. Zhao et al., The growth of GaMnAs films by molecular beam epitaxy using arsenic dimers, *J. Cryst. Growth* **251**, 311–316 (2003).

117. A. Richardella, P. Roushan, S. Mack, et al., Visualizing critical correlations near the metal-insulator transition in $Ga_{1-x}Mn_xAs$, *Science* **327**, 665–669 (2010).

118. B. C. Chapler, R. C. Myers, S. Mack et al., Infrared probe of the insulator-to-metal transition in $Ga_{1-x}Mn_xAs$ and $Ga_{1-x}Be_xAs$, *Phys. Rev. B* **84**, 081203 (2011).

119. A. M. Nazmul, T. Amemiya, Y. Shuto, S. Sugahara, and M. Tanaka, High temperature ferromagnetism in GaAs-based heterostructures with Mn-δ-doping, *Phys. Rev. Lett.* **95**, 017201 (2005).

120. P. W. Anderson, Antiferromagnetism. Theory of superexchange interaction, *Phys. Rev.* **79**, 350–356 (1950).

121. P. W. Anderson, Localized magnetic states in metals, *Phys. Rev.* **124**, 41–53 (1961).

122. P. W. Anderson and H. Hasegawa, Considerations on double exchange, *Phys. Rev.* **100**, 675–681 (1955).

123. J. K. Glasbrenner, I. Žutić, and I. I. Mazin, Theory of Mn-doped II–II–V semiconductors, *Phys. Rev. B* **90**, 140403(R) (2014).

124. T. O. Strandberg, C. M. Canali, and A. H. MacDonald, Magnetic interactions of substitutional Mn pairs in GaAs, *Phys. Rev. B* **81**, 054401 (2010).

125. A. W. Rushforth, N. R. S. Farley, R. P. Campion et al., Compositional dependence of ferromagnetism in (Al,Ga,Mn)As magnetic semiconductors, *Phys. Rev. B* **78**, 085209 (2008).

126. T. Jungwirth K. Y. Wang, J. Mašek et al., Prospects for high temperature ferromagnetism in (Ga,Mn)As semiconductors, *Phys. Rev. B* **72**, 165204 (2005).

127. M. Wang, K. W. Edmonds, B. L. Gallagher, et al., High Curie temperatures at low compensation in the ferromagnetic semiconductor (Ga,Mn)As, *Phys. Rev. B* **87**, 121301 (2013).

128. Y. Ohno, I. Arata, F. Matsukura, and H. Ohno, Valence band barrier at (Ga,Mn)As/GaAs interfaces, *Physica E* **13**, 521–524 (2002).

129. O. Thomas, O. Makarovsky, A. Patanè et al., Measuring the hole chemical potential in ferromagnetic $Ga_{1-x}Mn_xAs$/GaAs heterostructures by photoexcited resonant tunneling, *Appl. Phys. Lett.* **90**, 082106 (2007).

130. T. Tsuruoka, N. Tachikawa, S. Ushioda, F. Matsukura, K. Takamura, and H. Ohno, Local electronic structures of GaMnAs observed by cross-sectional scanning tunneling microscopy, *Appl. Phys. Lett.* **81**, 2800–2802 (2002).

131. B. I. Shklovskii and A. L. Efros, *Electronic Properties of Doped Semiconductors, Springer Series in Solid-State Sciences*, vol. 45, Springer–Verlag, Berlin, pp. 1–65, 1984.

132. T. Jungwirth, J. Sinova, A. H. MacDonald et al., Character of states near the Fermi level in (Ga,Mn)As: Impurity to valence band crossover, *Phys. Rev. B* **76**, 125206 (2007).

133. K. S. Burch, D. B. Shrekenhamer, E. J. Singley et al., Impurity band conduction in a high temperature ferromagnetic semiconductor, *Phys. Rev. Lett.* **97**, 087208 (2006).

134. K. Ando, H. Saito, K. C. Agarwal, M. C. Debnath, and V. Zayets, Origin of the anomalous magnetic circular dichroism spectral shape in ferromagnetic $Ga_{1-x}Mn_xAs$: Impurity bands inside the band gap, *Phys. Rev. Lett.* **100**, 067204 (2008).

135. J. M. Tang and M. E. Flatté, Magnetic circular dichroism from the impurity band in III–V diluted magnetic semiconductors, *Phys. Rev. Lett.* **101**, 157203 (2008).

136. D. Neumaier, M. Turek, U. Wurstbauer et al., All-electrical measurement of the density of states in (Ga,Mn)As, *Phys. Rev. Lett.* **103**, 087203 (2009).

137. T. Wojtowicz, W. L. Lim, X. Liu et al., Correlation of Mn lattice location, free hole concentration, and curie temperature in ferromagnetic GaMnAs, *J. Supercond. Nov. Magn.* **16**, 41–44 (2003).

138. S. J. Potashnik, K. C. Ku, R. F. Wang et al., Coercive field and magnetization deficit in $Ga_{1-x}Mn_xAs$ epilayers, *J. Appl. Phys.* **93**, 6784–6786 (2003).

139. K. Y. Wang, M. Sawiki, K. W. Edmonds et al., Control of coercivities in (Ga,Mn)As thin films by small concentrations of MnAs nanoclusters, *Appl. Phys. Lett.* **88**, 022510 (2006).

140. M. Adell, J. Adell, L. Ilver, J. Kanski, and J. Sadowski, Photoemission study of the valence band offset between low temperature GaAs and (GaMn)As, *Appl. Phys. Lett.* **89**, 172509 (2006).

141. S. Tiwari and D. J. Frank, Empirical fit to band discontinuities and barrier heights in III–V alloy systems, *Appl. Phys. Lett.* **60**, 630–632 (1992). If one takes into account explicitly the stress between (Ga,Mn)As and (In,Ga)As, this shift should correspond to the valence band offset between (Ga,Mn)As and the LH component of (In,Ga)As, that is preferentially transmitted through thick barriers.

142. S. Lodha, D. B. Janes, and N.-P. Chen, Unpinned interface fermi-level in Schottky contacts to n-GaAs capped with low-temperature-grown GaAs; experiments and modeling using defect state distributions, *J. Appl. Phys.* **93**, 2772–2779 (2003).

143. M. Malfait, J. Vanacken, W. V. Roy, G. Borghs, and V. Moshchalkov, Low-temperature annealing study of $Ga_{1-x}Mn_xAs$: Magnetic properties and Hall effect in pulsed magnetic fields, *J. Magn. Magn. Mater,* **290**, 1387–1390 (2005).

144. M. Sawiki, D. Chiba, A. Korbecka et al., Experimental probing of the interplay between ferromagnetism and localization in (Ga,Mn)As, *Nat. Phys.* **6**, 22–25 (2010).

145. Y. Nishitani, D. Chiba, M. Endo et al., Curie temperature *vs.* hole concentration in field-effect structures of (Ga,Mn) As, *Phys. Rev. B* **81**, 045208 (2010).

146. A. Koeder, S. Frank, W. Schoch et al., Curie temperature and carrier concentration gradients in epitaxy-grown $Ga_{1-x}Mn_xAs$ layers, *Appl. Phys. Lett.* **82**, 3278–3280 (2003).

147. K. Pappert, M. J. Schmidt, S. Humpfner et al., Magnetization-switched metal-insulator transition in a (Ga,Mn)As tunnel device, *Phys. Rev. Lett.* **97**, 186402 (2006).

148. K. I. Bolotin, F. Kuemmeth, and D. C. Ralph, Anisotropic magnetoresistance and anisotropic tunneling magnetoresistance due to quantum interference in ferromagnetic metal break junctions, *Phys. Rev. Lett.* **97**, 127202 (2006).

149. S. F. Shi, K. I. Bolotin, F. Kuemmeth, and D. C. Ralph, Temperature dependence of anisotropic magnetoresistance and atomic rearrangements in ferromagnetic metal break junctions, *Phys. Rev. B* **76**, 184438 (2007).

150. M. Viret, M. Gabureac, F. Ott et al., Giant anisotropic magneto-resistance in ferromagnetic atomic contacts, *Eur. Phys. J. B* **51**, 1–4 (2006).

151. A. Bernand-Mantel, P. Seneor, K. Bouzehouane et al., *Nat. Phys.* **5**, 920–924 (2009).

152. T. Uemura, Y. Imai, M. Harada, K. Matsuda, and M. Yamamoto, Tunneling anisotropic magnetoresistance in epitaxial CoFe/n-GaAs junctions, *Appl. Phys. Lett.* **94**, 182502 (2009).

153. M. Wimmer, M. Lobenhofer, J. Moser, et al., Orbital effects on tunneling anisotropic magnetoresistance in Fe/GaAs/Au junctions, *Phys. Rev. B* **80**, 121301(R) (2009).

154. J. Moser, A. Matos-Abiague, D. Schuh, W. Wegscheider, J. Fabian, and D. Weiss, Tunneling anisotropic magnetoresistance and spin-orbit coupling in Fe/GaAs/Au tunnel junctions, *Phys. Rev. Lett.* **99**, 056601 (2007).

155. L. Gao, X. Jiang, S. H. Yang et al., Bias voltage dependence of tunneling anisotropic magnetoresistance in magnetic tunnel junctions with MgO and $Al_2O_3$ tunnel barriers, *Phys. Rev. Lett.* **99**, 226602 (2007).

156. B. G. Park, J. Wunderlich, D. A. Williams et al., Tunneling anisotropic magnetoresistance in multilayer-(Co/Pt)/$AlO_x$/Pt structures, *Phys. Rev. Lett.* **100**, 087204 (2008).

157. A. N. Chantis, K. D. Belashchenko, E. Y. Tsymbal, and M. van Schilfgaarde, Tunneling anisotropic magnetoresistance driven by resonant surface states: First-principles calculations on an Fe(001) surface, *Phys. Rev. Lett.* **98**, 046601 (2007).

158. M. Zhu, M. J. Wilson, P. Mitra, P. Schiffer, and N. Samarth, Quasireversible magnetoresistance in exchange-spring tunnel junctions, *Phys. Rev. B* **78**, 195307 (2008).

159. M. J. Schmidt, K. Pappert, C. Gould, J. Schmidt, R. Oppermann, and L. W. Molenkamp, Bound-hole states in a ferromagnetic (Ga,Mn)As environment, *Phys. Rev. B* **76**, 035204 (2007).

160. J. M. Tang and M. E. Flatté, Multiband tight-binding model of local magnetism in $Ga_{1-x}Mn_xAs$, *Phys. Rev. Lett.* **92**, 047201 (2004).

161. J. M. Tang and M. E. Flatté, Spin-orientation-dependent spatial structure of a magnetic acceptor state in a zinc-blende semiconductor, *Phys. Rev. B* **72**, 161315(R) (2005).

162. P. Sankowski, P. Kacman, J. Majewski, and T. Dietl, Spin-dependent tunneling in modulated structures of (Ga,Mn)As, *Phys. Rev. B* **75**, 045306 (2007).

163. L. Thevenard, L. Largeau, O. Mauguin, A. Lemaitre, and B. Theys, Tuning the ferromagnetic properties of hydrogenated GaMnAs, *Appl. Phys. Lett.* **87**, 182506 (2005).

164. H. Ohno, N. Akira, F. Matsukura et al., Spontaneous splitting of ferromagnetic (Ga, Mn)As valence band observed by resonant tunneling spectroscopy, *Appl. Phys. Lett.* **73**, 363–365 (1998).

165. R. K. Hayden, D. K. Maude, L. Eaves et al., Probing the hole dispersion curves of a quantum well using resonant magnetotunneling spectroscopy, *Phys. Rev. Lett.* **66**, 1749–1752 (1991).

166. M. Tran, J. Peiro, H. Jaffrès, J.-M. George, and A. Lemaitre, Magnetization-controlled conductance in (Ga,Mn)As-based resonant tunneling devices, *Appl. Phys. Lett.* **95**, 172101 (2009).

167. J. Wunderlich, T. Jungwirth, B. Kaestner et al., Coulomb blockade anisotropic magnetoresistance effect in a (Ga,Mn)As single-electron transistor, *Phys. Rev. Lett.* **97**, 077201 (2006).

168. H. T. Dang, E. Erina, H. T. L. Nguyen, H. Jaffres, and H.-J. Drouhin, Spin-orbit assisted chiral-tunneling at semiconductor tunnel junctions: Study with advanced 30-band k.p methods, *Spintronics IX, Proc. SPIE* **9931**, 993127 (2016), doi:10.1117/12.2238796.

169. J.-M. Jancu, R. Scholz, E. A. de Andrada e Silva, and G. C. La Rocca, Atomistic spin-orbit coupling and k. p parameters in III–V semiconductors, *Phys. Rev. B* **72**, 193201 (2005).

170. S. Richard, F. Aniel, and G. Fishman, Energy-band structure of Ge, Si, and GaAs: A thirty-band k·p method, *Phys. Rev. B* **70**, 235204 (2004).

171. J. A. Katine, F. J. Albert, R. A. Buhrman, E. B. Myers, and D. C. Ralph, Current-driven magnetization reversal and spin-wave excitations in Co/Cu/Co pillars, *Phys. Rev. Lett.* **84**, 3149–3152 (2000).

172. S. I. Kiselev, J. Sankey, I. Krivorotov et al., Microwave oscillations of a nano-magnet driven by a spin-polarized current, *Nature* **425**, 380–383 (2003).

173. J. Grollier, V. Cros, A. Hamzic et al., Spin-polarized current induced switching in Co/Cu/Co pillars, *Appl. Phys. Lett.* **78**, 3663 (2001).

174. J. Grollier, V. Cros, H. Jaffrès et al., Field dependence of magnetization reversal by spin transfer, *Phys. Rev. B* **67**, 174402 (2003).

175. J. Sun, D. Monsma, D. Abraham, M. Rooks, and R. Koch, Batch-fabricated spin-injection magnetic switches, *Appl. Phys. Lett.* **81**, 2202 (2002).

176. S. Urazdhin, N. Birge, W. P. Pratt, and J. Bass, Switching current versus magnetoresistance in magnetic multilayer nanopillars, *Appl. Phys. Lett.* **84**, 1516 (2004).

177. Y. Huai, F. Albert, P. Nguyen, M. Pakala, and T. Valet, Observation of spin-transfer switching in deep submicron-sized and low-resistance magnetic tunnel junctions, *Appl. Phys. Lett.* **84**, 3118–3121 (2004).

178. Y. Liu, Z. Zhang, P. Freitas, and J. L. Martins, Current-induced magnetization switching in magnetic tunnel junctions, *Appl. Phys. Lett.* **82**, 2871 (2003).

179. G. D. Fuchs, N. C. Emley, I. N. Krivorotov et al., Spin-transfer effects in nanoscale magnetic tunnel junctions, *Appl. Phys. Lett.* **85**, 1205–1208 (2005).

180. J. Slonczewski, Currents, torques, and polarization factors in magnetic tunnel junctions, *Phys. Rev. B* **71**, 024411 (2005).

181. J. A. Katine and E. E. Fullerton, Device implications of spin-transfer torques, *J. Magn. Magn. Mater.* **320**, 1217–1226 (2008).

182. D. Chiba, F. Matsukura, and H. Ohno, Electrical magnetization reversal in ferromagnetic III–V semiconductors, *J. Phys. D Appl. Phys.* **39**, R215–R225 (2006).

183. A. Chernyshov, M. Overby, X. Liu, J. K. Furdyna, Y. Lyanda-Geller, and L. P. Rokhinson, Evidence for reversible control of magnetization in a ferromagnetic material by means of spin-orbit magnetic field, *Nat. Phys.* **5**, 656–659 (2009).

184. M. Tsoi, A. G. M. Jansen, J. Bass et al., Excitation of a magnetic multilayer by an electric current, *Phys. Rev. Lett.* **80**, 4281–4284 (1998).

185. M. D. Stiles and A. Zangwill, Anatomy of spin-transfer torque, *Phys. Rev. B* **66**, 014407 (2002).

186. A. Kalitsov, I. Theodonis, N. Kioussis, M. Chshiev, W. H. Butler, and A. Vedyayev, Spin-polarized current-induced torque in magnetic tunnel junctions, *J. Appl. Phys.* **99**, 08G501 (2006).

187. I. Theodonis, N. Kioussis, A. Kalitsov, M. Chshiev, and W. H. Butler, Anomalous bias dependence of spin torque in magnetic tunnel junctions, *Phys. Rev. Lett.* **97**, 237205 (2006).

188. A. Kalitsov, M. Chshiev, I. Theodonis, N. Kioussis, and W. H. Butler, Spin-transfer torque in magnetic tunnel junctions, *Phys. Rev. B* **79**, 174416 (2009).

189. Y. H. Tang, N. Kioussis, A. Kalitsov et al., Controlling the nonequilibrium interlayer exchange coupling in asymmetric magnetic tunnel junctions, *Phys. Rev. Lett.* **103**, 057206 (2009).

190. H. Kubota, A. Fukushima, K. Yakushiji et al., Quantitative measurement of voltage dependence of spin-transfer torque in MgO-based magnetic tunnel junctions, *Nat. Phys.* **4**, 37–41 (2007).

191. J. C. Sankey, Y. T. Cui, J. Z. Sun et al., Measurement of the spin-transfer-torque vector in magnetic tunnel junctions, *Nat. Phys.*, **4**, 67–71 (2007).

192. J. Sinova, T. Jungwirth, X. Liu et al., Magnetization relaxation in (Ga,Mn)As ferromagnetic semiconductors, *Phys. Rev. B* **69**, 085209 (2004).

193. S. Zhang, P. M. Levy, A. Marley, and S. S. P. Parkin, Quenching of magnetoresistance by hot electrons in magnetic tunnel junctions, *Phys. Rev. Lett.* **79**, 3744–3747 (1997).

194. A. M. Bratkovsky, Assisted tunneling in ferromagnetic junctions and half-metallic oxides, *Appl. Phys. Lett.* **72**, 2334–2336 (1998).

195. P. M. Levy and A. Fert, Spin transfer in magnetic tunnel junctions with hot electrons, *Phys. Rev. Lett.* **97**, 097205 (2006).

196. J. Schneider, U. Kaufmann, W. Wilkening, M. Baeumler, and F. Kohl, Electronic structure of the neutral manganese acceptor in gallium arsenide, *Phys. Rev. Lett.* **59**, 240–243 (1987).

197. A. O. Govorov, Optical probing of the spin state of a single magnetic impurity in a self-assembled quantum dot, *Phys. Rev. B* **70**, 035321 (2004).

198. G. Dresselhaus, Spin-orbit coupling effects in zinc blende structures, *Phys. Rev.* **100**, 580–586 (1955).

199. E. O. Kane, Band structure of indium antimonide, *J. Phys. Chem. Solids* **1**, 249–261 (1957).

200. E. O. Kane, The k.p method, in *Semiconductors and Semimetals*, R. K. Willardson and A. C. Beer (Eds.), Academic Press, New York, P. 1, 1966; E. O. Kane, Energy band theory, in *Handbook on Semiconductors*, T. S. Moss and W. Paul (Eds.), North-Holland, Amsterdam, P. 1, 1982.

201. G. F. Koster, J. O. Dimmock, R. G. Wheeler, H. Statz, *Properties of the Thirty-Two Point Groups*, M.I.T. Press, Cambridge, MA, 1963.

202. R. Winkler, *Spin-Orbit Coupling Effects in Two-Dimensional Electron and Hole Systems*, Springer–Verlag, Berlin, 2003.

203. J. M. Luttinger, Quantum theory of cyclotron resonance in semiconductors: General theory, *Phys. Rev.* **102**, 1030–1041 (1956).
204. M. Cardona, N. E. Christensen, and G. Fasol, Relativistic band structure and spin-orbit splitting of zinc-blende-type semiconductors, *Phys. Rev. B* **38**, 1806–1827 (1988).
205. E. L. Ivchenko, A. Yu. Kaminski, and U. Rössler, Heavy-light hole mixing at zinc-blende (001) interfaces under normal incidence, *Phys. Rev. B* **54**, 5852–5859 (1996).
206. M. V. Durnev, M. M. Glazov, and E. L. Ivchenko, Spin-orbit splitting of valence subbands in semiconductor nanostructures, *Phys. Rev. B* **89**, 075430 (2014).
207. D. Rideau, M. Feraille, L. Ciampolini et al., Strained Si, Ge, and $Si_{1-x}Ge_x$ alloys modeled with a first-principles-optimized full-zone k.p method, *Phys. Rev. B* **74**, 195208 (2006).
208. G. Fishman, *Semi-Conducteurs: Les Bases de la théorie k.p*, Les Editions de l'Ecole Polytechnique, Palaiseau, Paris, France, 2010.
209. S. Boyer-Richard, F. Raouafi, A. Bondi et al., 30-band k.p method for quantum semiconductor heterostructures, *Appl. Phys. Lett.* **98**, 251913 (2011).
210. N. A. Cukaric, M. Z Tadic, B. Partoens, and F. M. Peeters, 30-band k.p model of electron and hole states in silicon quantum wells, *Phys. Rev. B* **88**, 205306 (2013).
211. B. A. Foreman, Connection rules versus differential equations for envelope functions in abrupt heterostructures, *Phys. Rev. Lett* **80**, 3823–3826 (1998).
212. K. L. Kolokolov, J. Li, and C. Z. Ning, k.p Hamiltonian without spurious-state solutions, *Phys. Rev. B* **68**, 161308(R) (2003).
213. M. Cardona and F. Pollak, Energy-band structure of germanium and silicon: The k·p method, *Phys. Rev.* **142**, 530–543 (1966).
214. N. Cavassilas, F. Aniel, K. Boujdaria, and G. Fishman, Energy-band structure of GaAs and Si: A sps* k.p method, *Phys. Rev. B* **64**, 115207 (2001).
215. R. Eppenga, M. F. H. Schuurmans, and S. Colak, New k.p theory for $GaAs/Ga_{1-x}Al_xAs$-type quantum wells, *Phys. Rev. B* **36**, 1554–1564 (1987).
216. D. Y. K. Ko and J. C. Inkson, Matrix method for tunneling in heterostructures: Resonant tunneling in multilayer systems, *Phys. Rev. B* **38**, 9945–9951 (1988).
217. O. Krebs and P. Voisin, Giant optical anisotropy of semiconductor heterostructures with no common atom and the quantum-confined Pockels effect, *Phys. Rev. Lett.* **77**, 1829–1832 (1996).
218. S. Cortez, O. Krebs, and P. Voisin, Breakdown of rotational symmetry at semiconductor interfaces: a microscopic description of valence subband mixing, *Eur. Phys. J. B* **21**, 241–250 (2001).
219. J. Slonczewski, Conductance and exchange coupling of two ferromagnets separated by a tunneling barrier, *Phys. Rev. B* **39**, 6995–7002 (1989).
220. H. Jaffrès, D. Lacour, F. Nguyen Van Dau et al., Angular dependence of the tunnel magnetoresistance in transition-metal-based junctions, *Phys. Rev. B* **64**, 064427 (2001).
221. A. Braatas, G. E. W. Bauer, and P. J. Kelly, Non-collinear magnetoelectronics, *Phys. Rep.* **427**, 157–255 (2006).
222. J.-C. Rojas-Sanchez, N. Reyren, P. Laczkowski et al., Spin pumping and inverse spin hall effect in platinum: The essential role of spin-memory loss at metallic interfaces, *Phys. Rev. Lett.* **112**, 106602 (2014).
223. V. P. Amin and M. D. Stiles, Spin transport at interfaces with spin-orbit coupling: Phenomenology, *Phys. Rev. B* **94**, 104420 (2016).

# 5

# Spin Transport in Organic Semiconductors

**Valentin Dediu, Luis E. Hueso, and Ilaria Bergenti**

## 5.1 INTRODUCTION

Recently, spintronics has expanded its choice of materials to the organic semiconductors (OSCs) [1–3]. OSCs offer unique properties toward their integration with spintronics, the main one being the weakness of the spin-scattering mechanisms in OSC. Spin–orbit coupling is very small in most OSCs; carbon has a low atomic number (Z) and the strength of the spin–orbit interaction (SOI) is in general proportional to $Z^4$ [4]. Typical SOI values in OSC are less than about a few meV [5], well below any other characteristic energy, including main vibrational modes. Small spin–orbit coupling implies that the spin polarization of the carriers could be maintained for a very long time. Indeed, spin relaxation times in excess of 10 μs have been detected by various resonance techniques [6, 7], and these values compare very favorably (at least $10^3$ times larger) with those obtained with high-performing inorganic semiconductors, such as GaAs [8].

The field of OSCs nowadays combines considerable achievements in research and applications with challenging expectations. Over decades, the interest toward OSCs was mainly sustained by their exceptional optical properties, as the multicolor "offer" was backed by an enormous choice of available or easily modifiable molecules as well as by significant luminescence intensity. Indeed, these optical properties prompted a real breakthrough in the optoelectronics field where, along with the achievement of a deep basic knowledge, a variety of important applications have been developed. Display products based on hybrid light-emitting diodes with organic emitters (OLEDs) have become available to consumers, and organic photovoltaic devices are challenging existing commercial applications. These accomplishments are advancing a future strong demand for high quality control and guiding circuitry based on novel hybrid organic–inorganic devices. Organic field effect transistors (OFETs), in which considerable improvements have already been achieved, together with various elements based on magnetic and electrical resistance switching are foreseen as one of the most important candidates for these newly emerging applications.

Organic molecules are also considered as a promising contender for the future "beyond CMOS devices," offering 1–10 nm-sized building blocks with rich optical, electrical, and magnetic properties. OSCs have become a requisite of various roadmaps and research agendas, although many serious difficulties have still to be circumvented in order to meet these challenges. One of the biggest obstacles for commercial applications is the resistance contact problem [9], namely, the way to insert and extract electrically driven information, interfacing fragile organic blocks with hard metallic or oxide electrodes. However, it has to be noted, on the other hand, that the hybrid nature of most organic-based devices also has positive aspects, providing routes for large and fundamental modifications of the properties of the OSC, especially in the interface region.

# 5.2 ORGANIC SEMICONDUCTORS: GENERAL BACKGROUND

OSCs combine a strong intramolecular covalent carbon–carbon bonding with a weak van der Waals interaction between molecules. This combination of different interactions generates unusual materials whose optical properties are very similar to those of their constituent molecules, whereas their transport properties are instead firmly governed by the intermolecular interactions.

As a general rule, there are two very different classes of OSCs: small molecules and polymers. These two cases differ not only on their physical and chemical properties, but also on their technological processing. Small-molecule OSCs are rigid objects of about 1 nm size, with very different shapes and symmetries, but able to form highly ordered van der Waals layers and polycrystalline films. The small-molecule film growth involves either high or ultrahigh vacuum molecular beam deposition. A few of the main representatives of this kind of materials are the classes of oligothiophenes (4T, 6T, and others), metal chelates ($Alq_3$, $Coq_3$, and others), metal-phthalocyanines (CuPc, ZnPc, and others), acenes (tetracene, pentacene, and others), rubrene, and many more. Polymers, on the other hand, are long chains formed from a given monomer, and usually feature connection of oligo-segments of various lengths leading to a random conjugation length distribution. Polymer films are strongly disordered and generally cannot be produced in UHV conditions; instead, totally different processing technologies have to be applied, including ink-jet printing, spin coating, drop casting, and other extremely cheap methods. This potentially inexpensive fabrication constitutes one of the main technological advantages of the polymers. However, these growth approaches generally preclude a real interface control, as the electrode surface is not as clean as in UHV conditions and cannot be directly characterized by such important surface techniques as x-ray photoemission spectroscopy (XPS), ultraviolet photoemission spectroscopy (UPS), x-ray magnetic dichroism (XMCD), and others.

The electrical properties of OSC are usually depicted by two main components: charge injection and charge transport. The charge injection process from a metal into an OSC is conceptually different from the case of metals/inorganic semiconductors, mainly because of the vanishingly low density of intrinsic carriers—about $10^{10}$ $cm^{-3}$—in OSC. The external electrodes provide the electrical carriers responsible for the transport, as the OSC molecules can easily accommodate an extra charge, although this charge modifies considerably the molecular spatial conformation and creates strong polaronic effects. Carrier injection into OSC is best described [10] in terms of thermally and field-assisted charge tunneling across the inorganic–organic interface, followed by carrier diffusion into the bulk of OSC. The carriers propagate via a random site-to-site hopping between pseudo-localized states distributed within approximately a 0.1 eV energy interval. This interval describes the spatial distribution of the localized levels and is not to be confused with the bandwidth. OSCs are mainly families of π-conjugated molecules, which combine a considerable carrier delocalization inside the molecules with a weak van der Waals intermolecular interaction that limit considerably the carrier mobility. The electron mean free path is usually of about a molecule size, namely, about 1 nm, while standard mobilities are well below 1 cm²/V s. Two conducting channels are usually considered active: lowest unoccupied molecular level (LUMO) for *n*-type and highest occupied molecular level (HOMO) for *p*-type carriers. However, defects and interface states have to be considered depending on the material and its structural quality. These states usually emerge via one of the most important characteristics of OSC—trapping centers. Puzzlingly, in OSCs, the impurities do not induce additional carriers, as in inorganic semiconductors, but mainly generate trap states. Energetically distributed shallow (about/below 0.1 eV) and deep (about 0.5 eV, but even up to 1 eV) traps are known to characterize the electrical properties (especially the mobility) of many OSCs, their role being stronger in low-order materials like Alq₃, polymers, etc.

The mobility of OSC has a strong dependence on the electric field, and in the absence of traps, it is usually described by the so-called Poole–Frenkel dependence [11]

$$\mu(E) = \mu_0 \exp\left(\frac{\beta}{\sqrt{E}}\right) \tag{5.1}$$

where $\mu_0$ is the field-independent mobility. Including trapping complicates the equations and numerical solutions for different cases have to be calculated.

In order to complete this very brief description of electrical properties of OSC, we would like to add two more concepts.

The first concept highlights the difference between the two main transport regimes present for most organic-based devices: injection-limited current (ILC) and space charge-limited current (SCLC). In the first case, the interface resistance dominates the current–voltage curves and the flowing current is basically independent on the device thickness. However, the SCLC current ($J_{SCLC}$) requires at least one ohmic electrode and corresponds to the case when the injected carrier density reaches the charging capacity of the material. This SCLC current is characterized by a strong thickness ($d$) dependence, which for a trap-free OSC is given by [11]

$$J_{SCLC} \sim \frac{\mu(E)V^2}{d^3} \tag{5.2}$$

where $V$ is the voltage across the device and, at that given voltage, the electric field ($E$) value inside OSC depends as well on the thickness.

The second concept is the distinction between unipolar and bipolar transport in OSCs (the later featuring significant recombination effects). These two different transport regimes depend on the intrinsic properties of the OSC as well as on the matching of the electrode work function with either the LUMO or the HOMO levels. For more extensive reading on the electrical properties, we suggest the excellent reviews of Coropceanu et al. [12] and Brütting et al. [11].

The optical gap in many OSCs is about 2–3 eV, which conveniently fits very well in the visible range. The simultaneous injection of electrons and holes from two independent electrodes generates Frenkel excitons with a 3:1 triplet/singlet ratio. The $S1 \rightarrow S0$ optical transition is usually the only one allowed as a consequence of low spin–orbit coupling and Franck–Condon selection. The former rules out the singlet–triplet mixing, while the latter points out once again on the strong role of the polaronic effects in OSC [13, 14], which excludes the transitions where the equilibrium positions of the nuclei in both the ground and the excited states would be identical.

## 5.3 ELECTRICAL OPERATION: INJECTION

In the following discussion on the electrically operated organic spintronic devices (OSPDs), it is necessary to distinguish between two different cases: tunneling devices characterized by zero residence time of charges/spins in the organic barrier, and injection devices in which there is a net flow of spin-polarized carriers directly into the electronic levels of the OSC. While tunneling devices can feature large magnetoresistance (MR) values and are useful as 0–1 switching elements or magnetic sensors, injection devices are characterized by a finite lifetime of the generated spin polarization in OSC, and offer such opportunities as spin manipulation and control of the exciton statistics in OSC. These options may well open the way to magnetically controlled OFETs or OLEDs.

In the description of the spin injection process into OSC, we will need to mention numerous times two materials with exceptional importance in the field: manganites, mostly $La_{0.7}Sr_{0.3}MnO_3$ (LSMO), and $Alq_3$ (Tris(8-hydroxyquinoline) aluminum(III)). Importantly, about half of all the OSC spintronic devices available in the literature involve either one or both of these materials. This fact has some historical grounds, as the first successful experiments dealt with these materials, but it also indicates some outstanding spintronic properties of these materials, which have not been understood and developed in detail yet.

$La_{1-x}Sr_xMnO_3$ (with $0.2 < x < 0.5$) is a ferromagnetic material with 100% spin-polarized carriers at $T \ll T_c$. The Curie temperature is around 370 K in bulk samples, while thin films have $T_c$ around 330 K, changing with thickness and deposition method. The carrier hopping between $Mn^{3+}$ and $Mn^{4+}$ ions is governed by a strong on-site Hund interaction (nearly 1 eV), which makes $La_{1-x}Sr_xMnO_3$ an unusual metal, characterized by a narrow band (~1–2 eV) and small carrier density ($10^{21}$–$10^{22}$ cm$^{-3}$), very much like the high-temperature cuprate superconductors and some organic metals.

A critical issue for the spintronic applications of manganites is the difference between their bulk and surface magnetic properties: the latter, being responsible for the injected spin-polarized current, decays much faster with temperature than the former. This leads to the fact that in the room temperature region, the spin polarization is much smaller than the expected magnetization ratio $M(T)/M(0)$ [15]. Nevertheless, special surface treatments have succeeded in improving the thin film surface properties, and samples with reproducible XMCD signals at room temperature are currently routinely obtained [16]. Well before raising of organic spintronics, LSMO was already well-known for its spin injection properties, and has proved to be highly successful in diverse inorganic spintronic devices, such as tunnel junctions [17] and artificial grain boundaries devices [18].

Tris(8-hydroxyquinoline aluminum(III) or $Alq_3$ is a popular OSC from the early stages of xerographic and optoelectronics applications [19]. This material might look somewhat of a strange choice for transport-oriented applications, as $Alq_3$ has very low mobility values and a permanent electric dipole [20, 21]. These two factors are, in principle, able to compromise spin transport, but in spite of the expected problems with $Alq_3$, strong spin-mediated effects have been repeatedly reported by different groups.

The first experimental report on injection in OSPD featured a lateral device that combined LSMO electrodes and OSC conducting channels between 100 and 500 nm long (Figure 5.1a). Sexithiophene (6T), a rigid conjugated oligomer rod, a pioneer in organic thin film transistors [22], was chosen for the spin transport channel. A strong magnetoresistive response was recorded up to room temperature in 100 and 200 nm channels, and was explained as a result of the conservation of the spin polarization of the injected carriers. Using the time-of-flight approach, a spin relaxation time of the order of 1 µs was found.

This work introduced OSCs as very appealing materials for long-distance spin transport, although some important questions were left open. For example, the experiment did not provide a straightforward demonstration that the MR was related to the magnetization of the electrodes, and hence to its spin polarization, as no comparison between parallel and antiparallel magnetization of electrodes was available.

Following this report, an important step forward was the fabrication of a vertical spin-valve device consisting of LSMO and cobalt electrodes, sandwiching a thick (100–200 nm) layer of $Alq_3$ [23]. The spin valve showed MR up to a temperature around 200 K, and for voltages below 1 V. Unlike standard spin-valve devices, in which lower resistance corresponds to parallel electrode magnetization, the LSMO/$Alq_3$/Co devices showed lower resistance for an antiparallel magnetization configuration.

Starting from this work, the vertical geometry has acquired a prevailing role in organic spintronics, mainly because of its technological simplicity with respect to planar devices, requiring a substantial lithography involvement. Several groups have confirmed the so-called "inverse spin-valve effect" in LSMO/$Alq_3$/Co devices [24–28]. Nevertheless, its nature has not yet been clearly understood, although at least two different explanations have been proposed [23, 26]. One of these models requires the injection of a spin-down-oriented current from the cobalt electrode [23] (i.e. the spin polarization of carriers antiparallel to the applied magnetic field), in a serious disagreement with many experimental and theoretical reports [29, 30]. The second model

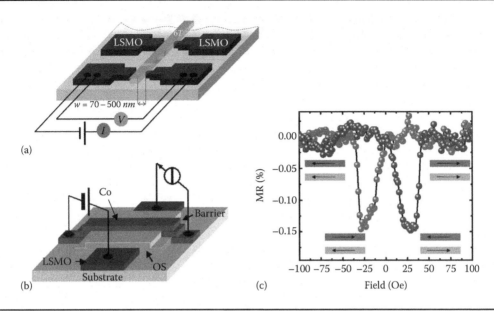

**FIGURE 5.1**    (a) Spin lateral device presented by Dediu et al. [32]: two LSMO electrodes obtained by electron beam lithography were connected by an organic thin film (T6). The channel length ranged between 70 and 500 nm. (b) Vertical spin-valve device consisting of LSMO electrode as bottom electrode and cobalt as the top one. The typical MR measurements are performed by applying a constant biasing voltage $V$ across the electrodes while measuring the resulting current as a function of an in-plane sweeping external magnetic field. (c) GMR loop of an LSMO (20 nm)/Alq$_3$ (100 nm)/AlO$_x$ (2 nm)/Co (20 nm) vertical spin-valve device measured at room temperature. The antiparallel configurations of the magnetizations correspond to the low resistance state.

introduces a very narrow hopping channel in OSCs, connecting at some given voltage the spin-up bands of one of the electrodes with the spin-down bands of the opposite one [26]. While in principle feasible, this model can be applied for the $n$-type carriers only and works mainly at voltages about 1 V or higher, that is, values at which spintronics effects in LSMO/Alq$_3$/Co devices generally vanish. This situation clearly indicates the limits of the overall comprehension of the main laws governing both the spin injection and transport in organic spintronics.

As explained above, Alq$_3$ is characterized by its very low mobility: about $10^{-5}$ cm$^2$/V s for electrons and less than $10^{-6}$ cm$^2$/V s for holes [31]; while the full relevance of the mobility has yet to be understood for OSPD, the wide use of Alq$_3$ is probably due to the good-quality thin films that can be grown on various ferromagnetic substrates by standard UHV evaporation.

Nevertheless, electrical detection of spin injection and transport has been actively pursued in a number of other OSCs (see Table 5.1). Thus, α-NPD (*N*,*N*-bis 1-naphtalenyl-*N*,*N*-bis phenylbenzidiane) and CVB (4, 4′-(Bis (9-ethyl-3-carbazovinylene)-1,1′-biphenyl) [24] were demonstrated to behave very similarly to Alq$_3$ with LSMO and Co set of magnetic electrodes: inverse spin-valve effect (around 10% at low temperatures) and fast decrease of the MR with increasing temperature, vanishing between 210 and 240 K. The same couple of electrodes was used to investigate spin injection in tetraphenylporphyrin (TTP)—the detected MR in this case was attributed,

**TABLE 5.1**
**Main Properties of OSCs Investigated in Organic Spintronics.**

| Oligoacenes | Rubrene | $\mu = 1\ cm^2/V\ s$ <br> p-type | Pentacene | $\mu = 10^{-1}\ cm^2/V\ s$ <br> p-type |
|---|---|---|---|---|
| Tiophenes | Sexitiophene | $\mu = 10^{-1}\ cm^2/V\ s$ <br> p-type | RRP3HT | $\mu = 10^{-1}\ cm^2/V\ s$ <br> p-type |
| Triarylamines | NPD | $\mu = 10^{-5}\ cm^2/V\ s$ <br> p-type | CVB | $\mu = 10^{-3}\ cm^2/V\ s$ |
| Porphyrines | TPP | $\mu = 10^{-5}\ cm^2/V\ s$ <br> n-type | | |
| Metal complexes | CuPC | $\mu = 10^{-2}\ cm^2/V\ s$ <br> n-type | Alq₃ | $\mu = 10^{-5}\ cm^2/V\ s$ <br> n-type |

however, to tunneling effects [28]. A planar geometry with two LSMO electrodes, similar to the first publication in the field [32], provided about 30% MR at low temperature in pentacene and BTQBT [33].

Moving from low to room temperature operation in vertical devices has required an advanced control of the interface quality. Thus, room temperature MR in LSMO/Alq₃/Al₂O₃/Co devices was demonstrated [26] by improving the quality of both injecting interfaces, especially the top one, in which a 2 nm thick Al₂O₃ insulating layer was added (see Figure 5.1b and c). Similarly, about 1% room temperature MR across thick Alq₃ and CuPc layers was achieved in vertical devices with both metallic electrodes and excellent interface quality (Figure 5.2c) [34]. Importantly, the substitution of the LSMO electrode by Fe has removed the inversion of the spin-valve effect [34], indicating that inversion is strongly caused by the manganite properties [26].

All the experiments described above were performed using UHV conditions for the OSC growth. Spin-coated polymers such as regio-random and regio-regular poly(3-hexylthiophene), (RRaP3HT and RRP3HT, respectively) have also shown noticeable MR [27, 35]. In these reports, the spin-valve effect was positive and a small value was recorded even at room temperature.

We can summarize the main findings obtained, thanks to the study of electrically operated injection devices. First, they unambiguously demonstrated both spin injection and spin transport in OSPDs. The available data confirms so far spin injection and transport only for voltages not exceeding approximately 1 V. Moreover, highly efficient MR are constrained in a narrower interval of about 0.1–0.2 V. Current MR measurements provide

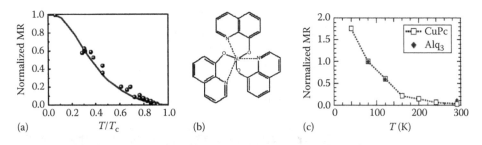

**FIGURE 5.2**    (a) Temperature dependence of normalized MR of vertical spin-valve devices consisting of LSMO and Co as electrodes and Alq$_3$ as spacer (dots) with $T_c = 325$ K. The line represents the LSMO surface magnetization (b) Alq$_3$ molecular structure. (c) Temperature dependence of maximum observed MR for two junctions based on Alq$_3$ and CuPc, as organic spacer and Fe and Co as metallic contact. From Liu et al., the MRs for each sample are normalized to their values at $T = 80$ K. (After Liu, Y. et al., *J. Appl. Phys.* 105, 07C708, 2009. With permission.)

mainly qualitative information and in the absence of well-defined models for spin injection and transport in OSC, it is difficult to extract the values of the spin polarization or its decay inside the organic material. The investigation of the thickness and temperature dependence of the MR in OSPDs provides important indications anyway for the basic physics underlying spintronics effects.

Surprisingly, very important information comes from the "confusing" set of data on the thickness dependence. Except for a few reported sequences where an expected decrease of MR upon increasing thickness was reported, many published [28, 36] as well as unpublished data (various conference discussions) indicate quite casual dependences.

A reasonable and, in our opinion, credible explanation for this comes, amazingly, from the analysis of the temperature behavior for the most archetypal LSMO-Alq$_3$-Co OSPD [24, 26]. Figure 5.2a presents the normalized MR versus temperature for a device with a 100 nm thick Alq$_3$ layer and a 2 nm thick Al$_2$O$_3$ barrier separating Alq$_3$ and top Co electrode [26]. Remarkably, the data extrapolate to zero at the Curie temperature of the manganite [26] (325 K in this particular case). Moreover, the temperature dependence of MR data agrees with that of the surface magnetization (SM) curve for LSMO (solid line in Figure 5.2a) obtained by Park et al. [15]. The SM represents the magnetization from the top 5 Å in a standard LSMO film, as determined by spin-polarized photoemission spectroscopy [15], and it is effectively the parameter of interest for spin injection in LSMO-based devices [37]. The results summarized above are in agreement with the previous claim that the temperature dependence of MR in Alq$_3$ spintronic devices is governed by the manganite electrode [24].

The temperature behavior described above emphasizes a very important statement: the LSMO-Alq$_3$-Co devices behave in an "injection limited"-like regime, as the electrode spin polarization nearly dominates the temperature dependence. The spin losses at the interface significantly prevail over the transport bulk losses. By analogy with the pure charge injection case, it seems coherent to expect a weak or even a random thickness dependence in such devices: for example, small fluctuations of the interface quality should influence the device behavior much more than a thickness variation. Also, we believe this regime can be expanded to other materials and interfaces, although there are no a priori reasons to suppose that all OSPDs should work in injection-limited mode.

## 5.4 ELECTRICAL OPERATION: TUNNELING

While the interest in spin injection and long-distance spin transport in OSCs was growing, several groups also started to explore the possibilities of these materials as spin tunnel barriers. Spin-polarized tunneling in OSPD was initially demonstrated in quite complex geometries. Petta et al. [38] fabricated Ni-octanethiol-Ni vertical tunneling devices in a nanopore geometry with an octanethiol self-assembled monolayer. These devices showed MR up to 16% at 4.2 K, although it vanished at about 30 K. The sign of the MR was observed to switch from positive to negative for different voltage values, but also from sample to sample at the same voltage. While evidence of the spin-polarized tunneling through a monolayer of organic was given, it has also been noticed that such devices are subject to strong intrinsic and extrinsic noises.

An interesting approach for the investigation of tunneling effects is the utilization of composite systems featuring magnetic nanoparticles embedded into an organic matrix. Depending on the separation between nanoparticles, a switching from interparticle tunneling to hopping conductivity between particles occurs. Such effects have been detected in $Co:C_{60}$ nanocomposite [39].

An important step forward for organic spin tunneling was the fabrication of simple devices via direct in situ UHV organic deposition with shadow masking. Vertical tunneling devices ($Co/Al_2O_3$-$Alq_3$/NiFe [40] and $Co/Al_2O_3$-rubrene/Fe [41]) made use of hybrid inorganic–organic ultrathin barriers to produce positive TMR when the electrode magnetizations were switched from parallel to antiparallel. TMR around 15% was recorded at 4.2 K [41], and, importantly, both OSC showed few percent TMR at room temperatures, opening the door to possible commercial applications. Another important aspect is that $Al_2O_3$-$Alq_3$ devices had the hybrid tunnel barrier well inside the typical quantum mechanical tunneling regime (0.6–1.6 nm). However, the $Al_2O_3$-rubrene hybrid barriers were considerably thicker, with rubrene layers ranged from 4.6 to 15 nm. These thicknesses are well over the tunneling regime and probably imply spin injection in rubrene layers. A transition from tunneling into injection regime was in principle tackled for the first time for organic–inorganic hybrid devices [41]. At the present state of the art, the organic spintronic community is not yet ready to provide a thickness value where such a transition can take place, nor is it ready to describe convincingly the system behavior at the two sides of transition. Further investigation is required in order to face these key fundamental questions. One has to keep firmly in mind, nevertheless, that OSCs of about 20–30 nm (e.g. $Alq_3$) are already good light emitters, indicating that this thickness is already well beyond tunneling limit.

The experiments reported above clearly show the possibility of achieving spin-polarized tunneling across organic or organic–inorganic hybrid barriers. Although the control of spin tunneling across organic layers is still relatively poor, it is also clear that there is a great margin for improvement, especially in the fields of interface control and engineering of single molecule studies.

## 5.5 ALTERNATIVE APPROACHES

In this section, we will outline different important experiments that have been reported recently in the field of organic spintronics, but not with standard electronic spin transport in devices. These alternative experiments are extremely important, as they allow us to have access to different physical parameters than the ones restricted by simple electronic transport data. Our overview here does not pretend to be systematic, but simply to highlight different recent approaches that are shedding light on different aspects of spin injection and transport in OSCs.

One important piece of information recently revealed is the spin polarization profile inside an OSC. Such information has deep implications, as it would allow the research community to evaluate the true potential of organic spintronics for carrying information at long distances. The data has been obtained by two powerful nonstandard techniques: two-photon photoemission spectroscopy and muon spin rotation technique.

Two-photon photoemission spectroscopy experiments allow us to obtain the spin polarization of injected electrons in the first monolayers of organic materials grown on top of a spin-polarized ferromagnetic material [42]. Spin-polarized electrons from the metal are excited by the first energy pulse into an intermediate energy state. Some of the electrons propagate into the OSC where, with a certain probability, they can be excited by the second photon causing the photoemission. The spin-polarized injection efficiency from cobalt into the first monolayer of the copper phthalocyanine was estimated to be close to 100% [42]. An obvious drawback in these measurements is that the energy requested for the electron injection process is much higher than the energy gap in the OSC. For those energies, the spin scattering is higher than for the (lower) voltages at which electron/spin injection usually takes place, making the comparison between photoemission spectroscopy and standard electron devices far from straightforward.

In a totally independent article, electrical injection was combined with low-energy muon spin rotation to study the spin diffusion length inside $Alq_3$ in a standard vertical spin-valve device [43]. The stopping distribution of the spin-polarized muons was varied in the range 3–200 nm by controlling the implantation energy, while a quantitative analysis determined the spin diffusion length ($l_S$) and its temperature dependence. On one hand, the diffusion length reached a low temperature value of 35 nm: a lower value than the one estimated from electrical measurements (>100 nm), a discrepancy that could nevertheless be explained, bearing in mind the imperfect injection efficiency and very small MR of the explored device. On the other hand, the weak temperature dependence of the $l_s$ is in good qualitative agreement with independent MR characterizations in $Alq_3$-based devices [26], and also in agreement with a recently proposed mechanism for the hyperfine-driven spin scattering in OSC.

In a different subset, we could position the articles that explore the spin transport properties in OSC in relation to their electroluminescence (EL) properties. In fact, EL has been key in the advancement of the OSC research, since it opens the door to multiple commercial applications, such as OLEDs. Similarly, the careful study of the emitted light could give us very useful information about the electron/hole recombination processes inside the organic. In OLEDs, due to low spin–orbit coupling, the dominant EL channel involves singlet–singlet transitions [44]. In OLED, the singlet/triplet ratio is determined by quantum statistics as 1:3 when considering a similar formation probability for one singlet and three triplet states (Figure 5.3). The injection of carriers with a controlled spin state could therefore enable the amplification of the singlet (triplet) exciton density, leading to a significant increase of efficiency for the EL [45], rather than a polarization in the emitted light as in other inorganic semiconductors (such as GaAs) [46–48]. This theoretically expected increase in the emitted light attracted many researchers into organic spintronics a few years ago, but unfortunately, the efficiency amplification has not been reported in the literature [49, 50]. The absence of amplification of light in spin-polarized OLED might be due to the fact that

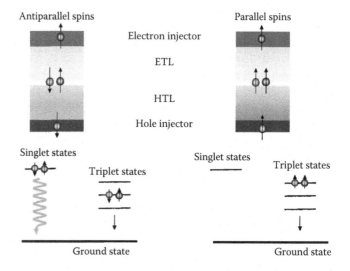

**FIGURE 5.3**  Schematic structure of a bilayered OLED. Electrons and holes are injected by cathode and anode, respectively, into the electron-transport layer (ETL) and the hole-transport layer (HTL). Exciton formation and recombination eventually take place at the interface. Spin statistic of carrier recombination in case of spin injection. Antiparallel spin states for electrons and holes generate singlet and triplet state in ratio 1:1. Parallel spin states for electrons and holes' spin state of carriers give rise only to the triplet state.

light emission can typically only be detected for applied voltages in excess of a few volts, while spin injection in organic materials has been demonstrated so far for voltages not exceeding 1 V [23, 45]. While this limitation does not seem to be a fundamental obstacle, it has yet to be understood whether it can be experimentally circumvented.

Irrespective of the current lack of successful results in light amplification in spin-polarized OLED, they are very useful anyway for the study of the electron spin conversion, that can be studied by electron spin resonance (ESR) [51]. In this chapter, the authors achieved a magnetically guided transfer between singlet ($PP_S$) and triplet ($PP_T$) polaron pairs caused by the spin precession. The two conversion channels give completely different inputs into the photoconductivity, and monitoring the later provides a way to control the PP dynamics. A periodical modification of the $PP_S \rightarrow PP_T$ transfer (Rabi oscillations) was observed in the conjugated polymer MEH-PPV, from which an extremely long spin coherence time above $0.5\,\mu s$ [51] was extracted.

## 5.6 INTERFACE EFFECTS

In a strict sense, for the MR in a spin-valve organic device to be present, both bottom and top ferromagnetic/OSC interfaces would have to be spin selective (such as, e.g. a tunnel barrier [52]). In this context, the perception of the complex nature of the interfaces between magnetic materials and OSCs is motivating several groups to undertake detailed studies of the organic–inorganic interfaces. Important issues such as the interfacial energy levels and the degree of chemical interaction or the different interfacial states are being clarified with this kind of research. In general, spectroscopic techniques are being used to study these systems. UPS/XPS are usually the preferred methods to study these interfaces.

Two groups reported almost parallel independent research on what, in principle, should be the optimum combination of materials, a high Curie temperature ferromagnetic material (cobalt) and a high-mobility OSC (pentacene, Pc) (see Figure 5.4a). However, it is important to remember that Co, as with any transition metal, is chemically very active, and this activity could result in a very complex interfacial structure. In general, interfacial dipoles of around 1 eV are observed between Co and Pc [53]; however, it is claimed that the Pc/Co interface (with the Co film deposited on top of the organic) behaves differently, showing vacuum-level alignment (no dipole) and highlighting again the importance of careful characterization of these complex devices [54].

The ubiquitous $LSMO/Alq_3/Co$ device has been fully investigated for interface electronic properties (Figure 5.4b). Detailed XPS and UPS investigations have revealed a 0.9 eV interfacial dipole at $LSMO/Alq_3$ interface [55] and a 1.5 eV dipole at direct $Co/Alq_3$ interfaces [56]. The latter is reduced from 1.5 to 1.3 eV when a 2 nm $Al_2O_3$ barrier is inserted between $Alq_3$ and Co layer [57]. This knowledge provides a complete set of barriers across the whole device, describing the current injection into LUMO and HOMO levels

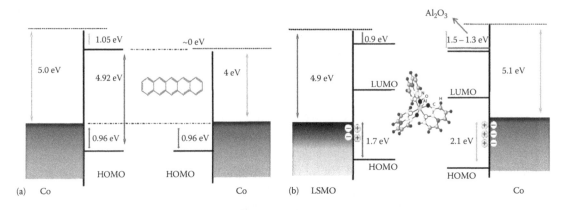

**FIGURE 5.4**   (a) Schematic band diagram for an LSMO/Alq$_3$ and Alq$_3$/Co structure presented by Zhang et al. [56, 57]. (b) Schematic band diagram for a layered Co/pentacene/Co structure presented by Popinciuc [54].

of OSC. This information is not, however, sufficient for the spin injection efficiency, as no clear data on the spin transfer is revealed. Thus, the necessary evolution of the interface studies consists in moving toward monitoring of the spin selectivity and injecting interfaces. For example, proximity effects at organic–inorganic interface can induce new states in the gap of OSCs with non-trivial spin polarization [58].

Recently, Grobosch et al. [59] have questioned whether interfacial energy values measured in UHV conditions can be used to explain the electronic transport behavior of actual devices measured in standard laboratory conditions.

## 5.7  MULTIFUNCTIONAL EFFECTS

A new field that is currently attracting much attention, and also one in which OSC could have a significant impact, is that of multifunctional or multipurpose devices. In general terms, these are devices that react either simultaneously or separately under different external stimuli, such as electric field, magnetic field, light irradiation, and pressure. In this area of research, the possibility of combining magnetic and electrical resistance switching has been realized recently [60], profiting from the unique combination of properties in OSPDs. Nonvolatile electrical memory effect in hybrid organic–inorganic materials has been extensively studied (see e.g. the extended recent review of Scott and Bozano [61]). Under the application of an electric field, resistance changes in excess of one order of magnitude, and retention times of tens of hours have been routinely detected. Following those successes, the International Technology Roadmap for Semiconductors has identified molecular-based memories among the emerging technology lines (http://www.itrs.net).

In organic spintronics, hybrid vertical devices with LSMO and Co as charge/spin-injecting electrodes show both spin-valve and nonvolatile electrical memory effects on a same chip (see Figure 5.5). This result was confirmed for two different OSCs, Alq$_3$ and 6T, and consists of the observation of

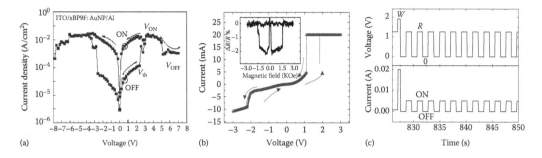

**FIGURE 5.5** (a) Typical current density–voltage characteristic for a nonvolatile memory cell based on composite of metallic nanoparticles and polymers embedded in an organic semiconducting host. (After Bozano, L.D. et al., *Adv. Funct. Mater.* 15, 1933–1939, 2005. With permission). (b) Current–voltage room temperature characteristics in T6-based vertical device. The irreversibility at both positive and negative voltages is clearly visible, and it is associated to the classical spin-valve effect presented in the inset. (c) Switching effect for the OrgSV memory cell. Writing (*W*), erasing (*E*), and reading (*R*) voltages have been used. Two very clear and reproducible ON and OFF states can be observed multiple times even if the voltage has been set to zero between different reading events.

10% negative spin-valve magnetic switching at voltages below 1 V, and 100% electrical switching at voltages above 2–3 V. Importantly, the resistance values achieved by electrical switching are not altered when the devices go back to zero voltage (nonvolatility; see Figure 5.5b and c). With an initial writing voltage pulse, the device moves to a resistance state (called 1), and this resistance state can be recovered multiple times with a reading voltage pulse (reading voltage < writing voltage), even if the voltage has been set to zero after the writing pulse. If we later apply a negative erasing pulse, the new resistance value at the reading voltage will be different (called 0), and again this value can be recovered multiple times even if the voltage is set to zero between reading pulses. This behavior, explained above, implies that a set resistance value (either 1 or 0) can be recovered even if the memory cell has been left without any voltage for a set amount of time, allowing an operation with much lower power and avoiding the need of constant refreshing of current RAM cells.

We believe this combination will become a subject of extensive investigations and can provide a basis for the generation of two bits/cell memory elements. As a future development, we envisage multifunctional devices with magnetic and electric field control of the resistance, and that in addition to these responses, will profit from the luminescent properties of the OSCs.

## 5.8 THEORETICAL MODELS

Many theoretical groups have done a considerable amount of work of different aspects of organic spintronics. Much of the current research is clearly progressing from previous works on inorganic devices; however, there are great new ideas and unique approaches that are helping the organic spintronics field to move forward. Probably one of the most active topics is the one related to the theoretical description of tunneling devices with a single (or a few) molecules placed between two spin-polarized electrodes [62–64]. Tunneling is perhaps the less-complicated organic spintronics process to describe, and very interesting results have been obtained by applying density functional theory (DFT) combined with nonequilibrium Green's functions to calculate the transport characteristics [64, 65]. Rocha et al. [64] considered two prototypical molecules, 8-alkane-dithiolate (octane-dithiolate) and 1,4-3-phenyl-dithiolate (tricene-dithiolate), sandwiched between nickel contacts. Very large MR values have been calculated for both cases: about 100% for the octane-dithiolate, and up to 600% for the tricene-dithiolate molecule (Figure 5.6).

In close contact with all the experimental research dealing with the characterization of organic–inorganic interfaces, many authors have developed different theoretical approaches to understand the optimum conditions for spin injection into an OSC. A few authors have indicated that an ohmic contact would be more efficient for spin injection into OSC [66]. By contrast, it has been suggested in many other papers that spin injection into OSC can only be achieved across an interfacial tunnel or Schottky barrier [67, 68]. In all the cases in which spin injection into OSC has been demonstrated, the organic–inorganic interfaces presented natural Schottky-like barriers, thus showing that experimentally this kind of interfacial contact seems more adequate for the realization of OSPD. A Su–Schrieffer–Heeger model was applied by Xie et al. [69] to study the importance of the electron–lattice interactions, considering interfaces featuring strong polaronic effects at both sides of them. The authors

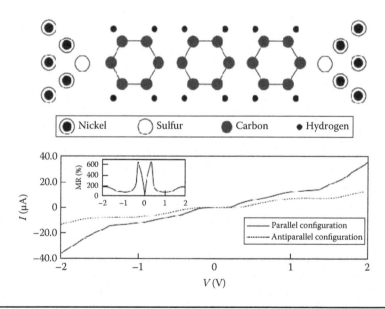

**FIGURE 5.6** Schematic representation of a tricene molecule attached to a (001) Ni surface from Rocha et al. (upper figure). Theoretical calculation of the current–voltage ($I$–$V$) characteristic of a tricene–nickel spin valve. The inset shows the calculated MR ratio with values up to 600%. (After Rocha, A.R. et al., *Nat. Mater.* 4, 335, 2005. With permission.)

showed that under certain biases, spin-polarized injection can be achieved for direct manganite–OSC interface, while substitution of the manganite with a conventional (non-polaronic) ferromagnetic metal resulted in the absence of a spin-polarized flow.

If the injection process has attracted much attention, it is clear as well that the decoherence mechanisms in organic are being subject of intense research. In a naive approach, it is clear that spin–orbit coupling in OSC is very small, but a detailed and predictive description has been lacking. Currently, alternative mechanisms for spin decoherence are being considered. Recently, Bobbert et al. have advanced the idea that the precession of the carrier spins in the random hyperfine fields of hydrogen nuclei [70] (around 5 mT) represents the dominant source for spin scattering in OSC. The authors have derived the spin diffusion length $l_S$ in a typical OSC and they obtain a very weak dependence of $l_S$ on temperature, in good agreement with recent experimental data [26]. Surprisingly, it was also shown that the spin-scattering strength is much less sensitive to disorder than the carrier mobility, a result that could shed light on the apparent discrepancy between low mobility and long spin diffusion lengths in OSC. It has to be noted that spin scattering by hyperfine fields is also considered as one of the most probable mechanisms [71] responsible for the so-called organic magnetoresistance (OMAR) effect in OSC [72], that is, MR without magnetic electrodes.

## 5.9 CONCLUSIONS

In spite of the important achievements described above, an understanding of the basic rules governing spin injection and transport in OSC is still lacking. It is also difficult to predict now the future evolution of organic spintronics, or any possible implementation into applications. Organic spintronics remains, so far, a fascinating scientific problem, which requires facing new physics conceptually and developing innovative technological procedures.

In addition to this stimulating fundamental interest, the results obtained in this novel field are already encouraging. It cannot be excluded that OSC will compete with other materials for the leadership in the spintronics field, or at least in some selected niche applications.

In order to link spintronics to such important organic applications like OLEDs and OFETs, it is necessary nevertheless to overcome some revealed limitations. Among these, we could cite the low voltages at which spin-polarized effects occur in electrically driven devices. While low operation voltages are required in order to reduce energy consumption, this limitation prevents the practical application of spin-polarized currents in OLEDs and OFETs, which are operated at $V > 1$ V (generally at few volts). The available data for spintronic effects in organic materials confirm so far spin injection and transport only for voltages not exceeding approximately 1 V. Expanding this working interval to about a few volts represents thus a challenging applicative goal. On the other hand, the available operation voltages do not prevent the utilization of OSPDs in sensor or memory applications.

As a final remark, we would like to point out on the immense potential of OSCs for the spintronics field. An almost infinite list of materials, easily modifiable interfaces, and cheap technologies for the growth of films on rigid and flexible substrates make us look optimistically toward the next steps of this young field.

## REFERENCES

1. V. A. Dediu, L. E. Hueso, I. Bergenti, and C. Taliani, Spin routes in organic semiconductors, *Nat. Mater.* **8**, 707–716 (2009).
2. F. Wang and Z. V. Vardeny, Organic spin valves: The first organic spintronics devices, *J. Mater. Chem.* **19**, 1685–1690 (2009).
3. W. J. M. Naber, S. Faez, and W. G. van der Wiel, Organic-spintronics, *J. Phys. D-Appl. Phys.* **40**, R205–R228 (2007).
4. D. S. McClure, Spin-orbit interaction in aromatic molecules, *J. Chem. Phys.* **20**, 682–686 (1952).
5. D. Beljonne, Z. Shuai, G. Pourtois, and J. L. Bredas, Spin-orbit coupling and intersystem crossing in conjugated polymers: A configuration interaction description, *J. Phys. Chem. A* **105**, 3899–3907 (2001).
6. C. B. Harris, R. L. Schlupp, and H. Schuch, Optically detected electron spin locking and rotary echo trains in molecular excited states, *Phys. Rev. Lett.* **30**, 1019–1022 (1973).
7. V. I. Krinichnyi, S. D. Chemerisov, and Y. S. Lebedev, EPR and charge-transport studies of polyaniline, *Phys. Rev. B* **55**, 16233–16244 (1997).
8. R. I. Dzhioev, K. V. Kavokin, V. L. Korenev et al., Low-temperature spin relaxation in n-type GaAs, *Phys. Rev. B* **66**, 245204 (2002).
9. I. Žutić, J. Fabian, and S. Das Sarma, Spintronics: Fundamentals and applications, *Rev. Mod. Phys.* **76**, 323–410 (2004).
10. V. I. Arkhipov, E. V. Emelianova, Y. H. Tak, and H. Bassler, Charge injection into light-emitting diodes: Theory and experiment, *J. Appl. Phys.* **84**, 848–856 (1998).
11. W. Brütting, S. Berleb, and A. G. Mückl, Device physics of organic light-emitting diodes based on molecular materials, *Org. Electron.* **2**, 1–36 (2001).
12. V. Coropceanu, J. Cornil, D. A. da Silva, Y. Olivier, R. Silbey, and J. L. Bredas, Charge transport in organic semiconductors, *Chem. Rev.* **107**, 926–952 (2007).
13. T. W. Hagler, K. Pakbaz, K. F. Voss, and A. J. Heeger, Enhanced order and electronic delocalization in conjugated polymers oriented by gel processing in polyethylene, *Phys. Rev. B* **44**, 8652–8666 (1991).
14. W. P. Su, J. R. Schrieffer, and A. J. Heeger, Solitons in polyacetylene, *Phys. Rev. Lett.* **42**, 1698–1701 (1979).
15. J.-H. Park, E. Vescovo, H.-J. Kim, C. Kwon, R. Ramesh, and T. Venkatesan, Magnetic properties at surface boundary of a half-metallic ferromagnet $La_{0.7}Sr_{0.3}MnO_3$, *Phys. Rev. Lett.* **81**, 1953–1956 (1998).
16. M. P. de Jong, I. Bergenti, V. A. Dediu, M. Fahlman, M. Marsi, and C. Taliani, Evidence for $Mn^{2+}$ ions at surfaces of $La_{0.7}Sr_{0.3}MnO_3$ thin films, *Phys. Rev. B* **71**, 014434 (2005).
17. J. M. De Teresa, A. Barthelemy, A. Fert, J. P. Contour, F. Montaigne, and P. Seneor, Role of metal-oxide interface in determining the spin polarization of magnetic tunnel junctions, *Science* **286**, 507–509 (1999).
18. J. E. Evetts, M. G. Blamire, N. D. Mathur et al., Defect-induced spin disorder and magnetoresistance in single-crystal and polycrystal rare-earth manganite thin films, *Philos. Trans. R. Soc. Lond. Ser. A-Math. Phys. Eng. Sci.* **356**, 1593–1613 (1998).

19. C. W. Tang and S. A. VanSlyke, Organic electroluminescent diodes, *Appl. Phys. Lett.* **51**, 913–915 (1987).

20. A. Curioni, M. Boero, and W. Andreoni, Alq(3): ab initio calculations of its structural and electronic properties in neutral and charged states, *Chem. Phys. Lett.* **294**, 263–271 (1998).

21. A. N. Caruso, D. L. Schulz, and P. A. Dowben, Metal hybridization and electronic structure of Tris(8-hydroxyquinolato) aluminum (Alq(3)), *Chem. Phys. Lett.* **413**, 321–325 (2005).

22. G. Horowitz, D. Fichou, X. Z. Peng, Z. G. Xu, and F. Garnier, A field-effect transistor based on conjugated alpha-sexithienyl, *Solid State Commun.* **72**, 381–384 (1989).

23. Z. H. Xiong, D. Wu, Z. V. Vardeny, and J. Shi, Giant magnetoresistance in organic spin-valves, *Nature* **427**, 821–824 (2004).

24. F. J. Wang, C. G. Yang, Z. V. Vardeny, and X. G. Li, Spin response in organic spin valves based on $La_{2/3}Sr_{1/3}MnO_3$ electrodes, *Phys. Rev. B* **75**, 245324 (2007).

25. A. Riminucci, I. Bergenti L. E. Hueso et al., Negative spin valve effects in manganite/organic based devices, arXiv:cond-mat/0701603v1 (2006).

26. V. Dediu, L. E. Hueso, I. Bergenti, et al., Room-temperature spintronic effect in $Alq_3$-based hybrid devices, *Phys. Rev. B* **78**, 115203 (2008).

27. S. Majumdar, H. S. Majumdar, R. Laiho, and R. Osterbacka, Comparing small molecules and polymer for future organic spin-valves, *J. Alloys Compd.* **423**, 169–171 (2006).

28. W. Xu, G. J. Szulczewski, P. LeClair et al., Tunneling magnetoresistance observed in $La_{0.67}Sr_{0.33}MnO_3$/organic molecule/Co junctions, *Appl. Phys. Lett.* **90**, 072506 (2007).

29. J. S. Moodera, L. R. Kinder, T. M. Wong, and R. Meservey, Large magnetoresistance at room-temperature in ferromagnetic thin-film tunnel-junctions, *Phys. Rev. Lett.* **74**, 3273–3276 (1995).

30. E. Y. Tsymbal, O. N. Mryasov, and P. R. LeClair, Spin-dependent tunnelling in magnetic tunnel junctions, *J. Phys. Condens. Matter* **15**, R109–R142 (2003).

31. S.-W. Liu, J.-H. Lee, C.-C. Lee, C.-T. Chen, and J.-K. Wang, Charge carrier mobility of mixed-layer organic light–emitting diodes, *Appl. Phys. Lett.* **91**, 142106 (2007).

32. V. Dediu, M. Murgia, F. C. Matacotta, C. Taliani, and S. Barbanera, Room temperature spin polarized injection in organic semiconductor, *Solid State Commun.* **122**, 181–184 (2002).

33. T. Ikegami, I. Kawayama, M. Tonouchi, S. Nakao, Y. Yamashita, and H. Tada, Planar-type spin valves based on low–molecular-weight organic materials with $La_{0.67}Sr_{0.33}MnO_3$ electrodes, *Appl. Phys. Lett.* **92**, 153304 (2008).

34. Y. Liu, T. Lee, H. E. Katz, and D. H. Reich, Effects of carrier mobility and morphology in organic semiconductor spin valves, *J. Appl. Phys.* **105**, 07C708 (2009).

35. N. A. Morley, A. Rao, D. Dhandapani, M. R. J. Gibbs, M. Grell, and T. Richardson, Room temperature organic spintronics, *J. Appl. Phys.* **103**, 07F306 (2008).

36. H. Vinzelberg, J. Schumann, D. Elefant, R. B. Gangineni, J. Thomas, and B. Büchner, Low temperature tunneling magnetoresistance on (La,Sr)$MnO_3$/Co junctions with organic spacer layers, *J. Appl. Phys.* **103**, 093720 (2008).

37. Y. Ogimoto, M. Izumi, A. Sawa et al., Tunneling magnetoresistance above room temperature in $La_{0.7}Sr_{0.3}MnO_3$/$SrTiO_3$/$La_{0.7}Sr_{0.3}MnO_3$ junctions, *Jpn. J. Appl. Phys. Part 2-Lett.* **42**, L369–L372 (2003).

38. J. R. Petta, S. K. Slater, and D. C. Ralph, Spin-dependent transport in molecular tunnel junctions, *Phys. Rev. Lett.* **93**, 136601 (2004).

39. S. Miwa, M. Shiraishi, S. Tanabe, M. Mizuguchi, T. Shinjo, and Y. Suzuki, Tunnel magnetoresistance of C-60-Co nanocomposites and spin-dependent transport in organic semiconductors, *Phys. Rev. B* **76**, 214414 (2007).

40. T. S. Santos, J. S. Lee, P. Migdal, I. C. Lekshmi, B. Satpati, and J. S. Moodera, Room-temperature tunnel magnetoresistance and spin-polarized tunneling through an organic semiconductor barrier, *Phys. Rev. Lett.* **98**, 016601 (2007).

41. J. H. Shim, K. V. Raman, Y. J. Park et al., Large spin diffusion length in an amorphous organic semiconductor, *Phys. Rev. Lett.* **100**, 226603 (2008).

42. M. Cinchetti, K. Heimer, J. P. Wustenberg et al., Deter-mination of spin injection and transport in a ferromagnet/organic semiconductor heterojunction by two-photon photoemission, *Nat. Mater.* **8**, 115–119 (2009).

43. A. J. Drew, J. Hoppler, L. Schulz et al., Direct measurement of the electronic spin diffusion length in a fully functional organic spin valve by low-energy muon spin rotation, *Nat. Mater.* **8**, 109–114 (2009).

44. R. H. Friend, R. W. Gymer, A. B. Holmes et al., Electroluminescence in conjugated polymers, *Nature* **397**, 121–128 (1999).

45. I. Bergenti, V. Dediu, E. Arisi et al., Spin polarised electrodes for organic light emitting diodes, *Org. Electron.* **5**, 309–314 (2004).

46. R. Fiederling, M. Keim, G. Reuscher et al., Injection and detection of a spin-polarized current in a light-emitting diode, *Nature* **402**, 787–790 (1999).

47. V. F. Motsnyi, P. Van Dorpe, W. Van Roy et al., Optical investigation of electrical spin injection into semiconductors, *Phys. Rev. B* **68**, 245319 (2003).

48. Y. Ohno, D. K. Young, B. Beschoten, F. Matsukura, H. Ohno, and D. D. Awschalom, Electrical spin injection in a ferromagnetic semiconductor heterostructure, *Nature* **402**, 790–792 (1999).

49. G. Salis, S. F. Alvarado, M. Tschudy, T. Brunschwiler, and R. Allenspach, Hysteretic electroluminescence in organic light-emitting diodes for spin injection, *Phys. Rev. B* **70**, 085203 (2004).

50. A. H. Davis and K. Bussmann, Organic luminescent devices and magnetoelectronics, *J. Appl. Phys.* **93**, 7358–7360 (2003).

51. D. R. McCamey, H. A. Seipel, S. Y. Paik et al., Spin Rabi-flopping in the photocurrent of a polymer light-emitting diode, *Nat. Mater.* **7**, 723–728 (2008).

52. E. I. Rashba, Theory of electrical spin injection: Tunnel contacts as a solution of the conductivity mismatch problem, *Phys. Rev. B* **62**, R16267–R16270 (2000).

53. M. V. Tiba, W. J. M. de Jonge, B. Koopmans, and H. T. Jonkman, Morphology and electronic properties of the pentacene on cobalt interface, *J Appl. Phys.* **100**, 093707 (2006).

54. M. Popinciuc, H. T. Jonkman, and B. J. van Wees, Energy level alignment symmetry at Co/pentacene/Co interfaces, *J. Appl. Phys.* **100**, 093714 (2006).

55. Y. Q. Zhan, I. Bergenti, L. E. Hueso, and V. Dediu, Alignment of energy levels at the Alq(3)/La$_{0.7}$Sr$_{0.3}$MnO$_3$ interface for organic spintronic devices, *Phys. Rev. B* **76**, 045406 (2007).

56. Y. Q. Zhan, M. P. de Jong, F. H. Li, V. Dediu, M. Fahlman, and W. R. Salaneck, Energy level alignment and chemical interaction at Alq(3)/Co interfaces for organic spintronic devices, *Phys. Rev. B* **78**, 045208 (2008).

57. Y. Q. Zhan, X. J. Liu, E. Carlegrim et al., The role of aluminum oxide buffer layer in organic spin-valves performance, *Appl. Phys. Lett.* **94**, 053301 (2009).

58. Y. Q. Zhan and M. Fahlman, Private communication (2009).

59. M. Grobosch, K. Dorr, R. B. Gangineni, and M. Knupfer, Energy level alignment and injection barriers at spin injection contacts between La$_{0.7}$Sr$_{0.3}$MnO$_3$ and organic semiconductors, *Appl. Phys. Lett.* **92**, 023302 (2008).

60. L. E. Hueso, I. Bergenti, A. Riminucci, Y. Q. Zhan, and V. Dediu, Multipurpose magnetic organic hybrid devices, *Adv. Mater.* **19**, 2639–2642 (2007).

61. J. C. Scott and L. D. Bozano, Nonvolatile memory elements based on organic materials, *Adv. Mater.* **19**, 1452–1463 (2007).

62. E. G. Emberly and G. Kirczenow, Molecular spintronics: Spin-dependent electron transport in molecular wires, *Chem. Phys.* **281**, 311–324 (2002).

63. R. Pati, L. Senapati, P. M. Ajayan, and S. K. Nayak, First-principles calculations of spin-polarized electron transport in a molecular wire: Molecular spin valve, *Phys. Rev. B* **68**, 100407(R) (2003).

64. A. R. Rocha, V. M. Garcia-Suarez, S. W. Bailey, C. J. Lambert, J. Ferrer, and S. Sanvito, Towards molecular spintronics, *Nat. Mater.* **4**, 335–339 (2005).

65. D. Waldron, P. Haney, B. Larade, A. MacDonald, and H. Guo, Nonlinear spin current and magnetoresistance of molecular tunnel junctions, *Phys. Rev. Lett.* **96**, 166804 (2006).

66. J. F. Ren, J. Y. Fu, D. S. Liu, L. M. Mei, and S. J. Xie, Spin polarized injection and transport in organic polymers, *Synth. Met.* **155**, 611–614 (2005).

67. P. P. Ruden and D. L. Smith, Theory of spin injection into conjugated organic semiconductors, *J. Appl. Phys.* **95**, 4898–4904 (2004).

68. J. H. Wei, S. J. Xie, L. M. Mei, J. Berakdar, and W. Yan, Conductance switching, hysteresis, and magnetoresistance in organic semiconductors, *Org. Electron.* **8**, 487–497 (2007).

69. S. J. Xie, K. H. Ahn, D. L. Smith, A. R. Bishop, and A. Saxena, Ground-state properties of ferromagnetic metal/conjugated polymer interfaces, *Phys. Rev. B* **67**, 125202 (2003).

70. P. A. Bobbert, W. Wagemans, F. W. A. van Oost, B. Koopmans, and M. Wohlgenannt, Theory for spin diffusion in disordered organic semiconductors, *Phys. Rev. Lett.* **102**, 156604 (2009).

71. P. A. Bobbert, T. D. Nguyen, F. W. A. van Oost, B. Koopmans, and M. Wohlgenannt, Bipolaron mechanism for organic magnetoresistance, *Phys. Rev. Lett.* **99**, 216801 (2007).

72. T. L. Francis, O. Mermer, G. Veeraraghavan, and M. Wohlgenannt, Large magnetoresistance at room temperature in semiconducting polymer sandwich devices, *New J. Phys.* **6**, 185 (2004).

73. L. D. Bozano, B. W. Kean, M. Beinhoff, K. R. Carter, P. M. Rice, and J. C. Scott, Organic materials and thin-film structures for cross-point memory cells based on trapping in metallic nanoparticles, *Adv. Funct. Mater.* **15**, 1933–1939 (2005).

# 6

# Spin Transport in Ferromagnet/III–V Semiconductor Heterostructures

**Paul A. Crowell and Scott A. Crooker**

The processes of electrical spin injection, transport, and detection in nonmagnetic semiconductors are inherently nonequilibrium phenomena. A nonequilibrium spin polarization, which we will refer to in this chapter as *spin accumulation*, may be generated and probed by a variety of optical and transport techniques. We will focus on spin accumulation in heterostructures of ferromagnetic metals (particularly Fe) and III–V semiconductors (particularly GaAs). Much of the essential physics of these systems for both metals and semiconductors has been addressed in other chapters in this volume. The basic phenomena of spin accumulation in metals have been covered in Chapter 5, Volume 1 and Chapter 9, Volume 3. This chapter will provide a review of a sequence of experiments that have advanced the understanding of spin transport in semiconductors through their sensitivity to electron spin dynamics. To a great extent, this has been a matter of careful experimental design and execution, and a good part of the discussion will address tests of well-known models. The benefit of this approach has been a reliable demonstration of all-electrical devices incorporating a ferromagnetic (FM) injector, semiconductor (SC) channel, and FM detector. In this context, we will show how quantitative electrical spin detection measurements can be made with these devices using both optical imaging and transport techniques, and we will conclude with some discussion of the new physics that can be addressed. Although the emphasis of the chapter is on the progression of experiments leading up to the demonstration of a III–V based lateral spin valve, some discussion of more recent developments as well as additional references are included in this revision.

## 6.1 SPIN TRANSPORT AND DYNAMICS IN III–V SEMICONDUCTORS

The basic physics of electron spin dynamics in III–V semiconductors was established through optical pumping experiments, which were reviewed authoritatively more than three decades ago [1]. Polarized photoluminescence measurements identified many of the important spin relaxation mechanisms in GaAs, with the significant caveat that most experiments were carried out on $p$-type, or compensated materials, in order to enhance the recombination rate of spin-polarized electrons. As a result, the electron spin dynamics observed in these experiments were influenced by the presence of holes and by recombination dynamics. The advent of absorption-based techniques, such as time-resolved Faraday rotation in the 1990s, brought particular attention to the case of $n$-type GaAs doped in close proximity to the metal-insulator transition (MIT) [2]. As will be evident below, the relatively long spin lifetime ($\sim 100\,\text{ns}$) of electrons [2–4] in this doping range and at low temperatures provides an enormous technical advantage *if* the semiconductor can be integrated with an appropriate source and detector of spin-polarized electrons. Before considering the role of the source and detector, however, it will be useful to discuss spin transport and dynamics in GaAs in the doping range near the MIT.

The interpretation of the spin transport measurements discussed below will be based on the standard spin drift-diffusion model, which provided the framework for the original proposal of electrical spin injection into semiconductors in the 1970s [5]. A detailed discussion of the elementary physics, elucidating some of the important differences with respect to spin transport in metals, is provided in Ref. [6]. An electron spin density **S** in a semiconductor is treated as a vector field evolving according to the dynamical equation

$$\frac{\partial \mathbf{S}}{\partial t} = D\nabla_r^2 \mathbf{S} + \mu(\mathbf{E}\cdot\nabla_r)\mathbf{S} + g_e\mu_B\hbar^{-1}(\mathbf{B}_{\text{eff}}\times\mathbf{S}) - \frac{\mathbf{S}}{\tau_s} + \mathbf{G}(\mathbf{r}) = 0, \qquad (6.1)$$

where:

   $D$   is the diffusion constant
   $\mathbf{E}$   is the electric field
   $\mathbf{B}_{\text{eff}}$   is the effective magnetic field acting on the electron spin
   $\tau_s$   is the spin lifetime
   $\mathbf{G}(\mathbf{r})$   is a term representing the spatial distribution of spins generated by an external source

The electron mobility is $\mu$, and we will use $g_e = -0.44$ for the Landé g-factor of an electron in GaAs. The corresponding gyromagnetic ratio can be expressed conveniently as $\gamma = |g_e|\mu_B/\hbar = 3.8\times10^6 \cdot \text{s}^{-1}\,\text{G}^{-1}$ or $\gamma/2\pi = 0.61\,\text{MHz/Gauss}$. Three features of semiconductors which make the spin dynamics non-trivial are (a) the relaxation mechanisms determining $\tau_s$, (b) the distinct contributions to the effective field $\mathbf{B}_{\text{eff}}$ (for example, applied fields, spin-orbit fields, or nuclear hyperfine fields), and (c) the relative importance of the drift term in the presence of internal electric fields.

In combination with ordinary diffusive transport theory, knowledge of the relevant spin lifetimes allows us to identify the critical length scales for a spin transport device. The relevant spin relaxation mechanisms are reviewed by Fabian in this volume. In the doping range of interest for us, the dominant channel of spin relaxation is the Dyakonov–Perel (DP) mechanism [7], which corresponds to the randomization of spin during diffusive transport by precession around the instantaneous spin-orbit field. Further discussion of DP and other spin relaxation mechanisms is given in Chapter 1, Volume 2. The bulk Dresselhaus spin-orbit coupling (due to the absence of inversion symmetry in GaAs), leads to a rapid decrease, proportional to $n_e^{-2}$, in the spin lifetime with increasing electron concentration $n_e$ [2, 4, 7]. The strong dependence on carrier density reflects the $k^3$ dependence in the spin-orbit Hamiltonian. A maximum in the spin lifetime of order 100–200 ns is typically observed near the MIT ($2 \times 10^{16} \text{cm}^{-3}$) at low temperatures [4], and even longer spin lifetimes (600ns) are found at elevated temperatures (10K) in samples doped slightly below the MIT [8, 9]. The characteristic length scales for spin transport in the semiconductor are the spin diffusion length $\lambda_s = \sqrt{D\tau_s}$ and drift length $l_d = \mu E \tau_s$. The characteristic mobility of $n$-GaAs at the MIT is of order 5000 cm²/V sec. Assuming a barely degenerate electron gas one finds that $\lambda_s$ is of the order of several microns.

We now consider the effective fields $\mathbf{B}_{\text{eff}}$ that couple to the spin through the torque term in Equation 6.1. In this case, the relevant magnetic field scale is determined by the gyromagnetic ratio and the spin lifetime. For example, the magnetic field required in order for a spin to precess a full cycle before it decays can be estimated from $2\pi / \gamma\tau_s \sim 17 \text{G}$ (for $\tau_s = 100 \text{ns}$). One immediate consequence of this relatively small field scale is that it is feasible to design a transport experiment that is extremely sensitive to the torque term in Equation 6.1. It then becomes possible to probe the effects of other "fields" that couple to spin, particularly those due to hyperfine and spin-orbit coupling. Hyperfine effects in GaAs near the MIT have been studied extensively using optical pumping [10, 11], and it suffices for now to note that the effective field acting on an electron spin due to dynamically polarized nuclei can easily exceed 1 kG. As will be seen below, this hyperfine field can influence the electron spin dynamics in a transport device profoundly. In contrast, the effective spin-orbit fields in bulk GaAs are small, and their effects on spin transport and dynamics are more subtle. The reader is referred to the monograph of Winkler [12] for a complete discussion. The only case that applies to the discussion of bulk GaAs in this chapter is the effect of strain [13], which leads to spin-orbit fields that depend linearly on momentum. We will not consider heterostructures, in which spin-orbit fields linear in momentum appear due to electric fields from structural inversion asymmetry (e.g. an applied gate voltage) [14, 15], or by modification of the Dresselhaus Hamiltonian by confinement [16].

For the case of a uniform electric field, the drift velocity $\mathbf{v}_d = \mu\mathbf{E}$, which for modest electric fields ($E \sim 1\,\text{V/cm}$) is on the order of $10^4$ cm/s. In this range, the spin drift length $v_d\tau_s$ is comparable to the spin diffusion length, and it is apparent that both the drift and diffusion terms in Equation 6.1 need to be considered. This is completely different from the situation in ordinary metals, where drift is negligible. One might guess that the drift length could become very large for transport near the MIT. In fact, however, $\tau_s$ drops precipitously due to impact ionization of donors [8] at modest electric fields above 10 V/cm. This indicates a maximum spin drift length in $n$-GaAs in the range of tens of microns. The physics near interfaces, where electric fields can be both large and nonuniform, is potentially much richer [6, 17].

## 6.2 SPIN INJECTION FROM FERROMAGNETIC METALS INTO SEMICONDUCTORS

As the previous section makes clear, the conditions for spin transport in bulk GaAs near the MIT are rather favorable. The length scales required are readily accessible with photolithography, and the magnetic fields required to probe spin dynamics are modest. Nonetheless, significant progress on the integration of ferromagnetic sources and detectors with semiconductors has occurred only since the turn of the century. The discussion here will focus on work since 2000. For a review of the status of the field at about that time, see Ref. [18].

As some stock was taken of the situation around 2000, a few important observations emerged. First, the standard description of spin transport in metals [19–22], when applied in a straightforward manner to semiconductors [23], appeared to be consistent with the absence of spin accumulation when electrons were injected from a ferromagnetic metal into a semiconductor. The essence of this argument was that the spin current flowing into the semiconductor is limited by the semiconductor's relatively large resistivity, while the low resistance of the ferromagnetic metal is itself an efficient sink for any spin accumulation at the interface.

It was immediately recognized that this obstacle to spin injection, often referred to as the "conductivity mismatch problem", could be circumvented by inserting a tunnel barrier between the ferromagnetic metal and the semiconductor [24–26]. In fact, this principle had already been exploited in the earliest demonstration of spin injection into GaAs using a scanning tunneling microscope with a ferromagnetic tip [27]. For FM/SC devices based on $Al_xGa_{1-x}As$, in which a Schottky contact is present anyway, the only problem would appear to be engineering a barrier of the appropriate thickness and height. Two general approaches have been followed: either using a pure "tunneling Schottky" barrier [28–30], or creating an additional artificial barrier by inserting an insulator at the FM/SC interface [31–33]. For the case of a high resistance barrier, ignoring the effects of dispersion, and assuming that that spin polarization of the ferromagnet at the interface with the semiconductor is equal to the bulk value, the polarization of the tunneling current should be limited only by the spin polarization of the ferromagnetic metal.

Although adopting a tunnel barrier for the injection contact was an obvious choice, it did not provide an immediate solution to the problem of electrical spin detection. At the same time, however, the community working on the problem of spin injection from magnetic *semiconductors* into their nonmagnetic counterparts developed a new version of the reliable luminescence polarization technique employed in optical pumping experiments. By making the magnetic semiconductor one contact in a *p-i-n* junction light-emitting diode (LED), it was possible to convert a spin-polarized current of either electrons [34] or holes [35] into polarized light. This so-called "spin-LED" (originally proposed in the 1970s [36]) has since been adapted by the community working on spin injection from FM metals and is discussed in greater detail in Chapter 2, Volume 2. By placing the Schottky tunnel barrier in series with the *p-i-n* diode, spin-polarized electrons flowing from the metal recombine with unpolarized holes. For a quantum well, the degree of polarization of the emitted light is equal to the spin polarization of the electron component of the recombination current, provided that the component of the spin being measured is along the growth axis. Using this principle, spin injection from Fe into GaAs was first reported by Zhu et al. in 2001 [28]. A significant improvement in the efficiency of the $Fe/Al_xGa_{1-x}As$ spin-LED was then reported by the NRL group using a graded Schottky barrier [29]. In parallel, the development of spin-LEDs based on insulating tunnel barriers inserted between the ferromagnet and semiconductor was being carried out by the IMEC and IBM groups [31, 32]. In all of these experiments, the extraction of the spin polarization of the tunneling current was based on rate equation arguments that we will not discuss here [37–39]. By 2005, the Schottky spin-LED was sufficiently well understood that it was being used for the evaluation of new potential spin injection materials, including Heusler alloys [40, 41], perpendicular anisotropy materials [42, 43], and others.

At the same time as spin-LEDs were being used to probe the polarization of electrons tunneling from a ferromagnet metal into a semiconductor, a different line of research was devoted to the study of spin accumulation due to unpolarized electrons incident on a FM/SC interface from the semiconductor. In 2001, Kawakami et al. [44] reported on spin accumulation at the interface between (100) GaAs and MnAs (a ferromagnetic semimetal) under optical pumping by linearly polarized light. This experiment as well as a subsequent one on Fe/GaAs [45] utilized time-resolved Faraday rotation, and hence could be carried out in simple bilayers of the ferromagnetic metal with an *n*-type GaAs epilayer. Ciuti, McGuire, and Sham interpreted this process using a tunneling approach [46, 47]. The spin accumulation was enhanced in experiments in which a forward electrical bias was applied while optically pumping [48], consistent with the tunneling hypothesis. The tunneling argument implied that it should be possible to generate a spin accumulation in semiconductors under forward bias even in the absence of optical pumping, as was then observed by Stephens et al. [49] in MnAs/GaAs bilayers.

For technical reasons, the majority of the spin-LED experiments discussed above probed the component of the spin accumulation parallel to an applied magnetic field. In contrast, the experiments under optical pumping and/or a forward-bias current probed the transverse component of the accumulated spin polarization, which could be detected unambiguously through its *precession* about the applied field. The exceptions among the spin-LED experiments were those carried out by the IMEC [31, 50, 51] and Minnesota groups [30, 37, 52], which were based on the precession of spins in the semiconductor after injection from the ferromagnet. Given the relatively short spin lifetimes in quantum wells (several hundred picoseconds), significant precession could be observed in these experiments only because of the large hyperfine contributions to the effective magnetic field. Although somewhat more complicated in its implementation, a steady-state measurement that is sensitive to electron spin precession allows for a much stronger quantitative analysis of the spin dynamics. We will call experiments of this type "Hanle measurements", as they are analogs of the Hanle effect commonly observed in photoluminescence polarization under optical pumping by circularly polarized light [1].

The experiments introduced in this section demonstrated that a spin accumulation could be generated in GaAs under either forward or reverse electrical bias across a FM/SC interface. There remained, however, significant limitations. First, the effects demonstrated were local. That is, the spin polarization was measured in the semiconductor at the location of the ferromagnetic source. Second, the detection techniques were strictly optical. Although there were some earlier attempts to detect optically induced spin accumulations electrically [53–55], they were not supported by the demonstration of a Hanle effect. A key point is that the tunnel barriers used in spin-LEDs, which were perfectly adequate for spin injection, were not particularly transparent to electrons at low bias voltages. The resistance-area product of a typical Schottky barrier of the design of Hanbicki et al. [29] is about $10^5 - 10^6 \, \Omega \, \mu m^2$, which is several orders of magnitude larger than the values achieved in typical magnetic tunnel junctions. This means that in simple two-terminal (FM-SC-FM) devices, the overall resistance is dominated by the interfaces. Except for the limit in which the entire semiconductor acts as a tunnel barrier [56, 57], the ordinary magnetoresistance $(R_{\uparrow\uparrow} - R_{\uparrow\downarrow}) / R_{\uparrow\uparrow}$ will be extremely small, as noted by Fert and Jaffrès [25]. This is in fact the other side of the two-edged sword of the "conductivity mismatch" argument. High resistivity barriers permit efficient spin injection, but the effectiveness of a such a barrier for spin detection is limited by its capacity for sinking the charge current. As detailed by Fert and Jaffrès, there is a narrow range of optimal barrier resistances for two-terminal FM-SC-FM structures, but these lie away from the practical values for Fe/GaAs devices with Schottky barriers.

In the next three sections, we address how these shortcomings have been addressed in lateral FM/GaAs devices, a process that has entailed three principal steps. The first of these was establishing that spin accumulations could be detected remotely from the ferromagnetic source (or drain) contact.

The experimental approach for this aspect of the project was similar to that used by Stephens et al. [49]. Scanning Kerr microscopy and local Hanle–Kerr effect studies were used to detect electron spins injected into a GaAs channel from Fe contacts under reverse bias, as expected from spin-LED experiments [9, 58, 59]. Also, spin accumulations generated by spin extraction at forward-biased contacts was demonstrated. As discussed below, these optical studies were used to directly measure the relevant spin transport parameters ($\tau_s, D, v_d$), and established that the drift-diffusion model of Equation 6.1 provides a satisfactory description of the spin transport.

The second step was to show that Fe/GaAs Schottky barriers, although highly resistive, could be used as electrical spin detectors [58, 60]. Electrical detection of both optically-injected as well as electrically-injected spin were demonstrated. These latter experiments were based on the serendipitous discovery that it was possible to electrically generate *and* detect a spin accumulation at a *single* FM/SC interface. As will be evident from the discussion below, this was possible only because the measurement was sensitive to the suppression of the spin accumulation by the Hanle effect.

The third step was the implementation of a completely electrical injection and detection scheme in a lateral geometry [61]. This experiment was based on the non-local approach in which spin and charge currents are separated in space. The overall philosophy follows that developed originally by Johnson and Silsbee for metals [19, 20], and implemented by many other groups in metallic nanostructures. The essential physics of the non-local measurement and its interpretation are covered in Chapter 5, Volume 1 and its application in more recent experiments on metals and graphene are covered in the chapters by Otani and van Wees. We will emphasize only those aspects that are of critical importance in semiconductors.

## 6.3 Fe/GaAs HETEROSTRUCTURES AND LATERAL SPIN-TRANSPORT DEVICES

In spite of the rapid refinement of spin-LEDs and the achievement of correspondingly high spin injection efficiencies, the appropriate modification of these devices for lateral transport experiments was not immediately apparent. For some period of time, the conductivity mismatch argument and its extension by Fert and Jaffrès [25] cast a pessimistic light over efforts to fabricate a lateral FM-SC-FM spin valve. In cases where successes were reported [62], the means for extracting information about dynamics (e.g. spin lifetimes) quantitatively were not demonstrated. In particular, no Hanle effect was observed.

Our own approach to this problem evolved from a purely technical question: is it possible to image an electrically injected spin accumulation in *n*-GaAs in a manner similar to that demonstrated for optical pumping experiments in the presence of an electric field [63–65]? In contrast to a spin-LED, in which electrons recombine with holes, luminescence is not a viable detection technique in this case. In contrast, Faraday or Kerr rotation are suitable probes of spin accumulation in *n*-type material. Given the time and length scales for spin transport in $n$-GaAs samples doped near the MIT, a relatively simple magneto-optical polar Kerr effect microscope is sufficient, provided that the injected spins can be tipped perpendicular to the sample plane (thereby enabling detection by the polar Kerr effect) without perturbing the magnetization of the Fe contacts. This can be accomplished by exploiting the large uniaxial anisotropy of the epitaxial Fe/GaAs (001) interface [37]. This anisotropy, derived from the surface electronic structure, leads to a magnetic easy axis along the [011] direction and allows for small "tipping" magnetic fields to be applied along the [01$\bar{1}$] (hard) direction without inducing significant rotation of the Fe magnetization. On the other hand, once injected into the semiconductor, an electron spin experiences a torque due to the small applied field and therefore precesses into the [100] direction, allowing for detection by polar Kerr rotation [49]. The spin accumulation, driven by lateral drift and diffusion, can then be imaged as it emerges from underneath a ferromagnetic contact. (An alternative approach avoids the use of precession by imaging the spin accumulation on the cleaved edge of the sample. This has been implemented by Kotissek et al. [66]).

A cross-section of a typical FM/SC heterostructure is shown in Figure 6.1a. The principal consideration in the design of these samples is the integration of a Fe/GaAs Schottky tunnel barrier with a conducting channel, while keeping the spin diffusion length long enough so that fabrication by photolithography and detection by optical Kerr microscopy are practical. The doping profile chosen for the Schottky barrier was originally perfected by Hanbicki et al. for spin LEDs [29], and the conduction band structure near the Fe/GaAs interface is shown in Figure 6.1b. The heterostructures we have used in all measurements described in this chapter are grown on semi-insulating GaAs (100) wafers and typically consist of (from bottom to top) a 2.5 μm thick $n$-doped channel (Si donor concentration $n \sim 3-7 \times 10^{16}\,\mathrm{cm^{-3}}$), a 15 nm transition region in which the doping is increased from the channel doping up to $5 \times 10^{18}\,\mathrm{cm^{-3}}$ and an additional 15 nm thick heavily doped ($5 \times 10^{18}\,\mathrm{cm^{-3}}$) region. After growth of the semiconductor layers, the sample

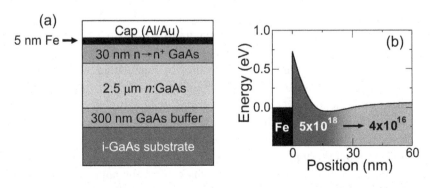

**FIGURE 6.1**    (a) A schematic cross-section of the heterostructures discussed in this chapter. See the text for a discussion of the dopings, which vary slightly from sample to sample. (b) Conduction band structure near the Fe/GaAs interface for the doping profile used in this work [29].

is cooled to approximately room temperature, and the epitaxial Fe layer is deposited. Typically, an As-rich $c(4\times4)$ reconstructed surface is used. Finally, the structures are capped with Al and Au. Transmission electron microscopy (TEM) studies have established that the interface between the Fe and GaAs is atomically abrupt (within one or two monolayers) [67], and all films have the required interfacial anisotropy, with the Fe easy magnetization axis along the [011] direction. The devices are prepared using standard wet and dry etching techniques. Gold vias are deposited over SiN isolation layers to contact the ferromagnetic electrodes. Micrographs of samples used in the three types of experiments discussed in this chapter are shown in Figure 6.2.

## 6.4 SCANNING KERR MICROSCOPY AND LOCAL HANLE EFFECT STUDIES

### 6.4.1 KERR SET-UP

Although any of these Fe/GaAs structures shown in Figure 6.2 can be (and have been) imaged with Kerr microscopy, the discussion in this section focuses first on the simple two-terminal device shown in Figure 6.2a. The "source" electrode is reverse-biased, meaning that electrons flow from the FM into the SC. The "drain" electrode at the opposite end of the $n$-GaAs channel is forward-biased. Figure 6.3 shows a schematic of the scanning Kerr microscopy experiment. The Fe/GaAs spin transport devices are mounted, nominally strain-free, on the vacuum cold finger of a small optical cryostat, which in turn is mounted on a $x$-$y$ positioning stage. The out-of-plane component of electron spin polarization in the semiconductor, $S_z$, is measured at a particular location via the polar Kerr rotation $\theta_K$ of a linearly-polarized probe beam that is focused onto (and reflected from) the sample at normal incidence. The probe beam is derived from a narrow-band continuous-wave Ti:sapphire ring laser, which is typically tuned just below the semiconductor bandgap in order to minimize the influence of the probe beam on the resident electrons in the $n$-GaAs while maintaining high sensitivity to spin polarization [59, 64]. The final probe focusing lens is also mounted on a positioning stage, and two-dimensional images of $S_z$ are obtained by raster-scanning either the cryostat or (more typically) the probe focusing lens.

Electrically-injected spins are studied by modulating, at kilohertz frequencies, the voltage bias applied to the Fe source/drain contacts, and then measuring the induced change in $\theta_K$ using lock-in amplifiers. As discussed above, the spins are injected along the magnetic easy axis of the Fe, which is the crystallographic [011] direction, or the $\hat{x}$-direction in Figure 6.3. They precess about the applied field **B**, and $\theta_K$ is proportional to $S_z$ at the position of the focused laser spot. External coils control the applied tipping fields $B_x$, $B_y$, and $B_z$. These coils are also used together with a fixed (local) probe beam to study the relaxation and dephasing of spins at a particular spot on the sample as a continuous function of an applied transverse magnetic field–the "local Hanle effect".

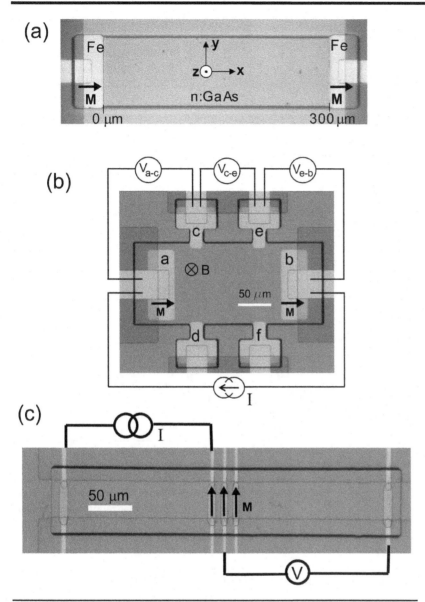

**FIGURE 6.2**    Micrographs of representative Fe/GaAs spin transport devices used for the experiments discussed in this chapter. (a) A simple two-terminal device used in the original Kerr imaging studies of Ref. [58, 59]. (b) Hall bars used for the detection of spin accumulation at a single Fe/GaAs interface [60]. Note how the source, drain, and channel contributions to the two-terminal voltage $V_{a-b}$ can be measured independently. (c) A typical non-local device, with a schematic of the connections to a current source and voltmeter [61]. Kerr imaging of non-local devices [9] is also discussed in the text.

The samples may also be held by a small cryogenic vise machined into the cold finger [64]. The uniaxial stress applied to the sample by the vise is uniform and can be varied *in situ* by a retractable actuator. For devices grown on GaAs (100) substrates and cleaved along the usual ⟨011⟩ crystal axes, this uniaxial stress along ⟨011⟩ leads to non-zero off-diagonal elements

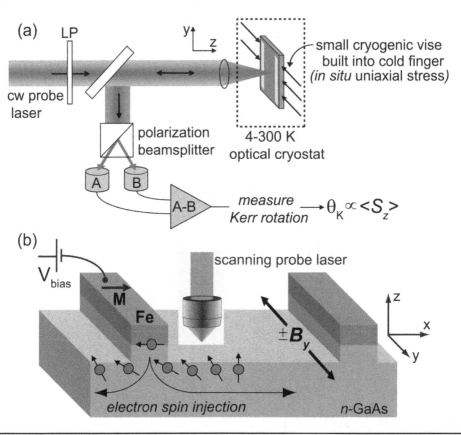

**FIGURE 6.3** (a) A schematic of the scanning Kerr microscope used to image electron spin transport in Fe/GaAs devices. The measured polar Kerr rotation $\theta_K$ of the reflected probe laser beam is proportional to the out-of-plane $\hat{z}$ component of the conduction electron spin polarization, $S_z$. External coils (not drawn) control the applied magnetic fields $B_x$, $B_y$, and $B_z$. (b) A representation of a Fe/GaAs spin transport structure in cross-section. In this drawing, electrons tunnel from Fe into the $n$-GaAs with initial spin polarization $S_0$ antiparallel to the Fe magnetization **M**. A small applied transverse magnetic field $\pm B_y$ is used to precess the injected spins out-of-plane (along $\pm\hat{z}$) so that they can be measured by the probe laser via the polar Kerr effect.

of the GaAs crystallographic strain tensor, $\varepsilon_{xy}$. For electrons moving in the $x$-$y$ sample plane, $\varepsilon_{xy}$ couples directly to electron spin $\sigma$ and momentum **k** via a spin-orbit coupling energy $\varepsilon_{xy}(\sigma_y k_x - \sigma_x k_y)$, which has the same form [13, 64] as the well-known Rashba spin-orbit Hamiltonian. The resulting effective magnetic field $\mathbf{B}_\varepsilon$ lies in-plane and is orthogonal to the electron momentum **k**.

## 6.4.2 Imaging Spin Injection, Local Hanle-Kerr Studies, and Modeling

In Figure 6.4, we show Kerr-rotation images and the associated local Hanle curves for electrically-injected spins. Figure 6.4a shows a spin transport device having rectangular Fe/GaAs source and drain contacts at either end of a 300 µm long $n$-GaAs channel. We image the $80 \times 80$ µm region shown by the dotted square, which includes part of the Fe injection contact and the

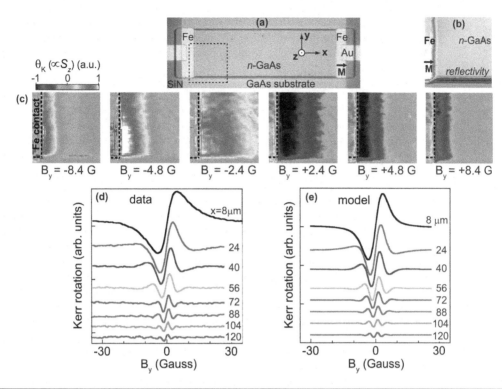

**FIGURE 6.4** (a) A lateral spin transport device consisting of a 300 μm long $n$-GaAs channel separating Fe/GaAs source and drain contacts. The dotted square shows the $80 \times 80$ μm region imaged. (b) An image of the reflected probe laser power, showing device features. (c) Images of electrical spin injection and transport in the $n$-GaAs channel. The electron current $I_e = 92$ μA. Electrons are injected with initial spin polarization $S_0 \parallel -\hat{x}$ and $B_y$ is varied from $-8.4$ to $+8.4$ G (left to right), causing spins to precess out of and into the page ($\pm\hat{z}$), respectively. (d) A series of local Hanle curves ($\theta_K$ vs. $B_y$) acquired with the probe laser fixed at different distances (8–120 μm) from the source contact (offset for clarity). (e) Simulated local Hanle curves using a 1D spin drift-diffusion model Equation 6.2, with $\tau_s = 125$ns, $v_d = 2.8 \times 10^4$ cm/s, and $D = 10$ cm$^2$/s.

bottom edge of the $n$-GaAs channel. An image of the reflected probe laser power (see Figure 6.4b) clearly shows these device features. With this contact under reverse voltage bias, so that spin-polarized electrons are injected from the Fe into the channel, Figure 6.4c shows a series of Kerr-rotation images as the applied field $B_y$ is varied from $-8.4$ G to $+8.4$ G. Injected electrons, initially polarized along the $-\hat{x}$ direction, precess into the $+\hat{z}$ direction when $B_y$ is oriented along $-\hat{y}$. When $B_y$ inverts sign and is oriented along $+\hat{y}$, the spins are tipped into the opposite direction (along $-\hat{z}$), reversing the sign of the measured Kerr rotation. These injected electrons flow down the channel with an average drift velocity $v_d$ that is the same in all the images. The drifting spins precess at a rate proportional to $|B_y|$; thus, the spatial period of the observed spin precession is shorter when $|B_y|$ is larger.

When the probe laser is fixed at a point in the $n$-GaAs channel and $S_z \propto \theta_K$ is measured as a continuous function of $B_y$, we obtain local Hanle curves having the characteristic antisymmetric lineshape shown in Figure 6.4d. These local Hanle curves are acquired in the $n$-GaAs channel at different distances from the Fe source contact (8–120 μm), as labeled.

With increasing distance, the amplitudes of the curves fall, their characteristic widths decrease, and they develop multiple oscillations associated with multiple precession cycles. The detailed structure of this family of Hanle curves contains considerable information about the dynamics of electron spin transport, including the electron spin lifetime $\tau_s$, diffusion constant $D$, and drift velocity $v_d$ [64, 68]. The Hanle curves invert when the magnetization **M** of the Fe contacts is reversed, as expected.

To model these local Hanle data we consider a simple one-dimensional picture of spin drift, diffusion, and precession in the GaAs channel. Spin-polarized electrons, injected with $S_0 = S_x$ (at $x = 0$ and $t = 0$), precess as they flow down the channel with a drift velocity $v_d$, arriving at the point of detection $x_0$ at a later time $t = x_0 / v_d$. At this point, $S_z = S_0 \exp(-t / \tau_s)\sin(\Omega_L t)$, where $\Omega_L = g_e \mu_B B_y / \hbar$ is the Larmor precession frequency. The actual signal is therefore computed by averaging the spin orientations of the precessing electrons over the Gaussian distribution of their arrival times (which has a half-width determined by diffusion).

$$S_z(B_y) = \int_{x_0}^{x_0+w} \int_0^{\infty} \frac{S_0}{\sqrt{4\pi Dt}} e^{-(x-v_dt)^2/4Dt} \times e^{-t/\tau_s} \sin(\Omega_L t) dt dx, \qquad (6.2)$$

where:

$v_d = \mu E$ is the drift velocity

$\mu$      is the electron mobility

$E$      is the electric field in the channel

The spatial integral accounts for the width $w$ of the source contact. This type of averaging is the basis of the Hanle effect observed in optical pumping experiments and in previous spin transport experiments in metals [20, 69] and semiconductors [30, 49, 50]. Using this model, Figure 6.4e shows that good overall agreement with the local Hanle data is obtained using $D = 10$ cm$^2$/s, $v_d = 2.8 \times 10^4$ cm/s, and $\tau_s = 125$ ns. The large drift velocity and spin lifetime in these devices allow access to a spatial regime far from the contacts and well beyond a spin diffusion length ($x_0 > \sqrt{D\tau_s}$), in which the average time-of-flight from the source to the point of detection, $T = (x_0 + w / 2)/ v_d$, determines the characteristic "age" of the measured spins. In this limit, the first peak in the data ($B_y = B_{peak}$) is the field at which electrons have precessed through one-quarter Larmor cycle, so that $T = \pi / (2\Omega_L) = \pi\hbar / (2 g_e \mu_B B_{peak})$. Thus, in Figure 6.4d, $B_{peak} = 1.05$ G when $x_0 = 88$ µm, indicating that $T \sim 380$ns and $v_d \sim 2.8 \times 10^4$ cm / s.

Simulating lateral spin flows in more complicated device geometries requires more sophisticated models of spin drift and diffusion. Two-dimensional models generally suffice for planar devices in which spin transport occurs in epilayers that are thinner than the characteristic spin diffusion lengths. For the case of applied in-plane magnetic and electric fields ($B_{x,y}, E_{x,y}$), the steady-state spin polarization **S**(**r**) can then be derived from the two-dimensional (2D) version of the spin drift-diffusion equation. Although Equation 6.1 in 2D can be solved analytically in certain limits [9], it is often easier to compute $S_z$ from a set of coupled spin drift-diffusion equations in two dimensions using numerical Fourier transform methods, particularly when including spin-orbit coupling terms [64, 68]. For in-plane electric and magnetic fields, and for spin-orbit coupling due to off-diagonal strain $\varepsilon_{xy}$ in bulk GaAs, the three equations determining the steady-state spin densities $S_{x,y,z}(\mathbf{r})$ are $O_1 S_x + O_2 S_z = -G_x$, $O_1 S_y + O_3 S_z = -G_y$, and $O_4 S_z - O_2 S_x - O_3 S_y = -G_z$, where

$$O_1 = D\nabla_r^2 + \mu\mathbf{E}\cdot\nabla_r - C_s^2 D - 1/\tau_s, \tag{6.3}$$

$$O_2 = g_e\mu_B B_y / \hbar + C_s(2D\nabla_x + \mu E_x), \tag{6.4}$$

$$O_3 = -g_e\mu_B B_x / \hbar + C_s(2D\nabla_y + \mu E_y), \tag{6.5}$$

$$O_4 = D\nabla_r^2 + \mu\mathbf{E}\cdot\nabla_r - 2C_s^2 D - 1/\tau_s. \tag{6.6}$$

$G_{x,y,z}(\mathbf{r})$ are the generation terms, and $C_s = C_3 m\varepsilon_{xy} / \hbar^2$ is the spin-orbit term that couples spin to the off-diagonal elements of the strain tensor in GaAs, $\varepsilon_{xy}$ [13, 68].

The presence of effective magnetic fields ($\mathbf{B}_{\text{eff}} = \mathbf{B}_\varepsilon$) due to spin-orbit coupling causes moving spins to precess, even in the absence of applied magnetic fields [64, 70]. The effective magnetic fields are directly revealed in local Hanle measurements. For example, for electrons flowing in the $\hat{x}$ direction as shown in Figure 6.4, $\mathbf{B}_\varepsilon$ is directed along $\pm\hat{y}$. $\mathbf{B}_\varepsilon$ therefore either directly *augments* or *opposes* the applied (real) magnetic field $B_y$ that is used to acquire the local Hanle curve. The net effect is therefore to shift the Hanle curve to the left or right, so that it is no longer antisymmetric with respect to $B_y$. An example of this is shown in Figure 6.5a, which shows local Hanle data acquired with increasing uniaxial stress applied to the device substrate along the [011] direction, which effectively turns on a Rashba-like spin-orbit interaction in the bulk $n$-GaAs channel of the device. The corresponding simulations shown in Figure 6.5b were performed using the numerical approach described above, and qualitatively reproduce the data when the applied strain $\varepsilon_{xy} = 0$, 1, and $2 \times 10^{-4}$.

## 6.4.3 IMAGING SPIN INJECTION AND EXTRACTION, BOTH UPSTREAM AND DOWNSTREAM OF THE CURRENT PATH

Scanning Kerr miroscopy can also be used to image the spin currents that exist when electrically-injected spins diffuse outside of the charge-current path. This is demonstrated on a "non-local" Fe/GaAs spin transport device of the type shown in Figure 6.2c, and again (with contact labels and wiring diagram) in Figure 6.6a. These devices have five Fe contacts in a $n$-GaAs channel, with easy-axis magnetization $\mathbf{M}$ along the $\hat{y}$ direction. We focus on spin injection and spin extraction at contact 4. The dotted square in Figure 6.6a shows the $65 \times 65$ μm region around contact 4 that was imaged. The reflected probe power in this region (Figure 6.6b) clearly shows contact 4,

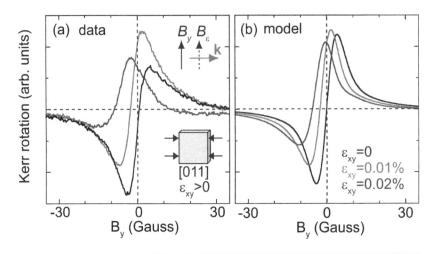

**FIGURE 6.5** (a) Local Hanle effect ($S_z$ vs. $B_y$) from electrically-injected spins in the presence of spin-orbit coupling and effective magnetic field $\mathbf{B}_\varepsilon$ due to strain. The device is the same as shown in Figure 6.4. These data are acquired at a point 4 μm from the source contact, for three values of uniaxial stress applied along the [011] substrate axis. Curves shift to the left with increasing stress, as the effective magnetic field $\mathbf{B}_\varepsilon \parallel \hat{y}$ increases due to the induced shear strain $\varepsilon_{xy}$. (b) Simulated local Hanle data, for $\varepsilon_{xy} = 0, 1, 2 \times 10^{-4}$.

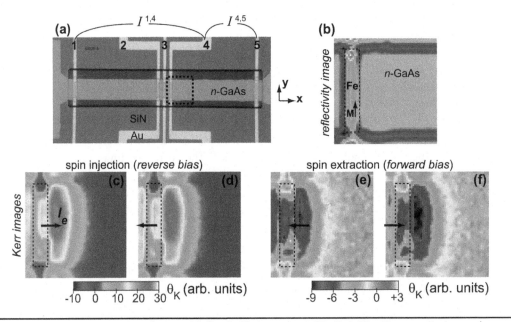

**FIGURE 6.6** (a) A spin transport device with five Fe/GaAs contacts (#1–5). A $65 \times 65$ μm region around contact 4 is imaged (dotted square). (b) The reflected probe power in this region, showing device features. (c, d) Images of the majority spin polarization that is electrically injected from contact 4 ($B_x = 2.4$ G). Arrows show the direction and magnitude of net electron current, $I_e$. In (c), spins are injected and drift/diffuse to the right ($I_e^{4\to5} = 25$ μA). In (d), spins are injected and drift to the *left* ($I_e^{4\to1} = 25$ μA); the spins to the right of contact 4 are diffusing outside of the current path. (e, f) Images of minority spin accumulation due to spin extraction at contact 4, both within and outside of the charge current path.

the $n$-GaAs channel, and the edges of the SiN layer. Under bias, the electron current $I_e$ flows *within* the imaged region when electrons flow between contacts 4 and 5. In this case the electric field in the channel, $E_x$, is non-zero in the imaged region and the drift velocity $v_d = \mu E_x$ may either augment or oppose electron diffusion away from contact 4, depending on the direction of $I_e$. Conversely, when $I_e$ flows between contacts 4 and 1, the current path is *outside* of the imaged region. In this case, $E_x$ is nominally zero in the imaged region regardless of $I_e$, and spin transport in this region should be purely diffusive.

Figure 6.6c and d show the imaged spin polarization for the case of spin injection (that is, the Fe/GaAs Schottky contact is reverse-biased and electrons flow from Fe into $n$-GaAs). In this device, majority spins having spin orientation antiparallel to the Fe magnetization **M** are injected for all reverse biases. The images show two cases, in which charge current path lies within or outside of the imaged region. For clarity, contact 4 in these images is outlined by a dotted rectangle, and the black arrows indicate the direction of the electron current $I_e$. In Figure 6.6c, injected electrons drift slowly to the right along $+\hat{x}$. The electron current $I_e^{4\rightarrow5} = 25$ μA and the voltage across the Fe/GaAs interface is $V_\text{int} \simeq 53$mV. All of these images show the *difference* between Kerr images acquired at $B_x = +2.4$ and $-2.4$ G. In this way, field-independent backgrounds are subtracted off, and only the signal that depends explicitly on electron spin precession remains. Figure 6.6c clearly shows a cloud of spin-polarized electrons emerging from and flowing away from contact 4, with an apparent spatial extent of order 15 μm in the $n$-GaAs channel. Note that the apparent extent of $S_z$ in these images depends on $|B_x|$, as shown earlier in Figure 6.4. In this case, both drift and diffusion drive a net flow of spins to the right.

A remarkably similar image, however, is observed in Figure 6.6d for the case in which spins are injected out of contact 4 under similar bias conditions ($I_e^{4\rightarrow1} = 25$ μA) but are drifting to the *left*, along $-\hat{x}$. In this case the current path is outside the imaged region, and $E_x$ in the imaged region is nominally zero. The cloud of spins seen to the right of contact 4 are those spins that have diffused out from under the contact, and which are now flowing to the right due to diffusion alone. The similarity of both the magnitude and the spatial extent of $S_z$ in Figure 6c and d indicates that diffusion, rather than drift, plays the major role in determining the overall transport of spins under these low injection bias conditions. These images show that robust spin currents, spatially separated from any charge currents, can be generated in the $n$-GaAs channel. It is precisely these "upstream" spin currents that we will measure using all-electrical non-local spin detection, as discussed later in the chapter.

In addition to electrical spin injection from a ferromagnet into a semiconductor, a spin polarization can also accumulate in a semiconductor when electrons flow from the semiconductor into the FM at a forward-biased Schottky barrier [49, 58]. Figure 6e and f show the corresponding set of images for the case of spin accumulation due to spin extraction out of contact 4 in this Fe/GaAs device. At this bias, the preferential tunneling of majority electrons from $n$-GaAs into contact 4 leaves behind an accumulation of *minority*-spin electrons in the channel, as evidenced by the opposite sign of the Kerr rotation signal. This accumulated minority spin polarization diffuses out from under the contact, either into the current path (where drift now *opposes* diffusion), or outside of the current path (where spins diffuse freely).

Figure 6e and f show spin accumulation due to spin extraction both within and outside of the current path under low forward bias conditions ($I_e^{5\to4} = I_e^{1\to4} = 25\,\mu\text{A}$). In both cases, the apparent spatial extent of the accumulated $S_z$ is similar to the case of purely diffusive spin transport. Local Hanle data confirm this, showing decay lengths of order 10 µm. Using these images and the associated local Hanle data (not shown), the important spin-transport parameters—$\tau_s, D$, and $v_d$—can be accurately determined as a function of temperature and bias conditions. Knowledge of these parameters is especially useful for interpreting and understanding all-electrical measurements of spin injection and spin detection in similar devices.

## 6.5 ELECTRICAL DETECTION OF SPIN ACCUMULATION

### 6.5.1 ELECTRICAL DETECTION OF OPTICALLY INJECTED SPINS

As shown by the Kerr imaging, a spin accumulation due to either spin injection or extraction forms at the Fe/GaAs interface under reverse or forward bias respectively. A rough calibration of the sensitivity of the Kerr microscope indicates that spin polarizations of 5–10% exist within a few microns of the injection electrode. We now turn to the question of how this spin accumulation can be detected electrically.

An obvious experiment to carry out along these lines is to inject spin-polarized electrons into GaAs by optical pumping [1] and then detect the correspond change in the conductance of a nearby FM/SC interface. As noted previously, related ideas had been pursued in earlier experiments on vertical FM/SC structures [53–55]. However, because in these studies the semiconductor was optically excited through the semitransparent FM contact, the observed signals were always comparable to background effects due to hot electrons, and from magneto-absorption of the pump light by the FM contact. Moreover, no Hanle effect was observed. These shortcomings can be overcome in lateral devices by injecting spin-polarized electrons into the $n$-GaAs channel, away from the FM contact [58, 71]. Doing so avoids magneto-absorption effects and allows the spins to cool before they drift and diffuse under the Fe contact. A schematic showing the implementation of this experiment is shown in Figure 6.7a, where we optically inject spin-polarized electrons, initially oriented along $\pm\hat{z}$, into the channel using circularly polarized modulated light as depicted. Spins are injected at a small spot approximately 40 µm from the edge of the forward-biased Fe drain contact. A Kerr

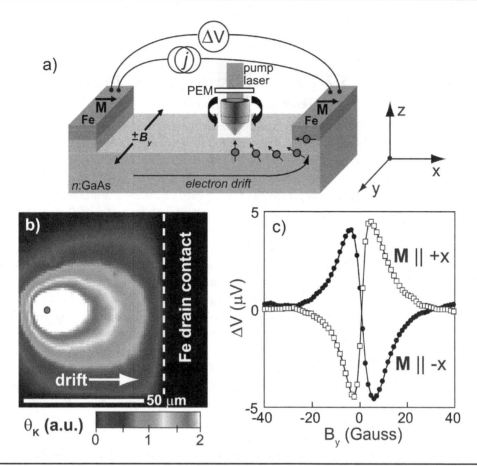

**FIGURE 6.7** (a) A schematic diagram of an experiment showing electrical detection of optically injected spins. Modulated circularly polarized light produced by passing a linearly polarized pump beam (50 μW at λ = 785 nm) through a photoelastic modulator (PEM) is incident on the semiconductor channel. The spin-dependent change in device voltage is measured using lock-in techniques. (b) A Kerr image of these optically-injected spins diffusing and drifting towards the drain contact. $B_y = 0$. The spins are generated in the small dot approximately 40 μm from the electrode. (c) Spin-dependent voltage change as a function of $B_y$, which induces precession of the optically injected spins parallel or antiparallel to **M**. Data for the two different magnetization directions of the detector are shown.

image of these spins in zero field is shown in Figure 6.7b. Small applied magnetic fields $B_y$ cause these spins to precess as they drift and diffuse towards the drain, tipping them into a direction parallel or antiparallel to the Fe magnetization **M**. The spin-dependent change in the conductance is then measured by lock-in techniques. Figure 6.7c shows, as a continuous function of $B_y$, the spin-dependent change in the voltage across the device that is explicitly due to the presence of the optically-injected and spin-polarized electrons at the drain contact. The signal, which reverses when **M** is reversed, shows the same antisymmetric local-Hanle lineshape observed in the imaging of electrical injection, as expected. These data demonstrate that these Schottky tunnel barrier contacts are sensitive to the spin polarization of the electrons in the GaAs, and therefore can be used as electrical spin detectors. In these studies, the spin-dependent voltages are on the order of microvolts, which represents a change of only a few parts per million of the total voltage across the device.

## 6.5.2 ELECTRICAL DETECTION BY THE THREE-TERMINAL TECHNIQUE

When the Kerr imaging experiments discussed above were originally being performed, the two-terminal resistance of these Fe/GaAs/Fe lateral devices was also measured, but we found that most of the magnetic field dependence was due to anisotropic magnetoresistance (AMR) in the Fe electrodes, as well as a dependence of the effective tunneling resistance on the orientation of the magnetization of the contact with respect to the crystalline axes. The latter effect, which is now known as tunneling anisotropic magnetoresistance (TAMR) [72, 73], could be as large as 1% of the device resistance. While attempting to characterize these backgrounds, we carried out measurements on devices with the electrodes several hundred microns apart, for which we reasoned that only the background AMR and TAMR effects should appear. A micrograph of one of these structures is shown in Figure 6.2b. When a small magnetic field was applied *perpendicular* to the sample, a peak in the two-terminal resistance appeared at zero magnetic field, as shown in Figure 6.8. At this field scale ($\sim 100$ G), the magnetization of the contact was unchanged, and so the effect could therefore be attributed to some effect in the semiconductor. Using the additional contacts on these devices, it was also possible to measure the voltage drops across the reverse-biased source contact, the forward biased drain contact, and the channel individually. In the vast majority of cases, the peak occurred only in a measurement across the forward-biased drain, and *never* appeared in the channel voltage.

Given the correspondence with the field scales observed in the Kerr imaging experiments, we hypothesized that the signature is due to dephasing of the spin accumulation at the Fe/GaAs interface, and subsequent experiments have verified that this is the case. The magnitude of the peak is the

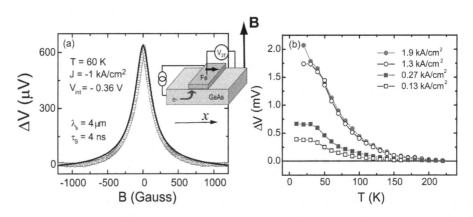

**FIGURE 6.8**    (a) The voltage measured across a forward-biased Fe/GaAs interface as a function of perpendicular magnetic field. A constant offset of 0.36 V has been subtracted from the data. The measurement geometry is shown in the inset. (b) The magnitude of the zero-field peak is shown as a function of temperature for several different forward bias current densities.

enhancement of the voltage drop across the tunnel junction due to the presence of a spin accumulation, which is destroyed (dephased) as the magnetic field is applied. This measurement and its interpretation are discussed in more detail in Ref. [60]. We represent the Fe/GaAs interface by a one-dimensional strip of length $w$. Spins polarized along $\hat{x}$ are generated at a point $x_1$ on the interface and diffuse to another point $x_2$, at which they are detected. The average steady-state spin accumulation is obtained by integrating the steady-state solutions of the drift diffusion equation over $x_1$ and $x_2$:

$$S_x(B_z) = \int_0^w \int_0^w \int_0^\infty \frac{S_0}{\sqrt{4\pi Dt}} e^{-(x_1-x_2-v_d t)^2/4Dt} \times e^{-t/\tau_s} \cos(\Omega_L t) dt dx_1 dx_2, \quad (6.7)$$

where $S_0$ is the spin generation rate (per unit contact length) and the other parameters are the same as in Equation 6.2. As with the imaging measurements, most of the parameters describing this Hanle curve can be obtained by independent transport or optical characterization. A typical fit is shown in Figure 6.8.

This single-contact measurement, often called a "three-terminal" experiment, might appear to be a special case, particularly when, as has now been demonstrated in many FM/GaAs heterostructures, measurements with a separate source and detector are possible. There are some cases, however, such as that discussed by Tran et al. [74], in which this approach allows the experimentalist to work with highly resistive tunnel barriers that cannot be probed by other techniques. Outside of the formalities of the drift-diffusion model, this type of measurement has a relatively simple interpretation in terms of two time scales: a residency time $\tau_r$ between spin generation and detection, as well as the spin lifetime $\tau_s$. Some simple statements can be made by direct analogy with a traditional optical Hanle effect experiment based on polarized photoluminescence [1] and the corresponding rate equations. The magnitude of the spin accumulation at zero field is proportional to $1/(1+\tau_r/\tau_s)$, while the width of the Hanle curve is proportional to the total relaxation rate $1/\tau_r + 1/\tau_s$. Expressed in this way, the tradeoff in the case of a resistive tunnel barrier is clear: the signal becomes small because $\tau_r$ is long, but the width of the Hanle curve can still be quite narrow (as it will depend only on $\tau_s$). This makes a small spin accumulation easier to measure in the sense that (a) it can still be inferred from the magnetic field dependence, even if its magnitude is small; and (b) the transport properties of the bulk semiconductor are not important provided that it does not function as a strong source of spin relaxation.

The three-terminal technique has continued to be applied to various other FM/SC heterostructures, due in large part to the fact that it is easy to implement. It has played a central, but controversial, role in the case of silicon-based devices with artificial tunnel barriers, in which observation of a Hanle effect at room temperature was first reported [75]. On the other hand, the observation of a peak in the three-terminal resistance in zero (perpendicular) field has now been observed in so many different cases that its utility as a probe of spin transport in the semiconductor channel has been called into question [76, 77, 78, 79, 80]. Some of this discussion has focused on the effect observed when the field is applied in the plane, for which the standard interpretation based only on the Hanle effect and a uniform in-plane magnetization does not apply [81]. In the case of silicon and germanium, similar spin lifetimes were extracted for p and n-type material [82, 83], which contradicts standard expectations. Nonetheless, the ubiquitous character of this effect, including its observation for junctions with and without artificial tunnel barriers as well as other hybrid structures such as FM/SrTiO$_3$ [84, 85] and FM/LaAlO$_3$/SrTiO$_3$ [79], has led to continued interest in the three-terminal approach as a simple means to probe spin-dependent transport, whether or not it is a signature of spin transport in the bulk of the semiconductor. In the case of Schottky barrier-based FM/III–V semiconductors, where non-local measurements of the type discussed below are now carried out routinely, and for which hyperfine effects (discussed below) serve as a fingerprint of bulk spin transport, the origin of the three-terminal signal is less controversial.

## 6.6  NON-LOCAL DETECTION OF SPIN ACCUMULATION

We now turn to the overall goal of a device in which a spin accumulation is generated at one FM electrode, transported, and then detected with a separate electrode. If the reader consults previous reviews of transport in ferromagnet-semiconductor heterostructures [18], a few clear themes emerge. First, backgrounds from local Hall effects, anisotropic magnetoresistance, and tunneling anisotropic magnetoresistance complicate virtually any measurement in which a charge current flows through a ferromagnetic detection electrode. Second, the most definitive way to establish the existence of a spin accumulation is to probe its dynamics. The first of these points can be addressed by separating the charge current in a device from the spin current. The second is addressed naturally by the observation of a Hanle effect.

The classic means by which spin and charge currents are separated is the non-local geometry of Johnson and Silsbee [19, 20], which is illustrated in Figure 6.9. Spins are introduced at a FM injector by an ordinary spin-polarized charge current. The charge current is sunk at an electrode at one end of the channel, but the spin accumulation created at the injector diffuses in both directions. A second FM contact, the detector, is located on the *opposite* branch of the channel, outside the path of the charge current. The lower part of Figure 6.9 shows a spatial profile of the spin-dependent chemical potentials for the ideal case in which no electric field (and hence no charge current) flows in the right-hand branch of the channel. The detector itself acts as a spin-dependent voltmeter, with chemical equilibrium established by the flow of a pure spin current between the FM detector and the channel. If the FM detector were a half-metal (with the Fermi level in a single spin band and therefore unity polarization), the electrochemical potentials for each of the two magnetization states ↑ and ↓ of the detector

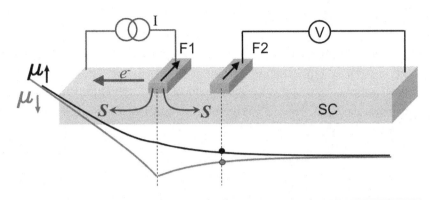

**FIGURE 6.9**   A schematic representation of a non-local spin transport device. F1 is the ferromagnetic injector, F2 is the detector, and SC is the semiconductor channel. Spin **S** diffuses in both directions from the injector contact. A representation of the spin-dependent chemical potentials $\mu_\uparrow$ and $\mu_\downarrow$ is shown for the ideal case in which the electric field vanishes to the right of the injector. The voltage $V$ is measured for both magnetization states of the detector.

would be equal to $\mu_\uparrow$ and $\mu_\downarrow$. The voltage difference $\Delta V = (\mu_\uparrow - \mu_\downarrow)/e$ would then provide a direct measurement of the spin accumulation. In practice, the polarization $P_{FM}$ of the FM at the Fermi level is less than one, and spin flips can occur as carriers flow back and forth across the detector interface. The latter process is accounted for by an efficiency factor $\eta$, which we can estimate from spin-LED measurements to be of order 0.5 for Fe/GaAs [38]. We make the assumption that the spin current flowing into (or out of) the detector does not impact the measurement, which corresponds to the limit of high-interface resistances. (In practice, the resistance of all of the devices discussed here are dominated by the interfaces. This is not necessarily the case in metallic devices, as discussed in Chapter 5, Volume 1).

The spin-dependent electrochemical potential is inferred from two non-local voltage measurements $V_{\uparrow\uparrow}$ and $V_{\uparrow\downarrow}$ obtained with the detector and injector magnetizations parallel and antiparallel, respectively. The difference $\Delta V_{NL} = V_{\uparrow\downarrow} - V_{\uparrow\uparrow}$ can be related to the spin-dependent chemical potential difference $\Delta\mu$ by

$$\Delta\mu = \frac{e\Delta V_{NL}}{\eta P_{FM}}, \tag{6.8}$$

where $P_{FM}$ is the spin polarization of the ferromagnet at the Fermi level. The electron spin polarization $P_{GaAs}$ in the semiconductor can be obtained from $\Delta\mu$, the carrier density $n$, and the density of states $\partial n/\partial\mu$:

$$P_{GaAs} = \frac{n_\uparrow - n_\downarrow}{n_\uparrow + n_\downarrow} = \frac{\Delta\mu}{n}\left(\frac{\partial n}{\partial\mu}\right). \tag{6.9}$$

Some indication of the overall efficiency is the *non-local resistance* $\Delta V_{NL}/I_{inj}$, which is proportional to the polarization per unit carrier flowing across the interface, which we denote $P_j$. We emphasize here that our samples are not truly degenerate semiconductors, and we therefore do not have direct knowledge of the density of states from the carrier density. One can, however, make a crude estimate of the order of magnitude of $\Delta V_{NL}$ if we assume a Pauli-like density of states for a degenerate semiconductor of carrier density $n \sim 3 \times 10^{16}\,\mathrm{cm}^{-3}$ and an effective mass $m^* = 0.07 m_e$. Given the injected polarizations from spin-LED measurements, and the approximate spatial dependence $P_{\mathrm{GaAs}}(x) \propto e^{-x/\lambda_s}$, a conservative estimate of the spin polarization at a detector several microns from the source is $P_{\mathrm{GaAs}} \sim 0.01$. We estimate the efficiency factor $\eta \sim 0.5$ from spin-LED measurements [38]. Using a literature value $P_{FM} = 0.4$ for the spin-polarization of iron [86], we find

$$\Delta V_{NL} \sim \frac{\eta P_{FM} \hbar^2 (3\pi^2 n)^{2/3}}{3 m^* e} P_{\mathrm{GaAs}} = 10\,\mu\mathrm{V}. \tag{6.10}$$

This is a large signal by the standards of a typical transport measurement. In our best devices, a noise floor of order $10\mathrm{nV}/\sqrt{\mathrm{Hz}}$ can be reached, although occasional devices will simply be too noisy to resolve the spin-dependent chemical potential shift. In practice, most of the difficulties in the experiments discussed below derive from the non-ideal aspects of the non-local measurement. In contrast to the cartoon representation of Figure 6.9, a significant voltage drop (up to several mV) will develop at the non-local detector due to spreading of the charge current from the source. This spin-independent background depends relatively strongly on magnetic field and temperature. Careful cryogenic practice, particularly temperature regulation, is necessary in order to obtain a stable background voltage throughout a measurement.

We now turn briefly to a discussion of the Hanle effect. Since the spin lifetimes and diffusion constants for these samples are known from the optical measurements discussed previously, it is possible to solve Equation 6.1 in steady-state, given the geometry of the source and detector. As was noted above, the typical field scale (width of a Hanle curve) is on the order of 100 G. For fields applied in the vertical direction, this will have no impact on the magnetizations of the contacts. It should therefore be possible to prepare the source and detector in either parallel or antiparallel states, and obtain Hanle curves in both cases. The large interfacial anisotropy of our samples facilitates this measurement, since the Fe contacts remain uniformly magnetized at remanence.

A micrograph of a typical non-local device is shown in Figure 6.2c. Photolithography and a combination of wet and dry etching are used to pattern a large mesa and a series of FM electrodes. Two of these are located over 150 μm from the center of the device and serve as the sink for the charge current, and the reference contact for the non-local voltage measurement. The bias current is generated with a current source and the non-local voltage $V_{NL}$ is measured with a nanovoltmeter. Data for a device with channel doping $5 \times 10^{16}\,\mathrm{cm}^{-3}$ obtained at 60 K are shown in Figure 6.10. Figure 6.10 shows raw data obtained in the longitudinal (or "spin-valve") geometry, in which the spin-dependent non-local voltage depends only on the relative orientations of the source and detector. The expected discontinuities when

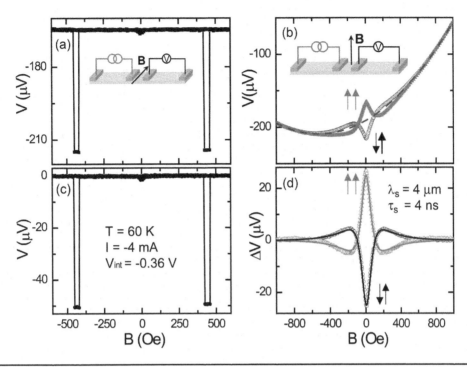

**FIGURE 6.10** Non-local electrical detection of spin accumulation at $T = 60$ K under a forward bias current of 4 mA ($J = 1$ kA / cm$^2$). The injection contact is $5 \times 80 \, \mu$m$^2$ and the detector contact is $5 \times 50 \, \mu$m$^2$. The center to center spacing of the contacts is 8 $\mu$m. (a) Raw data obtained in the longitudinal (or spin-valve) geometry. (b) Raw data obtained in the perpendicular (or Hanle) geometry for both parallel and antiparallel orientations of the detector and injector. The dashed line is a fit of the background obtained by a parabolic fit of the data at high fields. (c) The spin-valve data after subtraction of the spin-independent background. (d) Hanle data after subtraction of the parabolic background. The solid curves are fits to the drift-diffusion model. The lifetime $\tau_s = 4$ ns and the spin diffusion length $\lambda_s = 4 \, \mu$m.

the source and detector switch between parallel and antiparallel states are superimposed on a background voltage (approximately $-165 \, \mu$V in this case) due to the small amount of unwanted charge current spreading in the non-local geometry. This background is removed in Figure 6.10c. Data obtained in a perpendicular field (Figure 6.10b) show the Hanle effect when the source and detector are prepared in either the parallel or antiparallel states. In this case, there is a parabolic background due to the Lorentz forces acting on the carriers, but, significantly, the data for the two spin states merge at high fields, where the spins are completely dephased. After subtraction of the spin-independent background (dashed curve in Figure 6.10b), the Hanle curves of Figure 6.10d are obtained. As expected, these two curves are equal and opposite, and the difference in their peak values is equal to the magnitude of the jump in the spin-valve data. Curves showing fits to the drift-diffusion model are also shown in Figure 6.10d. In this case, we find a spin lifetime $\tau_s = 4$ nsec and a spin diffusion length $\lambda_s = 4 \, \mu$m.

It is also possible to determine the spin diffusion length by measuring the dependence of $\Delta V_{NL} = V_{\uparrow\downarrow} - V_{\uparrow\uparrow}$ on contact separation in multiterminal devices. By putting the detection contact in the current path (and subtracting out the additional background due to the ohmic voltage drop), it is

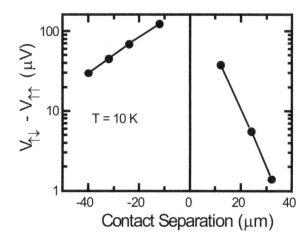

**FIGURE 6.11** The spin-dependent voltage $V_{\uparrow\downarrow} - V_{\uparrow\uparrow}$ is shown as a function of the separation between the injection and detection electrodes in the device with several injection/detection contacts. Positive separations correspond to the pure non-local geometry, in which the detection contact is outside the current path. For negative separations, the detector is located in the path of the charge current, and the signal is enhanced by drift.

also possible to measure the voltages $V_{\uparrow\downarrow}$ and $V_{\uparrow\uparrow}$ as well as the Hanle effect "down-stream" from the injector, where both diffusion and drift contribute to the spin transport. An example is shown in Figure 6.11, which shows $\Delta V_{NL}$ as a function of the injector/detector separation for a spin detector located outside the path of the charge current (contact separation $d > 0$), and in the current path ($d < 0$). The enhancement by drift in the latter case is clearly evident.

Demonstrations of the non-local detection of spin accumulation have been made by other groups in Fe/GaAs [87, 88] as well in other semiconductor-based systems, including MnAs/GaAs [89], Si [90], GaMnAs/GaAs [91], NiFe/InAs [92], and graphene [93, 94]. In many (but not all) cases, claims in these experiments have been supported by the observation of a Hanle effect, which has also been invoked extremely effectively in spin transport measurements on other spin-transport devices, particularly Si spin-valve transistors [95, 96]. We now turn to some aspects of these measurements that are less well-understood.

### 6.6.1 Bias Dependence of the Accumulated Spin Polarization

The overall phenomenology of the non-local measurements is in general agreement with expectations based on the drift-diffusion model for spin transport in the semiconductor. We now turn to aspects of the data that reflect the important and less understood role of the Fe/GaAs interface itself. Figure 6.12a shows a series of non-local spin-valve measurements as a function of longitudinal magnetic field for different injection currents. As expected, there are jumps at the transitions between parallel and

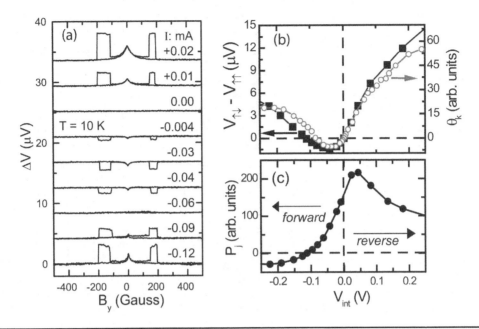

**FIGURE 6.12** (a) The non-local spin-valve voltage is shown as a function of the longitudinal magnetic field for several different bias voltages. Note that the signal inverts sign *twice*. (b) The spin-dependent non-local voltage $\Delta V_{NL} = V_{\uparrow\downarrow} - V_{\uparrow\uparrow}$ (closed squares, left axis) obtained from field sweeps is shown as a function of the interfacial voltage $V_{int}$ across the Schottky barrier injection contact. The Hanle–Kerr rotation signal $\theta_K$ (open circles, right axis) measured on this device as described in Section 4 is shown for the same bias conditions. (c) The polarization $P_j$ of the interfacial tunneling current is proportional to $\Delta V_{NL}(V_{int}) / I(V_{int})$, which is shown as a function of $V_{int}$. Note the change of sign at approximately −0.1 V.

antiparallel states of the injector and detector. (The additional peak at zero field is due to hyperfine interactions and will be discussed below). If the polarization of the tunneling current were fixed (e.g. by the bulk value of $P_{Fe}$), then the sign of the spin accumulation should change at zero bias. This is indeed the case, but the sign of $\Delta V_{NL}$ reverses again at a modest bias current of −60 μA (forward bias), equivalent to a voltage drop $V_{int} = -0.1$ V measured between the GaAs channel and the Fe injector. A more complete picture can be seen in Figure 6.12b, which shows $\Delta V_{NL}$ as a function of $V_{int}$. The same figure also shows the Kerr rotation $\theta_K$ measured in the same device (in the channel "upstream" of the injection contact) as a function of $V_{int}$ using the methods of Section 4. Up to a scale factor, the two measurements agree, confirming that the sign change at −0.1 V is due to a change in the polarity of the spin accumulation. The Kerr data also determine the absolute sign of the spin accumulation, which is positive (or majority) for positive values of $\theta_K$. Another representation of the data is provided in Figure 6.12c, in which we show the effective spin polarization of the current: $P_j = (j_\uparrow - j_\downarrow)/(j_\uparrow + j_\downarrow) \propto \Delta V_{NL}(V_{int}) / I(V_{int})$. In this form, it is evident that the spin polarization of the current depends very strongly on the injector bias voltage, with a peak near zero bias and a reversal in sign at −0.1 V.

The bias dependences of $\Delta V_{NL}$ and the Kerr rotation $\theta_K$ are shown for two additional samples in Figure 6.13, which illustrates that for a given sign of the injector bias voltage, the magnitude and sign of $\Delta V_{NL}$ (closed squares)

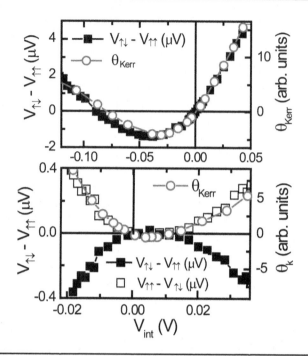

**FIGURE 6.13**   Bias dependences of $\Delta V_{NL}$ (closed squares, left axis) and Kerr rotation $\theta_K$ (open circles, right axis) for two different samples. As discussed in the text, the bias dependences of $\theta_K$ and $\Delta V_{NL}$ in (b) agree only after the *sign* of $\Delta V_{NL}$ is inverted (open squares).

can vary from heterostructure to heterostructure. In contrast, all devices grown from a single heterostructure (i.e. those based on a single growth of an epitaxial Fe/GaAs interface) will show approximately the same bias dependence. Interestingly, the bias dependence can be changed by annealing, which can even invert the sign of the non-local signal [97]. Regardless of annealing history, however, majority spin accumulation is observed for large forward and reverse bias currents in all cases. At first glance, this wide variation among heterostructures, particularly the overall sign, appears counterintuitive. In spin-LEDs, for example, the polarization of the current at reverse bias is always observed to have the correct sign for majority spin injection [38]. A first clue as to what is going on can be found in the Kerr rotation data of Figure 6.13. As was observed for the data of Figure 6.12b, the bias dependences of $\theta_K$ and $\Delta V_{NL}$ for the heterostructure of Figure 6.13b overlap nearly exactly, but for the sample of Figure 6.13c, $\theta_K$ overlaps with $-\Delta V_{NL}$. In other words, the sign of the non-local voltage for a given spin accumulation is reversed between the two samples. Otherwise, the $\Delta V_{NL}$ is proportional to the spin accumulation, as expected.

Focusing on the Kerr rotation, from which we can unambiguously determine the absolute sign of the spin accumulation, we always observe majority spin accumulation under large reverse bias (greater than +50 mV). In other words, a majority spin polarization flows from Fe into GaAs, as expected. Under forward bias, we observe majority spin accumulation for biases greater than (i.e. more negative than) –100 mV. This corresponds to the flow of *minority* spin from the semiconductor into the ferromagnet. In other words, the relative transmission probabilities for minority and majority spins always reverses near zero bias. This can be seen explicitly in Figure 6.14, which shows the polarization of the current $P_j \propto \Delta V_{NL}(V_{int})/I_{int}$ for the same two samples as in Figure 6.13. Based on the Kerr rotation measurements, we have chosen the sign of the polarization to be positive at large positive (reverse) bias voltages. Although the polarization of the current at zero bias can be either positive or negative, the polarization at large reverse bias is always positive, and it is always negative at large forward bias.

The absolute sign of the non-local signal, or, equivalently, the sign of $P_j$ at zero bias, is determined by the tunneling density of states along with the exact energy at which the Fermi level is pinned at the Fe/GaAs interface. The most direct access we have to this information comes from the Kerr rotation, which measures the spin accumulation directly (to the extent that $\theta_K \propto P_{GaAs}$), and hence the quantity $\delta P_{GaAs}/\delta V_{int}$ for a small change $\delta V_{int}$ in the bias. Hence, the slope of $\theta_K$ vs. $V_{int}$ at zero bias should serve as predictor for the signs of both $\Delta V_{NL}$ (which is always measured at zero bias across the *detector*) and $P_j$. This is in fact the case. Although this self-consistency is reassuring, the fact remains that we cannot predict the spin polarization at

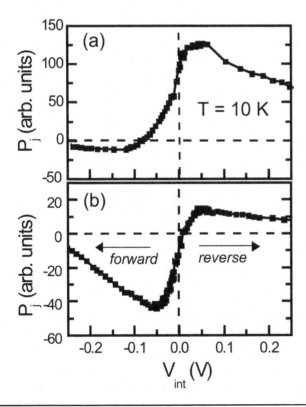

**FIGURE 6.14** The spin polarization of the tunneling current flowing from the semiconductor into the ferromagnet for the two samples of Figure 6.13. Note they have opposite sign at zero bias. Positive voltage (reverse bias) corresponds to electrons flowing from the ferromagnet into the semiconductor.

zero bias *a priori*. This remains a frustrating point in the materials physics of these systems, particularly because the interfaces appear to be remarkably well-controlled when evaluated by spin-LED or scanning transmission electron microscopy (STEM) measurements [67, 98].

While sample-specific effects may be regarded as problematic, it is important to emphasize the universal observation that for Fe/GaAs (100), $P_j$ always reverses sign over a narrow window ±0.1 V around zero bias [97]. Similar observations have been made by other groups [99]. For epitaxial Fe/GaAs (100), the minority spin current always dominates at large forward bias. Similar anomalous behavior (i.e. a crossover to minority-dominated tunneling) has been observed in Fe/GaAs/Fe tunnel junctions in which one interface is grown epitaxially [57]. In the case of metallic magnetic tunnel junctions, a reversal in the current polarization (and hence the sign of the tunneling magnetoresistance) has been observed in systems with crystalline barriers, as in Fe/MgO/Fe [100]. The association with crystalline barriers suggests that the effects of dispersion are important. On the other hand, the sign reversal is not necessarily observed for other ferromagnets grown epitaxially on GaAs. For example, Kotissek et al. found minority spin accumulation in $Fe_{0.32}Co_{0.68}$/GaAs structures under forward bias [66], and we find minority accumulation under forward bias in heterostructures (otherwise identical to those under discussion here) in which Fe is replaced by epitaxial $Fe_3Ga$. Experiments on Heusler alloys also show non-trivial injector bias dependence, depending on the degree of structural order [101].

Addressing the bias dependence requires moving beyond a simple Jullière-type model, in which the tunneling density of states and matrix elements are independent of energy. An extension of the Jullière model, accounting for the bulk Fe density of states, does produce a sign reversal, but only at much higher bias. The simplest approach to modifying the tunneling matrix elements, discussed by Smith and Ruden [102], accounts for the bias dependence of the Schottky barrier profile, while modeling the ferromagnet as an *s*-band metal with distinct Fermi wave-vectors for majority and minority spins. Other approaches emphasize different aspects of the electronic structure of the Fe/GaAs (001) Schottky barrier [103, 104, 105]. Chantis and coworkers [103] focus on the tunneling density of states at the Fe/GaAs interface. By calculating the surface density of states and then integrating the tunneling current for both majority and minority bands in *k*-space, they find an enhancement of the minority current at small forward bias, leading to a crossover from majority to minority transmission. Dery and Sham [104] focus on the conduction band near the surface of the semiconductor, arguing that the minority current is enhanced by tunneling from bound states formed in the shallow potential well just inside the Schottky barrier. It is impossible to distinguish between these models based only on existing data, although Li and Dery [106] have proposed inserting a barrier (of higher bandgap material) just inside the semiconductor. This would allow for an explicit test of the Dery–Sham model.

## 6.6.2 HYPERFINE EFFECTS

A second set of anomalies in the non-local measurements are associated with the magnetic field dependence and are illustrated in Figure 6.15. First, in the longitudinal field sweep at low temperatures (Figure 6.15a), a peak is observed at zero field in addition to the two expected jumps when the magnetizations of the source and detector are in the antiparallel state. (This peak can also be observed in the fields sweeps of Figure 6.12.) The zero-field peak depends very strongly on the field sweep rate and vanishes if the measurement time is sufficiently long (requiring wait times over a minute at each data point), or if the temperature is sufficiently high (typically above 60 K). Various checks can be used to ascertain that the magnetizations of the contacts are indeed fixed as the field is swept through zero, and so we conclude that it is due to some change in the spin dynamics in the semiconductor. Similar observations of a "zero-field peak" have been made by Salis et al. [87] and Ciorga et al. [91].

The second anomaly, illustrated in Figure 6.15b is the extreme sensitivity of the Hanle curves to the magnetic field orientation, an effect that becomes so strong at low temperatures that it is difficult to obtain a meaningful Hanle curve in an ordinary field sweep. As shown in Figure 6.15b, there is also a clear difference between the two different parallel configurations of the contacts. Instead of being identical, the Hanle curves for $V_{\uparrow\uparrow}$ and $V_{\downarrow\downarrow}$ are both distorted, and they appear to be mirror images with respect to inversion of the magnetic field axis. Unlike the zero-field peak in the longitudinal field sweeps, this effect persists in steady-state, and the distortion as well as the apparent shift with respect to zero field increase with the degree of field misalignment as well as with the bias current. Distortion of non-local Hanle curves has also been reported by the NRL and Regensburg groups [88, 91].

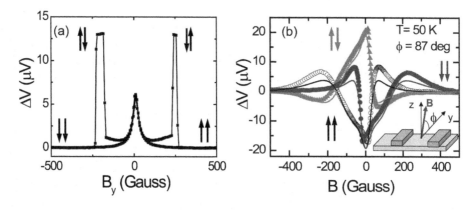

**FIGURE 6.15** Effects of hyperfine interactions. (a) Data from a longitudinal field sweep (field along easy axis of contacts) at 10 K. Note the peak at zero field in addition to the switching features. (b) Hanle data at 50 K obtained with the magnetic field oriented a few degrees away from vertical. Solid lines are a fit to the drift diffusion model with a constant hyperfine field with a sign determined by the magnetization of the injector. The samples and bias conditions are different for the two panels, and so the magnitudes of the signals should not be compared.

These effects, which are ubiquitous in Fe/GaAs devices, are due to the dynamic polarization of nuclei by electron spins localized on donor sites. This phenomenon was originally observed in optical pumping experiments [11] and has been reviewed extensively by Fleisher and Merkulov [10]. Discussions of similar effects in GaAs-based spin-LEDs appear in Refs. [37] and [51]. From the standpoint of *electron* spin transport and dynamics, the main consequence of the dynamic nuclear polarization (DNP) is an effective mean field $\mathbf{B}_N$ which acts on the electron spins. A proper calculation of $\mathbf{B}_N$ requires a self-consistent treatment of the electron and nuclear spin dynamics. In a relatively general form:

$$\mathbf{B}_N = b_n \frac{(\mathbf{B} + b_e \mathbf{S}) \cdot \mathbf{S}(\mathbf{B} + b_e \mathbf{S})}{(\mathbf{B} + b_e \mathbf{S})^2 + \xi B_l^2}, \tag{6.11}$$

where:

$b_n$ and $b_e$, which are both negative in GaAs [11], represent effective fields due to the polarized nuclei and electrons

$\mathbf{S}$ is the average electron spin ($|\mathbf{S}| = 1/2$ for electron spin polarization $P_{GaAs} = 1$)

$B_l$ is the local dipolar field experienced by the nuclei

$\xi$ parameterizes the assisting processes which allow energy to be conserved in mutual spin flips between electrons and nuclei. (We have simplified the notation of Paget [11] by incorporating prefactors into $b_n$ and $b_e$ that cannot be determined independently).

The electron spin precesses in a total field $\mathbf{B}_{tot} = \mathbf{B} + \mathbf{B}_N$, which can be much larger than the applied field $\mathbf{B}$.

If we ignore the exchange field (Knight field) $b_e$ acting on the nuclei, $\mathbf{B}_N$ is parallel to the applied field, but its magnitude and sign are determined by the electron spin accumulation. This is the origin of the distortion of the Hanle curves in Figure 6.15b as well as their dependence on the *sign* of the injected spins. A more extreme case is shown in Figure 6.16a, obtained for a misalignment of 15° in the direction of the applied field with respect to the sample normal. The satellite peak in the Hanle curve at ~300 G is due to the approximate cancelation of the applied field by the *z*-component of the hyperfine field. Given the dot product in Equation 6.11, the magnitude of the hyperfine field in this case is approximately 1000 G.

The sensitivity of the electron spin accumulation to $\mathbf{B}_N$ implied by Figure 6.16 can be regarded either positively or negatively depending on one's point of view, and it is important to note that it can be suppressed to a large extent by inverting the magnetization of the injector on a timescale that is fast relative to the timescale for the nuclei to become polarized, which is of the order of one second. This is feasible if small electromagnets are used. On the other hand, an optimist might be inclined to use the electron spin accumulation as a nuclear spin detector. For example, depolarizing a particular isotope (either $^{69}$Ga, $^{71}$Ga, or $^{75}$As) by resonant excitation suppresses its contribution to $\mathbf{B}_N$, resulting in a corresponding shift of $\mathbf{B}_{tot}$. As a result, the Hanle curve will shift along the field axis, and the value of

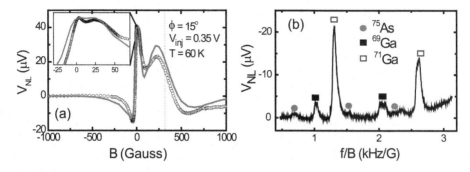

**FIGURE 6.16** (a) Non-local Hanle curve obtained with the field applied at 15° from the vertical direction. An expansion of the region around zero field is shown in the inset. The solid curve is obtained from the modified drift diffusion model with a nuclear field of the form in Equation 6.11. Note the double peak in both the model and the experimental data in the inset, which is due to the non-zero Knight field $b_e$. (b) NMR spectrum obtained in an applied field of 320 G for the same conditions in (a) as the frequency of the current in a coil over the sample is swept. The resonances for NMR transitions of $^{75}$As, $^{69}$Ga, and $^{71}$Ga are shown using circles, closed squares, and open squares, respectively.

the spin accumulation at a fixed field will change. This means of detecting nuclear magnetic resonance is illustrated in Figure 6.16b, which shows the Hanle signal measured at an applied field of 320 G as a function of the frequency of current passing through a small coil over the sample. The primary resonances for the three main isotopes are indicated by symbols.

Other manifestations of the hyperfine interaction are less amenable to simple interpretation. The zero-field feature in the longitudinal field sweeps has been studied carefully by Salis et al. [87], who were able to model it in terms of the formalism presented here by carrying out longitudinal field sweeps in a fixed perpendicular magnetic field. Its ubiquitous presence in longitudinal field sweeps is due to the rotation of $\mathbf{B}_N$ as the applied field is swept through zero.

It is evident that nuclear spins have a profound influence on electron spin dynamics. The influence of the electronic field $b_e$, also known as the Knight field, is more subtle, since it acts directly only on the nuclei. An effect on the electron spin accumulation occurs only because the exchange field $b_e\mathbf{S}$ changes the magnitude and direction of the effective field in which the nuclei precess. If, however, an experiment is designed to be sensitive to the Knight field, it then allows for a measurement of the magnitude and sign of the electron spin polarization *without requiring knowledge of the density of states in the semiconductor*. This comes about because the field $b_e\mathbf{S}$, which is proportional to the electron spin polarization, modifies the magnetic field dependence of the spin accumulation signal, whereas $\mathbf{S}$ alone only impacts the magnitude.

Chan et al. [107] have solved the problem of the coupled dynamics of the electron and nuclear spin system by incorporating the effective field of Equation 6.11 into the drift-diffusion model of Equation 6.1, solving the two equations self-consistently in one dimension. A model curve based on this result is shown in Figure 6.16a. The region around zero field is shown in the inset. The small double-peak near zero is due to the effective field $b_e\mathbf{S}$.

A more sensitive measurement of $b_e\mathbf{S}$ described in detail in Ref. [107], has allowed for a direct quantitative measurement of the spin accumulation $\mathbf{S}$ as a function of the injector bias. The resulting spin polarization $P_{GaAs}$ is shown in Figure 6.17 superimposed on the ordinary measurement of the spin-dependent non-local voltage $\Delta V_{NL} = V_{\uparrow\uparrow} - V_{\uparrow\downarrow}$. These two very different measurements show the same dependence on the bias voltage of the injector. As can be seen on the right-hand axis, the Knight field measurement gives an electron spin polarization of about 6% at the maximum injector bias. The sign corresponds to accumulation of majority spins, as expected based on the Kerr measurements of other structures at large forward bias. Given this polarization, the magnitude of $V_{NL}$ is in fact quite close to the very crude estimate of Equation 6.10, although the agreement is probably serendipitous. In subsequent work, the actual Knight shift in the electrically detected nuclear magnetic resonance frequency was also observed [108], and (within factors of order unity), the magnitudes of the shifts agreed with expectations based on the average Knight field $b_e\mathbf{S}$, as well as the magnitude of the spin accumulation as inferred from the non-local voltage. It is also possible to use the coupled electron-nuclear system to measure the nuclear spin relaxation time $T_1$. Kölbl et al. used spin valves to probe the electron-nuclear spin coupling down to temperature T ~ 0.1 K, observing a departure from the usual Korringa law ($T_1^{-1} \propto T$) at the lowest temperatures [109]. Uemura et al. have even developed a pulse sequence that allows for the observation of Rabi oscillations of the nuclear polarization in a $Co_2MnSi$/GaAs spin valve [110]. The success of these measurements reflects the extreme sensitivity of the polarized electron system to small changes in the hyperfine field. In effect, the FM/GaAs spin valves allow for the study of nuclear spin dynamics over a very small volume. On these scales, inhomogeneities of the hyperfine field also impact the electron spin dynamics [109], including the electron spin relaxation rate [111, 112].

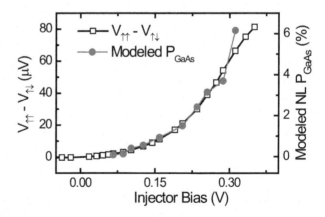

**FIGURE 6.17**    The non-local voltage $\Delta V_{NL} = V_{\uparrow\uparrow} - V_{\uparrow\downarrow}$ from spin-valve studies (open squares) compared with the spin polarization $P_{GaAs}$ determined from the measurement of $b_e$ (Knight field) as described in the text. Note the two independent $y$-scales.

### 6.6.3 DETECTOR BIAS DEPENDENCE

Given the difficulties described above in Section 6.1 in achieving a quantitative interpretation of the injector bias dependence, it might appear premature to consider applying an additional bias voltage to the detector. A biased detector would seem to remove some of the purported advantages of the non-local approach, particularly from the standpoint of traditional objections to two-terminal measurements [18]. In the context of the above measurements, however, we now have an excellent grasp of the behavior of spin-dependent signals (and artifacts) in Fe/GaAs devices. Furthermore, in the presence of the semiconductor channel, it is easy to bias the FM source and detector electrodes individually. The actual experimental phase space is therefore much larger than considered in some of the more pessimistic assessments of two-terminal devices with high-resistivity barriers [25, 113]. In fact [71], the application of separate injector and detector bias voltages allows one to exploit the advantages of semiconductor-based spin transport devices in ways that are not possible in similar metallic devices. The potential of multiterminal FM/SC devices has only begun to be explored, and the reader is referred to Chapter 17, Volume 3 for a discussion of some of the possible applications.

Given the complications addressed in Section 6.1, we separate into two contributions the effects of an additional detector bias voltage $V_d$ on the sensitivity of electrical spin detection [71]. The first contribution, discussed above, is simply the effect of $V_d$ on the polarization $P_j$ of the tunneling current that flows across the Fe/GaAs detector interface. Tuning the magnitude and/or sign of $P_j$ with $V_d$ will tune the magnitude and/or sign of the spin-detection sensitivity. The second contribution, which we exploit to further enhance the spin-dependent sensitivity of electrical spin detectors, originates from the electric fields in the semiconductor channel near the detector [6, 17] that arise *because* the detector is biased, including the enhancement of the spin transport by drift [61, 114], as well as the modification of the (spin-dependent) carrier densities near the detector interface.

We consider a situation where a bias voltage $V_d$ is applied between the FM detector contact and the SC channel. As a measure of the detector's spin-dependent sensitivity, we then consider the voltage change $\Delta V_d$ at the detector that is induced by a remotely-injected spin polarization (analogous to conventional non-local studies). If there is no electric field in the semiconductor (e.g. if the current is purely diffusive), simple reciprocity considerations require that for a given unit of spin-polarized current, the change $\Delta V_d$ must be proportional to the change $\Delta \mu$ in the spin accumulation that occurs when the same unit of spin-polarized current is *injected* from an identical contact at injection bias $V_{int}$. The sensitivity of the detector at a voltage $V$ is therefore proportional to $P_j = \Delta V_{NL}(V) / I(V)$. As noted in Section 6.1, this is proportional to the polarization $P_j$ of the current flowing through the injector. From a conventional non-local experiment (in which $V$ is the injector interface voltage), it is therefore possible to predict the electrical *detection* sensitivity of the contact *in the absence of electric field effects.*

In Ref. [71], we have addressed the reasoning of the previous paragraph by using optical pumping to generate a small spin-polarized current that is incident on a detector contact biased at a voltage $V_d$. The change $\Delta V_d$ in the detector voltage was measured using a standard modulation technique and compared with $\Delta V_{NL}(V)/I(V)$ for the same device. For detector bias voltages as small as 0.1 V, there are significant departures from the expectations of the simple reciprocity argument. In particular, the sensitivity of the detector at forward bias was observed to be dramatically enhanced. A simple model [115] showed that the enhancement is due to the modification of spin-dependent carrier densities $n_{\uparrow,\downarrow}$ (and their gradients) by the electric field at the interface.

The implication of this result is that the sensitivity of a non-local spin-valve can be tuned over a wide range by application of a bias voltage at the detector. Figure 6.18 shows that this is indeed the case. Figure 6.18a shows the magnitude of the spin-valve signal versus detector bias. We consider one of the sets of data (the black points) obtained at a *fixed* injector bias current (–1.5 mA). At zero detector bias—where non-local studies are usually performed—the spin-valve signal is quite small (<2 μV). However, when the non-local spin detector is forward-biased by –150 mV, the magnitude of the spin-valve signal increases by over an order of magnitude to approximately 20 μV. Furthermore, with the detection electrode under a small reverse bias, the spin valve signal actually inverts sign. Note that in all three cases the injected spin density remains unchanged. Thus, electrons in GaAs with a fixed spin polarization can be made to induce either positive or negative voltage changes at a detection electrode, simply by tuning the detector bias $V_d$ by a few tens of meV. The detector sensitivity in this voltage range is completely independent of the injector bias,

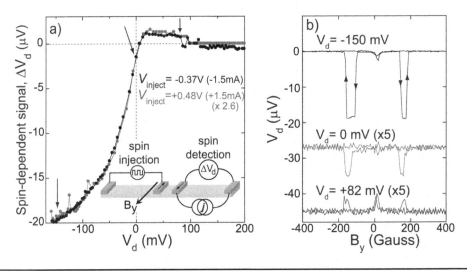

**FIGURE 6.18** (a) Inset: All-electrical lateral spin-valve setup with separately-biased injector and detector electrodes. Black (gray) points show the spin-valve signal $\Delta V_d$ versus detector bias $V_d$ for a forward (reverse) biased spin injector. (b) Raw spin-valve data at three detector biases [arrows in (a)], using a fixed –1.5 mA injector bias (curves offset). Note sign switching and ten-fold enhancement of the detected signal.

as demonstrated by the second set of data (gray points) in Figure 6.18a, which were obtained with a totally different injection current of +1.5 mA. After correcting for the different spin injection efficiencies, the two sets of data collapse onto a single curve. Figure 6.18b shows spin-valve data for these three scenarios ($V_d$ = −150, 0, and +82 mV), explicitly showing that spin-detection sensitivities are freely tunable in both sign and magnitude in all-electrical devices. This approach to enhancing the sensitivity of the detector has also been used to enable operation of FM/n-GaAs non-local spin valves at room temperature [116]. It is possible to enhance the detection sensitivity by over a factor of 10, while maintaining linearity of the output with respect to the injector bias current.

## 6.7 RECENT DEVELOPMENTS

### 6.7.1 SPIN-ORBIT EFFECTS

Another outstanding class of problems involves the detection of spin Hall effects (see Chapter 8, Volume 2) [117, 118, 119], an area which has seen activity over the last several years (since the previous edition of this chapter) on many different fronts [120]. The first obvious candidate is the electrical detection of the spin Hall effect in a manner analogous to the Hanle experiment of Kato et al. [118], who used Kerr microscopy to detect the spin accumulation at the edges of a Hall bar biased with an unpolarized current. In principle, this only requires placing ferromagnetic Hall contacts on a semiconductor channel. Given the small values of the spin accumulation (~0.1%), the technical requirements are more stringent than for the experiments introduced above. This experiment has been carried out [121], and it has allowed for a much more detailed comparison with theory than has been possible with optical probes. In particular, it was possible to measure the spin Hall conductivity (inferred from the spin accumulation and the drift-diffusion model) as a function of the ordinary charge conductivity. The skew and side jump scattering contributions to the spin Hall conductivity were extracted and compared with predictions for the case of ionized impurity scattering [122]. A similar experiment was subsequently carried out in $n$-GaAs based Esaki diodes using $Ga_{1-x}Mn_xAs$ as the spin detector [123]. It has also been possible to detect the transverse charge current generated in a FM/GaAs device by a spin current that is generated non-locally (the inverse spin Hall effect (ISHE)) [124]. The latter experiment is analogous to the original detection of the ISHE in a metallic non-local spin valve [125].

The classic Datta–Das spin-FET [15], in which the role of the magnetic field in a Hanle measurement is assumed by the effective spin-orbit magnetic field, requires a set of experiments, parallel to those discussed above, on a 2D electron system. This has been a remarkably difficult problem to address, in large part because the types of tunneling contacts developed for efficient spin injection into bulk GaAs have not been easily adaptable to two-dimensional electron gas (2DEG) heterostructures. Koo et al. did report on gate modulation of the spin valve signal in an InAs 2DEG [92], although no Hanle effect was observed. They later reported on a gate modulation of the ISHE in the same device [126]. A very clear demonstration of a 2D non-local spin valve was executed for the case of (Al,Ga)As/GaAs by Oltscher and coworkers [127, 128], who adapted their Esaki-diode approach in order to inject spin-polarized electrons into a 2DEG. These devices, however, were not gated, although, as in the bulk case [129], it has been possible to verify the spin accumulation using Kerr microscopy [130]. It remains to be seen whether all of the components of a spin-FET can be demonstrated in a FM/III–V device, although numerous alternatives are now being explored.

The presence of spin-orbit coupling, even in bulk III–V semiconductors, has been a key component of a series of experiments which have attempted to exploit spin pumping from a ferromagnet, in lieu of spin injection using a charge current. The essential idea of spin pumping is to drive the FM out of equilibrium using ferromagnetic resonance (FMR), and the system relaxes towards equilibrium through the flow of a spin current into an adjacent material [131]. This concept was demonstrated in metallic multilayers [132] and then non-local spin valves [133], but the measurement is much easier when the generated spin current can be detected using the ISHE [134]. In principle, the spin-pumping approach is complementary to spin injection by a charge current, in that it only requires an interface with a sufficiently high mixing conductance [131], and would hence appear to be a means to circumvent the conductivity mismatch problem [23]. Ando et al. tested this proposition by carrying out spin pumping measurements with ISHE detection on very simple devices consisting of permalloy ($Ni_{0.81}Fe_{0.19}$ sputtered on either $p$- or $n$-GaAs [135]. Remarkably, both electron and hole-doped systems showed ISHE signals on resonance, and these reversed sign with the reversal of the magnetization of the FM. Similar spin pumping signals have been observed in FM/Si [136, 137] and FM/Ge [138, 139] devices, and more recently in FM/GaN [140], for which the bulk spin-orbit coupling is weak. Shikoh et al. demonstrated that a spin current generated by spin pumping could be detected non-locally in Si [137], and Yamamoto et al. have carried out a similar experiment in GaAs [141]. In both of these experiments, the non-local detection was carried out using an ordinary heavy metal and the ISHE. Both ordinary magnetotransport (through the tunneling anistoropic magnetoresistance [72, 73]) and spin accumulation in FM/SC hybrids with tunnel barriers are sensitive to FMR [142, 143]. Given other difficulties in extracting spin transport parameters in traditional spin pumping devices [144], a direct comparison of the spin-pumping and non-local spin valve techniques would be a major step forward.

## 6.7.2 HEUSLER ALLOY-BASED HETEROSTRUCTURES

Since the early days of semiconductor spintronics, it has been realized that a large class of ternary alloys in the Heusler family are ferromagnets and lattice-matched to GaAs [145]. Moreover, many of these are predicted to be half-metallic, although most of these predictions have been made for bulk materials and not interfaces. Felser et al. [146] and Palmstrøm [145] have recently reviewed Heusler alloys in the context of spintronics. Although this chapter will not provide a review of the materials physics of the Heuslers (see, for example, Ref. [147]), it is important to note that their incorporation into III–V based devices has led to significant improvements, including operation at room temperature. Spin injection from Heusler alloys into GaAs was observed in spin-LED measurements [40, 148–150], and in recent years they have received considerably more attention in the context of non-local spin valves. Bruski and coworkers fabricated $Co_2FeSi/n$-GaAs spin valves demonstrating both the local and non-local transport signatures introduced above [151]. Although there has not been an explicit demonstration of half-metallicity of a Heusler/GaAs interface, there have now been several demonstrations of Heusler/III–V based lateral spin valves operating at room temperature [116, 152], and spin injection from $Co_2MnSi$ has been shown to be a particularly efficient source of dynamic nuclear polarization [110]. We anticipate additional improvements as the materials physics of this important class of ferromagnets, including their integration with semiconductors, is explored further.

## 6.8 CONCLUSIONS

Although this chapter has focused primarily on the early days of FM/III–V semiconductor heterostructures, devices based on this materials system have now advanced to the point that they can function as quantitative tools for studying spin transport. In particular, it is possible to probe spin transport and dynamics using combinations of optical and transport techniques that allow for detailed tests of theory. Although outstanding problems remain, particularly in the area of integration of FMs with quantum well heterostructures with strong spin-orbit coupling, the experiments described in this chapter have provided a foundation for this ongoing effort.

## ACKNOWLEDGMENTS

It is a pleasure to thank our collaborators who have made this work so enjoyable: Christoph Adelmann, Mun Chan, Athanasios Chantis, Kevin Christie, Madalina Furis, Chad Geppert, Eric Garlid, Qi Hu, Tsuyoshi Kondo, Changjiang Liu, Xiaohua Lou, Chris Palmstrøm, Sahil Patel, Tim Peterson, Madhukar Reddy, Darryl Smith, Gordon Stecklein, and Jianjie Zhang. The work discussed in this review has been supported by NSF under DMR-0804244 and DMR-1104951, the Office of Naval Research, the MRSEC Program of the National Science Foundation under Award Numbers DMR-0212302 and DMR-0819885, the Los Alamos LDRD program, the National High Magnetic Field Laboratory (which is supported by NSF DMR-1157490 and the State of Florida), the National Science Foundation NNIN and NNCI programs, and by C-SPIN, one of the six centers of STARnet, a Semiconductor Research Corporation program sponsored by MARCO and DARPA.

# REFERENCES

1. F. Meier and B. P. Zakharchenya (Eds.), *Optical Orientation*, North Holland, Amsterdam, 1984.
2. J. M. Kikkawa and D. D. Awschalom, Resonant spin amplification in $n$-type GaAs, *Phys. Rev. Lett.* **80**, 4313–4316 (1998).
3. R. I. Dzhioev, B. P. Zakharchenya, V. L. Korenev, and M. N. Stepanova, Spin diffusion of optically oriented electrons and photon entrainment in n-gallium arsenide, *Phys. Solid State* **39**, 1765–1768 (1997).
4. R. I. Dzhioev, K. V. Kavokin, V. L. Korenev et al., Low-temperature spin relaxation in n-type GaAs, *Phys. Rev. B* **66**, 245204 (2002).
5. A. G. Aronov and G. E. Pikus, Spin injection into semiconductors, *Sov. Phys. Semicond. - USSR* **10**, 698–700 (1976).
6. Z. G. Yu and M. E. Flatté, Spin diffusion and injection in semiconductor structures: Electric field effects, *Phys. Rev. B* **66**, 235302 (2002).
7. M. I. D'yakonov and V. I. Perel', Spin relaxation of conduction electrons in noncentrosymmetric semiconductors, *Sov. Phys. Solid State* **13**, 3023–3026 (1972).
8. M. Furis, D. L. Smith, S. A. Crooker, and J. L. Reno, Bias-dependent electron spin lifetimes in n-GaAs and the role of donor impact ionization, *Appl. Phys. Lett.* **89**, 102102 (2006).
9. M. Furis, D. L. Smith, S. Kos et al., Local Hanle-effect studies of spin drift and diffusion in $n$:GaAs epilayers and spin-transport devices, *New J. Phys.* **9**, 347 (2007).
10. V. G. Fleisher and I. A. Merkulov, Optical orientation of the coupled electron-nuclear spin system of a semiconductor, in *Optical Orientation*, F. Meier and B. P. Zakharchenya (Eds.), North Holland, Amsterdam, pp. 173–258, 1984.
11. D. Paget, G. Lampel, B. Sapoval, and V. I. Safarov, Low field electron-nuclear spin coupling in gallium arsenide under optical pumping conditions, *Phys. Rev. B* **15**, 5780–5796 (1977).
12. R. Winkler, *Spin-Orbit Coupling Effects in Two-Dimensional Electron and Hole Systems*, Springer-Verlag, Berlin, 2003.
13. G. E. Pikus and A. N. Titkov, Spin relaxation under optical orientation in semiconductors, in *Optical Orientation*, F. Meier and B. P. Zakharchenya (Eds.), North Holland, Amsterdam, pp. 73–132, 1984.
14. Y. A. Bychkov and E. I. Rashba, Oscillatory effects and the magnetic susceptibility of carriers in inversion layers, *J. Phys. C: Solid State Phys.* **17**, 6039–6045 (1984).
15. S. Datta and B. Das, Electronic analog of the electro-optic modulator, *Appl. Phys. Lett.* **56**, 665–667 (1990).
16. M. I. D'yakonov and V. Y. Kachorovskii, Spin relaxation of two-dimensional electrons in noncentrosymmetric semiconductors, *Sov. Phys. Semicond. - USSR* **20**, 110–112 (1986).
17. I. Žutić, J. Fabian and S. Das Sarma, Spin injection through the depletion layer: A theory of spin-polarized p-n junctions and solar cell, *Phys. Rev. B* **64**, 121201 (2001).
18. H. X. Tang, F. G. Monzon, F. J. Jedema, A. T. Filip, B. J. Van Wees, and M. L. Roukes, Spin injection and transport in micro- and nanoscale devices, in *Semiconductor Spintronics and Quantum Computation*, D. D. Awschalom, D. Loss, and N. Samarth (Eds.), Springer-Verlag, Berlin, pp. 31–87, 2002.
19. M. Johnson and R. Silsbee, Interfacial charge-spin coupling: Injection and detection of spin magnetization in metals, *Phys. Rev. Lett.* **55**, 1790–1793 (1985).
20. M. Johnson and R. H. Silsbee, Spin-injection experiment, *Phys. Rev. B* **37**, 5326–5335 (1988).

21. P. C. van Son, H. van Kempen, and P. Wyder, Boundary resistance of the ferromagnetic-nonferromagnetic metal interface, *Phys. Rev. Lett.* **58**, 2271–2273 (1987).

22. T. Valet and A. Fert, Theory of the perpendicular magnetoresistance in magnetic multilayers, *Phys. Rev. B* **48**, 7099–7113 (1993).

23. G. Schmidt, D. Ferrand, L. W. Molenkamp, A. T. Filip, and B. J. van Wees, Fundamental obstacle for electrical spin injection from a ferromagnetic metal into a diffusive semiconductor, *Phys. Rev. B* **62**, 4790–4793 (2000).

24. E. I. Rashba, Theory of electrical spin injection: Tunnel contacts as a solution of the conductivity mismatch problem, *Phys. Rev. B* **62**(24), 16267–16270 (2000).

25. A. Fert and H. Jaffrès, Conditions for efficient spin injection from a ferromagnetic metal into a semiconductor, *Phys. Rev. B* **64**, 184420 (2001).

26. D. L. Smith and R. N. Silver, Electrical spin injection into semiconductors, *Phys. Rev. B* **64**, 045323 (2001).

27. S. F. Alvarado and P. Renaud, Observation of spin-polarized-electron tunneling from a ferromagnet into GaAs, *Phys. Rev. Lett.* **68**, 1387–1390 (1992).

28. H. J. Zhu, M. Ramsteiner, H. Kostial, M. Wassermeier, H. P. Schönherr, and K. H. Ploog, Room temperature spin injection from Fe into GaAs, *Phys. Rev. Lett.* **87**, 016601 (2001).

29. A. T. Hanbicki, B. T. Jonker, G. Itskos, G. Kioseoglou, and A. Petrou, Efficient electrical injection from a magnetic metal/tunnel barrier contact into a semiconductor, *Appl. Phys. Lett.* **80**, 1240–1242 (2002).

30. J. Strand, B. D. Schultz, A. F. Isakovic, C. J. Palmstrøm, and P. A. Crowell, Dynamic nuclear polarization by electrical spin injection in ferromagnet-semiconductor heterostructures, *Phys. Rev. Lett.* **91**, 036602 (2003).

31. V. F. Motsnyi, J. De Boeck, J. Das et al., Electrical spin injection in a ferromagnet/tunnel barrier/semiconductor heterostructure, *Appl. Phys. Lett.* **81**, 265–267 (2002).

32. X. Jiang, R. Wang, S. van Dijken et al., Optical detection of hot-electron spin injection into GaAs from a magnetic tunnel transistor source, *Phys. Rev. Lett.* **90**, 256603 (2003).

33. X. Jiang, R. Wang, R. M. Shelby et al., Highly spin-polarized room-temperature tunnel injector for semiconductor spintronics using MgO(100), *Phys. Rev. Lett.* **94**, 056601 (2005).

34. R. Fiederling, M. Kelm, G. Reuscher et al., Injection and detection of a spin-polarized current in a light-emitting diode, *Nature* **402**, 787–789 (1999).

35. Y. Ohno, D. K. Young, B. Beschoten, F. Matsukura, H. Ohno, and D. D. Awschalom, Electrical spin injection in a ferromagnetic semiconductor heterostructure, *Nature* **402**, 790–792 (1999).

36. D. R. Scifres, B. A. Huberman, R. M. White, and R. S. Bauer, A new scheme for measuring itinerant spin polarizations, *Solid State Commun.* **13**, 1615–1617 (1973).

37. J. Strand, X. Lou, C. Adelmann et al., Electron spin dynamics and hyperfine interactions in Fe/$Al_{0.1}Ga_{0.9}As$/GaAs spin injection heterostructures, *Phys. Rev. B* **72**, 155308 (2005).

38. C. Adelmann, X. Lou, J. Strand, C. J. Palmstrøm, and P. A. Crowell, Spin injection and relaxation in ferromagnet-semiconductor heterostructures, *Phys. Rev. B* **71**, 121301 (2005).

39. G. Salis, R. Wang, X. Jiang et al., Temperature independence of the spin-injection efficiency of a MgO-based tunnel spin injector, *Appl. Phys. Lett.* **87**, 262503 (2005).

40. X. Y. Dong, C. Adelmann, J. Q. Xie et al., Spin injection from the Heusler alloy $Co_2MnGe$ into $Al_{0.1}Ga_{0.9}As$/GaAs heterostructures, *Appl. Phys. Lett.* **86**, 102107 (2005).

41. M. C. Hickey, C. D. Damsgaard, I. Farrer, D. A. Ritchie, R. F. Lee, G. A. C. Jones, M. Pepper, Spin injection between epitaxial $Co_{2.4}Mn_{1.6}Ga$ and an InGaAs quantum well, *Appl. Phys. Lett.* **86**, 252106 (2005).

42. N. C. Gerhardt, S. Hovel, C. Brenner et al., Electron spin injection into GaAs from ferromagnetic contacts in remanence, *Appl. Phys. Lett.* **87**, 032502 (2005).

43. C. Adelmann, J. L. Hilton, B. D. Schultz et al., Spin injection from perpendicular magnetized ferromagnetic δ-MnGa into (Al,Ga)As heterostructures, *Appl. Phys. Lett.* **89**, 112511 (2006).

44. R. K. Kawakami, Y. Kato, M. Hanson et al., Ferromagnetic imprinting of nuclear spins in semiconductors, *Science* **294**, 131–134 (2001).

45. R. J. Epstein, I. Malajovich, R. K. Kawakami et al., Spontaneous spin coherence in *n*-GaAs produced by ferromagnetic proximity polarization, *Phys. Rev. B* **65**, 121202(R) (2002).

46. C. Ciuti, J. P. McGuire, and L. J. Sham, Spin polarization of semiconductor carriers by reflection off a ferromagnet, *Phys. Rev. Lett.* **89**, 156601 (2002).

47. J. P. McGuire, C. Ciuti, and L. J. Sham, Theory of spin transport induced by ferromagnetic proximity on a two-dimensional electron gas, *Phys. Rev. B* **69**, 115339 (2004).

48. R. J. Epstein, J. Stephens, M. Hanson et al., Voltage control of nuclear spin in ferromagnetic Schottky diodes, *Phys. Rev. B* **68**, 41305 (2003).

49. J. Stephens, J. Berezovsky, J. P. McGuire, L. J. Sham, A. C. Gossard, and D. D. Awschalom, Spin accumulation in forward-biased MnAs/GaAs Schottky diodes, *Phys. Rev. Lett.* **93**, 097602 (2004).

50. V. F. Motsnyi, P. Van Dorpe, W. Van Roy et al., Optical investigation of electrical spin injection into semiconductors, *Phys. Rev. B* **68**, 245319 (2003).

51. P. Van Dorpe, W. Van Roy, J. De Boeck, and G. Borghs, Nuclear spin orientation by electrical spin injection in an $Al_xGa_{1-x}As$/GaAs spin-polarized light-emitting diode, *Phys. Rev. B* **72**, 035315 (2005).

52. J. Strand, A. F. Isakovic, X. Lou, P. A. Crowell, B. D. Schultz, and C. J. Palmstrøm, Nuclear magnetic resonance in a ferromagnet-semiconductor heterostructure, *Appl. Phys. Lett.* **83**, 3335–3337 (2003).

53. M. W. J. Prins, H. van Kempen, H. van Leuken, R. A. de Groot, W. Van Roy, and J. De Boeck, Spin-dependent transport in metal/semiconductor tunnel junctions, *J. Phys. Condens. Matter* **7**, 9447–9464 (1995).

54. A. Hirohata, Y. B. Xu, C. M. Guertler, J. A. C. Bland, and S. N. Holmes, Spin-polarized electron transport in ferromagnet/semiconductor hybrid structures induced by photon excitation, *Phys. Rev. B* **63**, 104425 (2001).

55. A. F. Isakovic, D. M. Carr, J. Strand, B. D. Schultz, C. J. Palmstrøm, and P. A. Crowell, Optical pumping in ferromagnet-semiconductor heterostructures: Magneto-optics and spin transport, *Phys. Rev. B* **64**, 161304 (2001).

56. S. Kreuzer, J. Moser, W. Wegscheider, D. Weiss, M. Bichler, and D. Schuh, Spin polarized tunneling through single-crystal GaAs(001) barriers, *Appl. Phys. Lett.* **80**, 4582–4584 (2002).

57. J. Moser, M. Zenger, C. Gerl, W. Wegscheider, and D. Weiss, Bias dependent inversion of tunneling magnetoresistance in Fe/GaAs/Fe tunnel junctions, *Appl. Phys. Lett.* **89**, 162106 (2006).

58. S. A. Crooker, M. Furis, X. Lou et al., Imaging spin transport in lateral ferromagnet/semiconductor structures, *Science* **309**, 2191–2195 (2005).

59. S. A. Crooker, M. Furis, X. Lou et al., Optical and electrical spin injection and spin transport in hybrid Fe/GaAs devices, *J. Appl. Phys.* **101**, 081716 (2007).

60. X. Lou, C. Adelmann, M. Furis, S. A. Crooker, C. J. Palmstrøm, and P. A. Crowell, Electrical detection of spin accumulation at a ferromagnet-semiconductor interface, *Phys. Rev. Lett.* **96**, 176603 (2006).

61. X. Lou, C. Adelmann, S. A. Crooker et al., Electrical detection of spin transport in lateral ferromagnet-semiconductor devices, *Nat. Phys.* **3**, 197–202 (2007).

62. P. R. Hammar and M. Johnson, Detection of spin-polarized electrons injected into a two-dimensional electron gas, *Phys. Rev. Lett.* **88**, 066806 (2002).

63. J. M. Kikkawa and D. D. Awschalom, Lateral drag of spin coherence in GaAs, *Nature* **397**, 139–141 (1999).
64. S. A. Crooker and D. L. Smith, Imaging spin flows in semiconductors subject to electric, magnetic, and strain fields, *Phys. Rev. Lett.* **94**, 236601 (2005).
65. M. Beck, C. Metzner, S. Malzer, and G. H. Döhler, Spin lifetimes and strain-controlled spin precession of drifting electrons in GaAs, *Europhys. Lett.* **75**, 597–603 (2006).
66. P. Kotissek, M. Bailleul, M. Sperl et al., Cross-sectional imaging of spin injection into a semiconductor, *Nat. Phys.* **3**, 872–877 (2007).
67. J. M. LeBeau, Q. O. Hu, C. J. Palmstrøm, and S. Stemmer, Atomic structure of postgrowth annealed epitaxial Fe/(001)GaAs interfaces, *Appl. Phys. Lett.* **93**, 121909 (2008).
68. M. Hruška, Š Kos, S. A. Crooker, A. Saxena, and D. L. Smith, Effects of strain, electric, and magnetic fields on lateral electron-spin transport in semiconductor epilayers, *Phys. Rev. B* **73**, 075306 (2006).
69. F. J. Jedema, H. R. Heersche, A. T. Filip, J. J. A. Baselmans, and B. J. van Wees, Electrical detection of spin procession in a metallic mesoscopic spin valve, *Nature* **416**, 713–716, (2002).
70. Y. Kato, R. C. Myers, A. C. Gossard, and D. D. Awschalom, Coherent spin manipulation without magnetic fields in strained semiconductors, *Nature* **427**, 50–53 (2004).
71. S. A. Crooker, E. S. Garlid, A. N. Chantis et al., Bias-controlled sensitivity of ferromagnet/semiconductor electrical spin detectors, *Phys. Rev. B* **80**, 041305 (2009).
72. C. Gould, C. Rüster, T. Jungwirth et al., Tunneling anisotropic magnetoresistance: A spin-valve-like tunnel magnetoresistance using a single magnetic layer, *Phys. Rev. Lett.* **93**, 117203 (2004).
73. J. Moser, A. Matos-Abiague, D. Schuh, W. Wegscheider, J. Fabian, and D. Weiss, Tunneling anisotropic magnetoresistance and spin-orbit coupling in Fe/GaAs/Au tunnel junctions, *Phys. Rev. Lett.* **99**, 056601 (2007).
74. M. Tran, H. Jaffrès, C. Deranlot et al., Enhancement of the spin accumulation at the interface between a spin-polarized tunnel junction and a semiconductor, *Phys. Rev. Lett.* **102**, 036601 (2009).
75. S. P. Dash, S. Sharma, R. S. Patel, M. P. de Jong, and R. Jansen, Electrical creation of spin polarization in silicon at room temperature, *Nature* **462**, 491–494 (2009).
76. Y. Song and H. Dery, Magnetic-field-modulated resonant tunneling in ferromagnetic-insulator-nonmagnetic junctions, *Phys. Rev. Lett.* **113**, 047205 (2014).
77. A. G. Swartz, S. Harashima, Y. Xie et al., Spin-dependent transport across Co/LaAlO$_3$/SrTiO$_3$ heterojunctions, *Appl. Phys. Lett.* **105**, 032406 (2014).
78. I. Appelbaum, H. N. Tinkey, and P. Li, Self-consistent model of spin accumulation magnetoresistance in ferromagnet/insulator/semiconductor tunnel junctions, *Phys. Rev. B* **90**, 220402 (2014).
79. H. Inoue, A. G. Swartz, N. J. Harmon et al., Origin of the magnetoresistance in oxide tunnel junctions determined through electric polarization control of the interface, *Phys. Rev. X* **5**, 041023 (2015).
80. O. Txoperena and F. Casanova, Spin injection and local magnetoresistance effects in three-terminal devices, *J. Phys. D: Appl. Phys.* **49**, 133001 (2016).
81. S. P. Dash, S. Sharma, J. C. Le Breton et al., Spin precession and inverted Hanle effect in a semiconductor near a finite-roughness ferromagnetic interface, *Phys. Rev. B* **84**, 054410 (2011).
82. K.-R. Jeon, B.-C. Min, Y.-H. Jo et al., Electrical spin injection and accumulation in CoFe/MgO/Ge contacts at room temperature, *Phys. Rev. B* **84**, 165315 (2011).
83. S. Iba, H. Saito, A. Spiesser et al., Spin accumulation in nondegenerate and heavily doped p-type germanium, *Appl. Phys. Express* **5**, 023003 (2012).

84. N. Reyren, M. Bibes, E. Lesne et al., Gate-controlled spin injection at $LaAlO_3$/$SrTiO_3$ interfaces, *Phys. Rev. Lett.* **108**, 186802 (2012).

85. W. Han, X. Jiang, A. Kajdos, S.-H. Yang, S. Stemmer, and S. S. P. Parkin, Spin injection and detection in lanthanum- and niobium-doped SrTiO3 using the Hanle technique, *Nat. Commun.* **4**, 2134 (2013).

86. R. J. Soulen Jr., J. M. Byers, M. S. Osofsky, C. T. Tanaka, J. Nowak, J. S. Moodera, A. Barry, J. M. D. Coey, Measuring the spin polarization of a metal with a superconducting point contact, *Science* **282**, 85–88 (1998).

87. G. Salis, A. Fuhrer, and S. F. Alvarado, Signatures of dynamically polarized nuclear spins in all-electrical lateral spin transport devices, *Phys. Rev. B* **80**, 115332 (2009).

88. C. Awo-Affouda, O. M. J. van't Erve, G. Kioseoglou et al., Contributions to Hanle lineshapes in Fe/GaAs nonlocal spin valve transport, *Appl. Phys. Lett.* **94**, 102511 (2009).

89. D. Saha, M. Holub, P. Bhattacharya, and Y. C. Liao, Epitaxially grown MnAs/GaAs lateral spin valves, *Appl. Phys. Lett.* **89**, 142504 (2006).

90. O. M. J. van't Erve, A. T. Hanbicki, M. Holub et al., Electrical injection and detection of spin-polarized carriers in silicon in a lateral transport geometry, *Appl. Phys. Lett.* **91**, 212109 (2007).

91. M. Ciorga, A. Einwanger, U. Wurstbauer, D. Schuh, W. Wegscheider, and D. Weiss, Electrical spin injection and detection in lateral all-semiconductor devices, *Phys. Rev. B* **79**, 165321 (2009).

92. H. C. Koo, J. H. Kwon, J. Eom, J. Chang, S. H. Han, and M. Johnson, Control of spin precession in a spin-injected field effect transistor, *Science* **325**, 1515–1518 (2009).

93. N. Tombros, C. Jozsa, M. Popinciuc, H. T. Jonkman, and B. J. van Wees, Electronic spin transport and spin precession in single graphene layers at room temperature, *Nature* **448**, 571–574 (2007).

94. W. Han, K. Pi, W. Bao et al., Electrical detection of spin precession in single layer graphene spin valves with transparent contacts, *Appl. Phys. Lett.* **94**, 222109 (2009).

95. I. Appelbaum, B. Q. Huang, and D. J. Monsma, Electronic measurement and control of spin transport in silicon, *Nature* **447**, 295–298 (2007).

96. B. Huang, D. J. Monsma, and I. Appelbaum, Coherent spin transport through a 350 micron thick silicon wafer, *Phys. Rev. Lett.* **99**, 177209 (2007).

97. Q. O. Hu, E. S. Garlid, P. A. Crowell, and C. J. Palmstrøm, Spin accumulation near Fe/GaAs (001) interfaces: The role of semiconductor band structure, *Phys. Rev. B* **84**, 085306 (2011).

98. T. J. Zega, A. T. Hanbicki, S. C. Erwin et al., Determination of interface atomic structure and its impact on spin transport using Z-contrast microscopy and density-functional theory, *Phys. Rev. Lett.* **96**, 196101 (2006).

99. G. Salis, S. F. Alvarado, and A. Fuhrer, Spin-injection spectra of CoFe/GaAs contacts: Dependence on Fe concentration, interface, and annealing conditions, *Phys. Rev. B* **84**, 041307 (2011).

100. C. Tiusan, J. Faure-Vincent, C. Bellouard, M. Hehn, E. Jouguelet, and A. Schuhl, Interfacial resonance state probed by spin-polarized tunneling in epitaxial Fe/MgO/Fe tunnel junctions, *Phys. Rev. Lett.* **93**, 106602 (2004).

101. P. Bruski, S. C. Erwin, J. Herfort, A. Tahraoui, and M. Ramsteiner, Probing the electronic band structure of ferromagnets with spin injection and extraction, *Phys. Rev. B* **90**, 245150 (2014).

102. D. L. Smith and P. P. Ruden, Spin-polarized tunneling through potential barriers at ferromagnetic metal/semiconductor Schottky contacts, *Phys. Rev. B* **78**, 125202 (2008).

103. A. N. Chantis, K. D. Belashchenko, D. L. Smith, E. Y. Tsymbal, M. van Schilfgaarde, and R. C. Albers, Reversal of spin polarization in Fe/GaAs (001) driven by resonant surface states: First-principles calculations, *Phys. Rev. Lett.* **99**, 196603 (2007).

104. H. Dery and L. J. Sham, Spin extraction theory and its relevance to spintronics, *Phys. Rev. Lett.* **98**, 046602 (2007).

105. S. Honda, H. Itoh, J. Inoue et al., Spin polarization control through resonant states in an Fe/GaAs Schottky barrier, *Phys. Rev. B* **78**, 245316 (2008).

106. P. Li and H. Dery, Tunable spin junction, *Appl. Phys. Lett.* **94**, 192108 (2009).

107. M. K. Chan, Q. O. Hu, J. Zhang, T. Kondo, C. J. Palmstrøm, and P. A. Crowell, Hyperfine interactions and spin transport in ferromagnet-semiconductor heterostructures, *Phys. Rev. B* **80**, 161206 (2009).

108. K. D. Christie, C. C. Geppert, S. J. Patel, Q. O. Hu, C. J. Palmstrøm, and P. A. Crowell, Knight shift and nuclear spin relaxation in Fe/$n$-GaAs heterostructures, *Phys. Rev. B* **92**, 155204 (2015).

109. D. Kölbl, D. M. Zumbühl, A. Fuhrer, G. Salis, and S. F. Alvarado, Breakdown of the Korringa law of nuclear spin relaxation in metallic GaAs, *Phys. Rev. Lett.* **109**, 086601 (2012).

110. T. Uemura, T. Akiho, Y. Ebina, and M. Yamamoto, Coherent manipulation of nuclear spins using spin injection from a half-metallic spin source, *Phys. Rev. B* **91**, 140410 (2015).

111. N. J. Harmon, T. A. Peterson, C. C. Geppert et al., Anisotropic spin relaxation in $n$-GaAs from strong inhomogeneous hyperfine fields produced by the dynamical polarization of nuclei, *Phys. Rev. B* **92**, 140201 (2015).

112. Y.-S. Ou, Y.-H. Chiu, N. J. Harmon et al., Exchange-driven spin relaxation in ferromagnet-oxide-semiconductor heterostructures, *Phys. Rev. Lett.* **116**, 107201 (2016).

113. R. Jansen and B. C. Min, Detection of a spin accumulation in nondegenerate semiconductors, *Phys. Rev. Lett.* **99**, 246604 (2007).

114. C. Jozsa, M. Popinciuc, N. Tombros, H. T. Jonkman, and B. J. Van Wees, Electronic spin drift in graphene field-effect transistors, *Phys. Rev. Lett.* **100**, 236603 (2008).

115. A. N. Chantis and D. L. Smith, Theory of electrical spin-detection at a ferromagnet/semiconductor interface, *Phys. Rev. B* **78**, 235317 (2008).

116. T. A. Peterson, S. J. Patel, C. C. Geppert et al., Spin injection and detection up to room temperature in Heusler alloy/$n$-GaAs spin valves, *Phys. Rev. B* **94**, 235309 (2016).

117. M. I. D'yakonov and V. I. Perel', Possibility of orienting electron spins with current, *JETP Lett.* **13**, 467–469 (1971).

118. Y. K. Kato, R. C. Myers, A. C. Gossard, and D. D. Awschalom, Observation of the spin Hall effect in semiconductors, *Science* **306**, 1910–1913 (2004).

119. J. Wunderlich, B. Kaestner, J. Sinova, and T. Jungwirth, Experimental observation of the spin-Hall effect in a two-dimensional spin-orbit coupled semiconductor system, *Phys. Rev. Lett.* **94**, 047204 (2005).

120. J. Sinova, S. O. Valenzuela, J. Wunderlich, C. H. Back, and T. Jungwirth, Spin Hall effects, *Rev. Mod. Phys.* **87**, 1213–1260 (2015).

121. E. S. Garlid, Q. O. Hu, M. K. Chan, C. J. Palmstrøm, and P. A. Crowell, Electrical measurement of the direct spin Hall effect in Fe/In$_x$Ga$_{1-x}$As heterostructures, *Phys. Rev. Lett.* **105**, 156602 (2010).

122. H.-A. Engel, B. I. Halperin, and E. I. Rashba, Theory of spin Hall conductivity in $n$-doped GaAs, *Phys. Rev. Lett.* **95**, 166605 (2005).

123. M. Ehlert, C. Song, M. Ciorga et al., All-electrical measurements of direct spin Hall effect in GaAs with Esaki diode electrodes, *Phys. Rev. B* 86, 205204 (2012).

124. K. Olejnk, J. Wunderlich, A. C. Irvine et al., Detection of electrically modulated inverse spin Hall effect in an Fe/GaAs microdevice, *Phys. Rev. Lett.* **109**, 076601 (2012).

125. S. O. Valenzuela and M. Tinkham, Direct electronic measurement of the spin Hall effect, *Nature* **442**, 176–179 (2006).

126. W. Y. Choi, H.-J. Kim, J. Chang, S. H. Han, H. C. Koo, and M. Johnson, Electrical detection of coherent spin precession using the ballistic intrinsic spin Hall effect, *Nat. Nanotechnol.* **10**, 666–670 (2015).

127. M. Oltscher, M. Ciorga, M. Utz, D. Schuh, D. Bougeard, and D. Weiss, Electrical spin injection into high mobility 2D systems, *Phys. Rev. Lett.* **113**, 236602 (2014).

128. M. Ciorga, Electrical spin injection and detection in high mobility 2DEG systems, *J. Phys. Condens. Matter* **28**, 453003 (2016).

129. B. Endres, M. Ciorga, R. Wagner et al., Nonuniform current and spin accumulation in a 1 μm thick n-GaAs channel, *Appl. Phys. Lett.* **100**, 092405 (2012).

130. M. Buchner, T. Kuczmik, M. Oltscher, C. Schüller, D. Bougeard, D. Weiss, and C. H. Back, Optical investigation of electrical spin injection into an inverted two-dimensional electron gas structure, *Phys. Rev. B* **95**, 035304 (2017).

131. Y. Tserkovnyak, A. Brataas, and G. E. W. Bauer, Enhanced Gilbert damping in thin ferromagnetic films, *Phys. Rev. Lett.* **88**, 117601 (2002).

132. R. Urban, G. Woltersdorf, and B. Heinrich, Gilbert damping in single and multilayer ultrathin films: Role of interfaces in nonlocal spin dynamics, *Phys. Rev. Lett.* **87**, 217204 (2001).

133. M. V. Costache, M. Sladkov, S. M. Watts, C. H. van der Wal, and B. J. van Wees, Electrical detection of spin pumping due to the precessing magnetization of a single ferromagnet, *Phys. Rev. Lett.* **97**, 216603 (2006).

134. E. Saitoh, M. Ueda, H. Miyajima, and G. Tatara, Conversion of spin current into charge current at room temperature: Inverse spin-Hall effect, *Appl. Phys. Lett.* **88**, 182509 (2006).

135. K. Ando, S. Takahashi, J. Ieda et al., Electrically tunable spin injector free from the impedance mismatch problem, *Nat. Mater.* **10**, 655–659 (2011).

136. K. Ando and E. Saitoh, Observation of the inverse spin Hall effect in silicon, *Nat. Commun.* 3, 629 (2012).

137. E. Shikoh, K. Ando, K. Kubo, E. Saitoh, T. Shinjo, and M. Shiraishi, Spin-pump-induced spin transport in p-type Si at room temperature, *Phys. Rev. Lett.* **110**, 127201 (2013).

138. M. Koike, E. Shikoh, Y. Ando et al., Dynamical spin injection into p-type germanium at room temperature, *Appl. Phys. Express* **6**, 023001 (2013).

139. J. C. Rojas-Sanchez, M. Cubukcu, A. Jain, A. Marty, L. Vila, J. P. Attane, E. Augendre, G. Desfonds, S. Gambarelli, H. Jaffrès, J. M. George, and M. Jamet, Spin pumping and inverse spin Hall effect in germanium, *Phys. Rev. B* **88**, 064403 (2013).

140. R. Adhikari, M. Matzer, A. T. Martin-Luengo, M. C. Scharber, and A. Bonanni, Rashba semiconductor as spin Hall material: Experimental demonstration of spin pumping in wurtzite *n*-GaN:Si, *Phys. Rev. B* **94**, 085205 (2016).

141. A. Yamamoto, Y. Ando, T. Shinjo, T. Uemura, and M. Shiraishi, Spin transport and spin conversion in compound semiconductor with non-negligible spin-orbit interaction, *Phys. Rev. B* **91**, 024417 (2015).

142. C. Liu, Y. Boyko, C. C. Geppert et al., Electrical detection of ferromagnetic resonance in ferromagnet/n-GaAs heterostructures by tunneling anisotropic magnetoresistance, *Appl. Phys. Lett.* **105**, 212401 (2014).

143. C. Liu, S. J. Patel, T. A. Peterson et al., Dynamic detection of electron spin accumulation in ferromagnet/semiconductor devices by ferromagnetic resonance, *Nat. Commun.* 7, 10296 (2016).

144. L. Chen, F. Matsukura, and H. Ohno, Direct-current voltages in (Ga,Mn)As structures induced by ferromagnetic resonance, *Nat. Commun.* 4, 2055 (2013).

145. C. Palmstrøm, Heusler compounds and spintronics, *Prog. Cryst. Growth Charact.* **62**, 371–397 (2016).

146. C. Felser, L. Wollmann, S. Chadov, G. H. Fecher, and S. S. P. Parkin, Basics and prospective of magnetic Heusler compounds, *APL Mater.* **3**, 041518 (2015).

147. C. Felser and A. Hirohata (Eds.), *Heusler Alloys–Properties, Growth, Applications*, Springer, Heidelberg, 2016.

148. M. C. Hickey, C. D. Damsgaard, S. N. Holmes et al., Spin injection from $Co_2MnGa$ into an InGaAs quantum well, *Appl. Phys. Lett.* **92**, 232101 (2008).

149. M. Ramsteiner, O. Brandt, T. Flissikowski et al., $Co_2FeSi$/GaAs/(Al,Ga)As spin light-emitting diodes: Competition between spin injection and ultrafast spin alignment, *Phys. Rev. B* **78**, 121303 (2008).

150. R. Farshchi and M. Ramsteiner, Spin injection from Heusler alloys into semiconductors: A materials perspective, *J. Appl. Phys.* **113**, 191101 (2013).

151. P. Bruski, Y. Manzke, R. Farshchi, O. Brandt, J. Herfort, and M. Ramsteiner, All-electrical spin injection and detection in the $Co_2FeSi$/GaAs hybrid system in the local and non-local configuration, *Appl. Phys. Lett.* **103**, 052406 (2013).

152. T. Saito, N. Tezuka, M. Matsuura, and S. Sugimoto, Spin injection, transport, and detection at room temperature in a lateral spin transport device with $Co_2FeAl_{0.5}Si_{0.5}$/n-GaAs Schottky tunnel junctions, *Appl. Phys. Express* **6**, 103006 (2013).

# Spin Polarization by Current

**Sergey D. Ganichev, Maxim Trushin, and John Schliemann**

## 7.1 INTRODUCTION

Spin generation and spin currents in semiconductor structures lie at the heart of the emerging field of spintronics and are a major and still growing direction of solid-state research. Among the plethora of concepts and ideas, current-induced spin polarization has attracted particular interest from both experimental and theoretical points of view; for reviews, see Refs. [1–14]. In nonmagnetic semiconductors or metals belonging to the gyrotropic point groups,* see Refs. [7, 16–20], *dc* electric current is generically accompanied by a non-zero average nonequilibrium spatially homogeneous spin polarization and vice versa. The latter phenomenon is referred to as the spin-galvanic effect observed in GaAs QWs [21] and other two-dimensional systems; see e.g. reviews [5, 7, 10, 11, 13, 14, 22–24, 58]. In low-dimensional semiconductor structures these effects are caused by asymmetric spin relaxation in systems with lifted spin degeneracy due to **k**-linear terms in the Hamiltonian, where **k** is the electron wave-vector. In spite of the terminological resemblance, spin polarization by electric current fundamentally differs from the spin Hall effect [4, 6, 8, 11, 12, 13, 25–31], which refers to the generation of a pure spin current transverse to the charge current, and causes spin accumulation at the sample edges. The distinctive features of the current-induced spin polarization are: that this effect can be present in gyrotropic media only, that it results in non-zero average spin polarization, and that it does not depend on the real-space coordinates. Thus, it can be measured in the whole sample under appropriate conditions. The spin Hall effect, in contrast, does not yield average spin polarization, and does not require gyrotropy, at least for the extrinsic spin Hall effect. Related discussion on spin Hall effect can be found in Chapter 8, Volume 2, and Chapter 7, Volume 3.

The ability of charge current to polarize spins in gyrotropic media was predicted more than thirty years ago by Ivchenko and Pikus [32]. The effect has been considered theoretically for bulk tellurium crystals, where, almost at the same time, it was demonstrated experimentally by Vorob'ev et al. [33]. In bulk tellurium, current-induced spin polarization is a consequence of the unique valence band structure of tellurium with hybridized spin-up and spin-down bands ("camel back" structure) and, in contrast to the spin polarization in quantum well structures, is not related to spin relaxation. In zinc-blende structure based QWs, this microscopic mechanism of the current-induced spin polarization is absent [34, 35]. Vas'ko and Prima [35], Aronov and Lyanda-Geller [36], and Edelstein [37] demonstrated that spin orientation by electric current is also possible in two-dimensional electron systems and is caused by asymmetric spin relaxation. Two microscopic mechanisms, namely scattering mechanism and precessional mechanism, based on Elliott–Yafet and D'yakonov–Perel spin relaxation, respectively, were developed. The first direct experimental proofs of this effect were obtained in semiconductor QWs by Ganichev et al. in (113)-grown p-GaAs/AlGaAs QWs [38, 39], by Silov et al.

---

* We remind readers that the gyrotropic point group symmetry makes no difference between certain components of polar vectors, like electric current or electron momentum, and axial vectors, like a spin or magnetic field, and is described by the gyration tensor [7, 15, 16]. Gyrotropic media are characterized by the linear in light or electron wave-vector **k** spatial dispersion resulting in optical activity (gyrotropy) or Rashba/Dresselhaus band spin-splitting in semiconductor structures [7, 16–20], respectively. Among 21 crystal classes lacking inversion symmetry, 18 are gyrotropic, from which 11 classes are enantiomorphic (chiral) and do not possess a reflection plane or rotation-reflection axis [7, 18, 19]. Three non-gyrotropic non-centrosymmetric classes are $T_d$, $C_{3h}$ and $D_{3h}$. We note that it is often, but misleadingly, stated that gyrotropy (optical activity) can be obtained only in non-centrosymmetric crystals having no mirror reflection plane. In fact seven non-enantiomorphic classes groups ($C_S$, $C_{2v}$, $C_{3v}$, $S_4$, $D_{2d}$, $C_{4v}$ and $C_{6v}$) are gyrotropic, also allowing spin orientation by the electric current.

in (001)-grown p-GaAs/AlGaAs heterojunctions [40, 41], by Sih et al. in (110)-grown n-GaAs/AlGaAs QWs [42], and by Yang et al. in (001)-grown InGaAs/InAlAs QWs [43], as well as in strained bulk (001)-oriented InGaAs and ZnSe epilayers by Kato et al. [44, 45], and Stern et al. [46], respectively. The experiments include the range of optical methods, such as Faraday rotation and linear-circular dichroism in transmission of terahertz radiation, time-resolved Kerr rotation, and polarized luminescence in near-infrared up to visible spectral range. We emphasize that current-induced spin polarization was observed even at room temperature [38, 39, 45]. It has been demonstrated that depending on the point group symmetry electric current in 2D system may result in the in-plane spin orientation, as in the case of e.g. (001)-grown structures [40, 41, 43–50], or may additionally align spins normal to the 2DEG's plane. The latter take place in [llh]- or [lmh]-oriented structures, e.g. [113]-, [110]- and [013]-grown QWs (see [10, 38, 39, 42, 51]). We also would like to note, that investigating spin injection from a ferromagnetic film into a two-dimensional electron gas, Hammar et al. [52, 53] used the concept of a spin orientation by current in a 2DEG (see also [54, 55]) to interpret their results. Though a larger degree of spin polarization was extracted, the experiment's interpretation is complicated by other effects [56, 57]. Later experiments on ferromagnet/(Ga,Mn)As bilayers [58], uniaxial (Ga,Mn)As epilayers [59], metallic interfaces and surfaces [60–65] and interfaces between ferromagnetic films and topological insulators [66, 67] clearly demonstrated the ability of spin polarization by electric current in ferromagnetics- and metal-based structures. Chapters 4 and 9, Volume 2 provide some background material on (Ga,Mn)As.

The experimental observation of current-induced spin polarization has given rise to extended theoretical studies of this phenomenon in various systems using various approaches and theoretical techniques. These include the semiclassical Boltzmann Equation [51, 68–73] derived from the quantum mechanical Liouville Equation [74] and other diffusion-type equations describing the dynamics of spin expectation values [6, 75–83]. Other authors have performed important and fruitful studies of the same issues using various Green's function techniques of many-body physics [2, 84–91]. The effect of external contacts to the system and boundaries was studied explicitly in Refs. [55, 92, 93], and in Refs. [94, 95] the spin response of an electron gas to a microwave radiation was calculated. The influence of four terminal geometry has been studied in Ref. [96] using a numerical Landauer-Keldysh approach, and weak localization corrections for current-induced spin polarization have been calculated in [97]. The current-induced spin polarization has also been investigated in hole systems [89, 98, 99]. A search for efficient spin generation and manipulation by all electrical means has given rise to theoretical analysis of current induced spin polarization in exotic regimes, like streaming caused by high electric fields [73, 100] or very weak electron-impurity interaction [101], in one-dimensional channels [102, 103] and combined structures with metal/insulator [104], ferromagnets/topological insulators [105–108], ferromagnets/graphene [109, 110], or metal/semiconductor [62, 111] interfaces. Magnetic heterostructures with topological insulators and graphene are discussed in Chapter 5, Volume 2, and Chapter 5, Volume 3.

Searching for publications on current-induced spin polarization, one can be confused by the fact that several different names are used to describe it. Besides current-induced spin polarization (CISP) this phenomenon is often referred to as the inverse spin-galvanic effect (ISGE), current-induced spin accumulation (CISA), the magnetoelectric effect (MEE), or kinetic magnetoelectric effect (KMEE) (this term was first used to describe the effect in nonmagnetic conductors by Levitov et al. [112]), or electric-field mediated in-plane spin accumulation. This variety of terms was extended further after the observation of the current-induced spin polarization at the interface between nonmagnetic metals [60]. The authors introduce the new term—Edelstein effect (EE)—which in following works was modified to Rashba–Edelstein effect (REE), as well as to its inversion (IEE) corresponding to the spin-galvanic effect. Despite the enormous diversity of labels, in all these cases we deal with one and the same microscopic effect: appearance of nonequilibrium spin polarization due to *dc* electric current in the gyrotropic media with Rashba/Dresselhaus spin splitting of the bands. In our chapter we will use two of these terms: current-induced spin polarization and the inverse spin-galvanic effect.

## 7.2 MODEL

Phenomenologically, the electron's averaged nonequilibrium spin **S** can be linked to an electric current **j** by

$$j_\lambda = \sum_\mu Q_{\lambda\mu} S_\mu, \tag{7.1}$$

$$S_\alpha = \sum_\gamma R_{\alpha\gamma} j_\gamma, \tag{7.2}$$

where **Q** and **R** are second rank pseudotensors. Equation 7.1 describes the spin-galvanic effect and Equation 7.2 represents the effect inverse to the spin-galvanic effect: an electron spin polarization induced by a *dc* electric current. We note the similarity of Equations 7.1 and 7.2 characteristic for effects due to gyrotropy: both equations linearly couple a polar vector with an axial vector. The phenomenological Equation 7.2 shows that the spin polarization can only occur for those components of the in-plane components of **j** which transform as the averaged nonequilibrium spin **S** for all symmetry operations. Thus the relative orientation of the current direction and the average spin is completely determined by the point group symmetry of the structure. This can most clearly be illustrated by the example of a symmetric (110)-grown zinc-blende QW where an electric current along $x \parallel [1\bar{1}0]$ results in a spin orientation along the growth direction $z$: see Figure 7.1a. These QWs belong to the point-group symmetry $C_{2v}$ and contain, apart from the identity and a $C_2$-axis, a reflection plane $m_1$ normal to the QW plane and $x$-axis, and a reflection plane $m_2$ being parallel to the interface plane, see Figure 7.1b. The reflection in $m_1$ transforms the current component $j_x$ and the average

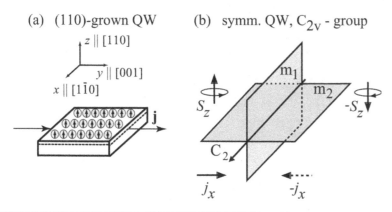

**FIGURE 7.1** (a) Electric current-induced spin polarization in symmetric (110)-grown zinc-blende structure based QWs. (b) symmetry elements of symmetrical QW grown in the $z \parallel [110]$ direction. Arrows in the sketch (b) show reflection of the components of polar vector $j_x$ and axial vector $S_z$ by the mirror reflection plane $m_1$. An additional reflection by the mirror reflection plane $m_2$ does not modify the components of an in-plane polar vector $j_x$ as well as not changing the polarity of an out-plane axial vector $S_z$. Thus, the linear coupling of $j_x$ and $S_z$ is allowed for structures of this symmetry.

spin component $S_z$ in the same way ($j_x \rightarrow -j_x$, $S_y \rightarrow -S_y$): see Figure 7.1b. Also the reflection in $m_2$ transforms these components in an equal way, and their sign remains unchanged for this symmetry operation. Therefore, a linear coupling of the in-plane current and the out-plane average spin is allowed, demonstrating that a photocurrent $j_x$ can induce the average spin polarization $S_z$. In asymmetric (110)-grown QWs or (113)-grown QWs, the symmetry is reduced to $C_s$ and additionally to the $z$-direction, spins can be oriented in the plane of QWs.* Similar arguments demonstrate that in (001)-grown zinc-blende structure based QWs, an electric current can result in an in-plane spin orientation *only*. In this case, the direction of spins depends on the relative strengths of the structure inversion asymmetry (SIA) [113] and bulk inversion asymmetry (BIA) [114], resulting in an anisotropy of the current-induced spin polarization.

A microscopic model of the current-induced spin polarization [38, 39] is sketched in Figure 7.2a. To be specific, we consider an electron gas in symmetric (110)-grown zinc-blende QWs. The explanation of the effect measured in structures of other crystallographic orientation or in a hole gas can be given in a similar way. In the simplest case, the electron's (or hole's) kinetic energy in a quantum well depends quadratically on the in-plane wave-vector components $k_x$ and $k_y$. In equilibrium, the spin degenerated $k_x$ and $k_y$ states are symmetrically occupied up to the Fermi energy $E_F$. If an external electric field is applied, the charge carriers drift in the direction of the resulting force. The carriers are accelerated by the electric field and gain kinetic energy until they are scattered. A stationary state forms where the energy gain and

---

* Note that for the lowest symmetry C1, which contains no symmetry elements besides identity, the relative direction between current and spin orientation becomes arbitrary.

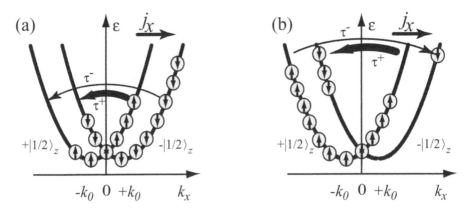

**FIGURE 7.2** Current-induced spin polarization (a) and spin-galvanic effect (b) due to spin-flip scattering in symmetric (110)-grown zinc-blende structure based QWs. In this case, only $\beta_{zx}\sigma_z k_x$ term are present in the Hamiltonian. The conduction subband is split into two parabolas with spin-up $|+1/2\rangle_z$ and spin-down $|-1/2\rangle_z$ pointing in the $z$-direction. In (a) biasing along the $x$-direction causes an asymmetric in **k**-space occupation of both parabolas. In (b) nonequilibrium spin orientation along $z$-direction causes an electric current in $x$-direction. (After Ganichev, S.D. et al., cond-mat/0403641, 2004; Ganichev, S.D. et al., *J. Magn. and Magn. Mater.* 300, 127, 2006. With permission.)

the relaxation are balanced, resulting in a non-symmetric distribution of carriers in **k**-space yielding an electric current. The electrons acquire the average quasimomentum

$$\hbar\Delta\mathbf{k} = e\mathbf{E}\tau_{ps} \tag{7.3}$$

where **E** is the electric field strength, $\tau_{ps}$ is the momentum relaxation time for a given spin split subband labeled by $s$, and $e$ is the elementary charge. As long as spin-up and spin-down states are degenerated in **k**-space the energy bands remain equally populated and a current is not accompanied by spin orientation. In QWs made of zinc-blende structure material like GaAs or strained bulk semiconductors, however, the spin degeneracy is lifted due to SIA and BIA [113, 114] and dispersion reads

$$E_{ks} = \frac{\hbar^2\mathbf{k}^2}{2m^*} + \beta_{lm}\sigma_l k_m \tag{7.4}$$

with the spin-orbit pseudotensor $\beta$, the Pauli spin matrices $\sigma_l$ and effective mass $m^*$ (note the similarity to Equations 7.1 and 7.2). The parabolic energy band splits into two subbands of opposite spin directions shifted in **k**-space symmetrically around $\mathbf{k} = 0$ with minima at $\pm k_0$. For symmetric (110)-grown zinc-blende structure based QWs, the spin-orbit interaction results in a Hamiltonian of the form $\beta_{zx}\sigma_z k_x$; the corresponding dispersion is sketched in Figure 7.2. Here the energy band splits into two subbands of $s_z = 1/2$ and $s_z = -1/2$, with minima symmetrically shifted in the **k**-space along the $k_x$ axis from the point $k = 0$ into the points $\pm k_0$, where $k_0 = m^*\beta_{z'x}/\hbar^2$. As long as the spin relaxation is switched off, the spin branches are equally populated and contribute equally to the current. Due to the band splitting, spin-flip relaxation processes $\pm 1/2 \rightarrow \mp 1/2$, however, become different because of the difference in quasimomentum transfer from initial to final states. In Figure 7.2a the **k**-dependent spin-flip scattering processes are indicated by arrows of different lengths and thicknesses. As a consequence, different amounts of spin-up and spin-down carriers contribute to the spin-flip transitions, causing a stationary spin orientation. In this picture, we assume that the origin of the current-induced spin orientation is, as sketched in Figure 7.2a, exclusively due to scattering and hence dominated by the Elliott–Yafet spin relaxation (scattering mechanism) [115]. The other possible mechanism resulting in the current-induced spin orientation is based on the D'yakonov–Perel spin relaxation [115] (precessional mechanism). In this case, the relaxation rate depends on the average $\Delta\mathbf{k}$-vector equal to $\bar{k}_{1/2} = -k_0 + \langle \Delta k_x \rangle$ for the spin-up and $\bar{k}_{-1/2} = k_0 + \langle \Delta k_x \rangle$ for the spin-down subband [114]. Hence, for the D'yakonov–Perel mechanism also, the spin relaxation becomes asymmetric in **k**-space, and a current through the electron gas causes spin orientation.

Figure 7.2b shows that not only phenomenological equations, but also microscopic models of the current-induced spin polarization and the spin-galvanic effect are similar. Spin orientation in the $x$-direction causes the unbalanced population in spin-down and spin-up subbands. As long as the carrier distribution in each subband is symmetric around the subband minimum at $k_{x_\pm}$, no current flows. The current flow is caused by **k**-dependent spin-flip relaxation processes [7, 21, 22]. Spins oriented in the $z$-direction are scattered along $k_x$ from the higher filled—e.g. spin subband $|+1/2\rangle_z$—to the less filled—e.g. spin subband $|-1/2\rangle_z$. The spin-flip scattering rate depends on the values of the wave-vectors of the initial and the final states [115]. Four quantitatively different spin-flip scattering events exist. They preserve the symmetric distribution of carriers in the subbands, and thus do not yield a current. While two of them have the same rates, the other two, sketched in Figure 7.2b by bent arrows, are inequivalent, and generate an asymmetric carrier distribution around the subband minima in both subbands. This asymmetric population results in a current flow along the $x$-direction. Within this model of elastic scattering, the current is not spin polarized, since the same number of spin-up and spin-down electrons move in the same direction with the same velocity. Like current-induced spin polarization, spin-galvanic effect can also result from the precessional mechanism [7] based on the asymmetry of the Dyakonov–Perel spin relaxation.

In order to foster the above phenomenological arguments and models, let us discuss a microscopic description of such processes as introduced in Ref. [71]. This theory is based on an analytical solution to the semiclassical Boltzmann equation, which does not include the spin relaxation time explicitly. Instead, one utilizes the so-called quasiparticle life time $\tau_0$, which is defined via the scattering probability alone. This is in contrast to the momentum relaxation time $\tau_{ps}$, which, in addition, depends on the distribution function itself and, therefore, should be self-consistently deduced from the Boltzmann equation written within the relaxation time approximation. In the framework of a simplest model dealing with the elastic delta-correlated scatterers, the quasiparticle life time $\tau_0$ depends neither on carrier momentum nor spin index, and in that way, it essentially simplifies our description.

Before we proceed, let us first discuss the applicability of the quasiclassical Boltzmann kinetic equation to the description of the spin polarization in quantum wells observed in [40, 41, 42, 43, 44, 45, 46]. There are two restrictions which are inherited by the Boltzmann equation due to its quasiclassical origin. The first one is obvious: the particle's de Broglie length must be much smaller than the mean free path. At low temperatures—compared to the Fermi energy—the characteristic de Broglie length $\lambda$ relates to the carrier concentration $n_e \sim \dfrac{m^* E_F}{\pi\hbar^2}$ approximately as $\lambda \sim \sqrt{2\pi / n_e}$, whereas the mean free path $l$ can be estimated as $l \sim \dfrac{\hbar}{m^*}\sqrt{2\pi n_e}\,\tau_0$. Thus, the first restriction can be written as

$$n_e \gg m^* / \hbar\tau_0. \tag{7.5}$$

The second one is less obvious and somewhat more specific to our systems here, but still it is important: the smearing of the spin-split subband due to the disorder $\hbar / \tau_0$ must be much smaller than the spin splitting energy $E_{+k} - E_{-k}$. The latter depends strongly on the quasimomentum, and therefore at the Fermi level, it is defined by the carrier concentration. As a consequence, the concentration must fulfill the following inequality

$$n_e \gg \hbar^2 / 8\pi\beta^2\tau_0^2, \tag{7.6}$$

where $\beta$ is the spin-orbit coupling parameter. This restriction can also be reformulated in terms of the mean free path $l$ and spin precession length $\lambda_s \sim \pi / k_0$. Namely, $l$ must be much larger than $\lambda_s$ so that an electron randomizes its spin orientation due to the spin-orbit precession between two subsequent scattering events. This restriction corresponds to the approximation which neglects the off-diagonal elements of the nonequilibrium distribution function in the spin space. Indeed, an electron spin cannot only be in one of two possible spin eigen states of the free Hamiltonian, but in an arbitrary superposition of them, the case described by the off-diagonal elements of the density matrix. The latter is not possible in classical physics where a given particle always has a definite position in the phase space. Thus, we could not directly apply Boltzmann equation in its conventional form for the description of the electron spin in 2DEGs with spin-orbit interactions as long as the inequality (7.6) is not fulfilled. Note, on the other hand, that all the quantum effects stemming from the quantum nature of the electron spin can be smeared out by a sufficient temperature larger than the spin splitting energy $E_{+k} - E_{-k}$. Thus, the room temperature $T_{\text{room}} = 25\text{meV}$ being much larger than the spin-orbit splitting energy of the order of 3meV (which is relevant for InAs quantum wells) makes the Boltzmann equation applicable for sure. In order to verify whether the conditions given by Equations 7.5 and 7.6 are indeed satisfied, one can deduce the quasiparticle life time $\tau_0$ from the mobility $\mu = e\tau_0 / m^*$. Then, using spin-orbit coupling parameters in the range usually found in experiments [116, 118, 119], the above inequalities turn out to be fulfilled. In a case when inequality (7.6) is not fulfilled, the theory based on spin-density matrix formalism yields the result for CISP degree, depending on the ratio between energy and spin relaxation times [51, 120]. If energy relaxation is slower than Dyakonov–Perel spin relaxation, then the electrically induced spin density is given by Equation 7.11 derived for well-split spin density subbands. This regime takes place at low temperatures. By contrast, spin density is twice as large in the opposite case of fast energy relaxation which is realized at moderate and high temperatures [51].

In general, the Boltzmann equation describes the time evolution of the particle distribution function $f(t,\mathbf{r},\mathbf{k})$, in the coordinate $\mathbf{r}$ and momentum $\mathbf{k}$ space. To describe the electron kinetics in presence of a small homogeneous electric field $\mathbf{E}$ in a steady state, one usually follows the standard procedure widely spread in the literature on solid state physics. The distribution function is then represented as a sum of an equilibrium $f_0(E_{sk})$ and nonequilibrium $f_1(s,\mathbf{k})$ contributions. The first one is just a Fermi-Dirac distribution, and the second one is a time and coordinate independent nonequilibrium correction linear in $\mathbf{E}$. This latter contribution should be written down as a solution of the kinetic equation; however, it might be also deduced from qualitative arguments in what follows.

Let us assume that the scattering of carriers is elastic, which means, above all, that spin flip is forbidden. Then, the average momentum $\Delta\mathbf{k}$ which the carriers gain due to the electric field can be estimated relying on the momentum relaxation time approximation from Equation 7.3. If the electric field is small (linear response), then to get the nonequilibrium term $f_1(s,\mathbf{k})$ one has to expand the Fermi–Dirac function $f_0(E_{s(k-\Delta k)})$ into the power series for small $\Delta\mathbf{k}$ up to the term linear in $\mathbf{E}$. Recalling $\hbar\mathbf{v} = -\partial_{\Delta k}E_{(sk-\Delta k)}|_{\Delta k=0}$, the nonequilibrium contribution $f_1(s,\mathbf{k})$ can be written as

$$f_1(s,\mathbf{k}) = -e\mathbf{E}\mathbf{v}\tau_{ps}\left[-\frac{\partial f^0(E_{sk})}{\partial E_{sk}}\right]. \tag{7.7}$$

Since $f_1(s,\mathbf{k})$ is proportional to the derivative of the step-like Fermi–Dirac distribution function, the nonequilibrium term substantially contributes to the total distribution function close to the Fermi energy only, and the sign of its contribution depends on the sign of the group velocity $\mathbf{v}$. Note that the group velocity $\mathbf{v}$ for a given energy and direction of motion is the same for both spin split subbands and, therefore, nonequilibrium addition to the distribution function would be the same for both branches as long as $\tau_{ps}$ were independent of the spin index. The latter is, however, not true, and $\tau_{ps}$ is different for two spin split subbands, as depicted in Figure 7.2. This is the microscopic reason why the nonequilibrium correction to the distribution function gives rise to the spin polarization.

The nonequilibrium correction can be also found as an analytical solution of the Boltzmann Equation [71], and has the form

$$g_1(s,\mathbf{k}) = -e\mathbf{E}\mathbf{k}\frac{\hbar\tau_0}{m^*}\left[-\frac{\partial f^0(E_{sk})}{\partial E_{sk}}\right]. \tag{7.8}$$

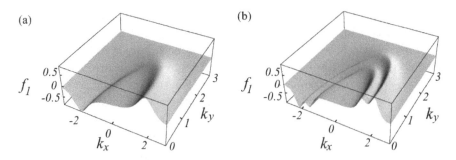

**FIGURE 7.3**   This plot shows schematically the nonequilibrium term of the distribution function vs. quasi-momentum for (a) spin-degenerate and (b) spin-split subbands. The electric field is directed along the x-axis. (a) Without spin splitting the nonequilibrium distribution function just provides a majority for right moving electrons at the expense of the ones with opposite momentum. (b) The spin-orbit splitting leads to the additional electron redistribution between two spin split subbands. As one can see from the plot, an amount of electrons belong to the outer spin split subband is larger then for the inner one. Thus, the spins from inner and outer subbands are not compensated with each other, and spin accumulation occurs.

In contrast to Equations 7.7 and 7.8 contains momentum **k** and quasiparticle life time as a prefactor.* Since the quasiparticle life time does not depend on the spin index, and the quasiparticle momentum calculated for a given energy is obviously different for two spin split subbands, Equation 7.8, immediately allows us to read out the very fact that the nonequilibrium contribution depends on the spin index; see also Figure 7.3. Thus, the spins are not compensated with each other as long as the system is out of equilibrium. Therewith one can see that the efficiency of the current-induced spin polarization is governed by the splitting between two subbands, i.e. it is directly proportional to the spin-orbit constants. One can also prove the later statements just calculating the net spin density

$$\langle S_{x,y,z}\rangle = \sum_{s}\int \frac{d^2k}{(2\pi)^2} S_{x,y,z}(k,s)g_1(s,\mathbf{k}),\qquad(7.9)$$

with $S_{x,y,z}$ being the spin expectation values.

To conclude this section, we would like to emphasize that the two Equations 7.7 and 7.8 are just two ways to describe the same physical content. One can think about spin accumulation either in terms of the momentum relaxation time $\tau_{ps}$ dependent on the spin split subband index, or one may rely on the exact solution Equation 7.8 which relates the spin accumulation to the quasimomentum difference between electrons with the opposite spin orientations.

## 7.3   ANISOTROPY OF THE INVERSE SPIN-GALVANIC EFFECT

Similar to the spin-galvanic effect [7, 116, 118] the current-induced spin polarization can be strongly anysotropic, due to the interplay of the Dresselhaus

---

* Note that the well-known textbook relation **v**=ħ**k**/m* holds only in the absence of spin-orbit coupling.

and Rashba terms. The relative strength of these terms is of general importance because it is directly linked to the manipulation of the spin of charge carriers in semiconductors, one of the key problems in the field of spintronics. Both Rashba and Dresselhaus couplings result in spin splitting of the band and give rise to a variety of spin-dependent phenomena which allow us to evaluate the magnitude of the total spin splitting of electron subbands [22, 23, 29, 115, 116, 118, 119, 121–130]. Dresselhaus and Rashba terms can interfere in such a way that macroscopic effects vanish, though the individual terms are large [115, 124, 125]. For example, both terms can cancel each other, resulting in a vanishing spin splitting in certain **k**-space directions [10, 14, 22, 125]. This cancellation leads to the disappearance of an antilocalization [122, 131], circular photogalvanic effect [118], magneto-gyrotropic effect [129, 130], spin-galvanic effect [116] and current-induced spin polarization [71], the absence of spin relaxation in specific crystallographic directions [115, 123], the lack of Shubnikov–de Haas beating [124], and has also given rise to a proposal for spin field-effect transistor operating in the nonballistic regime [125].

While the interplay of Dresselhaus and Rashba spin splitting may play a role in QWs of different crystallographic orientations, we focus here on anisotropy of the inversed spin-galvanic effect in (001)-grown zinc-blende structure based QWS. For (001)-oriented QWs linear in wave-vector part of Hamiltonian for the first subband reduces to

$$\mathcal{H}_k^{(1)} = \alpha(\sigma_{x_0} k_{y_0} - \sigma_{y_0} k_{x_0}) + \beta(\sigma_{x_0} k_{x_0} - \sigma_{y_0} k_{y_0}), \tag{7.10}$$

where the parameters $\alpha$ and $\beta$ result from the structure-inversion and bulk-inversion asymmetries, respectively, and $x_0$, $y_0$ are the crystallographic axes [100] and [010].

To study the anisotropy of the spin accumulation, one can just calculate the net spin density from Equation 7.9, where the integral over **k** can be taken easily making the substitution $\varepsilon = E(s,k)$ and assuming that $-\partial f^0(\varepsilon)/\partial\varepsilon = \delta(E_F - \varepsilon)$. The rest integrals over the polar angle can be taken analytically, and after some algebra we have

$$\langle \mathbf{S} \rangle = \frac{em^*\tau_0}{2\pi\hbar^3} \begin{pmatrix} \beta & \alpha \\ -\alpha & -\beta \end{pmatrix} \mathbf{E}. \tag{7.11}$$

This relation between the spin accumulation and electrical current can also be deduced phenomenologically applying Equation 7.2 (Figure 7.4).

The magnitude of the spin accumulation $\langle S \rangle = \sqrt{\langle S_x \rangle^2 + \langle S_y \rangle^2}$ depends on the relative strength of $\beta$ and $\alpha$ and varies after

$$\langle S \rangle = \frac{eEm^*\tau_0}{2\pi\hbar^3} \sqrt{\alpha^2 + \beta^2 + 2\alpha\beta\sin(2\widehat{\mathbf{E}}\mathbf{e}_x)}. \tag{7.12}$$

It is interesting to note, that $\langle S \rangle$ depends on the direction of the electric field (see Figure 7.4), i.e. the spin accumulation is anisotropic. This anisotropy

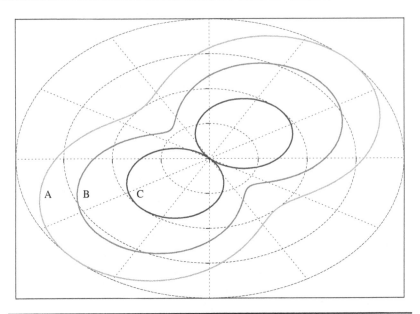

**FIGURE 7.4** Anisotropy of spin accumulation (in arbitrary units) vs. direction of the electric field in polar coordinates for different Rashba and Dresselhaus constants: A — $\alpha = 3\beta$, B — $\alpha = 2.15\beta$, C — $\alpha \sim \beta$. (After Žutić, I., *Nat. Phys.* 5, 630, 2009. With permission.) The curve B corresponds to the *n*-type InAs-based QWs investigated in [116]. (After Trushin, M. and Schliemann, J., *Phys. Rev.* B **75**, 155323, 2007. With permission.)

reflects the relation between Rashba and Dresselhaus spin–orbit constants. If either $\alpha = 0$ or $\beta = 0$ then the anisotropy vanishes. In the opposite case of $\alpha \sim \beta$, the anisotropy reaches its maximum. Note, that the case of $\alpha$ being exactly equal to $\beta$ requires special consideration, and the nonequilibrium distribution function turns out to be different from Equation 7.8. Here, the effective magnetic field does not depend on the direction of motion, and Dyakonov–Perel spin relaxation vanishes [10, 14, 123, 125]. A similar effect has also been found in rolled-up heterostructures [132], where Dyakonov–Perel spin relaxation also vanishes at a certain radius of curvature. In such situations, spin orientation by electric current becomes only possible due to other mechanisms of spin relaxation. Nevertheless, the anisotropy of spin relaxation will reflect the anisotropy of the band structure and in our case of (001)-oriented QW will qualitatively correspond to the curve C in Figure 7.4.

The simple Drude-like relation between electrical current and field allows us to write down a useful relation between the spin accumulation and charge current density

$$\langle \mathbf{S} \rangle = \frac{m^{*2}}{2\pi \hbar^3 e n_e} \begin{pmatrix} \beta & \alpha \\ -\alpha & -\beta \end{pmatrix} \mathbf{j}. \tag{7.13}$$

Note that, in the coordinate system with the coordinate axes parallel to the mirror crystallographic planes $x \parallel [1\bar{1}0]$ and $y \parallel [110]$, the Hamiltonian $\mathcal{H}_{\mathbf{k}}^{(1)}$

gets the form $\beta_{xy}\sigma_x k_y + \beta_{yx}\sigma_y k_x$ with $\beta_{xy} = \beta + \alpha$, $\beta_{yx} = \beta - \alpha$, and the relation between $\langle \mathbf{S} \rangle$ and $\mathbf{j}$ simplifies to

$$\langle \mathbf{S} \rangle = \frac{m^{*2}}{2\pi\hbar^3 e n_e} \begin{pmatrix} 0 & \beta + \alpha \\ \beta - \alpha & 0 \end{pmatrix} \mathbf{j}. \tag{7.14}$$

Hence, for the spin accumulation $\langle S_{[110]} \rangle$ and $\langle S_{[1\bar{1}0]} \rangle$ provided by the electrical current along [110] and [1$\bar{1}$0] crystallographic directions, respectively we obtain

$$\frac{\langle S_{[1\bar{1}0]} \rangle}{\langle S_{[110]} \rangle} = \frac{\alpha + \beta}{\beta - \alpha}. \tag{7.15}$$

From Equations 7.15 or 7.14, one can see that the spin accumulation is strongly anisotropic if the constants $\alpha$ and $\beta$ are close to each other. As was discussed earlier, the reason for such an anisotropy can be understood either from the spin–orbit splitting differently for different direction of the carrier motion, or from the anisotropic spin relaxation times [123].

Equations 7.14 and 7.15 show that by measuring the spin polarization for current flowing in particular crystallographic directions, one can map the spin–orbit constants and deduce their magnitudes. To find the absolute values of the spin–orbit constants, one needs to know the carrier effective mass $m^*$ and quasiparticle life time $\tau_0$. The latter can be roughly estimated from the carrier mobility. Indeed, though the band structure described by the Hamiltonian (7.10) is anisotropic, the electrical conductivity remains isotropic which can be verified directly computing the electrical current

$$\mathbf{j} = -e \sum_s \int \frac{d^2 k}{(2\pi)^2} \mathbf{v_s} g_1(s, \mathbf{k}).$$

Taking this integral in the same manner as in Equation 7.9, one can see that the anisotropy of the group velocity $\mathbf{v_s}$ is compensated by the one stemming from the distribution function $g_1(s, \mathbf{k})$. Therefore, the electrical conductivity is described by the Drude-like formula $\sigma = e^2 n_e \tau_0 / m^*$ with

$$n_e = \frac{m^* E_F}{\pi\hbar^2} + \left( \frac{m^*}{\hbar^2} \right)^2 \frac{\alpha^2 + \beta^2}{\pi}$$

being the carrier concentration. Using this Drude-like relation, one can estimate the quasiparticle life time as $\tau_0 \sim m^*\mu / e$. To give an example, an n-type InAs quantum well [116] with $\mu = 2\cdot10^4\,\text{cm}^2/(\text{V}\cdot\text{s})$ and $m^* \sim 0.032 m_0$, with $m_0$ being the bare electron mass, yields the quasiparticle life time $\tau_0 \sim 4\cdot10^{-13}\,\text{s}$.

We considered the regime linear in the electric current above. Stronger electric fields provide a needle-shaped electron distribution known as a "streaming" regime, in which each free charge carrier accelerates quasibal-listically in the "passive" region until it reaches the optical-phonon energy, then it emits an optical phonon and starts the next period of acceleration [117]. The inclusion of a spin degree of freedom into the streaming-regime kinetics gives rise to rich and interesting spin-related phenomena. In particular, the current-induced spin-orientation remarkably increases, reaching a high value $\simeq 2\%$ in the electric field $\sim 1\,\mathrm{kV/cm}$. The spin polarization enhancement is caused by squeezing of the electron momentum distribution in the direction of drift [73, 100]. Note that with further increase in the field spin polarization falls.

Finally, we would like to address another mechanism of the current-induced spin polarization and the spin-galvanic effect proposed in Refs. [69], [133] and [134], respectively. It does not require the spin splitting of electron spectrum, and is based on spin-dependent electron scattering by static defects or phonons which is asymmetric in the momentum space in gyrotropic structures [115, 135]. The spin polarization then occurs due to asymmetric spin-dependent scattering, followed by the processes of spin relaxation, which can be of both the Elliott–Yafet and the Dyakonov–Perel types, discussed further in Chapter 1, Volume 2. The scattering-related mechanism is expected to dominate at room temperature and/or electron gas of high density. It may be responsible for the observed current-induced spin polarization in $n$-doped InGaAs epilayers [49], where the anisotropy of both current-induced spin polarization and spin splitting, was studied in the same structure, and the smallest spin polarization was surprisingly observed for the electric current applied in the crystal directions corresponding to larger spin splitting. While it might be important in some particular cases, a detailed consideration of this mechanism is out of scope of our chapter. So far we have discussed spin- polarized electric transport in homogeneous two-dimensional structures. The family of these effects can be extended by using inhomogeneous structures. For example, spin solar cell and spin photodiode (spin-voltaic effect), proposed in Ref. [136], were recently observed in GaAs-based $p$-$n$ junctions [137]. These effects occurred due to illumination of a $p$-$n$ junction with circularly polarized interband light resulting in spin polarization of a charge current. Circular polarization generates spin-polarized electrons and holes. Due to the fast relaxation of hole spin polarization in the bulk and the long spin lifetime of electrons, the photocurrent becomes spin polarized.

At last but not least, so far we discussed spin polarized electric transport in homogeneous two-dimensional structures. The family of these effects can be extended by using inhomogeneous structure. An example are the spin solar cell and spin-voltaic effect suggested in [136] recently observed in n-GaAlAs/p-GaInAs/p-GaAs junction [137], see Chapter 8, Volume 2. These effects effect occur due to uniform illumination of a $p$-$n$ junction with circularly polarized inter-band light resulting in spin polarization of a charge current. Circular polarization generates spin-polarized electrons and holes. Due to the fast relaxation of hole spin polarization in the bulk and the long spin lifetime of electrons, the photocurrent becomes spin polarized.

## 7.4  CONCLUDING REMARKS

We have given an overview of current-induced spin polarization in gyrotropic semiconductor nanostructures. Such a spin polarization as a response to a charge current may be classified as the inverse of the spin-galvanic effect, and is sometimes called a magneto-electrical effect or Edelstein (Rashba–Edelstein) effect. Apart from reviewing the experimental status of affairs, we have provided a detailed theoretical description of both effects in terms of a phenomenological model of spin-dependent relaxation processes, and an alternative theoretical approach based on the quasiclassical Boltzmann equation. Two microscopic mechanisms of this effect, both involving **k**-linear band spin splitting, are known so far: scattering mechanism based on Elliott–Yafet spin relaxation and precessional mechanism due to D'yakonov–Perel spin relaxation. We note that recently, another mechanism of the current-induced spin polarization [133] and spin-galvanic effect [134] which may dominate at room temperature has been proposed. It is based on spin-dependent asymmetry of electron scattering, and is out of the scope of the present chapter. The relative direction between electric current and nonequilibrium spin for all mechanisms is determined by the crystal symmetry. In particular, we have discussed the anisotropy of the inverse spin-galvanic effect in (001)-grown zinc-blende structure based QWs in the presence of spin-orbit interaction of both the Rashba and the Dresselhaus type. Combined with the theoretical achievements derived from the Boltzmann approach, the precise measurement of this anisotropy allows in principle the determination of the absolute values of the Rashba and the Dresselhaus spin-orbit coupling parameter. Moreover, these interactions can be used for the manipulation of the magnitude and direction of electron spins by changing the direction of current, thereby enabling a new degree of spin control. We focused here on the fundamental physics underlying the spin-dependent transport of carriers in semiconductors, which persists up to room temperature [38, 39, 45], and, therefore, may become useful in future semiconductor spintronics.

As of December 2018, the field still attracts a considerable amount of attention as reflected by the number of recent papers devoted to the problem of current induced spin polarization. The major theoretical efforts have been recently applied towards understanding the interfaces between ferromagnetic films [138, 139, 140, 141], leave alone a few Keldysh-based descriptions of the current-induced spin polarization [142, 143, 144] and a unified description of current-induced spin orientation, spin-galvanic, and related spin-Hall effects [7, 145, 146], see Chapter 4, Volume 2. The current induced spin polarization has also been found in a few novel materials including two-dimensional transition metal dichalcogenides [147], oxide interfaces [148], $p$-doped tellurium [149, 150, 151], and topological insulators [152].

## ACKNOWLEDGMENTS

We thank E.L. Ivchenko, V.V. Bel'kov, D. Weiss, L.E. Golub, and S.A. Tarasenko for helpful discussions. This work is supported by the DFG (SFB 689 and 1277) and the Elite Network of Bavaria (K-NW-2013-247). M.T. is supported by the Director's Senior Research Fellowship from CA2DM at NUS (NRF Singapore Medium Sized Centre Programme R-723-000-001-281).

## REFERENCES

1. R. H. Silsbee, Spin-orbit induced coupling of charge current and spin polarization, *J. Phys. Condens. Matter* **16**, R179 (2004).
2. E. I. Rashba, Spin currents, spin populations, and dielectric function of non-centrosymmetric semiconductors, *Phys. Rev. B* **70**, 161201(R) (2004).
3. E. I. Rashba, Spin dynamics and spin transport, *J. Supercond. Incorporating Novel Magn.* **18**, 137 (2005).
4. E. I. Rashba, Semiconductor spintronics: Progress and challenges, in *Future Trends in Microelectronics. Up to Nano Creek*, S. Luryi, J. M. Xu, and A. Zaslavsky (Eds.), Wiley-Interscience, Hoboken, pp. 28–40, 2007, arXiv:cond-mat/0611194.
5. R. Winkler, *Spin-Dependent Transport of Carriers in Semiconductors, in Handbook of Magnetism and Advanced Magnetic Materials*, vol. 5, H. Kronmuller and S. Parkin (Eds.), John Wiley & Sons, New York, 2007, arXiv:cond-mat/0605390.
6. I. Adagideli, M. Scheid, M. Wimmer, G. E. W. Bauer, and K. Richter, Extracting current-induced spins: Spin boundary conditions at narrow Hall contacts, *New J. Phys.* **9**, 382 (2007).
7. E. L. Ivchenko and S. D. Ganichev, *Spin Photogalvanics, in Spin Physics in Semiconductors* M. I. Dyakonov (Ed.), Springer, Berlin, pp. 245–277, 2008, second edition pp. 281–328 (2017).
8. Y. K. Kato and D. D. Awschalom, SPECIAL TOPICS, Advances in spintronics, electrical manipulation of spins in nonmagnetic semiconductors, *J. Phys. Soc. Jpn.* **77**, 031006 (2008).
9. S. D. Ganichev, Spin-galvanic effect and spin orientation by current in non-magnetic semiconductors – A review, *Int. J. Mod. Phys. B* **22**, 1 (2008).
10. S. D. Ganichev and L. E. Golub, Interplay of Rashba/Dresselhaus spin splittings probed by photogalvanic spectroscopy – A review, *Phys. Stat. Solidi B* **251**, 1801 (2014).
11. J. Sinova, S. O. Valenzuela, J. Wunderlich, C. H. Back, and T. Jungwirth, Spin Hall effects, *Rev. Mod. Phys.* **87**, 1213 (2015).
12. A. Manchon, H. C. Koo, J. Nitta, S. M. Frolov, and R. A. Duine, New perspectives for Rashba spin-rbit coupling, *Nat. Mater.* **14**, 871 (2015).
13. T. Jungwirth, X. Marti, P. Wadley, and J. Wunderlich, Antiferromagnetic spintronics, *Nat. Nanotechnol.* **11**, 231 (2016).
14. J. Schliemann, Persistent spin textures in semiconductor nanostructures, *Rev. Mod. Phys.* **89**, 011001 (2017).
15. L. D. Landau, E. M. Lifshits, and L. P. Pitaevskii, *Electrodynamics of Continuous Media*, vol. 8, Elsevier, Amsterdam, 1984.
16. J. F. Nye, *Physical Properties of Crystals: Their Representation by Tensors and Matrices*, Oxford University Press, Oxford, 1985.
17. V. M. Agranovich and V. L. Ginzburg, *Crystal Optics with Spatial Dispersion, and Ex-citons in Springer Series*, in Solid-State Sciences, vol. 42, Springer, Berlin, 1984.
18. V. A. Kizel', Yu. I. Krasilov, and V. I. Burkov, Experimental studies of gyrotropy of crystals, *Usp. Fiz. Nauk* **114**, 295 (1974) [*Sov. Phys. Usp.* **17**, 745 (1975)].
19. J. Jerphagnon and D. S. Chemla, Experimental studies of gyrotropy of crystals, *J. Chem. Phys.* **65**, 1522 (1976).
20. B. Koopmans, P. V. Santos, and M. Cardona, Optical activity in semiconductors: stress and confinement effects, *Phys. Stat. Sol.* (B) **205**, 419 (1998).
21. S. D. Ganichev, E. L. Ivchenko, V. V. Bel'kov et al., Spin-galvanic effect, *Nature* **417**, 153 (2002).

22. S. D. Ganichev and W. Prettl, Spin photocurrents in quantum wells, *J. Phys. Condens. Matter* **15**, R935 (2003).

23. S. D. Ganichev and W. Prettl, *Intense Terahertz Excitation of Semiconductors*, Oxford University Press, Oxford, 2006.

24. E. L. Ivchenko, *Optical Spectroscopy of Semiconductor Nanostructures*, Alpha Science International, Harrow, UK, 2005.

25. J. Schliemann, *Int. J. Mod. Phys. B* **20**, 1015 (2006).

26. H.-A. Engel, E. I. Rashba, and B. I. Halperin, in *Handbook of Magnetism and Advanced Magnetic Materials*, H. Kronmüller and S. Parkin (Eds.), John Wiley, 2007.

27. E. M. Hankiewicz and G. Vignale, Spin-Hall effect and spin-Coulomb drag in doped semiconductors, *J. Phys. Condens. Matter* **21**, 253202 (2009).

28. M. I. Dyakonov and V. I. Perel, Possibility of orienting spins with current, *Pis'ma Zh. Eksp. Teor. Fiz.* **13**, 657 (1971) [*JETP Lett.* **13**, 467 (1971)].

29. J. Fabian, A. Matos-Abiague, C. Ertler, P. Stano, and I. Žutić, Semiconductor spintronics, *Acta Phys. Slov.* **57**, 565 (2007), arXiv:cond-mat/0711.1461.

30. J. Sinova, and A. MacDonald, Theory of spin-orbit effects in semiconductors, in *Spintronics*, vol. 82, T. Dietl, D.D. Awschalom, M. Kaminska and H. Ohno, (Eds.), Semiconductors and Semimetals, Elsevier, Amsterdam, pp. 45–89, 2008.

31. M. I. Dyakonov and A. V. Khaetskii, Spin Hall effect, in *Spin Physics in Semiconductors*, M. I. Dyakonov (Ed.), Springer, Berlin, pp. 211–244, 2008, second edition pp. 241–280.

32. E. L. Ivchenko and G. E. Pikus, New photogalvanic effect in gyrotropic crystals, *Pis'ma Zh. Eksp. Teor. Fiz.* **27**, 640 (1978) [*JETP Lett.* **27**, 604 (1978)].

33. L. E. Vorob'ev, E. L. Ivchenko, G. E. Pikus, I. I. Farbstein, V. A. Shalygin, and A. V. Sturbin, Optical activity in tellurium induced by a current, *Pis'ma Zh. Eksp. Teor. Fiz.* **29**, 485 (1979) [*JETP Lett.* **29**, 441 (1979)].

34. A. G. Aronov, Yu. B. Lyanda-Geller, and G. E. Pikus, Spin polarization of electrons by an electric current, *Zh Eksp. Teor. Fiz.* **100**, 973 (1991) [*Sov. Phys. JETP* **73**, 537 (1991)].

35. F. T. Vas'ko and N. A. Prima, Spin splitting of the spectrum of two-dimensional electrons, *Fiz. Tverdogo Tela* **21**, 1734 (1979) [*Sov. Phys. Solid State* **21**, 994 (1979)].

36. A. G. Aronov and Yu. B. Lyanda-Geller, Nuclear electric resonance and orientation of carrier spins by an electric field, *Pis'ma Zh. Eksp. Teor. Fiz.* **50**, 398 (1989) [*JETP Lett.* **50**, 431 (1989)].

37. V. M. Edelstein, Spin polarization of conduction electrons induced by electric current in two-dimensional asymmetric electron systems, *Solid State Commun.* **73**, 233 (1990).

38. S. D. Ganichev, S. N. Danilov, P. Schneider et al., Can electric current orient spins in quantum wells? cond-mat/0403641 (2004).

39. S. D. Ganichev, S. N. Danilov, P. Schneider et al., Electric current induced spin orientation in quantum well structures, *J. Magn. Magn. Mater.* **300**, 127 (2006).

40. A. Yu. Silov, P. A. Blajnov, J. H. Wolter, R. Hey, K. H. Ploog, and N. S. Averkiev, Currentinduced spin polarization at a single heterojunction, *Appl. Phys. Lett.* **85**, 5929 (2004).

41. N. S. Averkiev, and A. Yu. Silov, Circular polarization of luminescence caused by the current in quantum wells, *Semiconductors* **39**, 1323 (2005).

42. V. Sih, R. C. Myers, Y. K. Kato, W. H. Lau, A. C. Gossard, and D. D. Awschalom, Spatial imaging of the spin Hall effect and current-induced polarization in two-dimensional electron gases, *Nat. Phys.* **1**, 31 (2005).

43. C. L. Yang, H. T. He, L. Ding et al., Spin photocurrent and converse spin polarization induced in a InGaAs/InAlAs two-dimensional electron gas, *Phys. Rev. Lett.* **96**, 186605 (2006).

44. Y. K. Kato, R. C. Myers, A. C. Gossard, and D. D. Awschalom, Current-induced spin polarization in strained semiconductors, *Phys. Rev. Lett.* **93**, 176601 (2004).

45. Y. K. Kato, R. C. Myers, A. C. Gossard, and D. D. Awschalom, Current-induced spin polarization in strained semiconductors, *Appl. Phys. Lett.* **87**, 022503 (2005).

46. N. P. Stern, S. Ghosh, G. Xiang, M. Zhu, N. Samarth, and D. D. Awschalom, Current-induced polarization and the spin Hall effect at room temperature, *Phys. Rev. Lett.* **97**, 126603 (2006).

47. S. Kuhlen, K. Schmalbuch, M. Hagedorn et al., Electric field-driven coherent spin reorientation of optically generated electron spin packets in InGaAs, *Phys. Rev. Lett.* **109**, 146603 (2012).

48. I. Stepanov, S. Kuhlen, M. Ersfeld, M. Lepsa, and B. Beschoten, All-electrical timeresolved spin generation and spin manipulation in n-InGaAs, *Appl. Phys. Lett.* **104**, 062406 (2014).

49. B. M. Norman, C. J. Trowbridge, D. D. Awschalom, and V. Sih, Current-induced spin polarization in anisotropic spin-orbit fields, *Phys. Rev. Lett.* **112**, 056601 (2014).

50. F. G. G. Hernandez, S. Ullah, G. J. Ferreira, N. M. Kawahala, G. M. Gusev, and A. K. Bakarov, Macroscopic transverse drift of long current-induced spin coherence in two-dimensional electron gases, *Phys. Rev. B* **94**, 045305 (2016).

51. L. E. Golub and E. L. Ivchenko, Spin orientation by electric current in (110) quantum wells, *Phys. Rev. B* **84**, 115303 (2011).

52. P. R. Hammar, B. R. Bennett, M. J. Yang, and M. Johnson, Observation of spin injection at a ferromagnet-semiconductor interface, *Phys. Rev. Lett.* **83**, 203 (1999).

53. P. R. Hammar, B. R. Bennett, M. J. Yang, and M. Johnson, A Reply to the Comment by F. G. Monzon, H. X. Tang, and M. L. Roukes, and also to the Comment by B. J. van Wees, *Phys. Rev. Lett.* **84**, 5024 (2000).

54. M. Johnson, Theory of spin-dependent transport in ferromagnet-semiconductor heterostructures, *Phys. Rev. B* **58**, 9635 (1998).

55. R. H. Silsbee, Theory of the detection of current induced spin polarization in a two-dimensional electron gas, *Phys. Rev. B* **63**, 155305 (2001).

56. F. G. Monzon, H. X. Tang, and M. L. Roukes, Magnetoelectronic phenomena at a ferromagnet-semiconductor interface, *Phys. Rev. Lett.* **84**, 5022 (2000).

57. B. J. van Wees, *Phys. Rev. Lett.* **84**, 5023 (2000).

58. T. D. Skinner, K. Olejník, L. K. Cunningham et al., Complementary spin-Hall and inverse spin-galvanic effect torques in a ferromagnet/semiconductor bilayer, *Nat. Commun.* **6**, 6730 (2015).

59. K. Olejník, V. Novák, J. Wunderlich, and T. Jungwirth, Electrical detection of magnetization reversal without auxiliary magnets, *Phys. Rev. B* **91**, 180402(R) (2015).

60. J. C. Rojas Sánchez, L. Viva, G. Desfonds et al., Spin-to-charge conversion using Rashba coupling at the interface between non-magnetic materials, *Nat. Commun.* **4**, 2944 (2013).

61. H. J. Zhang, S. Yamamoto, Y. Fukaya et al., Current-induced spin polarization on metal surfaces probed by spin-polarized positron beam, *Sci. Rep.* **4**, 4844 (2014).

62. W. Zhang, M. B. Jungfleisch, W. Jiang, J. E. Pearson, and A. Hoffmann, Spin pumping and inverse Rashba-Edelstein effect in NiFe/Ag/Bi and NiFe/Ag/Sb, *J. Appl. Phys.* **117**, 17C727 (2015).

63. S. Sangiao, J. M. De Teresa, L. Morellon, I. Lucas, M. C. Martinez-Velarte, and M. Viret, Control of the spin to charge conversion using the inverse Rashba-Edelstein effect, *Appl. Phys. Lett.* **106**, 172403 (2015).

64. A. Nomura, T. Tashiro, H. Nakayama, and K. Ando, Temperature dependence of inverse Rashba-Edelstein effect at metallic interface, *Appl. Phys. Lett.* **106**, 212403 (2015).

65. M. Isasa, M. C. Martínez-Velarte, E. Villamor et al., Origin of inverse Rashba-Edelstein effect detected at the Cu/Bi interface using lateral spin valves, *Phys. Rev. B* **93**, 014420 (2016).

66. Y. Shiomi, K. Nomura, Y. Kajiwara et al., Spin-electricity conversion induced by spin injection into topological insulators, *Phys. Rev. Lett.* **113**, 196601 (2014).

67. A. R. Mellnik, J. S. Lee, A. Richardella et al., Spin-transfer torque generated by a topological insulator, *Nature* **511**, 449 (2014).

68. M. G. Vavilov, Giant magneto-oscillations of electric-field-induced spin polarization in a two-dimensional electron gas, *Phys. Rev. B* **72**, 195327 (2005).

69. S. A. Tarasenko, Scattering induced spin orientation and spin currents in gyrotropic structures, *JETP Lett.* **84**, 199 (2006).

70. H.-A. Engel, E. I. Rashba, and B. I. Halperin, Out-of-plane spin polarization from in plane electric and magnetic fields, *Phys. Rev. Lett.* **98**, 036602 (2007).

71. M. Trushin and J. Schliemann, Anisotropic current-induced spin accumulation in the two-dimensional electron gas with spin-orbit coupling, *Phys. Rev. B* **75**, 155323 (2007).

72. O. E. Raichev, Frequency dependence of induced spin polarization and spin current in quantum wells, *Phys. Rev. B* **75**, 205340 (2007).

73. L. E. Golub and E. L. Ivchenko, Spin-dependent phenomena in semiconductors in strong electric fields, *New J. Phys.* **15**, 125003 (2013).

74. D. Culcer and R. Winkler, Generation of spin currents and spin densities in systems with reduced symmetry, *Phys. Rev. Lett.* **99**, 226601 (2007).

75. A. A. Burkov, A. S. Nunez, and A. H. MacDonald, Theory of spin-charge-coupled transport in a two-dimensional electron gas with Rashba spin-orbit interactions, *Phys. Rev. B* **70**, 155308 (2004).

76. O. Bleibaum, Spin diffusion equations for systems with Rashba spin-orbit interaction in an electric field, *Phys. Rev. B* **73**, 035322 (2006).

77. V. L. Korenev, Bulk electron spin polarization generated by the spin Hall current, *Phys. Rev. B* **74**, 041308 (2006).

78. P. Kleinert and V. V. Bryksin, Spin polarization in biased Rashba –Dresselhaus two-dimensional electron systems, *Phys. Rev. B* **76**, 205326 (2007).

79. M. Duckheim and D. Loss, Loss Mesoscopic fluctuations in the spin-electric susceptibility due to Rashba spin-orbit interaction, *Phys. Rev. Lett.* **101**, 226602 (2008).

80. P. Kleinert and V. V. Bryksin, Electric-field-induced long-lived spin excitations in two-dimensional spin-orbit coupled systems, *Phys. Rev. B* **79**, 045317 (2009).

81. M. Duckheim, D. Loss, M. Scheid, K. Richter, İ. Adagideli, and P. Jacquod, Spin accumulation in diffusive conductors with Rashba and Dresselhaus spin-orbit interaction, *Phys. Rev. B* **81**, 085303 (2010).

82. Ka Shen, G. Vignale, and R. Raimondi, Microscopic theory of the inverse Edelstein effect, *Phys. Rev. Lett.* **112**, 096601 (2014).

83. Ka Shen, R. Raimondi, and G. Vignale, Theory of coupled spin-charge transport due to spin-orbit interaction in inhomogeneous two-dimensional electron liquids, *Phys. Rev. B* **90**, 245302 (2014).

84. F. T. Vasko and O. E. Raichev, *Quantum Kinetic Theory and Application: Electrons, Photons, Phonons*, Springer, Berlin, pp. 662–663, 2005.

85. A. V. Chaplik, M. V. Entin, and L.I. Magarill, *Phys. E* **13**, 744 (2002).

86. J. Inoue, G. E. W. Bauer, and L. W. Molenkamp, Diffuse transport and spin accumulation in a Rashba two-dimensional electron gas, *Phys. Rev. B* **67**, 033104 (2003).

87. Y. J. Bao and S. Q. Shen, Electric-field-induced resonant spin polarization in a two-dimensional electron gas, *Phys. Rev. B* **76**, 045313 (2007).

88. R. Raimondi, C. Gorini, M. Dzierzawa, and P. Schwab, Current-induced spin polarization and the spin Hall effect: A quasiclassical approach, *Solid State Commun.* **144**, 524 (2007).

89. C. X. Liu, B. Zhou, S. Q. Shen, and B.-f. Zhu, Current-induced spin polarization in a two-dimensional hole gas, *Phys. Rev. B* **77**, 125345 (2008).

90. C. Gorini, P. Schwab, M. Dzierzawa, and R. Raimondi, Spin polarizations and spin Hall currents in a two-dimensional electron gas with magnetic impurities, *Phys. Rev. B* **78**, 125327 (2008).

91. R. Raimondi and P. Schwab, Tuning the spin hall effect in a two-dimensional electron gas, *Europhys. Lett.* **87**, 37008 (2009).

92. I. Adagideli, G. E. W. Bauer, and B. I. Halperin, Detection of current-induced spins by ferromagnetic contacts, *Phys. Rev. Lett.* **97**, 256601 (2006).

93. Y. Jiang and L. Hu, Kinetic magnetoelectric effect in a two-dimensional semiconductor strip due to boundary-confinement-induced spin-orbit coupling, *Phys. Rev. B* **74**, 075302 (2006).

94. J. H. Jiang, M. W. Wu, and Y. Zhou, Kinetics of spin coherence of electrons in n-type InAs quantum wells under intense terahertz laser fields, *Phys. Rev. B* **78**, 125309 (2008).

95. M. Pletyukhov and A. Shnirman, Spin density induced by electromagnetic waves in a two-dimensional electron gas with both Rashba and Dresselhaus spin-orbit coupling, *Phys. Rev. B* **79**, 033303 (2009).

96. M.-H. Liu, S.-H. Chen, and C.-R. Chang, Current-induced spin polarization in spin-orbit-coupled electron systems, *Phys. Rev. B* **78**, 165316 (2008).

97. D. Guerci, J. Borge, and R. Raimondi, Spin polarization induced by an electric field in the presence of weak localization effects, *Phys. E.* **75**, 370 (2016).

98. A. Zakharova, I. Lapushkin, K. Nilsson, S. T. Yen, and K. A. Chao, Spin polarization of an electron-hole gas in InAs/GaSb quantum wells under a dc current, *Phys. Rev. B* **73**, 125337 (2007).

99. I. Garate and A. H. MacDonald, Influence of a transport current on magnetic anisotropy in gyrotropic ferromagnets, *Phys. Rev. B* **80**, 134403 (2009).

100. L. E. Golub and E. L. Ivchenko, *Advances in Semiconductor Research: Physics of Nanosystems, Spintronics and Technological Applications*, D. Persano Adorno and S. Pokutnyi (Eds.), Nova Science Publishers, pp 93–104, 2014.

101. G. Vignale and I. V. Tokatly, Theory of the nonlinear Rashba-Edelstein effect: The clean electron gas limit, *Phys. Rev. B* **93**, 035310 (2016).

102. I. A. Kokurin and N. S. Averkiev, Orientation of electron spins by the current in a quasi-one-dimensional system, *JETP Lett.* **101**, 568 (2015).

103. M. P. Trushin and A. L. Chudnovskiy, A curved one-dimensional wire with Rashba coupling as a spin switch, *JETP Lett.* **83**, 318 (2006).

104. I. V. Tokatly, E. E. Krasovskii, and G. Vignale, Current-induced spin polarization at the surface of metallic films: A theorem and an ab-initio calculation, *Phys. Rev. B* **91**, 035403 (2015).

105. I. Garate and M. Franz, Inverse spin-galvanic effect in the interface between a topological insulator and a ferromagnet, *Phys. Rev. Lett.* **104**, 146802 (2010).

106. T. Yokoyama, J. Zang, and N. Nagaosa, Theoretical study of the dynamics of magnetization on the topological surface, *Phys. Rev. B* **81**, 241410(R) (2010).

107. T. Yokoyama, Current-induced magnetization reversal on the surface of a topological insulator, *Phys. Rev. B* **84**, 113407 (2011).

108. W. Luo, W. Y. Deng, H. Geng et al., Perfect inverse spin Hall effect and inverse Edelstein effect due to helical spin-momentum locking in topological surface states, *Phys. Rev. B* **93**, 115118 (2016).

109. H. Li, X. Wang, and A. Manchon, Valley-dependent spin-orbit torques in two-dimensional hexagonal crystals, *Phys. Rev. B* **93**, 035417 (2016).

110. A. Dyrdał, J. Barnaś, and V. K. Dugaev, Current-induced spin polarization in graphene due to Rashba spin-orbit interaction, *Phys. Rev. B* **89**, 075422 (2014).

111. J. Borge, C. Gorini, G. Vignale, and R. Raimondi, Spin Hall and Edelstein effects in metallic films: From two to three dimensions, *Phys. Rev. B* **89**, 245443 (2014).

112. L. S. Levitov, Y. V. Nazarov, and G. M. Eliashberg, Magnetoelectric effects in conductors with mirror isomer symmetry, *Zh Eksp. Teor. Fiz.* **88**, 229 (1985) [*Sov. Phys. JETP* **61**, 133 (1985)].

113. Y. A. Bychkov and E. I. Rashba, Properties of a 2D electron gas with lifted spectral degeneracy, *Pis'ma Zh. Eksp. Teor. Fiz.* **39**, 66 (1984) [*JETP Lett.* **39**, 78 (1984)].

114. M. I. D'yakonov and V. Yu. Kachorovskii, Spin relaxation of two-dimensional electrons in noncentrosymmetric semiconductors, *Fiz. Tekh. Poluprovodn.* **20**, 178 (1986) [*Sov. Phys. Semicond.* **20**, 110 (1986)].

115. N. S. Averkiev, L. E. Golub, and M. Willander, Spin relaxation anisotropy in two-dimensional semiconductor systems, *J. Phys. Condens. Matter* **14**, R271 (2002).

116. S. D. Ganichev, V. V. Bel'kov, L. E. Golub et al., Experimental separation of Rashba and Dresselhaus spin-splittings in semiconductor quantum wells, *Phys. Rev. Lett.* **92**, 256601 (2004).

117. I. I. Vosilyus and I. B. Levinson, Galvanomagnetic effects in strong electric fields during nonelastic electron scattering, *Sov. Phys. JETP* **25**, 672 (1967).

118. S. Giglberger, L. E. Golub, V. V. Bel'kov et al., Rashba and dresselhaus spin-splittings in semiconductor quantum wells measured by spin photocurrents, *Phys. Rev. B* **75** 035327 (2007).

119. L. Meier, G. Salis, E. Gini, I. Shorubalko, and K. Ensslin, Two-dimensional imaging of the spin-orbit effective magnetic field, *Phys. Rev. B* **77**, 035305 (2008).

120. L. E. Golub, Spin transport in heterostructures, *Phys.-Usp.* **55**, 814 (2012).

121. J. Luo, H. Munekata, F. F. Fang, and P. J. Stiles, Effects of inversion asymmetry on electron energy band structures in GaSb/InAs/GaSb quantum wells, *Phys. Rev. B* **41**, 7685 (1990).

122. W. Knap, C. Skierbiszewski, A. Zduniak et al., Weak antilocalization and spin precession in quantum wells, *Phys. Rev. B* **53**, 3912 (1996).

123. N. S. Averkiev and L. E. Golub, Giant spin relaxation anisotropy in zinc-blende heterostructures, *Phys. Rev. B* **60**, 15582 (1999).

124. S. A. Tarasenko and N. S. Averkiev, Interference of spin splittings in magneto-oscillation phenomena in two-dimensional systems, *Pis'ma Zh. Èksp. Teor. Fiz.* **75**, 669 (2002) [*JETP Lett.* **75**, 552 (2002)]; N. S. Averkiev, M. M. Glazov, and S. A. Tarasenko, *Solid State Commun.* **133**, 543 (2005).

125. J. Schliemann, J. C. Egues and D. Loss, Nonballistic spin-field-effect transistor, *Phys. Rev. Lett.* **90**, 146801 (2003).

126. R. Winkler, *Spin-Orbit Coupling Effects in Two-Dimensional Electron and Hole Systems*, vol. 191, Springer Tracts in Modern Physics, Springer, Berlin, 2003.

127. W. Zawadzki and P. Pfeffer, Spin splitting of subbands energies due to inversion asymmetry in semiconductor heterostructures, *Semicond. Sci. Technol.* **19**, R1 (2004).

128. I. Žutić, J. Fabian, and S. Das Sarma, Spintronics: Fundamentals and applications. *Rev. Mod. Phys.* **76**, 323 (2004).

129. V. V. Bel'kov, P. Olbrich, S. A. Tarasenko et al., Symmetry and spin dephasing in (110)-grown quantum wells, *Phys. Rev. Lett.* **100**, 176806 (2008).

130. V. Lechner, L. E. Golub, P. Olbrich et al., Tuning of structure inversion asymmetry by the δ-doping position in (001)-grown GaAs quantum wells, *Appl. Phys. Lett.* **94**, 242109 (2009).

131. M. Kohda, V. Lechner, Y. Kunihashi et al., Gate-controlled persistent spin helix state in InGaAs quantum wells, *Phys. Rev. B* **86**, R081306 (2012).

132. M. Trushin and J. Schliemann, Spin dynamics in rolled-up two-dimensional electron gase, *New J. Phys.* **9**, 346 (2007).

133. S. A. Tarasenko, Spin orientation of free carriers by dc and high-frequency electric field in quantum wells, *Phys. E* **40**, 1614 (2008).

134. L. E. Golub, *JETP Lett.* **85**, 393 (2007).

135. E. L. Ivchenko and S. A. Tarasenko, New mechanism of the spin-galvanic effect, *JETP* **99**, 379 (2004).

136. I. Žutić, J. Fabian, and S. Das Sarma, Spin-polarized transport in inhomogeneous magnetic semiconductors: Theory of magnetic/nonmagnetic *p–n* junctions, *Phys. Rev. Lett.* **88**, 066603 (2002); I. Žutić, J. Fabian, and S. Das Sarma, Spin injection through the depletion layer: A theory of spin-polarized *p–n* junctions and solar cells, *Phys. Rev. B* **64**, 121201 (2001).

137. B. Endres, M. Ciorga, M. Schmid, M. Utz, D. Bougeard, D. Weiss, G. Bayreuther, and C.H. Back, Demonstration of the spin solar cell and spin photodiode effect, *Nat. Commun.* **4**, 2068 (2013).

138. V. P. Amin and M. D. Stiles, Spin transport at interfaces with spin-orbit coupling: Formalism, *Phys. Rev. B* **94**, 104419 (2016).

139. V. P. Amin and M. D. Stiles, Spin transport at interfaces with spin-orbit coupling: Phenomenology, *Phys. Rev. B* **94**, 104420 (2016).

140. T. H. Dang, D. Q. To, E. Erina, T. L. Hoai Nguyen, V. I. Safarov, H. Jaffrès, H.-J. Drouhin, Theory of the anomalous tunnel Hall effect at ferromagnet-semiconductor junctions, *J. Magn. Magn. Mater.* **459**, 37 (2018).

141. S. Tölle, U. Eckern, and C. Gorini, Spin-charge coupled dynamics driven by a time-dependent magnetization, *Phys. Rev. B* **95**, 115404 (2017).

142. C. Gorini, A. M. Sheikhabadi, K. Shen, I. V. Tokatly, G. Vignale, and R. Raimondi, Theory of current-induced spin polarization in an electron gas, *Phys. Rev. B* **95**, 205424 (2017).

143. A. M. Sheikhabadi, I. Miatka, E. Y. Sherman, and R. Raimondi, Theory of the inverse spin galvanic effect in quantum wells, *Phys. Rev. B* **97**, 235412 (2018).

144. I. V. Tokatly, Usadel equation in the presence of intrinsic spin-orbit coupling: A unified theory of magnetoelectric effects in normal and superconducting systems, *Phys. Rev. B* **96**, 060502(R) (2017).

145. D. S. Smirnov and L. E. Golub, Electrical spin orientation, spin-galvanic, and spin-Hall effects in disordered two-dimensional systems, *Phys. Rev. Lett.* **118**, 116801 (2017).

146. A. V. Shumilin, D. S. Smirnov, and L. E. Golub, Spin-related phenomena in the two-dimensional hopping regime in magnetic field, *Phys. Rev. B* **98**, 155304 (2018).

147. X.-T. An, M. W.-Y. Tu, V. I. Fal'ko, and W. Yao, Realization of valley and spin pumps by scattering at nonmagnetic disorders, *Phys. Rev. Lett.* **118**, 096602 (2017).

148. G. Seibold, S. Caprara, M. Grilli, and R. Raimondi, Theory of the spin galvanic effect at oxide interfaces, *Phys. Rev. Lett.* **119**, 256801 (2017).

149. V. A. Shalygin, A. N. Sofronov, L. E. Vorob'ev, and I. I. Farbshtein, Current-induced spin polarization of holes in tellurium, *Phys. Solid State* **54**, 2362 (2012).

150. C. Sahin, J. Rou, J. Ma, and D. A. Pesin, Pancharatnam-Berry phase and kinetic magnetoelectric effect in trigonal tellurium, *Phys. Rev. B* **97**, 205206 (2018).

151. S. S. Tsirkin, P. A. Puente, and I. Souza, Gyrotropic effects in trigonal tellurium studied from first principles, *Phys. Rev. B* **97**, 035158 (2018).

152. J. Tian, S. Hong, I. Miotkowski, S. Datta, and Y. P. Chen, Observation of current-induced, long-lived persistent spin polarization in a topological insulator: A rechargeable spin battery, *Sci. Adv.* **3** (4), e1602531 (2017).

# 8

# Anomalous and Spin-Injection Hall Effects

**Jairo Sinova, Jörg Wunderlich, and Tomás Jungwirth**

## 8.1  INTRODUCTION

This Chapter provides a valuable introduction to a growing family of Hall effects and the importance of the spin-orbit coupling. Subsequent developments in Hall effects, spurred by the discovery of novel two-dimensional materials (Chapter 5, Volume 3) and topological insulators (Chapter 14, Volume 2), are discussed throughout this book. For example, recent advances in fundamental phenomena and possible applications of Hall effects in a wide class of materials are addressed in Chapter 9, Volume 1, Chapters 4 and 15, Volume 2, and Chapters 7–9, Volume 3. Throughout this book we have learned about many aspects of spin dynamics in both metal and semiconducting systems involving the diagonal transport of spin and charge and their interactions, in some cases, with the collective ferromagnetic-order parameter driven out of equilibrium. In this chapter, we venture into another aspect of spin-charge dynamics, in which an applied electric field induces a spin or a spin-charge response transverse to the field. These are the spin-dependent Hall effects.

In metallic spintronics, many studies have centered around the injection of spin and its manipulation via external magnetic fields, or nonequilibrium current effects such as spin-torque effects. Within these studies, the spin–orbit (SO) coupling in the system becomes an important source of spin dephasing and, in many instances, it is important to minimize it in order to enhance the observed effects. On the other hand, in semiconductor spintronics, SO coupling comes to the forefront as a source of spin manipulation and spin generation, becoming a necessary element in physical phenomena of the spin-dependent Hall effects, such as the anomalous Hall effect, spin Hall effect, inverse spin Hall effect, and spin-injection Hall effect. These spin-dependent Hall effects are, of course, governed by common mechanisms that we describe in this chapter.

We focus primarily on the anomalous Hall effect and the spin-injection Hall effect. The former is the primary spin-dependent Hall effect that has been studied for many decades, but it is one that continues to fascinate, and is at the center of many theoretical and experimental studies. The latter is one of the recent members of the spintronics Hall family [1], and its phenomenology combines many interesting physics observed in semiconducting spintronics, and can be a good playground to study in detail these mechanisms and other spin-dynamics effects controlled by particular choices of SO-coupling strengths that can be tuned experimentally. The other principal spin-dependent Hall effects, the spin Hall effect and the inverse spin Hall effect, have also been intensively studied over the past five years and their recent progress has been presented in several reviews. We will therefore only provide a simple and short qualitative description of these reviews. Other chapters that are pertinent to these physics are Chapter 5, Volume 1 on spin-injection and detection through non-local measurements by Mark Johnson, and Chapter 7, Volume 3 on spin Hall effects in metallic system by Takahashi and Maekawa.

This chapter is hence structured as follows: In the remainder of this chapter, we introduce the different spin-dependent Hall effects, which are most closely related. In the discussion of the anomalous Hall effect (Section 8.2.1), we introduce the basic mechanisms that give rise to these effects within the metallic systems (Section 8.2.2). In Section 8.2.5, we discuss the experimental phenomenology of the spin-injection Hall effect, and in Section 8.4, we discuss the theoretical aspects of this phenomenology, which incorporates anomalous Hall effect physics as well as persistent spin-helix physics. Throughout, we focus more on qualitative descriptions rather than detailed derivations, and refer to the list of references for further details.

## 8.2 SPIN-DEPENDENT HALL EFFECTS IN SPIN–ORBIT-COUPLED SYSTEMS

Within the family of spin-dependent Hall effects, the anomalous Hall effect, the spin Hall effect, the inverse spin Hall effect, and the spin-injection Hall effect are among the most closely related, as sketched in Figure 8.1. The more extensively studied of these is the anomalous Hall effect from which many of the mechanisms have been identified. Several extensive reviews have been written over the years on these subjects. The anomalous Hall effect has been recently reviewed by Nagaosa et al. [2], on which we base part of our chapter. Within the spin Hall effect physics, several recent reviews have appeared, such as the one by Schliemann [3], focusing on some aspect of the intrinsic spin Hall effect [4,5], and Engel et al. [6] and Hankiewicz and Vignale [7], focusing on the extrinsic spin Hall effect and quantum spin Hall effect. This particular field is evolving rapidly, generating many new fields of interest such as the spin Hall insulators following the studies of the quantum spin Hall effect [8–10], and extending even to spin thermodynamic Hall effects.

### 8.2.1 Anomalous Hall Effect

In 1879, Edwin Hall ran a current through a gold foil and discovered that a transverse voltage was induced when the film was exposed to a perpendicular magnetic field. The ratio of this Hall voltage to the current is the Hall resistivity. For paramagnetic materials, the Hall resistivity is proportional to the applied magnetic field, and Hall measurements can be shown

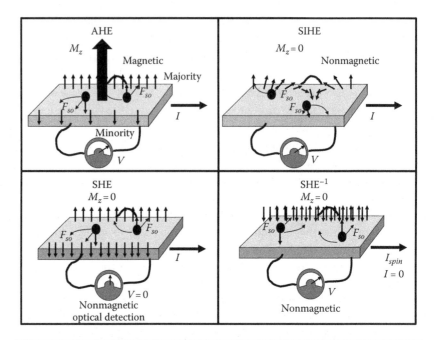

**FIGURE 8.1** Schematic of the family of spin-dependent Hall effects: Anomalous Hall effect (AHE), spin-injection Hall effect (SIHE), spin Hall effect (SHE), and inverse spin Hall effect (SHE$^{-1}$). A common thread is their existence at zero external magnetic field (i.e. no ordinary Hall effect present), with the key difference on the nature of the currents involved and the method of detection. Within the spin-injection Hall effect, the Hall response is overlaid on a new type of spin excitation termed persistent spin-helix.

to be equivalent to a measurement of the concentration of free carriers and determination whether they are holes or electrons. Magnetic films were found to exhibit both this ordinary Hall response and an extraordinary or anomalous Hall effect response that may persist at zero magnetic fields, which is proportional to the internal magnetization perpendicular to the plane:

$$R_{Hall} = R_o H + R_s M_z, \qquad (8.1)$$

where

$R_{Hall}$ is the Hall resistance

$R_o$ and $R_s$ are the ordinary and anomalous Hall coefficients

$M_z$ is the magnetization perpendicular to the plane

$H$ is the applied magnetic field

Based on this phenomenology, the anomalous Hall effect has been an important tool to measure electrically the magnetization in many systems. The key player in this effect is the SO coupling, without which one would not observe the effect in a homogeneously magnetized system. The role of the broken time-reversal symmetry is mainly to create an asymmetric spin population that leads to a charge Hall voltage, but would not produce a measurable anomalous Hall signal without the SO coupling.

The field of the anomalous Hall effect and its recent progress have been recently reviewed by Nagaosa et al. [2], focusing particularly on the new understanding of the underlying physics based on the topological nature of the effect. Let us first embark on a qualitative description of the different mechanisms.

## 8.2.2 Mechanism of the Spin-Dependent Hall Effects

The usual descriptions of the mechanisms of the anomalous Hall effect are shown in Figure 8.2. Many of these, at least as explained in many texts, are based on semiclassical descriptions of the different scattering and deflection mechanisms. Although we are going to retain the historical labeling of each contribution, in what follows, we recast their definition in a more physical and phenomenological way, anticipating their correct connection to the actual microscopic physics and to a precise experimental definition of each.

A natural classification of contributions to the anomalous Hall effect and other spin-dependent Hall effects, which is guided by experiment and by the microscopic theory of metals, is to separate them according to their dependence on the Bloch state transport lifetime $\tau$, which is equivalent to a $1/k_F l$ expansion, where $k_F$ is the Fermi wave-vector and $l$ is the mean free

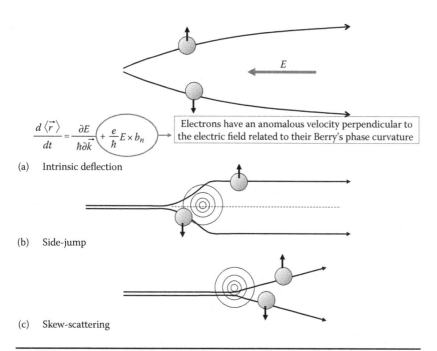

$$\frac{d\langle \vec{r} \rangle}{dt} = \frac{\partial E}{\hbar \partial \vec{k}} \left( + \frac{e}{\hbar} E \times b_n \right)$$

Electrons have an anomalous velocity perpendicular to the electric field related to their Berry's phase curvature

(a)    Intrinsic deflection

(b)    Side-jump

(c)    Skew-scattering

**FIGURE 8.2**    Schematic description of the mechanisms of the anomalous Hall effect. (a) Intrinsic deflection arises from interband coherence induced by an external electric field, giving rise to a velocity contribution perpendicular to the field direction. (b) Side-jump arises from the fact that electron velocity is deflected in opposite directions by the opposite electric fields experienced upon approaching and leaving an impurity. (c) Skew-scattering arises from asymmetric scattering due to the effective SO coupling of the electron or the impurity.

path. It is relatively easy to identify contributions to the anomalous Hall conductivity, $\sigma_{xy}^{AH}$, which vary as $\tau^1$ and as $\tau^0$. In experiment, a similar separation can sometimes be achieved by plotting $\sigma_{xy}^{AH}$ versus the longitudinal conductivity $\sigma_{xx} \propto \tau$, when $\tau$ is varied by altering disorder or varying temperature. More commonly (and equivalently), the Hall resistivity is separated into contributions proportional to $\rho_{xx}$ and $\rho_{xx}^2$. Since in almost all cases $\sigma_{xx} \gg \sigma_{xy}^{AH}$, it means that an anomalous Hall resistivity, $\rho_{xy}^{AH} = \sigma_{yx}^{AH}/\left(\sigma_{xx}^2 + \sigma_{xy}^{AH^2}\right)$, proportional to $\rho_{xx}^2$ implies a dominating mechanism that is scattering independent, that is, $\sigma_{xy}^{AH} \approx \rho_{yx}/\rho_{xx} \sim \sigma_{xx}^0$, and an anomalous Hall resistivity proportional to $\rho_{xx}$ implies a dominating mechanism that is very dependent on scattering, that is, $\sigma_{xy}^{AH} \approx \rho_{yx}^{AH}/\rho_{xx} \sim \sigma_{xx}^1$. The illustration of this scaling is shown in Figure 8.3a and b. This already highlights the counterintuitive nature of the anomalous Hall effect, which implies that a "dirtier" system will be dominated by scattering-independent mechanisms whereas a "cleaner" system will be dominated by impurity-scattering mechanisms.

This partitioning seemingly gives only two contributions to $\sigma_{xy}^{AH}$, one $\sim\tau$ and the other $\sim\tau^0$. The first contribution is known as the *skew-scattering* contribution, $\sigma_{xy}^{AH-skew}$. Note that in this parsing of anomalous Hall effect contributions, it is the dependence on $\tau$ (or $\sigma_{xx}$) that defines it, not a particular mechanism linked to a microscopic or semiclassical theory. The second contribution proportional to $\tau^0$ (or independent of $\sigma_{xx}$) is further separated into two parts: *intrinsic* and *side-jump*. It has long been believed that this partitioning is only a theoretical exercise. However, although these two contributions cannot be separated experimentally by dc measurements, they *can* be separated experimentally by defining the intrinsic contribution, $\sigma_{xy}^{AH-int}$, as the extrapolation of the ac-interband Hall conductivity to zero frequency in the limit of $\tau \to \infty$, with $1/\tau \to 0$ faster than $\omega \to 0$. This then leaves a unique definition for the third and last contribution, termed side-jump, as $\sigma_{xy}^{AH-sj} \equiv \sigma_{xy}^{AH} - \sigma_{xy}^{AH-skew} - \sigma_{xy}^{AH-int}$. This possible experimental extraction of $\sigma_{xy}^{AH-sj}$ and $\sigma_{xy}^{AH-int}$ is illustrated in Figure 8.3c.

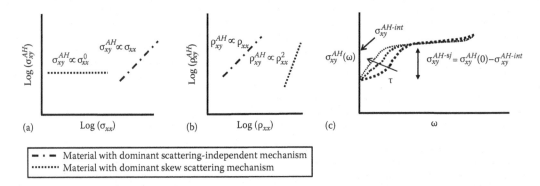

**FIGURE 8.3** Scaling of anomalous Hall conductivity (a) and resistivity (b) versus diagonal conductivity and resistivity, respectively, for a material whose anomalous Hall effect is dominated by skew scattering (dot-dashes) and for a material dominated by scattering-independent mechanisms. (c) Illustration of the scaling of ac-anomalous Hall conductivity for samples with decreasing impurity concentration (increasing $\tau$) to determine $\sigma_{xy}^{AH-int}$ and, from this, obtain $\sigma_{xy}^{AH-sj}$.

We examine these three contributions in the following. It is important to note that the above definitions have not relied on identifications of semi-classical processes such as side-jump scattering [11] or skew scattering from asymmetric contributions to the semiclassical scattering rates [12] identified in earlier theories. Not surprisingly, the contributions defined earlier contain these semiclassical processes. However, it is now understood that other contributions are present in the fully generalized semiclassical theory, which were not precisely identified previously, and are necessary to be fully consistent with microscopic theories [2, 13, 14].

## 8.2.2.1 Intrinsic Contribution, $\sigma_{xy}^{AH\text{-}int}$

We have defined the intrinsic contribution microscopically as the *dc* limit of the interband *ac* conductivity, a quantity that is not zero in ferromagnets when the SO-coupling interactions are included. There is, however, a direct link to semiclassical theory in which the induced interband coherence is captured by a momentum space Berry phase–related contribution to the anomalous velocity. We show this equivalence in the following.

Historically, this contribution to the anomalous Hall effect was first derived by Karplus and Luttinger [15] but its topological nature was not fully appreciated until recently [16–18]. The concept of intrinsic anomalous Hall effect was emphasized by Jungwirth et al. [17], motivated by the experimental importance of the anomalous Hall effect in ferromagnetic semiconductors and also by the earlier analysis of the relationship between momentum space Berry phases and anomalous velocities in semiclassical transport theory by Niu and coworkers [18, 19]. The frequency-dependent interband Hall conductivity, which reduces to the intrinsic anomalous Hall conductivity in the *dc* limit, had been evaluated earlier for a number of materials by Mainkar et al. [20] and by Guo and Ebert [21].

The intrinsic contribution to the conductivity is dependent only on the band structure of the perfect crystal. It can be calculated directly from the simple Kubo formula for the Hall conductivity for an ideal lattice, given the eigenstates $|n, \mathbf{k}\rangle$ and eigenvalues $\varepsilon_n(\mathbf{k})$ of a Bloch Hamiltonian $H$:

$$\sigma_{ij}^{AH\text{-}int} = e^2\hbar \sum_{n \neq n'} \int \frac{d\mathbf{k}}{(2\pi)^d} \left[ f(\varepsilon_n(\mathbf{k})) - f(\varepsilon_{n'}(\mathbf{k})) \right]$$
$$\times \frac{\text{Im}\left[ \langle n,\mathbf{k}|v_i(\mathbf{k})|n',\mathbf{k}\rangle\langle n',\mathbf{k}|v_j(\mathbf{k})|n,\mathbf{k}\rangle \right]}{(\varepsilon_n(\mathbf{k}) - \varepsilon_{n'}(\mathbf{k}))^2}, \tag{8.2}$$

and the velocity operator is defined by

$$v(\mathbf{k}) = \frac{1}{i\hbar}[\mathbf{r}, H(\mathbf{k})] = \frac{1}{\hbar}\nabla_k H(\mathbf{k}), \tag{8.3}$$

where $H$ is the $\mathbf{k}$-dependent Hamiltonian for the periodic part of the Bloch functions. Note the restriction $n \neq n'$ in Equation 8.2.

What makes this contribution quite unique is that it is proportional to the integration over the Fermi sea of the Berry's curvature of each occupied band, or equivalently [22, 23] to the integral of Berry phases over cuts of Fermi surface segments. This result can be derived by noting that

$$\left\langle n, \mathbf{k} \middle| \nabla_k \middle| n', \mathbf{k} \right\rangle = \frac{\left\langle n, \mathbf{k} \middle| \nabla_k H(\mathbf{k}) \middle| n', \mathbf{k} \right\rangle}{\varepsilon_{n'}(\mathbf{k}) - \varepsilon_n(\mathbf{k})}. \tag{8.4}$$

Using this expression, Equation 8.2 reduces to

$$\sigma_{ij}^{AH\text{-}int} = -\varepsilon_{ijl} \frac{e^2}{\hbar} \sum_n \int \frac{d\mathbf{k}}{(2\pi)^d} f(\varepsilon_n(\mathbf{k})) b_n^l(\mathbf{k}), \tag{8.5}$$

where:

$\varepsilon_{ijl}$   is the antisymmetric tensor
$\mathbf{a}_n(\mathbf{k})$ is the Berry-phase connection $\mathbf{a}_n(\mathbf{k}) = i\langle n, \mathbf{k} | \nabla_k | n, \mathbf{k} \rangle$,
$\mathbf{b}_n(\mathbf{k})$ is the Berry-phase curvature

$$\mathbf{b}_n(\mathbf{k}) = \nabla_k \times \mathbf{a}_n(\mathbf{k}) \tag{8.6}$$

corresponding to the states $\{|n, \mathbf{k}\rangle\}$.

This same linear response contribution to the anomalous Hall effect conductivity can be obtained from the semiclassical theory of wave-packet dynamics [18, 19, 24]. It can be shown that the wave-packet group velocity has an additional contribution in the presence of an electric field: $\dot{r}_c = \partial E_n(\mathbf{k})/\hbar\partial \mathbf{k} - (E/\hbar) \times \mathbf{b}_n(\mathbf{k})$. The intrinsic Hall conductivity formula, Equation 8.5, is obtained simply by summing the second (anomalous) term over all occupied states.

One of the motivations for identifying the intrinsic contribution $\sigma_{xy}^{AH\text{-}int}$ is that it can be evaluated accurately, even for relatively complex materials, using first-principles electronic structure theory techniques. In many materials that have strongly SO-coupled bands, the intrinsic contribution seems to dominate the anomalous Hall effect.

### 8.2.2.2 Skew-Scattering Contribution, $\sigma_{xy}^{AH\text{-}skew}$

This contribution is simply the component of $\sigma_{xy}^{AH}$ proportional to the Bloch state transport lifetime. It therefore tends to dominate in nearly perfect crystals. It is the only contribution to the anomalous Hall effect that appears within the confines of traditional Boltzmann transport theory in which interband coherence effects are completely neglected. Skew-scattering is due to chiral features that appear in the disorder scattering of SO-coupled ferromagnets [12, 25].

Treatments of semiclassical Boltzmann transport theory found in textbooks often appeal to the principle of detailed balance, which states that the transition probability $W_{n \to m}$ from $n$ to $m$ is identical to the transition probability in the opposite direction ($W_{m \to n}$). Although these two transition probabilities are identical in a Fermi's golden rule approximation, since $W_{n \to n'} = (2\pi/\hbar) \left| \langle n | V | n' \rangle \right|^2 \delta(E_n - E_{n'})$, where $V$ is the perturbation inducing the transition, detailed balance in this microscopic sense is not generic. In the presence of SO coupling, either in the Hamiltonian of the perfect crystal or in the disorder Hamiltonian, a transition that is right-handed with respect to the magnetization direction has a different transition probability than the corresponding left-handed transition. When the transition rates are evaluated perturbatively, asymmetric chiral contributions appear first at third order. In simple models, the asymmetric chiral contribution to the transition probability is often assumed to have the form [12, 25]

$$W_{kk'}^{A} = -\tau_{A}^{-1} k \times k' \cdot M_{s}. \tag{8.7}$$

When this asymmetry is inserted into the Boltzmann equation, it leads to a current proportional to the longitudinal current driven by $E$ and perpendicular to both $E$ and $M_s$. When this mechanism dominates, both the anomalous Hall conductivity $\sigma_{xy}^{AH}$ and the conductivity $\sigma_{xx}$ are approximately proportional to the transport lifetime $\tau$, and the anomalous Hall resistivity $\rho^{AH\text{-}skew} = \sigma^{AH\text{-}skew}\rho_{xx}^{2}$ is therefore proportional to the longitudinal resistivity $\rho_{xx}$.

There are several specific mechanisms for skew scattering. Evaluation of the skew-scattering contribution to the Hall conductivity or resistivity requires simply that the conventional linearized Boltzmann equation be solved using a collision term with accurate transition probabilities, since these will generically include a chiral contribution. We emphasize that skew-scattering contributions to $\sigma_{xy}^{AH}$ are present not only because of SO coupling in the disorder Hamiltonian, but also because of SO coupling in the perfect crystal Hamiltonian combined with purely scalar disorder. Either source of skew scattering could dominate $\sigma_{xy}^{AH\text{-}skew}$, depending on the host material and also on the type of impurities, and, as we will see, it dominates the physics of the present experiments in spin-injection Hall effect.

One important thing to note is that we have not defined the skew-scattering contribution to the anomalous Hall effect as the sum of *all* the contributions arising from the asymmetric scattering rate present in the collision term of the Boltzmann transport equation. We know from microscopic theory that this asymmetry also makes an anomalous Hall effect contribution of order $\tau^{0}$. There exists a contribution from this asymmetry, which is actually present in the microscopic theory treatment associated with the so-called ladder diagram corrections to the conductivity, and therefore of order $\tau^{0}$. In our experimentally practical parsing of anomalous Hall effect contributions, we do not associate this contribution with skew scattering but place it under the umbrella of side-jump scattering, even though it does not physically originate from any sidestep type of scattering.

### 8.2.2.3 Side-Jump Contribution $\sigma_{xy}^{AH\text{-}sj}$

Given the sharp definition we have provided for the intrinsic and skew-scattering contributions to the anomalous Hall effect conductivity, the equation

$$\sigma_{xy}^{AH} = \sigma_{xy}^{AH\text{-}int} + \sigma_{xy}^{AH\text{-}skew} + \sigma_{xy}^{AH\text{-}sj} \tag{8.8}$$

defines the side-jump contribution as the difference between the full Hall conductivity and the two simpler contributions. In using the term side-jump for the remaining contribution, we are appealing to the historically established taxonomy. Establishing this connection mathematically has been the most controversial aspect of anomalous Hall effect theory, and the one that has taken the longest to clarify from a theory point of view. Although this classification of Hall conductivity contributions is often useful, it is not generically true that the only correction to the intrinsic and skew contributions can be physically identified with the side-jump process defined as in the earlier studies of the anomalous Hall effect [26].

The basic semiclassical argument for a side-jump contribution can be stated straightforwardly: when considering the scattering of a Gaussian wave-packet from a spherical impurity with SO interaction ($H_{SO} = (1/2m^2c^2)$ $(r^{-1}\partial V/\partial r)S_z L_z$), a wave-packet with incident wave-vector $\mathbf{k}$ will suffer a displacement transverse to $\mathbf{k}$ equal to $(1/6)k\hbar^2/m^2c^2$. This type of contribution was first noticed, but discarded, by Smit [25] and reintroduced by Berger [26], who argued that it was the key contribution to the anomalous Hall effect. This kind of mechanism clearly lies outside the bounds of traditional Boltzmann transport theory in which only the probabilities of transitions between Bloch states appear, and not microscopic details of the scattering processes. This contribution to the conductivity ends up being independent of $\tau$ and therefore contributes to the anomalous Hall effect at the same order as the intrinsic contribution in an expansion in powers of scattering rate. The separation between intrinsic and side-jump contributions, which cannot be distinguished by their dependence on $\tau$, has been perhaps the most argued aspect of anomalous Hall effect theory [2].

Side-jump and intrinsic contributions have different dependences on more specific system parameters, especially in systems with complex band structures [14]. Some of the initial controversy that surrounded side-jump theories was associated with the physical meaning ascribed to quantities that were plainly gauge dependent, like the Berry's connection, which in early theories is typically identified as the definition of the side step upon scattering. Studies of simple models, for example, models of semiconductor conduction bands, also gave results in which the side-jump contribution seemed to be the same size, but opposite in sign compared to the intrinsic contribution [27]. We now understand [13] that these cancelations are unlikely, except in models with a very simple band structure, for example, one with a constant Berry's curvature. It is only by a careful comparison between fully microscopic linear response calculations, such as Keldysh (nonequilibrium Green's function) or Kubo formalisms, and the systematically developed semiclassical theory that the specific contribution due to the side-jump mechanism can be separately identified with confidence [13].

A practical approach, which is followed at present for materials in which $\sigma_{xy}^{AH}$ seems to be independent of $\sigma_{xx}$, is to first calculate the intrinsic contribution to the anomalous Hall effect. If this explains the observation (and it appears that it usually does), then it is deemed that the intrinsic mechanism dominates. If not, we can take some comfort from understanding on the basis of simple model results, that there can be other contributions to $\sigma_{xy}^{AH}$, which are also independent of $\sigma_{xx}$ and can for the most part be identified with the side-jump mechanism. It seems extremely challenging, if not impossible, to develop a predictive theory for these contributions, partly because they require many higher-order terms in the perturbation theory that must be summed, but more fundamentally because they depend sensitively on details of the disorder in a particular material, which are normally unknown and difficult to model accurately.

## 8.2.3 RECENT DEVELOPMENTS ON THE ANOMALOUS HALL EFFECT

The earlier description of the different mechanisms of the anomalous Hall effect in ferromagnetic metals has been refined over the past few years, with some of the recent research focused on unifying the different theories that disagreed. We finalize this section by pointing out some of the key recent developments, which are important to know when discussing the anomalous Hall effect:

1. When the anomalous Hall conductivity, $\sigma_{xy}^{AH}$, is independent of $\sigma_{xx}$, the anomalous Hall effect can often be understood in terms of the geometric concepts of Berry phase and Berry curvature in momentum space. This anomalous Hall effect mechanism is responsible for the intrinsic anomalous Hall effect (see following text). In this regime, the anomalous Hall current can be thought of as the unquantized version of the quantum Hall effect.

2. Three broad regimes have been identified when surveying a large body of experimental data for diverse materials [2]: (i) A high-conductivity regime ($\sigma_{xx} > 10^6$ ($\Omega$ cm)$^{-1}$) in which a linear contribution to $\sigma_{xy}^{AH} \sim \sigma_{xx}$ due to skew scattering dominates $\sigma_{xy}^{AH}$. In this regime, the normal Hall conductivity contribution can be significant and even dominate $\sigma_{xy}$. (ii) An intrinsic or scattering-independent regime in which $\sigma_{xy}^{AH}$ is roughly independent of $\sigma_{xx}$ ($10^4$ ($\Omega$ cm)$^{-1} < \sigma_{xx} < 10^6$ ($\Omega$ cm)$^{-1}$). (iii) A bad-metal or insulating regime ($\sigma_{xx} < 10^4$ ($\Omega$ cm)$^{-1}$) in which $\sigma_{xy}^{AH}$ decreases with decreasing $\sigma_{xx}$ at a rate faster than linear. This regime remains still a challenge, theoretically.

3. The band structure of the ferromagnetic materials plays a key role in these systems. In particular, the band (anti-)crossings near the Fermi energy has been identified using first-principles Berry curvature calculations as a mechanism that can lead to a large intrinsic anomalous Hall effect. The effect of scattering on these crossings is still not well understood.

4. A semiclassical treatment by a generalized Boltzmann equation taking into account the Berry curvature and coherent interband mixing effects due to band structure and disorder has been formulated.

This theory provides a clearer physical picture of the anomalous Hall effect than early theories by identifying correctly all the semiclassically defined mechanisms [13, 14]. This generalized semiclassical picture has been verified by comparison with controlled microscopic linear response treatments for identical models.

### 8.2.4 SPIN HALL EFFECT

A natural progression of the physics of the anomalous Hall effect to systems with time-reversal symmetry and with SO coupling is the spin Hall effect. In describing the different mechanisms giving rise to anomalous Hall effect, the exchange field is only important in creating a spin population imbalance that will give rise to a finite voltage that one can measure. Hence, since the mechanisms that give rise to the anomalous Hall effect are also present in these systems, the "spin" separation will supposedly take place, but it will not be directly measurable through a voltage perpendicular to the flow of the current in the sample, since there is no exchange field that induces the population asymmetry between spin-up and spin-down states. However, consistent with the global time-reversal symmetry, a spin current will be generated perpendicular to an applied charge current, which in turn should lead to spin accumulations with opposite magnetization at the edges. This is the spin Hall effect first predicted over three decades ago by simply invoking the phenomenology of the anomalous Hall effect in ferromagnets and applying it to semiconductors [28].

Motivated by studies of intrinsic anomalous Hall effect in strongly SO-coupled ferromagnetic materials, where scattering contributions do not seem to dominate, the possibility of an intrinsic spin Hall effect was put forward by Murakami et al. [4] and by Sinova et al. [5], with scattering playing a minor role. This proposal generated an extensive theoretical debate motivated by its potential as a spin-injection tool in low dissipative devices. The interest in the spin Hall effect has also been dramatically enhanced by experiments by two groups reporting the first observations of the spin Hall effect in n-doped semiconductors [29, 30] and in 2D hole gases (2DHG) [31]. These observations, based on optical experiments, have been followed by metal-based experiments in which both the spin Hall effect and the inverse spin Hall effect have been observed [32–36], and other experiments in semiconductors at other regimes, even at room temperature [37–40].

In the case of the inverse spin Hall effect, it is a spin current that generates a charge current in the so-called non-local measurements. These spin currents can be utilized in the spin-torque experiments and are being very actively studied. Given the extent of detail and still unsettled debate in this field, and the several reviews already mentioned in the introduction, we will not go further into the experiment and theory of spin Hall effect and inverse spin Hall effect. Instead, we will focus on the new phenomenology of the spin-injection Hall effect that combines several effects, is at present well understood in the diffusive regime, and is perhaps the most promising from a technology standpoint.

## 8.2.5 SPIN-INJECTION HALL EFFECT

The spin-injection Hall effect was observed in a photovoltaic cell device that opens direct application potential in opto-spintronics, for example, as a solid-state polarimeter. Importantly, the spin-injection Hall effect can be also used to measure spin dynamics of the electrons propagating in the semiconducting channel, in particular the spin-helix effect. The spin-injection Hall effect can, therefore, be used as an efficient spin-detection tool for basic studies of SO coupling phenomena in semiconductors and can also be directly incorporated in a number of spintronic devices, including the spin-field-effect transistor of the Datta–Das type [63].

The electrical detection of spin-polarized transport in semiconductors is a key requisite for the possible incorporation of spin in semiconductor microelectronics. In the spin-injection Hall effect, a polarized current is injected (optically in the present case) and its polarization can be detected by transverse electrical signals, via the anomalous Hall effect, directly along the semiconducting channel. One key practical feature is that this is achieved without disturbing the spin-polarized current or using magnetic elements, thus opening the door for possible spin-based sequential logic schemes in logic circuits.

Spin-polarized transport phenomena in semiconductors have been studied by a range of techniques used in ferromagnets and novel approaches developed specifically for semiconductor spintronics. The techniques include magneto-optical scanning probes [29, 37, 41–43], spin-polarized electroluminescence [31, 44–47], and magnetoelectric measurements utilizing spin-valve effects, magnetization dependence of nonequilibrium chemical potentials, and SO coupling phenomena [32, 42, 43, 48–54]. Two of these spintronic research developments have been particularly important for the observation of the spin-injection Hall effect. First, it has been demonstrated that electrons carrying electrical current in a ferromagnet align their spins with the local direction of magnetization, and that the resulting electrical signals due to the anomalous Hall effect can be used as an alternative to magneto-optical scanning probes to measure the local magnetization [55, 56]. The second is the observation of the spin Hall effect that verified that the physics of the anomalous Hall effect translate directly to semiconducting systems without broken time-reversal symmetry, that is, the spin Hall effect. Combining these two facts, one can surmise that injecting spin-polarized electrical currents into nonmagnetic semiconductors should also generate a Hall effect, which, as long as the spins of the charge carriers remain coherent, yields transverse charge accumulation and is therefore detectable electrically. In Wunderlich et al. [1], the device diode is operated in the reverse regime as a photocell in order to inject spin-polarized electrical currents into the semiconductor from a spatially confined region.

The effect, as discussed in the theory section, is well understood by microscopic calculations that reflect spin dynamics induced by an internal SO field. The spin-injection Hall effect is observed up to high temperatures

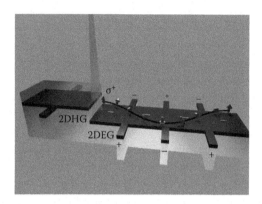

**FIGURE 8.4**   Sketch of the spin Hall effect device. Polarized light injects spin-polarized electron–hole pairs. The coplanar p–n junction [31] is connected in reverse bias, and the electrons, moving down the semiconductor channel in the $(1\bar{1}0)$ direction, experience a precessing motion in their spin, which is recorded via Hall probes. (After Wunderlich, J. et al., *Nat. Phys.* 5, 675, 2009. With permission.)

and the devices can also be thought of as a realization of a nonmagnetic spin-photovoltaic polarimeter, which directly converts polarization of light into transverse voltage signals (Figure 8.4).

## 8.3  SPIN-INJECTION HALL EFFECT: EXPERIMENT

In this section, we first describe the experimental findings of Wunderlich et al. [1] and follow it up with the theoretical analysis in the subsequent and final section of this chapter.

Figure 8.5a shows lateral micrographs of the planar 2D electron-hole gas (2DEG-2DHG) photodiodes with the p-region and n-region patterned into unmasked 1 μm wide Hall bars (similar experiments were performed in masked samples) in the $(1\bar{1}0)$ direction. The effective width of each Hall contact is 50–100 nm and separation between two Hall crosses is 2 μm. The sketch of the device and its experimental setup is also illustrated in Figure 8.1. As in the spin Hall effect experiments [31], the semiconductor wafer consists of a modulation p-doped AlGaAs/GaAs heterojunction on top of the structure separated by 90 nm of intrinsic GaAs from an n-doped AlGaAs/GaAs heterojunction underneath. Without etching, the top heterojunction in the wafer is populated by the 2DHG while the 2DEG at the bottom heterojunction is depleted. The n-side of the coplanar p–n junction is formed by removing the p-doped surface layer from a part of the wafer through wet etching. The removal of this top heterojunction results in populating the 2DEG. At zero or reverse bias, the device is not conductive in dark, due to charge depletion of the p–n junction. Counterpropagating electron and hole currents can be generated by illumination at sub-gap wavelengths. Because of the optical selection rules, the spin-polarization of injected electrons and holes is determined by the sense and degree of the circular polarization of vertically incident light. The optical spin-injection area is controlled by bias-dependent p–n junction depletion and, additionally, by

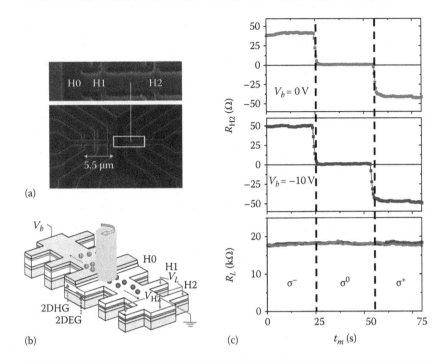

(a)

(b)

(c)

**FIGURE 8.5**   Devices, schematics of the experiment, and basic observation of the spin-injection Hall effect. (a) Micrograph of the coplanar p–n junction device. (b) Schematic diagram of the wafer structure of the 2DEG-2DHG p–n diode and of the spin-injection Hall effect measurement setup. (c) Steady-state spin-injection Hall effect signals changing sign for opposite helicities ($\sigma^-$ and $\sigma^+$) of the incident light beam, that is, for opposite spin-polarizations of the injected electrons. $R_{H2}$ is the second Hall bar resistance and $R_L$ is the longitudinal resistance. For linearly polarized light ($\sigma^0$), the injected electron current is spin-unpolarized and the spin-injection Hall effect vanishes. Measurements were performed at the 2DEG Hall cross H2 at a temperature of 4 K. (After Wunderlich, J. et al., *Nat. Phys.* 5, 675, 2009. With permission.)

the position and focus of the laser spot, or by including metallic masks on top of the Hall bars [1].

Figure 8.5c shows the measurements at the Hall cross bar H2 in the n-channel at 4 K, laser wavelength of 850 nm, and for 0 and –10 V reverse bias with the laser spot fixed on top of the p–n junction and focused on approximately 1 μm in diameter. While the longitudinal resistance $R_L$ is insensitive to the polarization, the transverse signal $R_H$ is observed only for polarized spin-injection into the electron channel and reverses sign upon reversing the polarization. With the laser spot focused on the p–n junction, the transverse voltage is only weakly bias dependent. Note that the large signals of tens of microvolts, corresponding to transverse resistances of tens of ohms, are detected outside the spin-injection area at a Hall cross separated by 3.5 μm from the p–n junction.

Figure 8.6a shows the simultaneous electrical measurements at Hall cross H0, which is wider and partially overlaps with the injection area, and at remote Hall crosses H1, H2, and H3. To confirm that the transverse signals

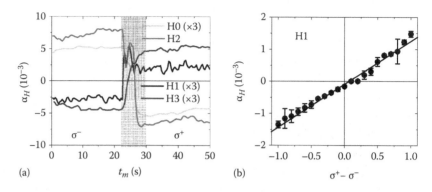

**FIGURE 8.6**    (a) Spin-injection Hall effect Hall probe signals are plotted for Hall crosses H0H3 in the 2DEG channel. Gray region corresponds to the manual rotation of the $\lambda/2$ wave plate, which changes the helicity of the incident light and, therefore, the spin-polarization of injected electrons. (b) Hall angles measured at n-channel Hall crosses H1 (similar results with opposite sign occur for H2). The spin-injection Hall effect angles are linear in the degree of polarization, hence behaving phenomenologically, as in the anomalous Hall effect. The laser spot is focused on the p–n junction and the bias voltage, laser wavelength, and measurement temperature are also the same as in Figure 8.5. (After Wunderlich, J. et al., *Nat. Phys.* 5, 675, 2009. With permission.)

do not result from spurious effects but originate from the polarized spin-injection, the helicity of the incident light is reversed by manually rotating a $\lambda/2$ wave plate by 45°. The signals are recalculated to Hall angles, $\alpha_H = R_H/R_L$, whose magnitude $10^{-3}$–$10^{-2}$ is comparable to the anomalous Hall effect in conventional metal ferromagnets. The spatial variation of the sign of the Hall signals observed suggests a modulated spin dynamics in the length scale of microns. The scattering mean free path in this system is on the order of ~10–100 nm and the typical SO-coupling length to ~1 μm. The first length scale determines the onset of the anomalous Hall effect–like transverse charge accumulation due to skew scattering off SO-coupled impurity potentials (see theory section). The second length scale governs the spin precession about the internal SO field, which in this configuration tends to point in-plane and perpendicular to the electron momentum. In quasi 1D channels, electrons injected with out-of-plane spins would precess coherently about this internal SO field, but in wider channels (diffusive transport), the spin-coherence length is expected to be much shorter. As we will see in the next section, the coherence can also exceed micrometer scales in wider channels due to the additional Dresselhaus SO field originating from inversion asymmetry of GaAs. In Figure 8.6b, the spin-injection Hall effect as a function of the degree of polarization at Hall cross H1 is shown. The linear dependence of the Hall angle on the polarization of injected electrons is analogous to the linear dependence on magnetization of the anomalous Hall effect in a ferromagnet.

Within the experiment, other important checks on the Hall symmetries were made in order to discard external effects for the observed phenomenology. Masked samples (only injection area open) show the same

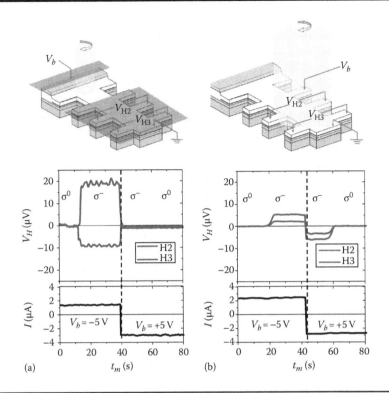

**FIGURE 8.7**    (a) Spin-injection Hall effect measurements in a masked sample with linearly polarized light and circularly polarized light of a fixed helicity for opposite polarities of the optical current. Schematics of each experimental setup are shown in the upper panel. Middle panel shows that spin-injection Hall effect voltages are detected only at negative bias when spin-polarized electrons move from the illuminated aperture toward the measured Hall crosses H2 and H3 in the n-channel. The optical current is plotted in the lower panel. (b) Complementary measurements to (a) in an unmasked sample with only the n-channel biased and Hall crosses H2 and H3 directly illuminated with a 10× stronger light intensity as compared to (a). Weak anomalous Hall effect signals are detected in this case, which are antisymmetric with respect to the polarity of the current. Lower panel shows the optically generated part of the current. (After Wunderlich, J. et al., *Nat. Phys.* 5, 675, 2009. With permission.)

phenomenology (shown in Figure 8.7a as the unmasked ones. Hence, the possibility that the observed effects arises from stray light from the focused polarized beam seems unlikely. Complete Hall symmetries of the transverse signals measured in our devices are confirmed by the experiments presented in Figure 8.7. Figure 8.7a shows data recorded in a sample with Hall crosses H2 and H3 fully covered by an insulating thin film and a metallic mask. The spin-injection Hall effect is observed only at reverse bias when electrons move from the illuminated aperture toward the n-channel Hall crosses. At forward bias, optically generated electrons are accelerated in the opposite direction and no Hall signals are detected independent of polarization.

The distinct features of the spin-injection Hall effect are highlighted by comparison to complementary data shown in Figure 8.7b. After etching the surface layers in the n-side of the p–n diode, Wunderlich et al. [1] were able to select wafers with residual sub-gap optical activity in an unmasked n-channel. The dark current generates zero Hall voltage at crosses H2 and H3 while clear (albeit weak) Hall signals are detected upon directly illuminating the crosses with intense circularly polarized light. The signals are attributed to the anomalous Hall effect since they occur inside the spin-generation area and, as expected, they are antisymmetric with respect to the current polarity. Also consistent with the anomalous Hall effect, they observe in these experiments Hall voltages with polarity depending only on current orientation and spin-polarization, but with the same signs on all irradiated Hall crosses measured. It contrasts the spin-injection Hall effect measurements of spin-injection from the p–n junction in which the sign can alternate among the Hall crosses.

It is also interesting that when measuring within the p-region, consistent with expectations, no clear Hall signal was measured at the first p-channel Hall cross with the laser spot focused on top of the p–n junction when the measurements were taken simultaneously with the n-channel measurements. This confirms that it is the spin-decoherence of propagating holes rather than an inherent absence of the Hall effect in the 2DHG, which explains the negative result of measurements.

Wunderlich et al. [1] also found that the measurements in a sample showed rectifying p–n junction characteristics at temperatures up to 240 K, hence demonstrating that the spin-injection Hall effect is readily detectable at high temperatures. Together with the zero-bias operation shown in Figure 8.5c and linearity of the spin-injection Hall effect in the degree of circular polarization of the incident light, these characteristics represent the realization of the spin-photovoltaic effect in a nonmagnetic structure and demonstrate the utility of the device as an electrical polarimeter [52, 57, 58].

## 8.4 THEORY OF SPIN-INJECTION HALL EFFECT

The previous section was limited to presenting the experimental phenomenology of the spin-injection Hall effect. As may have become apparent to the reader, it is rather surprising that such large spin-coherence lengths are also associated with an apparent coherent precession across enormous length scales beyond what is expected in normal 2DEG. On the other hand, because of the particular parameters chosen in the experiments, this is a system that we can model rather precisely and, as we will see in the following text, it is highly anisotropic in this coherent behavior.

The theoretical approach is based on the observation that the micrometer length scale governing the spatial dependence of the nonequilibrium spin-polarization is much larger than the ~10–100 nm mean free path in our 2DEG, which governs the transport coefficients. This allows us to disregard the Hall signal at first instance and first calculate the steady-state spin-polarization profile along the channel, and only afterward consider the spin-injection Hall effect as a response to the local out-of-plane component of the polarization. Note that this would not be possible when these length scales are of similar order, as they will be when mobilities increase and we exit the diffusive regime.

Our starting point is the description of GaAs near the $\Gamma$ point with the effect of the valence bands taken into account through a two-band effective Hamiltonian; this can be achieved through the so-called Lowin transformation. In the presence of an electric potential $V(r)$, the corresponding 3D SO-coupling Hamiltonian reads

$$
\begin{aligned}
H_{3D\text{-}SO} = & \left[\lambda^\star \, \sigma \cdot (k \times \nabla V(r))\right] \\
& + \left[\mathcal{B}k_x\left(k_y^2 - k_z^2\right)\sigma_x + \text{cyclic permutations}\right],
\end{aligned}
\tag{8.9}
$$

where:

$\quad \sigma \quad$ are the Pauli spin matrices
$\quad k \quad$ is the momentum of the electron
$\quad \mathcal{B} \approx 10\,\text{eV}\,\text{Å}^3$
$\quad \lambda^* = 5.3\,\text{Å}^2$ for GaAs [59, 60]

Equation 8.9 together with the 2DEG confinement yields an effective 2D Rashba and Dresselhaus SO-coupled Hamiltonian [61, 62],

$$
\begin{aligned}
H_{2DEG} = & \frac{\hbar^2 k^2}{2m} + \alpha(k_y\sigma_x - k_x\sigma_y) + \beta(k_x\sigma_x - k_y\sigma_y) \\
& + V_{dis}(r) + \lambda^\star \, \sigma \cdot (k \times \nabla V_{dis}(r)),
\end{aligned}
\tag{8.10}
$$

where:

$\quad m = 0.067 m_e$
$\quad \beta = -\mathcal{B}\langle k_z^2 \rangle \approx -0.02\,\text{eV Å}$
$\quad \alpha = e\lambda^* E_z \approx 0.01 - 0.03\ \text{eV Å}$ for the strength of the confining electric field
$\quad eE_z \approx 2\text{--}5 \times 10^{-3}$ eV/Å pointing along the [001] direction

The strength of the confinement is obtained from a self-consistent Poisson–Schrödinger simulation of the conduction band profile of our GaAs/AlGaAs heterostructure [31]. Typically $\alpha$ can be tuned whereas $\beta$ is a material-dependent parameter fixed by the choice of growth direction and, to a lesser extent, the degree of confinement.

### 8.4.1 NONEQUILIBRIUM POLARIZATION DYNAMICS ALONG THE [1$\bar{1}$0] CHANNEL

The realization of the original Datta–Das device concept in a purely Rashba SO-coupled system has been unsuccessful until recently due to spin-coherence issues; that is, in the required length scales in which transport is diffusive, no oscillating persistent precession states are present [63]. However, in a 2DEG where both Rashba and Dresselhaus SO coupling have similar strengths, a long-lived precessing excitation of the system has been shown to exist along a particular direction [37, 62]. When $\alpha$ and $\beta$ are equal in magnitude, the component of the spin along the [110] direction for $\alpha = -\beta$ or along the [1$\bar{1}$0] direction for $\alpha = \beta$ is a conserved quantity [61], as well as a precessing spin-wave (spin-helix) of wavelength $\lambda_{spin\text{-}helix} = \pi\hbar^2/(2m\alpha)$ in the direction perpendicular to the conserved spin component [62]. This spin-helix state has been observed through optical transient spin-grating experiments [37].

In the strong scattering regime of the structure considered, with $\alpha k_F$ and $\beta k_F \sim 0.5$ meV, much smaller than the disorder scattering rate $\hbar/\tau \sim 5$ meV, the system obeys a set of spin-charge diffusion equations [62] for arbitrary ratio of $\alpha$ and $\beta$:

$$\partial_t n = D\nabla^2 n + B_1\partial_x + S_{x-} - B_2\partial_{x-}S_{x+},$$

$$\partial_t S_{x+} = D\nabla^2\partial_t S_{x+} - B_2\partial_{x-}n - C_1\partial_{x+}S_z - T_1 S_{x+},$$

$$\partial_t S_{x-} = D\nabla^2\partial_t S_{x-} + B_1\partial_{x+}n - C_2\partial_{x-}S_z - T_2 S_{x-},$$

$$\partial_t S_z = D\nabla^2\partial_t S_z + C_2\partial_{x-}S_{x-} + C_1\partial_{x+}S_{x+} - (T_1 + T_2)S_z,$$

where:

$x_+$ and $x_-$ correspond to the [110] and [1$\bar{1}$0] directions
$B_{1/2} = 2(\alpha \mp \beta)^2(\alpha \pm \beta)k_F^2\tau^2$
$T_{1/2} = (2/m)(\alpha \pm \beta)^2(k_F^2\tau/\hbar^2)$
$D = v_F^2\tau/2$
$C_{1/2}^2 = 4DT_{1/2}$

For the present device, where $\alpha \approx -\beta$, the 2DEG channel is patterned along the [1$\bar{1}$0] direction, which corresponds to the direction of the spin-helix propagation. Within this direction, the dynamics of $S_{x-}$ and $S_z$ couple through the diffusion equations above. Seeking steady-state solutions of the form exp $[qx_-]$ yields the transcendental equation $(Dq^2 + T_2)(Dq^2 + T_1 + T_2) - C_2^2 q^2 = 0$, which can be reduced to $q^4 + (\tilde{Q}_1^2 - 2\tilde{Q}_2^2)q^2 + \tilde{Q}_1^2\tilde{Q}_2^2 + \tilde{Q}_2^4 = 0$, where $\tilde{Q}_{1/2} \equiv \sqrt{T_{1/2}/D} = 2m|\alpha \pm \beta|/\hbar^2$. This yields a physical solution for $q = |q|\exp[i\theta]$ of

$$|q| = \left(\tilde{Q}_1^2\tilde{Q}_2^2 + \tilde{Q}_2^4\right)^{1/4} \text{ and } \theta = \frac{1}{2}\arctan\left(\frac{\sqrt{2\tilde{Q}_1^2\tilde{Q}_2^2 - \tilde{Q}_1^4/4}}{\tilde{Q}_2^2 - \tilde{Q}_1^2/2}\right). \tag{8.11}$$

The resulting damped spin precession of the out-of-plane polarization component for the parameter range of the device, where we have set $\beta = -0.02$ eV Å and varied $\alpha$ from $-0.5\beta$ to $-1.5\beta$, is shown in Figure 8.8 in the main text. These results are in agreement with Monte Carlo calculations on similar systems (modeling an InAs 2DEG), but with higher applied biases [64]. In the Monte Carlo calculations, longer decaying lengths were observed at higher biases. However, the present device is well within the linear regime with very low driving fields; this results in shorter decay length scales of the oscillations as compared to Ref. [64].

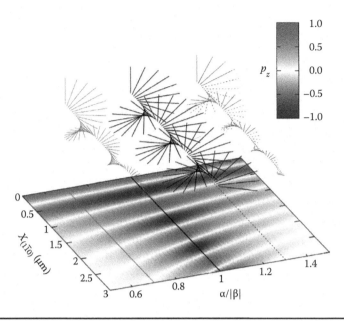

**FIGURE 8.8**  Steady-state solution of the spin precession for $\alpha/|\beta|$=0.7, 1.0, 1.3. The bottom panel indicates the precession of the polarization out-of-plane component for the parameter range of our device, where we have $\beta = -0.02$ eV Å and varied $\alpha$ from $-0.5\beta$ to $-1.5\beta$.

These results show good agreement with the steady-state variations (changes in the length scale of 1–2 μm) in the out-of-plane nonequilibrium polarization of our system observed indirectly through the Hall signals. We note that Monte Carlo simulations including temperature broadening of the quasiparticle states confirm the validity of the analytical results up to the high temperatures used in the experiment.

## 8.4.2 HALL EFFECT

The Hall effect signal can be understood within the theory of the anomalous Hall effect. The contributions to the anomalous Hall effect in SO-coupled systems with non-zero polarization can be classified into two types: the first type arises from the SO-coupled quasiparticles interacting with the spin-independent disorder and the electric field, and the second type arises from the non-SO-coupled part of the quasiparticles scattering from the SO-coupled disorder potential (last term in Equation 8.10). The contributions of the first type have recently been studied in 2DEG ferromagnetic systems with Rashba SO coupling [65–68]. These have shown that, within the regime applicable to our present devices, the anomalous Hall effect contribution due to the intrinsic and side-jump mechanisms vanish in the presence of even moderate disorder. In addition, the skew-scattering contribution from this type of contribution is also small (with respect to the contribution shown in the following) and furthermore is not linear in polarization [65]. Hence, we can disregard the contributions of the first type within our devices.

This is not surprising, since in our devices $\alpha k_F, \beta k_F \ll \hbar/\tau$, and hence these contributions arising from the SO coupling of the bands are not expected to dominate. Instead, the observed signal originates from contributions of the second type, that is, from interactions with the SO-coupled part of the disorder [27, 69]. Within this type, one contribution is due to the anisotropic scattering, the extrinsic skew scattering, and is obtained within the second Born approximation treatment of the collision integral in the semiclassical linear transport theory [27, 69]:

$$\left|\sigma_{xy}\right|^{skew} = \frac{2\pi e^2 \lambda^*}{\hbar^2} V_0 \tau n (n_\uparrow - n_\downarrow), \tag{8.12}$$

where $n = n_\uparrow + n_\downarrow$ is the density of photoexcited carriers with polarization $p_z = (n_\uparrow - n_\downarrow)/(n_\uparrow + n_\downarrow)$. Using the relation for the mobility $\mu = e\tau/m$ and the relation between $n_i$, $V_0$, and $\tau$, $\hbar/\tau = n_i V_0^2 m/\hbar^2$, the extrinsic skew-scattering contribution to the Hall angle, $\alpha_H \equiv \rho_{xy}/\rho_{xx} \approx \sigma_{xy}/\sigma_{xx}$, can be written as

$$\alpha_H^{skew} = 2\pi \lambda^* \sqrt{\frac{e}{\hbar n_i \mu}} n p_z (x_{[1\bar{1}0]})$$

$$= 2.44 \times 10^{-4} \frac{\lambda^*[\mathring{A}^2](n_- - n_\downarrow)\left[10^{11}\,\mathrm{cm}^{-2}\right]}{\sqrt{\mu\left[10^3\,\mathrm{cm}^2/\mathrm{Vs}\right]n_i\left[10^{11}\,\mathrm{cm}^{-2}\right]}} \tag{8.13}$$

$$\sim 1.1 \times 10^{-3} p_z,$$

where we have used $n = 2 \times 10^{11}\,\mathrm{cm}^{-2}$, $p_z$ is the polarization, $\mu = 3 \times 10^3\,\mathrm{cm}^2/\mathrm{V\,s}$, and $n_i = 2 \times 10^{11}\,\mathrm{cm}^{-2}$; the last estimate is introduced to give a lower bound to the Hall angle contribution within this model. In addition to this contribution, there also exists a side-jump scattering contribution in the SO-coupled disorder term given by

$$\alpha_H^{s-j} = \frac{2e^2 \lambda^*}{\hbar \sigma_{xx}}(n_\uparrow - n_\downarrow) \tag{8.14}$$

$$= 3.0 \times 10^{-4} \frac{\lambda^*\left[\mathring{A}^2\right]}{\mu\left[10^3\mathrm{cm}^2/\mathrm{Vs}\right]} p_z \sim 5.3 \times 10^{-4} p_z. \tag{8.15}$$

As expected, this is an order of magnitude lower than the skew-scattering contribution within this system.

We can then combine this result with the results from the previous section to predict, in this diffusive regime, the resulting theoretical $\alpha_H$ along the $[1\bar{1}0]$ direction for the relevant range of Rashba and Dresselhaus parameters corresponding to the experimental structure [1]. This is shown in Figure 8.9. We have assumed a donor impurity density $n_i$ on the order of the equilibrium density $n_{2DEG} = 2.5 \times 10^{11}\,\mathrm{cm}^{-2}$ of the 2DEG in dark, which is an upper bound for the strength of the impurity scattering in our modulation-doped heterostructure and, therefore, a lower bound for the Hall angle. For the mobility of the injected electrons in the 2DEG channel, one can consider

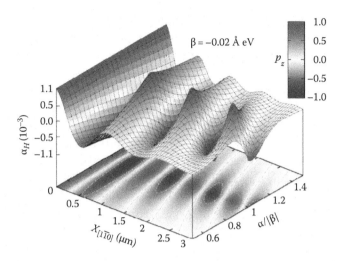

**FIGURE 8.9** Microscopic theory of the spin-injection Hall effect assuming SO-coupled band structure parameters of the experimental 2DEG system. The calculated spin-precession and spin-coherence lengths and the magnitude of the Hall angles are consistent with experiment. The grayscale surface shows the proportionality between the Hall angle and the out-of-plane component of the spin-polarization. (After Wunderlich, J. et al., *Nat. Phys.* 5, 675, 2009. With permission.)

the experimental value determined from ordinary Hall measurements without illumination, $\mu = 3 \times 10^3$ cm$^2$/V s. The density of photoexcited carriers of $n \approx 2 \times 10^{11}$ cm$^{-2}$ was obtained from the measured longitudinal resistance between successive Hall probes under illumination, assuming constant mobility.

The theory results shown in Figure 8.9 provide a semiquantitative account of the magnitude of the observed spin-injection Hall effect angle ($\sim 10^{-3}$), and explain the linear dependence of the spin-injection Hall effect on the degree of spin-polarization of injected carriers. The calculations are also consistent with the experimentally inferred precession length on the order of a micrometer, and the spin coherence exceeding micrometer length scales. We emphasize that the 2DEG in the strong disorder, weak SO-coupling regime realized in our experimental structures is a particularly favorable system for establishing theoretically the presence of the spin-injection Hall effect. In this regime, and for the simple band structure of the archetypal 2DEG, the spin-diffusion equations and the leading skew-scattering mechanism of the SO-coupling-induced Hall effect are well understood areas of the physics of quantum-relativistic spin-charge dynamics.

### 8.4.3 COMBINED SPIN–ORBIT AND ZEEMAN EFFECTS IN EXTERNAL MAGNETIC FIELDS

The above-mentioned theoretical considerations help explain the observed phenomenology and predict other expectations, such as higher damped oscillations observed in the orthogonal direction for the given device.

Another aspect that will affect these oscillations is the application of an external magnetic field whose consequences we consider next.

The spin dynamics at high temperatures and in external magnetic field can be studied theoretically by Monte Carlo–Boltzmann simulations, assuming the additional Zeeman term in the Hamiltonian, $g_e \mu_B \mathbf{B} \cdot \sigma/2$, where $g_e$ is the $g$-factor of GaAs. The orbital effects of the magnetic field are included on a classical level by considering the Lorentz force acting on injected electrons between scattering events. To model the device, the system is chosen to have an $L \times W = 3\,\mu\text{m} \times 0.5\,\mu\text{m}$ rectangular 2DEG connected to a source at $x = 0$ and a drain at $x = L$. The $x$-axis is along the $[1\bar{1}0]$ direction and the $y$-axis is along the $[110]$ direction. The electron density is $n_{2DEG} = 2.5 \times 10^{11}$ cm$^{-2}$. Fermi energy is $E_F = 9$ eV and Fermi momentum is $k_F = 0.125$ nm$^{-1}$. The Dresselhaus parameter $\beta = -0.02$ eV Å with the temperature of injected electrons is $T = 300$ K. Note that temperature enters our calculations only through the Fermi distribution of electrons; we do not include temperature-dependent scattering effects like, for example, electron-phonon scattering. The results we obtain are only very weakly temperature-dependent when comparing simulations at 1 and 300 K. Electrons are injected along the $[1\bar{1}0]$ direction, with initial spin-polarization along the $[001]$ axis. In the simulations, only point-like impurities such that the mean free path is $L_m f = 40$ nm were considered. The details of the Monte-Carlo–Boltzmann calculations can be found in Ref. [1].

Figure 8.10 shows the calculated spin-polarizations in the channel in magnetic fields oriented along the $[001]$, $[1\bar{1}0]$, and $[110]$ directions for the Rashba coupling parameter $\alpha = -0.5\beta$ (the effects of the magnetic field for $\alpha = -\beta$ are even less apparent than for the case of $|\alpha| \neq |\beta|$ shown in the figure). Since the calculations do not include quantum orbital effects of the magnetic field, their validity is limited to fields $\lesssim 1$ T. In agreement with expectations for the present 2DEG with relatively large SO-coupling strength, the Zeeman term has only a minor effect on the spin-polarization profile along the channel even at ~1 T fields. For the out-of-plane and $[1\bar{1}0]$-oriented in-plane field, the Zeeman coupling causes a weak suppression of the spin coherence, while for the $[110]$-oriented field, it leads to a weak suppression of the spin coherence, and a weak reduction or extension of the precession length depending on the sign of the magnetic field. The variations at magnetic fields $\lesssim 1$ T are not large enough to provide practical means for controlling the spin dynamics in the 2DEG channels. At higher magnetic fields, we expect an increasing role of the quantum orbital effects, which, due to the SO coupling, cannot be simply disentangled from the spin effects, and their description requires more sophisticated theoretical modeling. Furthermore, the field dependence of the spin-current generation characteristics of the p–n junctions is expected to be significant at fields of several Tesla.

## 8.4.4 Prospectives of Spin-Injection Hall Effect

The spin-injection Hall effect in nonmagnetic semiconductors and the ability to tune independently the strengths of disorder and SO coupling in semiconductor structures open up new opportunities for resolving

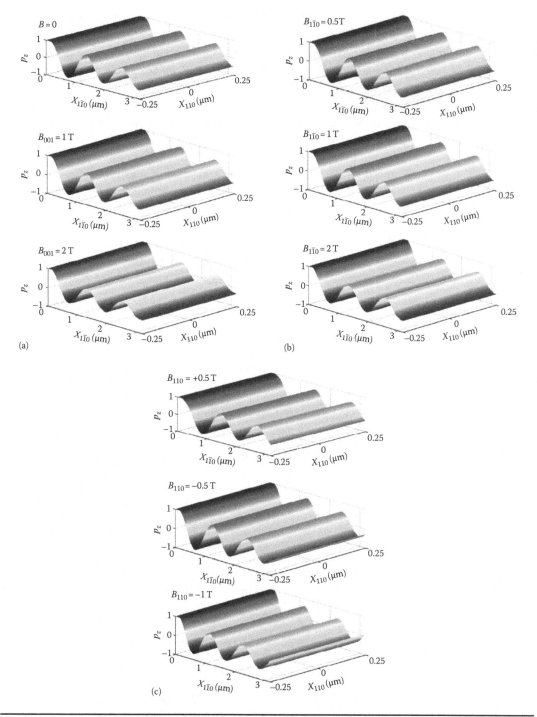

**FIGURE 8.10** Monte Carlo modeling of the out-of-plane component of the spin-polarization in the channel for $\alpha = -0.5\beta$ at 300 K. (a) Zero magnetic field and magnetic field oriented perpendicular to the plane of the 2DEG. (b) Magnetic field along the in-plane direction (the 2DEG channel direction). (c) Magnetic field along the [1$\bar{1}$0] in-plane direction (the 2DEG channel direction). Magnetic field causes a weak suppression of the spin coherence for these two field orientations. (After Wunderlich, J. et al., *Nat. Phys.* 5, 675, 2009. With permission.)

long-standing debates on the nature of spin-charge dynamics in the intriguing strong SO-coupling, weak disorder regime [66]. From the application perspective, the spin-injection Hall effect devices can be directly implemented as spin-photovoltaic cells and polarimeters, switches, invertors, and, due to the non-destructive nature of the spin-injection Hall effect, also as interconnects. We also foresee application of the spin-injection Hall effect in the Datta–Das [63], and other proposed spintronic transistor concepts [70]. An important next step toward a practical implementation in nanotechnology, as pointed out in Ref. [71], is the replacement of optical spin-injection by other solid-state means of spin-injection. These lightless devices utilizing the spin-injection Hall effect can be fabricated in a broad range of materials, including indirect-gap SiGe semiconductors [72]. Since the magnitude of the spin-injection Hall effect scales linearly with the SO-coupling strength, we expect ~100× weaker signals in the SiGe 2DEGs as compared to our measurements in the GaAs/AlGaAs, which is still readily detectable.

## REFERENCES

1. J. Wunderlich, A. C. Irvine, J. Sinova et al., Spin-injection Hall effect in a planar photovoltaic cell, *Nat. Phys.* **5**, 675 (2009).
2. N. Nagaosa, J. Sinova, S. Onoda, A. H. MacDonald, and N. P. Ong, Anomalous Hall effect, *Rev. Mod. Phys.* **82**, 1539 (2010).
3. J. Schliemann, Spin Hall effect, *Int. J. Mod. Phys. B* **20**, 1015 (2006).
4. S. Murakami, N. Nagaosa, and S. C. Zhang, Dissipationless quantum spin current at room temperature, *Science* **301**, 1348 (2003).
5. J. Sinova, D. Culcer, Q. Niu, N. A. Sinitsyn, T. Jungwirth, and A. H. MacDonald, Universal intrinsic spin Hall effect, *Phys. Rev. Lett.* **92**, 126603 (2004).
6. H.-A. Engel, E. I. Rashba, and B. I. Halperin, Theory of spin Hall effects in semiconductors, in *Handbook of Magnetism and Advanced Magnetic Materials*, vol. 5, H. K. S. Parkin (Ed.), John Wiley & Sons Ltd., Chichester, UK, p. 2858, 2007.
7. E. M. Hankiewicz and G. Vignale, Spin-Hall effect and spin-Coulomb drag in doped semiconductors, *J. Phys. Condens. Matter* **21**, 253202 (2009).
8. M. Koenig, H. Buhmann, L. W. Molenkamp et al., The quantum spin Hall effect: Theory and experiment, *J. Phys. Soc. Jpn.* **77**, 310 (2008).
9. B. Bernevig and S. Zhang, Quantum spin Hall effect, *Phys. Rev. Lett.* **96**, 0680 (2006).
10. C. L. Kane and E. J. Mele, Quantum spin hall effect in graphene, *Phys. Rev. Lett.* **95**, 226801 (2005).
11. L. Berger, Side-jump mechanism for the Hall effect of ferromagnets, *Phys. Rev. B* **2**, 4559 (1970).
12. J. Smit, The spontaneous Hall effect in ferromagnetics-I, *Physica* **21**, 877 (1955).
13. N. A. Sinitsyn, A. H. MacDonald, T. Jungwirth, V. K. Dugaev, and J. Sinova, Anomalous Hall effect in a two-dimensional Dirac band: The link between the Kubo-Streda formula and the semiclassical Boltzmann equation approach, *Phys. Rev. B* **75**, 045315 (2007).
14. N. A. Sinitsyn, Semiclassical theories of the anomalous Hall effect, *J. Phys. Condens. Matter* **20**, 023201 (2008).
15. R. Karplus and J. M. Luttinger, Hall effect in ferromagnetics, *Phys. Rev.* **95**, 1154 (1954).
16. M. Onoda and N. Nagaosa, Topological nature of anomalous Hall effect in ferromagnets, *J. Phys. Soc. Jpn.* **71**, 19 (2002).

17. T. Jungwirth, Q. Niu, and A. H. MacDonald, Anomalous Hall effect in ferro-magnetic semiconductors, *Phys. Rev. Lett.* **88**, 207208 (2002).

18. G. Sundaram and Q. Niu, Wave-packet dynamics in slowly perturbed crystals: Gradient corrections and Berry-phase effects, *Phys. Rev. B* **59**, 14915 (1999).

19. M. Chang and Q. Niu, Berry phase, hyperorbits, and the Hofstadter spectrum: Semi-classical dynamics in magnetic Bloch bands, *Phys. Rev. B* **53**, 7010 (1996).

20. N. Mainkar, D. A. Browne, and J. Callaway, First-principles LCGO calculation of the magneto-optical properties of nickel and iron, *Phys. Rev. B* **53**, 3692 (1996).

21. G. Y. Guo and H. Ebert, Band-theoretical investigation of the magneto-optical Kerr effect in Fe and Co multilayers, *Phys. Rev. B* **51**, 12633 (1995).

22. F. D. M. Haldane, Berry curvature on the Fermi surface: Anomalous Hall effect as a topological Fermi-liquid property, *Phys. Rev. Lett.* **93**, 206602 (2004).

23. X. Wang, D. Vanderbilt, J. R. Yates, and I. Souza, Fermi-surface calculation of the anomalous Hall conductivity, *Phys. Rev. B* **76**, 195109 (2007).

24. M. P. Marder, *Condensed Matter Physics*, Wiley, New York, 2000.

25. J. Smit, The spontaneous Hall effect in ferromagnetics–II, *Physica* **24**, 39 (1958).

26. L. Berger, Influence of spin–orbit interaction on the transport processes in ferromagnetic nickel alloys, in the presence of a degeneracy of the 3d band, *Physica* **30**, 1141 (1964).

27. P. Nozieres and C. Lewiner, Simple theory of anomalous Hall effect in semi-conductors, *J. Phys. (Paris)* **34**, 901 (1973).

28. M. I. D'yakonov and V. I. Perel', Possibility of orienting electron spins with cur-rent, *Sov. Phys. JETP* 467 (1971).

29. Y. K. Kato, R. C. Myers, A. C. Gossard, and D. D. Awschalom, Observation of the spin Hall effect in semiconductors, *Science* **306**, 1910 (2004).

30. V. Sih, R. Myers, Y. Kato, W. Lau, A. Gossard, and D. Awschalom, Spatial imag-ing of the spin Hall effect and current-induced polarization in two-dimen-sional electron gases, *Nat. Phys.* **1**, 31 (2005).

31. J. Wunderlich, B. Kaestner, J. Sinova, and T. Jungwirth, Experimental observa-tion of the spin-Hall effect in a two dimensional spin–orbit coupled semicon-ductor system, *Phys. Rev. Lett.* **94**, 047204 (2005).

32. S. Valenzuela and M. Tinkham, Electrical detection of spin polarized currents: The spin-current induced Hall effect, *J. Appl. Phys.* **101**, 09B103 (2007).

33. S. Valenzuela and M. Tinkham, Direct electronic measurement of the spin Hall effect, *Nature* **442**, 176 (2006).

34. S. Valenzuela, Nonlocal spin detection, spin accumulation and the spin Hall effect, *Int. J. Mod. Phys. B* **23**, 2413 (2009).

35. E. Saitoh, M. Ueda, H. Miyajima, and G. Tatara, Conversion of spin current into charge current at room temperature: Inverse spin-Hall effect, *Appl. Phys. Lett.* **88**, 8250 (2006).

36. T. Kimura, Y. Otani, T. Sato, S. Takahashi, and S. Maekawa, Room temperature reversible spin Hall effect, *Phys. Rev. Lett.* **98**, 156601 (2007).

37. C. P. Weber, J. Orenstein, B. A. Bernevig, S.-C. Zhang, J. Stephens, and D. D. Awschalom, Non-diffusive spin dynamics in a two-dimensional electron gas, *Phys. Rev. Lett.* **98**, 076604 (2007).

38. N. P. Stern, S. Ghosh, G. Xiang, M. Zhu, N. Samarth, and D. D. Awschalom, Current-induced polarization and the spin Hall effect at room temperature, *Phys. Rev. Lett.* **97**, 126603 (2006).

39. N. P. Stern, D. W. Steuerman, S. Mack, A. C. Gossard, and D. D. Awschalom, Drift and diffusion of spins generated by the spin Hall effect, *Phys. Rev. Lett.* **91**, 6210 (2007).

40. B. Liu, J. Shi, W. Wang et al., Experimental observation of the inverse spin Hall effect at room temperature, arXiv:cond-mat/0610150 (2006).

41. J. M. Kikkawa and D. D. Awschalom, Lateral drag of spin coherence in gallium arsenide, *Nature* **397**, 139 (1999).

42. S. A. Crooker, M. Furis, X. Lou et al., Imaging spin transport in lateral ferromagnet/semiconductor structures, *Science* **309**, 2191 (2005).

43. X. Lou, C. Adelmann, S. A. Crooker et al., Electrical detection of spin transport in lateral ferromagnet-semiconductor devices, *Nat. Phys.* **3**, 197 (2007).

44. R. Fiederling, M. Keim, G. Reuscher et al., Injection and detection of a spin-polarized current in a light-emitting diode, *Nature* **402**, 787 (1999).

45. H. Ohno, Properties of ferromagnetic III–V semiconductors, *J. Magn. Magn. Mater.* **200**, 110 (1999).

46. H. J. Zhu, M. Ramsteiner, H. Kostial, M. Wassermeier, H. P. Schönherr, and K. H. Ploog, Room-temperature spin injection from Fe into GaAs, *Phys. Rev. Lett.* **87**, 016601 (2001).

47. X. Jiang, R. Wang, R. M. Shelby et al., Highly spin-polarized room-temperature tunnel injector for semiconductor spintronics using MgO(100), *Phys. Rev. Lett.* **94**, 056601 (2005).

48. J. N. Chazalviel, Spin-dependent Hall effect in semiconductors, *Phys. Rev. B* **11**, 3918 (1975).

49. H. Ohno, H. Munekata, T. Penney, S. von Molmar, and L. L. Chang, Magnetotransport properties of p-type (In,Mn)As diluted magnetic III–V semiconductors, *Phys. Rev. Lett.* **68**, 2664 (1992).

50. J. Cumings, L. S. Moore, H. T. Chou et al., Tunable anomalous Hall effect in a nonferromagnetic system, *Phys. Rev. Lett.* **96**, 196404 (2006).

51. M. I. Miah, Observation of the anomalous Hall effect in GaAs, *J. Phys. D: Appl. Phys.* **40**, 1659 (2007).

52. S. D. Ganichev, E. L. Ivchenko, S. N. Danilov et al., Conversion of spin into directed electric current in quantum wells, *Phys. Rev. Lett.* **86**, 4358 (2001).

53. P. R. Hammar and M. Johnson, Detection of spin-polarized electrons injected into a two-dimensional electron gas, *Phys. Rev. Lett.* **88**, 066806 (2002).

54. B. Huang, D. J. Monsma, and I. Appelbaum, Coherent spin transport through a 350 micron thick silicon wafer, *Phys. Rev. Lett.* **99**, 177209 (2007).

55. J. Wunderlich, D. Ravelosona, C. Chappert et al., Influence of geometry on domain wall propagation in a mesoscopic wire, *IEEE Trans. Magn.* **37**, 2104 (2001).

56. M. Yamanouchi, D. Chiba, F. Matsukura, and H. Ohno, Current-induced domain-wall switching in a ferromagnetic semiconductor structure, *Nature* **428**, 539 (2004).

57. I. Žutić, J. Fabian, and S. D. Sarma, Spin-polarized transport in inhomogeneous magnetic semiconductors: Theory of magnetic/nonmagnetic p–n junctions, *Phys. Rev. Lett.* **88**, 066603 (2002).

58. T. Kondo, J. ji Hayafuji, and H. Munekata, Investigation of spin voltaic effect in a pn heterojunction, *J. Appl. Phys.* **45**, L663 (2006).

59. W. Knap, C. Skierbiszewski, A. Zduniak et al., Weak antilocalization and spin precession in quantum wells, *Phys. Rev. B* **53**, 3912 (1996).

60. R. Winkler, *Spin–Orbit Coupling Effects in Two-Dimensional Electron and Hole Systems*, Springer-Verlag, New York, 2003.

61. J. Schliemann, J. C. Egues, and D. Loss, Nonballistic spin-field-effect transistor, *Phys. Rev. Lett.* **90**, 146801 (2003).

62. B. A. Bernevig, T. L. Hughes, and S.-C. Zhang, Quantum spin Hall effect and topological phase transition in HgTe quantum wells, *Science* **314**, 1757 (2006).

63. S. Datta and B. Das, Electronic analog of the electro-optic modulator, *Appl. Phys. Lett.* **56**, 665 (1990).

64. M. Ohno and K. Yoh, Datta-Das-type spin-field-effect transistor in the nonballistic regime, *Phys. Rev. B* **77**, 045323 (2008).

65. A. A. Kovalev, K. Vyborny, and J. Sinova, Hybrid skew scattering regime of the anomalous Hall effect in Rashba systems: Unifying Keldysh, Boltzmann, and Kubo formalisms, *Phys. Rev. B* **78**, 41305 (2008).

66. M. Borunda, T. Nunner, T. Luck et al., Absence of skew scattering in two-dimensional systems: Testing the origins of the anomalous Hall effect, *Phys. Rev. Lett.* **99**, 066604 (2007).

67. T. S. Nunner, N. A. Sinitsyn, M. F. Borunda et al., Anomalous Hall effect in a two-dimensional electron gas, *Phys. Rev. B* **76**, 235312 (2007).

68. S. Onoda, N. Sugimoto, and N. Nagaosa, Quantum transport theory of anomalous electric, thermoelectric, and thermal Hall effects in ferromagnets, *Phys. Rev. B* **77**, 165103 (2008).

69. A. Crépieux and P. Bruno, Theory of the anomalous Hall effect from the Kubo formula and the Dirac equation, *Phys. Rev. B* **64**, 014416 (2001).

70. I. Žutić, J. Fabian, and S. Das Sarma, Spintronics: Fundamentals and applications, *Rev. Mod. Phys.* **76**, 323 (2004).

71. I. Žutić, Spins take sides, *Nat. Phys.* **5**, 630 (2009).

72. I. Žutić, J. Fabian, and S. C. Erwin, Spin injection and detection in silicon, *Phys. Rev. Lett.* **97**, 026602 (2006).

# Section V
# Magnetic Semiconductors, Oxides and Topological Insulators

# 9

# Magnetic Semiconductors
## *III–V Semiconductors*

**Carsten Timm**

## 9.1 INTRODUCTION

The present chapter discusses the properties of III–V diluted magnetic semiconductors (DMS) and their theoretical understanding. By DMS, we mean semiconducting compounds doped with magnetic ions and showing ferromagnetic order below a Curie temperature $T_C$. Due to the semiconducting properties of DMS, the carrier concentration can be changed significantly by doping, and even *in situ* by the application of a gate voltage, or by optical excitations. Since the carriers couple strongly to the magnetic moments, this leads to unique properties, and makes DMS promising for spintronics. In particular, magnetotransport and spin transfer in DMS-based devices are discussed in Chapter 4, Volume 2, while DMS quantum dots are discussed in Chapter 6, Volume 3. Electric-field control of magnetism in DMS has also revealed important opportunities in ferromagnetism metals, a discussed in Chapter 13 of Volume 2. The strong spin-orbit coupling in most DMS is responsible for the anomalous Hall effect discussed in Chapter 8, Volume 2.

Of the many excellent reviews on III–V DMS we only mention a few. Jungwirth et al. [1] focus on the theory of III–V DMS, in particular for the prototypical compound (Ga,Mn)As, and cover both *ab initio* calculations and the Zener kinetic-exchange theory discussed below. More recently, Dietl [2], as well as Dietl and Ohno [3], have reviewed experimental and theoretical aspects of DMS in general, where the latter article focuses mostly on (Ga,Mn)As and also discusses device concepts. Jungwirth et al. [4] discuss the physical properties of (Ga,Mn)As in view of spintronics applications. A book chapter by Jungwirth [5] more generally discusses Mn-doped III–V DMS and spintronics. Sato et al. [6] review *ab initio* calculations as well as model-based calculations and simulations for III–V and II-VI DMS. In a contribution for a broader audience, Zunger et al. [7] review theoretical approaches, contrasting model-based and *ab initio* calculations, and discussing potential pitfalls. More specialized reviews discuss wide-gap III–V DMS and dilute oxides [8, 9], disorder effects in DMS [10], ferromagnetic resonance (FMR) in (Ga,Mn)As [11], carrier localization in DMS [12], optical properties of III–V DMS [13], synthesis and epitaxial growth [14], the anomalous Hall effect [15], and the nanomorphology of DMS [16]. The basic picture put forward in Ref. [13] has been challenged by Jungwirth et al. [17], as discussed below.

Two classes of III–V DMS have emerged: On the one hand, compounds with narrow to intermediate gaps, such as antimonides and arsenides, which show Curie temperatures below room temperature, and on the other wide-gap materials, such as nitrides and phosphides, often with higher apparent $T_C$. The first class appears to be much better understood, and will be discussed first and in more detail. At least some of the wide-gap materials are more similar to the dilute magnetic oxides discussed in the following Chapter 10, and we will restrict ourselves to a number of comments pertinent to wide-gap III–V compounds.

## 9.2 III–V DMS WITH NARROW TO INTERMEDIATE GAPS

The equilibrium solubility of Mn in DMS with narrow to intermediate gaps lies below one percent of cations, too low for ferromagnetic order to occur. This led to efforts to achieve higher Mn concentrations by means of nonequilibrium molecular beam epitaxy (MBE) at low substrate temperatures. With this technique, ferromagnetism in (In,Mn)As with $T_C \approx 7.5\,\text{K}$ was achieved in 1992 [18], and in (Ga,Mn)As with $T_C \approx 60\,\text{K}$ in 1996 [19]. The Curie temperature of (Ga,Mn)As has been pushed above 100 K [20] and recently up to around 190 K [21–24] by improvements in the MBE technique and careful post-growth annealing, which allow higher Mn concentrations up to nominally $x \approx 16\%$ per cation site [3, 21–25]. In GaAs δ-doped with Mn, i.e. containing an essentially two-dimensional Mn-rich layer, $T_C \approx 250\,\text{K}$ has been reached [26]. (Ga,Mn)As with large Mn concentration has also been produced by Mn-ion implantation of GaAs followed by pulsed-laser annealing [27]. The magnetization curves and the temperature-dependent resistivity are comparable to annealed MBE-grown samples, while the values of $T_C$ are somewhat lower, but still exceeding 100 K [27]. Meanwhile, $T_C$ for (In,As)Mn has been increased to 90 K [28]. Recently, Tu et al. have observed room-temperature ferromagnetism with $T_C \approx 340\,\text{K}$ in p-type (Ga,Fe)Sb thin films [29] and with $T_C \approx 335\,\text{K}$ in n-type (In,Fe)Sb thin films [30], all grown by low-temperature MBE.

(Ga,Mn)As is the most extensively studied III–V DMS, not the least because the host material GaAs is technologically important and well-understood. It is known that ferromagnetism in (Ga,Mn)As requires Mn concentrations of $x \gtrsim 1\%$ [20, 31, 32]. There is a zero-temperature insulator-to-metal transition at a somewhat higher concentration of about $x = 1.5\%$ for well annealed samples [31–33]. The narrow-gap antimonides (Ga,Mn)Sb and (In,Mn)Sb show lower $T_C$ [34–36], but are otherwise similar to the arsenides. Recently, ferromagnetism with Curie temperatures of up to 40 K for $x \approx 5\%$ Mn concentration has been observed in Mn-ion-implanted and pulsed laser annealed InP [37], which is insulating for all concentrations. InP is of interest, since its band gap and hole effective masses are similar to those of GaAs, while the Mn acceptor level is much deeper [38].

At present, two central questions regarding (Ga,Mn)As remain partially open. It is clear that for relatively light doping, bound acceptor states form a narrow impurity band (IB) in the gap. It is also obvious that for heavy doping, this IB will become very broad, merge with the valence band (VB), and lose its identity [39]. There is an ongoing debate whether in high-$T_C$ (Ga,Mn)As one can still identify an IB. A related, but separate, question is whether the states close to the Fermi energy, which are responsible for charge transport, and for the carrier-mediated magnetic interaction, are localized or extended. In the following sections, we discuss the physics of (III,Mn)V DMS, progressing from light to heavy Mn-doping.

### 9.2.1 PROPERTIES OF ISOLATED MN DOPANTS: SINGLE-ELECTRON PICTURE

Manganese in narrow and intermediate-gap III–V semiconductors forms $Mn^{2+}$ ions at cation sites of the zinc blende lattice, as seen in electron paramagnetic resonance (EPR) [40, 41], and FMR [42]. The lattice with substitutional Mn impurities is depicted for the example of (Ga,Mn)As in Figure 9.1a. Since $Mn^{2+}$ replaces a 3+ ion, manganese provides one fewer electron or, equivalently, donates a hole and thus acts as an acceptor. Since it has a negative charge relative to the cation sublattice, it can bind the hole. The neutral acceptor-hole complex has been observed in infrared (IR) spectroscopy [43, 44] and EPR [45]. The hole binding energy is found to be 112.4 meV [43–45], in agreement with the thermal activation energy seen in transport [46]. Mn acceptors close to the surface have been imaged and switched between the ionized (no bound hole) and neutral states with a scanning tunneling microscope (STM) at room temperature [47].

In contrast to simple shallow acceptors, $Mn^{2+}$ has a half-filled $d$-shell. Photoemission experiments [48, 49] find the maximum weight of occupied $d$-orbitals about 4.5 eV below the Fermi energy and do not find evidence for a mixed-valence state, suggesting that the unoccupied $d$-levels also do not lie close to the Fermi energy. The absence of $d$-level signatures in the optical conductivity shows that they lie above the conduction-band (CB) minimum. The $d$-shell thus mostly acts as a local spin $S = 5/2$, consistent

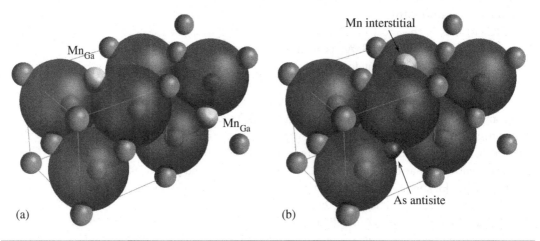

(a)                                    (b)

**FIGURE 9.1** (a) Zinc blende lattice structure of GaAs with two substitutional Mn impurities. $Ga^{3+}$ ions are shown as small red spheres, $As^{3-}$ as large, semitransparent blue spheres, and $Mn^{2+}$ as small green spheres. The gray cube denotes the conventional unit cell. The ionic radii are used in the plot, which is an exaggeration, since the bonds are partially covalent. (b) Lattice structure of GaAs with a $Mn^{2+}$ interstitial in the As-coordinated tetrahedral position and an As antisite defect (small blue sphere with the $As^{5+}$ ionic radius).

with EPR experiments [45]. It is useful to notionally divide the Mn impurity into a simple acceptor and the half-filled $d$-shell. Ignoring the $d$-shell for the moment, the attractive potential should lead to hydrogenic energy levels for the hole. A spherical approximation for the band structure of GaAs gives a binding energy of 25.7 meV [50, 51]. Binding energies of typical hydrogenic acceptors are larger, for example 30.6 meV for $Zn^{2+}$ [52], since the dielectric screening of the impurity charge becomes less effective close to the impurity. This can be taken into account by a central-cell correction [51], i.e. an additional short-range attractive potential.

A more complex picture emerges if we take into account that the hole also experiences the periodic potential of the host lattice. The bound acceptor states are derived from small wave-vector states close to the VB top, which mostly consist of anion $p$-orbitals. The point group at the impurity (cation) site in zinc blende semiconductors is the tetrahedral group $T_d$ [53]. Neglecting spin-orbit coupling, the anion $p$-orbitals correspond to the $\Gamma_5$ irreducible representation of $T_d$, i.e. they have $t_2$ symmetry [53]. Since $\Gamma_5$ is three-dimensional, the bound acceptor state is six-fold degenerate, including the spin degeneracy.

In the presence of spin-orbit coupling, the theory of double groups [53] shows that the level splits into a $\Gamma_7$ ($e''$) doublet and a $\Gamma_8$ ($u'$) quartet. The doublet can be thought of as resulting from the split-off band, which for GaAs is separated from the heavy-hole and light-hole bands by $\Delta_{SO} = 341$ meV at the $\Gamma$ point. The doublet is therefore expected to form a resonance in the VB. The remaining four-fold degeneracy can be understood in terms of a total angular momentum $j = 3/2$ resulting from orbital angular momentum $l = 1$ and spin $s = 1/2$. STM experiments [33, 47, 54, 55] find a pronounced spin-dependent spatial anisotropy of these states, in agreement with theory [47, 54–57].

We now turn to the Mn $d$-electrons. In the $T_d$ crystal field, the five $d$-orbitals split into three orbitals with $t_2$ character and two with $e$ character. All five are singly occupied and the electron spins are aligned (high-spin state), showing that the Hund's-rule coupling is larger than the crystal-field splitting. Importantly, the $t_2$-type $d$-orbitals hybridize with the $p$-orbitals of the neighboring anions. This hybridization strongly affects the bound acceptor states, which are also of $t_2$ type, if spin-orbit coupling is neglected, and have a large amplitude at the neighboring sites [58, 59]. The occupied $d$-orbitals, which all have the same spin, can only hybridize with the bound acceptor states with the same spin. Consequently, level repulsion shifts the occupied $d$-levels down in energy and the bound acceptor states with the same spin up. The resulting bound acceptor states, called the *dangling-bond hybrid* [58, 59], contain some $d$-level admixture, and are deeper in the gap, and thus more strongly bound than for a hydrogenic impurity [51, 60]. The occupied $d$-levels with some admixture of anion $p$-orbitals are called the *crystal-field resonance* [58, 59]. The resulting spin-up single-particle levels are sketched in the left-hand part of Figure 9.2.

On the other hand, the bound acceptor state with spin opposite to the Mn $d$-shell hybridizes with the unoccupied $t_2$ orbitals and is shifted

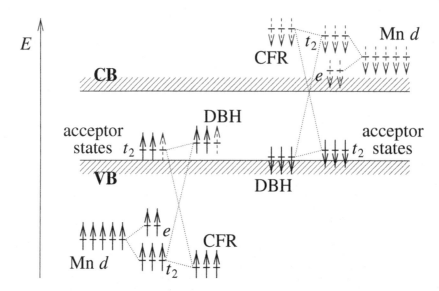

**FIGURE 9.2**   Sketch of single-particle states introduced by substitutional Mn acceptors in GaAs. The spin of the Mn $d$-shell is assumed to point upward. Spin-orbit coupling is ignored. Bold (dashed) arrows indicate occupied (unoccupied) states. The VB and CB edges are indicated. Left: level repulsion between spin-up $d$-states of $t_2$ symmetry and bound acceptor states also of $t_2$ symmetry leads to the occurrence of a spin-up dangling-bond hybrid (DBH) in the gap and a crystal-field resonance (CFR) deep in the VB. Right: for spin-down states, the unoccupied $d$-states are high above the Fermi energy, and the DBH is shifted downward in energy. In the ground state, the single hole introduced by the Mn acceptor occupies one of the most strongly bound spin-up DBHs.

downward, as shown in the right-hand part of Figure 9.2. This leads to spin splitting of the bound state, which depends on the orientation of the Mn spin [58, 59, 61]. In the ground state, a hole will occupy the most strongly bound state, which has the same spin as the $d$-electrons. Thus an electron with this spin is missing, and the total spin of the system is reduced. In this sense, the exchange coupling $J_{pd}$ between the Mn $d$-shell and the VB-derived states is antiferromagnetic.

## 9.2.2 PROPERTIES OF ISOLATED MN DOPANTS: MANY-PARTICLE PICTURE

For several reasons, the single-particle picture is not sufficient for the understanding of Mn dopants [17]: First, the splitting between occupied and unoccupied Mn $d$-levels is large, mostly due to the on-site Coulomb repulsion $U \approx 3.5$ eV [49]. Second, there is a large Hund's-rule coupling $J_H \approx 0.55$ eV [62]. The resulting strong correlations are not well-captured by density-function theory (DFT) using standard approximations such as the local (spin) density approximation (LDA, LSDA). There has been progress in incorporating correlations either by phenomenologically introducing a Hubbard-$U$ (LDA+$U$, LSDA+$U$) [6, 63, 64] or by correcting for self-interaction artifacts [6, 65, 66]. These approaches generically push Mn $d$-orbital weight away from the Fermi energy, leading to a smaller Mn $d$-orbital weight in the bound acceptor states. Since the hybridization between the anion $p$-orbitals and the Mn $d$-orbitals is small compared to their energy separation, it can be described perturbatively. Canonical perturbation theory [67, 68] maps the $d$-electrons onto a local spin and a short-range potential scatterer. The latter is a typically negligible correction to the Coulomb attraction. One finds an antiferromagnetic $pd$-exchange interaction $J_{pd}(\mathbf{r})$ between the hole and Mn spins, which falls off on the length scale of the anion-cation separation. In reciprocal space, this leads to a $\mathbf{k}$-dependent interaction $J_{pd}(\mathbf{k})$ [68]. The exchange coupling quoted in the literature is the limit $J_{pd} = J_{pd}(\mathbf{k} \to 0)$. Third, each Mn acceptor can bind a single hole, but not a second one [59]. Thus the Coulomb repulsion $U_{\mathrm{acc}}$ in the bound acceptor state is at least of the same order as the binding energy. Impurity states in doped semiconductors are strongly correlated systems [69]. Naive single-particle approaches predict too many bound acceptor states, seen for example in Figure 9.2.

The emerging picture describes an impurity state containing a single hole with total angular momentum $j = 3/2$ antiferromagnetically coupled to a local spin $S = 5/2$. In the four-dimensional subspace spanned by the $j = 3/2$ impurity states, one finds the operator identity $\mathbf{s} = \mathbf{j}/3$ [1, 51], which allows one to write the exchange interaction either as $\varepsilon \mathbf{j} \cdot \mathbf{S}$ or as $3\varepsilon \mathbf{s} \cdot \mathbf{S}$. Taking the factor of three into account, one can express the exchange constant $\varepsilon$ in terms of the exchange coupling $J_{pd}$ and the envelope function $f(\mathbf{r})$ describing the spatial extent of the bound acceptor state [1, 51],

$$\varepsilon = \frac{1}{3} \int d^3 r J_{pd}(\mathbf{r}) | f(\mathbf{r}) |^2 \approx \frac{1}{3} J_{pd} \overline{| f(\mathbf{r}) |^2}. \tag{9.1}$$

The approximation in the second step is valid if the envelope function changes much more slowly in space than the exchange interaction. The average is taken over the range of $J_{pd}(\mathbf{r})$. From IR spectroscopy [44], a value of $\varepsilon \approx 5$ meV is obtained. Photoemission experiments [48] independently give $J_{pd} \approx 54$ meV nm$^3$. (In the literature, $J_{pd}$ is often denoted by $\beta$ and $J_{pd}n_{Mn}$ by $\beta N_0$, where $n_{Mn}$ is the Mn concentration [1]). The exchange interaction can be rewritten as $\varepsilon \mathbf{j} \cdot \mathbf{S} = (\varepsilon / 2)[F(F + 1) - j(j + 1) - S(S + 1)]$, where $\mathbf{F} = \mathbf{j} + \mathbf{S}$ is the total angular momentum. The splitting between the ground state triplet ($F = 1$) and the first excited quintet ($F = 2$) is thus $2\varepsilon \approx 10$ meV [44].

## 9.2.3 OTHER ISOLATED DEFECTS

III–V DMS contain other defects apart from substitutional magnetic dopants. For (Ga,Mn)As, the most important ones are As ions at Ga sites (arsenic antisites), and Mn ions in interstitial positions. The positions of these two defects are shown in Figure 9.1b. Vacancy defects are present in much smaller concentrations, but can strongly enhance the mobilities of other defects [70].

Even undoped GaAs grown by low-temperature MBE can contain a relatively high concentration of As antisites [71], which act as double donors, as expected for As$^{5+}$ cations. In (Ga,Mn)As, antisites should thus lead to partial compensation: There are fewer holes than Mn acceptors. Variation of the As:Ga ratio during growth has been found to lead to a corresponding change in the hole concentration, confirming that antisites are also present in (Ga,Mn)As [72]. Cracking $As_4$ clusters to $As_2$ during growth can substantially reduce the compensation [73], probably by reducing the concentration of antisites. The mobility of antisites in GaAs is likely small at relevant temperatures [74, 75] so that their distribution is not affected by post-growth annealing. However, due to their Coulomb attraction, antisite donors could preferentially be incorporated close to Mn acceptors during growth [76].

During low-temperature growth of (Ga,Mn)As, some $Mn^{2+}$ ions are incorporated in interstitial positions. Channeling Rutherford backscattering experiments [77] have provided evidence for about 17% of Mn residing in interstitial positions for as-grown samples with a Mn concentration of $x \approx 7\%$. DFT calculations find a lower formation energy of the As-coordinated tetrahedral Mn interstitial than of the Ga-coordinated one [59, 78, 79], as expected for a cation, but the predicted energy differences do not agree well. Channeling Rutherford backscattering cannot distinguish between these two positions.

$Mn^{2+}$ interstitials in GaAs are double donors contributing to compensation. Since the mobility of Mn interstitials at typical growth and annealing temperatures around 250°C is substantial [74, 77, 78, 80, 81], they can lower their electrostatic energy by assuming positions close to substitutional Mn acceptors during annealing, not just during growth [76, 82]. Mn interstitials also diffuse to the film surface during annealing, where they form a magnetically inactive layer [78, 80, 81, 83–85]. This is thought to be the main origin of the observed decrease of compensation during annealing [70, 83, 86] and of the higher $T_C$ of thinner films [87, 88]. This is nicely demonstrated by capping the surface with a thin GaAs layer, which strongly suppresses the effect of annealing [83–85]. $Mn^{2+}$ interstitials also carry a spin $S = 5/2$. For a Mn interstitial and a substitutional Mn dopant in nearest-neighbor positions, one expects a strong direct antiferromagnetic exchange interaction [79]. X-ray magnetic circular dichroism experiments [89] support this picture. Substitutional-interstitial pairs can thus form strongly bound spin singlets, effectively removing active spins. The contribution of these pairs is suppressed by annealing [89]. On the other hand, the question of the exchange interaction between VB holes and Mn-interstitial spins is still open, with DFT calculations giving conflicting results [79, 82].

Since As antisites and Mn interstitials are double donors, the electrostatic interaction favors their formation for higher concentrations of substitutional Mn acceptors. This *self-compensation* is a limiting factor in the growth of heavily Mn-doped (Ga,Mn)As. Yuan et al. [61] have reported that ion implantation of Mn in GaAs and InAs followed by pulsed laser melting leads to samples that are essentially free of both As antisite defects and Mn interstitials.

A promising idea for how to overcome this problem as well as the limitation of the same Mn dopants supplying both magnetic moments and carriers is to dope III–V semiconductors with iron [29, 30, 90]. Iron mostly replaces the cation in the $Fe^{3+}$ state so that it only contributes magnetic moments, while the carrier are introduced by independent doping. This allows to grow both p-type and n-type DMS and room-temperature ferromagnetism has been reported for the representative compounds (Ga,Fe)Sb [29] and (In,Fe) Sb [30], respectively.

## 9.2.4 EFFECTS OF WEAK MN DOPING

A non-zero concentration of Mn impurities does not have a strong effect on the exchange interaction $J_{pd}$, since $J_{pd}$ is mainly due to hybridization between Mn and neighboring anion orbitals. In contrast, it does have important effects on the electronic system. For small Mn concentrations, the total potential seen by the holes consists of the sum of the Coulomb potentials of the individual acceptors. The impurity potentials fall off like $1/r$ in the absence of metallic screening, whereas the bound-state wave function falls off exponentially. Hence, the overlap of the potentials remains relevant for arbitrarily small concentrations, unlike the overlap of the bound states. This situation of very weak doping is sketched in Figure 9.3a. The acceptor binding energy in this regime is reduced, since an acceptor can be ionized by exciting the hole to roughly the *average* potential. This leads to a reduction of the binding energy proportional to $x^{1/3}$ [46, 91], which is observed in Mn-doped GaAs [17]. In this regime, IR spectroscopy shows a broad mid-IR peak in the ac conductivity [43, 44, 92]. This peak is broad in spite of the sharp energy of bound acceptor states, since the bound impurity wave functions contain many wave vector components, so that transitions from large-$\mathbf{k}$ states deep within the VB to the impurity states are possible [17].

With increasing Mn doping $x$, the bound acceptor states start to overlap. Let us for the moment assume a periodic superlattice of Mn dopants,

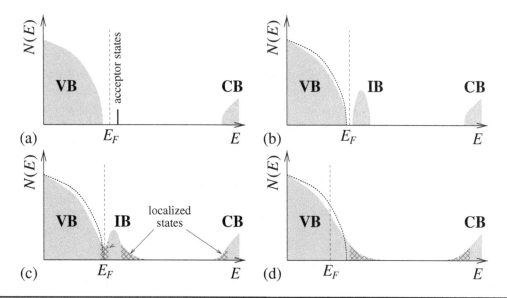

**FIGURE 9.3**   Density of states of III–V DMS with narrow to intermediate gaps. The Fermi energy $E_F$ is in all cases shown assuming no compensation. (a) For very weak doping, the overlap and thus the hybridization of bound acceptor states is negligible. (b) For larger doping, but assuming a regular superlattice of dopants, a sharply delimited IB emerges. If the doping level is not too high, there is a true gap between IB and VB. For zero compensation, $E_F$ falls into this gap. (c) A quite different picture emerges if disorder is included. The bands develop tails of localized states (shown cross-hatched), and the IB becomes asymmetric. Note that it is at present unknown whether extended states exist in the center of the IB before it merges with the VB, as shown here, or not. (d) For heavy doping, VB and IB merge completely, and the Fermi energy lies deep in the region of extended states.

ignoring disorder. Tunneling of electrons between the acceptor states leads to a narrow IB. Consequently, the mid-IR conductivity peak broadens, and shifts slightly to lower energies [92]. The total weight of the IB increases linearly with $x$, whereas the tunneling matrix elements, and thus the band width, increase exponentially. Thus the density of states does not become very large. This might explain why the Stoner enhancement of ferromagnetism, which is governed by the product of local Coulomb repulsion $U_{acc}$ and density of states, remains small [93–96].

Since we have so far ignored disorder, the density of states shows sharp edges of IB and VB and a true gap between the bands. A density of states of this type is predicted by theories that effectively average over disorder [60, 97, 98]. The doping also affects the host VB, for example since the impurity states take away spectral weight from the VB. The density of states in this regime is sketched in Figure 9.3b.

Due to the random positions of Mn dopants, the system is not periodic. Hence, band theory is not strictly valid, whereas the density of states remains well-defined. Due to disorder, the band edges in the density of states develop tails [56, 99], which always overlap, so that there is no true gap. However, it is reasonable to say that the IB still significantly affects the physics, if there is a clear minimum in the density of states in the overlap region with the VB. In addition, since the tunneling matrix elements decay exponentially with separation, the density of states of the IB becomes highly asymmetric [100]. The disorder also leads to strong (Anderson–Mott) localization of the states in the band tails, which are separated from extended states by mobility edges [12, 101–103]. This situation is sketched in Figure 9.3c.

*Ab initio* calculations suggest that Mn dopants have a relatively large attractive interaction [3, 104]. Hence, they are likely incorporated in positively correlated positions. They might further cluster during annealing but only in the presence of vacancies [74]. While the qualitative picture is not changed by correlated disorder, for a quantitative description detailed modeling of dopant positions is desirable. With compensation, $E_F$ can shift to the extended states close to the IB center. While it is controversial up to which Mn concentration $x$ this picture applies in (Ga,Mn)As, for (In,Mn)P, which has a similar electronic structure but a much larger acceptor binding energy of 220 meV, all studied samples with $x \leq 5\%$ are expected to be in the IB regime [37, 38].

It remains true that the IB contains one state per dopant, whereas single-particle approaches taking the angular momentum $j = 3/2$ into account find four states. Such approaches predict the Fermi energy $E_F$ to lie deep within the IB, even for vanishing compensation, leading to a relatively large density of states at $E_F$ [105, 106]. However, taking many-particle effects into account, $E_F$ falls into the minimum of the density of states between IB and VB, leading to very different physics. A simplistic model with a separate IB containing too many states can evidently mimic the large density of states at $E_F$ found if the bands have merged.

## 9.2.5 Effects of Heavy Mn Doping

For further increased Mn doping, the IB peak in the density of states grows in weight and width. Consequently, the minimum in the density of states between IB-type and VB-type states is gradually filled in. Krstajić et al. [39] estimate that the energy of the IB center minus half its width—a rough measure of its lower edge—crosses the unperturbed VB top already for a Mn concentration below 1%.

In the heavily doped limit, the IB completely merges with the VB. Due to disorder, the states in the band tail are expected to be localized, see Section 9.2.7, but the hole concentration is relatively high so that $E_F$ lies deep in the band, unless compensation is nearly complete. Both the impurity potential and the Coulomb repulsion between the holes are screened by the itinerant holes close to $E_F$. In this regime, it is reasonable to start from the unperturbed VB with the chemical potential located in the band and to incorporate disorder perturbatively in the second step [1]. The density of states for this case is sketched in Figure 9.3d.

The picture emerging for heavily doped DMS—holes in a weakly perturbed VB coupled to local spins by an antiferromagnetic exchange interaction—is the Zener kinetic-exchange or *pd*-exchange model [1, 39, 93, 94, 101, 107, 108]. Taking the *pd*-exchange interaction to be local, the model is characterized by a Hamiltonian of the form

$$H = H_{\text{holes}} + J_{pd} \sum_i \mathbf{S}_i \cdot \mathbf{s}_i, \tag{9.2}$$

where

$$H_{\text{holes}} = \sum_{n\mathbf{k}\sigma} \varepsilon_{n\mathbf{k}\sigma} c^\dagger_{n\mathbf{k}\sigma} c_{n\mathbf{k}\sigma} \tag{9.3}$$

describes non-interacting holes in the VB and $\mathbf{S}_i$ and $\mathbf{s}_i$ are the impurity and electronic spins, respectively, at Mn position $\mathbf{R}_i$. In this basic form, the Coulomb potentials due to acceptors and donors are neglected. Since properties of the host band structure are thought to be crucial, a realistic description is called for. The band structure of GaAs is sketched in Figure 9.4. In DFT one obtains the (Kohn–Sham) band structure as a matter of course, which is in fair agreement with the real (Landau-quasiparticle) bands if spin-orbit coupling, which is strong in arsenides and antimonides, is included. The main difference is the well-understood underestimation of the band gap. Adding a local approximation to the Coulomb potential of the Mn acceptors, $V_{\text{Mn}} \sum_i n_i$, where $n_i$ is the electronic charge density at Mn position $\mathbf{R}_i$, one arrives at the so-called *V-J* model [109, 110]. As discussed further in Section 9.2.8, Bouzerar and Bouzerar [109, 110] obtain reasonable results from a *V-J* model with $H_{\text{holes}}$ describing holes on a simple cubic lattice, suggesting that while the Coulomb disorder potential is crucial, details of the band structure are not as important as is often believed.

There are two main model-based approaches to a realistic band structure $\varepsilon_{n\mathbf{k}\sigma}$. On the one hand, in Luttinger–Kohn $\mathbf{k} \cdot \mathbf{p}$ theory [57, 95,

111–113], one restricts the Hilbert space to the relevant bands (usually the heavy-hole, light-hole, and split-off valence bands, perhaps adding the conduction band) and expands $\varepsilon_{nk\sigma}$ in the wave vector $\mathbf{k}$ around the $\Gamma$ point ($\mathbf{k} = 0$). The approach incorporates spin-orbit coupling and the non-spherical $\mathbf{k}$-dependence of $\varepsilon_{nk\sigma}$ and includes the correct effective masses at $\Gamma$. The bands become increasingly inaccurate away from $\mathbf{k} = 0$. If only the light-hole and heavy-hole bands are taken into account and the non-spherical $\mathbf{k}$-dependence is averaged over all directions, one arrives at the spherical approximation, which is described by the first-quantized $4 \times 4$ Hamiltonian

$$H_{\text{sph}}(\mathbf{k}) = \frac{\hbar^2}{2m}\left[\left(\gamma_1 + \frac{5}{2}\gamma_2\right)k^2 - 2\gamma_2(\mathbf{k}\cdot\mathbf{j})^2\right] \qquad (9.4)$$

in terms of the $4 \times 4$ total-angular-momentum operator $\mathbf{j}$ with $j = 3/2$. $\gamma_1$ and $\gamma_2$ are parameters. The last term stems from spin-orbit coupling, and is responsible for the splitting between the light-hole and heavy-hole bands away from the $\Gamma$ point.

The other approach, Slater–Koster tight-binding theory [56, 68, 99, 114–116], is formulated in real space by specifying on-site energies and hopping amplitudes for the relevant orbitals. Spin-orbit coupling is included [115]. These parameters are chosen so as to match the real band structure at high-symmetry points in the Brillouin zone. Therefore, the overall shape of the band structure agrees well with experiments—see Figure 9.4—but local properties, such as effective masses, are not well described. To resolve this problem, Yildirim et al. [117] have introduced a tight-binding model that reduces to the $\mathbf{k}\cdot\mathbf{p}$ Hamiltonian for small $\mathbf{k}$.

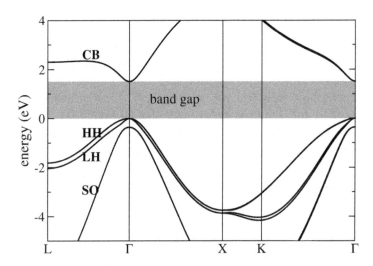

**FIGURE 9.4** Sketch of the band structure of GaAs (more precisely, the result of a tight-binding calculation using parameters from Ref. [115]). The band structure of other zinc blende semiconductors is topologically identical. The conduction band (CB) and the heavy-hole (HH), light-hole (LH), and spin-off (SO) valence bands are indicated.

Returning to the controversy regarding the range of applicability of the VB picture, model-based theories tend to predict that the IB and VB have merged at Mn concentrations $x$ of a few percent in GaAs [1, 39, 60, 100, 118]. In particular, Mašek et al. [118] discuss various proposed models with separate IBs, and argue that none of them can be derived from a plausible microscopic model. The VB picture has been very successful in explaining many experiments, for example, magnetization curves, magnetic anisotropies, magnetic stiffness, Gilbert damping, magnetotransport properties, the anomalous Hall effect, photoreflectance spectroscopy, Faraday and Kerr effect, and Raman spectroscopy [1, 3, 15, 119]. Experiments that appear to be inconsistent with this picture are discussed in the following [1, 3, 13, 119].

(1) Some angle-resolved photoemission spectroscopy (ARPES) experiments show a flat (momentum-independent) band merging with the VB top close to the $\Gamma$ point for (Ga,Mn)As with $x \approx 3.5\%$ [120] and $x \approx 2.5\%$ [121]. (The flat band has not been observed in other ARPES experiments with $x \approx 3\%$ [122]). The flat band is naturally interpreted as an IB that in the density of states has merged with the VB. However, it can also be described in terms of strongly localized—and thus momentum non-conserving—states at the top of the VB. Recent off- and on-resonance photoemission for $x \approx 0.1\%$, 1%, and 6% [123], and hard x-ray photoemission experiments for both $x \approx 1\%$ and $x \approx 13\%$ [124], also show merged IB and VB. The latter work [124] does not show any dip in the density of states even at the low concentration $x \approx 1\%$, in agreement with photoemission results at larger doping. Hence, the IB picture as understood in Section 9.2.4 does not apply. Care should be taken when interpreting the additional density of states close to the Fermi energy as being due to the merged IB. These states derive from the DBH, i.e. they mostly have anion $p$-orbital character with some Mn $d$-orbital admixture (see Section 9.2.1). It is thus misleading to speak of a "Mn $3d$-derived impurity band" [121]. Off- and on-resonance difference photoemission spectroscopy [123] is sensitive to the $d$-orbital weight. According to this work, $d$-orbital weight exists up to the merged VB edge and holes are predominantly located close to the Mn acceptors, as expected for DBHs. Nevertheless, the states at the Fermi energy have only about 1% Mn $d$-orbital character [123]. We note in passing that the comparison between experiment and theory [123] suggests that correlation effects have to be taken into account to achieve quantitative agreement. This can be understood in terms of a Coulomb anomaly at the Fermi energy [3].

(2) The effects of intentionally increased disorder are interpreted in terms of the IB picture. Stone et al. [125] find that in the quaternary DMS (Ga,Mn)(As,P) and (Ga,Mn)(As,N) already a concentration of phosphorus or nitrogen in the percent range makes the samples insulating and significantly reduces $T_C$. This is in agreement with estimates based on the IB picture. Furthermore, Sinnecker et al. [126] have irradiated highly compensated (Ga,Mn)As with Li ions to induce disorder, finding that the transport and magnetic properties are rather robust. This is argued to indicate that transport takes place in an IB, since the VB should be much more strongly affected by the disorder [126]. Note that high and low sensitivity of the DMS against disorder are both cited as support for the IB picture [125, 126], indicating

that further work is needed. Recently, Zhou et al. [127] have performed He ion implantation for nearly uncompensated (Ga,Mn)As. The magnetization curves remain Brillouin-function like but $T_C$ decreases continuously with increasing He fluence, which is interpreted in terms of the VB picture.

(3) Various IR spectroscopy experiments are interpreted as favoring the IB picture. There are contradictory results from ellipsometry concerning the shift of critical points with Mn concentration $x$ [128–130]. These experiments are sensitive to the evolution of the VB-CB band gap in various parts of the Brillouin zone. Gluba et al. [130] also perform photoreflectance experiments, which have a higher energy resolution, and find, in agreement with Ref. [129] but contradicting Ref. [128], no significant blueshift of the critical points. While the blueshift reported in Ref. [128] was interpreted in terms of the IB picture, its absence is argued to favor the VB picture [129, 130]. It should be noted that no detailed comparison with theory has been made.

Singley et al. [131] and Burch et al. [132] have studied (Ga,Mn)As with $x = 1.7\% - 7.3\%$ using IR absorption. Except for $x = 1.7\%$, the samples are metallic and show a broad mid-IR peak in the ac conductivity [131, 132]. This peak increases in weight for increasing $x$ and also with annealing. The peak maximum lies at around 250 meV for $x = 2.8\%$ and shifts to lower energies (redshifts) for increasing $x$ and with annealing. In a newer study over a broader doping range [25], the mid-IR peak is argued to develop adiabatically from insulating to very heavily doped ($x = 16\%$) samples. The main tenet of Refs. [13, 25, 131, 132] is to interpret this peak in terms of VB-to-IB transitions. The redshift with increasing $x$ is then naturally explained in terms of increased electronic screening, which moves the IB closer to the VB, and growing IB width [13, 131, 132]. These features are incorporated in a recent theoretical study [106], which treats the disorder by exact diagonalization in a supercell. The authors consider $x \leq 1.5\%$, mostly outside of the controversial region. Note that the calculation is based on a single-particle picture and therefore overestimates the number of states in the IB. The calculated peak shows a slight redshift or blueshift with increasing $x$ depending on the compensation [106].

A systematic experimental study of the mid-IR peak vs. Mn concentration $x$ for samples with low compensation [133] mostly shows a blueshift with increasing $x$. This agrees with the VB picture under the assumption of weak disorder [134]. In this picture, the mid-IR peak is due to transitions between the light-hole and heavy-hole bands. For increasing $x$, the Fermi energy moves deeper into the VB, the typical **k** vectors of these transitions grow, and the energy difference between light-hole and heavy-hole bands increases. Jungwirth et al. [17] have challenged the IB interpretation of IR experiments showing a redshift [13, 131, 132]. They point out that the maximum of the mid-IR peak is at a higher energy than expected from extrapolating the weak-doping data [92], and that the VB picture can also explain a redshift of the mid-IR peak, if disorder is treated beyond the first-order Born approximation (see Section 9.2.7). The main idea is that transitions into localized states in the tail of the merged VB do not have to conserve momentum, and can therefore contribute with large weight at relatively high energies [17]. For increasing $x$, the samples become more metallic and the high-energy contribution from localized states decreases, which could lead to a redshift. This view is supported by calculations by Yang et al. [135] within the VB picture, which incorporate the electron–electron interaction in the Hartree approximation, and the disorder by exact diagonalization for supercells.

For some of the samples, the ac conductivity also shows a smaller maximum at zero frequency [131, 132], which is interpreted as a Drude peak. By fitting a two-component model to the data and applying the conductivity sum rule to the Drude part alone, Burch et al. [132] extract the effective mass of the carriers. The resulting large effective mass is interpreted in terms of strong electron–electron interactions, as expected for an IB [132]. The analysis should be viewed with caution, since it relies on a single sample and neglects that the Drude peak can loose weight by disorder scattering [17]. Hot-electron photoluminescence experiments [136] are interpreted by the authors in terms of transitions of hot electrons from the CB into an IB. However, Jungwirth et al. [17] argue that a VB picture including disorder can explain the results, along with the shift of the mid-IR peak. Detailed calculations based on the VB or IB picture do not exist, however.

Finally, magnetic circular dichroism experiments [137] on (Ga,Mn)As with $x \lesssim 3\%$ have been interpreted in terms of multiple IBs. The data and in particular the positive sign of the dichroism signal are claimed to be inconsistent with the VB picture [137]. A calculation that is based on a tight-binding model for the VB, and includes disorder in real space [138], can explain the positive signal. This model still predicts a clear IB, only slightly overlapping with the VB for $x = 2\%$ [139]. Turek et al. [139] show that the positive-circular-dichroism result in Ref. [138] mainly comes from VB-to-CB transitions, not from VB-to-IB transitions. They also find that a tight-binding model based on *ab initio* calculations [140], which does not contain an IB, predicts a positive dichroism signal over most of the energy range. Note that IR magneto-optical experiments have been explained within the VB picture [141].

(4) Resonant tunneling spectroscopy [142, 143] for (Ga,Mn)As and (In,Ga,Mn)As suggests that Mn doping affects the VB only very weakly and that the Fermi energy is pinned above the VB edge, presumably in an IB. This interpretation has been challenged by Dietl and Sztenkiel [3, 144], who explain the observations in terms of the VB picture for (Ga,Mn)As and invoking the formation of hole subbands in the GaAs:Be electrode. They also point out that the tunneling spectra do not show tunneling into the purported IB. The alternative interpretation has been rejected by the original authors [145]. Detailed calculations for heterostructures should help to resolve the issue.

On the other hand, several rather general arguments support the VB picture for not too low Mn concentrations. First, for carrier-mediated magnetic interactions, a minimum in the density of states as sketched in Figure 9.3c would lead to a suppression of $T_C$ at weak compensation. Experimentally, there is no sign of this for metallic (Ga,Mn)As [1, 21, 22, 127, 146–150]. Indeed, this question was specifically addressed in a study by Wang et al. [150] employing hydrogen-co-doping. The authors find a monotonic increase in $T_c$ with carrier concentration, and observe the largest values of $T_c$ for the least compensated samples, in agreement with the VB picture. Furthermore, in the He-implantation study [127], the main effect of irradiation appears to be the reduction of the hole concentration through the creation of deep traps. The monotonic decrease of $T_c$ for increasing fluence is then inconsistent with the IB picture.

Second, the temperature-dependent dc conductivity should show an upturn when thermal activation of holes from the Fermi energy, supposed to lie in an IB, to extended states in the VB becomes possible. Indeed, this is observed for weakly Mn-doped GaAs [17, 151]. In metallic (Ga,Mn)As with $x \gtrsim 2\%$, no activated behavior is observed [17]. Neumaier et al. [152] have extracted the density of states from the temperature-dependent correction to the conductivity due to the electron–electron interaction [153] for samples of various geometries with $x \approx 4\%$ and $x \approx 6\%$. They do not find any evidence for a dip in the density of states and do find effective masses on the order of the bare mass. Their results are close to the prediction of a tight-binding model including disorder [56], and suggest merged bands and itinerant holes at $E_F$. Taken together, these results indicate that there is no pronounced dip in the density of states and no mobility gap, i.e. an energy range with localized states.

### 9.2.6 CARRIER-MEDIATED MAGNETIC INTERACTION

For heavily doped (Ga,Mn)As, we have arrived at the Zener *kinetic-exchange* model [107]. This is different from the Zener *double-exchange* model [154], which is concerned with the interaction between mixed-valence ions. The two pictures are opposite limiting cases on a continuum, though. For heavy doping, superexchange between neighboring Mn dopants might also be relevant [63, 96, 155, 156]. For $Mn^{2+}$ in III–V DMS, superexchange is antiferromagnetic [157]. Dietl et al. [101] have argued that it is small. Within standard perturbation theory, the superexchange interaction is indeed smaller than the RKKY interaction discussed below by a factor on the order of $E_F$ divided by the Coulomb repulsion $U$ in the $d$-shell [68].

There are a number of smoking-gun experiments showing that the magnetic interaction is indeed carrier-mediated: in field-effect transistors based on (In,Mn)As [158] and (Ga,Mn)As [146, 148, 149, 159–161], (In,Fe)As [162], as well as on GaAs $\delta$-doped with Mn [163], the hole concentration can be changed by a gate voltage (see also Chapter 13, Volume 2). A scheme of the device is shown in Figure 9.5. A positive (negative) voltage, which repels (attracts) the holes, leads to a reversible decrease (increase) of the magnetization and of $T_c$, as expected for a hole-mediated interaction. A particularly large change in $T_c$ of 42% has been achieved by using a thin (In,Fe)As layer within a broader InAs quantum well [162]. The effect of the electric field is large since it tunes the overlap of the relevant electronic wave functions with the DMS layer, while the carrier concentration is nearly unaffected, realizing an earlier theoretical proposal [164]. A change of the magnetic anisotropy by a gate voltage has also been demonstrated [159, 161]. This mechanism can be used to electrically switch the magnetization direction. The reverse effect, i.e. using a magnetic field close to $T_c$ to change the magnetization, and thereby

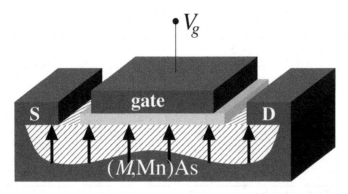

**FIGURE 9.5**  Field-effect transistor with (*M*,Mn)As, *M* = Ga,Mn, active region (schematically). By applying a voltage $V_g$ to the gate electrode, which is electrically isolated from the (*M*,Mn)As channel, the hole concentration in the DMS can be changed, leading to a change in the magnetic interactions and thus in the magnetization. Source (S) and drain (D) electrodes can be used to measure transport properties of the gated DMS film.

the carrier concentration [165] has not yet been observed. In an nonuniform magnetic field, this should lead to a voltage drop in the direction of the field gradient [165].

Furthermore, Koshihara et al. [166] have shown that irradiation of originally paramagnetic (In,Mn)As/GaSb heterostructures with deep red to IR light induces magnetic hysteresis. The results are easily understood in terms of photoinduced holes transferred from the GaSb layer to the (In,Mn)As layer. In ferromagnetic (Ga,Mn)As, Wang et al. [167] have observed an ultrafast enhancement of ferromagnetism after irradiation. The light pulse is thought to generate additional holes in the VB, which increase the magnetic interaction.

Measurements of the anomalous Hall effect in (Ga,Mn)As [15, 168] also support the picture of carrier-mediated magnetic interactions, since the magnetization inferred from the anomalous Hall effect agrees well with the one obtained from direct magnetometer measurements (see Chapter 8, Volume 2). In addition, experiments show a large magnetic anisotropy in (In,Mn)As and (Ga,Mn)As films [11, 169–172], which for example leads to reorientation transitions of the easy axis as a function of carrier concentration, temperature etc. In view of the tiny spin-orbit splitting [173] and the resulting tiny magnetic anisotropy of the Mn $d^5$ configuration, this would be hard to understand based on direct exchange interactions.

The kinetic-exchange model predicts a Ruderman–Kittel–Kasuya–Yosida (RKKY) interaction [174] between the Mn spins. The interaction is essentially proportional to $J_{pd}^2 \chi(\mathbf{r}_1 - \mathbf{r}_2)$, where $\chi$ is the non-local magnetic susceptibility of the carriers. The Stoner mechanism due to carrier–carrier interactions leads to an enhancement of the susceptibility by on the order of 20% [3, 93, 94]. If the Zeeman splitting of the VB is small, as is the case close to $T_C$, it is sufficient to assume unpolarized bands. The RKKY interaction calculated for realistic bands [68, 105, 175–177] is quite different from the textbook expression for a parabolic band [174]. Using a tight-binding band structure of GaAs, Timm and MacDonald [68] find an RKKY interaction that

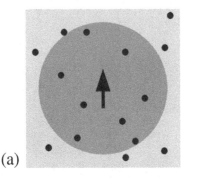 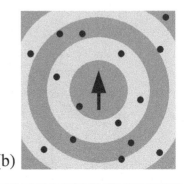

(a)　　　　　　　　　　　(b)

**FIGURE 9.6** Sketch of regions with ferromagnetic (pink) and antiferromagnetic (blue) RKKY interaction with regard to a given impurity spin, denoted by the arrow in the center. In (a) several neighboring impurity spins are ferromagnetically coupled, favoring ferromagnetic order, whereas in (b) the nearest-neighbor interaction is essentially random in sign.

is highly anisotropic, both in real space, and in spin space. The interaction is ferromagnetic and large at small separations, and shows the characteristic Friedel oscillations with period $\pi/k_F$ at larger separations, albeit with anisotropic Fermi wavenumber $k_F$. In compensated samples, typically, several neighboring impurities lie within the first ferromagnetic maximum, as sketched in Figure 9.6a, favoring ferromagnetic order [93, 105, 175, 176, 178]. Interactions at larger distances have random signs, but are small and average out for most impurity spins. However, new high-quality samples are only weakly compensated so that the period of the Friedel oscillations is comparable to the impurity separation, as shown in Figure 9.6b. In this case it might be important that the first minimum in the (110) direction does not reach negative—i.e. antiferromagnetic—values [68], e.g. in Figure 9.6b, the innermost antiferromagnetic (blue) region is absent in some directions. Similar "avoided frustration" is also found in DFT calculations [155, 179].

Kitchen et al. [54] have performed STM experiments for Mn ions on the (110) surface of Zn-doped p-type GaAs. The observed effective magnetic interactions are also highly anisotropic in real space. They agree well with tight-binding calculations using the bulk GaAs band structure [54].

The effective interaction between impurity spins has also been calculated within DFT from the total energy vs. Mn spin orientation and the corresponding torques [155, 179, 180]. The DFT and model-based results agree in that they show strong anisotropy in real space. DFT results typically lead to an overestimation of $T_C$, while model-based theories underestimate $T_C$ [68]. Probably contributing to this discrepancy are the neglect of the Coulomb potential of the acceptors in the model calculations, the neglect of spin-orbit coupling in the DFT results, and the unphysically large weight of Mn $d$-orbitals close to the Fermi energy in DFT. Indeed, LDA+$U$ usually leads to a reduction of the predicted $T_C$ [63]. Disorder is also expected to play a major role, as discussed below.

The conceptually simplest treatment of the kinetic-exchange model involves two distinct mean-field-type approximations: First, the $pd$-exchange interaction is decoupled at the mean-field (MF) level, and, second, the random distribution of impurity spins is replaced by a smooth magnetization (virtual-crystal approximation, VCA). Note that the same result for $T_C$ is obtained by first calculating the RKKY interaction between impurity spins, and then making the VCA/MF approximations [93]. The result reads [1, 93, 94, 178, 181, 182]

$$k_B T_C = \frac{S(S+1)}{3} \frac{J_{pd}^2 n_{Mn} \chi_{Pauli}}{g^2 \mu_B^2}, \qquad (9.5)$$

where:

$n_{Mn}$   is the concentration of Mn impurities

$g$   is their g-factor

$\mu_B$   is the Bohr magneton

$\chi_{Pauli}$ is the Pauli susceptibility of the carriers

For a single parabolic band, one obtains $k_B T_C = S(S+1)N(E_F)J_{pd}^2 n_{Mn}/6$ [178], where $N(E_F)$ is the density of states per spin direction at $E_F$. This leads to $T_C \propto a_0^{-6} p^{1/3} x^{4/3}$, where $a_0$ is the lattice constant, $x$ is the Mn doping level related to the Mn concentration by $n_{Mn} = 2.21 \times 10^{22}\,\mathrm{cm}^{-3} x$, and $p$ is the number of carriers per impurity. For increasing $x$, the Curie temperature should thus increase without any saturation. An increase is indeed observed for high-quality samples [1, 21, 22, 146–150]. The present model does not predict a reduction of $T_C$ close to vanishing compensation, $p = 1$, in contrast to the IB picture.

Dietl et al. [101] have predicted the Curie temperatures of 5% Mn-doped III–V and II–VI DMS based on the $\mathbf{k} \cdot \mathbf{p}$ Hamiltonian. The main trend is a strong increase of $T_C$ for lighter elements [101], which can already be understood from the parabolic-band model: Assuming that $J_{pd}$ is approximately independent of the host semiconductor [101], the strong dependence $T_C \propto a_0^{-6}$ favors high $T_C$ for hosts with small lattice constant $a_0$. This dependence stems, in equal parts, from the Mn concentration and from the electronic density of states by way of the Pauli susceptibility in Equation 9.5. Reinforcing this trend, the electronic effective mass is typically larger for lighter elements, further increasing the density of states.

Several routes have been taken to go beyond the MF approximation. We here restrict ourselves to equilibrium properties, spin dynamics is discussed in Chapter 1, Volume 2. A common approach is to first map the system onto a (Mn) spin-only model [155, 177, 183]. While Gaussian fluctuations (non-interacting spin waves) do not affect $T_C$, higher-order fluctuations can appreciably reduce it. This has been shown within the Tyablikov (random-phase) approximation combined with a treatment of random spin positions in a supercell [177, 183], called the "selfconsistent local random-phase approximation" (LRPA) [110, 177]. The results for $T_C(x)$ agree quantitatively with experiments for high-quality, annealed (Ga,Mn)As.

Calculations of the spin-wave stiffness find a much higher energy of spin waves for realistic band structures incorporating spin-orbit coupling compared to a parabolic band [1, 108]. This is partly due to the splitting between heavy-hole and light-hole bands, which can be understood as magnetic anisotropy of the total angular momentum **j**. The large stiffness stabilizes ferromagnetism and is consistent with FMR experiments [184]. The theoretical situation is not quite clear, though, since the spin-only model mentioned above [177, 183] does not include spin-orbit coupling, but nevertheless within the LRPA yields a stiffness in good agreement with experiments [185]. In addition, a calculation going beyond RPA but neglecting disorder [186] finds a strong enhancement of the spin-wave stiffness compared to the RPA. The relative importance of spin-orbit coupling, disorder, and higher-order many-particle corrections for the spin-wave stiffness is thus not clear.

### 9.2.7 DISORDER AND TRANSPORT

The resistivity of high-$T_C$ (Ga,Mn)As generically shows a finite residual value at low temperatures, sometimes with a small upturn for $T \to 0$ [20, 21, 27, 70, 86, 187, 188]. A finite limit of the resistivity is the defining property of a metal. (Ga,Mn)As is metallic due to the heavy doping. In addition, there is a broad and high peak in the resistivity in the vicinity of $T_C$ [17, 20, 21, 27, 70, 86, 131, 187, 188], also seen in (In,Mn)As [189]. In high-quality samples, its slope on the $T > T_C$ side is relatively flat [17, 21, 86].

DMS are highly disordered, since the magnetic dopants and the compensating defects are randomly distributed. Due to the strong interactions between charged defects, they are expected to be incorporated in correlated positions, likely as defect clusters, or to achieve such positions during annealing [76, 82, 96, 190–196]. Random-telegraph noise in the charge transport in (Ga,Mn)As [197] is interpreted in terms of the random switching of the magnetic moments of small impurity clusters.

There are two main contributions to the disorder felt by the carriers. First, the Mn acceptors attract the holes. In metallic DMS, the long-range part of this attraction is screened by itinerant holes, but screening is not very efficient due to the small hole concentration. The strong short-range attraction, which includes the central-cell correction, is not appreciably screened. The acceptor potentials, together with the repulsive potentials from Mn interstitials and antisites, form the Coulomb disorder potential. Clustering of defects of course has a large effect on this potential [76, 195]. Second, the *pd*-exchange interaction leads to disorder. Kyrychenko and Ullrich [196] have found that dynamical spin fluctuations are important for transport. Most theoretical approaches have nevertheless assumed the Mn spins to be frozen on the time scale relevant for electronic transport. In this limit, they act like a partially random Zeeman term, which is small compared to the Coulomb disorder [1]. The metallic behavior of high-quality samples suggests that disorder can be treated perturbatively in the first-order Born approximation [198, 199], which leads to a finite lifetime of quasiparticle states and to the exponential decay of the electronic Green function on the length scale of the elastic mean free path $l$. This causes a finite conductivity at $T = 0$. For annealed (Ga,Mn)As, the product $k_F l$, which is a measure for the strength of disorder, is experimentally found to increase from $k_F l \approx 1$ for $x \approx 1.5\%$ to $k_F l \approx 3-5$ for $x \approx 7\%$ [17, 132], consistent with a bad metal.

Going beyond the weak-disorder limit, disorder can lead to localized electronic states [12, 153, 199–201]: Electrons close to band minima tend to be trapped in the lowest depressions of the random potential landscape. The analog holds for holes close to band maxima. If disorder is not too strong, only states in the band tails are localized, whereas states in the band center are extended, with *mobility edges* separating the two types (see Figures 9.3c and d). If the states in the vicinity of $E_F$ are localized, conduction occurs through thermal activation into extended states, leading to an Arrhenius form of the conductivity, $\sigma \propto \exp(-\Delta E / T)$, where $\Delta E$ is the difference between $E_F$ and the nearest mobility edge [202]. In (Ga,Mn)As, $E_F$ seems to pass through a mobility edge for $x \approx 1.5\%$ [17, 31–33], corresponding to an insulator-to-metal transition. Note that there is one conflicting report of metallic conduction in an IB in (Ga,Mn)As with $x \approx 0.3\%$ [203]. A higher dopant concentration is required to make the system metallic compared to GaAs with shallow dopants, since the potential of a single Mn acceptor, and consequently the disorder, are stronger [17].

Moca et al. [188] find that measurements of the magnetoresistivity of (Ga,Mn)As with $x \lesssim 6.7\%$ and small $k_F l \sim 1$ are reasonably well-fitted by expressions derived from the scaling theory of localization [200]. The comparison shows that the drop in resistivity below $T_C$ is strongly correlated with the magnetization and suggests that the upturn of the resistivity at low $T$ [20, 21, 70, 86, 187, 188] results from disorder scattering. STM and scanning tunneling spectroscopy experiments [33] exhibit a rich spatial structure of the density of states at the surface of (Ga,Mn)As close to the metal-insulator transition. The correlations of the density of states at $E_F$, but not at other energies, decay as a power law with distance [33]. If this power-law behavior is indeed pinned to $E_F$ and not to the mobility edge, this indicates that the electron–electron interaction plays a crucial role [33]. It is interesting in this context that photoemission experiments also suggest strong correlation effects [3, 123]. How the pronounced spatial dependence [33] can be reconciled with the homogeneous ferromagnetism observed by muon spin relaxation (μSR) [204] is not fully clear.

As noted above, the question of localization of states is separate from the question of the survival of the IB [12]. In particular, an insulator-to-metal transition has been observed in p-doped GaAs quantum wells with $E_F$ in an IB [205]. Metallic conduction in an IB is also proposed for (Ga,Mn)As with $x \approx 0.3\%$ [203], contradicting other transport measurements [31, 32]. A picture of localized and extended states within an IB, as sketched in Figure 9.3c, is corroborated by exact-diagonalization studies [105, 106]. At present, it is unclear whether such a scenario pertains to any III–V DMS.

In the regime of weak doping, where an IB exists and the states at $E_F$ are localized, it is reasonable to start from a picture of holes hopping between bound acceptor states [105, 106, 191, 206–210]. The bound-hole spin is exchange-coupled to one or more impurity spins, forming a hole dressed with a spin-polarization cloud, called a *bound magnetic polaron* (BMP) [207, 208]. The exchange coupling to the same hole spins leads to an effective ferromagnetic coupling of the impurity spins in the BMP. The confinement of carriers in DMS quantum dots increases the tendency of BMP formation (see Chapter 6, Volume 3).

In metallic DMS, where strong localization is not relevant, disorder can still lead to subtle effects, in particular in reduced dimensions. Closed multiple-scattering paths such as the ones sketched in Figure 9.7 are important here: in the absence of a magnetic field and of spin-orbit coupling, the wave functions of carriers traversing a closed loop in opposite directions accumulate the same quantum phase, and therefore interfere constructively. This increases the probability for carriers to return to the same point and, consequently, decreases the conductivity. This *weak localization* [153, 198, 199] leads to a logarithmic increase in the resistivity for $T \to 0$, as observed in (Ga,Mn)As nanowires [211].

If a magnetic field is applied, the quantum phases accumulated for the two directions become different by an amount proportional to the magnetic flux $\Phi$ enclosed by the loop (see Figure 9.7). For a single loop, the conductivity shows Aharonov–Bohm oscillations as a function of $\Phi$. In typical nanoscopic samples, loops of different sizes contribute, which have different oscillation periods. This leads to a characteristic quasi-random dependence of the conductivity on magnetic field with universal amplitude (universal conductance fluctuations) [212]. Both the Aharonov–Bohm effect and universal conductance fluctuations have been observed in nanoscopic (In,Mn)As [189] and (Ga,Mn)As [211, 213, 214] devices. The reason why these effects can occur at all is that DMS are both diluted ferromagnets and bad metals, as discussed by Garate et al. [215]. In normal metallic ferromagnets, they are suppressed by the stronger internal magnetic field. In larger samples, so many loops of different sizes contribute that the net effect is the suppression of weak localization by a magnetic field, leading to negative magnetoresistance. Negative

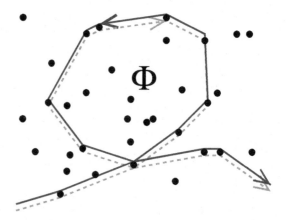

**FIGURE 9.7** Scattering paths traversed in opposite directions. The black dots represent random scatterers. The constructive interference of the wave functions of carriers following the solid and dashed paths leads to weak localization. A magnetic flux $\Phi$ threading the loop leads to a phase difference between the two paths.

magnetoresistance has been attributed to weak localization for (Ga,Mn)As films with up to $x \approx 8\%$ Mn and non-optimal $T_C$ [216]. The effect was observed at low temperatures, $T \lesssim 3\,\mathrm{K}$, and in weak magnetic fields on the order of 20 mT [216].

Since spin-orbit coupling in (Ga,Mn)As is strong, its effect on weak localization has to be taken into account. For a parabolic band with strong spin-orbit coupling one expects *positive* magnetoresistance (weak antilocalization) [199, 215]. Rokhinson et al. [216] argue that their observation of *negative* magnetoresistance in (Ga,Mn)As with $x \approx 8\%$ therefore indicates that $E_F$ is not located in the VB, but in an IB. However, it is unclear why spin-orbit effects should be significantly smaller in an IB derived from VB states. Garate et al. [215] show that within a $\mathbf{k} \cdot \mathbf{p}$ approach, the complex band structure leads to negative magnetoresistance and weak localization, as observed. Weak antilocalization has also been invoked to explain the broad resistivity peak around $T_C$ [217]. However, it is now thought that the peak maximum is located a few Kelvin above $T_C$, and that the peak is mostly due to temperature-dependent changes in the band structure and the scattering rate [1, 188]. On the other hand, a singularity in the temperature derivative of the resistivity, $d\rho/dT$, has been observed precisely at $T_C$ [21, 24]. This singularity is coupled to the one observed in the specific heat [218] and is consistent with the Fisher–Langer theory [219] for the resistive anomaly in metallic ferromagnets [4]. Deviations from Fisher–Langer theory are more likely to be observed if the resistivity at $T_C$ is dominated by disorder scattering [217].

## 9.2.8 DISORDER AND MAGNETISM

Disorder affects the magnetic properties in two principal ways. First, the carrier-mediated magnetic interaction is sensitive to carrier scattering. Second, the Mn local moments occupy random positions in the DMS lattice. We will discuss these mechanisms in turn.

The Curie temperature of (Ga,Mn)As decreases with decreasing $x$, but remains non-zero in the doping range $1\% \lesssim x \lesssim 1.5\%$, where the material is insulating at low temperatures. It has been suggested that an RKKY description remains valid in the presence of strong localization [93, 101–103]. In this regime, the RKKY interaction between two typical impurity spins decays exponentially on the scale of the localization length [102, 103]. Ferromagnetic order is still possible, as long as the localization length is larger than the separation between impurity spins. While the RKKY paradigm might hold in this regime, disorder effects are very strong, precluding a VCA/MF description. Indeed, Bouzerar et al. [109, 110, 220] find that the RKKY interaction calculated including electronic disorder in the coherent-potential approximation strongly deviates from the clean case and leads to enhanced $T_C$. Myers et al. [72] have measured the dependence of $T_C$ on the number $p$ of holes per Mn in this regime by systematically varying the As:Ga ratio across their samples during growth. They indeed find a strong deviation from the prediction $T_C \propto p^{1/3}$ of VCA/MF theory [72].

In this regime, a BMP picture may be more appropriate [105, 106, 191, 206–209]. For strong compensation, a single BMP comprises a hole and several ferromagnetically aligned impurity spins. For larger impurity concentrations, the BMPs overlap, leading to a ferromagnetic interaction between impurity spins within the same cluster of overlapping BMPs. At zero temperature, ferromagnetic long-range order emerges if the BMPs percolate. At finite temperatures, thermal fluctuations destroy the alignment between weakly coupled clusters. Kaminski and Das Sarma [207] have calculated $T_C$ based on this picture. A more recent theory based on variable-range hopping [209] is similar in spirit. Above the percolation transition at $T_C$, rare but large ordered regions have been predicted to lead to Griffiths singularities [221], which would result in characteristic scaling behavior of the susceptibility and specific heat. Such ordered regions have been observed in (In,Mn)Sb [222], but Griffiths singularities have not yet been found in III–V DMS.

Carrier scattering has another important consequence: One cannot simply grow quaternary DMS such as (Ga,Mn)(As,P) and hope to achieve properties interpolating between the end points. The alloying introduces additional disorder, which dramatically reduces the mean free path and $T_C$ [125].

We now turn to the effects of disorder due to the random distribution of magnetic dopants. In the extreme case, certain regions of the sample may be effectively decoupled from the rest and thus not take part in the spontaneous magnetization. There is evidence that such regions are present even in high-quality samples [148], whereas a recent study using μSR as the local probe found homogeneous ferromagnetism in the metallic regime and even through the metal-insulator transition [204]. A random distribution can lead to non-collinear ground states, since the RKKY interaction changes sign as a function of separation, which leads to frustration, and is anisotropic both in real space and in spin space [10, 68, 105, 175, 176, 223, 224]. The resulting reduction of the low-temperature magnetization has been estimated to be small [1, 68, 225].

In various theoretical approaches, disorder is generically found to lead to an anomalous temperature dependence of the magnetization, characterized by an upward curvature or nearly linear behavior in a broad intermediate temperature range [103, 117, 178, 191, 206, 224, 226–228]. Magnetization curves of this type have been observed for insulating (Ga,Mn)As [229–231] and also for early metallic samples [168, 229, 230]. Note that an upward curvature can also result from a temperature-induced change of the easy axis [1]. New high-quality samples show normal, Brillouin function-like magnetization curves [21].

Simultaneously treating both effects of disorder—the carrier scattering and the spatial disorder of spins—in a realistic calculation is computationally demanding. Extending earlier work [191, 232, 233], Yildirim et al. [117] perform large-scale Monte Carlo simulations of (Ga,Mn)As, where the local spins are treated classically, and an electronic tight-binding Hamiltonian is diagonalized for each spin configuration. For reasonably large $x$ and weak compensation, the magnetization curves are fairly

Brillouin function-like, in agreement with experiments [21]. However, to achieve realistic values of $T_C$, $J_{pd}$ had to be assumed to be much smaller than usually thought [234]. Moreover, the Coulomb disorder potential was neglected, although it is likely the larger contribution to the disorder seen by the holes [1]. Its inclusion would probably reduce $T_C$, counteracting the effect of a larger $J_{pd}$.

The disorder due to the random distribution of spins has also been studied in spin-only models within the LRPA [110, 177, 183], which includes disorder by employing a large supercell. As noted above, the results for $T_C(x)$ are in good agreement with experiments. Disorder is crucial for this. Remarkably, $T_C(x)$ calculated based on a spin-only model derived from a V-J model using a simple, unrealistic electronic band agrees well with DFT calculations [110]. This suggests that details of the band structure are not crucial for certain magnetic properties. However, this is not the case for properties relying on anisotropic exchange.

## 9.3 III–V DMS WITH WIDE GAPS

Wide-gap III–V DMS such as (Ga,Mn)N are interesting, since for several of these compounds Curie temperatures above room temperature have been reported [235–244]. Dietl [103, 147] has discussed several scenarios for apparent ferromagnetic response of DMS. For quasi-uniform DMS with merged VB and IB, the kinetic-exchange model can be applied, leading to an RKKY-type magnetic interaction. Indeed, both the arguments based on a parabolic band given in Section 9.2.6 and more sophisticated $\mathbf{k} \cdot \mathbf{p}$ theory predict high $T_C$ values [101]. Deep Mn-derived levels in the gap likely invalidate the underlying kinetic-exchange model for wide-gap DMS, though. Ferromagnetic response can also result from precipitates of known magnetic compounds [103, 147]. In this case, the semiconducting and magnetic properties result from different parts of the sample. A more interesting possibility is chemical nanoscale phase separation into regions rich and poor in magnetic ions [103, 147]. In contrast to precipitates, the crystal structure of the rich regions is in this case determined by the host, which can stabilize phases that are unstable in the bulk [103, 147]. Nanoscale phase separation has been observed in (Ga,Mn)N [245] and (Ga,Fe) [242] as well as in the intermediate-gap DMS (Ga,Mn)As [246]. Note that magnetic and transport properties can still be strongly coupled in such systems. In both cases of phase separation, an apparently ferromagnetic response is seen below a blocking temperature $T_B$: For $T < T_B$, the thermal energy is too small to overcome the size-dependent magnetic anisotropy energy of the magnetic regions, which therefore show frozen magnetic moments. This superparamagnetic behavior has been observed for Gd-ion-implanted GaN [247] and for (Ga,Fe)N [242]. Ney et al. [248] have pointed out that a large fraction of the apparently ferromagnetic signals could result from metastable states. Careful analysis is required to check that static, quasi-uniform ferromagnetism is actually present in a wide-gap DMS. In the following, we discuss such a still partially hypothetical state.

We concentrate on (Ga,Mn)N. Like in (Ga,Mn)As, substitutional $Mn^{2+}$ introduces one hole, which is attracted by the $Mn^{2+}$ core. Neglecting hybridization, one would find a hydrogenic acceptor state. However, the crucial difference between (Ga,Mn)N and (Ga,Mn)As is that in the nitride, the $d^5 \rightarrow d^4$ ($Mn^{2+} \rightarrow Mn^{3+}$) transition lies in the band gap [65, 66, 249]. It is probably deeper than the bound acceptor state, becoming a $d^4 \rightarrow d^5$ transition. Due to the smaller energy difference, the effect of hybridization, and, in particular, the $pd$-exchange interaction $J_{pd}$ are larger, as seen in photoemission as well as in optical and x-ray spectroscopy for (Ga,Mn)N [250, 251]. $J_{pd}$ may even become ferromagnetic [252, 253]. Hybridization and level repulsion lead to an unoccupied level with more than 50% $d$-orbital character deep in the gap, and an occupied level with more than 50% VB character, which may occur as a resonance in the VB.

The strong $pd$-hybridization agrees with LDA+$U$ and LSDA+$U$ calculations [63, 64, 254], which find large Mn $d$-orbital weight at the Fermi energy in (Ga,Mn)N unlike in (Ga,Mn)As [63, 64]. Employing self-interaction corrected LSDA, Schulthess et al. [65, 66] also find a large Mn $d$-orbital weight in the gap, close to $E_F$, and a predominantly $d^4$ state of the Mn dopants for both (Ga,Mn)N and (Ga,Mn)P, but not for (Ga,Mn)As. This picture is supported by optical spectroscopy [255], x-ray spectroscopy [250], and transport and magnetization measurements [256]. However, in other x-ray spectroscopy experiments [257], (Ga,Mn)P appears to be more similar to (Ga,Mn) As, suggesting that its properties are intermediate between (Ga,Mn)As and (Ga,Mn)N.

The overlap between the strongly $d$-orbital-like states in the gap is small. The IB formed for given Mn doping $x$ is thus narrower than in (Ga,Mn)As. Furthermore, it is much deeper in the gap. Therefore, a significant overlap, let alone merging, of IB and VB is unlikely [258]. The density of states is sketched in Figure 9.8. Dynamical mean-field theory calculations assuming local Coulomb and exchange interactions between holes and Mn impurities support this view [60]. The kinetic-exchange picture with the Fermi energy in a weakly perturbed VB is thus not applicable Figure 9.8.

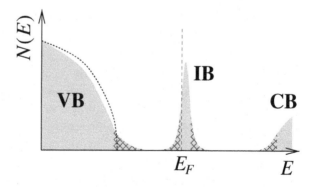

**FIGURE 9.8** Sketch of the density of states for a typical wide-gap III–V DMS. Compared to Figure 9.3c, the IB is both narrower and deeper in the gap. The Fermi energy is shown for partial compensation.

The high observed $T_C$ [235, 240, 241] at concentrations of magnetic dopants below the percolation threshold suggests a carrier-mediated mechanism. The suppression of the remanent magnetization in (Ga,Mn)N below about 10 K, accompanied by an increase of the resistivity attributed to localization supports a carrier-mediated scenario [241]. This picture is also favored by the observation that exposure to atomic hydrogen has a strong effect on the magnetic order [259], presumably by changing the compensation. Since in the absence of compensation, the narrow IB would be completely empty, a carrier-mediated mechanism requires partial compensation [65]. A high concentration of nitrogen vacancies, which are donors with one weakly bound electron [260], is typical for GaN and thus provides a likely source of compensation. Transport and x-ray spectroscopy experiments [261] on (Ga,Mn)N with $x \lesssim 1\%$, grown by metal-organic chemical vapor deposition, suggest a strong increase in the concentration of nitrogen vacancies with $x$ (self-compensation). Due to the large $d$-orbital weight in the IB, strong compensation leads to strong charge fluctuations in the Mn $d$-shell. This mixed-valence nature of Mn might lead to interactions through the double-exchange mechanism [64, 154, 240, 241, 254, 262], which can lead to high $T_C$ for large $x$.

Dhar et al. have observed ferromagnetic order in MBE-grown [237] and ion-implanted [239] GaN with very low concentrations of Gd, $n_{Gd}$, down to $7 \times 10^{15}\,\mathrm{cm}^{-3}$. Dividing the measured magnetization by $n_{Gd}$, they obtain magnetic moments per Gd of up to 5000 $\mu_B$. This, of course, implies that most of the magnetization is not carried by the Gd dopants. Giant moments per Gd have been confirmed for Gd-implanted GaN by Wang et al. [244]. The natural candidates are *defects* [263]. This interpretation is supported by the observations that (a) similar magnetic properties are found if the bare sapphire substrate or Si wavers are implanted with Gd, showing that the effect is not material specific [244], and (b) the magnetic signal vanishes if the implanted sample is kept at room temperature for a few weeks, presumably due to removing part of the defects by annealing [244]. Since the concentrations of unwanted impurities are small [237], native defects, most likely nitrogen vacancies [260], are probably responsible. Electrons weakly bound to nitrogen-vacancy donors [260] would each contribute a spin 1/2. Indeed, the total magnetization is found to depend only weakly on $n_{Gd}$ for small concentrations, followed by a crossover to a roughly linear behavior for larger $n_{Gd}$ with a coefficient consistent with the $f^7$ configuration of $Gd^{3+}$ [237]. However, it is not yet understood how very weak Gd doping could induce magnetic ordering of the defect spins. It has been found that in some Gd-doped GaN samples the magnetic effects are metastable [248], perhaps due to electrons trapped by defects. Magnetic moments of defects have also been invoked to explain magnetic signals in other semiconductors, including samples without transition-metal or rare-earth dopants [264–267]. This so-called $d^0$ magnetism [264], which is particularly relevant for oxides, is discussed in Chapter 10, Volume 2.

# REFERENCES

1. T. Jungwirth, J. Sinova, J. Mašek, J. Kučera, and A. H. MacDonald, Theory of ferromagnetic (III,Mn)V semiconductors, *Rev. Mod. Phys.* **78**, 809–864 (2006).

2. T. Dietl, A ten-year perspective on dilute magnetic semiconductors and oxides, *Nat. Mater.* **9**, 965–974 (2010).

3. T. Dietl and H. Ohno, Dilute ferromagnetic semiconductors: Physics and spintronic structures, *Rev. Mod. Phys.* **86**, 187–251 (2014).

4. T. Jungwirth, J. Wunderlich, V. Novák et al., Spin-dependent phenomena and device concepts explored in (Ga,Mn)As, *Rev. Mod. Phys.* **86**, 855–896 (2014).

5. T. Jungwirth, III–V based magnetic semiconductors, *Handbook of Spintronics: Part V.*, Y. Xu, D. D. Awschalom, and J. Nitta (Eds), Springer, Dordrecht, pp. 465–521, (2016).

6. K. Sato, L. Bergqvist, J. Kudrnovský et al., First-principles theory of dilute magnetic semiconductors, *Rev. Mod. Phys.* **82**, 1633–1690 (2010).

7. A. Zunger, S. Lany, and H. Raebiger, Trend: The quest for dilute ferromagnetism in semiconductors: Guides and misguides by theory, *Physics* **3**, 53 (2010).

8. S. J. Pearton, C. R. Abernathy, M. E. Overberg et al., Wide band gap ferromagnetic semiconductors and oxides, *J. Appl. Phys.* **93**, 1–12 (2003).

9. C. Liu, F. Yun, and H. Morkoç, Ferromagnetism of ZnO and GaN: A review, *J. Mater. Sci. Mater. Electron.* **16**, 555–597 (2005).

10. C. Timm, Disorder effects in diluted magnetic semiconductors, *J. Phys. Condens. Matter* **15**, R1865–R1896 (2003).

11. X. Liu and J. K. Furdyna, Ferromagnetic resonance in $Ga_{1-x}Mn_xAs$ dilute magnetic semiconductors, *J. Phys. Condens. Matter* **18**, R245–R279 (2006).

12. T. Dietl, Interplay between carrier localization and magnetism in diluted magnetic and ferromagnetic semiconductors, *J. Phys. Soc. Jpn.* **77**, 031005 (2008).

13. K. S. Burch, D. D. Awschalom, and D. N. Basov, Optical properties of III–Mn–V ferromagnetic semiconductors, *J. Magn. Magn. Mater.* **320**, 3207–3228 (2008).

14. A. Bonanni and T. Dietl, A story of high-temperature ferromagnetism in semiconductors, *Chem. Soc. Rev.* **39**, 528–539 (2010).

15. N. Nagaosa, J. Sinova, S. Onoda, A. H. MacDonald, and N. P. Ong, Anomalous hall effect, *Rev. Mod. Phys.* **82**, 1539–1592 (2010).

16. T. Dietl, K. Sato, T. Fukushima et al., Spinodal nanodecomposition in semiconductors doped with transition metals, *Rev. Mod. Phys.* **87**, 1311–1377 (2015).

17. T. Jungwirth, J. Sinova, A. H. MacDonald et al., Character of states near the Fermi level in (Ga,Mn)As: Impurity to valence band crossover, *Phys. Rev. B* **76**, 125206 (2007).

18. H. Ohno, H. Munekata, T. Penney, S. von Molnár, and L. L. Chang, Magnetotransport properties of p-type (In,Mn)As diluted magnetic III–V semiconductors, *Phys. Rev. Lett.* **68**, 2664–2667 (1992).

19. H. Ohno, A. Shen, F. Matsukura et al., (Ga,Mn)As: A new diluted magnetic semiconductor based on GaAs, *Appl. Phys. Lett.* **69**, 363–365 (1996).

20. F. Matsukura, H. Ohno, A. Shen, Y. Sugawara, Transport properties and origin of ferromagnetism in (Ga,Mn)As, *Phys. Rev. B* **57**, R2037–R2040 (1998).

21. V. Novák, K. Olejnk, J. Wunderlich et al., Curie point singularity in the temperature derivative of resistivity in (Ga,Mn)As, *Phys. Rev. Lett.* **101**, 077201 (2008).

22. M. Wang, R. P. Campion, A. W. Rushforth, K. W. Edmonds, C. T. Foxon, and B. L. Gallagher, Achieving high curie temperature in (Ga,Mn)As, *Appl. Phys. Lett.* **93**, 132103 (2008).

23. L. Chen, S. Yan, P. F. Xu et al., Low-temperature magnetotransport behaviors of heavily Mn-doped (Ga,Mn)As films with high ferromagnetic transition temperature, *Appl. Phys. Lett.* **95**, 182505 (2009).

24. P. Němec, V. Novák, N. Tesařová et al., The essential role of carefully optimized synthesis for elucidating intrinsic material properties of (Ga,Mn)As, *Nat. Commun.* **4**, 1422 (2013).

25. B. C. Chapler, R. C. Myers, S. Mack et al., Infrared probe of the insulator-to-metal transition in $Ga_{1-x}Mn_xAs$ and $Ga_{1-x}Be_xAs$, *Phys. Rev. B* **84**, 081203(R) (2011).

26. A. M. Nazmul, T. Amemiya, Y. Shuto, S. Sugahara, and M. Tanaka, High temperature ferromagnetism in GaAs-based heterostructures with Mn δ-Doping, *Phys. Rev. Lett.* **95**, 017201 (2005).

27. M. A. Scarpulla, R. Farshchi, P. R. Stone et al., Electrical transport and ferromagnetism in $Ga_{1-x}Mn_xAs$ synthesized by ion implantation and pulsed-laser melting, *J. Appl. Phys.* **103**, 073913 (2008).

28. T. Schallenberg and H. Munekata, Preparation of ferromagnetic (In,Mn)As with a high Curie temperature of 90 K, *Appl. Phys. Lett.* **89**, 042507 (2006).

29. N. T. Tu, P. N. Hai, L. D. Anh, and M. Tanaka, High-temperature ferromagnetism in heavily Fe-doped ferromagnetic semiconductor (Ga,Fe)Sb, *Appl. Phys. Lett.* **108**, 192401 (2016).

30. N. T. Tu, P. N. Hai, L. D. Anh, and M. Tanaka, High-temperature ferromagnetism in new n-type Fe-doped ferromagnetic semiconductor (In,Fe)Sb, *Appl. Phys. Express* **11**, 063005 (2018).

31. S. J. Potashnik, K. C. Ku, R. Mahendiran et al., Saturated ferromagnetism and magnetization deficit in optimally annealed $Ga_{1-x}Mn_xAs$ epilayers, *Phys. Rev. B* **66**, 012408 (2002).

32. R. P. Campion, K. W. Edmonds, L. X. Zhao et al., The growth of GaMnAs films by molecular beam epitaxy using arsenic dimers, *J. Cryst. Growth* **251**, 311–316 (2003).

33. A. Richardella, P. Roushan, S. Mack et al., Visualizing critical correlations near the metal-insulator transition in $Ga_{1-x}Mn_xAs$, *Science* **327**, 665–669 (2010).

34. E. Abe, F. Matsukura, H. Yasuda, Y. Ohno, and H. Ohno, Molecular beam epitaxy of III–V diluted magnetic semiconductor (Ga,Mn)Sb, *Phys. E* **7**, 981–985 (2000).

35. T. Wojtowicz, G. Cywi n'ski, W. L. Lim et al., $In_{1-x}Mn_xSb$—a narrow-gap ferromagnetic semiconductor, *Appl. Phys. Lett.* **82**, 4310–4312 (2003).

36. M. Csontos, G. Mihály, B. Jankó, T. Wojtowicz, X. Liu, and J. K. Furdyna, Pressure-induced ferromagnetism in (In,Mn)Sb dilute magnetic semiconductor, *Nat. Mater.* **4**, 447–449 (2005); G. Mihály, M. Csontos, S. Bordács et al., Anomalous hall effect in the (In,Mn)Sb dilute magnetic semiconductor, *Phys. Rev. Lett.* **100**, 107201 (2008).

37. M. Khalid, E. Weschke, W. Skorupa, M. Helm, and S. Zhou, Ferromagnetism and impurity band in a magnetic semiconductor: InMnP, *Phys. Rev. B* **89**, 121301(R) (2014); M. Khalid, K. Gao, E. Weschke et al., A comprehensive study of the magnetic, structural, and transport properties of the III–V ferromagnetic semiconductor InMnP, *J. Appl. Phys.* **117**, 043906 (2015).

38. R. Bouzerar, D. May, U. Löw, D. Machon, P. Melinon, and G. Bouzerar, *Effective Spin Models and Critical Temperatures for Diluted Magnetic Semiconductors*, Verhandlungen der DPG, Frühjahrstagung Regensburg, HL 96.3 (2016).

39. P. M. Krstajič, F. M. Peeters, V. A. Ivanov, V. Fleurov, and K. Kikoin, Double-exchange mechanisms for Mn-doped III–V ferromagnetic semiconductors, *Phys. Rev. B* **70**, 195215 (2004).

40. N. Almeleh and B. Goldstein, Electron paramagnetic resonance of manganese in gallium arsenide, *Phys. Rev.* **128**, 1568–1570 (1962).

41. J. Szczytko, A. Twardowski, K. Świątek et al., Mn impurity in $Ga_{1-x}Mn_xAs$ epilayers, *Phys. Rev. B* **60**, 8304–8308 (1999).

42. Y. Sasaki, X. Liu, J. K. Furdyna, M. Palczewska, J. Szczytko, and A. Twardowski, Ferromagnetic resonance in GaMnAs, *J. Appl. Phys.* **91**, 7484–7486 (2002).

43. R. A. Chapman and W. G. Hutchinson, Photoexcitation and photoionization of neutral manganese acceptors in gallium arsenide, *Phys. Rev. Lett.* **18**, 443–445 (1967).

44. M. Linnarsson, E. Janzén, B. Monemar, M. Kleverman, and A. Thilderkvist, Electronic structure of the GaAs:Mn$_{Ga}$ center, *Phys. Rev. B* **55**, 6938–6944 (1997).

45. J. Schneider, U. Kaufmann, W. Wilkening, M. Baeumler, and F. Köhl, Electronic structure of the neutral manganese acceptor in gallium arsenide, *Phys. Rev. Lett.* **59**, 240–243 (1987).

46. J. S. Blakemore, W. J. Brown Jr., M. L. Stass, and D. A. Woodbury, Thermal activation energy of manganese acceptors in gallium arsenide as a function of impurity spacing, *J. Appl. Phys.* **44**, 3352–3354 (1973).

47. A. M. Yakunin, A. Yu. Silov, P. M. Koenraad, W. Van Roy, J. De Boeck, and J. H. Wolter, Imaging of the (Mn$^{2+}$3d$^5$ + hole) complex in GaAs by cross-sectional scanning tunneling microscopy, *Phys. E* **21**, 947–950 (2004); A. M. Yakunin, A. Yu. Silov, P. M. Koenraad et al., Spatial structure of an individual Mn acceptor in GaAs, *Phys. Rev. Lett.* **92**, 216806 (2004).

48. J. Okabayashi, A. Kimura, O. Rader et al., Core-level photoemission study of Ga$_{1-x}$Mn$_x$As, *Phys. Rev. B* **58**, R4211–R4214 (1998).

49. J. Okabayashi, A. Kimura, T. Mizokawa, A. Fujimori, T. Hayashi, and M. Tanaka, Mn 3d partial density of states in Ga$_{1-x}$Mn$_x$As studied by resonant photoemission spectroscopy, *Phys. Rev. B* **59**, R2486–R2489 (1999).

50. A. Baldereschi and N. O. Lipari, Spherical model of shallow acceptor states in semiconductors, *Phys. Rev. B* **8**, 2697–2709 (1973).

51. A. K. Bhattacharjee and C. B. à la Guillaume, Model for the Mn acceptor in GaAs, *Solid State Commun.* **113**, 17–21 (2000).

52. R. F. Kirkman, R. A. Stradling, and P. J. Lin-Chung, An infrared study of the shallow acceptor states in GaAs, *J. Phys. C* **11**, 419–433 (1978).

53. M. S. Dresselhaus, G. Dresselhaus, and A. Jorio, *Group Theory: Application to the Physics of Condensed Matter*, Springer, Berlin, 2008.

54. D. Kitchen, A. Richardella, J.-M. Tang, M. E. Flatté, and A. Yazdani, Atom-by-atom substitution of Mn in GaAs and visualization of their hole-mediated interactions, *Nature* **442**, 436–439 (2006).

55. C. Çelebi, J. K. Garleff, A. Yu. Silov et al., Surface induced asymmetry of acceptor wave functions, *Phys. Rev. Lett.* **104**, 086404 (2010).

56. J.-M. Tang and M. E. Flatté, Multiband tight-binding model of local magnetism in Ga$_{1-x}$Mn$_x$As, *Phys. Rev. Lett.* **92**, 047201 (2004).

57. M. J. Schmidt, K. Pappert, C. Gould, G. Schmidt, R. Oppermann, and L. W. Molenkamp, Bound-hole states in a ferromagnetic (Ga,Mn)As environment, *Phys. Rev. B* **76**, 035204 (2007).

58. A. Zunger in H. Ehrenreich and D. Turnbull (eds), *Solid State Physics*, Vol. 39. Orlando: Academic (1986), p. 275.

59. P. Mahadevan and A. Zunger, Ferromagnetism in Mn-doped GaAs due to substitutional-interstitial complexes, *Phys. Rev. B* **68**, 075202 (2003).

60. F. Popescu, C. Şen, E. Dagotto, and A. Moreo, Crossover from impurity to valence band in diluted magnetic semiconductors: Role of Coulomb attraction by acceptors, *Phys. Rev. B* **76**, 085206 (2007).

61. Y. Yuan, C. Xu, R. Hübner et al., Interplay between localization and magnetism in (Ga,Mn)As and (In,Mn)As, *Phys. Rev. Mater.* **1**, 054401 (2017).

62. L. Craco, M. S. Laad, and E. Müller-Hartmann, *Ab initio* description of the diluted magnetic semiconductor Ga$_{1-x}$Mn$_x$As: Ferromagnetism, electronic structure, and optical response, *Phys. Rev. B* **68**, 233310 (2003).

63. L. M. Sandratskii, P. Bruno, and J. Kudrnovský, On-site Coulomb interaction and the magnetism of (GaMn)N and (GaMn)As, *Phys. Rev. B* **69**, 195203 (2004).

64. M. Wierzbowska, D. Sánchez-Portal, and S. Sanvito, Different origins of the ferromagnetic order in (Ga,Mn)As and (Ga,Mn)N, *Phys. Rev. B* **70**, 235209 (2004).

65. T. C. Schulthess, W. M. Temmerman, Z. Szotek, W. H. Butler, and G. M. Stocks, Electronic structure and exchange coupling of Mn impurities in III–V semiconductors, *Nat. Mater.* **4**, 838–844 (2005).

66. T. C. Schulthess, W. M. Temmerman, Z. Szotek, A. Svane, and L. Petit, First-principles electronic structure of Mn-doped GaAs, GaP, and GaN semiconductors, *J. Phys. Condens. Matter* **19**, 165207 (2007).

67. J. R. Schrieffer and P. A. Wolff, Relation between the Anderson and Kondo Hamiltonians, *Phys. Rev.* **149**, 491–492 (1966); K. A. Chao, J. Spałek, and A. M. Oleś, Canonical perturbation expansion of the Hubbard model, *Phys. Rev. B* **18**, 3453–3464 (1978); B. E. Larson, K. C. Hass, H. Ehrenreich, and A. E. Carlsson, Theory of exchange interactions and chemical trends in diluted magnetic semiconductors, *ibid.* **37**, 4137–4154 (1988).

68. C. Timm and A. H. MacDonald, Anisotropic exchange interactions in III–V diluted magnetic semiconductors, *Phys. Rev. B* **71**, 155206 (2005).

69. E. Nielsen and R. N. Bhatt, Search for ferromagnetism in doped semiconductors in the absence of transition metal ions, *Phys. Rev. B* **82**, 195117 (2010).

70. S. J. Potashnik, K. C. Ku, S. H. Chun, J. J. Berry, N. Samarth, and P. Schiffer, Effects of annealing time on defect-controlled ferromagnetism in $Ga_{1-x}Mn_xAs$, *Appl. Phys. Lett.* **79**, 1495–1497 (2001).

71. M. Luysberg, H. Sohn, A. Prasad et al., Electrical and structural properties of LT-GaAs: Influence of As/Ga flux ratio and growth temperature, *Mater. Res. Soc. Symp. Proc.* **442**, 485–490 (1997).

72. R. C. Myers, B. L. Sheu, A. W. Jackson et al., Antisite effect on hole-mediated ferromagnetism in (Ga,Mn)As, *Phys. Rev. B* **74**, 155203 (2006).

73. R. P. Campion, K. W. Edmonds, L. X. Zhao et al., High-quality GaMnAs films grown with arsenic dimers, *J. Cryst. Growth* **247**, 42–48 (2003).

74. B. Tuck, *Atomic Diffusion in III–V Semiconductors*, IOP Publishing, Bristol, 1988.

75. M. Stellmacher, R. Bisaro, P. Galtier, J. Nagle, K. Khirouni, and J. C. Bourgoin, Defects and defect behaviour in GaAs grown at low temperature, *Semicond. Sci. Technol.* **16**, 440–446 (2001).

76. C. Timm, F. Schäfer, and F. von Oppen, Correlated Defects, Metal-Insulator Transition, and Magnetic order in ferromagnetic semiconductors, *Phys. Rev. Lett.* **89**, 137201 (2002).

77. K. M. Yu, W. Walukiewicz, T. Wojtowicz et al., Effect of the location of Mn sites in ferromagnetic $Ga_{1-x}Mn_xAs$ on its Curie temperature, *Phys. Rev. B* **65**, 201303(R) (2002).

78. K. W. Edmonds, P. Bogusławski, K. Y. Wang et al., Mn interstitial diffusion in (Ga,Mn)As, *Phys. Rev. Lett.* **92**, 037201 (2004).

79. J. Mašek and F. Máca, Interstitial Mn in (Ga,Mn)As: Binding energy and exchange coupling, *Phys. Rev. B* **69**, 165212 (2004).

80. V. Holý, Z. Matěj, O. Pacherová et al., Mn incorporation in as-grown and annealed (Ga,Mn)As layers studied by x-ray diffraction and standing-wave fluorescence, *Phys. Rev. B* **74**, 245205 (2006).

81. G. S. Chang, E. Z. Kurmaev, L. D. Finkelstein et al., Post-annealing effect on the electronic structure of Mn atoms in $Ga_{1-x}Mn_xAs$ probed by resonant inelastic x-ray scattering, *J. Phys. Condens. Matter* **19**, 076215 (2007).

82. J. Blinowski and P. Kacman, Spin interactions of interstitial Mn ions in ferromagnetic GaMnAs, *Phys. Rev. B* **67**, 121204 (2003).

83. M. B. Stone, K. C. Ku, S. J. Potashnik, B. L. Sheu, N. Samarth, and P. Schiffer, Capping-induced suppression of annealing effects on $Ga_{1-x}Mn_xAs$ epilayers, *Appl. Phys. Lett.* **83**, 4568–4570 (2003).

84. J. Adell, I. Ulfat, L. Ilver, J. Sadowski, K. Karlsson, and J. Kanski, Thermal diffusion of Mn through GaAs overlayers on (Ga,Mn)As, *J. Phys. Condens. Matter* **23**, 085003 (2011).

85. L. Horák, J. Matějová, X. Mart et al., Diffusion of Mn interstitials in (Ga,Mn)As epitaxial layers, *Phys. Rev. B* **83**, 245209 (2011).

86. K. W. Edmonds, K. Y. Wang, R. P. Campion et al., Hall effect and hole densities in $Ga_{1-x}Mn_xAs$, *Appl. Phys. Lett.* **81**, 3010–3012 (2002).

87. B. S. Sørensen, J. Sadowski, S. E. Andresen, and P. E. Lindelof, Dependence of Curie temperature on the thickness of epitaxial (Ga,Mn)As film, *Phys. Rev. B* **66**, 233313 (2002).

88. K. C. Ku, S. J. Potashnik, R. Wang et al., Highly enhanced Curie temperature in low-temperature annealed [Ga,Mn]As epilayers, *Appl. Phys. Lett.* **82**, 2302–2304 (2003).

89. K. W. Edmonds, N. R. S. Farley, T. K. Johal et al., Ferromagnetic moment and antiferromagnetic coupling in (Ga,Mn)As thin films, *Phys. Rev. B* **71**, 064418 (2005).

90. P. N. Hai, L. D. Anh, S. Mohan et al., Growth and characterization of n-type electron-induced ferromagnetic semiconductor (In,Fe)As, *Appl. Phys. Lett.* **101**, 182403 (2012); N. T. Tu, P. N. Hai, L. D. Anh, and M. Tanaka, (Ga,Fe)Sb: A p-type ferromagnetic semiconductor, *ibid.* **105**, 132402 (2014)..

91. G. L. Pearson and J. Bardeen, Electrical properties of pure silicon and silicon alloys containing boron and phosphorus, *Phys. Rev.* **75**, 865–883 (1949).

92. W. J. Brown Jr. and J. S. Blakemore, Transport and photoelectrical properties of gallium arsenide containing deep acceptors, *J. Appl. Phys.* **43**, 2242–2246 (1972).

93. T. Dietl, A. Haury, and Y. Merle d'Aubigné, Free carrier-induced ferromagnetism in structures of diluted magnetic semiconductors, *Phys. Rev. B* **55**, R3347–R3350 (1997).

94. T. Jungwirth, W. A. Atkinson, B. H. Lee, and A. H. MacDonald, Interlayer coupling in ferromagnetic semiconductor superlattices, *Phys. Rev. B* **59**, 9818–9821 (1999).

95. T. Dietl, H. Ohno, and F. Matsukura, Hole-mediated ferromagnetism in tetrahedrally coordinated semiconductors, *Phys. Rev. B* **63**, 195205 (2001).

96. T. Jungwirth, K. Y. Wang, J. Mašek et al., Prospects for high temperature ferromagnetism in (Ga,Mn)As semiconductors, *Phys. Rev. B* **72**, 165204 (2005).

97. A. Chattopadhyay, S. Das Sarma, and A. J. Millis, Transition temperature of ferromagnetic semiconductors: A dynamical mean field study, *Phys. Rev. Lett.* **87**, 227202 (2001).

98. C. Santos and W. Nolting, Ferromagnetism in the Kondo-lattice model, *Phys. Rev. B* **65**, 144419 (2002); J. Kienert and W. Nolting, Magnetic phase diagram of the Kondo lattice model with quantum localized spins, *ibid.* **73**, 224405 (2006).

99. M. Turek, J. Siewert, and J. Fabian, Electronic and optical properties of ferromagnetic $Ga_{1-x}Mn_xAs$ in a multiband tight-binding approach, *Phys. Rev. B* **78**, 085211 (2008).

100. C. Timm, F. Schäfer, and F. von Oppen, Comment on "Effects of disorder on ferromagnetism in diluted magnetic semiconductors", *Phys. Rev. Lett.* **90**, 029701 (2003).

101. T. Dietl, H. Ohno, F. Matsukura, J. Cibert, and D. Ferrand, Zener model description of ferromagnetism in zinc-blende magnetic semiconductors, *Science* **287**, 1019–1022 (2000).

102. J. A. Sobota, D. Tanasković, and V. Dobrosavljević, RKKY interactions in the regime of strong localization, *Phys. Rev. B* **76**, 245106 (2007).

103. T. Dietl, Origin of ferromagnetic response in diluted magnetic semiconductors and oxides, *J. Phys. Condens. Matter* **19**, 165204 (2007).

104. M. van Schilfgaarde and O. N. Mryasov, Anomalous exchange interactions in III–V dilute magnetic semiconductors, *Phys. Rev. B* **63**, 233205 (2001).

105. G. A. Fiete, G. Zaránd, and K. Damle, Effective hamiltonian for $Ga_{1-x}Mn_xAs$ in the dilute limit, *Phys. Rev. Lett.* **91**, 097202 (2003); G. A. Fiete, G. Zaránd, K. Damle, and C. P. Moca, Disorder, spin-orbit, and interaction effects in dilute $Ga_{1-x}Mn_xAs$, *Phys. Rev. B* **72**, 045212 (2005).

106. C. P. Moca, G. Zaránd, and M. Berciu, Theory of optical conductivity for dilute $Ga_{1-x}Mn_xAs$, *Phys. Rev. B* **80**, 165202 (2009).

107. C. Zener, Interaction between the *d* shells in the transition metals, *Phys. Rev.* **81**, 440–444 (1951).

108. J. König, H.-H. Lin, and A. H. MacDonald, Theory of diluted magnetic semiconductor ferromagnetism, *Phys. Rev. Lett.* **84**, 5628–5631 (2000); J. König, T. Jungwirth, and A. H. MacDonald, Theory of magnetic properties and spin-wave dispersion for ferromagnetic (Ga,Mn)As, *Phys. Rev. B* **64**, 184423 (2001).

109. R. Bouzerar, G. Bouzerar, and T. Ziman, Non-perturbative $V$-$J_{pd}$ model and ferromagnetism in dilute magnets, *EPL* **78**, 67003 (2007); R. Bouzerar and G. Bouzerar, Unified picture for diluted magnetic semiconductors, *ibid.* **92**, 47006 (2010).

110. G. Bouzerar and R. Bouzerar, Unraveling the nature of carrier-mediated ferromagnetism in diluted magnetic semiconductors, *Comptes. Rendus. Phys.* **16**, 731 (2015).

111. J. M. Luttinger and W. Kohn, Motion of electrons and holes in perturbed periodic fields, *Phys. Rev.* **97**, 869–883 (1955).

112. I. Vurgaftman, J. R. Meyer, and L. R. Ram-Mohan, Band parameters for III–V compound semiconductors and their alloys, *J. Appl. Phys.* **89**, 5815–5875 (2001).

113. M. Abolfath, T. Jungwirth, J. Brum, and A. H. MacDonald, Theory of magnetic anisotropy in III$_{1-x}$Mn$_x$V ferromagnets, *Phys. Rev. B* **63**, 054418 (2001).

114. J. C. Slater and G. F. Koster, Simplified LCAO method for the periodic potential problem, *Phys. Rev.* **94**, 1498–1524 (1954); G. F. Koster and J. C. Slater, Wave functions for impurity levels, *ibid.* **95**, 1167–1176 (1954).

115. D. J. Chadi, Spin-orbit splitting in crystalline and compositionally disordered semiconductors, *Phys. Rev. B* **16**, 790–796 (1977).

116. T. Jungwirth, J. Mašek, J. Sinova, and A. H. MacDonald, Ferromagnetic transition temperature enhancement in (Ga,Mn)As semiconductors by carbon codoping, *Phys. Rev. B* **68**, 161202 (2003).

117. Y. Yildirim, G. Alvarez, A. Moreo, and E. Dagotto, Large-scale monte carlo study of a realistic lattice model for Ga$_{1-x}$Mn$_x$As, *Phys. Rev. Lett.* **99**, 057207 (2007).

118. J. Mašek, F. Máca, J. Kudrnovská et al., Microscopic analysis of the valence band and impurity band theories of (Ga,Mn)As, *Phys. Rev. Lett.* **105**, 227202 (2010).

119. H. Ohno and T. Dietl, Spin-transfer physics and the model of ferromagnetism in (Ga,Mn)As, *J. Magn. Magn. Mater.* **320**, 1293–1299 (2008).

120. J. Okabayashi, A. Kimura, O. Rader, et al., Angle-resolved photoemission study of Ga$_{1-x}$Mn$_x$As, *Phys. Rev. B* **64**, 125304 (2001).

121. M. Kobayashi, I. Muneta, Y. Takeda et al., Unveiling the impurity band induced ferromagnetism in the magnetic semiconductor (Ga,Mn)As, *Phys. Rev. B* **89**, 205204 (2014).

122. A. X. Gray, J. Minár, S. Ueda et al., Bulk electronic structure of the dilute magnetic semiconductor Ga$_{1-x}$Mn$_x$As through hard X-ray angle-resolved photoemission, *Nat. Mater.* **11**, 957–962 (2012).

123. I. Di Marco, P. Thunström, M. I. Katsnelson et al., Electron correlations in Mn$_x$Ga$_{1-x}$As as seen by resonant electron spectroscopy and dynamical mean field theory, *Nat. Commun.* **4**, 2645 (2013).

124. J. Fujii, B. R. Salles, M. Sperl et al., Identifying the electronic character and role of the Mn States in the valence band of (Ga,Mn)As, *Phys. Rev. Lett.* **111**, 097201 (2013).

125. P. R. Stone, K. Alberi, S. K. Z. Tardif et al., Metal-insulator transition by isovalent anion substitution in Ga$_{1-x}$Mn$_x$As: Implications to ferromagnetism, *Phys. Rev. Lett.* **101**, 087203 (2008).

126. E. H. C. P. Sinnecker, G. M. Penello, T. G. Rappoport et al., Ion-beam modification of the magnetic properties of Ga$_{1-x}$Mn$_x$As epilayers, *Phys. Rev. B* **81**, 245203 (2010).

127. S. Zhou, L. Li, Y. Yuan et al., Precise tuning of the Curie temperature of (Ga,Mn)As-based magnetic semiconductors by hole compensation: Support for valence-band ferromagnetism, *Phys. Rev. B* **94**, 075205 (2016).

128. K. S. Burch, J. Stephens, R. K. Kawakami, D. D. Awschalom, and D. N. Basov, Ellipsometric study of the electronic structure of $Ga_{1-x}Mn_xAs$ and low-temperature GaAs, *Phys. Rev. B* **70**, 205208 (2004).

129. T. D. Kang, G. S. Lee, H. Lee, D. Koh, and Y. J. Park, Optical properties of $Ga_{1-x}Mn_xAs$ ($0 \leq x \leq 0.09$) studied using spectroscopic ellipsometry, *J. Korean Phys. Soc.* **46**, 482–486 (2005).

130. L. Gluba, O. Yastrubchak et al., On the nature of the Mn-related states in the band structure of (Ga,Mn)As alloys via probing the $E_1$ and optical transitions, *Appl. Phys. Lett.* **105**, 032408 (2014); O. Yastrubchak, J. Sadowski, H. Krzyżanowska et al., Electronic- and band-structure evolution in low-doped (Ga,Mn)As, *J. Appl. Phys.* **114**, 053710 (2013).

131. E. J. Singley, R. Kawakami, D. D. Awschalom, and D. N. Basov, Infrared probe of itinerant ferromagnetism in $Ga_{1-x}Mn_xAs$, *Phys. Rev. Lett.* **89**, 097203 (2002); E. J. Singley, K. S. Burch, R. Kawakami, J. Stephens, D. D. Awschalom, and D. N. Basov, Electronic structure and carrier dynamics of the ferromagnetic semiconductor, *Phys. Rev. B* **68**, 165204 (2003).

132. K. S. Burch, D. B. Shrekenhamer, E. J. Singley et al., Impurity band conduction in a high temperature ferromagnetic semiconductor, *Phys. Rev. Lett.* **97**, 087208 (2006).

133. T. Jungwirth, P. Horodyská, N. Tesařová et al., Systematic study of Mn-doping trends in optical properties of (Ga,Mn)As, *Phys. Rev. Lett.* **105**, 227201 (2010).

134. J. Sinova, T. Jungwirth, S.-R. E. Yang, J. Kučera, and A. H. MacDonald, Infrared conductivity of metallic (III,Mn)V ferromagnets, *Phys. Rev. B* **66**, 041202 (2002).

135. S.-R. E. Yang, J. Sinova, T. Jungwirth, Y. P. Shim, and A. H. MacDonald, Nondrude optical conductivity of (III,Mn)V ferromagnetic semiconductors, *Phys. Rev. B* **67**, 045205 (2003).

136. V. F. Sapega, M. Moreno, M. Ramsteiner, L. Däweritz, and K. H. Ploog, Polarization of valence band holes in the (Ga,Mn)As diluted magnetic semiconductor, *Phys. Rev. Lett.* **94**, 137401 (2005); V. F. Sapega, M. Ramsteiner, O. Brandt, L. Däweritz, and K. H. Ploog, Hot-electron photoluminescence study of the (Ga,Mn)As diluted magnetic semiconductor, *Phys. Rev. B* **73**, 235208 (2006).

137. K. Ando, H. Saito, K. C. Agarwal, M. C. Debnath, and V. Zayets, Origin of the anomalous magnetic circular dichroism spectral shape in ferromagnetic $Ga_{1-x}Mn_xAs$: Impurity bands inside the band gap, *Phys. Rev. Lett.* **100**, 067204 (2008).

138. J.-M. Tang and M. E. Flatté, Magnetic circular dichroism from the impurity band in III–V diluted magnetic semiconductors, *Phys. Rev. Lett.* **101**, 157203 (2008).

139. M. Turek, J. Siewert, and J. Fabian, Magnetic circular dichroism in $Ga_{1-x}Mn_xAs$: Theoretical evidence for and against an impurity band, *Phys. Rev. B* **80**, 161201(R) (2009).

140. J. Mašek, J. Kudrnovský, F. Máca et al., Mn-doped Ga(As,P) and (Al,Ga)As ferromagnetic semiconductors: Electronic structure calculations, *Phys. Rev. B* **75**, 045202 (2007).

141. G. Acbas, M.-H. Kim, M. Cukr et al., Electronic structure of ferromagnetic semiconductor $Ga_{1-x}Mn_xAs$ probed by subgap magneto-optical spectroscopy, *Phys. Rev. Lett.* **103**, 137201 (2009).

142. S. Ohya, I. Muneta, P. N. Hai, and M. Tanaka, Valence-band structure of the ferromagnetic semiconductor GaMnAs studied by spin-dependent resonant tunneling spectroscopy, *Phys. Rev. Lett.* **104**, 167204 (2010); S. Ohya, K. Takata, and M. Tanaka, Nearly non-magnetic valence band of the ferromagnetic semiconductor GaMnAs, *Nat. Phys.* **7**, 342–347 (2011).

143. S. Ohya, I. Muneta, Y. Xin, K. Takata, and M. Tanaka, Valence-band structure of ferromagnetic semiconductor (In,Ga,Mn)As, *Phys. Rev. B* **86**, 094418 (2012).

144. T. Dietl and D. Sztenkiel, Reconciling results of tunnelling experiments on (Ga,Mn)As, arXiv:1102.3267 (2011).

145. S. Ohya, K. Takata, I. Muneta, P. N. Hai, and M. Tanaka, Comment on [140], arXiv:1102.4459 (2011).

146. D. Chiba, F. Matsukura, and H. Ohno, Electric-field control of ferromagnetism in (Ga,Mn)As, *Appl. Phys. Lett.* **89**, 162505 (2006).

147. T. Dietl, Origin and control of ferromagnetism in dilute magnetic semiconductors and oxides, *J. Appl. Phys.* **103**, 07D111 (2008).

148. M. Sawicki, D. Chiba, A. Korbecka et al., Experimental probing of the interplay between ferromagnetism and localization in (Ga,Mn)As, *Nat. Phys.* **6**, 22–25 (2010).

149. Y. Nishitani, D. Chiba, M. Endo et al., Curie temperature versus hole concentration in field-effect structures of $Ga_{1-x}Mn_xAs$, *Phys. Rev. B* **81**, 045208 (2010).

150. M. Wang, K. W. Edmonds, B. L. Gallagher et al., High Curie temperatures at low compensation in the ferromagnetic semiconductor (Ga,Mn)As, *Phys. Rev. B* **87**, 121301(R) (2013).

151. D. A. Woodbury and J. S. Blakemore, Impurity conduction and the metal-nonmetal transition in manganese-doped gallium arsenide, *Phys. Rev. B* **8**, 3803–3810 (1973).

152. D. Neumaier, M. Turek, U. Wurstbauer et al., All-electrical measurement of the density of states in (Ga,Mn)As, *Phys. Rev. Lett.* **103**, 087203 (2009).

153. P. A. Lee and T. V. Ramakrishnan, Disordered electronic systems, *Rev. Mod. Phys.* **57**, 287–337 (1985).

154. C. Zener, Interaction between the d-Shells in the Transition Metals. II. Ferromagnetic compounds of manganese with perovskite structure, *Phys. Rev.* **82**, 403–405 (1951).

155. J. Kudrnovský, I. Turek, V. Drchal, F. Máca, P. Weinberger, and P. Bruno, Exchange interactions in III–V and group–IV diluted magnetic semiconductors, *Phys. Rev. B* **69**, 115208 (2004).

156. G. Bouzerar, R. Bouzerar, and O. Cépas, Superexchange induced canted ferromagnetism in dilute magnets, *Phys. Rev. B* **76**, 144419 (2007).

157. J. B. Goodenough, An interpretation of the magnetic properties of the perovskite-type mixed crystals $La_{1-x}Sr_xCoO_{3-\lambda}$, *J. Phys. Chem. Solids* **6**, 287–297 (1958); J. Kanamori, Superexchange interaction and symmetry properties of electron orbitals, *ibid.* **10**, 87–98 (1959).

158. H. Ohno, D. Chiba, F. Matsukura et al., Electric-field control of ferromagnetism, *Nature* **408**, 944–946 (2000); D. Chiba, M. Yamanouchi, F. Matsukura, and H. Ohno, Electrical manipulation of magnetization reversal in a ferromagnetic semiconductor, *Science* **301**, 943–945 (2003).

159. D. Chiba, M. Sawicki, Y. Nishitani, Y. Nakatani, F. Matsukura, and H. Ohno, Magnetization vector manipulation by electric fields, *Nature* **455**, 515–518 (2008); D. Chiba, T. Ono, F. Matsukura, and H. Ohno, Electric field control of thermal stability and magnetization switching in (Ga,Mn)As, *Appl. Phys. Lett.* **103**, 142418 (2013).

160. M. H. S. Owen, J. Wunderlich, V. Novák et al., Low-voltage control of ferromagnetism in a semiconductor p-n junction, *New J. Phys.* **11**, 023008 (2009).

161. T. Niazi, M. Cormier, D. Lucot et al., Electric-field control of the magnetic anisotropy in an ultrathin (Ga,Mn)As/(Ga,Mn)(As,P) bilayer, *Appl. Phys. Lett.* **102**, 122403 (2013).

162. L. D. Anh, P. N. Hai, Y. Kasahara, Y. Isawa, and M. Tanaka, Modulation of ferromagnetism in (In,Fe)As quantum wells via electrically controlled deformation of the electron wave functions, *Phys. Rev. B* **92**, 161201(R) (2015).

163. A. M. Nazmul, S. Kobayashi, S. Sugahara, and M. Tanaka, Electrical and optical control of ferromagnetism in III–V semiconductor heterostructures at high temperature (~100K), *Jpn. J. Appl. Phys.* Part 2, **43**, L233–L236 (2004).

164. E. Dias Cabral, M. A. Boselli, R. Oszwałdowski, I. Žutić, and I. C. da Cunha Lima, Electrical control of magnetic quantum wells: Monte Carlo simulations, *Phys. Rev. B* **84**, 085315 (2011)

165. C. Timm, Charge and magnetization inhomogeneities in diluted magnetic semiconductors, *Phys. Rev. Lett.* **96**, 117201 (2006); Possible magnetic-field-induced voltage and thermopower in diluted magnetic semiconductors, *Phys. Rev. B* **74**, 014419 (2006).

166. S. Koshihara, A. Oiwa, M. Hirasawa et al., Ferromagnetic order induced by photogenerated carriers in magnetic III–V semiconductor heterostructures of (In,Mn)As/GaSb, *Phys. Rev. Lett.* **78**, 4617–4620 (1997); H. Munekata, T. Abe, S. Koshihara et al., Light-induced ferromagnetism in III–V-based diluted magnetic semiconductor heterostructures, *J. Appl. Phys.* **81**, 4862–4864 (1997).

167. J. Wang, I. Cotoros, K. M. Dani, X. Liu, J. K. Furdyna, and D. S. Chemla, Ultrafast enhancement of ferromagnetism via photoexcited holes in GaMnAs, *Phys. Rev. Lett.* **98**, 217401 (2007).

168. H. Ohno, Properties of ferromagnetic III–V semiconductors, *J. Magn. Magn. Mater.* **200**, 110–129 (1999); H. Ohno and F. Matsukura, A ferromagnetic III–V semiconductor: (Ga,Mn)As, *Solid State Commun.* **117**, 179–186 (2001).

169. X. Liu, Y. Sasaki, and J. K. Furdyna, Ferromagnetic resonance in $Ga_{1-x}Mn_xAs$: Effects of magnetic anisotropy, *Phys. Rev. B* **67**, 205204 (2003); X. Liu, W. L. Lim, L. V. Titova et al., External control of the direction of magnetization in ferromagnetic InMnAs/GaSb heterostructures, *Phys. E* **20**, 370–373 (2004).

170. M. Sawicki, F. Matsukura, A. Idziaszek et al., Temperature dependent magnetic anisotropy in (Ga,Mn)As layers, *Phys. Rev. B* **70**, 245325 (2004); M. Sawicki, K.-Y. Wang, K. W. Edmonds et al., In-plane uniaxial anisotropy rotations in (Ga,Mn)As thin films, *ibid.* **71**, 121302 (2005).

171. S. C. Masmanidis, H. X. Tang, E. B. Myers et al., Nanomechanical measurement of magnetostriction and magnetic anisotropy in (Ga,Mn)As, *Phys. Rev. Lett.* **95**, 187206 (2005).

172. K. W. Edmonds, G. van der Laan, N. R. S. Farley et al., Strain dependence of the Mn anisotropy in ferromagnetic semiconductors observed by x-ray magnetic circular dichroism, *Phys. Rev. B* **77**, 113205 (2008).

173. Y. Wan-Lun and T. Tao, Theory of the zero-field splitting of $^6S(3d^5)$ -state ions in cubic crystals, *Phys. Rev. B* **49**, 3243–3252 (1994).

174. K. Yosida, *Theory of Magnetism*. Berlin: Springer (1996).

175. G. Zaránd and B. Jankó, $Ga_{1-x}Mn_xAs$: A frustrated ferromagnet, *Phys. Rev. Lett.* **89**, 047201 (2002).

176. L. Brey and G. Gómez-Santos, Magnetic properties of GaMnAs from an effective Heisenberg Hamiltonian, *Phys. Rev. B* **68**, 115206 (2003).

177. G. Bouzerar, T. Ziman, and J. Kudrnovský, Calculating the Curie temperature reliably in diluted III–V ferromagnetic semiconductors, *Europhys. Lett.* **69**, 812–818 (2005); Compensation, interstitial defects, and ferromagnetism in diluted ferromagnetic semiconductors, *Phys. Rev. B* **72**, 125207 (2005).

178. S. Das Sarma, E. H. Hwang, and A. Kaminski, Temperature-dependent magnetization in diluted magnetic semiconductors, *Phys. Rev. B* **67**, 155201 (2003).

179. P. Mahadevan, A. Zunger, and D. D. Sarma, Unusual directional dependence of exchange energies in GaAs diluted with Mn: Is the RKKY description relevant?, *Phys. Rev. Lett.* **93**, 177201 (2004).

180. H. Ebert and S. Mankovsky, Anisotropic exchange coupling in diluted magnetic semiconductors: *Ab initio* spin-density functional theory, *Phys. Rev. B* **79**, 045209 (2009); S. Mankovsky, S. Polesya, S. Bornemann et al., Spin-orbit coupling effect in (Ga,Mn)As films: Anisotropic exchange interactions and magnetocrystalline anisotropy, *ibid.* **84**, 201201(R) (2011).

181. A. A. Abrikosov and L. P. Gor'kov, On the nature of impurity ferromagnetism, *Sov. Phys. JETP* **16**, 1575–1577 (1963) [*Zh. Eksp. Teor. Fiz.* **43**, 2230–2233 (1962)].

182. T. Jungwirth, J. König, J. Sinova, J. Kučera, and A. H. MacDonald, Curie temperature trends in (III,Mn)V ferromagnetic semiconductors, *Phys. Rev. B* **66**, 012402 (2002).

183. S. Hilbert and W. Nolting, Magnetism in (III,Mn)-V diluted magnetic semiconductors: Effective Heisenberg model, *Phys. Rev. B* **71**, 113204 (2005).

184. S. T. B. Goennenwein, T. Graf, T. Wassner et al., Spin wave resonance in $Ga_{1-x}Mn_xAs$, *Appl. Phys. Lett.* **82**, 730–732 (2003).

185. G. Bouzerar, Magnetic spin excitations in diluted ferromagnetic systems: The case of $Ga_{1-x}Mn_xAs$, *EPL* **79**, 57007 (2007).

186. M. D. Kapetanakis and I. E. Perakis, Spin dynamics in (III,Mn)V ferromagnetic semiconductors: The role of correlations, *Phys. Rev. Lett.* **101**, 097201 (2008).

187. A. Van Esch, L. Van Bockstal, J. De Boeck et al., Interplay between the magnetic and transport properties in the III–V diluted magnetic semiconductor $Ga_{1-x}Mn_xAs$, *Phys. Rev. B* **56**, 13103–13112 (1997).

188. C. P. Moca, B. L. Sheu, N. Samarth, P. Schiffer, B. Jankó, and G. Zaránd, Scaling theory of magnetoresistance and carrier localization in $Ga_{1-x}Mn_xAs$, *Phys. Rev. Lett.* **102**, 137203 (2009).

189. S. Lee, A. Trionfi, T. Schallenberg, H. Munekata, and D. Natelson, Mesoscopic conductance effects in InMnAs structures, *Appl. Phys. Lett.* **90**, 032105 (2007).

190. J. Mašek, I. Turek, V. Drchal, J. Kudrnovský, and F. Máca, Correlated doping in semiconductors: The role of donors in III–V diluted magnetic semiconductors, *Acta Phys. Pol. A* **102**, 673–678 (2002); J. Mašek, I. Turek, J. Kudrnovský, F. Máca, and V. Drchal, Compositional dependence of the formation energies of substitutional and interstitial Mn in partially compensated (Ga,Mn)As, *ibid.* **105**, 637–644 (2004).

191. G. Alvarez, M. Mayr, and E. Dagotto, Phase diagram of a model for diluted magnetic semiconductors beyond mean-field approximations, *Phys. Rev. Lett.* **89**, 277202 (2002).

192. G. Alvarez and E. Dagotto, Clustered states as a new paradigm of condensed matter physics, *J. Magn. Magn. Mater.* **272–276**, 15–20 (2004).

193. P. Mahadevan, J. M. Osorio-Guillén, and A. Zunger, Origin of transition metal clustering tendencies in GaAs based dilute magnetic semiconductors, *Appl. Phys. Lett.* **86**, 172504 (2005).

194. H. Raebiger, M. Ganchenkova, and J. von Boehm, Diffusion and clustering of substitutional Mn in (Ga,Mn)As, *Appl. Phys. Lett.* **89**, 012505 (2006).

195. F. V. Kyrychenko and C. A. Ullrich, Enhanced carrier scattering rates in dilute magnetic semiconductors with correlated impurities, *Phys. Rev. B* **75**, 045205 (2007).

196. F. V. Kyrychenko and C. A. Ullrich, Transport and optical conductivity in dilute magnetic semiconductors, *J. Phys. Condens. Matter* **21**, 084202 (2009); Temperature-dependent resistivity of ferromagnetic $Ga_{1-x}Mn_xAs$: Interplay between impurity scattering and many-body effects, *Phys. Rev. B* **80**, 205202 (2009).

197. M. Zhu, X. Li, G. Xiang, and N. Samarth, Random telegraph noise from magnetic nanoclusters in the ferromagnetic semiconductor (Ga,Mn)As, *Phys. Rev. B* **76**, 201201(R) (2007).

198. H. Bruus and K. Flensberg, *Many-Body Quantum Theory in Condensed Matter Physics*, Oxford University Press, Oxford, 2004.

199. J. Rammer, *Quantum Transport Theory*, Westview Press, Boulder, 2004.

200. E. Abrahams, P. W. Anderson, D. C. Licciardello, and T. V. Ramakrishnan, Scaling theory of localization: Absence of quantum diffusion in two dimensions, *Phys. Rev. Lett.* **42**, 673–676 (1979).

201. B. I. Shklovskii and A. L. Efros, *Electronic Properties of Doped Semiconductors*, Springer, Berlin, (1984).

202. D. M. Basko, I. L. Aleiner, and B. L. Altshuler, Metal-insulator transition in a weakly interacting many-electron system with localized single-particle states, *Ann. Phys.* **321**, 1126–1205 (2006); On the problem of many-body localization, cond-mat/0602510 (2006).

203. T. Słupiński, J. Caban, and K. Moskalik, Hole transport in impurity band and valence bands studied in moderately doped GaAs:Mn single crystals, *Acta Phys. Pol. A* **112**, 325–330 (2007).

204. S. R. Dunsiger, J. P. Carlo, T. Goko et al., Spatially homogeneous ferromagnetism of (Ga,Mn)As, *Nat. Mater.* **9**, 299–303 (2010).

205. N. V. Agrinskaya, V. I. Kozub, D. V. Poloskin, A. V. Chernyaev, and D. V. Shamshur, Transition from strong to weak localization in the split-off impurity band in two-dimensional p-GaAs/AlGaAs structures, *JETP Lett.* **80**, 30–34 (2004) [*P. Zh. Eksp. Teoret. Fiz.* **80**, 36–40 (2004)]; Crossover from strong to weak localization in the split-off impurity band in two-dimensional p-GaAs/AlGaAs structures, *Phys. Stat. Sol. (C)* **3**, 329–333 (2006).

206. M. Berciu and R. N. Bhatt, Effects of disorder on ferromagnetism in diluted magnetic semiconductors, *Phys. Rev. Lett.* **87**, 107203 (2001); Mean-field approach to ferromagnetism in (III,Mn)V diluted magnetic semiconductors at low carrier densities, *Phys. Rev. B* **69**, 045202 (2004).

207. A. Kaminski and S. Das Sarma, Polaron percolation in diluted magnetic semiconductors, *Phys. Rev. Lett.* **88**, 247202 (2002); Magnetic and transport percolation in diluted magnetic semiconductors, *Phys. Rev. B* **68**, 235210 (2003).

208. A. C. Durst, R. N. Bhatt, and P. A. Wolff, Bound magnetic polaron interactions in insulating doped diluted magnetic semiconductors, *Phys. Rev. B* **65**, 235205 (2002).

209. B. L. Sheu, R. C. Myers, J.-M. Tang et al., Onset of ferromagnetism in low-doped $Ga_{1-x}Mn_xAs$, *Phys. Rev. Lett.* **99**, 227205 (2007).

210. A. N. Andriotis and M. Menon, The synergistic character of the defect-induced magnetism in diluted magnetic semiconductors and related magnetic materials, *J. Phys. Condens. Matter* **24**, 455801 (2012).

211. D. Neumaier, K. Wagner, S. Geißler et al., Weak localization in ferromagnetic (Ga,Mn)As nanostructures, *Phys. Rev. Lett.* **99**, 116803 (2007).

212. P. A. Lee, A. D. Stone, and H. Fukuyama, Universal conductance fluctuations in metals: Effects of finite temperature, interactions, and magnetic field, *Phys. Rev. B* **35**, 1039–1070 (1987).

213. K. Wagner, D. Neumaier, M. Reinwald, W. Wegscheider, and D. Weiss, Dephasing in (Ga,Mn)As nanowires and rings, *Phys. Rev. Lett.* **97**, 056803 (2006).

214. L. Vila, R. Giraud, L. Thevenard et al., Universal conductance fluctuations in epitaxial GaMnAs ferromagnets: Dephasing by structural and spin disorder, *Phys. Rev. Lett.* **98**, 027204 (2007).

215. I. Garate, J. Sinova, T. Jungwirth, and A. H. MacDonald, Theory of weak localization in ferromagnetic (Ga,Mn)As, *Phys. Rev. B* **79**, 155207 (2009).

216. L. P. Rokhinson, Y. Lyanda-Geller, Z. Ge et al., Weak localization in $Ga_{1-x}Mn_xAs$: Evidence of impurity band transport, *Phys. Rev. B* **76**, 161201(R) (2007).

217. C. Timm, M. E. Raikh, and F. von Oppen, Disorder-induced resistive anomaly near ferromagnetic phase transitions, *Phys. Rev. Lett.* **94**, 036602 (2005).

218. S. Yuldashev, K. Igamberdiev, S. Lee et al., Specific heat study of GaMnAs, *Appl. Phys. Express* **3**, 073005 (2010).

219. M. E. Fisher and J. S. Langer, Resistive anomalies at magnetic critical points, *Phys. Rev. Lett.* **20**, 665–668 (1968).

220. G. Bouzerar, J. Kudrnovský, and P. Bruno, Disorder effects in diluted ferromagnetic semiconductors, *Phys. Rev. B* **68**, 205311 (2003).

221. V. M. Galitski, A. Kaminski, and S. Das Sarma, Griffiths phase in diluted magnetic semiconductors, *Phys. Rev. Lett.* **92**, 177203 (2004).

222. A. Geresdi, A. Halbritter, M. Csontos et al., Nanoscale spin polarization in the dilute magnetic semiconductor (In,Mn)Sb, *Phys. Rev. B* **77**, 233304 (2008).

223. J. Schliemann and A. H. MacDonald, Noncollinear Ferromagnetism in (III,Mn)V Semiconductors, *Phys. Rev. Lett.* **88**, 137201 (2002); J. Schliemann, Disorder-induced noncollinear ferromagnetism in models for (III,Mn)V semiconductors, *Phys. Rev. B* **67**, 045202 (2003).

224. G. A. Fiete, G. Zaránd, B. Jankó, P. Redliński, and C. P. Moca, Positional disorder, spin-orbit coupling, and frustration in $Ga_{1-x}Mn_xAs$, *Phys. Rev. B* **71**, 115202 (2005).

225. T. Jungwirth, J. Mašek, K. Y. Wang et al., Low-temperature magnetization of (Ga,Mn)As semiconductors, *Phys. Rev. B* **73**, 165205 (2006).

226. M. Mayr, G. Alvarez, and E. Dagotto, Global versus local ferromagnetism in a model for diluted magnetic semiconductors studied with Monte Carlo techniques, *Phys. Rev. B* **65**, 241202(R) (2002).

227. D. J. Priour Jr., E. H. Hwang, and S. Das Sarma, Disordered RKKY lattice mean field theory for ferromagnetism in diluted magnetic semiconductors, *Phys. Rev. Lett.* **92**, 117201 (2004).

228. K. Aryanpour, J. Moreno, M. Jarrell, and R. S. Fishman, Magnetism in semiconductors: A dynamical mean-field study of ferromagnetism in $Ga_{1-x}Mn_xAs$, *Phys. Rev. B* **72**, 045343 (2005).

229. B. Beschoten, P. A. Crowell, I. Malajovich et al., Magnetic circular dichroism studies of carrier-induced ferromagnetism in $(Ga_{1-x}Mn_x)As$, *Phys. Rev. Lett.* **83**, 3073–3076 (1999).

230. R. Mathieu, B. S. Sørensen, J. Sadowski et al., Magnetization of ultrathin (Ga,Mn)As layers, *Phys. Rev. B* **68**, 184421 (2003).

231. W. Limmer, A. Koeder, S. Frank et al., Waag, Electronic and magnetic properties of GaMnAs: annealing effects, *Phys. E* **21**, 970–974 (2004).

232. J. Schliemann, J. König, and A. H. MacDonald, Monte Carlo study of ferromagnetism in (III,Mn)V semiconductors, *Phys. Rev. B* **64**, 165201 (2001).

233. G. Alvarez and E. Dagotto, Single-band model for diluted magnetic semiconductors: Dynamical and transport properties and relevance of clustered states, *Phys. Rev. B* **68**, 045202 (2003).

234. G. Bouzerar and R. Bouzerar, Comment on "Large-Scale Monte Carlo Study of a Realistic Lattice Model for $Ga_{1-x}Mn_xAs$", *Phys. Rev. Lett.* **100**, 229701 (2008); A. Moreo and E. Dagotto, Reply, *ibid.* **100**, 229702 (2008).

235. S. Sonoda, S. Shimizu, T. Sasaki, Y. Yamamoto, and H. Hori, Molecular beam epitaxy of wurtzite (Ga,Mn)N films on sapphire(0001) showing the ferromagnetic behaviour at room temperature, *J. Cryst. Growth* **237–239**, 1358–1362 (2002); T. Sasaki, S. Sonoda, Y. Yamamoto et al., Magnetic and transport characteristics on high Curie temperature ferromagnet of Mn-doped GaN, *J. Appl. Phys.* **91**, 7911–7913 (2002).

236. N. Theodoropoulou, A. F. Hebard, M. E. Overberg et al., Unconventional carrier-mediated ferromagnetism above room temperature in ion-Implanted (Ga,Mn)P:C, *Phys. Rev. Lett.* **89**, 107203 (2002).

237. S. Dhar, O. Brandt, M. Ramsteiner, V. F. Sapega, and K. H. Ploog, Colossal magnetic moment of Gd in GaN, *Phys. Rev. Lett.* **94**, 037205 (2005); S. Dhar, L. Pérez, O. Brandt et al., Gd-doped GaN: A very dilute ferromagnetic semiconductor with a Curie temperature above 300 K, *Phys. Rev. B* **72**, 245203 (2005).

238. R. Rajaram, A. Ney, G. Solomon, J. S. Harris Jr., R. F. C. Farrow, and S. S. P. Parkin, Growth and magnetism of Cr-doped InN, *Appl. Phys. Lett.* **87**, 172511 (2005).

239. S. Dhar, T. Kammermeier, A. Ney et al., Ferromagnetism and colossal magnetic moment in Gd-focused ion-beam-implanted GaN, *Appl. Phys. Lett.* **89**, 062503 (2006).

240. S. Sonoda, I. Tanaka, H. Ikeno et al., Coexistence of $Mn^{2+}$ and $Mn^{3+}$ in ferromagnetic GaMnN, *J. Phys. Condens. Matter* **18**, 4615–4621 (2006).

241. S. Yoshii, S. Sonoda, T. Yamamoto et al., Evidence for carrier-induced high-$T_c$ ferromagnetism in Mn-doped GaN film, *EPL* **78**, 37006 (2007).

242. A. Bonanni, M. Kiecana, C. Simbrunner et al., Paramagnetic GaN:Fe and ferromagnetic (Ga,Fe)N: The relationship between structural, electronic, and magnetic properties, *Phys. Rev. B* **75**, 125210 (2007).

243. J. M. Zavada, N. Nepal, C. Ugolini et al., Optical and magnetic behavior of erbium-doped GaN epilayers grown by metal-organic chemical vapor deposition, *Appl. Phys. Lett.* **91**, 054106 (2007).

244. X. Wang, C. Timm, X. M. Wang et al., Metastable giant moments in Gd-Implanted GaN, Si, and Sapphire, *J. Supercond. Novel Magn.* **24**, 2123–2128 (2011).

245. G. Martnez-Criado, A. Somogyi, S. Ramos et al., Mn-rich clusters in GaN: Hexagonal or cubic symmetry? *Appl. Phys. Lett.* **86**, 131927 (2005).

246. M. Yokoyama, H. Yamaguchi, T. Ogawa, and M. Tanaka, Zinc-blende-type MnAs nanoclusters embedded in GaAs, *J. Appl. Phys.* **97**, 10D317 (2005).

247. S. Y. Han, J. Hite, G. T. Thaler et al., Effect of Gd implantation on the structural and magnetic properties of GaN and AlN, *Appl. Phys. Lett.* **88**, 042102 (2006).

248. A. Ney, R. Rajaraman, T. Kammermeier et al., Metastable magnetism and memory effects in dilute magnetic semiconductors, *J. Phys. Condens. Matter* **20**, 285222 (2008).

249. P. Mahadevan and A. Zunger, First-principles investigation of the assumptions underlying model-Hamiltonian approaches to ferromagnetism of 3$d$ impurities in III–V semiconductors, *Phys. Rev. B* **69**, 115211 (2004).

250. J. I. Hwang, Y. Ishida, M. Kobayashi et al., High-energy spectroscopic study of the III–V nitride-based diluted magnetic semiconductor $Ga_{1-x}Mn_xAs$, *Phys. Rev. B* **72**, 085216 (2005).

251. S. Marcet, D. Ferrand, D. Halley et al., Magneto-optical spectroscopy of (Ga,Mn)N epilayers, *Phys. Rev. B* **74**, 125201 (2006).

252. T. Dietl, Hole states in wide band-gap diluted magnetic semiconductors and oxides, *Phys. Rev. B* **77**, 085208 (2008).

253. W. Pacuski, P. Kossacki, D. Ferrand et al., Observation of strong-coupling effects in a diluted magnetic semiconductor $Ga_{1-x}Fe_xN$, *Phys. Rev. Lett.* **100**, 037204 (2008).

254. B. Sanyal, O. Bengone, and S. Mirbt, Electronic structure and magnetism of Mn-doped GaN, *Phys. Rev. B* **68**, 205210 (2003).

255. T. Graf, M. Gjukic, M. S. Brandt, M. Stutzmann, and O. Ambacher, The $Mn^{3+/2+}$ acceptor level in group III nitrides, *Appl. Phys. Lett.* **81**, 5159–5161 (2002).

256. M. A. Scarpulla, B. L. Cardozo, R. Farshchi et al., Ferromagnetism in $Ga_{1-x}Mn_xP$: Evidence for inter-Mn exchange mediated by localized holes within a detached impurity band, *Phys. Rev. Lett.* **95**, 207204 (2005).

257. P. R. Stone, M. A. Scarpulla, R. Farshchi et al., Mn $L_{3,2}$ x-ray absorption and magnetic circular dichroism in ferromagnetic $Ga_{1-x}Mn_xP$, *Appl. Phys. Lett.* **89**, 012504 (2006).

258. L. Kronik, M. Jain, and J. R. Chelikowsky, Electronic structure and spin polarization of $Mn_xGa_{1-x}N$, *Phys. Rev. B* **66**, 041203 (2002).

259. S. Sonoda, I. Tanaka, F. Oba et al., Awaking of ferromagnetism in GaMnN through control of Mn valence, *Appl. Phys. Lett.* **90**, 012504 (2007).

260. J. Neugebauer and C. G. Van de Walle, Atomic geometry and electronic structure of native defects in GaN, *Phys. Rev. B* **50**, 8067–8070 (1994); P. Bogusławski, E. L. Briggs, and J. Bernholc, Native defects in gallium nitride, *ibid.* **51**, 17255–17258 (1995); S. Limpijumnong and C. G. Van de Walle, Diffusivity of native defects in GaN, *ibid.* **69**, 035207 (2004).

261. X. L. Yang, Z. T. Chen, C. D. Wang et al., Effects of nitrogen vacancies induced by Mn doping in (Ga,Mn)N films grown by MOCVD, *J. Phys. D: Appl. Phys.* **41**, 125002 (2008).

262. K. Sato and H. Katayama-Yoshida, First principles materials design for semiconductor spintronics, *Semicond. Sci. Technol.* **17**, 367–376 (2002).

263. G. M. Dalpian and S.-H. Wei, Electron-induced stabilization of ferromagnetism in $Ga_{1-x}Gd_xN$, *Phys. Rev. B* **72**, 115201 (2005).

264. J. M. D. Coey, M. Venkatesan, and C. B. Fitzgerald, Donor impurity band exchange in dilute ferromagnetic oxides, *Nat. Mater.* **4**, 173–179 (2005).

265. T. Dubroca, J. Hack, R. E. Hummel, and A. Angerhofer, Quasiferromagnetism in semiconductors, *Appl. Phys. Lett.* **88**, 182504 (2006).

266. D. M. Edwards and M. I. Katsnelson, High-temperature ferromagnetism of sp electrons in narrow impurity bands: Application to $CaB_6$, *J. Phys. Condens. Matter* **18**, 7209–7225 (2006).
267. J. M. D. Coey, K. Wongsaprom, J. Alaria, and M. Venkatesan, Charge-transfer ferromagnetism in oxide nanoparticles, *J. Phys. D: Appl. Phys.* **41**, 134012 (2008).

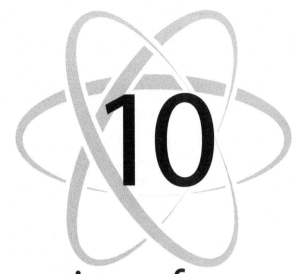

# 10

# Magnetism of
# Dilute Oxides

**J.M.D. Coey**

Reports of Curie temperatures well above room temperature for wide-bandgap oxides doped with a few percent of transition-metal cations have triggered intense interest in these materials as potential magnetic semiconductors. The origin of the magnetism is debated; in some systems, the ferromagnetism can be attributed to nanoparticles of a ferromagnetic secondary phase, but in others, properties are found that are incompatible with any secondary phase, and an intrinsic origin related to structural defects is implicated. The magnetic interactions in these materials are qualitatively different from those in magnetically concentrated compounds, and the standard "$m$-$J$ paradigm" of localized magnetism is unable to explain their behavior. In this chapter, we first summarize the established view of magnetism in the oxides, before reviewing data on a selection of the most widely studied materials; then we consider the physical models that have been advanced to explain the magnetism, and raise issues that still have to be addressed. As noted in Chapter 9, Volume 2, some of the concepts also apply to magnetically doped wide-gap III–V compounds.

## 10.1 INTRODUCTION

Transition-metal oxides have long been regarded as an important and well-understood class of magnetic materials. We have built up an excellent working knowledge of the spinel ferrites $TFe_2O_4$, garnets $M_3Fe_5O_{12}$, and hexagonal ferrites $MFe_{12}O_{19}$, which have widespread industrial applications, and represent about 40% of the global market for hard and soft magnets. This knowledge can be regarded as a familiar landmark in our understanding of magnetism in solids, epitomized by the award of the 1970 Nobel Prize in physics to Louis Néel.

The crystal structures of these oxides are usually based on a close-packed lattice of oxygen anions where the $3d$ cations mostly occupy octahedral interstices with six oxygen neighbors or tetrahedral interstices with four oxygen neighbors. The larger $3d$ cations show some preference for the octahedral sites. Rare earth or alkali ions such as Ba or Sr have a greater coordination number, and may sometimes replace an anion in the close-packed oxygen lattice. A list of the ionic radii of common $3d$ cations is given in Table 10.1, based on a radius of 140 pm for $O^{2-}$.

An ionic picture of the bonding, with formal cation valences of 2+, 3+, or 4+ and an oxygen valence of 2– leading to a $2p^6$ closed shell, is the traditional starting point for looking at the electronic structure. The number of electrons per cation is an integer, and the magnetic moments are well-localized. The oxides are usually insulators or wide-gap semiconductors.

**TABLE 10.1**
**Charge State, Electronic Configuration, and Ionic Radius in Octahedral Coordination for Common 3d Cations in Oxides**

| Cation | Charge State | Configuration | Ionic Radius (pm) |
|--------|--------------|---------------|-------------------|
| Sc | $3+$ | $3d^0$ | 83 |
| Ti | $3+/4+$ | $3d^0/3d^1$ | 61/69 |
| V | $2+$ | $3d^3$ | 72 |
| Cr | $3+$ | $3d^3$ | 64 |
| Mn | $2+/3+/4+$ | $3d^5/3d^4/3d^3$ | 83/65/53 |
| Fe | $2+/3+$ | $3d^6/3d^5$ | 82/65 |
| Co | $2+/3+$ | $3d^7/3d^6$ | 82/61 |
| Ni | $2+/3+$ | $3d^8/3d^7$ | 78/69 |
| Cu | $2+$ | $3d^9$ | 72 |

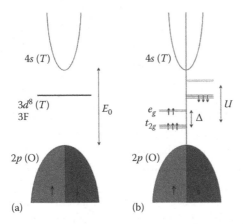

(a)    (b)

**FIGURE 10.1**  Electronic structure diagrams for a $3d$ metal oxide (a) shows the $3d$ electrons as a highly correlated level, (b) shows the $3d$ one-electron states in the crystal field. The example illustrated is a $3d^8$ cation with $S = 1$ in an octahedral site.

They owe their colors—green for yttrium iron garnet, brown for barium hexaferrite—to electronic transitions involving the $3d$ levels, which normally lie in the primary bandgap $E_0$ between a valence band with largely oxygen $2p$ character and a conduction band formed from unoccupied metal $4s$ orbitals. The magnitude of $E_0$ decreases on moving along the $3d$ series from 8.1 eV for $TiO_2$ or 5.7 eV for $Sc_2O_3$, to 4.9 eV for $Ga_2O_3$ or 3.4 eV for ZnO. The measured bandgap $E_g$ can be much less than $E_0$ because of the possibility of $p–d$, or $d–s$ transitions.

According to Hund's rules, the $3d$ ions are in either an A ($d^5$), D ($d^1$, $d^4$, $d^6$, $d^9$), or F ($d^2$, $d^3$, $d^7$, $d^8$) orbital ground state, depending on the number of $d$ electrons they possess. The ground-state levels are split by the crystal/ligand field interaction [1, 2] and there are higher energy levels, which can be excited optically. A schematic representation of the electronic structure is given in Figure 10.1a.

It is often helpful to think of one-electron $d$ states, even though the $d$ shell is strongly correlated. The five $d$ orbitals split into a group of three

$(d_{xy}, d_{yz}, d_{zx})$ and a group of two $(d_{x^2-y^2}, d_{z^2})$ in cubic symmetry—undistorted octahedral and tetrahedral sites have cubic symmetry. These are labeled as $t_{2g}$ and $e_g$ states in octahedral symmetry, and as $t_2$ and $e$ states in tetrahedral symmetry where there is no inversion center. The $e_g$ and $t_2$ overlap and hybridize strongly with the oxygen $2p$ ligand orbitals $(p_x, p_y, p_x)$ but the $t_{2g}$ and $e$ do not. This produces the ligand field splitting of the triplet and the doublet, which reinforces the purely electrostatic crystal field effect of the oxygen charge on the $d$ orbitals to give a splitting $\Delta$. The $d$–$d$ correlations are roughly taken into account in this picture by a separation of the ↑ and ↓ levels by an energy $U$, as shown in Figure 10.1b. When $U > \Delta$, Hund's first rule applies to the ion, and it is in a *high-spin* state. Otherwise, when $U < \Delta$, it is in a *low-spin* state, with a spin moment determined by the lowest orbital occupancy. The crystal field quenches the orbital moments for $3d$ ions in solids, so their magnetic moments are, to a first approximation, spin-only moments of just one Bohr magneton, $\mu_B$, per unpaired electron. $S$ is the good quantum number for the $3d$ ion. The one-electron picture works well for D-state ions, where there is a single electron or hole outside a filled or half-filled shell. It is not nearly as good for the F-state ions.

Three other points complete the beginner's toolkit for $3d$ ions in oxides. (a) Filled shells and half-filled shells are particularly stable. In other cases, a crystal field stabilization energy can be calculated for the ion. For $Cr^{3+}$ $d^3$, for example, this is $3 \times (2/5)\Delta$ in octahedral sites. (The crystal field splitting of the energies of the five $d$ orbitals preserves their center of gravity). (b) Jahn–Teller distortion of the octahedron or tetrahedron may stabilize the energy of the ion. The effect is strongest for a singly occupied $e_g$ or $e$ level—in other words, $d^4$ or $d^9$ ions in octahedral coordination, and $d^1$ and $d^6$ ions in tetrahedral coordination. (c) There is a tendency for the $d$-levels to become more stable as we move across the series from Sc to Zn because of the increasing nuclear charge. For example, the $d$ levels of Ti are close to the conduction band, whereas those of Cu may overlap the $2p$ valence band.

Electrons or holes, introduced into oxides by doping or non-stoichiometry, are often trapped. If they are at all mobile, they tend to form *polarons* with a large effective mass. Electron doping can populate states in the $4s$ conduction band, or $3d^{n+1}$ states in the primary gap, but the extra electrons do not part easily from their dopant atom. Tin in $In_2O_3$ is an exception, and indium tin oxide (ITO) is a good example of a transparent conducting oxide. Hole doping, by oxygen substoichiometry, for example, usually creates donor states near the bottom of the conduction band, which also tend to remain localized. The insulating character of the oxides effectively resists tinkering with the stoichiometry. Transport properties are quite different from those of the classical covalent semiconductors, where electrons or holes diffuse freely in the conduction or valence bands. The best oxide semiconductors are those with the most covalent bonding, especially those with the $3d$ cations in tetrahedral coordination—ZnO and $CuAlO_2$, for example.

Electronic structure calculations based on density functional theory have become very popular. These calculations work well for metals, and can accurately predict the magnetic moments. There are difficulties in applying them to oxides, where the primary gap $E_0$ and the unoccupied electronic levels may be difficult to reproduce.

The sources of the magnetism in oxides are the metal cations, which bear a moment due to unpaired electrons in their $3d$ shells. Any $3d$ cation with $1 \le n \le 9$ has a spin moment, with very few exceptions—$3d^6$ cations in octahedral coordination and $3d^4$ cations in tetrahedral coordination may be in a nonmagnetic low-spin state. The wave functions of the $3d$ cations have exponentially decaying tails, which ensure that there is negligible overlap between cations that are not nearest-neighbors. These neighbors share one or more common oxygen ligands. Hence, the magnetic superexchange coupling essentially involves a threesome of two nearest-neighbor cations and an intervening oxygen, as shown schematically for the $Mn^{2+}$–$O^{2-}$–$Mn^{2+}$ bond in Figure 10.2. The Mn has a half-filled $3d$ shell, so only minority-spin electron transfer from the oxygen to the manganese is possible. A $2p$ electron with $\downarrow$ is transferred to the Mn on the left, leaving a $2p$ hole that can be filled by an electron from the other Mn, provided it is $\downarrow$. Superexchange theory leads to a Heisenberg-type Hamiltonian

$$\mathcal{H} = -2J\mathbf{S_i} \cdot \mathbf{S} \tag{10.1}$$

where the exchange constant $J = J_0 \cos^2 \theta$, where $\theta$ is the superexchange bond angle, and $J_0 = -t^2/U'$, where $t$ is the M–O transfer integral and $U'$ is the on-site Coulomb interaction for adding an electron to the M ion. Typically, $t \approx 0.1$ eV and $U \approx 5$ eV. Hence, the order of magnitude of the exchange constant is 2 meV or −20 K.

The expression for the Curie temperature in mean-field theory is

$$T_C = \frac{2Z_0 J S(S+1)}{3k_B} \tag{10.2}$$

where $k_B$ is the Boltzmann constant and S is the spin. Hence, if $S = 5/2$, the largest value possible in the $3d$ series, and the cation coordination number $Z_0 = 8$, we find $T_C \approx 800$ K. In practice, the superexchange interactions are often frustrated—the antiparallel coupling of all nearest-neighbors cannot be achieved for geometric reasons imposed by the topology of the lattice. Furthermore, Equation 10.2 is known to overestimate $T_C$ by about 30%, because it takes no account of spin-wave excitations. In practice, the magnetic-ordering temperatures of oxides almost never exceed 1000 K. Hematite—$\alpha Fe_2O_3$—has a Néel temperature of 960 K. In most cases, the ordering temperatures are a few hundred Kelvin, at best, as indicated in

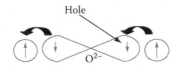

Hole

$O^{2-}$

**FIGURE 10.2**   Superexchange interaction between ions with more than half-filled $d$ shells *via* an $O^{2-}$ anion.

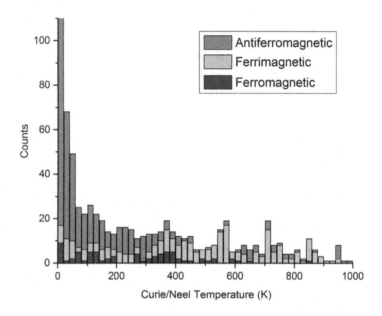

**FIGURE 10.3** Histogram of the magnetic-ordering temperature of a selection of magnetic oxides. (After Ackland, K. et al., *Phys. Rep.* 746, 1, 2018. With permission.)

Figure 10.3. When the oxide is diluted with nonmagnetic cations, the coordination number $Z = Z_0 x$ is reduced, and $T_C$ varies as $x$ above the percolation threshold as shown in Figure 10.4.

The validity of the Heisenberg model in oxides has been amply demonstrated by fitting the exchange constants to the spin-wave dispersion relations determined by inelastic neutron scattering. Complete data sets have been obtained for $Fe_2O_3$, $Cr_2O_3$, and $Fe_3O_4$, among others [3–5]. The values of the exchange constants range from $-30$ to $+6\,K$ in $\alpha$-$Fe_2O_3$. Negative

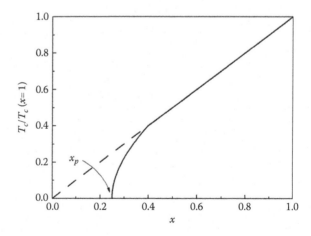

**FIGURE 10.4** Sketch of the magnetic-ordering temperature in a dilute magnetic oxide with nearest-neighbor superexchange interactions. The dashed line is the prediction of mean-field theory for long-range interactions, $x_p$ is the percolation threshold.

(antiferromagnetic) interactions are for superexchange bonds involving the same ions, with large bond angles. Positive (ferromagnetic) values are for bonds involving ions in different valence states, or near-90° bonds between ions in the same valence state.

During the 1960s, a wealth of information was accumulated on the nature of the exchange interactions between different cations for different superexchange bond angles. An extensive set of rules was formulated by Goodenough [6] and Kanamori [7]. These rules were simplified by Anderson [8], as follows:

1. 180° exchange between half-filled orbitals is strong and antiferromagnetic.
2. 90° exchange between half-filled orbitals is ferromagnetic, and rather weak.
3. Exchange between a half-filled orbital and an empty orbital of different symmetry is ferromagnetic and rather weak.

The stronger interactions tend to be negative (antiferromagnetic), coupling the two cation spin moments antiparallel, and leaving no net moment on the intervening oxygen. The three main families of magnetic insulators mentioned in the opening paragraph are all ferrimagnetic, with dominant nearest-neighbor antiferromagnetic exchange between ions on different, unequal sublattices.

Not all transition-metal oxides are ferrimagnetic or antiferromagnetic, and not all are insulators. The extended $3d$ orbitals of the elements at the beginning of the series can form band states. There exist a few ferromagnetic oxides, such as $CrO_2$, which is the only simple oxide that is a ferromagnetic metal ($T_C = 390$ K).

In an effort to increase the magnetization of oxides, beginning in the 1950s, ways were sought to create strong ferromagnetic coupling. A successful approach, implemented in mixed-valence manganites, was to impose two different charge states on the manganese on the octahedral sites of the perovskite structure by including cations on the cubic sites, which had well-defined, but different charge states [9]. A famous example is $(La_{0.7}Sr_{0.3})MnO_3$. Formally, 70% of the manganese is $Mn^{3+}$ $3d^4$, while 30% is $Mn^{4+}$ $3d^3$. The fourth manganese $d$ electrons can hop around among the $3d^3$ ion cores, preserving their spin memory and providing ferromagnetic coupling via a mechanism known as *double exchange* [10]. The delocalization of the $d$ electrons is conditional on parallel alignment of the ion cores, which gives rise to the characteristic colossal magnetoresistance in these oxides near the Curie point, which can be as high as 370 K. Another route to ferromagnetism is illustrated by the double perovskite $Sr_2FeMoO_6$, where the spins of the iron $3d^5$ ion cores are coupled antiparallel to the spins of the electrons in a spin-polarized $4d^1$ molybdenum band, giving ferromagnetic alignment of the iron.

To summarize, concentrated magnetic oxides are well described in terms of the "*m*-J paradigm": there are localized magnetic moments $m$ on the cations, and superexchange or double-exchange interactions to couple them. The paradigm provides a good account of both the magnetic order and the spin waves. Exchange interactions are normally antiferromagnetic, and magnetic-ordering temperatures do not exceed 1000 K.

## 10.2 DILUTE OXIDES

So much for magnetic general knowledge. Our concern in this chapter is dilute magnetic oxides with general formula

$$(M_{1-x}T_x)O_n \tag{10.3}$$

where:

$M$    is a nonmagnetic cation
$T$    is a transition-metal cation
$n$    is an integer or rational fraction
$x < 0.1$ (10%)
Typical values of $x$ are a few percent.

An important limit is $x_p$, the percolation threshold [11], where continuous nearest-neighbor paths first appear, which link $M$ cations throughout the crystal. We are interested in magnetic bond percolation, where cations are regarded as neighbors when their spins are coupled via a superexchange bond. Below $x_p$ there are only isolated cations, and small clusters of nearest-neighbor pairs, triplets, etc. Above $x_p$ there is a bulk cluster, which encompasses most of the magnetic cations. Provided the exchange interactions only involve nearest neighbors, there is no possibility of long-range order below $x_p$, which is approximately $2/Z_0$ [12], where $Z_0$ is the cation coordination number. Depending on the structure, $Z_0$ lies between 6 and 12, which means that $x_p$ is in the range 16%–33%. Our dilute oxides fall below the percolation threshold, and should not be expected to order magnetically. A random distribution of magnetic dopant ions is illustrated schematically in Figure 10.5a. At low concentrations, most of the dopants are isolated, with no magnetic nearest neighbors, and a Curie-law susceptibility is expected for them:

$$\chi_1 = \frac{x_1 C}{T} \tag{10.4}$$

Here $x_1$ is the fraction of isolated ions, and $C$ is the Curie constant $\mu_0 N g^2 \mu_B^2 S(S+1)/3k_B$, where $N$ is the number of cations per unit volume, and $g = 2$ for spin-only moments. As the concentration increases, there will be an increasing proportion of dimers and larger groups. The dimers (and other even-membered groups) have a Curie–Weiss susceptibility

$$\chi_2 = \frac{x_2 C}{(T - \theta)} \tag{10.5}$$

where:

$x_2$    is the fraction ions present as dimers
$\theta$    is a negative constant of order −10 to −100 K, which is proportional to the antiferromagnetic exchange coupling $J_0$

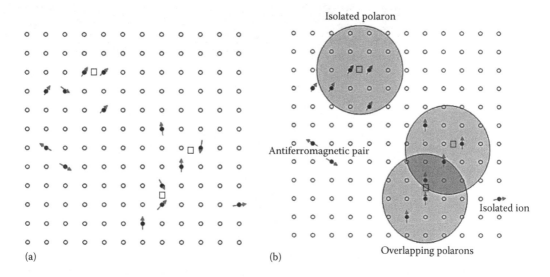

**FIGURE 10.5**    (a) Schematic representation of distribution of dopant ions in a DMS. (b) The same, but with donor defects (□) that create magnetic polarons where the dopant ions are coupled ferromagnetically.

The odd-membered groups with a net moment contribute a modified Curie-law susceptibility. With all this, deviations from a linear response to an applied field of up to 1 T, for example, are only perceptible below about 10 K. The room-temperature susceptibility is of order $10^{-3}x$ ($x$ is defined in the formula (10.3)). This is the behavior expected for a dilute magnetic oxide, and indeed it is exactly what is found in bulk crystalline materials and in well-crystallized, defect-free thin films. There are numerous examples in the literature, for example, Refs. [13–17].

The conventional picture of magnetism in insulators or magnetic semi-conductors is at a loss to account for high-temperature ferromagnetism in dilute magnetic oxides. The Curie or Néel temperatures of oxides were plotted on a histogram in Figure 10.3. None exceeded the Néel point of $\alpha Fe_2O_3$, which is 960 K. The ordering temperatures in solid solutions vary as $x$ or as $x^{1/2}$, depending on the nature of the interactions [18]. No order is expected at room temperature when $x < 10\%$, or in insulators at any temperature when $x < x_p$. The magnetic behavior we would normally expect from the $m$-$J$ paradigm is a superposition of Curie-law paramagnetism for isolated ions and Curie–Weiss behavior for the small, antiferromagnetically coupled groups of two, three, or more ions. Indeed, this is precisely what is found in well-crystallized, bulk material [15].

It therefore came as a surprise when, at the beginning of 2001, Matsumoto et al. published their claim that anatase $TiO_2$ doped with 7% cobalt was ferromagnetically ordered at room temperature. There followed a deluge of similar reports on other materials. The first samples were all thin films, but soon reports of high-temperature ferromagnetism were appearing for nanoparticles and nanocrystalline material as well. Table 10.2 lists some of these early reports. There are now well over 2000 publications on dilute ferromagnetic oxides, about half of them on ZnO, and new ones continued

**TABLE 10.2**
**Early Reports of Ferromagnetic Oxide Thin Films with $T_C$ above Room Temperature**

| Material | $E_g$ (eV) | Doping | Moment ($\mu_B$) | $T_C$(K) | Reference |
|---|---|---|---|---|---|
| TiO$_2$ | 3.2 | V—5% | 4.2 | >400 | Hong et al. [37] |
| | | Co—7% | 0.3 | >300 | Matsumoto et al. [38, 39] |
| | | Co—7% | 1.4 | >650 | Shinde et al. [40] |
| | | Fe—2% | 2.4 | 300 | Wang et al. [41] |
| SnO$_2$ | 5.5 | Fe—5% | 1.8 | 610 | Coey et al. [42] |
| | | Co—5% | 7.5 | 650 | Ogale et al. [43] |
| HfO$_2$ | 3.2 | Co—2% | 9.4 | >400 | Coey et al. [44] |
| | | Fe—1% | 2.2 | >400 | Hong et al. [45] |
| CeO$_2$ | | Co—3% | 6.1 | 725 | Tiwari et al. [46] |
| In$_2$O$_3$ | 2.9 | Mn—5% Sn | 0.8 | >400 | Philip et al. [47] |
| | | Fe—5% | 1.4 | >600 | He et al. [48] |
| | | Cr—2% | 1.5 | 900 | Philip et al. [49] |
| ZnO | 3.4 | V—15% | 0.5 | >350 | Saeki et al. [50] |
| | | Mn—2.2% | 0.16 | >300 | Sharma et al. [51] |
| | | Fe 5%, Cu 1% | 0.75 | 550 | Han et al. [52] |
| | | Co—10% | 2.0 | 280–300 | Ueda et al. [53] |
| | | Ni—0.9% | 0.06 | >300 | Radovanovic and Gamelin [54] |

to appear every month. Many of these are electronic structure calculations for dilute magnetic oxides, with various crystalline defects.

The experimental results were astonishing for three reasons:

1. The magnetism appears in oxide materials where the doping is far below $x_p$.
2. The magnetism appears to be ferromagnetic, whereas superexchange in oxides is usually antiferromagnetic.
3. The samples appear to be ferromagnetic at room temperature and above, although no dilute magnetic material had ever been found to be magnetically ordered at room temperature.

In these circumstances, the claims of ferromagnetism were met with deep skepticism, and the conviction that the high-temperature ferromagnetism must somehow be associated with a segregated, magnetically concentrated impurity phase, or else it was simply a measurement artifact. Otherwise, one is led to believe that the high-temperature ferromagnetic-like behavior of dilute oxides is a new and significant magnetic phenomenon. Although the field is fraught with irreproducibility, and impurity phases are often implicated, such is the view of the present author. It is difficult to digest the flood of often-contradictory experimental information, that is now beginning to slacken. There are some helpful reviews, for example by Ogale (2010) [19], Fukumura and Kawasaki (2013) [20], and Hong (2016) [21]. Our approach here is to present experimental data rather than electronic structure calculations on six main oxide systems, and then to focus on physical models that could account for important aspects of the phenomena.

## 10.3 MAGNETIC MEASUREMENTS

Before considering the experimental evidence, some words of caution are in order. We are dealing mainly with films, about 100 nm thick. If $x = 1\%$, there may the equivalent of just four layers of ferromagnetic cations in the film. The ferromagnetic volume fraction of a typical $5 \times 5$ mm$^2$ sample on its substrate, which is usually about 0.5 mm thick, is just two parts per million. Assuming 1 $\mu_B$ per cation, the ferromagnetic moment is approximately $1.5 \times 10^{-8}$ A m$^2$ ($1.5 \times 10^{-5}$ emu). While this is comfortably above the sensitivity of a super-conducting quantum interference device magnetometer (SQUID), an alternating-gradient force magnetometer (AGFM), or even a good vibrating-sample magnetometer (VSM), it does represent a *very small moment*. A 0.1 μg speck of iron or magnetite has a moment of this magnitude. These contaminants are ubiquitous, and some care is needed to be sure that what is being measured is the sample, and not magnetic contamination. Co nanoparticles are difficult to detect, but Fe and $Fe_3O_4$ are readily picked up by Mössbauer spectroscopy. Element-specific x-ray absorption is also useful. Secondary phases are an important consideration, and are probably the complete explanation of the ferromagnetism in certain cases. Other high-temperature ferromagnets are spinel phases, where the antiferromagnetic interactions lead to ferromagnetic structures where the moment does not exceed 1.5 $\mu_B$ per magnetic cation. Low concentrations of nanoscale ferromagnetic inclusions may elude detection by x-ray diffraction, but should be detectable by careful transmission-electron microscopy. In magnetic measurements, they are expected to behave super-paramagnetically, with the characteristic signature of anhysteretic magnetization curves that superpose perfectly when plotted as a function of $H/T$.

Several publications have emphasized how bogus magnetic signals may arise from markers, kapton tape, silver paint, contamination from steel tweezers, unsuitable substrate holders, and other sources of external pollution [22–27]. It is essential, as part of the experimental protocol, to run blank, control samples that have been treated in exactly the same way as those containing the magnetic cations. In order to make a case that magnetic cations or unpaired electrons in the oxide film are the source of the observed magnetism, precautions in sample preparation and handling are imperative, to do justice to the sensitivity of the magnetometer.

The problems for nanocrystalline material are different. Here the moments reported are of order $2 \times 10^{-4}$ A m$^2$ kg$^{-1}$ (2 10$^{-4}$ emu g$^{-1}$). The issue now is sample purity; the presence of ferromagnetic metals or oxides at the ppm level has to be discounted. The specific moments of Fe, Co, Ni, and $Fe_3O_4$ are, respectively, 217, 162, 55, and 92 A m$^2$ kg$^{-1}$ (emu g$^{-1}$ is an equivalent unit); their magnetizations are 1710, 1440, 490, and 480 kA m$^{-1}$ (emu cc$^{-1}$ is an equivalent unit).

For thin films deposited on substrates, the huge mismatch between the mass of the thin film sample—typically some tens of micrograms—and that of the substrate—which is about a thousand times greater—presents a challenge. Commonly used substrates, sapphire ($Al_2O_3$), MgO, $SrTiO_3$, $LaAlO_3$, and Si, are all diamagnetic with susceptibilities of −4.8, −3.1, −1.3, −2.7, and $−1.5 \times 10^{-9} \, m^3 kg^{-1}$. The problem is illustrated in Figure 10.6. The substrate gives a diamagnetic signal, which is perhaps reproducible to within 1%, given the uncertainties in positioning, and centering the substrate in the magnetometer used to measure the hysteresis loop. It is therefore practically impossible to determine the diamagnetic susceptibility of the undoped film. A Curie-law signal due to paramagnetic dopant ions can be readily detected at low temperatures, but it is difficult to measure any high-field slope that may be associated with the thin film.

The measured response for a ferromagnetic thin film is shown in Figure 10.6b. The magnetization appears to saturate in a field of order 1 T, and the high-field slope becomes that of the substrate, within experimental error. The procedure is then to suppose the magnetization is indeed saturated, and subtract the high-field diamagnetic slope as the substrate correction, yielding the magnetization curve for the film shown in Figure 10.6c. Typically, these 'ferromagnetic' hysteresis loops for dilute magnetic oxides exhibit little or no coercivity ($\approx 10 \, mT$), and a small remanence ratio ($M_r/M_s \approx 5\%–10\%$). Trapped flux in a SQUID, or the remanence of an iron-cored electromagnet create fields of this magnitude, so it is important to measure precisely the field actually acting on the sample. A flaw in the additivity argument of the moments of thin film and substrate is that it ignores the possible emergence of a new electronic or magnetic state on one side or the other of the interface [28]. Best known is the example of a thin film of $LaAlO_3$ on (100) $SrTiO_3$ when a two-dimensional electron gas

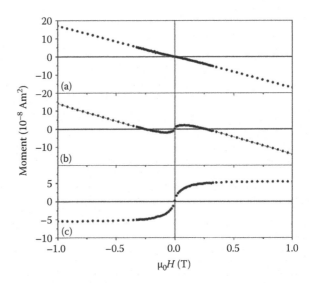

**FIGURE 10.6** Data reduction for magnetization measurements on a thin film of a typical dilute magnetic oxide: (a) diamagnetic substrate, (b) ferromagnetic thin film with diamagnetic substrate, (c) ferromagnetic component isolated from (b).

that shows evidence of magnetic order at liquid-helium temperatures forms in the $SrTiO_3$ provided the thickness of the $LaAlO_3$ overlayer exceeds four unit cells [29–31]. The common explanation is charge transfer of about 0.5 electrons per unit cell into $3d(Ti)$ states just below the interface, although oxygen vacancies in $SrTiO_3$ just below the interface may also contribute [32, 33]. The charge transfer tends to occur towards the oxide with the lower bandgap, so in films of $LaMnO_3$ on $SrTiO_3$, the charge buildup in the manganite lead to an abrupt switch from antiferromagnetic to ferromagnetic order when the cap layer thickness 6 unit cells [34].

A more recent study of polished (100), (110) and (111) $SrTiO_3$ crystal slices from four different suppliers has revealed a new magnetic phenomenon related to oxygen vacancies near the surface. This is quite different from the effects of the magnetic impurities that are present to a greater or lesser degree these substrates — paramagnetic $3d$ impurities, usually present in ppm or sub ppm amounts, that give a characteristic Curie-law upturn in the susceptibility at low temperature and hysteretic ferromagnetic Fe-Ni inclusions in the surface, probably introduced by the polishing process. The new effect is a ferromagnetic-like response with little or no hysteresis or temperature dependence, at least between 4 K and 400 K. The effect is increased by heating the substrate under vacuum, and it is increased by heating the substrate under vacuum, and reduced by treating the surface with a Ti-specific electron-donor catechol molecule [35]. These studies illustrate the potential difficulties in disentangling the magnetic response of a thin film from that of the substrate on which it rests. However $SrTiO_3$ may be an extreme case. Other substrates such as c- or r-cut sapphire appear to be magnetically inert.

Units of $\mu_B$ $nm^{-2}$ are useful when the surface or interface is the source of the magnetic signal. The oxides may even be blank substrates [35, 36, 204]. In $SrTiO_3$, for example, anhysteretic, saturating magnetic moments of up to 60 $\mu_B$ $nm^{-2}$ are found [35].

At first, it seemed reasonable to quote the magnetization of the samples in units of Bohr magnetons per $3d$ cation, normalizing the measured magnetic moment by the number of transition-metal cations thought to be in the sample. This unit is useful when comparing different doping concentrations and also when comparing different host materials, provided the transition-metal concentration is accurately known. However, the practice is misleading when the oxides turn out not to be dilute magnetic semiconductors (DMSs), and the dopant cations are not the primary source of the magnetic order. Units of $\mu_B$ $nm^{-2}$ are useful when the surface or interface is the source of the magnetic signal. The oxides may even be blank substrates [34–36]. In $SrTiO_3$, for example, anhysteretic, saturating magnetic moments of up to 30 $\mu_B$ $nm^{-2}$ are found [36].

We now present experimental results on a selection of dilute oxides, which can exhibit ferromagnetic behavior when they are in thin film or nanocrystalline form. The crystal structures of host materials are presented in Figure 10.7. Some structural details are included in Table 10.3, including the percolation threshold $x_p$, the cation site coordination, and bandgap $E_g$. The systems are now discussed in turn.

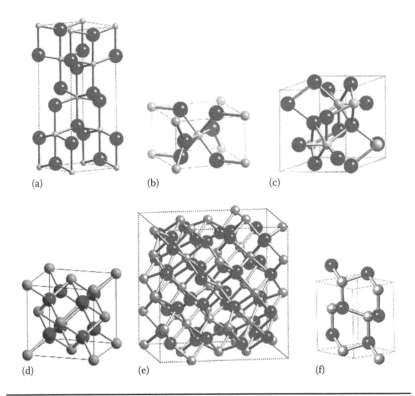

**FIGURE 10.7**   Crystal structures of host semiconductors: (a) $TiO_2$ (anatase), (b) $SnO_2$, $TiO_2$ (rutile), (c) $HfO_2$, (d) $CeO_2$ (fluorite), (e) $In_2O_3$ (bixbyite), and (f) ZnO (wurtzite), oxygen are the large dark ions.

| TABLE 10.3 Parameters for Some Oxides | | | | | |
|---|---|---|---|---|---|
| **Material** | **Structure** | $\varepsilon$ | **m\*/m** | **Coordination** | **xp** |
| $TiO_2$ | Anatase | 37 | 0.4 | Octahedral | 0.25 |
| $SnO_2$ | Rutile | 25 | 0.4 | Octahedral | 0.25 |
| $HfO_2$ | Monoclinic | 26 | 0.5 | Sevenfold | 0.18 |
| $CeO_2$ | Fluorite | 9 | 0.3 | Cubic | 0.18 |
| $In_2O_3$ | Bixbyite | 8 | 0.3 | Octahedral | 0.18 |
| ZnO | Wurtzite | 9 | 0.3 | Tetrahedral | 0.18 |

$\varepsilon$, static dielectric constant; m\*/m, effective electron mass

## 10.4  RESULTS ON SPECIFIC OXIDES

### 10.4.1  DIOXIDES

#### 10.4.1.1  $TiO_2$

Anatase with 7% Co-doping was the first example of high-temperature ferromagnetism in a dilute oxide film [38]. The same group reported ferromagnetism in rutile films [39]. The solubility of Co in these oxides is low, and there is the possibility of cobalt-metal clustering [55, 56]. But films with

1%–2% of Co can be prepared, which are apparently free of clusters, and exhibit ferromagnetism with a moment of $\approx 1\,\mu_B$/Co up to 700 K [40]. Other evidence that Co-doped $TiO_2$ can be intrinsically ferromagnetic, and carrier-mediated include: electric-field modulation of the magnetization [57], observation of band-edge optical dichroism [58], and anomalous Hall effect [59], as well as tunnel magnetoresistance [60].

Other cation dopants reported to make $TiO_2$ ferromagnetic include V–Ni [37], Cr [16], Mn [41], Fe [61], and Cu [62, 63]. In some highly perfect, Fe-doped films, an $Fe_3O_4$ secondary phase forms at the surface [64]. It has been shown that highly perfect Cr-doped $TiO_2$ films are not ferromagnetic, whereas films with structural defects can be [65]. The Cr is trivalent in $TiO_2$ films, and the magnetization of (001) anatase films grown on $LaAlO_3$ can be highly anisotropic [66], but the magnetism again disappears as the structural quality improves [67]. The order of magnitude of the magnetization of the doped $TiO_2$ films is 10–20 kA m$^{-1}$.

$TiO_2$, unlike ZnO or $SnO_2$, is usually quite insulating in thin film form. Griffin et al. [68–70] have established that free carriers are not necessary for the high-temperature ferromagnetic order in Co-$TiO_2$, and that it does not depend critically on the uniformity of the cobalt distribution. The materials may be dilute magnetic dielectrics, rather than DMSs. Nevertheless, Fukumura et al. [71] concluded from a detailed study of magnetization, anomalous Hall effect, and magnetic circular dichroism of epitaxial films with different cobalt contents and carrier densities, that the ferromagnetism *is* related to the electron density. Many studies indicate that a critical requirement for ferromagnetism in $TiO_2$ films is an abundance of oxygen vacancies, $O_v$. These form easily in titanium oxide, as is evident from the existence of a sequence of Magneli phases $Ti_nO_{2n-1}$.

Giant moments of $4.2\,\mu_B$/V were reported for 5% V-doped $TiO_2$ [37], and moments as large as $22.9\,\mu_B$/Co have been reported for 0.15% Co-doping [72]. These numbers imply that the transition-metal cation cannot be the carrier of the moment. Ferromagnetism with a magnetization of $\approx 10$–20 kA m$^{-1}$ at temperatures up to 880 K has also been reported in 200–300 nm thick films of undoped anatase-$TiO_2$ deposited on $LaAlO_3$ [73, 74]. It has been shown by Zhou et al. [75] that a $TiO_2$ crystal irradiated with a dose of $5 \times 10^{15}$ cm$^{-2}$ of 2 MeV oxygen ions develops a tiny magnetization of 2 A m$^{-1}$. The induced defects are $Ti^{3+}$–$O_v$ complexes. Oxygen vacancies are also thought to be responsible for the magnetism of $TiO_2$ nanoribbons [76]. Other procedures that have the effect of introducing cation vacancies by nitrogen doping (1 μB/N) [77], Ta doping [78], or simply by Ti substoichiometry in anatase [79] all produce material with room-temperature ferromagnetism, in the absence of substitution of any magnetic 3d cation for Ti. These results may be interpreted in terms of the magnetism of strongly-correlated 2p holes [80].

### 10.4.1.2 $SnO_2$

First reports of high-temperature ferromagnetism in $SnO_2$ films came from Ogale et al. [43], who found a Curie temperature of 650 K, and an extraordinary moment of 7.5 Bohr magnetons per cobalt atom for films doped with

5% Co. High-temperature ferromagnetism was subsequently observed for films doped with V [81], Cr [82], Fe [42], Mn [83], and elements from Cr–Ni. [84]; in some of these films, the moments again exceed the cation spin-only values [18]. In the case of V-doping, for example, the results depend rather critically on the choice of substrate [85, 86]. Co-doped films exhibit Faraday rotation of 570° cm$^{-1}$, and show some sign of an anomalous Hall effect [87]. The magnetization of the $3d$ doped $SnO_2$ films is typically 20 kA m$^{-1}$.

There is a remarkable anisotropy of the magnetization when the films grow with a (101) texture, which is found in films with V, Mn, Fe, and Co-doping [84, 86], and even in undoped films [88]. The effect cannot be explained by magnetocrystalline anisotropy, because the ferromagnetic moments measured in different directions do not converge in high fields. The anisotropy becomes very pronounced at low temperature, and it has been modeled in terms of orbital current loops [89].

There is much evidence that the magnetism is somehow associated with defects related to the oxygen stoichiometry. It was originally suggested that these could be F-centers—an oxygen vacancy that traps an electron, which serves to couple the magnetic dopant ions [42]. The defects are not necessarily adjacent to the cation dopant, which are usually octahedrally coordinated in substitutional sites. $SnO_2$ forms dilute magnetic oxides not only in thin film form, but also in nanoparticles [90–92] and nanocrystalline ceramics [93, 94]. A significant experiment by Archer and Gamelin [95] demonstrated that the magnetism of colloidal, 2 nm Ni-doped $SnO_2$ nanoparticles could be switched on and off as they aggregate under gentle thermal annealing at 100°C. The ferromagnetism is controlled by grain-boundary oxygen defects, as was shown in similar experiments on colloidal particles of Ni-doped ZnO [54] and Co-doped $TiO_2$ [96], which become strongly ferromagnetic when spin coated into thin films. Magnetic force microscopy of Fe-doped $SnO_2$ thin films shows magnetic contrast at the grain boundaries [84].

There are also reports of magnetic moments in undoped $SnO_2$ films [88], nanoparticles [97], and nanowires [98]. Magnetization is of order 10 kAm$^{-1}$, 10 Am$^{-1}$ and 10 kAm$^{-1}$, respectively. In the latter case, ferromagnetism is enhanced by UV irradiation. Cation doping with Li in films [99], or nanoparticles [100], or with Mg in films [101] produces magnetism that is associated with $2p$ holes. Magnetization is of order 10 kA m$^{-1}$ and 10 A m$^{-1}$, respectively.

### 10.4.1.3 $HfO_2$

A landmark in the study of ferromagnetism in oxide thin films was the report by Venkatesan et al. [102] of high-temperature ferromagnetism in undoped $HfO_2$ films. These results helped shift the focus from dopants to defects, and led to coining the term $d^0$ *magnetism* for the weak, anhysteretic high-temperature magnetism of a wide range of materials that are not normally expected to be magnetic, because they contain no $3d$ ions [18]. A subsequent investigation [44] of more than 40 $HfO_2$ films of different thickness, deposited on different substrates by pulsed-laser deposition, with and without doping, established that moments were largest on c-cut sapphire and smallest on Si; they showed no systematic variation with film thickness (except that the thinnest films were not magnetic). Dopants, whether magnetic or nonmagnetic, tended to reduce the moment. The moments were widely scattered around an average of ~200 $\mu_B$ nm$^{-2}$, but they were unstable, decaying over times of order six months. Magnetization of the thinnest films was comparable to that of Ni (400 kA m$^{-1}$), and in some samples, it was very anisotropic, being larger when the field was applied perpendicular to the film, than when it was applied parallel. The results were substantially confirmed in a series of papers by Hong and coworkers [74, 45, 103, 104] who looked at HfO$_2$ undoped, or with Fe or Ni dopants. Oxygen annealing reduced the moment, which was associated with defects near the film/substrate interface. However, other studies could find no intrinsic ferromagnetic signal in similarly prepared HfO$_2$ films, irrespective of oxygen pressure during deposition [105, 106]. It was shown by Abraham et al. [22] that similar signals could be observed in samples manipulated with stainless steel tweezers, and it is now standard practice to manipulate thin film samples with plastic or wooden tools.

A study of colloidal HfO$_2$ nanorods 2.6 nm in diameter [107] found small magnetic moments ~100 A m$^{-1}$. Furthermore, a magnetization of comparable magnitude can be reversibly induced by heating pure HfO$_2$ powder in vacuum or air at 750°C [44]. Oxygen vacancy-related magnetism has also been identified in nanorods [108], and amorphous HfO$_2$ nanostructures [109]. It has been shown that magnetism and oxygen vacancy content can be controlled by magnetic field annealing [110]. However, a detailed study by Hildebrandt et al. of monoclinic HfO$_2$-x grown on c-cut sapphire by reactive molecular beam epitaxy for a range of oxygen stoichiometry found no sign of room-temperature ferromagnetism, even in films that exhibit p-type conductivity [111]. Oxygen vacancies may be necessary, but they cannot be the whole story.

### 10.4.1.4 CeO$_2$

The first report of ferromagnetism in CeO$_2$ was by Tiwari et al. [46] who found a moment of 6.1 $\mu_B$/Co in 3% Co-doped films that increased with temperature up to 725 K; it corresponds to a magnetization of 40 kA m$^{-1}$, and is too big to be explained by the presence of Co metal in the films. The magnetization depends on the substrate and oxygen pressure during growth [112, 113]. Very much smaller magnetization has been found in CeO$_2$ nanoparticles that are undoped [97, 114–116] or doped with Ni [117] or Ca [118]. Values range from 1 to 5000 A m$^{-1}$, and magnetization is greatest in sub-20 nm particles with no 3d doping [114]. The high-temperature ferromagnetism in the nanoparticles is associated with Ce$^{3+}$–Ce$^{4+}$ surface pairs [116], rather than surface oxygen vacancies [115]. Trivalent rare-earth doping has a decided influence on the magnetism of CeO$_2$. Pr first increases the magnetism of thin films, and then decreases it when the Pr content > 10% [119]. In nanoparticles, the moment declines when the Pr content > 1% [120].

Cerium dioxide nanoparticles are perhaps the oxide materials that most reliably exhibit $d^0$ ferromagnetism, although the magnitude of the room-temperature moment depends on the form of the particles and the preparation method [121]. A thorough study of the magnetism of 4 nm CeO$_2$

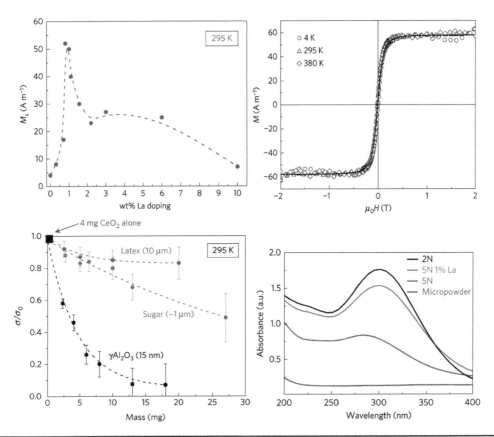

**FIGURE 10.8**   Magnetic properties of 4 nm CeO$_2$ nanoparticles. (a) Effect of La doping on the magnetization, (b) Temperature-independent anhysteretic magnetization curves of a sample containing 1 wt% La, (c) Effect of physical dilution of a 4 mg sample with similar masses of different magnetically-inert powders, showing a characteristic length scale of order 100 nm for the appearance of magnetism in agglomerates of the CeO$_2$ nanoparticles, and (d) Comparison of UV absorption spectra of magnetic (upper two curves) and nonmagnetic (lower two curves) CeO$_2$ powders. (After Coey, J.M.D. et al., *Nature Phys.* 12, 694, 2016. With permission.)

nanoparticles [121] revealed two unexpected results. First was the need for a small amount of trivalent rare-earth doping to switch on the ferromagnetic-like response. The dopant does not have to be magnetic; 1% of La produces the best response (Figure 10.8a). More remarkable is the observation that the CeO$_2$ nanoparticles have to be in contact to form subassemblies greater than a certain size for the magnetism to appear. There is a *characteristic length scale* associated with the magnetism of about 300 nm, which is associated with the appearance of UV absorption around this wavelength in the magnetic samples (Figure 10.8d). The explanation proposed in terms of giant orbital paramagnetism is discussed in § 10.5.1. There is a comprehensive review of the room-temperature magnetism in CeO$_2$ [206].

## 10.4.2 Sesquioxides

### 10.4.2.1 In$_2$O$_3$

Indium oxide, In$_2$O$_3$, has the cubic bixbyite structure, with indium in undistorted 8*b* and highly distorted 24*d* octahedral sites (Figure 10.7e).

The structure is related to fluorite (Figure 10.7d) with a quarter of the oxygen atoms removed. The bandgap $E_0$ is 2.93 eV [122]. The oxide is usually oxygen-deficient, creating donor levels near the bottom of the conduction band. Electron concentrations of order $10^{24}–10^{26}$ m$^{-3}$ lead to resistivities of about 10 μΩ m, a high mobility, and an effective mass of $0.3m_e$. The electrons tend to accumulate at the oxide surface. Substituting some of the In by Sn is a convenient way of controlling the carrier concentration in $In_2O_3$. Indium tin oxide $(In_{1-x}Sn_x)_2O_3$ (ITO) with $x \approx 0.1$ is a widely used transparent $n$-type conducting oxide.

The first report of ferromagnetism in Mn-doped ITO by Philip et al. [47] showed a moment of about 1 $\mu_B$ per manganese, and evidence of an anomalous Hall signal. It was followed by indications of ferromagnetism in $In_2O_3$, for bulk material doped with Fe and Cu [123], and for oxygen-poor thin films doped with Fe [48, 124] and other dopants [125–127]. Subsequent investigations of ferromagnetic Mn- and Fe-doped ITO films [128] found that the manganese behaved paramagnetically at low temperature, and that the moment was not the primary source of the magnetism. The presence of paramagnetic manganese in mixed valence states ($Mn^{2+}$ and $Mn^{3+}$) [129] points to charge transfer ferromagnetism (§10.5.1). The anhysteretic magnetization curve, shown in Figure 10.9, may be regarded as typical of many ferromagnetic dilute magnetic oxide thin films. Hysteretic ferromagnetism of the iron-doped samples was due to an $Fe_3O_4$ secondary phase. Anhysteretic ferromagnetism was observed in Fe-doped films thought to be intrinsically ferromagnetic [124].

A study of Cr-doped $In_2O_3$ by Philip et al. [49] demonstrated that the ferromagnetism was related to carrier concentration, appearing at a concentration $n_e = 2 \times 10^{25}$ m$^{-3}$, which would correspond to the formation of an impurity band [130]. Moments of up to 1.5 $\mu_B$/Cr were found for 2% doping. Chromium is a good dopant, because there is little prospect of ferromagnetic secondary phases; $CrO_2$ is ferromagnetic, with $T_C = 390$ K, but $T_C$ was

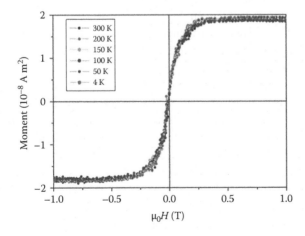

**FIGURE 10.9** Ferromagnetic magnetization curves of a thin film of ITO doped with 5% Mn, at different temperatures. The manganese in these films is paramagnetic. (After Venkatesan, M. et al., *J. Appl. Phys.* 103, 07D135, 2008. With permission.)

measured in this study to be $\approx 900\,\mathrm{K}$. Domains were observed by magnetic force microscopy (MFM). Further investigations by Panguluri et al. [131] on both undoped and Cr-doped films found paramagnetic Cr coexisting with a film moment of about $500\,\mathrm{A}\,\mathrm{m}^{-1}$. Both materials exhibited a spin polarization of 50% in measurements of point-contact Andreev reflection. The implication is that the moment of the $3d$ dopant is inessential for the ferromagnetism, which is associated with a narrow defect-related band. This picture is consistent with the dependence of the anomalous Hall effect on carrier concentration [127].

There are also reports of ferromagnetism of similar magnitude in undoped thin films [126], with smaller values in undoped nanoparticles [129, 205]. Sundaresan et al. [97] found very small moments in $12\,\mathrm{nm}$ nanoparticles of $In_2O_3$ ($<5\,\mathrm{A}\,\mathrm{m}^{-1}$), but in another study by Bérardan et al. [132], $In_2O_3$ nanoparticles prepared in a variety of different conditions were found to be completely diamagnetic, and transition-metal doped samples were paramagnetic. There is a recent summary of the magnetic properties of doped and undoped $In_2O_3$ and its derivatives [133].

## 10.4.3 Monoxides

### 10.4.3.1 ZnO

Although the first report of ferromagnetism in thin films of a dilute magnetic oxide was for $TiO_2$, it is on ZnO that the majority of the investigations have been carried out. Zinc oxide is a promising semiconductor material with a bandgap of $3.37\,\mathrm{eV}$. It crystallizes in the wurtzite structure (Figure 10.7f), where Zn occupies tetrahedral sites. The oxide has a natural tendency to be $n$-type on account of oxygen vacancies or interstitial zinc atoms. It can be doped n-type with atoms such as Al, and it has been possible to make nitrogen-doped $p$-type material, which opens the way to producing light-emitting diodes and laser diodes. Anion doping with Bi is a new approach to creating spin-polarized p-states in ZnO [134]. Extensive reviews of the semiconducting properties of ZnO are available [135, 136].

Various cations can replace zinc in the structure, and small dopings of all the $3d$ elements have somewhere been reported to make ZnO ferromagnetic. There is an admirable review of the voluminous early literature on doped ZnO by Pan et al. [137]. The largest moments are found for $Co^{2+}$. The $Co^{2+}$ ion gives a characteristic pattern of optical absorption in the bandgap, due to crystal field transitions of the $^4F$ ion in tetrahedral coordination. Studies of the magnetic properties of bulk material doped with Co [15, 138] or Mn [139] show only the paramagnetism expected for isolated ions, and small antiferromagnetically coupled nearest-neighbor clusters that arise statistically at a low doping level, as discussed above (Figure 10.5a).

The range of solid solubility of cations of the $3d$ series in ZnO films prepared by pulsed-laser deposition was established by Jin et al. [140]; solubility limits as high as 15% were found for Co, but most other cations could be introduced at the 5% level. None of these films turned out to be to be magnetic, but following Ueda et al. [53], who were the first to report high-temperature ferromagnetism in their films containing cobalt at the 10% level, there have been a great many reports of ferromagnetic moments in films doped with Co (see among others Janisch et al. [141], Prellier et al. [142], Pearton et al. [143], and references therein), as well as every other 3*d* dopant from Sc to Cu [50, 51, 54, 144]. The variation of magnetic moment for a series of films nominally doped with 5% of various transition-metal ions prepared by pulsed-laser deposition in the same conditions is shown in Figure 10.10. There are also numerous counterexamples where no room-temperature moment has been found in doped films, or an extrinsic origin has been established. Negative results tend to be underreported in the literature, so the observation of saturating magnetism in ZnO and other oxide films is probably less prevalent than one might imagine. It is very sensitive to deposition conditions such as substrate temperature and oxygen pressure [135, 137]. However, nanoparticles and nanorods of ZnO with cobalt and other dopants have also been found to be magnetic under certain process conditions [145], although the moments per cobalt atom are one or two orders of magnitude less than those found for the thin films, which are typically 0.1–1 $\mu_B$/Co.

Some progress has been made toward providing a systematic experimental account of the phenomenon. Narrow process windows have been delimited where the magnetism can be observed, which differ depending on the deposition method—pulsed-laser deposition, sputtering, evaporation, organometallic chemical vapor deposition, and others. With Cr and Mn, for example, it is difficult to obtain ferromagnetic moments in *n*-type material, whereas *p*-type samples can exhibit the symptoms [112] (Figure 10.11). The substrate temperature required for the magnetism is often around 400°C, where the films are not of the best crystalline quality. This implies a point-, line-, or planar-defect-related origin [146]. Other evidence in this sense comes from the appearance of magnetism in Zn-doped and oxygen-deficient

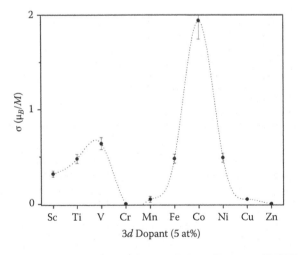

**FIGURE 10.10** Magnetic moment measured at room temperature for ZnO thin films with various 3*d* dopants (After Venkatesan, M. et al., *Appl. Phys. Lett.* 90, 242508, 2006. With permission.)

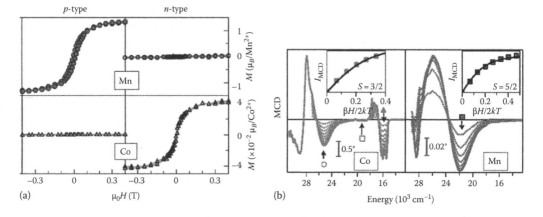

**FIGURE 10.11**   (a) Development of magnetism in *n*-type ZnO with Co or *p*-type ZnO with Mn. (After Kittilstved, K.R. et al., *Nat. Mater.* 5, 291, 2006. With permission.) (b) MCD spectra and the magnetic field dependence of the intensity of the MCD signal (insets) recorded at different energies in ZnO doped with Co (left) and Mn (right).

films [147, 148], and the creation of a large magnetization of ~200 kA m$^{-1}$ in the volume implanted by 100 keV Ar ions [149].

Evidence that Co-doped ZnO is an intrinsically ferromagnetic semiconductor is relatively sparse. No magnetoresistance has been observed at room temperature, despite the high $T_C$, nor are there clear signs of an anomalous Hall effect. Semiconductor-like magnetoresistance can, however, be observed below about 30 K [150]; there is a positive component, as expected in the classical two-band model, and a negative component due to ionized impurity scattering. There is also quite a large anisotropic magnetoresistance below 4 K (≈10%), which is a band effect in the semiconductor, not ferromagnetic AMR. Furthermore, there is band-edge magnetic dichroism with a paramagnetic dichroic response from the dilute Co$^{2+}$ ions in tetrahedral sites in the wurtzite lattice, as well as a component with a ferromagnetic response [151, 152]. The ferromagnetic dichroism has been observed for Ti, V, Mn, and Co-doping [152]. Magnetic circular dichroism experiments on single nanoparticles in the electron microscope by Zhang et al. [153] indicate that their particles are intrinsically ferromagnetic at room temperature when doped with Co, but not with Fe.

Some reports [154, 155] have related the carrier concentration to the existence of ferromagnetism, as would be expected if ZnO were a DMS. Results of Behan et al. [155], based on a systematic investigation of samples with 5% Co and carrier densities ranging from 10$^{16}$ to 10$^{21}$ cm$^{-3}$ that the material is a dilute magnetic insulator at low carrier concentrations and a DMS at high concentrations, with an intermediate concentration range around 10$^{19}$ cm$^{-3}$ where the ferromagnetism is quenched. However, the interaction between the conduction electrons and the Co spins seems to be rather weak, <20 K, and in other studies the magnetic properties are little influenced by changing the carrier concentration by Al doping, for example [156]. It has been shown that the sign of the weak Mn–Mn exchange in colloidal Mn–ZnO nanocrystals can be changed from antiferromagnetic to ferromagnetic, by exciting carriers optically [157], although the strength of the interaction is only a few Kelvin. It seems that there may be a correlation between carrier concentration and ferromagnetism, but this is not a cause and effect relation.

A problem in interpreting the magnetic data for some dopants is the tendency to form ferromagnetic impurity phases, which may escape detection by x-ray diffraction. For cobalt in particular, absence of evidence cannot be taken as evidence of absence. There is a well-documented tendency for nanometric coherent cobalt precipitates to appear in the ZnO films [158–161], and single crystals [162], which exhibit ferromagnetic properties, especially at doping levels above 5%. An impurity-phase-based explanation $(Mn_{2-x}Zn_xO_3)$ has been advanced also for Mn-doped material [163]. For other dopants, such as vanadium or copper, contamination by high-temperature ferromagnetic impurity phases is improbable.

In some cases, there are features of the data, which make an impurity-phase explanation untenable. These are (a) observation of a moment per transition-metal ion that is greater than that of any possible ferromagnetic impurity phase based on the dopant ion, and which may exceed the spin-only moment of the cation, and (b) observation of an anisotropy of the magnetization, measured in different directions relative to the crystal axes of the film, which is not a feature of any known ferromagnetic phase at room temperature [102, 144, 164]. An example is shown in Figure 10.12. Both the features have occasionally been reported in the other thin film oxide systems we have been considering.

A common feature of the ferromagnetism in ZnO and the other oxides is its tendency to decay over times of order of a few months [24, 164]. This has led to the name *phantom ferromagnetism*, for the fickle syndrome presented by dilute magnetic oxides.

The tunnel spin polarization of transition-metal-doped ZnO has been measured [165]. Although poorly reproducible, these results are in line with the observation of MCD at the band-edge as discussed above. The possibility to obtain and manipulate a spin current in ZnO could be of great interest, as the spin lifetime in ZnO is rather long [166].

There is much evidence to suggest that ferromagnetism in transition-metal-doped ZnO is unambiguously correlated with structural defects,

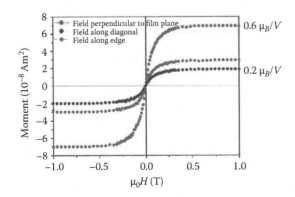

**FIGURE 10.12** Anisotropy of the magnetization of a thin film of $(Zn_{0.95}V_{0.05})O$.

which depend on the substrate temperature and oxygen pressure used during film deposition [26, 137]; the carriers are by-products of these defects. Ney et al. [17] have demonstrated that highly perfect Co-doped epitaxial films, for example, are purely paramagnetic, with cobalt occupying zinc sites, and no sign of anything other than weak antiferromagnetic Co–O–Co nearest-neighbor superexchange. Droubay et al. [167] show that Mn-doped epitaxial films are paramagnetic, with Mn substituting for Zn, and no moments on either oxygen or zinc. Barla et al. [168] showed that the cobalt in ferromagnetic Co-doped ZnO films was paramagnetic, and there was no moment on the zinc. A study by Tietze et al. [169] who investigated 5% Co-doped films produced by pulsed laser deposition (PLD), which exhibited a moment of ≈1 $\mu_B$/Co, corresponding to a magnetization of ≈20 kA m$^{-1}$ came to a remarkable conclusion: extensive x-ray magnetic circular dichroism (XMCD) measurements on the Co and O edges with a very good signal/noise ratio showed only the signatures of paramagnetic cobalt and oxygen. There is no sign of element-specific ferromagnetism. By a process of elimination, only oxygen vacancies remained as the possible intrinsic source of the ferromagnetism observed in the ZnO films. The influence of oxygen vacancies can be enhanced by doping ZnO with Co in mixed valence states [170], in line with the CTF model discussed below.

Thin films and tunnel junctions made of Cr-doped material, which is a magnetic insulator, show that defects near the surface rather than throughout the bulk are responsible for the magnetism [171]. The studies of many systems as a function of thickness, which indicate magnetic moments of 100–200 $\mu_B$ nm$^{-2}$ argue in the same sense [24, 164]. The idea then is that 3d ions may enhance the magnetism of ZnO, but that the magnetic moments are not the direct cause of it. Indeed, as in TiO$_2$ [172], the 3d dopants may not even be magnetically coupled to the ferromagnetic phase. [169, 173]. An optimized doping strategy uses one dopant to augment the concentration of defects, and another, which is in a mixed-valence state, to promote charge transfer to a spin-split impurity band, has produced magnetizaion as high as 20 kA/m [170]. As in the other oxides, there are reports of magnetism in undoped films [174–176], where the magnetization is greatest in the thinnest ZnO films, and is sensitive to strain [177]. The magnetism is very sensitive to the choice of substrate and processing conditions [177–179]. Interfaces are important, and the magnetism of ZnO may be modified in thin-film heterostructures [180, 181] and capped ZnO nanoparticles [182, 183], where the moment in organic-capped ZnO detected by Zn K-edge XMCD is found to reside within 0.8 nm of the interface [180]. Undoped nanocrystals [184, 185], nanorods [173], and other nanostructures [186] are also reported to be ferromagnetic, but with a very small magnetization ~1–100 A m$^{-1}$. A compilation of data from several groups by Straumal et al. [187] indicates that weak ferromagnetism is only found in undoped bulk ceramic material with a grain size below 60 nm, and in Mn-doped ceramic samples with a grain size of less than about 1.5 μm (Figure 10.13). The grain boundaries are interstitial amorphous layers surrounding 20 nm ZnO grains or 1000 nm Mn-doped ZnO grains [188] analysis of the nanocrystalline ceramic samples of undoped and

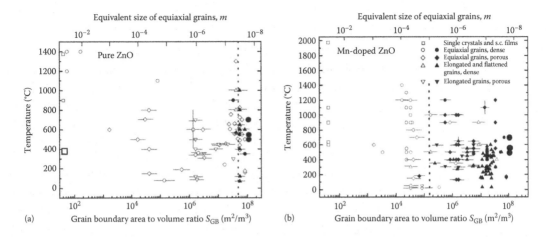

**FIGURE 10.13**  Plot of magnetic moment versus grain-boundary area for undoped (a) and Mn-doped (b) ZnO ceramics. (After Straumal, B.B. et al., *Phys. Rev. B* 79, 205206, 2009. With permission.)

Co- or Mn-doped ZnO by Straumal et al. [187] suggests that the magnetism is associated with the grain boundaries, and they envisage the ferromagnetism as a grain-boundary foam, which only occupies a few percent of the volume of the samples. This picture, supported by μsr measurements [189] captures the ideas that defects are somehow responsible, but the magnetism requires an extended structure of interfaces to appear.

Although there is as yet no consensus, the current view of transition-metal-doped ZnO is moving far away from the initial expectations that the materials are DMSs.

## 10.5  DISCUSSION

The experimental picture is confused by the difficulty in reproducing many of the results, the often ephemeral nature of the magnetism, the problem of characterizing defects and interface states in thin films and nanoparticles, as well as the difficulty in detecting small amounts ($\approx$1 wt%) of secondary phases in the films [26]. Despite the tendency for experiments where nothing unusual is found to pass unreported in the literature, there does seem to be a *prima facie* case that there is something to explain in the numerous reports of fickle ferromagnetism in dilute magnetic oxides, and $d^0$ systems which cannot be attributed to sloppy experimentation. If so, it seems better to seek a general explanation of the phenomenon, rather than to rely on a different *ad hoc* explanation for each material. For this reason, we will not discuss the numerous electronic structure calculations for specific defects in the six oxides, which we have been considering, but focus instead on generic physical models.

It is now quite clear that the anomalous ferromagnetism is not found in highly perfect films or well-crystallized bulk material. Those dilute magnetic oxides behave paramagnetically down to liquid helium temperatures, in the way that was described in Section 10.1, albeit with a nonstatistical distribution of dopants [167]. There is accumulating evidence, reviewed in Section 10.4, that the phenomenon is closely associated with defects of some sort, which are present in thin films and nanocrystals, but are absent in perfectly crystalline material [190]. The poor reproducibility of the experimental data may be due to the difficulty in precisely recapturing the process conditions that lead to a specific defect distribution and density. The samples as prepared are not in equilibrium, and they evolve with time and temperature.

At first, it was taken for granted that the magnetism was *spatially homogeneous*, and due to the *spin moments of the 3d ions*. Both these assumptions increasingly suspect. The difficulty with any spatially homogeneous explanation based on localized 3d moments is that the magnetic interactions have to be ferromagnetic, long-range, and extraordinarily strong—at least one or two orders of magnitude greater than anything previously encountered in oxides (Figure 10.3). The view [191] that dilute magnetic oxides were DMSs, similar to the canonical example of $(Ga_{1-x}Mn_x)As$, discussed in Chapter 9, Volume 2, was influential. There the manganese introduces occupied 3d states deep below the top of the As 4p valence band, and the ferromagnetism is mediated by 4p holes that are polarized by exchange with the $3d^5$ $Mn^{2+}$ ion cores. Unfortunately, the model does not work for most 3d dopants in oxides, because the 3d level lies above the top of the oxygen 2p valence band. Exchange via the 4s electrons in the conduction band introduced by donor doping, as in Gd-doped EuO, for example [192], provides weaker coupling than exchange via the 2d holes.

A third possibility illustrated in Figure 10.14 is to couple the 3d ion cores via electrons in a spin-polarized impurity band [193]. The impurity band in a semiconductor is made up of large Bohr-like orbitals associated with the dopant electrons. In dilute magnetic oxides, it could be derived from defect states such as F-centers: the idea is that the impurity band becomes spin-split, either spontaneously if it is narrow enough to satisfy the Stoner criterion, or else on account of s–d interaction with the magnetic dopant, when it is formed of overlapping magnetic polarons (Figure 10.5b). Calculations of $T_C$ in the molecular field approximation [193, 194] lead to the expression

$$T_C = \left[ \frac{S(S+1)s^2 x\delta}{3} \right]^{1/2} \frac{J_{sd}\omega_c}{k_B} \tag{10.6}$$

where:

$S$ and $s$ are the cation core spin and the donor spin, respectively
$\delta$ is the donor or acceptor defect concentration
$\omega_c \approx 8\%$ is the cation volume fraction in the oxide

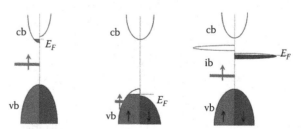

**FIGURE 10.14**   Schematic models for DMSs left; spin-split conduction band center; spin-split valence band and right; spin-split impurity band.

The difficulty is that, knowing the values of the parameters in the model, especially $J_{sd} \approx 1$ eV, the predicted Curie temperatures are of order 10 K, which is one or two orders of magnitude too low. The model can be modified by introducing hybridization and charge transfer from the donor orbitals to those of the dopants, but at some point in the dilute limit, it has to fail, and a different approach is needed. Small concentrations of conduction electrons do provide ferromagnetic coupling via the RKKY interaction, but an estimate of the magnitude of the magnetic-ordering temperature due to this interaction is [193]

$$T_C \approx \frac{Z_0 n^{5/3} m^* x^{1/3} \delta J_{sf}^2 S(S+1)}{(96\pi\hbar^2 n_c^{2/3} k_B)} \tag{10.7}$$

The Curie temperature at high carrier concentrations $n_c \sim 10^{21}$ cm$^{-3}$ does not exceed a few tens of Kelvin.

One way of increasing the exchange energy density is to look for an inhomogeneous distribution of magnetic ions. At one extreme, they may form a segregated secondary phase, which should have a high $T_C$ if it is to explain the magnetism. Phase segregation is expected when the solubility limit is exceeded, but the tolerance for dissolved ionic species may be greater in thin films or nanoparticles, which are not in thermodynamic equilibrium, than it is in the bulk. Magnetically ordered transition-metal oxides such as CoO or $Fe_3O_4$ have predominantly antiferromagnetic superexchange coupling, and are usually antiferromagnetic or ferrimagnetic. However, it is possible to find the reduced metal as a secondary phase, especially after deposition or thermal treatment in vacuum at high temperature. Fe, Co, and Ni are ferromagnetic metals, and the presence of segregated ferromagnetic metal or ferromagnetic oxide inclusions is undoubtedly the explanation of the ferromagnetism (with telltale hysteresis) of *some* of the dilute magnetic oxide samples. It is a pity that so much effort has been invested in studying these dopants, and especially cobalt, which is the ferromagnet with the highest Curie temperature of all. It is easier to make a case for an intrinsically new magnetic phenomenon when ferromagnetism is observed in oxides that are undoped, or doped with elements such as Sc, Ti, V, Cr, or Cu, which form no phases with high-temperature magnetic order.

The near-anhysteretic magnetization curves of dilute ferromagnetic oxides are practically temperature-independent, and emphatically do not imply an explanation in terms of superparamagnetism, where anhysteretic magnetization curves taken at different temperatures must superpose when plotted as a function of $H/T$. In the dilute ferromagnetic oxides, there is almost no difference between the curves measured at 4 and 300 K (Figure 10.9). It may be that in some cases the macrospin moments of cobalt or iron nanoinclusions are coupled by electrons in the host oxide. Such high-temperature-ferromagnetic conducting nanocomposites could possibly find applications in electronic devices.

Even when substituting for cations in the host oxide, the magnetic dopant ions may experience a tendency to cluster. Spontaneous chemical segregation in a solid solution is known as *spinodal decomposition*, and it often takes the form of a composition density wave with alternating regions rich and poor in one or other element. Spinodal decomposition has been suggested as a mechanism for high-temperature ferromagnetism in dilute oxides [195]. It has the virtue of segregating regions of the sample, which still looks like a solid solution, with a higher magnetic energy density than the average. Since these regions percolate as sheets throughout the material, there is no expectation of independent superparamagnetic clusters. The problem is to explain why the regions rich in transition-metal ions should be ferromagnetic, granted the propensity for $3d$ ions in oxides to couple antiferromagnetically by superexchange. There may be an unbalanced moment in a lamella related to certain antiferromagnetic arrangements of the spins, but this could not be a general explanation, especially for the $3d$ ions from the beginning of the series, which have a small spin moment, and do not form any oxide with a high magnetic-ordering temperature.

The accumulating evidence that defects play a central role in the magnetism of dilute magnetic oxides does not mean that the $3d$ cations are not involved, but it suggests that they do not contribute in the way at first assumed. Beyond a threshold thickness, the moments do not scale with film thickness, but seem to depend on surface area, with values of 100–400 $\mu_B$ nm$^{-2}$. Unphysically large moments have been inferred "per cation" at low doping levels in almost all the oxide systems. Then there are the reports of $d^0$ magnetism in some samples of undoped films and nanoparticles, and at some oxide crystal surfaces [35], and interfaces. Although they are particularly important for getting to the root of $d^0$ magnetism, these reports are rather uncommon. It is much more usual to find that the $x = 0$ end-member is nonmagnetic, and this then serves as a credibility check of the experimental protocol, establishing the $3d$ dopant's role in initiating the magnetism, and incidentally confirming that the laboratory is not irredeemably contaminated with airborne magnetic dust.

## 10.5.1 MODELS

Having discussed the messages that the experimental data seem to convey, we now summarize the generic models that have been proposed for ferromagnetism in dilute magnetic oxides. There are six of them. Five depend on whether the spin magnetic moments are localized or delocalized, and the nature of the electrons involved in the exchange. The sixth does not involve spin ferromagnetism, but collective orbital paramagnetism.

1. The DMS model [191]. There are well-defined local moments on the dopant cations, which are coupled ferromagnetically via $2p$ holes, or by RKKY interactions with $4s$ electrons.

2. The bound magnetic polaron (BMP) model [193]. Here again, there are well-defined local moments on the dopants, but they interact with electrons associated with defects, which form an impurity band. Each defect electron occupies a large orbit and interacts with several

dopant cations to form a magnetic polaron, and ferromagnetism sets in at the percolation threshold for these large objects.

3. A variant of the model (BMP), [131] dispenses with the magnetic dopants, but retains the idea of localized moments, which are now associated with electrons trapped at defects. Triplet pairs could form to give $S = 1$ moments, which are then coupled by electrons in an impurity or conduction band.

4. The spin-split impurity band (SIB) model [196], which is based on a local density of states associated with defects where the density of states at the Fermi level is sufficient to satisfy the Stoner criterion.

5. The charge-transfer ferromagnetism (CTF) model [172, 197] is a related model with a defect-based impurity band, but there is another charge reservoir in the system that allows for the facile transfer of electrons to or from the impurity band to create a filling that leads to spontaneous spin splitting. In dilute magnetic oxides, this reservoir is associated with the dopant ions.

6. Lastly, there is the possibility that the effects are not essentially related to collective spin magnetism at all, but that the saturating magnetic signal is due to giant orbital paramagnetism (GOP) associated with a new collective orbital state of the electrons. The magnetic moments are then *induced* rather than *aligned* by the field.

The first three of the models are Heisenberg models, with localized spin moments that fit the $m$-$J$ paradigm. The first two of them require roughly uniform ferromagnetism of the sample, since the moments are borne by the $3d$ dopants that are distributed more or less uniformly throughout the material. We have seen little proof that the dopants in the dilute magnetic oxides are magnetically ordered, and growing evidence that they are actually paramagnetic. The models also require long-range ferromagnetic exchange interactions of a strength not found in any concentrated magnetically ordered material. Curie temperatures of order 10–50 K might be possible for materials with a few percent doping, but not values of order 800 K. Neither model accounts for $d^0$ ferromagnetism in undoped, defective oxides. The third model associates the local moments with defects. They can occur in oxides with no $3d$ ions [80]. The defects could be distributed near interfaces or grain boundaries, and so the model can account for both inhomogeneous and $d^0$ ferromagnetism. The drawback is that, as with any Heisenberg model, spin-wave excitations are inevitable, and there is no reason to expect high Curie temperatures.

The next two models are Stoner models. Localized magnetic moments are not involved; all that is needed is a defect-related density of states with the right structure and occupancy. The magnetism is not homogeneous, but is confined to percolating regions such as surfaces, interfaces, or grain boundaries. Furthermore, Edwards and Katsnelson [196], who developed the SIB model for the $d^0$ ferromagnet $CaB_6$, show that spin waves are suppressed for the defect-related impurity band, so Curie temperatures may be very high. The main problem with the model is its lack of flexibility. There is no specific role for the $3d$ dopants. The number of electrons in the impurity band is determined by the defect structure of the sample, and it would be something of a fluke for the electron occupancy to coincide with a peak in the density of states where the Stoner criterion is satisfied:

$$I N(E_F) > 1 \qquad (10.8)$$

where $I$ is the Stoner exchange integral, which is of order 1 eV. The model can have ferromagnetic metallic, half-metallic, and insulating ground states.

The difficulty arising from a fixed number of electrons in the impurity band is resolved in the CTF model. In this model, there is a proximate electron reservoir, which can feed or drain the impurity band as required to meet the Stoner criterion. This makes it possible for a wider range of materials to become magnetically ordered. In the case of dilute magnetic oxides, the $3d$ dopants can serve as the reservoir, provided they are able to coexist in different valence states such as $Ti^{3+}/Ti^{4+}$, $Mn^{3+}/Mn^{4+}$, $Fe^{2+}/Fe^{3+}$, $Co^{2+}/Co^{3+}$, or $Cu^+/Cu^{2+}$. The point of doping with $3d$ cations is not that they carry a magnetic moment, but that their mixed valence allows them to act as the electron reservoir. The cations can remain paramagnetic. It is possible that they tend to migrate to the surfaces or grain boundaries, where the defects lie. The defect-based models can explain the inhomogeneous nature of magnetism. The magnetic regions are those that are rich in defects, which tend to concentrate at interfaces and grain boundaries (Figure 10.15).

It is a challenge to account for the observations that the dilute magnetic oxides may be insulating, semiconducting, or metallic. The Heisenberg models account for a nonconducting magnetic state, by supposing that the carriers that propagate the exchange are immobile; the Fermi energy in the conduction band lies below the mobility edge [155]. However, the Stoner models have the option of a half-filled SIB, which can arise from single-electron defect states. The CTF model has a rich electronic phase diagram, with insulating, half-metallic, and metallic ferromagnetic states, as well as nonmagnetic states (Figure 10.16). The undoped end-member is most often nonmagnetic, which shows that the doping is critically important. Ferromagnetic $d^0$ end-member oxide films remain very much an exception, rather than the rule.

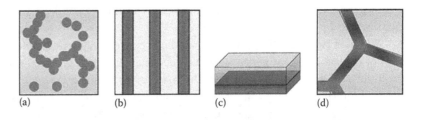

**FIGURE 10.15**   Inhomogeneous ferromagnetism in a dilute magnetic oxide. The ferromagnetic regions are distributed (a) at random, (b) in spinodally segregated regions, (c) at the surface/interface of a film, and (d) at grain boundaries.

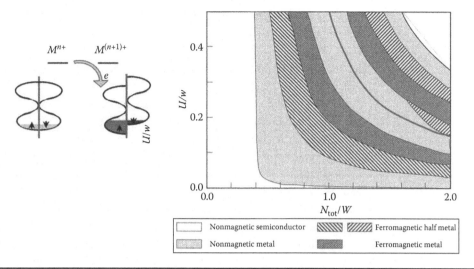

**FIGURE 10.16** Phase diagram for the CTF model. Electron transfer from the 3*d* charge reservoir into the defect-based impurity band (a "double-peak" structure is chosen) can lead to spin splitting, as shown on the left. Axes are the total number of electrons in the system $N_{tot}$ and the 3*d* coulomb energy $U_d$, each normalized by the impurity bandwidth *w*. The Stoner integral *I* is taken to be 0.6. In the unshaded nonmagnetic semiconductor regions, the impurity band is empty or full. The curved gray line marks half-filling. In the gray regions, it is partly filled, but not spin-split. In the blue regions, the Stoner criterion is satisfied, and it is ferromagnetic. (After Coey, J.M.D. et al., *New J. Phys.* 12, 053025, 2010. With permission.)

Finally, there are two features of the ferromagnetism, which have received less attention than they deserve. One is the magnetization process—the fact that the ferromagnetism exhibits very little hysteresis, and the magnetization curve is practically independent of temperature. Magnetization curves like that in Figure 10.9 are found in all the systems we have been discussing. It may be asked whether a material with a magnetization curve showing no hysteresis can really be ferromagnetic; where is the broken symmetry? But the magnetization curves of iron, permalloy, and other soft ferromagnets often fail to show hysteresis, at least at room temperature, because they form domains or other flux-closed structures to reduce their dipolar self-energy.

Hysteresis is normally related to magnetocrystalline anisotropy, which falls off rapidly with increasing temperature. The virtual lack of hysteresis below room temperature indicates that *dipolar interactions* are controlling the approach to saturation. A uniformly magnetized soft ferromagnetic film has a demagnetizing factor of zero when the magnetization lies in-plane and unity when the magnetization is perpendicular to the plane; the magnetization curve saturates in a small field in the first case and it is linear up to saturation at $H = M_0$ in the second. The measured magnetization curves are usually not like this; they saturate in almost the same way regardless of field direction, with a magnetization curve that can be approximated by a tanh function

$$M = M_0 \tanh\left(\frac{H}{H_d}\right) \tag{10.9}$$

where $H_d \approx 70$–$130$ kA m$^{-1}$. Isolated micron and submicron iron particles ($M_0 = 1720$ kA m$^{-1}$) behave similarly, but with $H_d \approx 300$ kA m$^{-1}$. The iron particles adopt a vortex magnetic configuration to minimize the demagnetizing energy, which progressively unwinds as the magnetization approaches saturation.

A spin-split defect-based s-like impurity band would not be expected to exhibit magnetocrystalline anisotropy. Assuming the ferromagnetic regions are thin layers following the grain boundaries, the ferromagnetic grain-boundary foam will tend to have a wandering ferromagnetic axis that lies in the plane of the grain boundary. On applying an increasing magnetic field, the wandering axis will unwind, and straighten up as saturation is approached. Modeling the magnetization process of an ensemble of macrospins subjected to a randomly oriented demagnetizing field gives a magnetization curve with $H_d \approx 0.16 M_0$ [198], where $M_0$ is the magnetization of the ferromagnetic regions. From the plot in Figure 10.17, based on the magnetization curves of 250 dilute ferromagnetic oxide films, and the summary in Table 10.4, it can be deduced that $M_0/M_s \approx 100$, where $M_s$ is the saturation magnetization of the sample. We infer that *the ferromagnetic volume fraction in the dilute magnetic oxides is about 1%.* When data on the nanocrystalline ceramics is analyzed in the same way, it turns out that the volume fraction there is around one part in $10^5$ or $10^6$.

The other point to flag is the anisotropy of the saturation magnetization of oriented films, either between parallel and perpendicular to the plane or within the plane. The effect has been reported for HfO$_2$ [44, 102, 144], TiO$_2$ [66], SnO$_2$ [84, 86, 88, 89], and ZnO [144, 164] (Figure 10.12); the anisotropy may be as large as a factor 10. Reports of this anisotropy of $d$-zero magnetism are intriguing, but rather uncommon. They merit further investigation.

The last model we consider, the GOP model, has an entirely different basis, namely that the temperature-independent "ferromagnetic" response

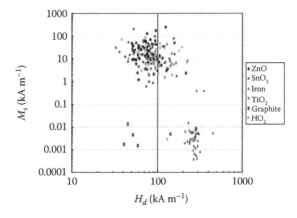

**FIGURE 10.17** Plot of magnetization $M_s$ versus internal field $H_d$ for thin films and nanoparticles of doped and undoped oxides. For films of the thin films of oxides, $M_s$ clusters around 10 kA m$^{-1}$, whereas for nanoparticles, it is three orders of magnitude lower. $H_d$ in both cases is about 100 kA m$^{-1}$. Also included on the scatter plot are data on graphite and dispersed iron nanoparticles.

| | $H_d$ (kA m$^{-1}$) |
|---|---|
| **TABLE 10.4** | |
| **Local Dipole Field H$_d$** | |
| TiO$_2$ | 125 (40) |
| SnO$_2$ | 79 (30) |
| HfO$_2$ | 94 (35) |
| ZnO | 83 (30) |
| Graphite | 68 (42) |
| Fe | 275 (40) |

to an applied field is not fundamentally related to any sort of collective, exchange-driven ferromagnetic spin ordering, but is entirely field-induced and originates from coherent mesoscopic conduction currents. In other words, the effect is a novel type of temperature-independent, saturating orbital paramagnetism. The possibility of surface orbital moments has been evoked in the context of Au nanoparticles [199, 200] and nanorods, and also of zinc oxide nanoparticles [201, 202]. A theory has recently been developed that envisages the possibility of coherent electronic states in quasi-two-dimensional systems with a large surface to volume ratio, due to coupling with zero-point fluctuations of the vacuum electromagnetic field [203]; the magnetic consequences of the theory have been developed in the context of CeO$_2$ nanoparticles [121]. The theoretical form for the magnetization curve is

$$M = M_s \, x / (1 + x^2)^{1/2} \tag{10.10}$$

where $x = C\,B$. electromagnetic field. The form of the saturation resembles Equation 10.9, but fitting experimental data gives the values of C, which yields the characteristic wavelength

$$\lambda = \left[ (C / M_s)(6\hbar c\, f_c) \right]^{1/4} \tag{10.11}$$

where $f_c$ is the volume fraction of the sample that is magnetically coherent.

In the case of CeO$_2$ nanoparticles, the analysis gives $\lambda = 300$ nm as the coherence length, which corresponds to a maximum in the UV absorption for magnetic nanoparticles at 4.1 eV. The remarkable disappearance of the magnetism when the nanoparticles are segregated by dilution (Figure 10.8c) is interpreted in terms of a minimum lengthscale of order needed to establish the coherent electronic state. The model may be applicable to a range of granular nanoscale oxides with conducting interfaces, and to oxide surfaces such as that of SrTiO$_3$ [35] or nanoporous amorphous alumina membranes [207]. We emphasize the essentially anhysteretic nature of the temperature-independent magnetic signal. By including the electron spin, and Rashba spin-orbit coupling of spin moments to the field-induced orbital moments in the GOP model, it may be possible to account for much of the behavior of nanocrystalline and nanogranular $d^0$ oxides, including the weak hysteresis that is occasionally observed.

## 10.6 CONCLUSIONS

The magnetism of dilute and undoped magnetic oxides remains one of the most puzzling and potentially important open questions in magnetism today. There is now sufficient evidence that the observations in many cases are not simply attributable to measurement artifacts or trivial impurities. There is definitely something to explain.

Two possible sources of the magnetism are defects and dopants. Both may be involved, but in the $d^0$ systems and perhaps in most of the dilute magnetic oxide systems that have been investigated, the defects suffice. The nature of the magnetic order and the coupling mechanism have to be better understood, but it seems that the $m$-J paradigm of magnetism in solids is unable to encompass these materials. On present evidence, the coupling between the $3d$ dopants is weak. They are not DMSs.

A new picture of high-temperature defect-related magnetism in oxides is emerging, where there are two contending explanations, based on quite different physical premises [208]. One model is Stoner spin ferromagnetism involving a spin-split defect-based impurity band. Charge transfer from the $3d$ cations can help stabilize the spin splitting. There is much to investigate in the CTF model, and to connect with specific materials systems – spin waves, micromagnetism, orbital moments, anisotropy, magnetotransport, and other spin-orbit coupling effects. The other model is giant orbital paramagnetism, where there is no symmetry-breaking magnetic order, but an unusual reversible orbital paramagnetic response of electrons that exhibit coherence on a mesoscopic scale as a result of a collective Lamb shift [203]. Here too there is much to investigate, especially in thin film structures artificially patterned on a mesoscopic scale.

The lack of reproducibility of magnetic properties between and within laboratories, and the lack of appropriate physical characterization tools that operate in ambient conditions for the aperiodic defects, such as oxygen vacancies and grain boundaries that appear to be important in many cases may be less of an obstacle than first thought. These hurdles can be circumvented to a great extent by experimental methods that create or destroy the d-zero magnetism in a particular sample. Specific chemical surface treatments with catechols are useful for strontium titanate surfaces [209] and nanoporous amorphous alumina membranes [207], $CeO_2$ nanoparticles come closest to a dilute magnetic 'fruitfly' system that is reproducible and easy to prepare [206].

The Co-doped $TiO_2$ or ZnO films produced by pulsed laser deposition that launched the field, have proved misleading. We need to tie down the correlation between optical, especially UV, absorption and magnetism, and better understand the significance of a critical size for the appearance of magnetism.

Device applications can follow by design or by serendipity. Once the basic spin or orbital character of the magnetism is firmly established, the challenge will be to generate stable and controllable defect structures where we can reap the benefit of this unusual high-temperature magnetism, whose occurrence is by no means limited to dilute magnetic oxides.

## ACKNOWLEDGMENT

This work was supported by Science Foundation Ireland, grant 16/IA/4534.

## REFERENCES

1. R. G. Burns, *Mineralogical Applications of Crystal Field Theory*, 2nd edn., Cambridge University Press, Cambridge, UK, 1993.
2. D. J. Newman and B. Ng, *Crystal Field Handbook*, Cambridge University Press, Cambridge, UK, p. 290, 2000.
3. E. J. Samuelsen and G. Shirane, Inelastic neutron scattering investigation of spin waves and magnetic interactions in alpha-$Fe_2O_3$, *Phys. Status Solidi* **42**, 241–256 (1970).
4. E. J. Samuelsen, M. T. Hutchings, and G. Shirane, Inelastic neutron scattering investigation of spin waves and magnetic interactions in $Cr_2O_3$, *Physica* **48**, 13–42 (1970).
5. H. Bourdonnay, A. Bousquet, J. P. Cotton et al., Experimental determination of exchange integrals in magnetite, *J. Phys.* 32(2–3), C1–C1182 (1970).
6. J. Goodenough, Theory of the role of covalence in the perovskite-type manganites [La, M(II)]$MnO_3$, *Phys. Rev.* **100**, 564–573 (1955).
7. J. Kanamori, Superexchange interaction and symmetry properties of electron orbitals, *J. Phys. Chem. Solids* **10**, 87–98 (1959).
8. P. W. Anderson, Theory of magnetic exchange interactions: Exchange in insulators and semiconductors, *Solid State Phys.* **14**, 99–214 (1963).
9. J. M. D. Coey, M. Viret, and S. von Molnar, Mixed-valence manganites, *Adv. Phys.* **48**, 167–293 (1999).
10. C. Zener, Interaction between the *d*-shells in the transition metals II: Ferromagnetic compounds of manganese with the perovskite structure, *Phys. Rev.* **82**, 403–405 (1951).
11. D. Stauffer, *Introduction to Percolation Theory*, Taylor & Francis, London, UK, 1985.
12. G. Deutscher, R. Zallen, and J. Adler (Eds), *Percolation Structures and Processes*, Adam Hilger, Bristol, UK, 1983.
13. P. Sati, R. Hayn, R. Kuzian et al., Magnetic anisotropy of $Co^{2+}$ as signature of intrinsic ferromagnetism in ZnO:Co, *Phys. Rev. Lett.* **96**, 017203 (2006).
14. W. Pacuski, D. Ferrand, J. Cibert et al., Effect of the s,p-d exchange interaction on the excitons in $Zn_{1-x}Co_xO$ epilayers, *Phys. Rev. B* **73**, 035214 (2006).
15. C. N. R. Rao and F. L. Deepak, Absence of ferromagnetism in Mn- and Co-doped ZnO, *J. Mater. Chem.* **15**, 573–578 (2005).
16. S. A. Chambers and R. F. Farrow, New possibilities for ferromagnetic semiconductors, *MRS Bull.* **28**, 729–733 (2003).
17. A. Ney, K. Ollefs, S. Ye et al., Absence of intrinsic ferromagnetic interactions with isolated and paired Co dopant atoms in $Zn_{1-x}Co_xO$ with high structural perfection, *Phys. Rev. Lett.* **100**, 157201 (2008).
18. J. M. D. Coey, $d^0$ ferromagnetism, *Solid State Sci.* **7**, 660–667 (2005).
19. S. B. Ogale, Dilute doping, defects and ferromagnetism in metal oxide systems, *Adv. Mater.* **22**, 3125–3155 (2010).
20. T. Fukumura and M. Kawasaki, Chapter 3: Magnetic oxide semiconductors: on the high-temperature ferromagnetism in $TiO_2$- and ZnO-based compounds, in *Functional Metal Oxides: New Science and Novel Applications*. S. B. Ogale, T. Venkatesen and M. Blamire (eds.), Wiley-VCH Verlag, Weinheim, Germany, 2013.
21. N. H. Hong, Magnetic oxide semiconductors, in *Handbook of Spintronics*, Y. Xu, D. D. Awschalom and J. Nitta (eds.), Springer, Dordrecht, the Netherlands, pp. 563–583, 2016.

22. D. W. Abraham, M. M. Frank, and S. Guha, Absence of magnetism in hafnium oxide films, *Appl. Phys. Lett.* **87**, 252502 (2005).

23. Y. Belghazi, G. Schmerber, S. Colis et al., Extrinsic origin of ferromagnetism in ZnO and $Zn_{0.9}Co_{0.1}O$ magnetic semiconductor films prepared by sol-gel technique, *Appl. Phys. Lett.* **89**, 122504 (2006).

24. J. M. D. Coey, Dilute magnetic oxides, *Curr. Opin. Solid State Mater. Sci.* **10**, 83–92 (2007).

25. M. A. Garcia, E. Fernandez Pinel, J. de la Venta et al., Sources of experimental errors in the observation of nanoscale magnetism, *J. Appl. Phys.* **105**, 013925 (2009).

26. K. Potzger and S. Q. Zhou, Non-DMS related ferromagnetism in transition-metal doped zinc oxide, *Phys. Status Solidi B* **246**, 1147–1167 (2009).

27. M. Khalid, A. Setzer, M. Ziese et al., Ubiquity of ferromagnetic signals in common diamagnetic oxides, *Phys. Rev. B* **81**, 214414 (2010).

28. J. M. D. Coey, Ariando, and W. E. Pickett, Magnetism at the edge; New phenomena at oxide interfaces, *MRS Bull.* **38**, 1040–1047 (2013).

29. A. Ohtomo, H. Y. Hwang, A high-mobility electron gas at the $LaAlO_3/SrTiO_3$ heterointerface, *Nature* **427**, 423–426 (2004).

30. N. Nakagawa, H. Y. Hwang, D. A. Muller, Why some interfaces cannot be sharp, *Nat. Mater.* **5**, 204–209 (2006).

31. G. Herranz, M. Basletic, M. Bibes et al., High mobility in $LaAlO_3/SrTiO_3$ heterostructures: Origin, dimensionality, and perspectives, *Phys. Rev. Lett.*, **98**, 216803 (2007).

32. Z. Q. Liu, C. J. Li, W. M. Lü et al., Origin of the two-dimensional electron gas at $LaAlO_3/SrTiO_3$ interfaces: The role of oxygen vacancies and electronic reconstruction, *Phys. Rev. X* **3**, 021010 (2013).

33. N. C. Bristowe, P. Ghosez, P. B. Littlewood, and E. Artacho, The origin of two-dimensional electron gases at oxide interfaces; Insights from theory, *J. Phys. Condens. Matter* **26**, 143201 (2014).

34. X. Renshaw Wang, C. J. Li, W. M. Lü et al., Imaging and control of ferromagnetism in $LaMnO_3/SrTiO_3$ heterostructures, *Science* **349**, 716–719 (2015).

35. J. M. D. Coey, M. Venkatesan, and P. Stamenov, Surface Magnetism of Strontium Titanate, *J. Phys. Condens. Matter* **28**, 485001 (2016).

36. M. Khalid, A. Setzer, M. Ziese et al., Ubiquity of ferromagnetic signals in common diamagnetic oxide crystals, *Phys. Rev. B* **81**, 214414 (2009).

37. N. H. Hong, J. Sakai, W. Prellier et al., Ferromagnetism in transition-metal doped $TiO_2$ thin films, *Phys. Rev. B* **70**, 195204 (2004).

38. Y. Matsumoto, M. Murakami, T. Shono et al., Room-temperature ferromagnetism in transparent transition metal-doped titanium dioxide, *Science* **291**, 854–856 (2001).

39. Y. Matsumoto, R. Takahashi, M. Murakami et al., Ferromagnetism in Co-doped $TiO_2$ rutile thin films grown by laser molecular beam epitaxy, *Jpn. J. Appl. Phys.* **40**, L1204–L1206 (2001).

40. S. R. Shinde, S. B. Ogale, S. Das Sarma et al., Ferromagnetism in laser deposited anatase $Ti_{1-x}Co_xO_{2-\delta}$ films, *Phys. Rev. B* **67**, 115211 (2003).

41. Z. J. Wang, J. K. Tang, Y. X. Chen et al., Room-temperature ferromagnetism in manganese-doped reduced rutile titanium dioxide thin films, *J. Appl. Phys.* **95**, 7384 (2004).

42. J. M. D. Coey, A. P. Douvalis, C. B. Fitzgerald, and M. Venkatesan, Ferromagnetism in Fe-doped $SnO_2$ thin films, *Appl. Phys. Lett.* **84**, 1332–1334 (2004).

43. S. B. Ogale, R. J. Choudhary, J. P. Buban et al., High temperature ferromagnetism with a giant magnetic moment in transparent Co-doped $SnO_{2-\delta}$, *Phys. Rev. Lett.* **91**, 077205 (2003).

44. J. M. D. Coey, M. Venkatesan, P. Stamenov et al., Magnetism in hafnium dioxide, *Phys. Rev. B* **72**, 024450 (2005).

45. H. N. Hong, N. Poirot, and J. Sakai, Evidence for magnetism due to oxygen vacancies in Fe-doped $HfO_2$ thin films, *Appl. Phys. Lett.* **89**, 042503 (2006).

46. A. Tiwari, V. M. Bhosle, R. Ramachandran et al., Ferromagnetism in Co-doped $CeO_2$: Observation of a giant magnetic moment with a high Curie temperature, *Appl. Phys. Lett.* **88**, 142511 (2006).

47. J. Philip, N. Theodoropoulou, G. Berera et al., High-temperature ferromagnetism in manganese-doped indium–tin oxide films, *Appl. Phys. Lett.* **85**(5), 777–779 (2004).

48. J. He, S. F. Xu, Y. K. Yoo et al., Room-temperature ferromagnetic n-type semiconductor $(In_{1-x}Fe_x)_2O_{3-\sigma}$, *Appl. Phys. Lett.* **87**, 052503 (2005).

49. J. Philip, A. Punnoose, B. I. Kim et al., Carrier-controlled ferromagnetism in transparent oxide semiconductors, *Nat. Mater.* **5**, 298–304 (2006).

50. H. Saeki, H. Tabata, and T. Kawai, Magnetic and -electric properties of vanadium doped ZnO films, *Solid State Commun.* **120**, 439–442 (2001).

51. P. Sharma, A. Gupta, K. V. Rao et al., Ferromagnetism above room temperature in bulk and transparent thin films of Mn-doped ZnO, *Nat. Mater.* **2**, 673–677 (2003).

52. S. J. Han, J. W. Song, C. H. Yang et al., A key to room-temperature ferromagnetism in Fe-doped ZnO: Cu, *Appl. Phys. Lett.* **81**, 4212–4214 (2002).

53. K. Ueda, H. Tabata, and T. Kawai, Magnetic and electric properties of transition-metal-doped ZnO films, *Appl. Phys. Lett.* **79**, 988–990 (2001).

54. P. V. Radovanovic and D. R. Gamelin, High-temperature ferromagnetism in $Ni^{2+}$-doped ZnO aggregates prepared from colloidal dilute magnetic semiconductor quantum dots, *Phys. Rev. Lett.* **91**, 157302 (2003).

55. J. Y. Kim, J. H. Park, B. G. Park et al., Ferromagnetism induced by clustered Co in Co-doped anatase $TiO_2$ thin films, *Phys. Rev. Lett.* **90**, 017401 (2003).

56. P. A. Stampe, R. J. Kennedy, Y. Xin, and J. S. Parker, Investigation of the cobalt distribution in the room temperature ferromagnet $TiO_2$:Co, *J. Appl. Phys.* **93**, 7864–7866 (2003).

57. T. Zhao, S. R. Shinde, S. B. Ogale et al., Electric field effect in diluted magnetic insulator anatase Co: $TiO_2$, *Phys. Rev. Lett.* **94**, 126601 (2005).

58. H. Toyosaki, T. Fukumura, Y. Yamada, and M. Kawasaki, Evolution of ferromagnetic circular dichroism coincident with magnetization and anomalous Hall effect in Co-doped rutile $TiO_2$, *Appl. Phys. Lett.* **86**, 182503 (2005).

59. H. Toyosaki, T. Fukumura, Y. Yamada et al., Anomalous Hall effect governed by electron doping in a room-temperature transparent ferromagnetic semiconductor, *Nat. Mater.* **3**, 221–224 (2004).

60. H. Toyosaki, T. Fukumura, K. Ueno et al., A ferromagnetic oxide semiconductor as spin injection electrode in a magnetic tunnel junction, *Jpn. J. Appl. Phys.* **44**, L896–L898 (2005).

61. Z. Wang, W. Wang, J. Tang et al., Extraordinary Hall effect and ferromagnetism in Fe-doped reduced rutile, *Appl. Phys. Lett.* **83**, 518–520 (2003).

62. S. Duhalde, M. F. Vignolo, F. Golmar et al., Appearance of room-temperature ferromagnetism in Cu-doped $TiO_{2-\delta}$ films, *Phys. Rev. B* **72**, 161313 (2005).

63. D. L. Hou, R. B. Zhao, H. J. Meng et al., Room-temperature ferromagnetism in Cu-doped $TiO_2$ thin films, *Thin Solid Films* **516**, 3223–3226 (2008).

64. Y. J. Kim, S. Thevuthasan, T. Droubay et al., Growth and properties of molecular beam epitaxially grown ferromagnetic Fe-doped $TiO_2$ rutile films on $TiO_2$(110), *Appl. Phys. Lett.* **84**, 3531–3533 (2004).

65. T. C. Kaspar, S. M. Heald, C. M. Wang et al., Negligible magnetism in excellent structural quality $Cr_xTi_{1-x}O_2$ anatase: Contrast with high-$T_C$ ferromagnetism in structurally defective $Cr_xTi_{1-x}O_2$, *Phys. Rev. Lett.* **95**, 217203 (2005).

66. J. Osterwalder, T. Droubay, T. Kaspar et al., Growth of Cr-doped $TiO_2$ films in the rutile and anatase structures by oxygen plasma assisted molecular beam epitaxy, *Thin Solid Films* **484**, 289–298 (2005).

67. S. A. Chambers, Ferromagnetism in doped thin film oxide and nitride semiconductors and dielectrics, *Surf. Sci. Rep.* **61**, 345–381 (2006).

68. K. A. Griffin, A. B. Pakhomov, C. M. Wang et al., Intrinsic ferromagnetism in insulating cobalt doped anatase $TiO_2$, *Phys. Rev. Lett.* **94**, 157204 (2005).

69. K. A. Griffin, A. B. Pakhomov, C. M. Wang et al., Cobalt-doped anatase $TiO_2$: A room-temperature dilute magnetic dielectric material, *J. Appl. Phys.* **97**, 10D320 (2006).

70. K. A. Griffin, M. Varela, S. J. Pennycook et al., Atomic-scale studies of cobalt distribution in Co-$TiO_2$ anatase $TiO_2$ thin films: Processing, microstructure and the origin of ferromagnetism, *J. Appl. Phys.* **97**, 10D320 (2006).

71. T. Fukumura, H. Toyosaki, K. Ueno et al., Role of charge -carriers for ferromagnetism in Co-doped rutile $TiO_2$, *New J. Phys.* **10**, 055018 (2008).

72. A. F. Orlov, L. A. Balagurov, A. S. Konstantinova et al., Giant magnetic moments in dilute magnetic semiconductors, *J. Magn. Magn. Mater.* **320**, 895–897 (2008).

73. S. D. Yoon, Y. Chen, A. Yang et al., Oxygen-defect-induced magnetism at 880 K in semiconducting anatase $TiO_{2-\delta}$ films, *J. Phys. Condens. Matter* **18**, L355–L361 (2006).

74. N. H. Hong, J. Sakai, N. Proirot et al., Room-temperature ferromagnetism observed in undoped semiconducting and insulating oxide thin films, *Phys. Rev. B* **73**, 132404 (2006).

75. Z. Q. Zhou, C. Cizmar, K. Potzger et al., Origin of magnetic moments in defective $TiO_2$ single crystals, *Phys. Rev. B* **79**, 113201 (2009).

76. B. Santora, P. K. Giri, K. Imakita and M. Fujii, Evidence of oxygen-vacancy-induced room-temperature ferromagnetism in solvothermally synthesised indoped $TiO_2$ nanoribbons, *Nanoscale* **5**, 5476–5480 (2013).

77. N. N. Bao, H. M. Fan, J. Ding, and J. B. Yi, Toom-temperature ferromagnetism in N-doped rutile $TiO_2$ films, *J. Appl. Phys.* **109**, 07C302 (2011).

78. A. Rusydi, S. Dhar, A. R. Barman et al., Cation-vacancy-induced room-temperature ferromagnetism in transparent, conducting anatase $Ti_{1-x}Ta_xO_2$ thin films, *Phil. Trans. Roy. Soc. A* **370**, 4926–4943 (2012).

79. S. Wang, L. Pan, J. J. Song et al., Titanium-defected undoped anatase $TiO_2$ with *p*-type conductivity, room-temperature ferromagnetism and remarkable photocatalytic properties, *J. Am. Chem. Soc.* **137**, 2975–2983 (2015).

80. I. S. Elfimov, S. Yunoki, and G. A. Sawatzky, Possible path to a new class of ferromagnetic and half-metallic ferromagnetic materials, *Phys. Rev. Lett.* **89**, 216403 (2002).

81. N. H. Hong and J. Sakai, Ferromagnetic V-doped $SnO_2$ thin films, *Phys. B* **358**, 265–268 (2005).

82. N. H. Hong, J. Sakai, W. Prellier, and A. Hassini, Transparent Cr-doped $SnO_2$ thin films: Ferromagnetism beyond room temperature with a giant magnetic moment, *J. Phys. Condens. Matter* **17**, 1697–1702 (2005).

83. K. Gopinadhan, S. C. Kashyap, D. K. Pandya et al., High temperature ferromagnetism in Mn-doped $SnO_2$, *J. Appl. Phys.* **102**, 113513 (2007).

84. C. B. Fitzgerald, M. Venkatesan, L. S. Dorneles et al., Magnetism in dilute magnetic oxide thin films based on $SnO_2$, *Phys. Rev. B* **74**, 11530 (2006).

85. N. H. Hong, J. Sakai, N. T. Huong et al., Role of defects in tuning ferromagnetism in diluted magnetic oxide thin films, *Phys. Rev. B* **72**, 045336 (2005).

86. J. Zhang, R. Skomski, L. P. Yue et al., Structure and magnetism of V-doped $SnO_2$ thin films: Effect of the substrate, *J. Phys. Condens. Matter* **19**, 256204 (2007).

87. H. S. Kim, L. Bi, G. F. Dionne et al., Structure, magnetic and optical properties, and Hall effect of Co- and Fe-doped $SnO_2$ films, *Phys. Rev. B* **77**, 214436 (2008).

88. N. H. Hong, N. Poirot, and J. Sakai, Ferromagnetism observed in pristine $SnO_2$ thin films, *Phys. Rev. B* **77**, 033205 (2008).

89. J. Zhang, R. Skomski, Y. F. Lu et al., Temperature-dependent orbital-moment anisotropy in dilute magnetic oxides, *Phys. Rev. B* **75**, 214417 (2007).

90. P. I. Archer, P. V. Radovanovic, S. M. Heald et al., Low-temperature activation and deactivation of high-curie-temperature ferromagnetism in a new diluted magnetic semiconductor: $Ni^{2+}$-doped $SnO_2$, *J. Am. Chem. Soc.* **127**, 14479–14487 (2005).

91. C. Van Komen, A. Thurber, K. M. Reddy et al., Structure-magnetic property relationship in transition metal (M = V, Cr, Mn, Fe, Co, Ni) doped $SnO_2$ nanoparticles, *J. Appl. Phys.* **103**, 07D141 (2008).

92. A. Thurber, K. M. Reddy, V. Shutthanandan et al., Ferromagnetism in chemically-substituted $CeO_2$ nanoparticles by Ni doping, *Phys. Rev. B* **76**, 165206 (2007).

93. C. B. Fitzgerald, M. Venkatesan, L. S. Dorneles et al., $SnO_2$ doped with Mn, Fe or Co; room-temperature dilute magnetic semiconductors, *J. Appl. Phys.* **95**, 7390–7395 (2004).

94. A. F. Cabrera, A. M. N. Mudarra, C. E. T. Rodriguez et al., Mechanosynthesis of Fe-doped $SnO_2$ nanoparticles, *Phys. B* **398**, 215–218 (2007).

95. P. I. Archer and D. R. Gamelin, Controlled grain-boundary defect formation and its role in the high-$T_c$ ferromagnetism on $Ni^{2+}$: $SnO_2$, *J. Appl. Phys.* **99**, 08M107 (2006).

96. J. D. Bryan, S. M. Heald, S. A. Chambers, and D. R. Gamelin, Strong room-temperature ferromagnetism in $Co^{2+}$-doped $TiO_2$ made from colloidal nanocrystals, *J. Am. Chem. Soc.* **126**, 11640–11647 (2004).

97. A. Sundaresan, R. Bhargavi, N. Rangarajan et al., Ferromagnetism as a universal feature of nanoparticles of otherwise nonmagnetic oxides, *Phys. Rev. B* **74**, 161306(R) (2006).

98. S. Bhaumik, A. K. Sinha, S. K. Ray, and A. K. Das, Defect-iduced room-temperature ferromagnetism in $SnO_2$ nanowires controlled by UV light irradiation, *IEEE Trans. Magn.* **50**, 2400206 (2014).

99. J. Wang, W. Zhou, and P. Wu, Band gap narrowig and $d^0$ ferromagnetism in epitaxial Li-doped $SnO_2$ films, *Appl. Surf. Sci.* **314**, 188–192 (2014)

100. N. Wang, W. Zhou, and P. Wu, Ferromagnetic order in $SnO_2$ nanoparticles with nonmagnetic Li doping, *J. Mater. Sci.* **26**, 4132–4137 (2015).

101. P. Wu, B. Zhou, and W. Zhou, Room-temperature ferromagnetism in epitaxial Mg-doped $SnO_2$ thin films, *Appl. Phys. Lett.* **100**, 182405 (2012).

102. M. Venkatesan, C. B. Fitzgerald, and J. M. D. Coey, Unexpected magnetism in a dielectric oxide, *Nature* **430**, 630 (2004).

103. H. N. Hong, Magnetism due to defects/oxygen vacancies in $HfO_2$ thin films, *Phys. Status Solidi C* **4**, 1270–1275 (2007).

104. N. H. Hong, J. Sakai, N. Poirot, and A. Ruyter, Laser ablated Ni-doped $HfO_2$ thin films: Room temperature ferromagnets, *Appl. Phys. Lett.* **86**, 242505 (2005).

105. M. S. R. Rao, D. C. Kundaliya, S. B. Ogale et al., Search for ferromagnetism in undoped and cobalt-doped $HfO_{2-\delta}$, *Appl. Phys. Lett.* **88**, 142505 (2006).

106. N. Hadacek, A. Nosov, L. Ranno et al., Magnetic properties of $HfO_2$ thin films, *J. Phys. Condens. Matter* **19**, 486206 (2007).

107. E. Tirosh and G. Markovich, Control of defects and magnetic properties in colloidal $HfO_2$ nanorods, *Adv. Mater.* **19**, 2608–2612 (2007).

108. X. Liu, Y. Chen, L. Wang, and D. L. Peng, Transition from paramagnetism to ferromagnetinm in $HfO_2$ nanorods, *J. Appl. Phys.* **113**, 076102 (2013).

109. Q. Xie, W. P. Mang, Z. Xie et al., Room-temperature ferromagnetism in undoped amorphous $HfO_2$ nanohelix arrays, *Chin. Phys. B* **24**, 057503 (2015).

110. Q. Xie, W. Wang, Z. Xie et al., High-magnetic field annealing effect on room-temperature ferromagnetism enhancement of undoped $HfO_2$ thin films, *App. Phys. A* **119**, 917921 (2015).

111. E. Hildebrandt, J. Kurian, and L. Alff, Physical properties and band structure of reactive molecular beam epitaxy grown oxygen engineered $HfO_{2-x}$, *J. Appl. Phys.* **112**, 114112 (2012).

112. Y. Q. Song, H. W. Zhang, Q. Y. Wen et al., Room-temperature ferromagnetism in Co-doped $CeO_2$ thin films on Si (111) substrates, *Chin. Phys. Lett.* **24**, 218–221 (2007).

113. R. Vodungbo, Y. Zheng, F. Vidal et al., Room-temperature ferromagnetism of Co-doped $CeO_{2-\delta}$ diluted magnetic oxide: Effect of oxygen and anisotropy, *Appl. Phys. Lett.* **90**, 162510 (2007).

114. Y. L. Liu, Z. Lockmann, A. Aziz et al., Size-dependent ferromagnetism in cerium oxide ($CeO_2$) nanostructures independent of oxygen vacancies, *J. Phys. Condens. Matter* **20**, 165201 (2008).

115. M. Y. Ge, H. Wang, E. Z. Liu et al., On the origin of ferromagnetism in $CeO_2$ nanocubes, *Appl. Phys. Lett.* **93**, 062505 (2008).

116. M. L. Li, S. H. Ge, W. Qiao et al., Relationship between the surface chemical states and magnetic properties of $CeO_2$ nanoparticles, *Appl. Phys. Lett.* **94**, 152511 (2009).

117. A. Thurber, K. M. Reddy, and A. Punnoose, Influence of oxygen level on structure and ferromagnetism in $Sn_{0.95}Fe_{0.05}O_2$ nanoparticles, *J. Appl. Phys.* **105**, 07E706 (2009).

118. X. B. Chen, G. S. Li, Y. G. Su et al., Synthesis and room--temperature ferromagnetism of $CeO_2$ nanocrystals with nonmagnetic $Ca^{2+}$ doping, *Nanotechnology* **20**, 115600 (2009).

119. G. Niu, E. Hildebrandt, M. A. Schubert et al., Oxygen vacancy induced room temperature ferromagnetism in Pr-doped $CeO_2$ thin films on silicon, *ACS Appl. Mater. Interfaces* **4**, 17496–17505 (2014).

120. N. Paunovic, Z. Dohčević-Mitrović et al., Suppression of inherent ferromagnetism in Pr-doped $CeO_2$ Nanocrystals, *Nanoscale* **4**, 5469–5476 (2012).

121. J. M D. Coey, K. Ackland, M. Venkateran, and S. Sen, Collective magnetic response of $CeO_2$ nanoparticles, *Nat. Phys.* **12**, 694–699 (2016).

122. P. D. C. King, T. D. Veal, F. Fuchs et al., Band gap, electronic structure and surface electron accumulation in cubic and rhombohedral $In_2O_3$, *Phys. Rev. B* **79**, 205211 (2009).

123. Y. K. Yoo, Q. Z. Xue, H. C. Lee et al., Bulk synthesis and high-temperature ferromagnetism of $(In_{1-x}Fe_x)_2O_{3-s}$ with Cu co-doping, *Appl. Phys. Lett.* **86**, 042506 (2005).

124. X. H. Xu, F. X. Jiang, J. Zhang et al., Magnetic properties of *n*-type Fe-doped $In_2O_3$ ferromagnetic thin films, *Appl. Phys. Lett.* **94**, 212510 (2009).

125. G. Peleckis, X. L. Wang, and S. X. Dou, Room-temperature ferromagnetism in Mn and Fe codoped $In_2O_3$, *Appl. Phys. Lett.* **88**, 132507 (2006).

126. N. H. Hong, J. Sakai, N. T. Huong et al., Magnetism in transition-metal-doped $In_2O_3$ thin films, *J. Phys. Condens. Matter* **18**, 6897–6905 (2006).

127. Z. G. Yu, J. He, S. F. Xu et al., Origin of ferromagnetism in semiconducting $(In_{1-x-y}Fe_xCu_y)_2O_{3-\sigma}$, *Phys. Rev. B* **74**, 165321 (2006).

128. M. Venkatesan, R. D. Gunning, P. Stamenov et al., Room-temperature ferromagnetism in Mn- and Fe-doped indium tin oxide thin films, *J. Appl. Phys.* **103**, 07D135 (2008).

129. S. S. Farvid, T. Sabergharesou, L. N. Hutfluss et al., Evidence of charge-transfer ferromagnetism in transparent diluted magnetic oxide nanocrystals: Switching the mechanism of magnetic interactions, *J. Am. Chem. Soc.* **136**, 7669–7679 (2014).

130. H. Raebiger, S. Lam, and A. Zunger, Control of ferromagnetism via electron doping in $In_2O_3$: Cr, *Phys. Rev. Lett.* **101**, 027203 (2008).

131. R. P. Panguluri, P. Kharel, C. Sudakar et al., Ferromagnetism and spin-polarized charge carriers in $In_2O_3$ thin films, *Phys. Rev. B* **79**, 165208 (2009).

132. D. Bérardan, E. Guilmeau, and D. Pelloquin, Intrinsic magnetic properties of $In_2O_3$ and transition-metal doped $In_2O_3$, *J. Magn. Magn. Mater.* **320**, 983–989 (2008).

133. S. H. Babu, S. Kaleemulla, N. M. Rao, and C. Krishnamoorthi, Indium oxide: A transparent, conducting ferromagnetic semiconductor for spintronic applications, *J. Magn. Magn. Mater.* **416**, 66–74 (2016).

134. J. Lee, N. G. Subramaniam, I.A. Kowalik et al., Towards a new class of heavy ion doped magnetic semiconductors for room temperature applications, *Sci. Rep.* **5**, 17053 (2015).

135. U. Ozgur, Y. I. Alivov, C. Liu et al., A comprehensive review of ZnO materials and devices, *J. Appl. Phys.* **98**, 041301 (2005).

136. A. Janotti and C. G. van de Walle, Fundamentals of Zinc oxide as a semiconductor, *Rep. Prog. Phys.* **76**, 126501 (2009).

137. F. Pan, C. Song, X. J. Liu, Y. C. Yang, and F. Zeng, Ferromagnetism and possible application in spintronics of transition-metal doped ZnO films, *Mater. Sci. Eng. R* **62**, 1–35 (2008).

138. M. Bouloudenine, N. Viart, S. Colis et al., Antiferro-magnetism in bulk $Zn_{1-x}Co_xO$ magnetic semiconductors prepared by the coprecipitation technique, *Appl. Phys. Lett.* **87**, 052501 (2005).

139. J. Alaria, P. Turek, M. Bernard et al., No ferromagnetism in Mn-doped ZnO semiconductors, *Chem. Phys. Lett.* **415**, 337–341 (2005).

140. Z. Jin, T. Fukumura, M. Kawasaki et al., High throughput fabrication of transition-metal-doped epitaxial ZnO thin films: A series of oxide-diluted magnetic semiconductors and their properties, *Appl. Phys. Lett.* **78**(24), 3824–3826 (2001).

141. R. Janisch, P. Gopal, and N. A. Spaldin, Transition metal-doped $TiO_2$ and ZnO-present status of the field, *J. Phys. Condens. Matter* **17**, R657–R689 (2005).

142. W. Prellier, A. Fouchet, and B. Mercey, Oxide-diluted magnetic semiconductors: A review of the experimental status, *J. Phys. Condens. Matter* **15**, R1583–R1601 (2003).

143. S. J. Pearton, W. H. Heo, M. Ivill et al., Dilute magnetic semiconducting oxides, *Semicond. Sci. Technol.* **19**, R59–R74 (2004).

144. M. Venkatesan, C. B. Fitzgerald, J. G. Lunney, and J. M. D. Coey, Anisotropic ferromagnetism in substituted zinc oxide, *Phys. Rev. Lett.* **93**, 177206 (2004).

145. B. Martinez, F. Sandiumenge, L. Balcells et al., Role of the microstructure on the magnetic properties of Co-doped ZnO nanoparticles, *Appl. Phys. Lett.* **86**, 103113 (2005).

146. N. Khare, M. J. Kappers, M. Wei et al., Defect induced ferromagnetism in Co-doped ZnO, *Adv. Mater.* **18**, 1449–1452 (2006).

147. L. E. Halliburton, N. C. Giles, N. Y. Garces et al., Production of native donors in ZnO by annealing at high temperature in Zn vapor, *Appl. Phys. Lett.* **87**, 172108 (2005).

148. D. A. Schwartz and D. R. Gamelin, Reversible 300 K ferromagnetic ordering in a diluted magnetic semiconductor, *Adv. Mater.* **16**, 2115–2119 (2004).

149. R. P. Borges, R. C. da Silva, S. Magalhaes et al., Magnetism in Ar-implanted ZnO, *J. Phys. Condens. Matter* **19**, 476207 (2007) (see also ibid 20, 429801).

150. P. Stamenov, M. Venkatesan, L. S. Dornales et al., Magnetoresistance of Co-doped ZnO thin films, *J. Appl. Phys.* **99**, 08M124 (2006).

151. K. R. Kittilstved, W. K. Liu, and D. R. Gamelin, Electronic origins of polarity-dependent high $T_C$ ferromagnetism in oxide diluted magnetic semiconductors, *Nat. Mater.* **5**, 291 (2006).

152. J. R. Neal, A. J. Behan, R. M. Ibrahim et al., Room temperature magneto-optics of ferromagnetic transition-metal doped ZnO thin films, *Phys. Rev. Lett.* **96**, 107208 (2006).

153. Z. H. Zhang, X. F. Wang, J. B. Xu et al., Evidence of intrinsic ferromagnetism in individual dilute magnetic semiconducting nanostructures, *Nat. Nanotechnol.* **4**, 523–527 (2009).

154. X. H. Xu, H. J. Blythe, M. Ziese et al., Carrier-induced ferromagnetism in ZnMnAlO and ZnCoAlO thin films at room temperature, *New J. Phys.* **8**, 135–144 (2006).

155. A. J. Behan, H. Mokhtari, H. Blythe et al., Two magnetic regimes in doped ZnO corresponding to a dilute magnetic semiconductor and a dilute magnetic insulator, *Phys. Rev. Lett.* **100**, 047206 (2008).

156. M. Venkatesan, R. D. Gunning, P. Stamenov et al., Magnetic, magnetotransport and optical properties in Al doped $Zn_{0.95}Co_{0.05}$ O thin films, *Appl. Phys. Lett.* **90**, 242508 (2006).

157. S. T. Oschenbein, Y. Feng, K. M. Whittaker et al., Charge-controlled magnetism in colloidal doped semiconductor nanocrystals, *Nat. Nanotechnol.* **4**, 681–687 (2009).

158. J. H. Park, M. G. Kim, H. M. Jang et al., Co metal clustering as the origin of ferromagnetism in Co-doped ZnO thin films, *Appl. Phys. Lett.* **84**, 1338–1341 (2004).

159. L. S. Dorneles, M. Venkatesan, R. Gunning et al., Magnetic and structural properties of Co-doped ZnO thin films, *J. Magn. Magn. Mater.* **310**, 2087–2088 (2007).

160. K. Rode, R. Mattana, A. Anane et al., Magnetism of (Zn,Co)O thin films probed by x-ray absorption spectroscopies, *Appl. Phys. Lett.* **92**, 012509 (2008).

161. M. Ivill, S. J. Pearton, S. Rawal et al., Structure and magnetism of cobalt-doped ZnO thin films, *New J. Phys.* **10**, 065002 (2008).

162. D. P. Norton, M. E. Overberg, S. J. Pearton et al., Ferromagnetism in VCo-implanted ZnO, *Appl. Phys. Lett.* **83**, 5488–5490 (2003).

163. D. C. Kundaliya, S. B. Ogale, S. E. Lofland et al., On the origin of high-temperature ferromagnetism in the low-temperature processed Mn–Zn–O system, *Nat. Mater.* **3**, 709–712 (2004).

164. A. Zhukova, A. Teiserskis, S. van Dijken et al., Giant moment and magnetic anisotropy in Co-doped ZnO films grown by pulse-injection metal organic chemical vapor deposition, *Appl. Phys. Lett.* **89**, 232503 (2006).

165. K. Rode, Contribution à l'étude des semiconducteurs ferromagnétiques: Cas des films minces d'oxyde de zinc dopé au cobalt, PhD thesis, Université Paris XI, France, 2006.

166. S. Ghosh, V. Sih, W. H. Lau et al., Room-temperature spin coherence in ZnO, *Appl. Phys. Lett.* **86**, 232507 (2005).

167. T. C. Droubay, D. J. Keavney, T. C. Kaspar et al., Correlated substitution in paramagnetic $Mn^{2+}$-doped ZnO epitaxial films, *Phys. Rev. B* **79**, 155203 (2009).

168. A. Barla, G. Schmerber, E. Beaurepaire et al., Paramagnetism of the Co sublattice in ferromagnetic $Zn_{1-x}Co_xO$ films, *Phys. Rev. B* **76**, 125201 (2007).

169. T. Tietze, M. Gacic, G. Schütz et al., XMCD studies on Co and Li doped ZnO magnetic semiconductors, *New J. Phys.* **10**, 055009 (2008).

170. J. J. Beltrán, C. A. Barrero, and A. Punnoose, Combination of defects plus mixed valence of transition metals: A strong strategy for ferromagnetic enhancement in ZnO Nanoparticles, *J. Phys. Chem. C* **120**, 8969–8978 (2016).

171. B. K. Roberts, A. B. Pakhomov, P. Voll et al., Surface scaling of magnetism in Cr-ZnO dilute magnetic dielectric thin films, *Appl. Phys. Lett.* **92**, 162511 (2008).

172. J. M. D. Coey, R. D. Gunning, M. Venkatesan, P. Stamenov, and K. Paul, Magnetism in defect-ridden oxides, *New J. Phys.* **12**, 053025 (2010).

173. M. V. Limaye, S. B. Singh, R. Das, P. Poddar, and S. K. Kulkarni, Room temperature ferromagnetism in undoped and Fe doped ZnO nanorods: Microwave-assisted synthesis *J. Solid State Chem.* **184**, 391–400 (2011).

174. Q. Y. Xu, H. Schmidt, S. Q. Zhou et al., Room temperature ferromagnetism in ZnO films due to defects, *Appl. Phys. Lett.* **92**, 082508 (2008).

175. N. H. Hong, E. Chikoidze, and Y. Dumont, Ferromagnetism in laser-ablated ZnO and Mn-doped ZnO thin films: A comparative study from magnetization and Hall-effect measurements, *Phys. B* **404**, 3978–3981 (2009).

176. H. N. Hong, A. Barla, J. Sakai et al., Can undoped semiconducting oxides be ferromagnetic? *Phys. Status Solidi C* **4**(12), 4461–4466 (2007).

177. C. S. Ong, T. S. Herng, X. L. Huang, Y. P. Feng, and J. Ding, Strain-induced ZnO spinterfaces, *J. Phys. Chem. C* **116**, 610–617 (2012).

178. M. Khalid, M. Ziese, A. Setzer et al., Defect-induced magnetic order in pure ZnO films, *Phys. Rev. B* **80**, 035331(2009).

179. P. Zhan, W. P. Wang, Z. Xie et al., Substrate effect on the room-temperature ferromagnetism in undoped ZnO films, *Appl. Phys. Lett.* **101**, 031913 (2012).

180. D. Gao, Z. Zhang, Y. Li et al., Abnormal room temperature ferromagnetism in CuO–ZnO heterostructures: Interface related or not? *Chem. Commun* **51**, 1151–1153 (2015).

181. T. Taniguchi, K. Yamaguchi, A. Shigeta, Y. Matsuda, and S. Hayami, Enhanced and engineered $d^0$ ferromagnetism in molecularly-thin zinc oxide nanosheets, *Adv. Func. Mater.* **13**, 2140–2145 (2013).

182. J. Chaboy, R. Boada, C. Piquer et al., Evidence of intrinsic magnetism in capped ZnO nanoparticles, *Phys. Rev. B* **82**, 064411 (2010).

183. C. Guglieri, M. A. Laguna-Marco, M. A. García, N. Carmona, and E. Céspedes, XMCD Proof of Ferromagnetic Behavior in ZnO Nanoparticles, *J. Phys Chem. C* **116**, 6608–6614 (2012).

184. A. Sundaresan and C. N. R. Rao, Implications and consequences of ferromagnetism universally exhibited by inorganic nanoparticles, *Solid State Commun.* **140**, 1197–2000 (2009).

185. X. Y. Xu, C. X. Xu, J. Dai et al., Size dependence of defect-induced room temperature ferromagnetism in undoped ZnO nanoparticles. *J. Phys. Chem. C* **116**, 8813–8818 (2012).

186. X. F. Bie, C. Z. Wang, H. Ehrenberg, Y. J. Wei, and G. Chen, Room-temperature ferromagnetism in pure ZnO nanoflowers, *Solid State Sci.* **12**, 1365–1367 (2010).

187. B. B. Straumal, A. A. Mazilkin, S. G. Protasova et al., Magnetization study of nanograined pure and Mn-doped ZnO films: Formation of a ferromagnetic grain-boundary foam, *Phys. Rev. B* **79**, 205206 (2009).

188. B. B. Straumal, A. A. Mazilkin, S. G. Protasova et al., Ferromagnetism of nanostructured zinc oxide films, *Phys. Metal. Metallog.* **113**, 1244–1256 (2012).

189. T. Tietze, P. Audehm, Y. C. Chen et al., Interfacial dominated ferromagnetism in nanograined ZnO: A μSR and DFT study, *Sci. Rep.* **5**, 8871 (2015).

190. J. M. D. Coey, High-temperature ferromagnetism in dilute magnetic oxides, *J. Appl. Phys.* **97**, 10D313 (2004).

191. T. Dietl, H. Ohno, F. Matsukura et al., Zener model description of ferromagnetism in zinc-blende magnetic semiconductors, *Science* **287**, 1019 (2000).

192. S. Methfessel and D. C. Mattis, Magnetic semiconductors, in *Handbuch der Physik*, vol. 18(1), S. Flügge (Ed.), Springer-Verlag, Berlin, 1968.

193. J. M. D. Coey, M. Venkatesan, and C. B. Fitzgerald, Donor impurity band exchange in dilute ferromagnetic oxides, *Nat. Mater.* **4**, 173–179 (2005).

194. D. J. Priour Jr. and S. Das Sarma, Clustering in disordered ferromagnets: The Curie temperature in diluted magnetic semiconductors, *Phys. Rev. B* **73**, 165203 (2006).

195. T. Dietl, Origin and control of magnetism in dilute magnetic semiconductors and oxides, *J. Appl. Phys.* **103**, 07D111 (2008).

196. D. M. Edwards and M. I. Katsnelson, High-temperature ferromagnetism of sp electrons in narrow impurity bands: Application to $CaB_6$, *J. Phys. Condens. Matter* **18**, 7209–7225 (2006).

197. J. M. D. Coey, K. Wongsaprom, J. Alaria et al., Charge transfer ferromagnetism in oxide nanoparticles, *J. Phys. D: Appl. Phys.* **41**, 134012 (2008).

198. J. M. D. Coey, J. T. Mlack, M. Venkatesan, and P. Stamenov, Magnetization process in dilute magnetic oxides, *IEEE Trans. Magn.* **46**, 2501–2504 (2010).

199. R. Gréget, G. L. Nealon, B. Vileno et al., Magnetic properties of gold nanoparticles: A room-temperature quantum effect, *Chem. Phys. Chem.* **13**, 3092–3097 (2012).

200. G. L. Nealon, B. Donnio, R. Gréget et al., Magnetism in gold nanoparticles, *Nanoscale* **4**, 5244–5258 (2012).

201. A. Hernando, P. Crespo, and M. A. Garcia, Origin of orbital ferromagnetism and giant magnetic anisotropy on the nanoscale, *Phys. Rev. Lett.* **96**, 057206 (2006).

202. A. Hernando, P. Crespo, M. A. Garcia, M. Coey, A. Ayuela, and P. M. Eschnique, Revisiting magnetism of capped Au and ZnO nanoparticles: Surface band structure and atomic orbitals with giant magnetic moment, *Phys. Status Solidi B* **248**, 2352–2360 (2011).

203. S. Sen, K. S. Gupta, and J. M. D. Coey, Mesoscopic structure formation in condensed matter due to vacuum fluctuations, *Phys. Rev. B* **92**, 155115 (2015).

204. M. Venkatesan, P. Kavle, S. B. Porter, K. Ackland, and J. M. D. Coey, Magnetic analysis of polar and nonpolar oxide substrates, *IEEE Trans. Magn.* **50**, 2201704 (2014).

205. S. K. S. Patel, K. Dewangan, S. K. Srivastav, and N. S. Gajbhiye, Synthesis of monodisperse $In_2O_3$ nanoparticles and their $d^0$ ferromagnetism, *Curr. Appl. Phys.* **14**, 905–908 (2014).

206. K. Ackland, and J. M. D. Coey, Room-temperature magnetism in $CeO_2$ – a review. *Phys. Rep.* **746**, 1–40 (2018).

207. A. S. Esmaeily, M. Venkatesan, S. Sen and J. M. D. Coey, d-zero magnetism in nanoporous amorphous alumina membranes, *Phys. Rev. Mater.* 2, 054404 (2018).

208. J. M. D. Coey, d-zero magnetism in oxides, *Nat. Mater.* **18** (2019), in press.

209. J. M. D. Coey, M. Venkatesan, and P. Stamenov, Surface magnetism of strontium titanate, *J. Phys.: Condens. Matter* **28**, 485001 (2016).

# 11

# Magnetism of
# Complex Oxide
# Interfaces

**Satoshi Okamoto, Shuai Dong, and Elbio Dagotto**

## 11.1 INTRODUCTION

Controlling the magnetic properties of magnetic materials in bulk crystals and at interfaces is crucial for spintronic applications and related phenomena. In this chapter, we will review theoretical studies focused on interfaces or heterostructures of complex oxides such as perovskite transition-metal oxides (TMOs). TMOs have been a focus of materials science for decades because of their exotic behavior originating from strong electron-electron or electron-lattice interactions, including high critical temperature ($T_c$) superconductivity in cuprates [1] and colossal magnetoregistance effects in manganites [2, 3]. In fact, after the discovery of high-$T_c$ cuprates, crystal growth and measurement techniques have been improved dramatically, and a variety of novel phenomena have been discovered, such as novel metal-insulator transitions and spin charge orbital ordering [4, 5]. Further, the recent progress in thin-film growth techniques led to the discovery of metallic states at interfaces between dissimilar insulators [6, 7]. Thus, while bulk complex TMOs can provide interesting functionalities, heteroengineering of TMOs could also provide additional useful and novel functionalities for applications.

One potential application of such TMOs is a tunneling magneto-resistance (TMR) [8, 9] device using ferromagnetic metallic manganites. Currently, popular materials used for such devices are ferromagnetic metallic Fe with MgO as tunneling barriers [10–14] (see also Chapter 12, Volume 1 of this book). Replacing Fe with manganites such as $La_{1-x}Sr_xMnO_3$ is expected to improve the TMR performance because of the higher spin polarization. While such efforts have been reported [15–17], manganite-based TMR devices still do not outperform Fe-MgO-based devices [18–21]; while the TMR ratio at low temperatures could exceed more than 1000%, it becomes extremely small at room temperature. A possible mechanism for such a rapid decrease in the TMR ratio for manganite-based devices is that the interface magnetization decreases more rapidly than that in the bulk region [22, 23]. Thus, to improve the TMR performance of manganite-based devices, it is necessary to carefully engineer their interfaces [19].

This principle is applied to all other TMOs; for practical applications, we need to have full understanding of their interface properties. However, this task is not easy because strong correlation effects sometimes render the interfacial properties incompatible with the standard density functional theory (DFT), which is widely used to describe a wide range of materials properties. Therefore, one should take multiple approaches, such as advanced theoretical techniques to solve model Hamiltonians, phenomenological calculations, DFT calculations, and the combination of some of them. This chapter describes examples of such theoretical studies of TMO interfaces employing a variety of methods.

This Chapter is organized as follows: in Section 11.2, the band diagram of many TMOs is presented. From experimental estimations of the work-functions, we are able to align the bands of several complex oxides allowing for estimations of the direction of charge flow when they are combined in heterostructures. Section 11.3 describes theoretical work at the interfaces of a ferroelectric and a magnetic manganite. We provide rules to judge the type of coupling between the spins at opposite sides of the interfaces. We also study how the electric field of the ferroelectric component affects the electronic density of the other component, typically a metallic manganite, thus altering its transport properties and sometimes even inducing a phase transition. In Section 11.4, manganite superlattices are studied. Depending on the type of manganite components used in the superlattices and their thickness, a variety of states can be found, sometimes involving metal to insulator transitions. The possibility of a novel type of spin frustration only present at interfaces is also discussed. In Section 11.5, the puzzling exchange bias effect is addressed for the interfaces involving a ferromagnet on one side and a G-type (staggered) antiferromagnetic state on the other. A rationalization of this effect based on the spin-orbit coupling is provided. In Section 11.6, cobaltite thin films are studied, with emphasis on the dimensionality dependence of the spin state that can be high or low, depending on the geometry. Finally, in Section 11.7, the intriguing growth of interfaces along the unusual (111) direction is described. It is argued that, by this procedure, a physical realization of the quantum spin Hall state at room temperature can be achieved.

Overall, this chapter is presented in such a manner that the most important physical aspects of the many problems addressed are described, while the technicalities are only briefly mentioned. We encourage readers to consult the literature cited for more details.

## 11.2 BAND DIAGRAM

For the success of the emergent field of oxide heterostructures, it is crucial to properly determine the relative work functions of the materials involved, since these work functions control the bending of the valence and conduction bands (VB and CB) of the constituent materials at the interfaces, and ultimately the carrier concentration at those interfaces. Work functions of conventional metals and semiconductors have been studied for decades, establishing the fundamental background for current electronics. As a consequence, for the next-generation electronic devices utilizing the complex properties of correlated-electron systems, such as high-$T_c$ cuprates and CMR manganites, determining the relative work functions of a variety of transition-metal oxides is equally important.

In this section, we provide a crude estimation of the band diagram of perovskite transition-metal oxides using experimental data currently available. This is achieved by combining information from chemical potential shifts obtained using photoemission spectroscopy (PES) with diffusion voltage measurements on heterostructures.

PES is a powerful technique in this context and has provided fundamental information to uncover the properties of complex oxides. The workfunction of a target material could be directly measured by this technique. On the other hand, the diffusion voltage $V_d$ (or built-in potential) measurement requires a junction consisting of two materials. By measuring the diffusion voltage $V_d$ (or built-in potential) of the junction, one can also extract the chemical potential differences between the constituents. If there are no extra effects, such as interface polarities, impurities, or lattice reconstructions, the diffusion voltage $V_d$ is equivalent to the work-function difference between the two members of the junction. If the work function of one of the materials is known, the work function of the other can be estimated. In Ref. [24], we started investigating this interesting subject by constructing the band diagram for the case of cuprates and manganites.

In addition to cuprates and manganites, iron-based oxides, ruthenium-based oxides, and aluminum oxides are also important materials for spintronics applications. In the cases of $LaFeO_3$ and $BiFeO_3$, the $Fe^{+3}$ ion has a large moment, formally $5\mu_B$, and $BiFeO_3$ is the only known material showing multiferroic behavior at room temperature [25, 26]. $SrRuO_3$ is an itinerant ferromagnet with $T_C \sim 160$ K [27]. Finally, $LaAlO_3$ is a nonmagnetic wide band gap insulator. This material has been used as a tunneling barrier for devices for several years. Moreover, since the discovery of high-mobility electron gases at an interface between $LaAlO_3$ and $SrTiO_3$, considerable attention has been paid to $LaAlO_3$ [7]. In this section, we will complement the work that started in Ref. [24] by adding these other oxides into the previously established band diagram with a brief summary of experimental results.

In Ref. [28, 29], the workfunction of cubic LSMO was estimated to be ~4.8 eV. This value is slightly smaller than that of $SrRuO_3$ ~5.2 eV [30]. In Ref. [31], the diffusion voltage, i.e. the workfunction difference, between $SrRuO_3$ and both $LaFeO_3$ and $BiFeO_3$ was studied. The electron affinity and the band gap were determined to be ~3.3 eV and ~2.7 eV, respectively, for $LaFeO_3$ [32] and ~3.3 eV and ~2.8 eV, respectively, for $BiFeO_3$ [31, 33, 34]. From optical absorption experiments, the band gap of $LaAlO_3$ has been estimated to be 5.6 eV [35]. Furthermore, the electron affinity was deduced to be 2.5eV [36]. From these numbers, we can construct the band diagram covering cuprates, manganites, ferrites, $SrRuO_3$, and $LaAlO_3$ as shown in Figure 11.1. It is remarkable that all of the valence band maxima and the conduction band minima of the listed compounds are located inside the band gap of $LaAlO_3$. This is why $LaAlO_3$ can work as a nice tunneling barrier for many materials, as long as the polarity discontinuity induces interfacial conduction electron gases [7]. It should be also noted that the Fermi level of LSMO is expected to lie in the middle of the band gap of $LaFeO_3$ and $BiFeO_3$. Therefore, we do not expect a considerable charge transfer between the two compounds when an interface is formed between them.

In summary, band diagrams as discussed in this section are important to guide in the fabrication of complex oxide interfaces, particularly with regard to the possibility of charge transfer between the components that may induce novel states at those interfaces.

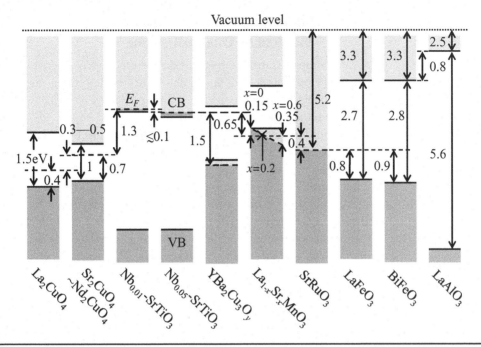

**FIGURE 11.1**    Schematic band diagrams of $La_2CuO_4$, $Sm_2CuO_4$ ($Nd_2CuO_4$), $Nb_{0.01}$-$SrTiO_3$, $Nb_{0.05}$-$SrTiO_3$, $YBa_2Cu_3O_y$, $La_{1-x}Sr_xMnO_3$, $SrRuO_3$, $LaFeO_3$, $BiFeO_3$, and $LaAlO_3$ based on diffusion voltage measurements [30, 31, 37–40], photoemission spectroscopy [28, 29, 41–43], and optical absorption [32, 34, 35]. Tops of valence bands (VB) and bottoms of conduction bands (CB) are indicated by solid lines, while chemical potentials are indicated by dashed lines. The vacuum level is indicated by a dotted line. Dark (light) hatched regions are occupied (unoccupied) states.

## 11.3 FERROELECTRIC-MANGANITE HETEROSTRUCTURES

As in the case of ferromagnetism, the presence of ferroelectricity is also an important property of functional materials. Moreover, the combination of ferromagnetism and ferroelectricity gives rise to the so-called "multiferroicity", which is physically interesting due to the coexistence of two types of orders. Multiferroic materials are important for novel applications because electric fields that are usually associated with ferroelectricity, may be used to manipulate magnetic properties if the two orders (ferromagnetism and ferroelectricity) are coupled as expected in multiferroics [44, 45]. However, in general for single phase materials, it is very hard to host these two ferroic orders simultaneously [46]. For this reason, heterostructures constructed by combining strong ferroelectric and strong ferromagnetic (FM) oxides have received considerable attention, because they are good candidates to realize the above-mentioned magnetoelectric functions, namely, electric field control of magnetism or vice versa. To pursue these magnetoelectric applications, it is necessary to perform both theoretical and experimental studies involving the interfaces between ferroelectric and FM materials, such as $BiFeO_3/La_{1-x}Sr_xMnO_3$, and $Pb(Zr,Ti)O_3/La_{1-x}Sr_xMnO_3$.

## 11.3.1 INTERFACE MAGNETIC INTERACTION: MOLECULAR ORBITAL PICTURE

Several complex transition metal oxide are intrinsically magnetic. Therefore, when different compounds are combined to form interfaces, novel magnetic ordering patterns could emerge. In this section, we describe how different TMOs interact when an interface is formed.

In bulk magnetic materials, magnetic ordering is induced through a variety of interactions. It is well-established that, in strongly correlated Mott insulators, local moments interact with each other via superexchange (SE) interactions [47]. The sign of exchange interaction between two magnetic sites would depend on the angle formed by the magnetic sites and ligand ions [48, 49] or orbital ordering [50]. In the presence of itinerant electrons, the interaction between localized spins is mediated by those electrons. Such interactions include the double-exchange (DE) interaction [51] and the RKKY interaction [52–55].

At an interface between different materials, the symmetry is lowered, and its natural consequence is the appearance of the Dzyaloshinskii–Moriya interaction [55, 56]. In addition, constituent materials could interact with each other via interfacial magnetic interactions. As discussed in Section 2, different TMOs have different chemical potentials in general. Therefore, electrons could be transferred across interfaces and a purely localized picture may not be entirely true, even if the constituent TMOs of the heterostructure are Mott insulators. For such a complicated situation, a simple description based on molecular orbitals turned out to be useful [57]. While it is qualitative, this provides a physically transparent picture by which both experiment and theory may greatly benefit.

Here we describe a general consideration for the magnetic interaction at an interface involving manganites [57]. We focus on the interfacial interaction derived by $d_{3z^2-r^2}$ orbitals which have the largest hybridization along the $z$ layer-stacking direction. We see that the sign of the magnetic interaction via the $d_{3z^2-r^2}$ orbitals is naturally fixed based on the molecular orbitals formed at the interface and the generalized Hund's rule. The molecular orbitals effectively lift the degeneracy between the $d_{3z^2-r^2}$ and $d_{x^2-y^2}$ orbitals by an energy of order the hopping amplitude. The $t_{2g}$ electrons are assumed to be electronically inactive and considered as localized spins when a finite number of electrons occupy the $t_{2g}$ orbitals; for example, $S = 3/2$ in a manganite component. The TM region is specified by the number of electrons occupying a $d_{3z^2-r^2}$ orbital. Reference [57] also discussed how to generalize the molecular-orbital-based argument for more complicated situations including $t_{2g}$ orbitals.

$(d_{3z^2-r^2})^0$ system: Let us start from the simplest case, an interface between Mn and a $(d_{3z^2-r^2})^0$ system (Figure 11.2, top figure). In this case, the bonding (B) orbital is occupied by an electron whose spin is parallel to the localized $t_{2g}$ spin of Mn. When there are other unpaired electrons in the $(d_{3z^2-r^2})^0$ system, their spins align parallel to that of the electron in the B orbital, due to the Hund coupling. Thus, the FM coupling is generated between Mn and $(d_{3z^2-r^2})^0$ systems. When the orbital $d_{x^2-y^2}$ is much lower in energy than $d_{3z^2-r^2}$ in the interfacial Mn layer, the Mn and $(d_{3z^2-r^2})^0$ systems are virtually decoupled. Thus, the magnetic coupling becomes antiferromagnetic (AFM) due to the superexchange (SE) interaction between $t_{2g}$ electrons.

$(d_{3z^2-r^2})^{1,2}$ normal: This simple consideration can be easily generalized to $(d_{3z^2-r^2})^1$ and $(d_{3z^2-r^2})^2$ systems. First, we consider that the $d_{3z^2-r^2}$ and $d_{x^2-y^2}$ orbitals are nearly degenerate in the interfacial Mn layer, and the unoccupied $d_{x^2-y^2}$ level in the TM region is much higher than the $d_{3z^2-r^2}$ level (Figure 11.2, second top figures). We call this configuration "normal" (N) configuration. In the lowest energy configuration, the B orbitals and the Mn $d_{x^2-y^2}$ orbital are occupied by electrons. For the $(d_{3z^2-r^2})^1$ system, the FM interaction is favored, as in the $(d_{3z^2-r^2})^0$ system. On the other hand, for the $(d_{3z^2-r^2})^2$ system, the down electron orbital is hybridized with the minority band in the Mn region. Thus, the "down" B orbital is higher in energy and has larger weight on the TM than the "up" B orbital. Because of the Hund

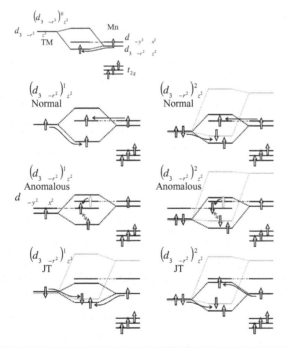

**FIGURE 11.2**    Molecular orbitals formed by $d_{3z^2-r^2}$ orbitals on Mn and the $(d_{3z^2-r^2})^{0,1,2}$ systems. In the normal (anomalous) configurations, $d_{3z^2-r^2}$ and $d_{x^2-y^2}$ orbitals are nearly degenerate in the interfacial Mn, and the unoccupied $d_{x^2-y^2}$ orbital in the neighboring TM is higher in energy (lower in energy than the occupied Mn $d_{x^2-y^2}$). In the Jahn–Teller case, the $d_{3z^2-r^2}$ level is significantly lower than the $d_{x^2-y^2}$ level. Black (light) lines indicate the level of majority (minority) spins. The up level and down level are exchange split, resulting in the level scheme as indicated. The minority levels are neglected where these are irrelevant. (After Okamoto, S., *Phys. Rev. B* 82 024427, 2010. With permission.)

coupling with the "down" electron in the B orbital, other unpaired electrons, if they exist, tend to be antiparallel to the Mn spin.

$(d_{3z^2-r^2})^{1,2}$ anomalous: When the $d_{x^2-y^2}$ level in the TM region becomes lower than the Mn $d_{x^2-y^2}$ level, the charge transfer occurs. We shall call this configuration "anomalous" (AN) (Figure 11.2, third figures). The electron transferred to the TM $d_{x^2-y^2}$ orbital has the same spin as the higher energy B orbital due to the Hund coupling (indicated by arrows). Therefore, the sign of the magnetic coupling between the Mn and $(d_{3z^2-r^2})^{1,2}$ systems is unchanged.

The magnetic interactions discussed so far are insensitive to the electronic density in the interfacial Mn because the interactions are mainly derived from the virtual electron excitation from the occupied $d_{3z^2-r^2}$ orbital in the TM region to the unoccupied counterpart in the Mn region.

Next, we consider the case where the $d_{3z^2-r^2}$ level is much lower than the $d_{x^2-y^2}$ in the interfacial Mn layer due to either the local Jahn–Teller distortion, or compressive strain originating from the substrate (Figure 11.2, lower figures denoted by "JT").

$(d_{3z^2-r^2})^1$ JT: When the Mn $d_{3z^2-r^2}$ density is close to 1, the superexchange interaction between the occupied $d_{3z^2-r^2}$ orbitals becomes AFM. On the other hand, when the density is much less than 1, the FM interaction between $(d_{3z^2-r^2})^0$ configuration on Mn and $(d_{3z^2-r^2})^1$ becomes dominant.

$(d_{3z^2-r^2})^2$ JT: When the Mn $d_{3z^2-r^2}$ occupancy is close to 1, "up" electrons are localized on both sites while "down" electrons can be excited from the $(d_{3z^2-r^2})^2$ system to the Mn minority level, i.e. the "down" electron density is virtually reduced in the TM region. As a result, unpaired spins, if they exist, become parallel to the "up" spin, i.e. FM coupling. When the Mn $d_{3z^2-r^2}$ density becomes much less than 1, the "up" AB orbital becomes less occupied while keeping the occupancy of B orbitals relatively unchanged. Eventually, the "down" density in the TM region becomes larger than the "up" density, and the magnetic coupling between the Mn and $(d_{3z^2-r^2})^2$ regions becomes AFM.

It is instructive to see the relation between the molecular-orbital picture described here, and other known mechanisms, although those are for bulk magnetic interactions. An interaction with a $(d_{3z^2-r^2})^0$ system with a $t_{2g}$ spin is ferromagnetic. This corresponds to the double exchange interaction originally proposed by Zener [51]. When the $e_g$ degeneracy is lifted by the JT effect and there is one $e_g$ electron per site, one should expect the antiferromagnetic superexchange interaction [47]. This situation corresponds to the $(d_{3z^2-r^2})^1$ Jahn–Teller case. Indeed, we have antiferromagnetic interaction in this case. On the other hand, when the $e_g$ degeneracy remains, one could expect the ferromagnetic interaction with antiferro-type orbital ordering, i.e. $d_{3z^2-r^2}/d_{x^2-y^2}$ [50]. This situation corresponds to the $(d_{3z^2-r^2})^1$ normal case. Indeed, we have ferromagnetic interaction in this situation.

For what systems can we apply the above discussion? In Ref. [58], the orbital and magnetic states at a cuprate/manganite interface were studied. In cuprates, the $d_{3z^2-r^2}$ orbital is completely filled while the $d_{x^2-y^2}$ orbital is less-than-half filled. As discussed in Section 2, cuprates have larger work-function than manganites, and the electron transfer from the latter to the former is expected. Thus, this situation corresponds to $\left(d_{3z^2-r^2}\right)^2$ anomalous. In agreement with these observations the antiferromagnetic spin alignment was observed in Ref. [58]. Such antiferromagnetic coupling between cuprate and manganite causes an interesting consequence called inverse spin switch behavior, the increase in the superconducting transition temperature, in manganite/cuprate/manganite trilayer systems [59, 60]. In Ref. [61], the magnetic states at a $BiFeO_3/(La,Sr)MnO_3$ interface were analyzed. Since an $Fe^{3+}$ ion has the high spin state with $S = 5/2$ and the minimal electron transfer is expected across such an interface according to the band diagram shown in Section 11.2, this situation corresponds to $\left(d_{3z^2-r^2}\right)^1$ normal. While we expect ferromagnetic coupling across the interface, experimental results show the antiferromagnetic spin arrangement. Reference [57] argued that, because of the presence of a robust $d_{x^2-y^2}$ orbital ordering at an interface Mn layer, this layer and the second Mn layer develop an antiferromagnetic coupling, resulting in the antiferromagnetic spin arrangement between an interface Fe layer and the bulk manganite region.

## 11.3.2 INTERFACES WITH FERROELECTRICS

In addition to the famous case of $BiFeO_3/La_{1-x}Sr_xMnO_3$ briefly addressed in the previous subsection, several other ferroelectrics have been used at interfaces, as described in this subsection. As a general motivation, and as described before, note that the most striking effect in these ferroelectric/ferromagnetic heterostructures is expected to be the ferroelectric field effect. Physically, the presence of a ferroelectric polarization can be mimicked by the presence of charge at the surface or interface. For example, a polarization of 10 µC/cm² equals to 0.1 elementary charge per $4\,\text{Å}\times 4\,\text{Å}$ surface. To screen this surface charge, mobile electrons or holes will be attracted to or repelled away from the interface, due to the bending of energy bands by electrostatic potentials, as sketched in Figure 11.3a.

Considering the large polarization of $Pb(Zr,Ti)O_3$ (up to ~80 µC/cm²), this screening effect and effective charge accumulation at the interface is a strong force to modulate the carrier density of ferromagnetic materials on the other side of that same interface, which may alter their physical properties if correlated electrons systems are used as the ferromagnetic component. For example, it is well-known that $La_{1-x}Sr_xMnO_3$ has the best ferromagnetic metallic behavior when $x = 3/8$. If the interfacial carrier density is modulated away from this region, the ferromagnetism and transport will become worse [62, 63]. In contrast, if the original doping $x$ is not in the optimal region, a proper modulation by field effect may enhance the ferromagnetism as well as transport [62, 63], as sketched in Figure 11.3b. In fact, there

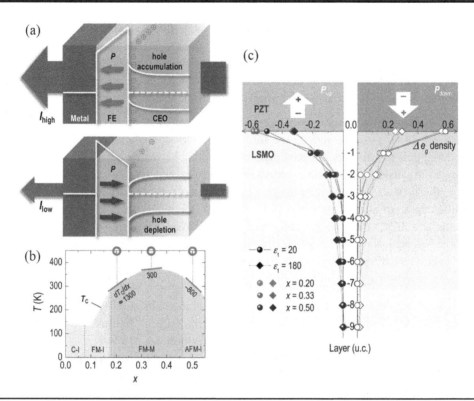

**FIGURE 11.3** Illustrations and results corresponding to a metal-ferroelectric-correlated electron oxide (CEO) heterostructure studied both experimentally and theoretically. (a) Schematic of tunneling current and energy bands modulated by ferroelectric polarization. The carrier population is controlled by the direction of the polarization, which yields either a hole accumulated (top) or depleted (bottom) state in the CEO layer. (b) Phase diagram of a typical CEO: $La_{1-x}Sr_xMnO_3$. The field effect modulation will give different results for $x = 0.20$, 0.33, and 0.50 according to this phase diagram. (c) The theoretical depth profile showing the changes in the $e_g$ electronic density for upward and downward polarizations, at different electronic compositions. Two different dielectric constants ($\varepsilon$) have been used to compare the degree of ferroelectric control. These idealized calculations show that the electronic density modifications in the first and second layers can indeed be very large, compatible with the experimental results. (After Jiang, L. et al., *Nano Letters*. 13, 5837, 2013. With permission.)

have been many experimental and first-principles calculations addressing this important issue [64-69]. Because of the rich phase diagrams of manganites, it is expected that the modifications in the carrier density at the interface induced by the ferroelectric effective interfacial charge may even induce complete phase transitions from metals to insulators on the manganite side, restricted to the interfacial zone, vastly increasing the modifications in transport properties. Readers are referred to related topical reviews for additional details on these interesting ideas [70–73].

Model studies as described before can also be applied to investigate the field effect modulated magnetism in ferroelectric-manganite heterostructures. By solving self-consistently the Poisson equation for the electrostatic potential, the distribution of carrier density can be finally obtained, as shown in Figure 11.3c for example. The modifications in the magnetic properties can also be simulated, which only occurs within one or two unit cells of the interface, because of the short screening length in metallic systems such as $La_{1-x}Sr_xMnO_3$ [75]. Such short range screening limits the magnetoelectricity effects to occurring within one or two unit cells [68, 69, 75], although the tunneling current across the interface can still be significantly modified [62]. To overcome this barrier and improve the magnetoelectricity, a theoretical design was recently proposed based on a manganite bilayer [74], as sketched in Figure 11.4a. The asymmetric interfaces ($n$-type *vs* $p$-type terminations) will break the symmetry between the two layers of Mn, then the ferroelectric polarization can tune this asymmetry, as shown in Figure 11.4b. Accompanying the tuning of the charge density, a magnetic phase transition may occur between ferromagnetism and A-type antiferromagnetism. Indeed, both the model simulation and first-principles calculations confirmed the possibility of ferroelectric field tuned magnetic transition [74]. The ideal on/off ratio of magnetization can be as high as 93%.

Summarizing this section for the non-expert readers, it is clear that the wide variety of possible combinations of complex oxides at interfaces, with both magnetic and ferroelectric components, can lead to a plethora of spin arrangements, charge redistributions, phase transitions, and other interesting effects. This is why oxide interfaces are considered a "playground" to create new artificial materials. Even if the potential promise of new functionalities and oxide electronics are not realized in the future, at a minimum these interfaces can provide fertile ground to create novel states of correlated electronic systems that do not appear in bulk transition metal oxides.

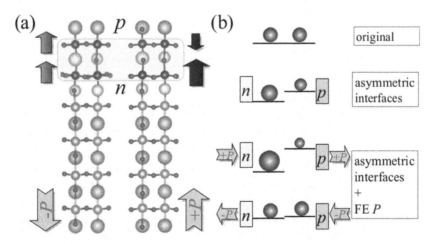

**FIGURE 11.4**  Schematic of a ferroelectric (e.g. $BaTiO_3$) and bilayer manganite (e.g. $La_{0.75}Sr_{0.25}MnO_3$) heterostructure. (a) Crystal structure. Green = Ba; red = O; cyan = Ti; purple = Mn; yellow = $La_{0.75}Sr_{0.25}$. The $n$-/$p$-type interfaces are indicated. Left/right are the negative/positive polarization cases, with switched magnetic orders. (b) The cases of three $e_g$ electron densities (spheres) and potential (bars) modulated by asymmetric interfaces (bricks) and ferroelectric polarization (arrows). (After Dong, S. et al., *Phys. Rev. B* 88, 140404(R), 2013. With permission.)

## 11.4 MANGANITE SUPERLATTICES AND INTERFACES

As typical correlated electronic materials, the manganese oxides widely known as manganites have received considerable attention in the past two decades for their unique physical properties, especially the colossal magnetoresistance where relatively small magnetic fields induce very large modifications in the resistivity, usually by several orders of magnitude [76–78]. Moreover, these materials are known to have a very rich phase diagram when varying the chemical composition and the electronic bandwidth, and after introducing external fields. Several insulating phases arise, with antiferromagnetic spin states, including the A-type state in the undoped compound and the CE-type state at half doping with spin, charge, and orbital order. In the past decade, these materials have received even more attention because of remarkable experimental advances that have allowed for the precise control of the digital growth of manganite heterostructures, leading to layer-by-layer grown manganite ultrathin films and/or superlattices [79–84]. Interesting physics emerges in these digital superlattices, such as particular metal-insulator transitions, that do not appear in the materials when in bulk form.

Manganite heterostructures were theoretically studied using the famous two-orbital double-exchange model [85, 86], which has been repeatedly confirmed to be a successful description of manganites [76, 77]. Both the kinetic double-exchange hopping term, the antiferromagnetic superexchange among the localized $t_{2g}$ sping, and the electron-lattice coupling involving Jahn–Teller modes, have been properly incorporated in the model Hamiltonian. In addition, the electrostatic potential $V_{Mn}$, which raises from the polar discontinuity and charge transfer across the interface, should also be taken into account in the theoretical study of heterostructures. As a first order approximation, $V_{Mn}$ can be assumed to be a function of the eight surrounding A-site ions [87]. By adopting this approximation, Dong and collaborators studied the $(LaMnO_3)_{2n}/(SrMnO_3)_n$ superlattices [85]. Despite the well-known physical properties of the bulk alloy-mixed $La_{2/3}Sr_{1/3}MnO_3$, with a dominant state that is ferromagnetic and metallic, these superlattices display a novel $n$-dependent metal-insulator transition and a spatial modulation of the magnetization [81, 82, 84]. The model studied by Dong and coworkers can successfully reproduce the observations reported in experiments. More specifically, these superlattices are metallic when $n < 3$, as it occurs for the material in bulk from, but the artificial structures become insulating once $n \geq 3$, as sketched in Figure 11.5. When $n \geq 3$, the $LaMnO_3$ layers and interfacial layers are ferromagnetic, while the inner portion of the $SrMnO_3$ component is antiferromagnetic, which also agrees with intuitive expectations (note that for large enough $n$, each building block has to display properties resembling their bulk counterparts) as well as with neutron measurements [84]. In addition, first-principles calculations also revealed the presence of layer-modulated magnetism in $(LaMnO_3)_{2n}/(SrMnO_3)_n$ [88–90].

A subsequent experiment now on $(LaMnO_3)_n/(SrMnO_3)_{2n}$ (note the sub-indexes $n$ and $2n$ are switched with regard to the previous case) revealed

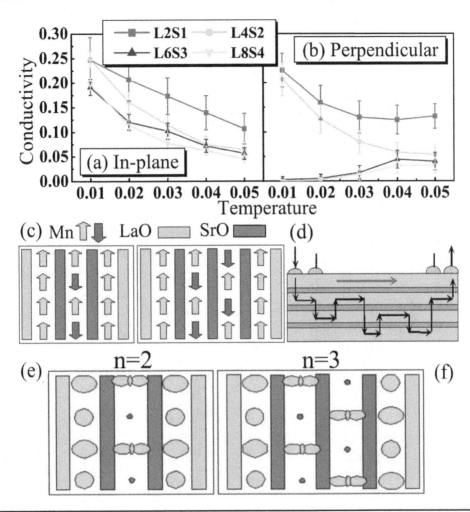

**FIGURE 11.5**   Theoretical results for $(LaMnO_3)_{2n}/(SrMnO_3)_n$ superlattices. (a-b) The calculated in-plane and out-of-plane conductivity, respectively. L2S1 denotes the $n = 1$ case and other cases are indexed following the same rule. (c) Sketch of spin (denoted by arrows) arrangements found at interfaces for the cases $n = 2$ (left) and $n = 3$ (right). (d) Sketch of the in-plane current flowing through superlattices. (e) The orbital orders expected to be found at the interfaces. (After Dong, S. et al., *Phys. Rev. B* 78, 201102(R), 2008. With permission.)

even more exotic phenomena, more specifically a much enhanced Néel temperature ($T_N$) for the cases $n = 1$ or 2 as compared with the corresponding results for bulk $La_{1/3}Sr_{2/3}MnO_3$ [91]. This enhanced A-type antiferromagnetism looks non-trivial, since the strained $LaMnO_3$ has a ferromagnetic tendency [85, 92], while $SrMnO_3$ displays G-type (i.e. spin staggered) antiferromagnetism. In fact, the A-type antiferromagnetic ordering in doped manganites has been known for $La_{1-x}Sr_xMnO_3$ [93], $Pr_{1-x}Sr_xMnO_3$ [93, 94], and $Nd_{1-x}Sr_xMnO_3$ [95, 96]. Early theoretical studies for bulk manganites incorporating strong correlation effects already clarified for this magnetic ordering the importance of $x^2 - y^2$ orbital ordering by which itinerant $e_g$ electrons could gain the kinetic energy [97–99]. The aforementioned double-exchange model with an on-site potential modulation (accompanying the A-site ions) was also adopted to study the superlattice $(RMnO_3)_n/(AMnO_3)_{2n}$ ($R$: rare earth; $A$: alkaline earth) [100]. As shown in Figure 11.6b–c, the A-type antiferromagnetism was found to be the most favorable ground state phase in a particular region of the phase diagram (namely, when varying model parameters), corresponding to the case of large-bandwidth manganites (e.g. $R$ = La and $A$ = Sr). The A-type antiferromagnetic region is much larger for the $n$ = 1 case (Figure 11.6b) than for the $n$ = 2 one (Figure 11.6c), in agreement with the experimental observation that $T_N$ is higher in the $n$ = 1 superlattice. The theoretical study emphasized the crucial role of the $Q_3$ mode of Jahn–Teller distortions, imposed by the tensile strain that is induced by the substrate material $SrTiO_3$ into the $SrMnO_3$ layers. This $Q_3$ mode of the Jahn–Teller distortions prefers the $x^2 - y^2$ orbital ordering (Figure 11.6f–g), which enhances/suppresses the double-exchange tendencies in-plane/out-of-plane, and thus leads to the enhanced A-type antiferromagnetism.

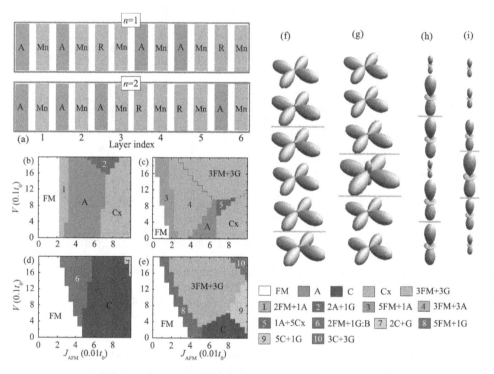

**FIGURE 11.6** Theoretical results for the $(RMnO_3)_n/(AmnO_3)_{2n}$ superlattices. (a) Sketch of layers in a finite cluster with six Mn-oxide layers (gray). "Mn" stands for the $MnO_2$ layers, "R" for the $RO$ layers, and "A" for the $AO$ layers. The only active layers for the mobile electrons are the Mn layers, but the other layers influence the Mn layers by their electrostatic potential. (b)–(e) Possible ground state phase diagrams arising from the theoretical calculations (see text): (b)–(c) Using a $SrTiO_3$ substrate; (d)–(e) Using a $LaAlO_3$ substrate; (b) and (d): $n$ = 1. (c) and (e): $n$ = 2. The magnetic configurations can be simple (e.g. A-type or C-type antiferromagnetism), or complex (combination of different spin patterns): see Ref. [100] for more details. The vertical axis is the on-site potential modulation by the La and Sr layer structure. The horizontal axis is the superexchange intensity between Mn's $t_{2g}$ spins: the narrower the bandwidth, the larger the superexchange. As the Jahn–Teller coupling constant, $\lambda Q = 0.9t$ is used. (f)–(g) Orbital ordering patterns using $SrTiO_3$ for the $n$ = 1 and $n$ = 2 cases, respectively. (h)–(i) Orbital ordering patterns using $LaAlO_3$ for the $n$ = 1 and $n$ = 2 cases, respectively. The cyan bars denote the positions of the SrO sheets. (After Dong, S. et al., *Phys. Rev. B* 86, 205121, 2012. With permission.)

Not only the $Q_3$ mode or the tensile strain, but also the lower dimensionality could stabilize the A-type antiferromagnetic ordering [101, 102]. This effect is intensively studied in layered manganites such as $La_{2-2x}Sr_{1+2x}Mn_2O_7$. It was established that the A-type antiferromagnetic ordering is stabilized in a wider-range of hole concentration than in the cubic manganites [103, 104]. Interestingly, the C-type antiferromagnetic ordering at large hole concentrations also appears [104], as predicted in Ref. [102]. In this ordering, ferromagnetically-ordered Mn chains lie along an $x$ or $y$ direction, not along the $z$ direction, which requires compressive strain as discussed below. While the heteroepitaxy of layered compounds is more difficult than that of cubic perovskite systems, the idea of "dimensionality" is another useful route to control the electronic property of complex oxides. We will review such a study in Section 6.

An interesting extension of this physical mechanism is provided by the opposite $Q_3$ mode, namely, by using compressive strain that arises from small lattice constant substrates such as $LaAlO_3$. In this case, the calculation predicted a different result; in particular, the ground state is probably given by the C-type antiferromagnetism, with $3z^2 - r^2$ orbital ordering as shown in Figure 11.6h–i. This type of orbital ordering prefers the ferromagnetic double-exchange to be out-of-plane, but it displays antiferromagnetic coupling in-plane. Thus, both the A-type antiferromagnetic and C-type antiferromagnetic order of $(LaMnO_3)_n /(SrMnO_3)_{2n}$ are basically caused by the substrate strains. Similar conclusions were also obtained by using first-principles calculations [105].

Contrary to the most studied wide-bandwidth case of $La_{1-x}Sr_xMnO_3$, the narrow-bandwidth manganites in bulk form, such as for the case of $Pr_{1-x}Ca_xMnO_3$, usually show more complex magnetic/charge/orbital orders [76, 78]. This situation also occurs in manganite superlattices. Still considering the case of $(RMnO_3)_n/(AMnO_3)_{2n}$ as an example, the ground state for narrow bandwidth manganites (e.g. $R$ = Pr and $A$ = Ca) could become quite different from the above described simple cases of A-type or C-type antiferromagnetism. According to the phase diagram Figure 11.6b–e, the possible ground state can be much complex, with spatial modulation of magnetic orders, which certainly need experimental confirmation.

A theoretical study using a single interface between narrow bandwidth manganites can reveal the true characteristics of a non-trivial interfacial state [86]. In this particular study, the electrostatic potential was solved self-consistently accompanying the charge transfer, using the Poisson equation (as opposed to a phenomenological potential increasing the difficulty of the calculation). The magnetic ground state was obtained by optimizing all classical spins and lattice distortions of the double exchange model. The magnetism emergent from this highly non-trivial calculation revealed a layer-dependent evolution, from the original A-type antiferromagnetism on the $RMnO_3$ side, through an exotic (canted) CE-type charge/orbital-ordered antiferromagnetism in the vicinity of the interface, to a normal collinear CE-type state, and finally to the G-type antiferromagnetism in the $AMnO_3$ end side of the superstructure, as summarized in Figure 11.7. This calculation predicts a new state, the canted CE state, and moreover provides a

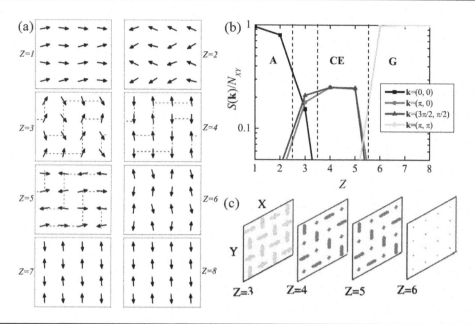

**FIGURE 11.7** The case of a narrow bandwidth $RMnO_3/AMnO_3$ interface to illustrate a possible new source of spin frustration. $Z$ labels layers in the vicinity of an interface: $Z \leq 4$ is nominally $RMnO_3$, while $Z \geq 5$ is $AMnO_3$. (a) Optimized spin configurations near the interface. The dashed lines highlight the spin zigzag chains that are characteristic of CE states with ferromagnetic order within each zigzag chain. (b) Layer dependence of the spin structure factor for the optimized spin configurations. (c) Layer dependence of the real-space orbital pattern. (After Yu, R. et al., *Phys. Rev. B* 80, 125115, 2009. With permission.)

concrete example of a novel type of frustration that emerges in superlattices: for layers "sandwiched" between very stable states, such as the A-type and CE-type states, the spin arranges in new patterns that are not stabilized in manganites when in bulk form. This theoretically proposed novel form of frustration deserves more work, and for concrete predictions to be contrasted against experiments.

## 11.5 EXCHANGE BIAS EFFECTS ACROSS THE FERROMAGNETIC/G-TYPE ANTIFERROMAGNETIC INTERFACE

Exchange bias is a widely observed unusual effect corresponding to ferromagnetic-antiferromagnetic interfaces [106]. In principle, the pinning effect by the antiferromagnetic interface is expected to bias the hysteresis loop of the attached ferromagnetic layer, as sketched in Figure 11.8a. However, the origin of this effect is still under much discussion. For example, for a fully compensated antiferromagnetic interface, such as the (001) surface of a G-type antiferromagnetic perovskite (Figure 11.8b–c), it intuitively appears that the antiferromagnetic moments are symmetrically distributed with respect to their orientations, and thus they cannot bias the neighboring ferromagnetic moments. Several possible mechanisms have been proposed to understand the exchange bias effect. Extrinsic factors are often

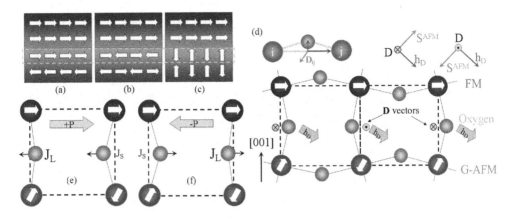

**FIGURE 11.8** Sketches that illustrate new ideas proposed to understand exchange bias effects. (a)–(c) Schematic of various ferromagnetic-antiferromagnetic configurations at an interface. Local magnetic moments are denoted by the arrows. (a) corresponds to fully uncompensated antiferromagnetic interface, as it occurs for example at the (001) surface of an A-type antiferromagnetic state or at the (111) surface of a G-type antiferromagnetic state. (b) and (c) are fully compensated antiferromagnetic interfaces, as they occur in a G-type antiferromagnetic state. In (a) and (b), the ferromagnetic and antiferromagnetic magnetic moments are collinear. In (c), the magnetic moments are non-collinear, namely, the magnetic easy axes of the ferromagnetic and antiferromagnetic materials are different. ((a)–(c) After Dong, S. et al., *Phys. Rev. B* 84, 224437, 2011. With permission.) (d) The (mutually perpendicular) relationship between the $M_i$-O-$M_j$ bond, oxygen displacement, and $\vec{D}_{i,j}$ vector. (e) Schematic of the interface between ferromagnetic and G-type antiferromagnetic perovskites, including the oxygen octahedral tilting. The staggered directions of the $\vec{D}_{ij}$ vectors at the interface are marked as in- and out-arrows, while the uniform $\vec{h}_D$ vectors are also shown near the oxygens. (f) The resulting uniform $\vec{h}_D$ should be perpendicular to $\vec{S}^{AFM}$ and $\vec{D}$. (g) Ferroelectric-polarization-driven asymmetric bond angles and modulated normal superexchange at the interface. (h) A switch of the FE polarization will switch the biased field $\vec{h}_J$. ((d)–(g) After Dong, S. et al., *Phys. Rev. Lett.* 103, 127201, 2009. With permission.)

considered, such as interface roughness, spin canting near the interface, as well as frozen interfacial and domain pinning, but there is no universally accepted explanation for this exotic phenomenon [107–111]. Recent experiments demonstrated the presence of exchange bias in $BiFeO_3/La_{0.7}Sr_{0.3}MnO_3$ heterostructures [61, 112], as well as in other $BiFeO_3$/ferromagnetic alloys [113]. More interestingly, it was observed that this exchange-bias can be affected by the switching of the ferroelectric polarization of $BiFeO_3$. This new ingredient cannot be well understood by any of the traditional theories of exchange bias.

As an alternative, Dong and collaborators proposed two other related mechanisms to enlarge the existing possible theories of exchange bias [115]. These new mechanisms are based on the Dzyaloshinskii–Moriya interaction, and on the ferroelectric polarization. The latter mechanism is only active in those heterostructures that involve multiferroics (e.g. $BiFeO_3$). To understand these new ideas, consider that the spin-spin interaction in perovskites can be described by a simplified effective Hamiltonian:

$$H = \sum_{\langle ij \rangle} [J_{ij}\vec{S}_i \cdot \vec{S}_j + \vec{D}_{ij} \cdot (\vec{S}_i \times \vec{S}_j)], \qquad (11.1)$$

where $J_{ij}$ is the standard superexchange coupling between two nearest-neighbor spins, while $\vec{D}_{ij}$ is the Dzyaloshinskii–Moriya interaction vector, which, at a more fundamental level, arises from the spin-orbit coupling. In perovskites, $\vec{D}_{ij}$ is determined by the bending of the $M_i$-O-$M_j$ bond ($M$: metal cation), which is induced by the oxygen octahedral rotations and tilting. Moreover, the vector is perpendicular to the $M_i$-O-$M_j$ bond, as shown in Figure 11.8d. Since the tilting and rotations are collective, the nearest-neighbor oxygens in the same direction will move away from midpoint in opposite directions, namely, the nearest-neighbor displacements are themselves staggered, as shown in Figure 11.8e. As a consequence, the $\vec{D}_{ij}$ vectors between nearest-neighbor bonds along the same direction are also staggered. But note that the spins in a G-type antiferromagnetic state $S^{AFM}$ are staggered as well. Combining these two staggered components, namely $\vec{D}_{ij}$ and $\vec{S}^{AFM}$, will give rise to a *uniform* Dzyaloshinskii–Moriya effect at the interface, as shown in Figure 11.8d–e, which can be described by an effective Hamiltonian:

$$H_{DM}^{\text{interface}} = \sum_{<ij>} \vec{D}_{ij} \cdot (\vec{S}_i^{FM} \times \vec{S}_j^{AFM}) = -\vec{h}_D \cdot \sum_i \vec{S}_i^{FM}, \qquad (11.2)$$

where $\vec{h}_D = \vec{D} \times \vec{S}_j^{AFM}$ can be regarded as the effective biasing magnetic field which is fixed by the field-cooling process, and then assumed to be frozen at low temperatures during the hysteresis loop measurement.

Furthermore, if one component has a ferroelectric polarization which induces a uniform displacement between cations and anions, the bond angles at the interface become no longer symmetric. Since the magnitude of normal superexchange coupling depends on the bond angle, these modulated bond angles induce staggered interfacial superexchange couplings, which are denoted as $J_L$ and $J_S$, as shown in Figure 11.8f. As in the previous example, once again the staggered superexchange couplings at the interface will also induce a *uniform* bias field when in the presence of the G-type antiferromagnetic spin order, which can be described by:

$$H_{DM}^{\text{interface}} = \sum_{<ij>} \vec{J}_{ij} \cdot (\vec{S}_i^{FM} \cdot \vec{S}_j^{AFM}) = -\vec{h}_J \cdot \sum_i \vec{S}_i^{FM}, \qquad (11.3)$$

where $\vec{h}_J = \dfrac{(J_S - J_L)}{2} \vec{S}_j^{AFM}$ arises from the modulation of the superexchange $J$, which can be switched by polarization (Figure 11.8g).

These new ideas emphasize the interactions between the lattice distortion and magnetism, rather than continuing to rely on the existence of uncompensated antiferromagnetic moments, which are often used in other models, as previously explained. Using the first principles theory, Dong and collaborators chose the $SrRuO_3/SrMnO_3$ system to verify these two proposals for the exchange bias effect [114]. Needless to say, it is important to continue exploring these and other new ideas to explain the exchange bias effect, and to design experiments to confirm them.

## 11.6 COBALTITE THIN FILMS

In this section, we discuss a more complicated situation where both $e_g$ states and $t_{2g}$ states are electronically active, as realized in perovskite cobalbite LaCoO$_3$ (LCO) [116, 117]. While LCO has not been actively utilized for spintronics applications, as discussed below, this could be useful for controllable spin filter devices.

LCO is a unique perovskite compound in which two quantum many-body states are close in energy; in bulk, a low spin (LS) state ($t_{2g}^6$; $S = 0$) is the ground state and a high spin (HS) state ($t_{2g}^4 e_g^2$; $S = 2$) is the first excited state. The pure LS state is realized below $T \lesssim 100\,$K and a mixed spin state of HS and LS is realized at $T \gtrsim 100\,$K, due to thermal excitation [118]. This indicates that the LS and HS states are in a delicate balance [119], and tuning between these two states with a small energy cost would be possible, for example, by applying strain. It is also important to note that the electrical resistivity shows an insulating behavior at all temperature regime, while it becomes small at $T \gtrsim 400\,$K, indicating the importance of correlation effects.

In Ref. [120], an alternative approach was taken to control the spin state in LSO, i.e. the dimensionality, by fabricating superlattices of LCO and LaAlO$_3$, as schematically shown in Figure 11.9a. Since LaAlO$_3$ is a large bandgap insulator, SLs with thin LCO could be regarded as a two-dimensional limit. As shown in Figure 11.9b, the orbital level scheme in Co$^{3+}$ is expected to be changed by reducing the dimensionality.

The spin-state transition in LCO thin films are experimentally suggested from optical conductivity and x-ray absorption spectroscopy (XAS) measurements. For the bulk LCO, the optical conductivity $\sigma(\omega)$ consists of three features: strong absorption features at $\hbar\omega \sim 3\,$eV and ~1.5 eV and a weak feature at $\hbar\omega \sim 0.5\,$eV. By reducing the thickness of LCO in SLs, the weak feature at $\hbar\omega \sim 0.5\,$eV is systematically diminished and the strong feature at ~1.5 eV showed a blue shift. In XAS measurements, bulk LCO shows two features: a strong absorption peak at $\hbar\omega \sim 529\,$eV, and a relatively weak shoulder at $\hbar\omega \sim 528\,$eV. With reducing dimensionality, the weak shoulder feature is systematically diminished, and the strong peak shows a blue shift.

These features are consistent with the dynamical mean-field theory (DMFT) calculations [121], which support the spin-state transition in the LCO/LAO superlattices. Numerical results for the orbital-resolved density of states (DOS) for the 3D and 2D systems are presented in Figure 11.9c. The energy distribution of the orbital states shows broad Co $e_g$ (narrow Co $t_{2g}$) bands located between +1 and +4 eV (1 and +1 eV). O $p$ bands are mostly located below –2 eV. For 3D LCO, a portion of the $e_g$ orbital is occupied near $E = 0.8$ eV and a portion of the $t_{2g}$ orbital is unoccupied near $E = 0.5\,$eV. This configuration manifests a mixed spin state in 3D LCO at room temperature. By contrast, for 2D LCO, the $e_g$ ($3z^2 - r^2$ and $x^2 - y^2$) bands are empty, and the $t_{2g}$ ($xy$, $yz$ and $xz$) bands are fully occupied, suggesting an electron transfer from the $e_g$ orbitals to the $t_{2g}$ orbitals by reducing the dimensionality. Based these results, the characteristic features in $\sigma(\omega)$ for 3D LAO can be attributed to a O p–Co $d$ charge transfer transition (labeled

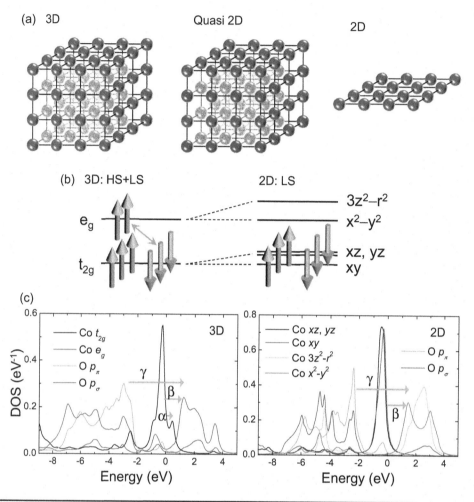

**FIGURE 11.9** Dimensional crossover of the spin state of cobaltites. Schematic geometry of LaCoO$_3$/LaAlO$_3$ superlattices, bulk cubic (3D) cobaltite, artificial superlattice of cobaltite having quasi 2D structure, and monolayer of cobaltite which is purely 2D. Only B sites of perovskite structure ABO$_3$ are shown; red circles are Co sites and blue circles are Al. (b) Spin states. In 3D, a mixture of the high-spin (HS) state and the low-spin (LS) state is realized, while in 2D the pure LS state is realized. (c) The DMFT results for the orbital-resolved spectral function for 3D cobaltite and 2D cobaltite. (After Jeong, D.W. et al., *Scientific Reports.* 4, 6124, 2014. With permission.)

$\gamma$ with $\hbar\omega \sim 3\,\text{eV}$), a Co $t_{2g} - e_g$ transition (labeled $\beta$ with $\hbar\omega \sim 1.5\,\text{eV}$), and a Co $t_{2g} - t_{2g}$ transition (labeled $\alpha$ with $\hbar\omega \sim 0.5\,\text{eV}$). For 2D LCO, because the $t_{2g}$ orbitals are fully filled, the weak absorption feature $\alpha$ seen for 3D LCO at $\hbar\omega \sim 0.5\,\text{eV}$ is absent. The optical gap is thus characterized by the lowest excitations $\alpha$ mode for 3D and $\beta$ mode for 2D. Two features in XAS, $\hbar\omega \sim 528\,\text{eV}$ and $\sim 529\,\text{eV}$, are ascribed to the electron excitations from core states to unoccupied $t_{2g}$ states at $\hbar\omega \sim 0.5\,\text{eV}$ and to $e_g$ states with the low-energy peak at $\hbar\omega \sim 1.5\,\text{eV}$. So, the absence of the low energy absorption edge at $\hbar\omega \sim 528\,\text{eV}$ for 2D LCO is naturally understood.

By reducing the dimensionality from 3D to 2D, the insulating nature is changed from a Mott insulator to a band insulator with the gap amplitude increased from $\lesssim 0.5\,eV$ to $\sim 1\,eV$. Since a $Co^{3+}$ ion has finite spin in the HS+LS mixed state for 3D LCO, it is expected to be easy to align the Co moments by an external magnetic field or attaching a ferromagnet. Therefore, when LCO is prepared at the critical regime between the mixed HS-LS state and the LS state, and used as a tunneling barrier between magnetic metals, this would work as a efficient and controllable spin filter; for the HS state, the band gap for electrons with parallel spin arrangement to the Co moment is smaller than those with antiparallel spin, allowing one of two spins to transmit, and for the LS state, the band gap is large and independent of spin, prohibiting the spin transport across the barrier.

## 11.7 THEORETICAL PREDICTIONS FOR INTERFACES GROWN IN THE (111) DIRECTION

In this section, we will discuss interesting novel phenomena, the topological insulators, that could be realized in oxide heterostructures.

The study of novel electronic states driven by the non-trivial band topology of electrons has been a major subject of interest in condensed matter physics since the discovery of the integer quantum Hall effect (QHE) [122, 123]. After the proposal by Haldane [124] that the QHE could be realized in an electronic system with a honeycomb lattice without Landau levels, Kane and Mele proposed that graphene could show the spin Hall effect when the spin-orbit coupling (SOC) strength is strong [125]. While the SOC in graphene turned out to be too small to realize the spin Hall effect at non-zero temperatures, the spin Hall effect or topological insulator (TI) state in HgTe quantum wells was theoretically proposed [126] and experimentally verified [127]. Further theoretical predictions [128, 129] and experimental realizations [130, 131] have established on a firm ground the area of TIs in materials with strong SOC. For potential applications, and for exploring further novel phenomena, inducing magnetism by doping magnetic ions or interfacing with magnetic and/or superconducting systems has also been proposed [61, 132–134] Once realized in real materials, these phenomena could lead to entirely new device paradigms for spintronics and quantum computing. However, until now most efforts on TIs have focused on narrow bandgap semiconductors involving heavy elements such as Hg and Bi where the electronic properties are dominated by $s$ and $p$ orbitals.

Here, we will consider another class of materials—heterostructures of transition-metal oxides (TMOs) involving $d$ electrons. In terms of magnetism and superconductivity, TMOs are ideal playgrounds as they have already demonstrated a wide variety of symmetry breaking effects arising from strong correlations between electrons. However, only a few studies addressing TIs based on the transition-metal oxides have appeared so far [135]. As we have discussed in early sections, experimentally the quality of oxide-based heterostructures is becoming extremely high. Recent highlights of TMO heterostructures research include the realization of the integer quantum Hall effect [136] and the fractional quantum Hall effect [137] in Zn oxide heterostructures, as well as the integer quantum Hall effect in δ-doped $SrTiO_3$ [138]. We believe that our results may open new directions focusing on topological phenomena in the rapidly growing field of oxide electronics.

We will focus on two-dimensional (2D) TIs realized using cubic perovskite TMOs. Our design principle starts from the simple observation that, when viewed from the [111] crystallographic axis, threefold symmetry is realized in cubic perovskite structure (Figure 11.10a). Furthermore, when we focus on a bilayer, it forms a buckled honeycomb lattice, which resembles graphene (Figure 11.10b). Therefore, Dirac dispersions are naturally expected. One could also adjust the Fermi level at those Dirac points by choosing appropriate elements. Finally, when the SOC is turned on, the system would become a 2D TI.

In order to find possible candidate systems, Ref. [139] started from a tight-binding model for either $t_{2g}$ or $e_g$ electrons and examined the $z_2$ topological index [140] and edge spectra using finite-thick slabs. For $t_{2g}$ systems, this study revealed that $t_{2g}^1$, $t_{2g}^2$, $t_{2g}^3$ and $t_{2g}^5$ TMOs are possible candidates to realize TIs when the SOC is much larger than the crystal field splitting between the $a_{1g}$ level and $e_g'$ level, and $t_{2g}^2$, $t_{2g}^4$ and $t_{2g}^5$ are candidates when the SOC is much weaker than the crystal field. From similar analyses for $e_g$ electron systems, $e_g^1$, $e_g^2$ and $e_g^3$ systems were found to be all possible candidates for 2D TIs. One might find this rather strange, because the SOC is supposed to be quenched in the $e_g$ multiplet. However, even in the $e_g$ multiplet, the SOC is found to be active because of the symmetry lowering from octahedral ($O_h$) to trigonal ($C_{3v}$) in our (111) bilayers.

In practice, the (111) bilayer of perovskite TMO $ABO_3$ has to be stabilized by being sandwiched between insulating TMO A'B'O$_3$. Based on the tight-binding results, Ref. [139] examined the $t_{2g}^4$ system $LaReO_3$, $t_{2g}^5$ systems $LaRuO_3$, $LaOsO_3$, $SrRhO_3$, $SrIrO_3$, and $e_g^2$ systems $LaCuO_3$, $LaAgO_3$ and $LaAuO_3$ using DFT methods. For A = La, $LaAlO_3$ is used for an insulator A'B'O$_3$, and for A = Sr, $SrTiO_3$ is used. In this research it was found that, in the cases of $LaOsO_3$, $SrIrO_3$, $LaAgO_3$ and $LaAuO_3$, the Fermi level is located inside the gap. Therefore, (111) bilayers of these materials would become two-dimensional topological insulators. However, for $LaReO_3$, $SrRhO_3$, $LaRuO_3$ and $LaCuO_3$, the Fermi level crosses more than one band. Furthermore, in $LaCuO_3$, antiferromagnetic ordering is realized within DFT. As a consequence, the $LaReO_3$, $SrRhO_3$, $LaRuO_3$ (111) bilayers are classified as topological metals rather than TIs, and in particular $LaCuO_3$ is an antiferromagnetic trivial metal. The dispersion relation of the $SrIrO_3$ (111) and $LaAuO_3$ (111) bilayers are shown in Figure 11.10c–f, respectively. Figure 11.10c,e are for bulk, and Figure 11.10d,f are for finite-thick zigzag slabs computed by using Wannier parametrization. The gapless edge modes crossing the Fermi level ($E = 0$) in the latter two panels indicate that these are indeed 2D TIs.

It is remarkable that the non-trivial gap amplitude of the $LaAuO_3$ (111) bilayer is about 150 meV. For this reason, this system should remain a 2D TI even at room temperature. However, since the extension of the electron wave functions shrinks when moving from $5d$ TMOs to $4d$ TMOs, and then

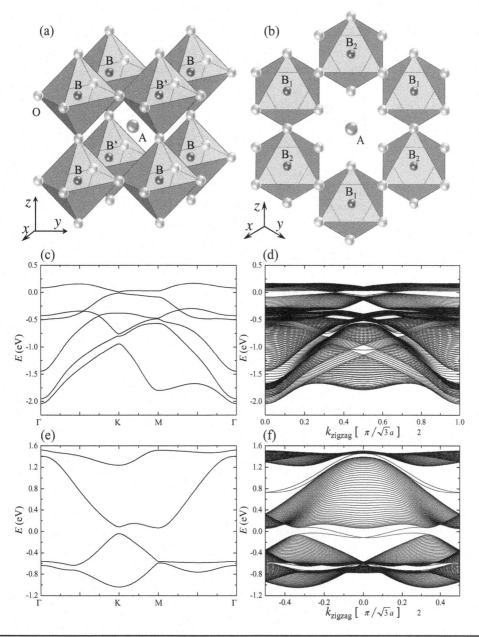

**FIGURE 11.10**  Crystal structure of perovskite (111) bilayer of $ABO_3$ sandwiched by $AB'O_3$, conventional view (a). (b) (111) bilayer of $ABO_3$ viewed from the [111] direction. Wannier band structure of $SrIrO_3$ (111) bilayer for bulk (c) and finite-thick zigzag slab (d) Wannier band structure of $LaAuO_3$ (111) bilayer for bulk (e) and finite-thick zigzag slab (f). (After Okamoto, S. et al., *Phys. Rev. B* 89, 195121, 2014. With permission.)

to $3d$ TMOs, some instability caused by stronger local Coulomb interactions could appear. Reference [141] reexamined the stability of the 2D TI states of $SrIrO_3$ and $LaAuO_3$ (111) bilayers using DFT+DMFT [121]. It was found that the $SrIrO_3$ (111) bilayer is indeed unstable against an antiferromagnetic trivial insulating state, result consistent with the strong coupling approach [142]. Recently, Hirai and coworkers successfully fabricated (111) bilayers of perovskite iridate, $Ca_{0.5}Sr_{0.5}IrO_3$, and found that it is strongly insulating with magnetic ordering (the particular magnetic order pattern remains unclear) [143]. This result appears to be consistent with the DFT+DMFT results. On the other hand, the $LaAuO_3$ (111) bilayer remains a 2D TI even after including correlation effects, at least within the DFT+DMFT approximation. The band gap is reduced somewhat, but remains larger than 100 meV. Therefore, if synthesized properly, the $LaAuO_3$ (111) bilayer is expected to become a physical realization of the quantum spin Hall effect at room temperature.

The growth direction [111] described in this section opens a new subfield of research within complex oxide interfaces, providing yet another interesting variable to produce novel materials and states. In particular, topological insulators could be realized by this procedure. While the theoretical basis for these ideas appears to be on a firm ground, the next natural challenge is to realize these ideas experimentally in real systems exploiting the [111] crystalographic axis.

## 11.8 CONCLUSION

In this chapter, we have reviewed theoretical studies involving transition metal oxide interfaces and heterostructures that might have potential relevance for spintronic applications of complex oxide materials. Because of the subtleties associated with strong correlation effects, including phase competition among many spin, charge, orbital, and lattice ordered states, a variety of approaches must be considered depending on the situation being analyzed. In the many sections of this chapter, we have described (phenomenological) models for bulk and interfacial oxides, advanced numerical calculations using model Hamiltonians, and the implementation of some of them together with DFT methodologies. We envision that, from the combination of all these points of view, we could help to understand or predict novel phenomena in TMO heterostructures. We are also convinced that TMOs have a great potential for novel spintronic applications not only within magnetism, as intensively discussed here, but also in the context of non-trivial band topology and correlations. It is clear that the field of oxide interfaces provides a new and exciting playground to create new states that cannot be observed in bulk compounds. We hope this article motivates further research to realize efficient electronic devices based on TMOs.

## ACKNOWLEDGMENTS

S.O. and E.D. were supported by the U.S. Department of Energy, Office of Science, Basic Energy Sciences, Materials Sciences and Engineering Division. S.D. was supported by the National Natural Science Foundation of China (Grant No. 11274060).

## REFERENCES

1. J. G. Bednorz and K. A. MÜller, Possible high *Tc* superconductivity in the BaLaCuO system, *Z. fur Phys. B Condens. Matter* **64**, 189 (1986).

2. S. Jin, T. H. Tiefel, M. McCormack, R. A. Fastnacht, R. Ramesh, and L. H. Chen, Thousandfold change in resistivity in magnetoresistive La-Ca-Mn-O films, *Science* **264**, 413 (1994).

3. Y. Tokura, A. Urushibara, Y. Moritomo et al., Giant magnetotransport phenomena in filling-controlled kondo lattice system: La1-*x* SrxMnO3, *J. Phys. Soc. Jpn.* **63**, 3931 (1994).

4. M. Imada, A. Fujimori, and Y. Tokura, Metal-insulator transitions, *Rev. Mod. Phys.* **70**, 1039 (1998).

5. Y. Tokura and N. Nagaosa, Orbital physics in transition-metal oxides, *Science* **288**, 462 (2000).

6. A. Ohtomo, D. A. Muller, J. L. Grazul, and H. Y. Hwang, Artificial charge-modulation in atomic-scale perovskite titanate superlattices, *Nature* **419**, 378 (2000).

7. A. Ohtomo and H. Y. Hwang, A high-mobility electron gas at the LaAlO3/SrTiO3 heterointerface, *Nature* **427**, 423 (2004).

8. M. Julliere, Tunneling between ferromagnetic films, *Phys. Lett.* **54A**, 225 (1975).

9. S. Maekawa and U. Gafvert, Electron tunneling between ferromagnetic films, *IEEE Trans. Magn.* **18**, 707 (1982).

10. W. H. Buttler, X.-G. Zhang, T. C. Schulthess, and J. M. MacLaren, Spin-dependent tunneling conductance of Fe/MgO/Fe sandwiches, *Phys. Rev. B* **63**, 054416 (2001).

11. J. Mathon and A. Umerski, Theory of tunneling magnetoresistance of an epitaxial Fe/MgO/Fe (001) junction, *Phys. Rev. B* **63**, 220403 (2001).

12. M. Bowen, V. Cros, F. Petroff et al., Large magnetoresistance in Fe/MgO/FeCo(001) epitaxial tunnel junctions on GaAs(001), *Appl. Phys. Lett.* **79**, 1655 (2001).

13. S. Yuasa, T. Nagahama, A. Fukushima, Y. Suzuki, and K. Ando, Giant room-temperature magnetoresistance in single-crystal Fe/MgO/Fe magnetic tunnel junctions, *Nat. Mater.* **3**, 868 (2004).

14. S. S. P. Parkin, C. Kaiser, A. Panchula et al., Giant tunnelling magnetoresistance at room temperature with MgO (100) tunnel barriers, *Nat. Mater.* **3**, 862 (2004).

15. M. Viret, M. Drouet, J. Nassar, J. P. Contour, C. Fermon, and A. Fert, Low-field colossal magnetoresistance in manganite tunnel spin valves, *Europhys. Lett.* **39**, 545 (1997).

16. M.-H. Jo, N. D. Mathur, N. K. Todd, and M. G. Blamire, Very large magnetoresistance and coherent switching in half-metallic manganite tunnel junctions, *Phys. Rev. B* **61**, R14905(R) (2000).

17. M. Bowen, M. Bibes, A. Barthelemy, J.-P. Contour, Y. Lemaitre, and A. Fert, Nearly total spin polarization in $La_{2/3}Sr_{1/3}MnO_3$ from tunneling experiments, *Appl. Phys. Lett.* **82**, 233 (2003).

18. Y. Ogimoto, M. Izumi, A. Sawa et al., Tunneling magnetoresistance above room temperature in $La_{0.7}Sr_{0.3}MnO_3/SrTiO_3/La_{0.7}Sr_{0.3}MnO_3$ junctions, *Jpn. J. Appl. Phys.* **42**, L369 (2003).

19. H. Yamada, Y. Ogawa, Y. Ishii et al., Engineered interface of magnetic oxides, *Science* **305**, 646 (2004).

20. Y. Ishii, H. Yamada, H. Sato et al., Improved tunneling magnetoresistance in interface engineered (La,Sr)MnO3 junctions, *Appl. Phys. Lett.* **89**, 042509 (2006).

21. M. Bibes and A. Barthelemy, Oxide spintronics, *IEEE Trans. Electron Devices* **54**, 1003 (2007).

22. J.-H. Park, E. Vescovo, H.-J. Kim, C. Kwon, R. Ramesh, and T. Venkatesan, Magnetic properties at surface boundary of a half-metallic ferromagnet $La_{0.7}Sr_{0.3}MnO_3$, *Phys. Rev. Lett.* **81**, 1953 (1998).

23. J. J. Kavich, M. P. Warusawithana, J. W. Freeland et al., Nanoscale suppression of magnetization at atomically assembled manganite interfaces: XMCD and XRMS measurements, *Phys. Rev. Lett.* **76**, 014410 (2007).

24. S. Yunoki, A. Moreo, E. Dagotto, S. Okamoto, S. S. Kancharla, and A. Fujimori, Electron doping of cuprates via interfaces with manganites, *Phys. Rev. B* **76**, 064532 (2007).

25. G. A. Smolenskii and I. E. Chupis, Ferroelectromagnets, *Soviet Physics Uspekhi* **25**, 475 (1982).

26. J. Wang, J. B. Neaton, H. Zheng et al., Epitaxial $BiFeO_3$ multi-ferroic thin film heterostructures, *Science* **299**, 1719 (2003).

27. A. Callaghan, C. W. Moeller, and R. Ward, Magnetic interactions in ternary ruthenium oxides, *Inorg. Chem.* **5**, 1572 (1966).

28. M. P. de Jong, V. A. Dediu, C. Taliani, and W. R. Salaneck, Electronic structure of $La_{0.7}Sr0_3MnO_3$ thin films for hybrid organic/inorganic spintronics applications, *J Appl. Phys.* **94**, 7292 (2003).

29. M. Minohara, I. Ohkuho, H. Kumigashira, and M. Oshima, Band diagrams of spin tunneling junctions $La_{0.6}Sr_{0.4}MnO_3$:Nb:$SrTiO_3$ and $SrRuO_3$Nb:$SrTiO_3$ determined by in situ photoemission spectroscopy, *Appl. Phys. Lett.* **90**, 2123 (2007).

30. C. Yoshida, A. Yoshida, and H. Tamura, Nanoscale conduction modulation in $AuPb(Zr,Ti)O_3$/$SrRuO_3$ heterostructure, *Appl. Phys. Lett.* **75**, 1449 (1999).

31. A. Tsurumaki-Fukuchi, H. Yamada, and A. Sawa, Resistive switching artificially induced in a dielectric/ferroelectric composite diode, *Appl. Phys. Lett.* **103**, 152903 (2013).

32. M. D. Scafetta, Y. J. Xie, M. Torres, J. E. Spanier, and S. J. May, Optical absorption in epitaxial $La_{1-x}SrxFeO_3$ thin films, *Appl. Phys. Lett.* **102**, 081904 (2013).

33. S. J. Clark and J. Robertson, Band gap and Schottky barrier heights of multiferroic $BiFeO_3$, *Appl. Phys. Lett.* **90**, 132903 (2007).

34. J. F. Ihlefeld, N. J. Podraza, Z. K. Liu et al., Optical band gap of $BiFeO_3$ grown by molecular-beam epitaxy, *Appl. Phys. Lett.* **92**, 142908 (2008).

35. S. G. Lim, S. Kriventsov, T. N. Jackson et al., Dielectric functions and optical bandgaps of high-K dielectrics for metal-oxide-semiconductor field-effect transistors by far ultraviolet spectroscopic ellipsometry, *J. Appl. Phys.* **91**, 4500 (2002).

36. P. W. Peacock and J. Robertson, Band offsets and Schottky barrier heights of high dielectric constant oxides, *J. Appl. Phys.* **92**, 4712 (2002).

37. Y. Muraoka, T. Muramatsu, J. Yamaura, and Z. Hiroi, Photogenerated hole carrier injection to $YBa_2Cu_3O_{7-x}$ in an oxide heterostructure, *Appl. Phys. Lett.* **85**, 2950 (2004).

38. T. Muramatsu, Y. Muraoka, and Z. Hiroi, Photocarrier injection and current voltage characteristics of $La_{0.8}Sr_{0.2}MnO_3$/$SrTiO_3$:Nb heterojunction at low temperature, *Jpn. J. Appl. Phys.* **44**, 7367 (2005).

39. M. Nakamura, A. Sawa, H. Sato, H. Akoh, M. Kawasaki, and Y. Tokura, Optical probe of electrostatic-doping in an n-type Mott insulator, *Phys. Rev. B* **75**, 155103 (2007).

40. T. Fujii, M. Kawasaki, A. Sawa, Y. Kawazoe, H. Akoh, and Y. Tokura, Electrical properties and colossal electroresistance of heteroepitaxial $SrRuO_3$/$SrTi_{1-x}Nb_xO_3$ ($0.0002 \leq x \leq 0.02$) Schottky junctions, *Phys. Rev. B* **75**, 165101 (2007).

41. H. Namatame, A. Fujimori, Y. Tokura et al., Resonant-photoemission study of $Nd_{2-x}Ce_xCuO_4$, *Phys. Rev. B* **41**, 7205 (1990).

42. A. Fujimori, A. Ino, J. Matsuno, T. Yoshida, K. Tanaka, and T. Mizokawa, Core-level photoemission measurements of the chemical potential shift as a probe of correlated electron systems, *J. Electron Spectrosc. Relat. Phenom.* **124**, 127 (2002).

43. J. Matsuno, A. Fujimori, Y. Takeda, and M. Takanoi, Chemical potential shift in $La_{1-x}Sr_xMnO_3$: Photoemission test of the phase separation scenario, *Europhys. Lett.* **59**, 252 (2002).

44. S.-W. Cheong and M. Mostovoy, Multiferroics: A magnetic twist for ferroelectricity, *Nat. Mater.* **6**, 13–20 (2007).

45. S. Dong, J. M. Liu, S. W. Cheong, and Z. F. Ren, Multiferroic materials and magnetoelectric physics: Symmetry, entanglement, excitation, and topology, *Adv. Phys.* **64**, 519–626 (2015).

46. N. A. Hill, Why are there so few magnetic ferroelectrics? *J. Phys. Chem. B* **104**, 6694–6709 (2000).

47. P. W. Anderson, Antiferromagnetism. Theory of superexchange interaction, *Phys. Rev.* **79**, 350 (1950).

48. J. B. Goodenough, *Magnetism and the Chemical Bond*, Interscience, New York, 1963.

49. J. Kanamori, Superexchange interaction and symmetry properties of electron orbitals, *J. Phys. Chem. Solids* **10**, 87 (1959).

50. K. I. Kugel and D. I. Khomskii, The Jahn–Teller effect and magnetism: Transition metal compounds, *Soviet Physics Uspekhi* **25**, 231 (1982).

51. C. Zener, Interaction between the d-Shells in the transition metals. II. Ferromagnetic compounds of manganese with perovskite structure, *Phys. Rev.* **82**, 403 (1951).

52. M. A. Ruderman and C. Kittel, Indirect exchange coupling of nuclear magnetic moments by conduction electrons, *Phys. Rev.* **96**, 99 (1954).

53. T. Kasuya, A theory of metallic ferro- and antiferromagnetism on zener's model, *Prog. Theor. Phys.* **16**, 45 (1956).

54. K. Yoshida, Magnetic properties of Cu-Mn alloys, *Phys. Rev.* **106**, 893 (1957).

55. I. Dzyaloshinskii, A thermodynamic theory of "weak" ferromagnetism of antiferromagnetics, *J. Phys. Chem. Solids* **4**, 241 (1958).

56. T. Moriya, Anisotropic superexchange interaction and weak ferromagnetism, *Phys. Rev.* **120**, 91 (1960).

57. S. Okamoto, Magnetic interaction at an interface between manganite and other transition-metal oxides, *Phys. Rev. B* **82**, 024427 (2010).

58. J. Chakhalian, J. W. Freeland, H.-U. Habermeier et al., Orbital reconstruction and covalent bonding at an oxide interface, *Science* **318**, 1114 (2007).

59. N. M. Nemes, M. Garcia-Hernandez, S. G. E. te Velthuis et al., Origin of the inverse spin-switch behavior in manganite/cuprate/manganite trilayers, *Phys. Rev. B* **78**, 094515 (2008).

60. J. Salafranca and S. Okamoto, Unconventional proximity effect and inverse spin-switch behavior in a model manganite-cuprate-manganite trilayer system, *Phys. Rev. Lett.* **105**, 256804 (2010).

61. P. Yu, J.-S. Lee, S. Okamoto et al., Interface ferromagnetism and orbital reconstruction in $BiFeO_3$-$La_{0.7}Sr_{0.3}MnO_3$ heterostructure, *Phys. Rev. Lett.* **105**, 027201 (2010).

62. L. Jiang, W. S. Choi, H. Jeen et al., Tunneling electroresistance induced by interfacial phase transitions in ultrathin oxide heterostructures, *Nano Lett.* **13**, 5837–5843 (2013).

63. L. Jiang, W. S. Choi, H. Jeen, T. Egami, and H. N. Lee, Strongly coupled phase transition in ferroelectric/correlated electron oxide heterostructures, *Appl. Phys. Lett.* **101**, 042902 (2012).

64. C. H. Ahn, J. M. Triscone, N. Archibald et al., Ferroelectric field effect in epitaxial thin film oxide $SrCuO_2$/$Pb(Zr_{0.52}Ti_{0.48})O_3$ heterostructures, *Science* **269**, 373–376 (1995).

65. J. D. Burton and E. Y. Tsymbal, Prediction of electrically induced magnetic reconstruction at the manganite/ferroelectric interface, *Phys. Rev. B* **80**, 174406 (2009).

66. J. D. Burton and E. Y. Tsymbal, Giant tunneling electroresistance effect driven by an electrically controlled spin valve at a complex oxide interface, *Phys. Rev. Lett.* **106**, 157203 (2011).

67. J. Hoffman, X. A. Pan, J. W. Reiner et al., Ferroelectric field effect transistors for memory applications, *Adv. Mater.* **22**, 2957–2961 (2010).

68. H. J. A. Molegraaf, J. Hoffman, C. A. F. Vaz et al., Magnetoelectric effects in complex oxides with competing ground states, *Adv. Mater.* **21**, 3470 (2009).

69. C. A. F. Vaz, J. Hoffman, Y. Segal et al., Origin of the magnetoelectric coupling effect in $Pb(Zr_{0.2} Ti_{0.8})O_3/La_{0.8}Sr_{0.2}MnO_3$ multiferroic heterostructures, *Phys. Rev. Lett.* **104**, 127202 (2010).

70. C. H. Ahn, A. Bhattacharya, M. Di Ventra et al., Electrostatic modification of novel materials, *Rev. Mod. Phys.* **78**, 1185–1212 (2006).

71. C. A. F. Vaz, Electric field control of magnetism in multiferroic heterostructures, *J. Phys. Condens. Matter* **24**, 333201 (2012).

72. J. P. Velev, S. S. Jaswal, and E. Y. Tsymbal, Multi-ferroic and magnetoelectric materials and interfaces, *Philos. Trans. R. Soc. A-Math. Phys. Eng. Sci* **369**, 3069–3097 (2011).

73. X. Huang and S. Dong, Ferroelectric control of magnetism and transport in oxide heterostructures, *Mod. Phys. Lett. B* **28**(23), 1430010 (2014).

74. S. Dong and E. Dagotto, Full control of magnetism in a manganite bilayer by ferroelectric polarization, *Phys. Rev. B* **88**, 140404(R) (2013).

75. S. Dong, X. T. Zhang, R. Yu, J. M. Liu, and E. Dagotto, Microscopic model for the ferroelectric field effect in oxide heterostructures, *Phys. Rev. B* **84**, 155117 (2011).

76. E. Dagotto, *Nanoscale Phase Separation and Colossal Magnetoresistance*, Springer, Berlin, 2003.

77. E. Dagotto, T. Hotta, and A. Moreo, Colossal magnetoresistant materials: The key role of phase separation, *Phys. Rep. Rev. Sect. Phys. Lett.* **344**, 1–153 (2001).

78. Y. Tokura, Critical features of colossal magnetoresistive manganites, *Rep. Prog. Phys.* **69**, 797–851 (2006).

79. S. Smadici, B. B. Nelson-Cheeseman, A. Bhattacharya, and P. Abbamonte, Interface ferromagnetism in a $SrMnO_3/LaMnO_3$ superlattice, *Phys. Rev. B* **86**, 174427 (2012).

80. A. Bhattacharya, X. Zhai, M. Warusawithana, J. N. Eckstein, and S. D Bader, Signatures of enhanced ordering temperatures in digital superlattices of $(LaMnO_3)m/(SrMnO_3)_{2m}$, *Appl. Phys. Lett.* **90**, 222503 (2007).

81. A. Bhattacharya, S. J. May, S. G. E. te Velthuis et al., Metal-insulator transition and its relation to magnetic structure in $(LaMnO_3)_{2n}/(SrMnO_3)_n$ superlattices, *Phys. Rev. Lett.* **100**, 257203 (2008).

82. C. Adamo, C. A. Perroni, V. Cataudella, G. De Filippis, P. Orgiana, and L. Maritato, Tuning the metal-insulator transitions of $(SrMnO_3)_n/(LaMnO_3)_{2n}$ superlattices: Role of interfaces, *Phys. Rev. B* **79**, 045125 (2009).

83. C. Adamo, X. Ke, P. Schiffer et al., Electrical and magnetic properties of $(SrMnO_3)_n/(LaMnO_3)_{2n}$ superlattices, *Appl. Phys. Lett.* **92**, 112508 (2008).

84. S. J. May, A. B. Shah, S. G. E. te Velthuis et al., Magnetically asymmetric interfaces in a $LaMnO_3/SrMnO_3$ superlattice due to structural asymmetries, *Phys. Rev. B* **77**, 174409 (2008).

85. S. Dong, R. Yu, S. Yunoki, G. Alvarez, J. M. Liu, and E. Dagotto, Magnetism, conductivity, and orbital order in $(LaMnO_3)_{2n}/(SrMnO_3)_n$ super-lattices, *Phys. Rev. B* **78**, 201102 (2008).

86. R. Yu, S. Yunoki, S. Dong, and E. Dagotto, Electronic and magnetic properties of $RMnO_3/AMnO_3$ heterostructures, *Phys. Rev. B* **80**, 125115 (2009).

87. G. Bouzerar and O. Cepas, Effect of correlated disorder on the magnetism of double exchange systems, *Phys. Rev. B* **76**, 020401 (2007).

88. B. R. D. Nanda and S. Satpathy, Effects of strain on orbital ordering and magnetism at perovskite oxide interfaces: $LaMnO_3/SrMnO_3$, *Phys. Rev. B* **78**, 054427 (2008).

89. B. R. K. Nanda and S. Satpathy, Effects of strain on orbital ordering and magnetism at perovskite oxide interfaces: $LaMnO_3/SrMnO_3$, *Phys. Rev. B* **78**, 054427 (2008).

90. B. R. K. Nanda and S. Satpathy, Polar catastrophe, electron leakage, and magnetic ordering at the $LaMnO_3/SrMnO_3$ interface, *Phys. Rev. B* **81**, 224408 (2010).

91. S. J. May, P. J. Ryan, J. L. Robertson et al., Enhanced ordering temperatures in antiferromagnetic manganite superlattices, *Nat. Mater.* **8**, 892–897 (2009).

92. Y. S. Hou, H. J. Xiang, and X. G. Gong, Intrinsic insulating ferromagnetism in manganese oxide thin films, *Phys. Rev. B* **89**, 064416 (2014).

93. O. Chmaissem, B. Dabrowski, S. Kolesnik, J. Mais, J. D. Jorgensen, and S. Short, Structural and magnetic phase diagrams of $La_{1-x}Sr_xMnO_3$ and $Pr_{1-y}Sr_yMnO_3$, *Phys. Rev. B* **67**, 094431 (2003).

94. Z. Jirak, J. Hejtmanek, E. Pollert et al., Magnetic ground states in $Pr_{1-x}Sr_xMnO_3$ (x = 0.48 – 0.75), *J. Appl. Phys.* **89**, 7404 (2001).

95. H. Kuwahara, T. Okuda, Y. Tomioka, A. Asamitsu, and Y. Tokura, Two-dimensional charge-transport and spin-valve effect in the layered anti-ferromagnet $Nd_{0.45}Sr_{0.55}MnO_3$, *Phys. Rev. Lett.* **82**, 4316 (1999).

96. R. Kajimoto, H. Yoshizawa, H. Kawano et al., Hole-concentration-induced transformation of the magnetic and orbital structures in $Nd_{1-x}Sr_xMnO_3$, *Phys. Rev. B* **60**, 9506 (1999).

97. R. Maezono, S. Ishihara, and N. Nagaosa, Orbital polarization in manganese oxides, *Phys. Rev. B* 57, R13993 (1998).

98. R. Maezono, S. Ishihara, and N. Nagaosa, Phase diagram of manganese oxides, *Phys. Rev. B* **58**, 11583 (1998).

99. S. Okamoto, S. Ishihara, and S. Maekawa, Phase transition in perovskite manganites with orbital degree of freedom, *Phys. Rev. B* **61**, 14647 (2000).

100. S. Dong, Q. F. Zhang, S. Yunoki, J. M. Liu, and E. Dagotto, Magnetic and orbital order in $(RMnO_3)_n/(AMnO_3)_{2n}$ superlattices studied via a double-exchange model with strain, *Phys. Rev. B* **86**(20), 205121 (2012).

101. S. Okamoto, S. Ishihara, and S. Maekawa, Orbital structure and magnetic ordering in layered manganites: Universal correlation and its mechanism, *Phys. Rev. B* **63**, 104401 (2001).

102. S. Ishihara, S. Okamoto, and S. Maekawa, Roles of electron correlation and orbital degree of freedom in manganese oxides, *Trans. Mater. Res. Soc. Jpn.* **26**, 963 (2001).

103. M. Kubota, H. Fujioka, K. Ohoyama et al., Neutron scattering studies on magnetic structure of the double-layered manganite $La_{2-2x}Sr_{1+2x}Mn_2O_7$ ($0.30 \leq x \leq 0.50$), *J. Phys. Chem. Solids* **60**, 1161 (2000).

104. C. D. Ling, J. E. Millburn, J. F. Mitchell, D. N. Argyriou, J. Linton, and H. N. Bordallo, Interplay of spin and orbital ordering in the layered colossal magnetoresistance manganite $La_{2-2x}Sr_{1+2x}Mn_2O_7$($0.5 \leq x \leq 1.0$), *Phys. Rev. B* **62**, 15096 (2000).

105. Q. F. Zhang, S. Dong, B. L. Wang, and S. Yunoki, Strain-engineered magnetic order in $(LaMnO_3)_n/(SrMnO_3)_{2n}$ superlattices, *Phys. Rev. B* **86**, 094403 (2012).

106. J. Nogues, J. Sort, V. Langlais et al., Exchange bias in nanostructures, *Phys. Rep. Rev. Sect. Phys. Lett.* **422**, 65–117 (2005).

107. A. P. Malozemoff, Random-field model of exchange-anisotropy at rough ferromagnetic-antiferromagnetic interfaces, *Phys. Rev. B* **35**, 3679–3682 (1987).

108. N. C. Koon, Calculations of exchange bias in thin films with ferromagnetic/antiferromagnetic interfaces, *Phys. Rev. Lett.* **78**, 4865–4868 (1997).

109. M. Kiwi, J. Mejia-Lopez, R. D. Portugal, and R. Ramirez, Exchange-bias systems with compensated interfaces, *Appl. Phys. Lett.* **75**, 3995–3997 (1999).

110. T. C. Schulthess and W. H. Butler, Consequences of spin-flop coupling in exchange biased films, *Phys. Rev. Lett.* **81**, 4516–4519 (1998).

111. M. Kiwi, Exchange bias theory, *J. Magn. Magn. Mater.* **234**, 584–595 (2001).

112. S. M. Wu, S. A. Cybart, P. Yu et al., Reversible electric control of exchange bias in a multiferroic field-effect device, *Nat. Mater.* **9**, 756–761 (2010).

113. P. Borisov, A. Hochstrat, X. Chen, W. Kleemann, and C. Binek, Magnetoelectric switching of exchange bias, *Phys. Rev. Lett.* **94**, 117203 (2005).

114. S. Dong, Q. Zhang, S. Yunoki, J.-M. Liu, and E. Dagotto, Ab initio study of the intrinsic exchange bias at the $SrRuO_3/SrMnO_3$ interface, *Phys. Rev. B* **84**, 224437 (2011).

115. S. Dong, K. Yamauchi, S. Yunoki et al., Exchange bias driven by the Dzyaloshinskii-Moriya interaction and ferroelectric polarization at G-type antiferromagnetic perovskite interfaces, *Phys. Rev. Lett.* **103**, 127201 (2009).

116. M. Raccah and J. B. Goodenough, First-order localized-electron collective-electron transition in $LaCoO_3$, *Phys. Rev.* **155**, 932 (1967).

117. J. B. Goodenough, An interpretation of the magnetic properties of the perovskite-type mixed crystals $La_{1-x}Sr_xCoO_{3-\lambda}$, *J. Phys. Chem. Solids* **6**, 287 (1958).

118. Y. Tokura, Y. Okimoto, S. Yamaguchi, H. Taniguchi, T. Kimura, and H. Takagi, Thermally induced insulator-metal transition in $LaCoO_3$: A view based on the Mott transition, *Phys. Rev. B* **58**, R1699(R) (1998).

119. M. W. Haverkort, Z. Hu, J. C. Cezar et al., Spin state transition in $LaCoO_3$ studied using soft X-ray absorption spectroscopy and magnetic circular dichroism, *Phys. Rev. Lett.* **97**, 176405 (2006).

120. D. W. Jeong, W. S. Choi, S. Okamoto et al., Dimensionality control of d-orbital occupation in oxide superlattices, *Sci. Rep.* **4**, 6124 (2014).

121. A. Georges, G. Kotliar, W. Krauth, and M. J. Rozenberg, Dynamical mean-field theory of strongly correlated fermion systems and the limit of infinite dimensions, *Rev. Mod. Phys.* **68**, 13 (1996).

122. R. E. Prange and S. M. Girvin (Eds.), *The Quantum Hall Effect*, Springer–Verlag, New York, 1987.

123. D. J. Thouless, M. Kohmoto, M. P. Nightingale, and M. den Nijs, Quantized hall conductance in a two-dimensional periodic potential, *Phys. Rev. Lett.* **49**, 405 (1982).

124. F. D. M. Haldane, Model for a quantum Hall effect without Landau levels: Condensed-matter realization of the "parity anomaly", *Phys. Rev. Lett.* **61**, 2015 (1988).

125. C. L. Kane and E. J. Mele, Z2 topological order and the quantum spin hall effect, *Phys. Rev. Lett.* **95**, 146802 (2005).

126. B. A. Bernevig, T. L. Hughes, and S.-C. Zhang, Quantum spin Hall effect and topological phase transition in HgTe quantum wells, *Science* **314**, 1757 (2006).

127. M. Konig, S. Wiedmann, C. BrÜne et al., Quantum spin Hall insulator state in HgTe quantum wells, *Science* **318**, 766 (2007).

128. J. E. Moore and L. Balents, Topological invariants of time-reversal-invariant band structures, *Phys. Rev. B* **75**, 121306 (2007).

129. L. Fu, C. L. Kane, and E. J. Mele, Topological insulators in three dimensions, *Phys. Rev. Lett.* **98**, 106803 (2007).

130. D. Hsieh, D. Qian, L. Wray et al., A topological Dirac insulator in a quantum spin Hall phase, *Nature* **452**, 970 (2008).

131. Y. Xia, D. Qian, D. Hsieh et al., Observation of a large-gap topological-insulator class with a single Dirac cone on the surface, *Nat. Phys.* **5**, 398 (2009).

132. X.-L. Qi, T. L. Hughes, and S.-C. Zhang, Topological field theory of time-reversal invariant insulators, *Phys. Rev. B* **78**, 195424 (2008).

133. L. Fu and C. L. Kane, Superconducting proximity effect and Majorana fermions at the surface of a topological insulator, *Phys. Rev. Lett.* **100**, 096407 (2008).

134. X.-L. Qi, R. Li, J. Zang, and S.-C. Zhang, Inducing a magnetic monopole with topological surface states, *Science* **323**, 1184 (2009).

135. A. Shitade, H. Katsura, J. Kunes, X.-L. Qi, S.-C. Zhang, and N. Nagaosa, Quantum spin Hall effect in a transition metal oxide $Na_2IrO_3$, *Phys. Rev. Lett.* **102**, 256403 (2009).

136. A. Tsukazaki, A. Ohtomo, T. Kita, Y. Ohno, H. Ohno, and M. Kawasaki, Quantum Hall effect in polar oxide heterostructures, *Science* **315**, 1388 (2007).

137. A. Tsukazaki, S. Akasaka, K. Nakahara et al., Observation of the fractional quantum Hall effect in an oxide, *Nat. Mater.* **9**, 889 (2010).

138. Y. Matsubara, K. S. Takahashi, M. S. Bahramy et al., Observation of the quantum Hall effect in 5-doped $SrTiO_3$, *Nat. Commun.* **7**, 11631 (2016).

139. D. Xiao, W. Zhu, Y. Ran, N. Nagaosa, and S. Okamoto, Interface engineering of quantum Hall effects in digital transition-metal oxide heterostructures, *Nat. Commun.* **2**, 596 (2011).

140. L. Fu and C. L. Kane, Time reversal polarization and a $z_2$ adiabatic spin pump, *Phys. Rev. B* **74**, 195312 (2006).
141. S. Okamoto, W. Zhu, Y. Nomura, R. Arita, D. Xiao, and N. Nagaosa, Correlation effects in (111) bilayers of perovskite transition-metal oxides, *Phys. Rev. B* **89**, 195121 (2014).
142. S. Okamoto, Doped Mott insulators in (111) bilayers of perovskite transition-metal oxides with the strong spin-orbit coupling, *Phys. Rev. Lett.* **110**, 066403 (2013).
143. D. Hirai, J. Matsuno, and H. Takagi, Fabrication of (111)-oriented $Ca_{0.5}Sr_{0.5}IrO_3/SrTiO_3$ superlattices A designed playground for honeycomb physics, *APL Mater.* **3**, 041508 (2015).

# 12

# LaAlO$_3$/SrTiO$_3$
## *A Tale of Two Magnetisms*

**Yun-Yi Pai, Anthony Tylan-Tyler, Patrick Irvin, and Jeremy Levy**

## 12.1 INTRODUCTION

Complex oxides have been a continual source of discoveries in solid state physics. Recent advances in thin-film growth techniques have enabled unit cell-level control over their composition, resulting in the creation of layered, complex oxide heterostructures. *Emergent phenomena*—properties or phases observed only within heterostructures and not found in the parent compounds—provide a recurring theme for this active interdisciplinary field that spans materials science, physics, chemistry, and engineering. One of the most striking, and controversial, examples of an emergent property is the observation of magnetism at the interface formed between the two complex oxides $LaAlO_3$ and $SrTiO_3$. Both materials are separately nonmagnetic, and yet there are many reports of magnetic phenomena associated with the $LaAlO_3/SrTiO_3$ interface.*

The properties of $SrTiO_3$ have fascinated researchers for generations. This simple perovskite material has been widely utilized as a substrate for the growth of other materials, but has its own wide-ranging properties that span "all of solid state physics"—*save for magnetism* [1] (see Figure 12.1). $SrTiO_3$ has also inspired important discoveries such as high-temperature superconductivity [2]. Interest in the fundamental properties of $SrTiO_3$ was renewed in the last decade due to the 2004 report by Ohtomo and Hwang of high-mobility electron transport at the interface between $LaAlO_3$ and $TiO_2$-terminated $SrTiO_3$ [3]. The discovery of magnetism at the $LaAlO_3/SrTiO_3$ interface was first reported by Brinkman et al. [4] in 2007. Following that initial report, a cascade of new signatures of magnetism at the $LaAlO_3/SrTiO_3$ interface were obtained using a variety of measurement techniques.

While most published findings are internally consistent, some reports apparently contradict the existence of interfacial magnetism, and some magnetic signatures have not always been reproduced under nominally identical conditions. The goal of this chapter is to help organize and classify the sometimes contradictory observations that have been made on $LaAlO_3/SrTiO_3$ heterostructures, and closely related systems, which have been investigated extensively, as well as explored theoretically. We argue that there is not a

**FIGURE 12.1**  Properties of $LaAlO_3/SrTiO_3$. Properties that have been observed for $SrTiO_3$ from superconductivity via doping, ferroelectricity, and piezoelectricity. Magnetism, however, is found at the interface of the $LaAlO_3/SrTiO_3$, even though neither of the two materials are magnetic. Quotes are from Ref. [1] and Ref. [101].

---

* It should be noted that there are also many complex-oxide heterostructures with magnetic properties that are derived from bulk behavior. For a more general discussion of magnetism at complex oxide interfaces, readers are referred to Chapter 11, Volume 2.

single form of magnetism, but rather that there are two main types of magnetic behavior with distinct properties and origins.

### 12.1.1 ORGANIZATION OF THIS CHAPTER

The purpose of this chapter is twofold. First, we will review the experimental evidence for magnetic phenomena at the LaAlO$_3$/SrTiO$_3$ interface. We will then argue that essentially all of the signatures of magnetism can be sorted into two distinct categories: (1) magnetic phases (e.g. ferromagnetic or Kondo) involving local magnetic moments and their coupling to itinerant electrons; (2) *metamagnetic* effects that are mediated by attractive electron-electron interactions that do not involve local moments. We will review possible candidates for the local moments that give rise to the ferromagnetic phases and focus on arguments for one potential source: oxygen vacancies. For the metamagnetic transport signatures, band-structure effects (e.g. Lifshitz transition) and strong attractive electron-electron interaction can help consolidate disparate experimental findings. The coexistence of magnetism and superconductivity is discussed briefly.

While the main focus will be on experimental evidence, we will also briefly summarize various theoretical approaches that have been taken so far. However, we argue that the starting point (assumptions) for many theories could be hindered by attempts to combine these two distinct (and weakly interacting) forms of magnetism. To make better progress theoretically, we argue that it will be simpler to initially restrict the domain of inquiry to one or the other phases. Toward the end of the article, we discuss open questions as well as some tests for this framework for understanding the two classes of magnetic phenomena.

## 12.2 EXPERIMENTAL SIGNATURES OF MAGNETISM

The first signature of magnetism at the LaAlO$_3$/SrTiO$_3$ interface was reported in 2007 by Brinkman et al. [4]. They reported a hysteresis loop in the magnetoresistance (Figure 12.2a) and a characteristic Kondo minimum (followed by a saturation at lower temperature) in the resistance versus temperature that was attributed to magnetic impurities (Figure 12.3a). The Kondo signature was later also observed by Hu et al. [5], as well as by Lee et al. [6] in electrolyte-gated SrTiO$_3$. Several other signatures of magnetism followed, including observations of anisotropic magnetoresistance [5, 7–13], anomalous Hall effect [11, 14–16], and direct observation of magnetization with tools from cantilever magnetometry [17], scanning SQUID magnetometry and susceptometry [18–20], magnetic force microscopy [21, 22], and x-ray magnetic circular dichroism [23, 24]. Below, we consider these findings in more detail.

### 12.2.1 MAGNETOMETRY

#### 12.2.1.1 SQUID Magnetometry

In 2011, Ariando et al. [9] reported a direct measurement of LaAlO$_3$/SrTiO$_3$ magnetization using SQUID magnetometry. As shown in Figure 12.3b, a

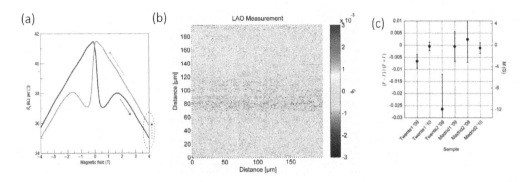

**FIGURE 12.2** Inconsistencies and null results. Even though the signatures of magnetism are reported by many, there have also been controversies. (a) A hysteresis loop in the magnetoresistance previously thought as purely magnetic, but later identified as magnetocaloric effect [55]. (After Brinkman, A. et al., *Nat. Mater.* 6, 493, 2007. With permission.) (b) A null result reported by Wijnands et al. [57]: scanning SQUID on samples nominally equally to those reported in Bert et al. [18]. No ferromagnetic patches are found on the area of the sample explored. (After Wijnands, T. *Scanning Superconducting Quantum Interference Device Microscopy*, Master's thesis, 2013. With permission.) (c) Using neutron spin-dependent reflectivity, where the magnetization is computed from asymmetry in the spin-dependent superlattice Bragg reflections as a function of wave vector transfer and neutron beam polarization, Fitzsimmons et al. [56] reported magnetization for LaAlO$_3$/SrTiO$_3$ superlattice close to the noise level through the presented six samples grown by two independent groups. (After Fitzsimmons, M.R. et al., *Phys. Rev. Lett.* 107, 217201, 2011. With permission.)

series of hysteresis loops of the magnetization in external magnetic field is observed. This hysteresis persists up to room temperature. It is worth noting that this ferromagnetic hysteresis was only reported for the samples grown at P(O$_2$) = 10$^{-2}$ mbar pressures, which exist at the boundary of the metal-insulator transition. A similar finding for Nb-doped SrTiO$_3$ has also been reported by Liu et al. [25].

### 12.2.1.2 Cantilever Magnetometry

Using cantilever-based torque magnetometry (Figure 12.3c), Li et al. [17] reported an in-plane magnetization of the 2DEG at LaAlO$_3$/SrTiO$_3$ with magnitude corresponding to ~0.3 $\mu_B$ per interface unit cell (if all the magnetization is assigned to the interface). The magnetization was found to be independent of temperature up to 40 K and persists beyond 200 K. The same sample was also found to be superconducting below 120 mK.

### 12.2.1.3 Scanning SQUID Microscopy

Scanning SQUID microscopy has been used to image inhomogeneous magnetism in LaAlO$_3$/SrTiO$_3$. With a 3 μm-diameter SQUID loop serving as a local sensor of magnetic flux, this technique can detect magnetic moments as small as $\sim 10^2 \mu_B / \sqrt{Hz}$ [26]. Bert et al. [18] reported dipole-like magnetic patches at the interface of LaAlO$_3$/SrTiO$_3$ (Figure 12.3d) [18]. Subsequent investigations by Kalisky et al. [19] determined that the patches were highly nonuniform in size, orientation, and physical placement, and only observed at or above a critical thickness for LaAlO$_3$ layer (3 u.c.), the same as the threshold for the insulator-to-metal transition in LaAlO$_3$/

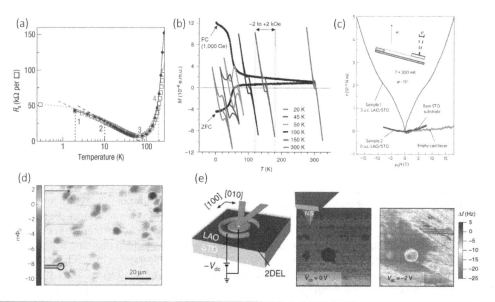

**FIGURE 12.3** Experimental evidence for ferromagnetism at the LaAlO$_3$/SrTiO$_3$ interface. (a) Kondo resistance minimum. (After Brinkman, A. et al., *Nat. Mater.* 6, 493, 2007. With permission.) (b) SQUID measurement: hysteresis loops taken at different temperature superimposed on top of the temperature dependence of magnetic moments. (After Ariando, X. et al., *Nat. Commun.* 2, 188, 2011. With permission.) (c) Cantilever-based magnetometry: the magnetic moment induced a torque under external magnetic field; the magnetization of the sample is inferred accordingly. (After Li, L. et al., *Nat. Phys.* 7, 762–766, 2011. With permission.) (d) With a micrometer-sized SQUID on a probe tip, microscopic magnetization can be imaged. Dipole-shaped patches are observed. (After Bert, J.A. et al., *Nat. Phys.* 7, 767–771, 2011. With permission.) (e) Electrically-controlled ferromagnetism observed with magnetic force microscope. The magnetism signal is observed only when the interface is insulating. (After Bi, F. et al., *Nat. Commun.* 5, 5019, 2014. With permission.)

SrTiO$_3$ [27]. By touching the LaAlO$_3$ surface with the SQUID sensor, it was discovered that the magnetic moment and orientation of the ferromagnetic patches could be manipulated (Figure 12.4f) [20]. The same scanning SQUID system can also simultaneously image diamagnetic susceptibility associated with superconductivity [18, 28]. Correlations between magnetic and superconducting order were not found within the spatial resolution of this technique.

### 12.2.1.4  β-NMR

Salman et al. [29] measured magnetic properties of LaAlO$_3$/SrTiO$_3$ using β-NMR. In this technique, spin-polarized radioactive $^8$Li atoms are shot at the LaAlO$_3$/SrTiO$_3$ sample. The spins of Li nuclei are inferred from the spins of electrons emitted via β decay. The presence of magnetic moments in the sample increases the spin decoherence rate. A "weak" ($\sim 1.8 \times 10^{-3} \mu_B$ / u.c.) magnetization is obtained for both 6 and 8 u.c samples. If a high degree of spatial nonuniformity is assumed, then a local density $\sim 10^{12} \mu_B$ / cm$^2$ may be present [29], which is on the same order of magnitude as Bert et al. [18].

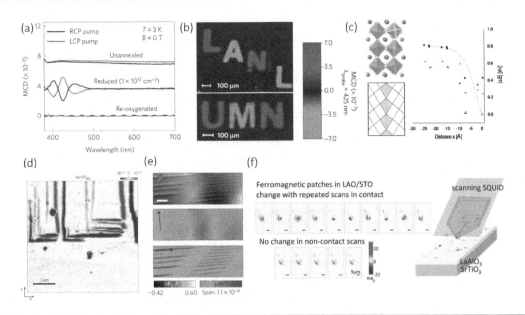

**FIGURE 12.4** Ferromagnetism, oxygen vacancies, and ferroelastic domains. (a) Using magnetic circular dichroism (MCD), Rice et al. [71] reported a persistent magnetic signal for oxygen deficient $SrTiO_3$ samples. The signal disappears upon further re-oxygenation and reappears after introduction of oxygen vacancies . (After Rice, W.D. et al., *Nat. Mater.* 13, 481, 2014. With permission.) (b) The persistent MCD signal created in real space. (After Rice, W.D. et al., *Nat. Mater.* 13, 481, 2014. With permission.) (c) With DFT calculation for ferroelectric perovskite $BaTiO_3$, Goncalves-Ferreira et al. [95] found oxygen vacancies do have lower energy when around twin wall of the ferroelastic domains. (After Goncalves-Ferreira, L. et al., *Phys. Rev. B.* 81, 024109, 2010. With permission.) (d) and (e) Experimentally observed ferroelastic domains using (d) a Scanning SET. (After Honig, M. et al., *Nat. Mater.* 12, 1112, 2013. With permission.) (e) scanning SQUID. (After Kalisky, B. et al., *Nat. Mater.* 12, 1091, 2013. With permission.) (f) Mechanical manipulation of the magnetization using the probe of scanning SQUID. (After Kalisky, F.B. et al., *Nano Lett.*, 2012, 12, 2012. With permission.)

## 12.2.1.5 Magnetic Force Microscopy (MFM)

Bi et al. [21] reported electronically-controlled magnetism at the interface of $LaAlO_3/SrTiO_3$ using magnetic force microscopy (Figure 12.3e). When an in-plane polarized magnetic tip in an atomic force microscope (AFM) is brought to the vicinity of the sample surface, the magnetization of the sample and its interaction with the MFM tip shifts the oscillation amplitude, phase, and frequency, which are recorded as a function of position. Magnetic interactions between the MFM tip and the interface take place through a thin layer of gold deposited on the surface of $LaAlO_3$ to serve as a top gate for the sample. The 2DEG at the interface can be tuned from insulating to conducting, which is monitored through the capacitance of the two-terminal device. Magnetic interactions are observed only when the tip is magnetized in-plane, and when the device is gated into the insulating regime. Related measurements show a pronounced magnetoelectric effect, which indicates that free carriers rapidly screen (on <10 μs time scales) ferromagnetic domains [21]. Follow-up measurements revealed that this electronically-controlled ferromagnetism is observed only within a thickness window 8–25 u.c. [22]. Outside this region, the samples cannot be gated into the insulating regime: samples with thinner LaAlO₃ layers are prone to leakage via direct tunneling, while samples with LaAlO₃ layer thicker than 25 u.c. are subjected to Zener tunneling.

## 12.2.2 TRANSPORT

### 12.2.2.1 Anisotropic Magnetoresistance

Anisotropy in the magnetotransport measurements is a widely reported signature of magnetism at the LaAlO₃/SrTiO₃ interface [5, 7–13]. When the magnetic field is (i) oriented out-of-plane, the magnetoresistance is positive [30]. When the magnetic field is (ii) in-plane and perpendicular to the direction of transport, the magnetoresistance is positive for small fields and eventually becomes negative at larger field [31]. When the magnetic field is (iii) parallel to the direction of transport, the magnetoresistance is negative [7, 9]. As the applied magnetic field rotates in plane, as shown in Figure 12.5a, a sinusoidal dependence of magnetoresistance as a function of the angle φ between the field and the direction of current (Figure 12.5b). However, when the carrier concentration $n_e$ is above a critical value $n_L$, a Lifshitz transition

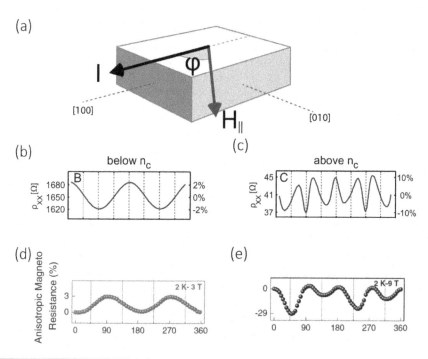

**FIGURE 12.5**   Transport metamagnetism: anisotropic magnetoresistance. (a) Setup for measuring in-plane anisotropy of the magnetoresistance and the metamagnetism transition critical field $B_p$. The anisotropy appears only when carrier concentration $n_e$ is greater than $n_L$ and the external magnetic field is greater than $B_p$ (which is also a function of $n_e$). A characteristic change in the magnetoresistance can be clearly seen from (b) to (c), when the carrier concentration is going above $n_L$, or (d) to (e), when the external field is exceeding $B_p$. ((b) and (c) after Joshua, A. et al., *Proc. Natl. Acad. Sci.* 110, 9633, 2013. With permission; (d) and (e) after Annadi, A. et al., *Phys. Rev. B* 87, 201102, 2013. With permission.)

occurs and this two-fold symmetry develops crystalline anisotropy with high harmonics (Figure 12.5c,e).

### 12.2.2.2 Non-Linear and Anomalous Hall Effects

Nonlinearity in the Hall coefficient is widely reported for $SrTiO_3$-based heterostructures [8, 11, 14–16, 30, 32–38]. Figure 12.6a [36] shows the most common nonlinearity in the Hall resistance $R_{xy}$, using a standard Hall geometry with an out-of-plane external magnetic field. This results in an overall parabolic shape of the Hall coefficient $R_H$ (the slope of the Hall resistance $R_{xy}$) in the non-linear regime, as shown in Figure 12.6b [16]. This feature is commonly attributed to the multiband nature of $SrTiO_3$ [11, 30, 38].

The attribution of non-linear Hall effect to the existence of multiple bands cannot explain the upturn of $R_H$ at smaller fields. Gunkel et al. [16] postulated an additional anomalous Hall effect (AHE) with Langevin-type dependence on the external magnetic field [16]. While the physical origin for this AHE was not identified, it was considered to be a manifestation of some magnetizable component in the system, e.g. due to the release and alignment of the spin degree of freedom.

Using a different configuration, namely a large in-plane magnetic field with small out-of-plane component, as shown in the inset of Figure 12.6c, Joshua et al. [15] reported a relatively sharp onset of an anomalous Hall effect a critical in-plane field $H_c^{\parallel}$ that depends strongly on carrier concentration $n_e$. This critical field diverges near the Lifshitz density $n_L$: $H_c^{\parallel} \sim \left(n_e - n_L\right)^{-1}$, increasing from 2 T to values as large as 14 T. This offset in Hall resistance was interpreted as some type of magnetization that has been "released" above a density-dependent critical magnetic field.

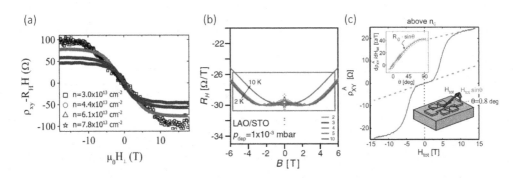

**FIGURE 12.6**    Transport metamagnetism: non-linear and anomalous Hall effect at the $LaAlO_3$/$SrTiO_3$ interface. (a) Commonly reported nonlinearity in the Hall resistance: with standard Hall geometry and an out-of-plane external magnetic field, the magnitude of the Hall coefficient $R_H$ (the slope of the Hall resistance $R_{xy}$) decreases with the increased field and settles at some value smaller than the zero field $R_H$. (After Ben Shalom, M. et al., *Phys. Rev. Lett.* 104, 126802, 2010. Copyright the American Physical Society. With permission.) (b) Hall coefficient $R_H$ as function of applied field. While the overall parabolic-like shape can be explained by the multiband nature of $SrTiO_3$, the small upturn around zero field cannot (adapted from [16]). (c) When applying a large in-plane field with a small out-of-plane field, the dependence of the Hall component on the Hall resistivity can be manipulated with the large in-plane field. There is a sudden release of the magnetic moments whenever the field exceeds a critical value. (After Joshue, A. et al., *Proc. Natl. Acad. Sci.* 110, 9633, 2013. With permission.)

### 12.2.2.3 Rashba Spin-Orbit Coupling

Strong gate-tunable Rashba spin-orbit coupling was first reported by Caviglia et al. [39] and Ben Shalom et al. [7]. Due to the polar nature of the LaAlO$_3$ layer, there is a built-in electric field in the direction perpendicular to the interface (along the c-axis for [001] the SrTiO$_3$). This field breaks inversion symmetry and introduces Rashba spin-orbit coupling (Figure 12.7) that mixes the $d_{xy}$ band and $d_{xz}$ / $d_{yz}$ bands and spin-splitting. The spin-splitting is at its largest (several meV) near the avoided crossings of the bands.

Rashba spin-orbit coupling was probed via magnetotransport. As the gating voltage increases, Caviglia et al. found that the magnetoconductance changes from positive (associated with weak localization and negligible Rashba spin-orbit coupling) to negative (associated with weak antilocalization and strong Rashba spin-orbit coupling). By fitting the magnetoconductance to the Maekawa–Fukuyama theory of spin-relaxation, the coupling

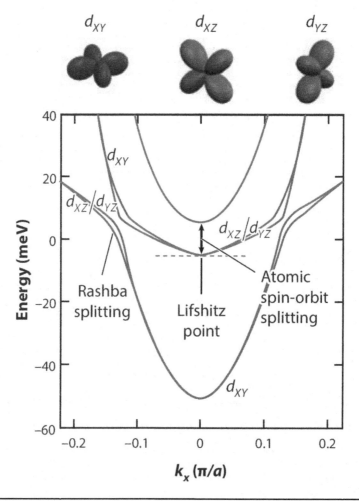

**FIGURE 12.7** Band structure of Ti 3d t$_{2g}$ orbitals. The 3d orbitals of Ti split into a less energetic triplet t$_{2g}$ ($d_{xy}$, $d_{yz}$, $d_{xz}$) and a more energetic doublet $e_g\left(d_{3z^2-r^2}, d_{x^2-y^2}\right)$. $d_{xy}$ has the lowest energy of the three t$_{2g}$ orbitals, and is populated first. As the carrier concentration increases, $d_{yz}$ / $d_{xz}$ start to be populated at Lifshitz point ($n_L$).

strength was found to be strongly dependent on gate, with a sharp rise near the gating voltage corresponding to insulator to superconductor transition of the same sample.

Ben Shalom et al. [36], on the other hand, reported a field-independent longitudinal resistance for in-plane magnetic fields greater than the superconducting upper critical field $H_{c2}^{\parallel}$ but below a larger critical field $H^*$, followed by a significant drop of the resistance above $H^*$. Ben Shalom et al. [36] associated $H^*$ with a spin-orbit field, and obtain a dependence of Rashba spin-orbit coupling on gate voltage opposite to that reported by Caviglia et al. [39].

A current-driven Rashba field was reported by Narayanapillai et al. [40]. Using strong Rashba spin-orbit coupling at $LaAlO_3/SrTiO_3$ for spin-charge conversion has been demonstrated recently by Lesne et al. [41]. The reported spin-charge conversion efficiency is comparable to non-2D materials such as Pt, Ta, W [41]. Additionally, a sign change of the Rashba coefficient is also reported [41].

Attempts to resolve the spin-dependent band structure using spin-polarized ARPES (SARPES) have been made, with conflicting reports. A large spin-splitting of 90 meV was reported by Santander-Syro et al. on doped $SrTiO_3$ [42], followed by a null result reported by Walker et al. [43], and a possible reconciliation of the two contradictory reports [44].

### 12.2.2.4 Electron Pairing Without Superconductivity

The superconducting state is associated with a variety of non-trivial magnetic properties. A conventional (type I) superconductor is a perfect diamagnet and the superconducting state consists of a condensate of Cooper pairs, which (in most cases) have a spin-singlet configuration. As will be described below, $LaAlO_3/SrTiO_3$ exhibits a robust phase in which electrons remain paired, but are not superconducting. The pairing transition itself involves significant changes in magnetization and orbital character that are associated with the complex metamagnetic behavior observed in this system.

Superconductivity in $SrTiO_3$ was first reported in 1964 by Schooley et al [45]. Apart from being one of a few semiconductors to exhibit superconducting behavior, it was suspected that its nature might be different from conventional BCS superconductivity. For example, the superconducting "dome" observed for $SrTiO_3$ [46] bears a striking resemblance to high-temperature superconductors, which were discovered two decades later [2]. One mysterious and controversial phase of high-temperature superconductors is the so-called "pseudogap" regime in which superconductivity is not observed, but a gap in the single-particle spectrum remains, as observed through tunneling experiments and other measurements [47]. While it is far from clear that $SrTiO_3$ shares a similar pairing mechanism—it lacks, for example, an antiferromagnetic parent phase that is believed to contribute to pairing in high-$T_c$ compounds—planar tunneling experiments on $LaAlO_3/SrTiO_3$ structures by Richter et al. [48] show a pseudogap feature past the boundary of the superconducting dome. The physical origin of the pseudogap phase in high-$T_c$ compounds continues to be debated. In particular, a seemingly straightforward question—whether the pseudogap phase is related to pre-formed Cooper pairs—has been challenging to resolve experimentally in the cuprates.

Cheng et al. [49] investigated the phenomenon of electron pairing using a single-electron transistor (SET) geometry (Figure 12.8a). A SET is a three-terminal device with a quantum dot (QD) that is tunnel-coupled to source and drain leads, with a gate that can change the number of carriers in the QD [50]. The conductance of a conventional SET is generally low if the energy of the lowest available state in the QD is not aligned with the chemical potential of the leads. When there is such an alignment, the energy of the state with $N$ and $N+1$ electrons in the QD becomes degenerate, and electrons can tunnel resonantly through the device. Cheng et al. created a SET device using conductive-AFM (c-AFM) at the LaAlO₃/SrTiO₃ interface (Figure 12.8a). They found that the differential conductance of the SET (Figure 12.8b) departed significantly from what one would expect for a SET formed from ordinary semiconductors. Three distinct regimes of behavior were observed as a function of applied magnetic field (out of plane): (i) at very low magnetic fields (below $|B| < \mu_0 H_{c2} \sim 0.1\,\text{T}$), a Josephson supercurrent (marked SC in Figure 12.8b) is found to flow through the device; (ii) at intermediate fields ($\mu_0 H_{c2} < |B| < B_P \sim 2-11\,\text{T}$), a series of vertical (magnetic field-independent) lines are observed; (iii) $|B| > B_P$, the vertical lines bifurcate, leading to a doubling of the number of conductance peaks which Zeeman shift as the magnetic field is further increased. Significantly, while $B_p$ varies from device to device, it generally is found to increase as the carrier

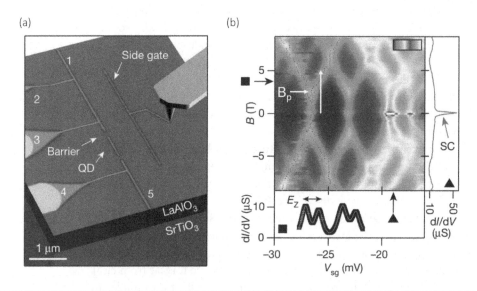

**FIGURE 12.8**  Electron pairing without superconductivity. Using a c-AFM lithography-defined single electron transistor, Cheng et al. [49] reported a phase of LaAlO₃/SrTiO₃ in which the electrons are paired, but not condensed into the superconducting phase. (a) Schematics for the c-AFM lithography process and the single electron transistor device, consisting of a quantum dot (QD) and a sidegate . (After Cheng, G. et al., *Nature* 521, 196, 2015. With permission.) (b) Evolution of the pairing transition as a function of external out-of-plane magnetic field and sidegate voltage. (After Cheng, G. et al., *Nature* 521, 196, 2015. With permission.)

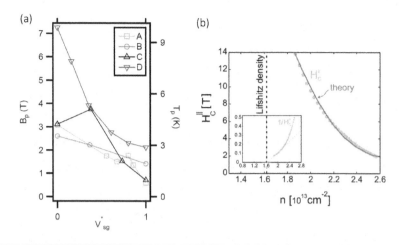

**FIGURE 12.9**    Metamagnetism and pairing transition. (a) Pairing field ($B_p$) as a function of voltage on sidegate ($V_{sg}$) (replotted from Ref [49]). (b) Density-dependent critical field $H_c^\parallel$ that marks the onset of AMR, AHE, and giant negative magnetoresistance. (After Joshue, A. et al., *Proc. Natl. Acad. Sci.* 110, 9633, 2013. With permission.)

density decreases (Figure 12.9a). Furthermore, this value is more than an order of magnitude greater than the upper critical field for the superconductivity, $\mu_0 H_{c2}$ (Figure 12.9a), and for temperatures that greatly exceed the highest observed superconducting state in SrTiO$_3$. That is to say, it is a state in which there is *electron pairing without superconductivity*.

## 12.2.3 Coexistence of Magnetism and Superconductivity

### 12.2.3.1 Superconductivity at LaAlO$_3$/SrTiO$_3$ Interface

The 2DEG exhibits superconductivity [51, 52], at low carrier concentrations ($n_e \sim 10^{13}$ cm$^{-2}$) and critical temperatures in the range $T \approx 200 - 300$ mK [51]. Like most other characteristics of this interface, the superconducting properties can be tuned by a backgate [52]. By tracing out the critical temperature as a function of gating voltage (and therefore carrier concentration), a dome-shaped outline enclosing the superconducting phase is seen, similar to high-$T_c$ superconductors as well as the doped bulk SrTiO$_3$.

### 12.2.3.2 Magnetic effects

While a true spatial coexistence of ferromagnetism and superconductivity at LaAlO$_3$/SrTiO$_3$ is not confirmed, coexistence at the sample level is reported by Li et al. [17] and Dikin et al. [53] and within the resolution of the scanning SQUID (~3 μm) by Bert et al. [18]. Dikin et al. [53] observed a clear hysteresis in magnetotransport measurements as a function of external magnetic field by mapping the critical temperature as a function of out-of-plane magnetic field while controlling the temperature of the sample to maintain a constant resistance near the superconducting transition. If, alternatively, the temperature is kept constant, hysteresis is seen in both the sheet resistance $R_s$ and the Hall resistance $R_{xy}$. The interplay of magnetic effects and superconducting state was later also probed by Ron et al. [54] in shadow-masked one dimensional nanowire with width ~50 nm, comparable to the superconducting coherence length. Ron reported a critical field $H_s$ (8mT to 12.5 mT) at the onset of a drop in the in-plane (both parallel and perpendicular to the current) magnetoresistance. Hysteresis loops in the magnetoresistance are reported for all directions with magnetization field close to $H_s$.

### 12.2.4 NULL RESULTS, INCONSISTENCIES AND ARTIFACTS

#### 12.2.4.1 Artifacts

The first report of magnetism at the LaAlO$_3$/SrTiO$_3$ interface [4], in addition to reporting a Kondo resistance minimum (Figure 12.9a), also shows a hysteretic magnetoresistance curve that depends on the direction of the sweeping of the external field (Figure 12.2a), as well as the sweep rate. The hysteresis is stronger at lower temperatures, or at higher sweep rate. After subsequent analyses, the authors determined that the hysteresis loops in those magnetoresistance curves can be ascribed to a magnetocaloric effect within the ceramic chip carrier used to hold the sample [55]. On the other hand, the other magnetic signature reported in [4], the Kondo resistance minimum, is not susceptible to this magnetocaloric effect.

#### 12.2.4.2 Null Results

While the 2DEG of LaAlO$_3$/SrTiO$_3$ is teeming with evidence of magnetism, there have been reports that are seemingly at odds, either quantitatively or qualitatively, with some of the primary reports. Using polarized neutron reflectometry, Fitzsimmons et al. [56] found essentially no evidence for magnetism in a variety of LaAlO$_3$/SrTiO$_3$ samples grown by two independent groups. The magnetization, which is calculated from the asymmetry in specular reflectivity as a function of wave-vector transfer and neutron beam polarization (Figure 12.2c), is close to the signal-to-noise limit of the measurement.

#### 12.2.4.3 Scanning SQUID

Follow-up scanning SQUID observations at the LaAlO$_3$/SrTiO$_3$ interface [18–20, 28] by Wijnands [57] failed to reproduce observed magnetic patches (Figure 12.2b). The experiments were performed using a scanning SQUID microscope of similar design and sensitivity, with samples prepared under nominally identical conditions, as Bert et al. [18].

## 12.3 THEORIES OF MAGNETISM AT LaAlO$_3$/SrTiO$_3$ INTERFACE

Theories of magnetism at the LaAlO$_3$/SrTiO$_3$ interface usually begin with a theory of the ferromagnetic phase. This phase was first attributed, by Pentcheva and Pickett [58], to the presence of oxygen vacancies at the interface that localize electrons in nearby Ti $3d$ states. Similar oxygen vacancies at the surface of SrTiO$_3$ may also introduce ferromagnetic order [59]. These localized states are located either in the Ti $d_{xy}$ orbitals [60, 61], which lie very close to the interface and are thus easily localized by interface defects, or the Ti $e_g$ orbitals by a restructuring of the Ti $3d$ orbitals near oxygen vacancies [62–64]. In the former case, the ferromagnetic state is assumed to arise from oxygen vacancies introducing localized magnetic moments, while in the latter case, the localized $e_g$ electrons interact with itinerant electrons leading to Stoner magnetism.

With the discovery of localized ferromagnetic patches separated by a paramagnetic phase, additional detail was necessary. By taking the view that oxygen vacancies lead to an orbital reconstruction in nearby Ti, Pavlenko et al. [64] showed that ferromagnetism induced by oxygen vacancies would only occur above a critical vacancy density. This would then allow for the phase separation of superconductivity and ferromagnetism [65], which has alternatively been explained by a spiral-spin state in the $d_{xz}$ / $d_{yz}$ bands by Banerjee et al. [66], while superconductivity is present in the $d_{xy}$ band. The patches [18–20, 28] are possibly the broken spirals due to defects [66, 67]. However, in this picture, the spiral states are made possible by the Rashba spin-orbit coupling, whose magnitude can be controlled via electric field effect [36, 39], the spiral states and, accordingly, ferromagnetic patches, are expected to be tunable via the electric field effect [66, 67], but this is not observed by Bert et al. [28].

Alternative explanations of the coexistence of superconductivity and ferromagnetism rely upon a similar phase-separation picture. Michaeli et al. [61] argue that Fulde–Ferrell–Larkin–Ovchinnikov (FFLO) pairing can occur in the $d_{xz}$ / $d_{yz}$ bands while the ferromagnetism arises from a Ruderman–Kittel–Kasuya–Yosida (RKKY) interaction amongst localized $d_{xy}$ electrons. Alternatively, Fidkowski et al. [68] argue that superconductivity in the interface is of a hybrid $s$- and $p$-wave pairing induced by the proximity effect with superconducting grains in the SrTiO$_3$ bulk, while the ferromagnetism arises from Kondo interactions between localized moments and itinerant electrons, to which the hybrid pairing is insensitive.

In addition to the coexistence of ferromagnetism and superconductivity, there have also been several theoretical attempts to explain anisotropic magnetic effects. Fischer et al. [69] shows that Rashba spin-orbit coupling in a system with Stoner magnetism can result in an anisotropic spin susceptibility, as well as a nematic phase arising from unequal occupation of the $d_{xz}$ / $d_{yz}$ bands. Also using spin-orbit coupling, Fete et al. [12] presented a model where anisotropic magnetoconductance arises from an orbital reconstruction in which, at low carrier densities, the $d_{xy}$ band is responsible for the electron transport, and, as density increases, the $d_{xz}$ / $d_{yz}$ bands begin to dominate and introduce anisotropic effects due to their nearly one-dimensional nature at the interface.

A recent attempt to explain many of these features simultaneously has been presented by Ruhman et al. [70]. In this theory, the magnetic features arise as a result of competition between Kondo screening in the $d_{xy}$ band and RKKY interactions in the $d_{xz}$ / $d_{yz}$ bands. In addition to having a paramagnetic (dominated by Kondo screening) and a ferromagnetic (dominated by localized magnetic moments/ordering of the RKKY interaction), there is also a spin-glass phase where the localized magnetic moments are insufficient to order the random RKKY interactions in the $d_{xz}/d_{yz}$ bands.

More recently, this many-body explanation of the magnetic effects in LaAlO$_3$/SrTiO$_3$ interfaces has been questioned due to the temperature and density dependence of the giant magnetoresistance. In the case of Kondo screening, this effect should be very sensitive to thermalization effects, but this is not observed experimentally. Thus, Diez et al. [31] put forward a model based upon the quasiclassical Boltzmann diffusion equation with spin-orbit coupling and band anisotropy to explain this phenomenon as a single-particle effect.

## 12.4 MAIN THESIS: TWO DISTINCT CLASSES OF MAGNETIC BEHAVIOR IN LaAlO$_3$/SrTiO$_3$

Here we argue that the collection of experimental evidence for magnetism at the LaAlO$_3$/SrTiO$_3$ interface can be meaningfully sorted into two categories. The first is a *ferromagnetic* or Kondo liquid phase derived from local magnetic moments that are coupled via exchange interactions with itinerant electrons (Figure 12.10b). The ferromagnetic phase is favored

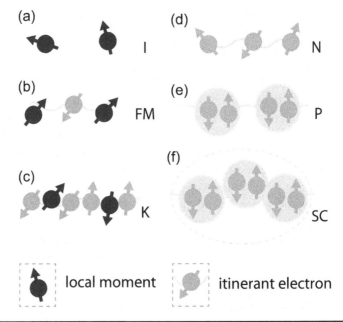

**FIGURE 12.10** Schematics for various phases for LaAlO$_3$/SrTiO$_3$. (a) Insulating state (I): the concentration of itinerant electrons is not high enough to support conduction at the interface. (b) Ferromagnetic (FM): local moments are antiferromagnetically coupled to the itinerant electrons. When the concentration of the itinerant electrons is high enough to support long-range ferromagnetic order, but not too high to screen the local moments. (c) Kondo-screened (K) phase: when the concentration of itinerant electrons is further increased, the spins of itinerant electrons are coupled to localized moment. (d) Normal state (N): when the itinerant electrons are sufficient to support conduction at the interface. (e) Paired state (P): where pairs of electrons are formed. (f) Superconducting (SC) phase.

when the interface is insulating, weakly conductive, or exhibiting locally insulating regions. The Kondo regime is favored when the density of itinerant electrons exceeds the density of local magnetic moments. We argue this is the magnetic phase that has been observed by scanning SQUID [18–20], SQUID [5, 9], cantilever magnetometry [17], MFM [21, 22], and MCD [71].

The second type of magnetism is a *metamagnetic* phase, resulting from an agglomeration of effects from the band structure of the $d$ manifold of the Ti atoms in $SrTiO_3$, spin-orbit coupling, and electronic pairing. This second type of magnetism manifests itself in a variety of magnetotransport signatures including magnetoresistance anisotropy [5, 7–12], sign changes of the magnetoresistance [15, 31], and anomalous Hall effects [11, 12, 14, 15]. Notably, these effects do not appear to interact with localized magnetic moments. Regarding interactions between ferromagnetism and superconductivity, we argue that the evidence suggests that they may indicate proximity effects, but not necessarily true coexistence [72].

## 12.5 FERROMAGNETIC AND KONDO REGIMES

Ferromagnetism at the $LaAlO_3/SrTiO_3$ interface results from localized moments and their antiferromagnetic (RKKY) exchange with itinerant electrons [73]. The detail of the exchange interaction depends on the concentration of the itinerant carriers, as well as the concentration of localized moments. Variations involving superexchange [66] and double exchange have also been described theoretically [74, 75]. As the local moments are antiferromagnetically exchange-coupled to itinerant electrons, ferromagnetic ordering between the local moments is expected [24, 76] as long as the density of local moments $n_m$ exceeds that of the itinerant electrons $n_e$. The ferromagnetic order disappears as the interface become conductive. The dynamic magnetic screening of magnetization by itinerant carriers was directly observed by Bi et al. [21] using top-gated $LaAlO_3/SrTiO_3$ structures [21]. In the regime where $n_e > n_m$, we expect Kondo physics to dominate (Figure 12.10c).

The fact that ferromagnetism can be tuned electronically is crucial for reconciling many of the apparent inconsistencies—including null results—described earlier. For example, the experiments of Fitzsimmons [56] did not actively control electron density at the interface (Figure 12.11a) and may have taken place in the Kondo regime. Inconsistencies in observations of ferromagnetic patches—either the patches themselves or sample-to-sample variations—can possibly be reconciled in two ways. If a sample is locally insulating, or simply has an excess of magnetic moments relative to the itinerant carrier density ($n_e < n_m$), then a ferromagnetic phase is favored. Local variations in the impurity density (or patches of magnetic moments) could fulfill this condition. Alternatively, local variations in the itinerant carrier density—with magnetic-impurity density presumably held constant—could produce a local ferromagnetic region. Surface adsorbates are known to strongly influence the local electron density [77–79].

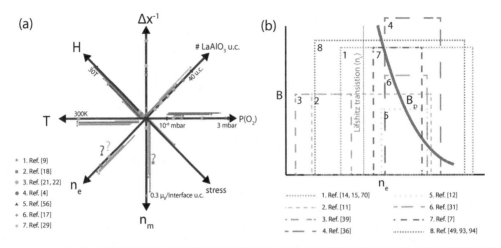

**FIGURE 12.11** Approximate parameter space and phase diagrams. The parameter space explored by various research groups discussed in the main text. (a) Ferromagnetism. The knobs commonly tuned, are: magnetic field (B), temperature (T), carrier concentration ($n_e$), moment concentration ($n_m$), mechanical stress, oxygen partial pressure for sample growth (P(O$_2$)), number of LaAlO$_3$ layers (# LaAlO$_3$ u.c.), and resolution ($\Delta\times^{-1}$). We argue that the null results reported by Fitzsimmons et al. [56] could potentially be reconciled if the carrier concentration is controlled. (b) Metamagnetism parameter space explored by various works discussed in the text, and the carrier concentration dependent critical field $B_p$ for the metamagnetic transition.

The magnetic patches were not tuned within the backgate range explored (–70 V to 390 V) by Bert et al. [28], suggesting that the local density of the itinerant electrons $n_e$ and magnetic moments $n_m$ inside the magnetic patches must be substantially different from the rest of the sample and are therefore difficult to control via backgating, unlike the gate-tuned superfluid density for the rest of the sample. This explanation, however, does not seem to be compatible with the fact that no clear phase competition between the superfluid density and ferromagnetic patches is observed [80]. The independence of the magnetic effect on the backgate is also reported by Ron et al. [54] in shadow-mask created 1D nanowires. One possible explanation is that magnetic ordering exists at the intrinsically insulating boundary of the nanowire, which makes it harder to manipulate using the electric field effect. However, the low critical temperature (1 K) reported by Ron et al. [54] suggests a different origin or regime altogether.

## 12.5.1 WHAT ARE THE LOCAL MOMENTS?

Perhaps one of the outstanding questions relates to the physical origin of the local moments. They can either be extrinsic in nature, e.g. derived from the interface, or simply magnetic impurities. Alternatively, they can have an intrinsic origin, such as defect states from vacancies of composite atoms, cation substitution, or interdiffusion [81]. Here we discuss both possibilities and then narrow the discussion to the case for oxygen vacancies.

### 12.5.1.1  Magnetic Impurities, Oxygen Vacancies (Vacancy Clusters), and F-Centers

Dilute magnetic impurities are generally found in $SrTiO_3$ substrates. Impurity levels in the order of one part per million have been reported [82–84], with large vendor-dependent variations [84]. For impurity concentrations of this level, it is possible to induce magnetism, as in the case of many dilute magnetic semiconductors [85]. However, observation of magnetic signals depends strongly on growth conditions, such as the oxygen partial pressure during growth of $LaAlO_3$ [9, 86], post-growth annealing temperature, and oxygen partial pressure [24, 25, 87]. This makes oxygen related defects such as vacancies and clusters the most common suspects. Both of these growth "knobs" can also tune the electron density. As mentioned, ferromagnetic hysteresis was only observed by Ariando [9] for samples with high oxygen pressure $(P(O_2) = 10^{-2}$ mbar), close to the metal-insulator transition. The more highly conductive samples grown at lower oxygen pressures did not show ferromagnetic hysteresis.

Oxygen vacancies are intrinsic, pervasive point defects for $SrTiO_3$. Oxygen vacancies and vacancy clusters modify the band structure and create deep, localized in-gap states [59, 63], at, or in the vicinity of, defect sites. Each oxygen vacancy, in principle, donates two electrons. Depending on the configuration of vacancy clustering, these electrons can remain localized on the vacancy site, or fill the $d$ orbitals of nearby Ti atoms. The defect sites with unpaired electrons are referred to as F-centers or color centers. Both F-centers and Ti 3d orbital can possess a local moment [88, 89].

Strong evidence linking magnetic moments to oxygen vacancies comes from a report by Rice et al. [71] which used magnetic circular dichroism (MCD) to optically excite long-lived magnetic states of reduced $SrTiO_3$ single crystals. The sample was magnetized using a circularly polarized pump (Figure 12.4a) and probed via differential absorption of right- and left-circularly polarized light. The observed magnetization persists after the pumping beam is blocked, enabling long-lived magnetic states to be imprinted (Figure 12.4b). Samples doped with niobium did not show magnetization induced by MCD until oxygen vacancies were subsequently induced. The samples with oxygen vacancies did not exhibit magnetization after being re-oxygenated. A similar dependence on oxygen vacancies was also reported by Liu et al. [25] using SQUID magnetometry.

Following Rice et al. [71], Choi et al. [59] provided DFT calculations for in-gap states induced by oxygen vacancies of different valence, $V_O$, $V_O^+$, and $V_O^{2+}$, consistent with Rice et al [71]. Moments due to unpaired spins of Ti are only induced from $V_O$ and $V_O^+$. Furthermore, $V_O^{2+}$ dominate the population in the bulk, suggesting that magnetic moments may cluster around interfaces or ferroelastic domain boundaries. Calculations by Cuong et al. [90] suggest that vacancies tend to cluster into apical divacancy pairs, creating in-gap state of $e_g$ character. Altmeyer et al. [91] discuss magnetic moments of Ti $3d$ as a function of distance between the Ti atom and single oxygen vacancy or apical divacancies.

**TABLE 12.1**
**Parameter Space Explored by the Works Discussed in the Text**

| | Max Field (T) | Temperature (K) | $n_e$ ($10^{13}$ cm$^{-2}$) | Backgate Voltage (V) | #LaAlO$_3$ Unit Cells (u.c.) | Device Size/ Dimensionality | P(O$_2$) (mbar) | Post Anneal? |
|---|---|---|---|---|---|---|---|---|
| Joshua et al. [15] | 14 T | 50 mK, 2 K, 4.2 K | 1–3 | −50 to 450 | 6, 10 | mm/2D | 10$^{-4}$ | o |
| Ben Shalom et al. [36] | 18 T | 20 mK-100 K | 0.1–5 | −50 to 50 | 8, 15 | mm/2D | 5 × 10$^{-4}$ – 10$^{-3}$ | x |
| Caviglia et al. [39] | 12 T | 30 mK - 20 K | 1.85–2.2 | −340 to 320 | 12 | mm/2D | 6 × 10$^{-5}$, 10$^{-4}$ | o |
| Fete et al. [12] | 7 T | 1.5 K | | | >4 | mm/2D | 10$^{-4}$ | o |
| Ariando et al. [9] | 9 T | 2 K–300 K | | | 10 | mm/2D | 10$^{-6}$ –5 × 10$^{-2}$ | x |
| Ron et al. [54] Maniv et al. [104] | 18 T | 20 mK, 60 mK | | −5 to 10 | 10, 6, 16 | 50 nm mm/2D | 10$^{-4}$ (Torr) | o |
| Cheng et al. [49] | 9 T | 50 mK | | | 3.4 | nm/1D | 10$^{-3}$ | o |

## 12.6 METAMAGNETISM

Here we describe a complementary set of effects, most of which are related to transport, that we label with a single term "metamagnetism". These effects generally take place well below room temperature (~35 K) but above the superconducting transition. Many different effects take place within the same range of tunable parameters (Figure 12.11b and Table 12.1); we argue below that these effects are manifestations of the same underlying phenomena.

### 12.6.1 LIFSHITZ TRANSITION

Among all the reports from giant negative magnetoresistance, anisotropy in magnetoresistance and the anomalous Hall effect, the Lifshitz transition appears to play a central role. Its importance within the context of magnetotransport phenomena was first stressed by Joshua et al. [14]. The critical (Lifshitz) density (referred as $n_L$ in the following text) is where the $d_{xy}$-derived branches of the $t_{2g}$ bands are partially filled, and the $d_{xz}$/$d_{yz}$-derived bands begin to be populated (Figure 12.7). Giant negative magnetoresistance, magnetotransport anisotropy, and anomalous Hall effect are all observed only above the Lifshitz transition. The magnetoresistance near the Lifshitz transition evolves from being weak and weakly-dependent on angle, to strong and highly anisotropic, with a crossover that depends on both magnetic field and electron density. The Lifshitz transition is also strongly linked to an observed tunable spin-orbit coupling [39] and takes place close to the maximum superconducting transition temperature [14].

### 12.6.2 ANISOTROPIC TRANSPORT VERSUS N, B

As discussed previously, the anisotropy of the magnetoresistance undergoes a vivid transition from sinusoidal 2-fold oscillation (Figure 12.5b) to a four-fold, or irregular oscillation (Figure 12.5c), when the carrier concentration $n_e$ is increased above the Lifshitz transition $n_L$. This transition between low and high anisotropy is also found to be a function of the external magnetic field (Figure 12.5d,e). The critical field for this transition depends on the carrier concentration, and a phase diagram (Figure 12.9b) was mapped out by Joshua et al. [15], showing that the critical magnetic field can vary from ~2 T to as large as 15 T as the carrier density is reduced to $n_L$. In separate measurements, Diez et al. [31] showed a large negative magnetoresistance that exhibits a strong backgate dependence, becoming much more pronounced at large backgate values (Figure 12.12b). In a separate magnetotransport study of Ben Shalom et al. [36] (Figure 12.12a), a large negative magnetoresistance effect takes place at a critical magnetic field $\mu_0 H^*$ that ranges from 2 T to >18 T as the carrier density is reduced from $n = 7.8 \times 10^{13} \, \text{cm}^{-2}$ to $3.0 \times 10^{13} \, \text{cm}^{-2}$. It is worth noting that all of these experiments are probing a similar range of magnetic field and carrier density, and all of them are observing two distinct phases that are separated by a curve similar to that shown in Figure 12.11b and Table 12.1.

### 12.6.3 ANOMALOUS HALL EFFECT

As mentioned previously, there is also an critical carrier density $n_L$ as well as a carrier $n_e$ dependent critical in-plane $H_c^{\parallel}$ and out-of-plane field $H_c^{\perp}$, as mapped out by Joshua et al. [15]. The (model-independent) interpretation is that the offset and increase in Hall angle represents an excess magnetization that is "released" above a critical magnetic field, indicating a "gate-tunable polarized phase" [15]. Now the question becomes: what is this magnetizable degree of freedom, and what causes the sudden release as the field is increased? This magnetization does not behave as a single-component Langevin-type paramagnetism, because it lacks the anticipated temperature dependence, i.e. *it is temperature-independent* [16]. As discussed, by adding an anomalous Hall term with Langevin-type paramagnetic field dependence, Gunkel et al. [16] was able to capture the small upturn in the Hall coefficient $R_H$ missed by the simple multiband model. However, Gunkel et al. [16] also pointed out that (i) a temperature-independent saturation field $B_C$ for the Hall coefficient $R_H$, and (ii) a temperature dependent $R_0^{AHE}$ deviates from the expected scaling of the Langevin-type spin-1/2 paramagnetic system.

### 12.6.4 CORRELATED TRANSPORT PROPERTIES

The anisotropic transport, anomalous Hall effect, and electron pairing without superconductivity all share features that change with carrier density and magnetic field in similar ways (see Figure 12.11b). Hence it is natural to consider the possibility that they are different manifestations of the same underlying phenomena. Here we discuss this possibility explicitly. First, let us consider the AHE and its connection to electron pairing without superconductivity in a SET reported by Cheng et al. [49]. Based on the stability of the paired state conductance peaks with respect to magnetic field (region $|B| < B_P \approx 2 \, \text{T}$ in Figure 12.8b) it is known that the electron pairs exist in a spin-singlet configuration, i.e. $S_{\text{tot}} = 0$. In the unpaired state (i.e. for $B > B_P$), electrons are now unpaired, with spin states $|S, S_z\rangle = |1/2, \pm 1/2\rangle$ that can subsequently Zeeman split and polarize in the excess magnetic field $B - B_p$. This polarization of unpaired spins should produce an anomalous Hall effect, similar to what is observed in bulk 2D experiments. The phase diagram for $H_C^{\parallel}$ from Joshua et al. [15] (Figure 12.9b) shows a monotonically decreasing dependence on the carrier density $n_e$. In the SET experiments, pairing fields as large as 5 T were reported; in follow-up experiments on ballistic quantum channels, the pairing transition has been observed at magnetic fields as large as 11 T [92]. The SET measurements only probe pairing at discrete values of carrier density, which itself is not directly measurable on the QD. For a single device, a few (2–5) data points are clearly observable. Figure 12.9a shows the pairing energy for a set of devices, plotted against a normalized gate voltage. While the scaling is not always monotonic, there is a general trend that favors larger pairing fields (in a particular device) with lower electron density. This behavior (Figure 12.9b) is similar to the crossover curve separating regions of large (anisotropic) and small (isotropic) magnetoresistance and anomalous Hall effect.

Regarding the transition between low and high anisotropy in transport, again, the collective evidence indicates that the paired state is most likely composed of $d_{xy}$ carriers, which, near the Lifshitz transition, is the preferred ground state due to a presumed attractive interaction shared only between $d_{xy}$ carriers. The $d_{xz}$ and $d_{yz}$ carriers, which have highly anisotropic band structure, become occupied when the pairing energy is overcome by sufficiently large magnetic fields. As the electron density increases above the Lifshitz point, the energy difference between the (unoccupied) $d_{xz}$ / $d_{yz}$ states and the (occupied) $d_{xy}$ states reduces and eventually vanishes. That is to say, a monotonically decreasing pairing field is expected, regardless of the physical origin of the electron pairing itself. The crossover between attractive ($U < 0$) and repulsive ($U > 0$) regimes was observed by Cheng et al. [93] in transport measurements on devices similar to those depicted in Figure 12.8a, but at higher carrier density.

Differences between the paired and unpaired phases are also manifested as negative magnetoresistance effects, notably the results from Ben Shalom et al. [36] (Figure 12.12a). In that experiment, a critical field $H^*$ was identified as the onset for giant negative magnetoresistance. The value of $H^*$ changes from <2 T to >18 T as the carrier density decreases toward what appears to be the Lifshitz transition (although about 2x higher than the value measured by Joshua et al. [15]). This behavior can also be incorporated into a framework in which changes in resistance may arise due to qualitative differences in transport between electron pairs and unpaired electrons. At sufficiently low carrier density, the electron pairs may propagate ballistically [94] or become localized.

## 12.7 FUTURE DIRECTIONS

### 12.7.1 EMPIRICAL TESTS

How would one test the proposed picture of two magnetisms? Regarding ferromagnetism and related phases, it is necessary to establish the origin of the magnetic moments. As discussed previously, oxygen vacancies are prime

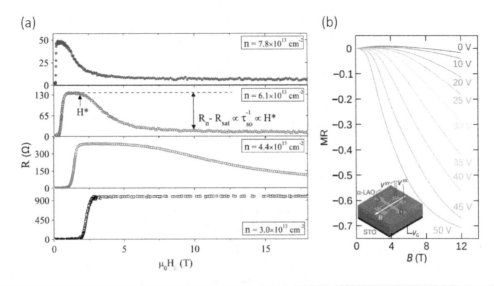

**FIGURE 12.12**   Giant negative longitudinal ($R_{xx}$) magnetoresistance. When an external magnetic field exceeds $H_c^{\parallel}$ and a carrier-concentration-dependent critical field $H^*$, the longitudinal magnetoresistance $R_{xx}$ drops significantly. (a) At $T = 20$ mK, the evolution of the sheet resistance is plotted as a function of in-plane magnetic field (parallel to the current) at different carrier concentration. The resistance is restored at $H_c^{\parallel}$, followed by a field-independent interval from $H_c^{\parallel}$ to $H^*$, and then a sudden drop at $H^*$. (After Shalom, B. et al., *Phys. Rev. Lett.* 104, 126802, 2010. With permission.) (b) At $T = 1.4$ K (above $T_c$), a drop of 70% in $R_{xx}$ is reported by Diez et al. (After Diez, M. et al., *Phys. Rev. Lett.* 115, 016803, 2015. With permission.)

suspects for producing the required magnetic moments. Engineering oxygen vacancies via growth conditions and/or annealing can control the concentration of local moments [87], but it is not clear where exactly they are stabilized and whether they are single vacancies or clusters. To capture the physics, more detailed experimental and theoretical studies need to be performed. Concerning experiments that probe ferromagnetism, it is important to be able to control the concentration of itinerant electrons, a parameter that has been shown to be critical to the stabilization of ferromagnetic phases [21]. Topgating, as opposed to backgating through the SrTiO$_3$ substrate, is more effective in controlling carrier density and reaching a properly insulating phase. Finally, spatially resolved measurements correlating oxygen vacancies and the magnetic moments, together with any long-range order they induce, could unequivocally establish the mechanism for dilute ferromagnetism at this interface.

## 12.7.2 WHERE ARE THE OXYGEN VACANCIES?

Here we turn specifically to the suspected role of oxygen vacancies. If they are indeed responsible for the magnetic moments, the question becomes: where exactly are these vacancies located? Are they pinned to the interfaces? Do they cluster around ferroelastic domain walls? While there is no direct experimental evidence showing that the oxygen vacancies are pinned within ferroelastic domain walls in SrTiO$_3$, in general, for perovskites, the movement of ferroelastic domain walls is restricted by defects, and could be easily pinned by them. Conversely, mobile oxygen vacancies may become trapped at ferroelastic domain walls, as pointed out by Goncalves-Ferreira et al. [95] for CaTiO$_3$ (Figure 12.4c), oxygen vacancies have lower energy when they are placed at a twin wall. Measurements using soundwaves, suggesting that domain walls are polar, has been reported by Scott et al. [96].

Ferroelastic domains in SrTiO$_3$ have been imaged in a variety of ways, including optical microscopy [97], scanning SET (Figure 12.4d), scanning SQUID (Figure 12.4e), and low temperature SEM [98]. Ferroelastic domains are found to modulate the flow of the current [99], as well as the critical temperature of the superconductivity [100], and they play an important role in the gating of essentially all LaAlO$_3$/SrTiO$_3$ heterostructures at low temperatures. Scanning SQUID experiments, (Figure 12.4f) in which the probe gently touches the LaAlO$_3$ surface, have been demonstrated to produce significant changes in the magnitude and direction of ferromagnetic patches [20]. The force magnitude (~1 µN) is too small to create damage or otherwise depart from the elastic regime; however, it is plausible that this kind of force may result in a rearrangement of ferroelastic domains and therefore the spatial arrangement of magnetic moments. Further experimentation would be necessary to establish a clearer connection and to relate the ferroelastic domain structure to the presence of stabilized magnetic moments.

### 12.7.3 WHAT IS THE ORIGIN OF ELECTRON PAIRING?

In all of the experiments related to electron pairing thus far, the picture appears to be that there is a strongly attractive pairing interaction (Hubbard $U < 0$) at low electron densities that is responsible for electron pairing and superconductivity. As the electron density increases, the sign of the electron-electron interaction changes ($U > 0$) and superconductivity is ultimately suppressed. The paired, non-superconducting phase, close to the Lifshitz transition, becomes the dominant mode of transport, and yet it is far from being a Fermi liquid. Perhaps the biggest open question relates to the underlying physical mechanism for electron pairing. So far, there is no clear experiment (or theory) that favors a particular mechanism. A deeper understanding of electron pairing (and, as a consequence, superconductivity) in SrTiO$_3$-based systems would represent an important milestone in our overall ability to describe the physics of one of the richest material systems known.

## ACKNOWLEDGMENTS

We would like to acknowledge Ariando, A. D. Caviglia, B. Kalisky, K. A. Moler, K. Nowack, and J. M. Triscone for helpful feedback on the manuscript. We thank NSF DMR (1124131, 1609519, 1124131) and ONR N00014-15-1-2847 for financial support.

# REFERENCES

1. M. L. Cohen, Superconductivity in low-carrier-density systems: Degenerate semiconductors, *Superconductivity: Part 1* **615** 1969.
2. J. G. Bednorz and K. A. Muller, Perovskite-type oxides-the new approach to High Tc superconductivity, *Nobel Lecture* (1987).
3. A. Ohtomo and H. Y. Hwang, A high-mobility electron gas at the $LaAlO_3$/$SrTiO_3$ heterointerface, *Nature* **427**, 423–426 (2004).
4. A. Brinkman, M. Huijben, M. Van Zalk et al., Magnetic effects at the interface between non-magnetic oxides, *Nat. Mater.* **6**, 493–496 (2007).
5. H.-L. Hu, R. Zeng, A. Pham et al., Subtle interplay between localized magnetic moments and itinerant electrons in $LaAlO_3$/$SrTiO_3$ heterostructures, *ACS Appl. Mater. Interfaces* 13630–13636 (2016).
6. M. Lee, J. R. Williams, S. Zhang, C. D. Frisbie, and D. Goldhaber-Gordon, Electrolyte gate-controlled Kondo effect in $SrTiO_3$, *Phys. Rev. Lett.* **107**, 256601 (2011).
7. M. Ben Shalom, C. W. Tai, Y. Lereah et al., Anisotropic magnetotransport at the $SrTiO_3$/$LaAlO_3$ interface, *Phys. Rev. B* **80**, 140403 (2009).
8. S. Seri and L. Klein, Antisymmetric magnetoresistance of the $SrTiO_3$/$LaAlO_3$ interface, *Phys. Rev. B* **80**, 180410 (2009).
9. X. Ariando, G. Wang, Z. Q. Baskaran et al., Electronic phase separation at the $LaAlO_3$/$SrTiO_3$ interface, *Nat. Commun.* **2**, 188 (2011).
10. X. Wang, W. M. Lu, A. Annadi et al., Magnetoresistance of two-dimensional and three-dimensional electron gas in $LaAlO_3$/$SrTiO_3$ heterostructures: Influence of magnetic ordering, interface scattering, and dimensionality, *Phys. Rev. B* **84**, 075312 (2011).
11. S. Seri, M. Schultz, and L. Klein, Interplay between sheet resistance increase and magnetotransport properties in $LaAlO_3$/$SrTiO_3$, *Phys. Rev. B* **86**, 085118 (2012).
12. A. Fête, S. Gariglio, A. D. Caviglia, J. M. Triscone, and M. Gabay, Rashba induced magnetoconductance oscillations in the $LaAlO_3$-$SrTiO_3$ heterostructure, *Phys. Rev. B* **86**, 201105 (2012).
13. A. Annadi, Z. Huang, K. Gopinadhan et al., Fourfold oscillation in anisotropic magnetoresistance and planar Hall effect at the $LaAlO_3$/$SrTiO_3$ heterointerfaces: Effect of carrier confinement and electric field on magnetic interactions, *Phys. Rev. B* **87**, 201102 (2013).
14. A. Joshua, S. Pecker, J. Ruhman, E. Altman, and S. Ilani, A universal critical density underlying the physics of electrons at the $LaAlO_3$/$SrTiO_3$ interface, *Nat. Commun.* **3**, 1129 (2012).
15. A. Joshua, J. Ruhman, S. Pecker, E. Altman, and S. Ilani, Gate-tunable polarized phase of two-dimensional electrons at the $LaAlO_3$/$SrTiO_3$ interface, *Proc. Natl. Acad. Sci.* **110**, 9633 (2013).
16. F. Gunkel, C. Bell, H. Inoue et al., Defect control of conventional and anomalous electron transport at complex oxide interfaces, *Phys. Rev. X* **6**, 031035 (2016).
17. L. Li, C. Richter, J. Mannhart, and R. C. Ashoori, Coexistence of magnetic order and two-dimensional superconductivity at $LaAlO_3$/$SrTiO_3$ interfaces, *Nat. Phys.* **7**, 762–766 (2011).
18. J. A. Bert, B. Kalisky, C. Bell et al., Direct imaging of the coexistence of ferromagnetism and superconductivity at the $LaAlO_3$/$SrTiO_3$ interface, *Nat. Phys.* **7**, 767–771 (2011).
19. B. Kalisky, J. A. Bert, B. B. Klopfer et al., Critical thickness for ferromagnetism in $LaAlO_3$/$SrTiO_3$ heterostructures, *Nat. Commun.* **3**, 922 (2012).
20. F. B. Kalisky, J. A. Bert, C. Bell et al., Scanning probe manipulation of magnetism at the $LaAlO_3$/$SrTiO_3$ heterointerface, *Nano Lett.* **12**, 4055–4059 (2012).
21. F. Bi, M. Huang, S. Ryu et al., Room-temperature electronically-controlled ferromagnetism at the $LaAlO_3$/$SrTiO_3$ interface, *Nat. Commun.* **5**, 5019 (2014).

22. F. Bi, M. Huang, H. Lee, C.-B. Eom, P. Irvin, and J. Levy, LaAlO$_3$ thickness window for electronically controlled magnetism at LaAlO$_3$/SrTiO$_3$ heterointerfaces, *App. Phys. Lett.* **107**, 082402 (2015).

23. J. S. Lee, Y. W. Xie, H. K. Sato et al., Titanium d$_{xy}$ ferromagnetism at the LaAlO$_3$/SrTiO$_3$ interface, *Nat. Mater.* **12**, 703–706 (2013).

24. M. Salluzzo, S. Gariglio, D. Stornaiuolo et al., Origin of interface magnetism in BiMnO$_3$/SrTiO$_3$ and LaAlO$_3$/SrTiO$_3$ heterostructures, *Phys. Rev. Lett.* **111**, 087204 (2013).

25. Z. Q. Liu, W. M. Lü, S. L. Lim et al., Reversible room-temperature ferromagnetism in Nb-doped SrTiO$_3$ single crystals, *Phys. Rev. B* **87**, 220405 (2013).

26. M. E. Huber, N. C. Koshnick, H. Bluhm et al., Gradiometric micro-SQUID susceptometer for scanning measurements of mesoscopic samples, *Rev. Sci. Instrum.* **79**, 053704 (2008).

27. S. Thiel, G. Hammerl, A. Schmehl, C. W. Schneider, and J. Mannhart, Tunable quasi-two-dimensional electron gases in oxide heterostructures, *Science* **313**, 1942–1945 (2006).

28. J. A. Bert, K. C. Nowack, B. Kalisky et al., Gate-tuned superfluid density at the superconducting LaAlO$_3$/SrTiO$_3$ interface, *Phys. Rev. B* **86**, 060503 (2012).

29. Z, Salman, O. Ofer, M. Radovic et al., Nature of weak magnetism in SrTiO$_3$/LaAlO$_3$ multilayers, *Phys. Rev. Lett.* **109**, 257207 (2012).

30. M. Ben Shalom, A. Ron, A. Palevski, and Y. Dagan, Shubnikov-De Haas oscillations in SrTiO$_3$/LaAlO$_3$ interface, *Phys. Rev. Lett.* **105**, 206401 (2010).

31. M. Diez, A. M. R. V. L. Monteiro, G. Mattoni et al., Giant negative magnetoresistance driven by spin-orbit coupling at the LaAlO$_3$/SrTiO$_3$ interface, *Phys. Rev. Lett.* **115**, 016803 (2015).

32. D. Stornaiuolo, C. Cantoni, G. M. De Luca et al., Tunable spin polarization and superconductivity in engineered oxide interfaces, *Nat. Mater.* **15**, 278–283 (2016).

33. A. Fête, C. Cancellieri, D. Li et al., Growth-induced electron mobility enhancement at the LaAlO$_3$/SrTiO$_3$ interface, *App. Phys. Lett.* **106**, 051604 (2015).

34. J. S. Kim, S. S. A. Seo, M. F. Chisholm et al., Nonlinear Hall effect and multichannel conduction in LaTiO$_3$/SrTiO$_3$ superlattices, *Phys. Rev. B* **82**, 201407 (2010).

35. P. Gallagher, M. Lee, T. A. Petach et al., A high-mobility electronic system at an electrolyte-gated oxide surface, *Nat. Commun.* **6**, 6437 (2015).

36. M. Ben Shalom, M. Sachs, D. Rakhmilevitch, A. Palevski, and Y. Dagan, Tuning spin-orbit coupling and superconductivity at the SrTiO$_3$/LaAlO$_3$ interface: A magnetotransport study, *Phys. Rev. Lett.* **104**, 126802 (2010).

37. Y. Lee, C. Clement, J. Hellerstedt et al., Phase diagram of electrostatically doped SrTiO$_3$, *Phys. Rev. Lett.* **106**, 136809 (2011).

38. C. Bell, S. Harashima, Y. Kozuka et al., Dominant mobility modulation by the electric field effect at the LaAlO$_3$/SrTiO$_3$ interface, *Phys. Rev. Lett.* **103**, 226802 (2009).

39. A. D. Caviglia, M. Gabay, S. Gariglio, N. Reyren, C. Cancellieri, and J. M. Triscone, Tunable Rashba spin-orbit interaction at oxide interfaces, *Phys. Rev. Lett.* **104**, 126803 (2010).

40. K. Narayanapillai, K. Gopinadhan, X. Qiu et al., Current-driven spin orbit field in LaAlO$_3$/SrTiO$_3$ heterostructures, *App. Phys. Lett.* **105**, 162405 (2014).

41. E. Lesne, Y. Fu, S. Oyarzun et al., Highly efficient and tunable spin-to-charge conversion through Rashba coupling at oxide interfaces, *Nat. Mater.* (advance online publication) (2016).

42. A. F. Santander-Syro, F. Fortuna, C. Bareille et al., Giant spin splitting of the two-dimensional electron gas at the surface of SrTiO$_3$, *Nat. Mater.* **13**, 1085–1090 (2014).

43. S. McKeown Walker, S. Riccò, F. Y. Bruno et al., Absence of giant spin splitting in the two-dimensional electron liquid at the surface of SrTiO$_3$ (001), *Phys. Rev. B* **93**, 245143 (2016).

44. A. C. Garcia-Castro, M. G. Vergniory, E. Bousquet, and A. H. Romero, Spin texture induced by oxygen vacancies in strontium perovskite (001) surfaces: A theoretical comparison between $SrTiO_3$ and $SrHfO_3$, *Phys. Rev. B* **93**, 045405 (2016).

45. J. F. Schooley, W. R. Hosler, and M. L. Cohen, Superconductivity in semiconducting $SrTiO_3$, *Phys. Rev. Lett.* **12**, 474–475 (1964).

46. C. S. Koonce, M. L. Cohen, J. F. Schooley, W. R. Hosler, and E. R. Pfeiffer, Superconducting transition temperatures of semiconducting $SrTiO_3$, *Phys. Rev.* **163**, 380–390 (1967).

47. T. Timusk and B. Statt, The pseudogap in high-temperature superconductors: an experimental survey, *Rep. Prog. Phys.* **62**, 61–122 (1999).

48. C. Richter, H. Boschker, W. Dietsche et al., Interface superconductor with gap behaviour like a high-temperature superconductor, *Nature* **502**, 528–531 (2013).

49. G. Cheng, M. Tomczyk, S. C. Lu et al., Electron pairing without superconductivity, *Nature* **521**, 196–199 (2015).

50. M. A. Kastner, The single-electron transistor, *Rev. Mod. Phys.* **64**, 849–858 (1992).

51. N. Reyren, S. Thiel, A. D. Caviglia et al., Superconducting interfaces between insulating oxides, *Science* **317**, 1196–1199 (2007).

52. A. D. Caviglia, S. Gariglio, N. Reyren et al., Electric field control of the $LaAlO_3$/$SrTiO_3$ interface ground state, *Nature* **456**, 624–627 (2008).

53. D. A. Dikin, M. Mehta, C. W. Bark, C. M. Folkman, C. B. Eom, and V. Chandrasekhar, Coexistence of superconductivity and ferromagnetism in two dimensions, *Phys. Rev. Lett.* **107**, 056802 (2011).

54. A. Ron, E. Maniv, D. Graf, J. H. Park, and Y. Dagan, Anomalous magnetic ground state in an $LaAlO_3$/$SrTiO_3$ interface probed by transport through nanowires, *Phys. Rev. Lett.* **113**, 216801 (2014).

55. V. K. Guduru, Surprising magnetotransport in oxide heterostructures, Master's thesis, Radboud University, Nijmegen, the Netherlands, 2014.

56. M. R. Fitzsimmons, N. W. Hengartner, S. Singh et al., Upper limit to magnetism in $LaAlO_3$/$SrTiO_3$ heterostructures, *Phys. Rev. Lett.* **107**, 217201 (2011).

57. T. Wijnands, Scanning superconducting quantum interference device microscopy, Master's thesis, University of Twente, Enschede, the Netherlands, 2013, http://essay.utwente.nl/62800/1/Master_Thesis_Tom_Wijnands_openbaar.pdf.

58. R. Pentcheva and W. E. Pickett, Charge localization or itineracy at $LaAlO_3$/$SrTiO_3$ interfaces: Hole polarons, oxygen vacancies, and mobile electrons, *Phys. Rev. B* **74**, 035112 (2006).

59. H. Choi, J. D. Song, K.-R. Lee, and S. Kim, Correlated visible-light absorption and intrinsic magnetism of $SrTiO_3$ due to oxygen deficiency: Bulk or surface effect? *Inorg. Chem.* **54**, 3759–3765 (2015).

60. R. Pentcheva and W. E. Pickett, Ionic relaxation contribution to the electronic reconstruction at the n-type $LaAlO_3$/$SrTiO_3$ interface, *Phys. Rev. B* **78**, 205106 (2008).

61. K. Michaeli, A. C. Potter, and P. A. Lee, Superconducting and ferromagnetic phases in $SrTiO_3$/$LaAlO_3$ oxide interface structures: Possibility of finite momentum pairing, *Phys. Rev. Lett.* **108**, 117003 (2012).

62. G. Chen and L. Balents, Ferromagnetism in itinerant two-dimensional $t_{2g}$ systems, *Phys. Rev. Lett.* **110**, 206401 (2013).

63. N. Pavlenko, T. Kopp, E. Y. Tsymbal, J. Mannhart, and G. A. Sawatzky, Oxygen vacancies at titanate interfaces: Two-dimensional magnetism and orbital reconstruction, *Phys. Rev. B* **86**, 064431 (2012).

64. N. Pavlenko, T. Kopp and J. Mannhart, Emerging magnetism and electronic phase separation at titanate interfaces, *Phys. Rev. B* **88**, 201104 (2013).

65. N. Pavlenko, T. Kopp, E. Y. Tsymbal, G. A. Sawatzky, and J. Mannhart, Magnetic and superconducting phases at the $LaAlO_3$/$SrTiO_3$ interface: The role of interfacial Ti 3d electrons, *Phys. Rev. B* **85**, 020407(R) (2012).

66. S. Banerjee, O. Erten, and M. Randeria, Ferromagnetic exchange, spin-orbit coupling and spiral magnetism at the LaAlO$_3$/SrTiO$_3$ interface, *Nat. Phys.* **9**, 626–630 (2013).

67. M. Gabay and J.-M. Triscone, Oxide heterostructures: Hund rules with a twist, *Nat. Phys.* **9**, 610–611 (2013).

68. L. Fidkowski, H. C. Jiang, R. M. Lutchyn, and C. Nayak, Magnetic and superconducting ordering in one-dimensional nanostructures at the LaAlO$_3$/SrTiO$_3$ interface, *Phys. Rev. B* **87** (2013).

69. M. H. Fischer, S. Raghu, and E. A. Kim, Spin-orbit coupling in LaAlO$_3$/SrTiO$_3$ interfaces: Magnetism and orbital ordering, *New J. Phys.* **15**, 023022 (2013).

70. J. Ruhman, A. Joshua, S. Ilani, and E. Altman, Competition between Kondo screening and magnetism at the LaAlO$_3$/SrTiO$_3$ interface, *Phys. Rev. B* **90**, 125123 (2013).

71. W. D. Rice, P. Ambwani, M. Bombeck et al., Persistent optically induced magnetism in oxygen-deficient strontium titanate, *Nat. Mater.* **13**, 481–487 (2014).

72. G. Cheng, J. P. Veazey, P. Irvin et al., Anomalous transport in sketched nanostructures at the LaAlO$_3$/SrTiO$_3$ interface, *Phys. Rev. X* **3**, 011021 (2013).

73. M. A. Ruderman and C. Kittel, Indirect exchange coupling of nuclear magnetic moments by conduction electrons, *Phys. Rev.* **96**, 99–102 (1954).

74. D. Odkhuu, S. H. Rhim, D. Shin, and N. Park, La displacement driven double-exchange like mediation in titanium dxy ferromagnetism at the LaAlO$_3$/SrTiO$_3$, *J. Phys. Soc. Jpn.* **85**, 043702 (2016).

75. F. Lechermann, L. Boehnke, D. Grieger, and C. Piefke, Electron correlation and magnetism at the LaAlO$_3$/SrTiO$_3$ interface: A DFT+DMFT investigation, *Phys. Rev. B* **90**, 085125 (2014).

76. M. Salluzzo in P. Mele et al. (Eds.), *Oxide Thin Films, Multilayers, and Nanocomposites*, Springer International Publishing, Switzerland, pp. 181–211, 2015.

77. Y. Xie, Y. Hikita, C. Bell, and H. Y. Hwang, Control of electronic conduction at an oxide heterointerface using surface polar adsorbates, *Nat. Commun.* **2**, 494 (2011).

78. K. A. Brown, S. He, D. J. Eichelsdoerfer et al., Giant conductivity switching of LaAlO$_3$/SrTiO$_3$ heterointerfaces governed by surface protonation, *Nat. Commun.* **7**, 10681 (2016).

79. W. Dai, S. Adhikari, A. C. Garcia-Castro et al., Tailoring LaAlO$_3$/SrTiO$_3$ interface metallicity by oxygen surface adsorbates, *Nano Lett.* **16**, 2739–2743 (2016).

80. J. A. Bert, Superconductivity in reduced dimensions, Doctoral thesis, Stanford University, CA, USA, 2012, http://www.stanford.edu/group/moler/theses/bert_thesis.pdf.

81. G. Salvinelli, G. Drera, A. Giampietri, and L. Sangaletti, Layer-resolved cation diffusion and stoichiometry at the LaAlO$_3$/SrTiO$_3$ heterointerface probed by X-ray photoemission experiments and site occupancy modeling, *ACS Appl. Mater. Interfaces* **7**, 25648–25657 (2015).

82. J. M. D. Coey and S. A. Chambers, Oxide dilute magnetic semiconductors—Fact or fiction? *MRS Bull.* **33**, 1053–1058 (2008).

83. M. A. Garcia, E. Fernandez Pinel, J. de la Venta et al., Sources of experimental errors in the observation of nanoscale magnetism, *J. Appl. Phys.* **105**, 013925 (2009).

84. J. M. D. Coey, M. Venkatesan, and P. Stamenov, surface magnetism of strontium titanate, *J. Phys. Condens. Matter* **28**, 485001 (2016).

85. T. Dietl and H. Ohno, Dilute ferromagnetic semiconductors: Physics and spintronic structures, *Rev. Mod. Phys.* **86**, 187–251 (2014).

86. M. Huijben, A. Brinkman, G. Koster, G. Rijnders, H. Hilgenkamp, and D. H. A. Blank, Structure-property relation of SrTiO$_3$/LaAlO$_3$ interfaces, *Adv. Mater.* **21**, 1665–1677 (2009).

87. Z. Q. Liu, L. Sun, Z. Huang et al., Venkatesan and Ariando, dominant role of oxygen vacancies in electrical properties of unannealed $LaAlO_3/SrTiO_3$ interfaces, *J. Appl. Phys.* **115**, 054303 (2014).

88. A. Lopez-Bezanilla, P. Ganesh, and P. B. Littlewood, Magnetism and metal-insulator transition in oxygen-deficient $SrTiO_3$, *Phys. Rev. B* **92**, 115112 (2015).

89. A. Lopez-Bezanilla, P. Ganesh, and P. B. Littlewood, Research update: Plentiful magnetic moments in oxygen deficient $SrTiO_3$, *APL Mater.* **3**, 100701 (2015).

90. D. D. Cuong, B. Lee, K. M. Choi, H.-S. Ahn, S. Han, and J. Lee, Oxygen vacancy clustering and electron localization in oxygen-deficient $SrTiO_3$: LDA+U study, *Phys. Rev. Lett.* **98**, 115503 (2007).

91. M. Altmeyer, H. O. Jeschke, O. Hijano-Cubelos et al., Magnetism, spin texture, and in-gap states: Atomic specialization at the surface of oxygen-deficient $SrTiO_3$, *Phys. Rev. Lett.* **116**, 157203 (2016).

92. Cheng et al., unpublished.

93. G. Cheng, M. Tomczyk, A. B. Tacla et al., Tunable electron-electron interactions in $LaAlO_3/SrTiO_3$ nanostructures, *Phys. Rev. X* **6**, 041042 (2016).

94. M. Tomczyk, G. Cheng, H. Lee et al., Micrometer-scale ballistic transport of electron pairs in $LaAlO_3/SrTiO_3$ nanowires, *Phys. Rev. Lett.* **117**, 096801 (2016).

95. L. Goncalves-Ferreira, S. A. T. Redfern, E. Artacho, E. Salje, and W. T. Lee, Trapping of oxygen vacancies in the twin walls of perovskite, *Phys. Rev. B* **81**, 024109 (2010).

96. J. F. Scott, E. K. H. Salje, and M. A. Carpenter, Domain wall damping and elastic softening in $SrTiO_3$: Evidence for polar twin walls, *Phys. Rev. Lett.* **109**, 187601 (2012).

97. Z. Erlich, Y. Frenkel, J. Drori et al., Optical study of tetragonal domains in $LaAlO_3/SrTiO_3$, *J. Supercond. Novel Magn.* **28**, 1017–1020 (2015).

98. H. J. H. Ma, S. Scharinger, S. W. Zeng et al., Local electrical imaging of tetragonal domains and field-induced ferroelectric twin walls in conducting $SrTiO_3$, *Phys. Rev. Lett.* **116**, 257601 (2016).

99. Y. Frenkel, N. Haham, Y. Shperber et al., Anisotropic transport at the $LaAlO_3/SrTiO_3$ interface explained by microscopic imaging of channel-flow over $SrTiO_3$ domains, *ACS Appl. Mater. Interfaces* **8**, 12514–12519 (2016).

100. H. Noad, E. M. Spanton, K. C. Nowack et al., Enhanced superconducting transition temperature due to tetragonal domains in two-dimensionally doped $SrTiO_3$, *Phys. Rev. B* **94**, 174516 (2016).

101. H. Kroemer, Nobel lecture: Quasielectric fields and band offsets: Teaching electrons new tricks, *Rev. Mod. Phys.* **73**, 783–793 (2001).

102. M. Honig, J. A. Sulpizio, J. Drori, A. Joshua, E. Zeldov, and S. Ilani, Local electrostatic imaging of striped domain order in $LaAlO_3/SrTiO_3$, *Nat. Mater.* **12**, 1112–1118 (2013).

103. B. Kalisky, E. M. Spanton, H. Noad et al., Locally enhanced conductivity due to the tetragonal domain structure in $LaAlO_3/SrTiO_3$ heterointerfaces, *Nat. Mater.* **12**, 1091–1095 (2013).

104. E. Maniv, M. B. Shalom, A. Ron et al., Strong correlations elucidate the electronic structure and phase diagram of $LaAlO_3/SrTiO_3$ interface, *Nat. Commun.* **6**, 8239 (2015).

# Electric-Field Controlled Magnetism

**Fumihiro Matsukura and Hideo Ohno**

## 13.1 INTRODUCTION

As learned in the previous chapters, especially Chapters 9, 10, and 17, Volume 1, many researchers are now studying intensively the interplay between electricity and magnetism. The electrical input to the magnets can induce magnetization dynamics, and even switch its direction via spin torque, spin-orbit torque, and electric-field-induced torque. This chapter focuses on the electric-field effects on magnetism, as well as the electric-field control of magnetization dynamics and direction. The effect was first demonstrated using ferromagnetic semiconductors, and is now observed in other material systems, e.g. metals and multiferroics. Here, we will restrict ourselves to describing the electric-field effect-related phenomena in magnetic semiconductors and metals. We will not cover the control of magnetic anisotropy in magnets through stress adjacent to the piezoelectric material to which an electric-field is applied. Readers who are interested in other material systems and electric-field-induced phenomena than those treated here may consult other reviews [1–4].

## 13.2 ELECTRIC FIELD EFFECTS IN MAGNETIC SEMICONDUCTORS

One of the most important characteristics of magnetic semiconductors is that the magnetic order is brought about by the presence of carriers. Hence, their magnetism is expected to be controlled by conventional means that vary the carrier concentration without changing temperature. This can be done electrostatically by applying an electric field onto the semiconductors using capacitor structures like a field effect transistor. The idea of electric field controlled magnetism was already proposed in the 1960s, in the context of research on rare earth magnetic semiconductors [5], and was demonstrated experimentally in 2000 for a III–V compound-based magnetic semiconductor, (In,Mn)As [6].

### 13.2.1 CARRIER-INDUCED FERROMAGNETISM IN SEMICONDUCTORS

As discussed in Chapters 9 and 10, Volume 2, many magnetic semiconductors are confirmed to exhibit carrier-induced ferromagnetism with the carrier concentration-dependent Curie temperature $T_C$ [7, 8]. In typical III–V magnetic semiconductors, (In,Mn)As and (Ga,Mn)As, the majority of Mn substituting for a III-group cation act as an acceptor, and thus one does not need additional carrier doping to induce the ferromagnetism [9, 10]. In II–VI counterparts, such as (Cd,Mn)Te and (Zn,Mn)Te, Mn ions are electrically neutral, and thus one needs additional hole doping to observe the carrier-induced ferromagnetism [11, 12]. There has been a long discussion on the position of the Fermi level in (Ga,Mn)As, i.e. whether it lies in the valence band or the impurity band [13, 14]. Recent experiments confirmed that it is in the valence band [15, 16], consistent with the electronic structure obtained from the first principles calculation [17, 18]. In such a case, the ferromagnetism is stabilized by the energy gain resulting from hole-repopulation in the valence band with spin splitting induced by the $p$-$d$ exchange interaction ($p$-$d$ Zener model) [13, 19]. The $p$-$d$ exchange interaction is the exchange interaction among $p$ electrons in the valence band of the host semiconductor and $d$ electrons of the guest magnetic ions. According to the model, $T_C$ of these materials for three-dimensional case is given by,

$$T_C = x_{eff} N_0 S(S+1)\beta^2 A_F \rho_s (T_C) / 12 k_B - T_{AF} \qquad (13.1)$$

where:

| | |
|---|---|
| $x_{eff}$ | is the effective Mn spin composition |
| $N_0$ | the cation density |
| $S$ | the Mn spin |
| $\beta$ | the $p$-$d$ exchange integral |
| $\rho_S$ | the spin density of states |
| $A_F$ | the Fermi-liquid parameter |
| $k_B$ | the Boltzmann constant |
| $T_{AF}$ | the reduction of the Curie temperature due to the short-range antiferromagnetic superexchange interaction among Mn spins |

One can take $T_{AF} = 0$ for Mn doped III–V compounds due to the formation of bound magnetic porlarons around close pairs of ionized Mn acceptors. By calculating $\rho_S$ from a $6 \times 6$ $k \cdot p$ matrix with $p$-$d$ exchange interaction, the experimentally observed $T_C$ can be reproduced reasonably well without any adjustable parameter. The magnitude of $T_C$ increases with the increase of hole concentration $p$, because $\rho_S$ increases with $p$. This carrier-induced nature of ferromagnetism is consistent with the experimental observation of higher $T_C$ of (Ga,Mn)As with higher conductivity [20]. The conductivity dependence of $T_C$ was also investigated by changing the conductivity through the co-doping of donors and the growth condition, and was shown to be consistent again with the carrier-induced mechanism of ferromagnetism [21, 22]. It is known that there exist a number of interstitial Mn in (Ga,Mn) As, which act as a double donor compensating holes, and whose spin couples antiferromagnetically with a spin of substitutional Mn [23, 24]. The post-growth annealing at ~200°C results in the diffusing out of interstitial Mn to the surface, and thus increases $p$ and magnetic moments [25, 26]. The systematic study utilizing the annealing effect confirmed also the monotonic increase of $T_C$ with increasing conductivity [25]. The *p-d* Zener model predicts higher $T_C$ for materials with wider bandgap due to their larger $\rho_S$ and *p-d* exchange interaction, if similar values of $x_{eff}$ and $p$ to those in (Ga,Mn) As are realized in them [13]. The prediction encouraged many researchers to work on related materials, and as a result, we now have a lot of information to help understand the physics of magnetic semiconductors from comprehensive points of view [27].

The *p-d* Zener model explains also the experimental observation of the strain-direction-dependent magnetic easy axis direction [28]. The magnetic anisotropy is brought about by the anisotropic carrier-mediated exchange interaction associated with the spin-orbit interaction (SOI) in the host materials. The model predicts that the direction and magnitude of the magnetic anisotropy are hole-concentration dependent [13, 19, 29].

## 13.2.2 ELECTRIC FIELD MODULATION OF THE CURIE TEMPERATURE

In this subsection, we introduce the electric field modulation of $T_C$ of ferromagnetic semiconductors, mainly focusing on the experimental results obtained for III–V compounds. Typical ferromagnetic III–V semiconductors contain a relatively large amount of holes, ranging from $10^{19}$ to $10^{21}$ cm$^{-3}$. Hence, the electric field screening length is the order of a nanometer, and thus most of experiments were done on thin films with several nanometers to observe sizable electric field effects.

The electric field modulation of $T_C$ was observed first in 2000 for (In,Mn) As thin layers in a field effect transistor as a channel layer [6]. The magnetization curves were probed by Hall resistance $R_{Hall}$ thanks to the anomalous Hall effect, as shown in schematically in Figure 13.1a. The increase of $p$ by the application of negative voltage $V_G$ to a gate electrode increases $T_C$ determined from the Arrott plots, and the decrease of $p$ by positive $V_G$ decreases $T_C$. The effect is reversible with the change in $p$, and consistent with the *p-d* Zener model. Similar observation of the modulation of $T_C$ has been reported with various kinds of ferromagnetic semiconductors, such as MnGe [30], (Cd,Mn)Te [31], (Ga,Mn)As [32], (Ga,Mn)Sb [33], Co doped $TiO_2$ [34], and so on. Among them, the effect in (Ga,Mn)As has been most extensively investigated.

For (Ga,Mn)As thin layers, the electric field modulation of $T_C$ was confirmed by both magnetization and magnetotransport measurements [32, 35, 36]. The magnetization measurements revealed that the electric fields modulate the magnitude of spontaneous magnetic moment, in addition to $T_C$ due to the change in the surface depletion layer thickness [35]. The Mn spins in the depleted region do not participate in the ferromagnetic order due to the absence of holes. The experimentally obtained relationship between $T_C$ and $p$ from transport measurements follows $T_C \propto p^\gamma$ with $\gamma \sim 0.2$ as shown in Figure 13.1b. On the other hand, Equation 13.1 gives $\gamma = 1/3$ for the free-electron model with quadratic band dispersion $\varepsilon = \hbar^2 k^2/(2m^*)$, where $\varepsilon$ is the kinetic energy, $k$ the wavenumber, and $m^*$ the effective mass of the carriers,

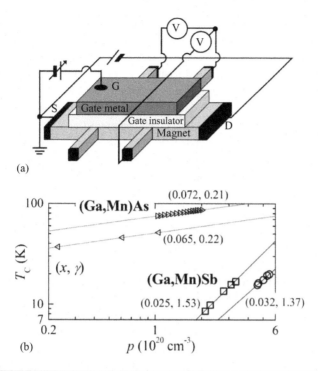

(a)

(b)

**FIGURE 13.1** (a) Schematic of electric field device and configuration for transport measurements. S, D, and G denote source, drain, and gate, respectively. (b) Logarithm plots of experimentally determined hole concentration $p$ dependence of the Curie temperature $T_C$ for (Ga,Mn)As and (Ga,Mn)Sb, where $p$ is modulated by the application of electric fields. Numbers in parentheses are nominal Mn composition $x$ and the exponent $\gamma$ in $T_C \propto p^\gamma$. Solid lines are linear fitting. (After Chang, H.W. et al., *Appl. Phys. Lett.* 103, 142402, 2013. With permission.)

and $\hbar$ the Dirac constant. More realistic valence-band dispersions with warps calculated by $6 \times 6$ $k \cdot p$ matrix give the value of $\gamma$ between 0.6 and 0.8 depending on $p$. This indicates that Equation 13.1 should be adapted for thin film cases with nonuniform hole distribution along the film normal ($z$ direction) due to the presence of depletion layers at the interfaces of (Ga,Mn)As and the short screening length of the electric field. The adapted expression for $T_C$ with $T_{AF} = 0$ is given by:

$$T_C = \left[ x_{eff} N_0 S(S+1)\beta^2 A_F \rho_{sheet} / 12k_B \right] \int \left( p(z) / p_{sheet} \right)^2 dz$$

$$= \int T_C^{3D}(z) dz \int \left( p(z) / p_{sheet} \right)^2 dz \tag{13.2}$$

where:

$\rho_{sheet}$ is sheet density of states given by $\rho_{sheet} = \int \rho_S dz$, $p(z)$ is $z$-dependent hole concentration, $p_{sheet}$ is the sheet hole concentration given by $p_{sheet} = \int p(z) dz$

$T_C^{3D}$ is the Curie temperature for three-dimensional case given by Equation 13.1 [35, 36]

The $p(z)$ is calculated from the one-dimensional Poisson equation, in which material parameters, the net concentration of Mn acceptors, the concentration of donor-like interface states at insulator/(Ga,Mn)As boundary, and the concentration of As antisite donors in a low-temperature grown GaAs buffer are treated as adjustable parameters. The calculated hole concentration $p$ dependence of $T_C$ by Equation 13.2, where $p$ is modulated by applying electric fields, reproduces well the experimental observation of $T_C \propto p^\gamma$ with $\gamma \sim 0.2$ [36]. The electric field dependence of the saturation moments is shown to be proportional to the thickness of a magnetically active layer (thickness of (Ga,Mn)As subtracted by the calculated depletion layer thickness), consistent with the nature of the carrier-induced ferromagnetism [35].

Because the value of $\gamma$ is related to the hole distribution in the channel, it is expected to depend on the channel materials in the electric field effect devices through the material-dependent Fermi-level pinning position at the interface. The pinning results in a surface depletion layer for p-GaAs, but an accumulation layer for p-GaSb [37]. The relation of $T_C \propto p^\gamma$ with $\gamma = 1.4$–$1.6$ was obtained for (Ga,Mn)Sb from the electric field effect measurements probed by $R_{Hall}$, as shown in Figure 13.1b. The larger value of $\gamma$ than that of (Ga,Mn)As was shown to be explained by the presence of the surface accumulation layer in (Ga,Mn)Sb [38].

When the magnetic ions are located in an only partial channel region along the film normal, one can change a degree of overlap of the magnetic ions with the wave function of carriers by the application of electric fields [39], which is expected to also result in the change in $T_C$ through the modulation of the carrier-mediated exchange coupling [6]. This scheme was utilized to demonstrate the electric field modulation of $T_C$ in heterostructures of Mn delta-doped GaAs/p-(Al,Ga)As, as well as InAs/(In,Fe)As/InAs quantum well [40, 41].

The modulation of $T_C$ of (Ga,Mn)As has been observed in a variety of field effect structures, which include a back gate structure with a p-n junction of (Ga,Mn)As/n-GaAs for low-voltage operation [42], ferroelectric-gate structure with nonvolatile functionality [43], and electric-double layer structure to apply higher electric fields [44]. It also demonstrated that the electric field formation of magnetic nanodots from a continuous (Ga,Mn)As layer using a meshed gate electrode, in which nanowindow regions without electric fields correspond to nanomagnets surrounded by the paramagnetic region created by depleting holes under the meshed gate [45].

### 13.2.3 ELECTRIC FIELD MODULATION OF MAGNETIZATION PROCESS AND ANISOTROPY

As described above, in the vicinity of $T_C$, one can change the magnetic phases either ferromagnetic or paramagnetic, by applying electric fields without changing temperature. In addition, one can change coercive force $H_C$ and magnetic anisotropy well below $T_C$.

The electric field control of $H_C$ was first demonstrated for (In,Mn)As channels with perpendicular magnetic easy axis in an electric field effect structure [46]. The increase of $p$ was found to result in the increase in $H_C$ below $T_C$. The same tendency was observed for (Ga,Mn)As [32]. The domain structure in (In,Mn)As, observed by magneto-optical Kerr microscope (MOKE), indicated that the magnetization process takes place through domain nucleation and its expansion [47], consistent with the magnetization process from the virgin state [46]. In addition, the dependence of $H_C$ on temperature $T$ and electric field $E$ was shown to follow the scaling relationship of $H_C \propto [1 - (T/T_C{}^*)]^\eta$ with $\eta \sim 2$, where $T_C{}^*$ is defined as the temperature at which $H_C$ becomes zero (Figure 13.2) [47]. These results suggest that the modulation of $H_C$ is related to the modulation of the exchange stiffness constant $A_S$. The electric field modulation of $H_C$ can be used for electric field-assisted magnetization switching; the reduction in $H_C$ enables us to induce the magnetization reversal by applying an electric field pulse in a small external magnetic field [46].

The presence of the SOI in ferromagnetic materials manifests itself as the presence of magnetic anisotropy. According to the $p$-$d$ Zener model, the magnetic anisotropy in (Ga,Mn)As is expected to depend on the strain (direction and magnitude), hole concentration, and the magnitude of the magnetization through SOI in the valence band. The expectation was corroborated by the experimental observation, the change of direction of the magnetic easy axis, either in-plane or perpendicular, by changing the direction of strain, annealing condition (thus hole concentration and magnetization), or temperature (thus magnetization) [28, 48, 49]. Hence, one expects that one can control the direction and magnitude of magnetic easy axis by applying electric fields.

The electric field modulation of the magnitude of the perpendicular magnetic anisotropy in (Ga,Mn)As with in-plane easy axis was detected through the anisotropic magnetoresistance (AMR) effect under a perpendicular magnetic field [50]. The saturation magnetic field of the AMR was shown to increase with increasing $p$ by applying a negative electric field. The result indicates that the in-plane easiness increases with increasing $p$, and is consistent qualitatively with the description of the $p$-$d$ Zener model for

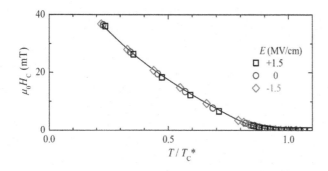

**FIGURE 13.2**  Dependence of coercivities $H_C$ on the reduced temperature $T/T^*$ as a function of applied electric fields $E$ for a 5-nm-thick $In_{0.937}Mn_{0.063}As$, where $T_C{}^*$ is defined as the temperature at which $H_C$ becomes zero. Line is guide for the eye. (After Chiba, D. et al., *J. Phys. D: Appl. Phys.* 39, R215, 2006. With permission.)

(Ga,Mn)As with compressive strain. The AMR effect is the phenomenon in which the resistance depends on a relative angle between the magnetization and current as a result of the presence of SOI. Clear AMR effect was observed for (Ga,Mn)As [51], and its strain-dependent sign was explained by the Boltzmann transport theory with the *p-d* Zener model [52, 53].

The electric field modulation of the in-plane magnetic anisotropy was also observed through the transverse component of AMR effect [50], sometimes referred as the planar Hall effect (PHE) [54]. The application of $E$ ranging from −4 to 4 MV/cm to a 4-nm-thick (Ga,Mn)As through a gate insulator modulates the in-plane magnetic field angle dependence of the planar Hall resistance, reflecting the modulation of the in-plane anisotropy. The (Ga,Mn)As possesses relatively complicated in-plane magnetic anisotropies; fourfold in-plane easy axis along ⟨100⟩, as well as two uniaxial easy axes along [110] or [$\bar{1}$10] and [100] or [010] [55]. The analysis of the angular dependence of the planar Hall resistance indicated that the sign and magnitude of the uniaxial anisotropy along [110] is modulated, and results in the change in in-plane magnetization direction by ~10° by applying $E$ in the absence of an external magnetic field [50]. The origin of the two uniaxial easy axes remains to be elucidated: one possible explanation for the [110] anisotropy is the formation of Mn dimers along [$\bar{1}$10] orientation during epitaxial growth [56]. The theoretical work indicated that the anisotropic Mn distribution can serve as effective shear strain [56], and thus results in the carrier-concentration dependent uniaxial anisotropy along [110] orientation [57].

The control of magnetic anisotropies by external means provides new schemes for magnetization reversal; the reversal through ratchet-type motion of magnetization by rotating the minima in the potential landscape, and the reversal through the magnetization precession induced by the abrupt change of the anisotropy direction [58–60]. Electric field-induced stochastic magnetization switching was observed for (Ga,Mn)As, in which the magnetocrystalline anisotropy energy barrier was reduced by $E$ to switch the magnetization through thermal fluctuation of magnetization [61].

## 13.2.4 ELECTRIC FIELD MODULATION OF OTHER MAGNETIC PARAMETERS

Because (Ga,Mn)As is in the vicinity of metal-insulator transition (MIT) [20], it contains two types of regions; one is populated by extended holes and the other by localized holes (two fluids model) [62]. The region populated by extended holes exhibits ferromagnetism, whereas that by localized holes exhibits superparamagnetic-like behavior. The presence of two-phase is hard to detect by transport measurements due to much higher conductivity in the ferromagnetic region than in the superparamagnetic region. However, it can be detected by history-dependent remanent magnetization measurements [35]. It was shown that the portion of the superparamagnetic-like region can be changed by the application of the electric fields; when $p$ in a (Ga,Mn)As layer is decreased, the portion increases [35].

Ferromagnetic resonance (FMR) spectra (resonance fields $H_R$ and spectral linewidths $\Delta H$) of (Ga,Mn)As were shown to be described well by the Landau–Lifshitz–Gilbert (LLG) equation with perpendicular and in-plane magnetic anisotropies, without considering extrinsic contributions such as magnetic anisotropy dispersion and two-magnon scattering [63]. The FMR spectra as a function of external magnetic-field angles $\theta_H$ were measured for a thin metallic (Ga,Mn)As film in a capacitor structure under $E$ between −4 and 4 MV/cm [64]. Hole concentration $p \sim 3.2 \times 10^{20}$ cm$^{-3}$ at $E = 0$ was modulated by ~25%, and thus the resistivity $\rho \sim 9$ m$\Omega$cm at $E = 0$ by ~25% by the application of $E$. The analysis of the angular dependence of $H_R$ indicated that the anisotropy of this particular (Ga,Mn)As is nearly independent of $E$. On the other hand, the analysis of the angular dependence of $\Delta H$ indicated that the damping constant $\alpha \sim 0.054$ at $E = 0$ is modulated by ~10% by $E$; the increase in $p$ results in the decrease of $\alpha$. The theoretical work based on the relaxation mechanism due to the SOI predicted the carrier-concentration dependent $\alpha$ [65], however, the observed tendency was opposite to the prediction. To understand the discrepancy, the relation among $\alpha$, $\rho$, and magnetic properties was investigated by changing $p$ through $E$ and the post annealing. As described above, the magnetic disorder defined as the ratio $M_{SP}/M_{tot}$ of the superparamagnetic-like magnetization component to total magnetization, increases with decreasing $p$. A clear correlation between $\alpha$ and $M_{SP}/M_{tot}$ was obtained as shown in Figure 13.3, suggesting strongly that the electric field modulation of $\alpha$ in (Ga,Mn)As is determined by the magnetic disorder rather than the SOI.

(Ga,Mn)As usually shows a positive anomalous Hall coefficient [66]. However, it was found that thin (Ga,Mn)As with relatively high conductivity $\sigma > 200$ S/cm sometimes shows a negative anomalous Hall coefficient. For such (Ga,Mn)As, the sign of the coefficient can be changed by the application of $E$ [67]. The origin of the behavior is not understood yet [68], but is probably

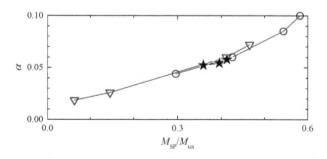

**FIGURE 13.3** Dependence of damping constant $\alpha$ of (Ga,Mn)As on the ratio $M_{SP}/M_{tot}$ of superparamagnetic-like component to total magnetization. Stars are the result for a 4-nm-thick (Ga,Mn)As with Mn composition $x$ of 0.13 in an electric field effect structure, in which the resistivity is changed by the application of electric fields. Triangles and circles are results for a 20-nm-thick metallic (Ga,Mn)As with $x = 0.068$ and a 200-nm-thick insulating (Ga,Mn)As with $x = 0.075$, respectively, for which the resistivity is changed by post annealing. (After Chen, L. et al., *Phys. Rev. Lett.* 115, 057204, 2015. With permission.)

related to the interplay between carrier confinement and topology of band structure.

The electric field modulation of the magnetic domain wall (DW) velocity was also reported for (In,Mn)As [69].

# 13.3 ELECTRIC FIELD EFFECTS IN FERROMAGNETIC METALS

Stimulated by studies on magnetic semiconductors, researchers began to investigate electric field effects in metals. The first observation of an electric field effect on magnetism in metals was reported for thin layers of FePt and FePd, where $H_C$ was modulated at room temperature using an electric double layer [70]. Prior to this study, it was assumed that a large electric field effect in metals would be difficult to observe owing to the associated short screening length. Subsequent theoretical work showed that electric field effects on surface magnetization and surface magnetocrystalline anisotropy can be sizable, even for metal ferromagnets such as Fe, Co, and Ni, because of the spin-dependent screening of the electric field [71]. The studies encouraged many to investigate electric field effects in a number of materials, including Fe(Co)/MgO system, and led to the observation of the electric field-induced change of magnetic anisotropy at room temperature, whose origin is attributed to the modulation of interfacial magnetic anisotropy [72, 73]. The effect is now being utilized to induce the magnetization switching [74, 75].

## 13.3.1 ELECTRIC FIELD MODULATION OF MAGNETIC ANISOTROPY AND OTHER PROPERTIES

Experimental works indicated the presence of the interfacial perpendicular magnetic anisotropy at a Fe/MgO, Co/AlO, and CoFeB/MgO interfaces [72, 73, 76–79]. These findings were followed by the fabrication of nanoscale CoFeB/MgO magnetic tunnel junctions (MTJs) with a perpendicular easy axis [80]. For the Fe/MgO junction, theoretical studies showed that the hybridization of Fe $3d$ and O $2p$ orbitals modifies the electronics structure at the interface, and the resultant occupancy of electrons among $3d$ orbitals with different magnetic quantum numbers determines the direction and magnitude of magnetic anisotropy through the SOI [81, 82]. This theoretical description was shown to be consistent with the results of x-ray magnetic circular dichroism (XMCD) measurement [83]. The formation of Fe(Co)-O bonds was also confirmed experimentally for the CoFeB/MgO interface [84]. Application of an electric field alters the Fermi energy position at the interface, which changes the occupancy of the orbitals and hence the interface magnetic anisotropy [81, 82]. The results of x-ray absorption (XAS) and XMCD for Fe/MgO suggested that an electric field can reversibly change the degree of oxidation of Fe at the interface, which may influence the magnetism [85]. However, XAS measurement of Fe/MgO revealed the presence of the electric field effect on the anisotropy without redox, indicating that the effect is associated with the electron doping and/or redistribution induced by the electric field at the interface [86]. Some experimental works showed that the efficiency and the polarity of the effect depend on the composition of materials and stack structure, as well as deposition and annealing condition [87–91]. Theoretical works predicted that they depend on the counter-interface material, strain, and the Rashba SOI [92–95]. The electric field modulation ratio $\xi$ is defined as $\xi = dK^{eff}t/dE$, where $K^{eff}$ is an effective perpendicular magnetic anisotropy density, and $t$ is the thickness of a ferromagnetic material. Typical reported values of $\xi$ for CoFeB/MgO range from ~10 to ~100 fJ/Vm, which is consistent with those obtained from the first principles calculations for (Co)Fe/MgO (or vacuum) systems [81, 82]. The applied electric field results in an atomic-scale inhomogeneous electric field at the metal/insulator interface due to electrostatic screening in the metal, which induces electric quadrupoles at the interface. The quadrupoles produce the magnetic dipole moments, which change the magnetic anisotropy through SOI. This mechanism based on the induced dipoles was confirmed by the combination of XMCD experiments and first principles calculations on an $L1_0$-FePt system [96]. The electric field-induced change in the anisotropy due to the magnetic dipole moments could be large, but is largely compensated by the anisotropy with the opposite sign, due to the electric field effect on the orbital moments in the FePt system [96].

The modulation of $T_C$ and magnetic moments have also been observed in ultrathin layers of Fe and Co at room temperature [97–99]. The observation may be related to the modulation of the exchange constant as indicated by the first principles calculation [100, 101]. The magnetic phase transition between ferromagnetic and antiferromagnetic phases was observed through the structural phase transition between body-centered and face-centered cubic phases induced by the application of the local electric field by the tip of a scanning tunneling microscope [102].

Ferromagnetic resonance (FMR) spectra were measured for Ta/CoFeB/MgO junctions with different CoFeB thickness $t$ ranging from 1.4 to 1.8 nm as functions of $E$ and external magnetic field angle $\theta_H$ measured from the film normal [103]. As shown in Figure 13.4 for $Co_{0.2}Fe_{0.6}B_{0.2}$ with $t = 1.5$

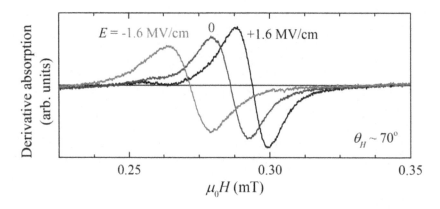

**FIGURE 13.4**   Ferromagnetic resonance spectra for a 1.5-nm-thick $Co_{0.2}Fe_{0.6}B_{0.2}$ sandwiched by Ta and MgO as a function of applied electric fields $E$.

nm, the resonance field $H_R$ is shifted by the application of $E$. The analysis of theangular dependence of the $H_R$ indicated that the first order effective perpendicular magnetic anisotropy field $H_{K1}^{\text{eff}}$ increases with decreasing $t$, indicating its interfacial origin, whereas the second order anisotropy field $H_{K2}$ is virtually independent of $E$. The magnitude of $H_{K1}^{\text{eff}}$ is modulated by $E$, while that of $H_{K2}$ is not. The magnitude of the modulation of the areal anisotropy energy density does not depend virtually on $t$, indicating that the electric field effect is the interfacial effect. The analysis of the FMR spectral linewidth $\Delta H$ indicated that the damping constant $\alpha$ increases with decreasing $t$, most probably due to the spin pumping at the Ta/CoFeB interface [104]. It was found that the FMR spectral linewidths $\Delta H$ can be also modulated by $E$. The analysis of the angular dependence indicated that the magnitude of $\alpha$ can be modulated by a few percent by $E$ of 1 MV/cm for CoFeB/MgO with thin CoFeB possessing perpendicular magnetic easy axis [103].

The electric field effect on magnetic domain structures was observed by a magneto-optical Kerr effect microscopy. It was shown that CoFeB/MgO junctions with perpendicular magnetic anisotropy show an in-plane isotropic maze domain pattern at demagnetized state [105]. The application of $E$ modulates the domain periods $D_P$ from 1.5 to 2.0 µm by applying $E$ of $\pm 1.1$ MV/cm for the particular 1.8-nm-thick CoFeB used in the study [106] (Figure 13.5). The electric field dependence of $D_P$ can be described well by the existing model with the modulation of $A_S$ by a few percent in addition to that of the magnetic anisotropy [107]. The electric field modulation of $D_P$ was observed also for Co [108], and the modulation can be used for magnetization reversal in a local area [109]. The modulation of $A_S$ was also shown by the modulation of the distance between the resonance fields between the Kittel and spin-wave modes in a nanoscale CoFeB/MgO MTJ detected by homodyne technique [110].

Electric field effect can also be used to control DW motion under an external magnetic field, as shown in Co wires. Here, the applied electric field modifies the creep velocity of the DW by over an order of magnitude, with changes most probably related to the modification of the magnetic anisotropy [111, 112]. By using patterned double-gate insulators, the DW pinning could

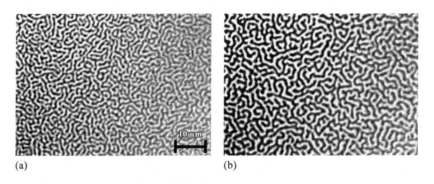

(a)                                        (b)

**FIGURE 13.5**   Domain structures at demagnetized state for a 1.5-nm-thick $Co_{0.2}Fe_{0.6}B_{0.2}$ sandwiched by Ta and MgO under (a) negative and (b) positive electric fields $E$. (After Dohi, T. et al., *AIP Adv.* 6, 075017, 2016. With permission.)

be controlled by an applied electric field, in which the electrode perimeter acts as a barrier for the DW motion after the application of the negative electric field. Relatively large pinning fields of 65 mT can be removed by applying a positive electric field. Once the electric field was applied, the DW pinning was maintained at zero electric field for several days, suggesting that the observed nonvolatile nature of the effect is related to the ionic transport and/or charge trap in the insulator [112]. The theoretical work predicted that the current-induced Walker breakdown can be suppressed by the application of the electric field through the spin flexoelectric interaction, leading to a faster maximum DW velocity [113]. There have also been many interesting topics on the electric field control of the DW motion in multiferroic structures [114].

Electric field effects on antiferromagnetic materials and the modulation of the anomalous Hall effect in a paramagnetic metal Pt were reported [115–117].

## 13.3.2 ELECTRIC FIELD-INDUCED MAGNETIZATION DYNAMICS

The electron spin resonance and nuclear magnetic resonance can be induced in nonmagnetic materials by applying a high-frequency electric field through the modulation of the SOI or quadrupole moments [118, 119]. It was shown that FMR in ferromagnetic metals can be excited also by the application of rf electric fields through the rf-modulation of the magnetic anisotropy [120, 121]. The FMR is usually excited in a free layer in an MTJ severed as a pseudocapacitor structure fabricated on a coplanar waveguide by applying rf electric fields. The FMR detection is done by homodyne technique, in which rectified dc voltage $V_{dc}$ is measured thanks to the tunnel magnetoresistance (TMR) effect (see Chapters 11–13, Volume 1), which results in the change in junction resistance $R$ synchrony with the input rf signal [122]. The homodyne-detected FMR is a powerful tool to determine magnetic parameters, such as anisotropy constant, its electric field modulation ratio, and damping constant, of nanomagnets in MTJs [123–126]. The excitation of the FMR by electric fields are confirmed from its spectral shape and magnetic field angle dependence, which are different from those for the FMR excited by spin-transfer torque (STT) [120]. One can observe electric field-induced non-linear FMR [127], in which spectral shape is largely distorted from anti-symmetric Lorentzian due to the foldover effect by increasing input rf power $P_{rf}$ as shown in Figure 13.6 [128]. From the rf power dependence of the FMR spectra, one can determine the value of ξ. The behavior was reproduced by micromagnetic simulations [127]. The electric field effect can also excite spin-wave resonance in nanoscale MTJs [110, 129]. The electric field-induced FMR was shown to be used as a microwave detector with higher sensitivity than that induced by STT [130].

The spin-wave propagation in magnetic insulator and metals can be modulated by the application of electric fields through the modification of the SOI or magnetic anisotropy [131, 132]. The analysis of spin-wave resonance frequency in an Au/Fe/MgO structure showed that the interfacial Dzyaloshinskii–Moriya interaction (DMI) is modulated by the electric

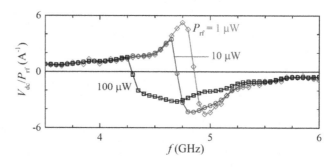

**FIGURE 13.6** Normalized homodyne-detected ferromagnetic resonance spectra $V_{dc}/P_{rf}$ for a CoFeB/MgO-based magnetic tunnel junction with perpendicular magnetic easy axis under in-plane magnetic field of 0.1 T as a function of input rf power $P_{rf}$. (After Hirayama, E. et al., *Appl. Phys. Lett.* 107, 132404, 2015. With permission.)

fields in addition to the modulation of the interfacial anisotropy [133]. The presence of the DMI sometimes stabilizes a topologically non-trivial spin texture, skyrmion [134]. Micromagnetic simulation showed that one can create and annihilate a skyrmion as well as switch its core polarity in a magnetic nanodisk by applying $E$ through the change in magnetic anisotropy [135].

### 13.3.3 ELECTRIC FIELD-INDUCED MAGNETIZATION SWITCHING

Electric field-induced magnetization switching that involves interface anisotropy was demonstrated in MTJs, which serve as a pseudocapacitor. The magnetization configuration is detected by the TMR effect. So far, two schemes have been demonstrated: one utilizes the electric field-induced change of coercivity [136], and is essentially the same as the electric field-assisted magnetization switching demonstrated for a single layer of a ferromagnetic semiconductor and ferromagnetic metal in a field effect structure [46, 137]; the other utilizes magnetization precession induced by a temporal change in the direction of magnetic anisotropy [74, 75]. The first scheme achieves unipolar switching under a constant magnetic field, and was demonstrated for a CoFeB/MgO MTJ in a small external magnetic field [136]. The field is needed to ensure that the potential wells—otherwise degenerate bistable magnetic states—are asymmetric, because the electric field does not break time-reversal symmetry. The application of an electric field reduces the coercivity of the CoFeB electrode, and switches the magnetization direction to align it with an applied magnetic field. The magnetization direction is detected by the junction resistance. STT switching was utilized to induce switching in the opposite direction; without STT, the polarity of the external magnetic field must be reversed to induce the electric field-assisted switching in the opposite direction.

In the second scheme, an initial study used a FeCo/MgO/Fe MTJ with an in-plane magnetic easy axis [74] was followed by a CoFeB/MgO/CoFeB with a perpendicular magnetic easy axis [75]. For the case of an MTJ with a perpendicular easy axis, the application of an electric field pulse temporarily aligns the easy axis in-plane, thus inducing magnetization precession in the free layer about the new easy axis. Applying an external constant magnetic field assists the change in direction for easy axis, fixes the precessional axis and changes the precessional period (the inverse of the Larmor frequency) as shown in Figure 13.7. The precessional period is determined by the magnitude of the effective magnetic field (sum of anisotropy, stray, and external fields). Magnetization switching takes place when the pulse duration is a half-integer multiple of the precession period. The obtained switching probabilities therefore show an oscillatory behavior with the pulse duration, which requires fine control of the pulse duration for switching. The amplitude of oscillation decays relatively fast as the pulse duration increases, which is due to thermal agitation enhanced by the distribution of effective fields [138]. The effect of thermal agitation results in randomization of the precessional phase, as confirmed by the real-time observation of the precession in Figure 13.8 [139]. The real-time observation was conducted by measuring the transmission voltage $V_T$ during the application of the bias voltage, which induces the precession in a free layer in an MTJ. The random phase shift becomes larger with time, and thus results in rapid decay in the probability. More reliable switching is possible by combining both the electric field-induced and STT-induced switching, with two successive voltage pulse applied to utilize the advantages of the two schemes [140]. The first pulse, $V_E$, induces the magnetization precession by the electric field effect, and the second pulse, $V_{STT}$, determines the final magnetization direction by STT. This combined scheme achieves higher probabilities than those of STT-induced or electric field-induced switching alone at shorter pulse duration regime, and gives it potential for applications.

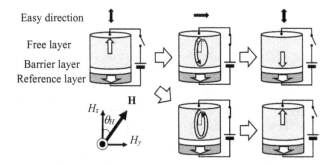

**FIGURE 13.7**  Schematics of operation scheme of electric field induced magnetization switching in a magnetic tunnel junction. Upper arrows show the direction of magnetic easy direction in a free layer. The application of an electric field pulse induces magnetization precession through the temporal change of the easy axis direction. Magnetization switching takes place when the pulse duration is a half-integer multiple of the precession period. The resultant direction is the same as the initial direction when the duration is an integer multiple of the period. A perpendicular component $H_z$ of an external magnetic field **H** compensates a perpendicular component of a stray field from a reference layer, and an in-plane component $H_y$ determines the precessional axis.

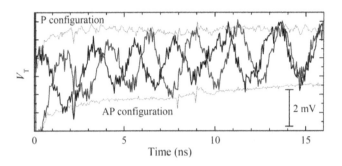

**FIGURE 13.8** Time evolution of transmitted voltage $V_T$ from a CoFeB/MgO-based magnetic tunnel junction under bias voltage. Upper curve and lower curves are $V_T$ for parallel and antiparallel magnetization configuration, respectively. Two middle oscillatory curves are for situations with magnetization precession from parallel and antiparallel magnetization configuration. (After Kanai, S. et al., *Jpn. J. Appl. Phys.* 58, 0802A3, 2017. With permission.)

Because the MTJ is a pseudocapacitor, tunnel current flows through the barrier layer during the application of electric fields. As a result, the switching energy is dominated by the Joule heating rather than the charging energy of the capacitor, although the STT is not important for the switching. In order to reduce the Joule energy, the electric field-induced switching was investigated for CoFeB/MgO-based MTJs with high junction resistance by increasing the MgO barrier-layer thickness [141, 142]. It was shown that the switching energy can be reduced to ~6 fJ/bit, which is the smallest reported value for the magnetization switching in MTJs, but is still dominated by the Joule heating. To reduce the switching energy further, one needs to improve the electric field modulation ratio ξ.

## 13.4 CONCLUSION

Curiosity about the interplay between electricity and magnetism brought about initial study of the electric field effect on magnetism. Nowadays, the effect is also drawing attention from a technological viewpoint, because of its compatibility with semiconductor device technology, and capability for realizing nonvolatile devices with low power consumption. To put the electric field devices to practical use, a number of challenges must be overcome; for instance, one needs to enhance the effect to reduce the operational energy, and improve the switching reliability, as well as to ensure thermal stability at small dimensions. Meeting these challenges will have a big impact on future nanoelectronics. In addition, the effect is vital as a powerful probe of condensed matter physics to explore the mechanism of cooperation phenomena.

## ACKNOWLEDGMENTS

The authors are grateful for many collaborations, especially with T. Dietl, D. Chiba, and L. Chen on ferromagnetic semiconductors, as well as with

S. Kanai, A. Okada, E. Hirayama, and T. Dohi on ferromagnetic metals. The authors acknowledge the support from MEXT, a Grant-in-Aid for Scientific Research (No. 26103002), and R&D Project for ICT Key Technology.

# REFERENCES

1. Y. Tokura, S. Seki, and N. Nagaosa, Multiferroics of spin origin, *Rep. Prog. Phys.* **77**, 075501 (2014).
2. F. Matsukura, Y. Tokura, and H. Ohno, *Nat. Nanotech.* **10**, 209–220 (2015).
3. K. Roy, Ultra-low-energy straintronics using multiferroic composites, *SPIN* **3**, 1330003 (2013).
4. M. Barangi and P. Mazumder, Straintronics, *IEEE Nanotech. Magazine*, 15–24 (2015).
5. S. Methfessel, Potential application of magnetic rare earth compounds, *IEEE Trans. Magn.* **1**, 144–155 (1965).
6. H. Ohno, D. Chiba, F. Matsukura et al., Electric field control of ferromagnetism, *Nature* **408**, 944–946 (2000).
7. F. Matsukura, H. Ohno, A. Shen, and Y. Sugawara, Transport properties and origin of ferromagnetism in (Ga,Mn)As, *Phys. Rev. B* **57**, R2037–R2040 (1998).
8. K. W. Edmonds, K. Y. Wang, R. P. Campion et al., High-Curie-temperature $Ga_{1-x}Mn_xAs$ obtained by resistance-monitored annealing, *Appl. Phys. Lett.* **81**, 4991–4993 (2002).
9. H. Ohno, H. Munekata, T. Penny, S. von Molnár, and L. L. Chang, Magnetotransport properties of p-type (In,Mn)As diluted magnetic III-V semiconductors, *Phys. Rev. Lett.* **68**, 2664–2667 (1992).
10. H. Ohno, A. Shen, F. Matsukura et al., (Ga,Mn)As: A new diluted magnetic semiconductor based on GaAs, *Appl. Phys. Lett.* **69**, 363–365 (1996).
11. A. Haury, A. Wasiela, A. Arnoult et al., Observation of ferromagnetic transition induced by two-dimensional hole gas in modulation-doped CdMnTe quantum wells, *Phys. Rev. Lett.* **79**, 511–514 (1997).
12. D. Ferrand, J. Cibert, A. Wasiela et al., Carrier-induced ferromagnetism in $p$-$Zn_{1-x}Mn_xTe$, *Phys. Rev. B* **63**, 085201 (2001).
13. T. Dietl, H. Ohno, F. Matsukura, J. Cibert, and D. Ferrand, Zener model description of ferromagnetism in zinc-blende magnetic semiconductors, *Science* **287**, 1019–1022 (2000).
14. M. Tanaka, S. Ohya, and P. N. Hai, Recent progress in III–V based ferromagnetic semiconductors: Band structure, Fermi level, and tunneling transport, *Appl. Phys. Rev.* **1**, 011102 (2014).
15. S. Souma. L. Chen. R. Oszwałdowski et al., Fermi level position, Coulomb gap, and Dresselhaus splitting in (Ga,Mn)As, *Sci. Rep.* **6**, 277266 (2016).
16. J. Kanski, L. Ilver, K. Karlsson et al., Electronic structure of (Ga,Mn)As revisited, *New J. Phys.* **19**, 023006 (2017).
17. J. H. Park, S. K. Kwon, and B. I. Min, Electronic structures of III–V based ferromagnetic semiconductors; Half-metallic phase, *Physica B* **281&282**, 703–704 (2000).
18. M. Toyoda, H. Akai, K. Sato, and H. Katayama-Yoshida, Curie temperature of GaMnN and GaMnAs from LDA-SIC electronic structure calculations, *Phys. Stat. Sol. (c)* **3**, 4155–4159 (2006).
19. T. Dietl, H. Ohno, and F. Matsukura, Hole-mediated ferromagnetism in tetrahedrally coordinated semiconductors, *Phys. Rev. B* **63**, 195205 (2001).
20. A. Oiwa, S. Katsumoto, A. Endo et al., Nonmetal-metal-nonmetal transition and large negative magnetoresistance in (Ga,Mn)As/GaAs., *Solid State Commun.* **103**, 209–213 (1997).
21. Y. Satoh, D. Okazawa, A. Nagashima, and J. Yoshino, Carrier concentration dependence of electronic and magnetic properties of Sn-doped GaMnAs, *Physica E*, **10**, 196–200 (2001).

22. H. Shimizu, T. Hayashi, T. Nishinaga, and M. Tanaka, Magnetic and transport properties of III–V based magnetic semiconductor (Ga,Mn)As: Growth condition dependence, *Appl. Phys. Lett.* **74**, 398–400 (1999).

23. K. M. Yu, W. Walukiewicz, T. Wojtowicz et al., Effect of the location of Mn sites in ferromagnetic $Ga_{1-x}Mn_xAs$ on its Curie temperature, *Phys. Rev B* **65**, 201303 (R) (2002).

24. J. Blinowski and P. Kacman, Spin interactions of interstitial Mn ions in ferromagnetic GaMnAs, *Phys. Rev. B* **67**, 121204 (R) (2003).

25. K. W. Edmonds, P. Bogusławski, K. Y. Wang et al., Mn interstitial diffusion in (Ga,Mn)As, *Phys. Rev. Lett.* **92**, 037201 (2004).

26. K. W. Wang, K. W. Edmonds, R. P. Campion et al., Influence of the Mn interstitial on the magnetic and transport properties of (Ga,Mn)As, *J. Appl. Phys.* **95**, 6512–6514 (2004).

27. T. Dietl, A ten-year perspective on dilute magnetic semiconductors and oxides, *Nat. Mater.* **9**, 965–974.

28. A. Shen, H. Ohno, F. Matsukura et al., Epitaxy of (Ga,Mn)As, a new diluted magnetic semiconductor based on GaAs, *J. Cryst. Growth* **175/176**, 1069–1074 (1997).

29. M. Abolfath, T. Jungwirth, J. Brum, and A. H. MacDonald, Theory of magnetic anisotropy in $III_{1-x}Mn_xV$ ferromagnets, *Phys. Rev. B* **63**, 054418 (2001).

30. Y. D. Park, A. T. Hanbicki, S. C. Erwin et al., A group IV ferromagnetic semiconductor; $Mn_xGe_{1-x}$, *Science* **295**, 651–654 (2002).

31. H. Boulari, P. Kossacki, M. Bertolini et al., Light and electric field control of ferromagnetism in magnetic quantum structures, *Phys. Rev. Lett.* **88**, 207204 (2002).

32. D. Chiba, F. Matsukura, and H. Ohno, Electric field control of ferromagnetism in (Ga,Mn)As, *Appl. Phys. Lett.* **89**, 162505 (2006).

33. Y. Nishitani, M. Endo, F. Matsukura, and H. Ohno, Magnetic anisotropy in a ferromagnetic (Ga,Mn)Sb thin film, *Physica E* **42**, 2681–2684 (2010).

34. Y. Yamada, K. Ueno, T. Fukumura et al., Electrically induced ferromagnetism at room temperature in cobalt-doped titanium dioxide, *Science* **332**, 1065–1067 (2011).

35. M. Sawicki, D. Chiba, A. Korbecka et al., Experimental probing of the interplay between ferromagnetism and localization in (Ga,Mn)As, *Nat. Phys.* **6**, 22–25 (2010).

36. Y. Nishitani, D. Chiba, M. Endo et al., Curie temperature versus hole concentration in field-effect structures of $Ga_{1-x}Mn_xAs$, *Phys. Rev. B* **81**, 045208 (2010).

37. H. Hasegawa, H. Ohno, and T. Sawada, Hybrid orbital energy for heterojunction band lineup, *Jpn. J. Appl. Phys. Part 2* **25**, L265–L268 (1986).

38. H. W. Chang, S. Akita, F. Matsukura, and H. Ohno, Hole concentration dependence of the Curie temperature of (Ga,Mn)Sb in a field-effect structure, *Appl. Phys. Lett.* **103**, 142402 (2013).

39. B. Lee, T. Jungwirth, and A. H. MacDonald, Theory of ferromagnetism in diluted magnetic semiconductor quantum wells, *Phys. Rev. B* **61**, 15606–15609 (2000).

40. A. M. Nazmul, S. Kobayashi, S. Sugahara, and M. Tanaka, Electrical and optical control of ferromagnetism in III–V semiconductor heterostructures at high temperature (~100 K), *Jpn. J. Appl. Phys. Part 2* **43**, L233–L236 (2004).

41. L. D. Anh, P. N. Hai, Y. Kasahara, Y. Iwasa, and M. Tanaka, Modulation of ferromagnetism (In,Fe)As quantum wells via electrically controlled deformation of the electron wave functions, *Phys. Rev. B* **92**, 161201(R) (2015).

42. K. Olejnik, M. H. S. Owen, V. Novák et al., Enhanced annealing, high Curie temperature, and low-voltage gating in (Ga,Mn)As; A surface oxide control study, *Phys. Rev. B* **78**, 054403 (2008).

43. I. Stolichnov, S. W. E. Riester, H. J. Trodahl et al., Non-volatile ferroelectric control of ferromagnetism in (Ga,Mn)As, *Nat. Mater.* **7**, 464–467 (2008).

44. M. Endo, D. Chiba, H. Shimotani, F. Matsukura, Y. Iwasa, and H. Ohno, Electric double layer transistor with a (Ga,Mn)As channel, *Appl. Phys. Lett.* **96**, 022515 (2000).

45. D. Chiba, F. Matsukura, and H. Ohno, Electrically defined ferromagnetic nanodots, *Nano Lett.* **10**, 4505–4508 (2010).

46. D. Chiba, M. Yamanouchi, F. Matsukura, and H. Ohno, Electrical manipulation of magnetization reversal in a ferromagnetic semiconductor, *Science* **301**, 943–945 (2003).

47. D. Chiba, F. Matsukura, and H. Ohno, Electrical magnetization reversal in ferromagnetic III–V semiconductors, *J. Phys. D: Appl. Phys.* **39**, R215–R225 (2006).

48. K. Takamura, F. Matsukura, D. Chiba, and H. Ohno, Magnetic properties of (Al,Ga,Mn)As, *Appl. Phys. Lett.* **81**, 2590–2592 (2002).

49. M. Sawicki, F. Matsukura, A. Idziaszek et al., Temperature dependent magnetic anisotropy in (Ga,Mn)As layers, *Phys. Rev. B* **70**, 245325 (2004).

50. D. Chiba, M. Sawicki, Y. Nishitani, Y. Nakatani, F. Matsukura, and H. Ohno, Magnetization vector manipulation by electric fields, *Nature* **455**, 515–518 (2008).

51. D. V. Baxter, D. Ruzumetov, J. Scherschiligh et al., Anisotropic magnetoresistance in $Ga_{1-x}Mn_xAs$, *Phys. Rev. B* **65**, 212407 (2002).

52. T. Jungwirth, M. Abolfath, J. Sinova, J. Kučera, and A. H. MacDonald, Boltzmann theory of engineered anisotropic magnetoresistance in (Ga,Mn)As, *Appl. Phys. Lett.* **81**, 4029–4031 (2002).

53. F. Matsukura, M. Sawicki, T. Dietl, D. Chiba, and H. Ohno, Magnetotransport properties of metallic (Ga,Mn)As films with compressive and tensile strain, *Physica E* **21**, 1032–1036 (2004).

54. H. X. Tang, R. K. Kawakami, D. D. Awschalom, and M. L. Roukes, Giant planar Hall effect in epitaxial (Ga,Mn)As devices, *Phys. Rev. Lett.* **90**, 107201 (2003).

55. K. Pappert, S. Hümpfner, J. Wenisch et al., Transport characterization of the magnetic anisotropy of (Ga,Mn)As, *Appl. Phys. Lett.* **90**, 062109 (2007).

56. M. Birowska, C. Śliwa, J. A. Majewski, and T. Dietl, Origin of bulk uniaxial anisotropy in zinc-blende dilute magnetic semiconductors, *Phys. Rev. Lett.* **108**, 237203 (2012).

57. M. Sawicki, K.-Y. Wang, K. W. Edmonds et al., In-plane uniaxial anisotropy rotations in (Ga,Mn)As thin films, *Phys. Rev. B* **71**, 121302(R) (2005).

58. Y. Iwasaki, Stress-driven magnetization reversal in magnetostrictive films with in-plane magnetocrystalline anisotropy, *J. Magn. Magn. Mater.* **240**, 395–397 (2002).

59. D. Chiba, Y. Nakatani, F. Matsukura, and H. Ohno, Simulation of magnetization switching by electric field manipulation of magnetic anisotropy, *Appl. Phys. Lett.* **96**, 192506 (2010).

60. P. Balestriere, T. Devolder, J. Wunderlich, and C. Chappert, Electric field induced anisotropy modification in (Ga,Mn)As: A strategy for the precessional switching of the magnetization, *Appl. Phys. Lett.* **96**, 142504 (2010).

61. D. Chiba, T. Ono, F. Matsukura, and H. Ohno, Electric field control of thermal stability and magnetization switching in (Ga,Mn)As, *Appl. Phys. Lett.* **103**, 142418 (2013).

62. T. Dietl, Interplay between carrier localization and magnetism in diluted magnetic and ferromagnetic semiconductors, *J. Phys. Soc. Jpn.* **77**, 031005 (2008).

63. L. Chen, F. Matsukura, and H. Ohno, Direct-current voltages in (Ga,Mn)As structures induced by ferromagnetic resonance, *Nat. Commun.* **4**, 2055 (2013).

64. L. Chen, F. Matsukura, and H. Ohno, Electric field modulation of damping constant in a ferromagnetic semiconductor (Ga,Mn)As, *Phys. Rev. Lett.* **115**, 057204 (2015).

65. J. Sinova, T. Jungwirth, X. Liu et al., Magnetization relaxation in (Ga,Mn)As ferromagnetic semiconductors, *Phys. Rev. B* **69**, 085209 (2004).

66. T. Jungwirth, Q. Niu, and A. H. MacDonald, Anomalous Hall effect in ferromagnetic semiconductors, *Phys. Rev. Lett.* **88**, 207208 (2002).

67. D. Chiba A. Werpachowska, M. Endo et al., Anomalous Hall effect in field-effect structures of (Ga,Mn)As, *Phys. Rev. Lett.* **104**, 106601 (2010).

68. A. Werpachowska and T. Dietl, Effect of inversion asymmetry on the intrinsic anomalous Hall effect in ferromagnetic (Ga,Mn)As, *Phys. Rev. B* **81**, 155205 (2010).

69. M. Yamanouchi, D. Chiba, F. Matsukura, and H. Ohno, Current-assisted domain wall motion in ferromagnetic semiconductors, *Jpn. J. Appl. Phys.* **45**, 3854–3859 (2006).

70. M. Weisheit, S. Fähler, A. Marty, Y. Souche, C. Poinsignon, and D. Givord, Electric field-induced modification of magnetism in thin-film ferromagnets, *Science* **315**, 349–351 (2007).

71. C.-G. Duan, J. P. Velev, R. F. Sabirianov et al., Surface magnetoelectric effect in ferromagnetic metal films, *Phys. Rev. Lett.* **101**, 137201 (2008).

72. T. Maruyama, Y. Shiota, T. Nozakim et al., Large voltage-induced magnetic anisotropy change in a few atomic layers of iron, *Nat. Nanotech.* **4**, 158–161 (2009).

73. M. Endo, S. Kanai, S. Ikeda, F. Matsukura, and H. Ohno, Electric field effects on thickness dependent magnetic anisotropy of sputtered $MgO/Co_{40}Fe_{40}B_{20}/$ Ta structures, *Appl. Phys. Lett.* **96**, 212503 (2010).

74. Y. Shiota, T. Nozaki, F. Bonell, S. Murakami, T. Shinjo, and Y. Suzuki, Induction of coherent magnetization switching in a few atomic layers of FeCo using voltage pulses, *Nat. Mater.* **11**, 39–43 (2012).

75. S. Kanai, M. Yamanouchi, S. Ikeda, Y. Nakatani, F. Matsukura, and H. Ohno, Electric field-induced magnetization reversal in perpendicular-anisotropy CoFeB-MgO magnetic tunnel junction, *Appl. Phys. Lett.* **101**, 122403 (2012).

76. T. Shinjo, S. Hine, and T. Takada, Mössbauer spectra of ultrathin Fe films coated by MgO, *J. Phys.-Paris* **40**, C2-86–87 (1979).

77. M. Klaua, D. Ulmann, J. Barthel et al., Growth, structure, electronic, and magnetic properties of MgO/Fe(001) bilayers and Fe/MgO/Fe(001) trilayers, *Phys. Rev. B* **64**, 134411 (2001).

78. A. Manchon, C. Ducret, L. Lombard et al., Analysis of oxygen induced anisotropy crossover in Pt/Co/MOx trilayers, *J. Appl. Phys.* **104**, 043914 (2008).

79. S. Yakata, H. Kubota, Y. Suzuki et al., Influence of perpendicular magnetic anisotropy on spin-transfer switching current in CoFeB/MgO/CoFeB magnetic tunnel junctions, *J. Appl. Phys.* **105**, 07D131 (2009).

80. S. Ikeda, K. Miura, H. Yamamoto et al., A perpendicular-anisotropy CoFeB-MgO magnetic tunnel junction, *Nat. Mater.* **9**, 721–724 (2010).

81. K. Nakamura, R. Shimabukuro, Y. Fujiwara, T. Akiyama, T. Ito, and A. Freeman, Giant modification of the magnetocrystalline anisotropy in transition-metal monolayers by an external electric field, *Phys. Rev. Lett.* **102**, 187201 (2009).

82. R. Shimabukuro, K. Nakamura, T. Akiyama, and T. Ito, Electric field effects on magnetocrystalline anisotropy in ferromagnetic Fe monolayers, *Physica E* **42**, 1014–1017 (2010).

83. S. Kanai, M. Tsujikawa, Y. Miura, M. Shirai, F. Matsukura, and H. Ohno, Magnetic anisotropy in Ta/CoFeB/MgO investigated by x-ray magnetic circular dichroism and first-principles calculation, *Appl. Phys. Lett.* **105**, 222409 (2014).

84. Z. Wang, M. Saito, K. P. McKenna et al., Atomic-scale structure and local chemistry of CoFeB-magnetic tunnel junctions, *Nano Lett.* **16**, 1530–1536 (2016).

85. F. Bonell, Y. T. Takahashi, D. D. Lam et al., Reversible change in the oxidation state and magnetic circular dichroism of Fe driven by an electric field at the FeCo/MgO interface, *Appl. Phys. Lett.* **102**, 152401 (2013).

86. S. Miwa, K. Matsuda, K. Tanaka et al., Voltage-controlled magnetic anisotropy in Fe|MgO tunnel junctions studied by x-ray absorption spectroscopy, *Appl. Phys. Lett.* **107**, 162402 (2015).

87. S. Kanai, M. Endo, S. Ikeda, F. Matsukura, and H. Ohno, Magnetic anisotropy modulation in Ta/CoFeB/MgO structure by electric fields, *J. Phys.: Conf. Ser.* **266**, 012092 (2011).

88. T. Nozaki, K. Yakushiji, S. Tamaru et al., Voltage-induced magnetic anisotropy changes in an ultrathin FeB layer sandwiched between two MgO layers, *Appl. Phys. Express* **6**, 073005 (2013).

89. T. Koyama, A. Obinata, Y. Hibino, and D. Chiba, Sign reversal of electric field on coercivity in MgO/Co/Pt system, *Appl. Phys. Express* **6**, 123001 (2013).

90. W. Skowroński, T. Nozaki, D. D. Lam et al., Underlayer material influence on electric field controlled perpendicular magnetic anisotropy in CoFeB/MgO magnetic tunnel junctions, *Phys. Rev. B* **91**, 188410 (2015).

91. Y. Hibino, T. Koyama, A. Obinata et al., Peculiar temperature dependence of electric field effect on magnetic anisotropy in Co/Pd/MgO system, *Appl. Phys. Lett.* **109**, 082403 (2016).

92. S. E. Barnes, J. Ieda, and S. Maekawa, Rashba spin-orbit anisotropy and the electric field control of magnetism, *Sci. Rep.* **4**, 4105 (2014).

93. D. Yoshikawa, M. Obata, Y. Taguchi, S. Haraguchi, and T. Oda, Possible origin of nonlinear magnetic anisotropy variation in electric field effect in a double interface system, *Appl. Phys. Express* **7**, 111305 (2014).

94. P. V. Ong, N. Kioussis, D. Odkhuu, P. K. Amiri, K. L. Wang, and G. P. Carman, Giant voltage modulation of magnetic anisotropy in strained heavy metal/magnet/insulator heterostructures, *Phys. Rev. B* **92**, 020407(R) (2015).

95. X. W. Guan, X. M. Cheng, T. Huang, S. Wang, K. H. Xue, and X. S. Mio, Effect of metal-to-metal interface states on the electric field modified magnetic anisotropy in MgO/Fe/non-magnetic metal, *J. Appl. Phys.* **119**, 133905 (2016).

96. S. Miwa, M. Suzuki, M. Tsujikawa et al., Voltage controlled interfacial magnetism through platinum orbits, *Nat. Commun.* **8**, 15848 (2017).

97. D. Chiba, S. Fukami, K. Shimamura, N. Ishiwata, and T. Ono, Electrical control of the ferromagnetic phase transition in cobalt at room temperature, *Nat. Mater.* **10**, 853–856 (2011).

98. K. Shimamura, D. Chiba, S. Ono et al., Electrical control of Curie temperature in cobalt using an ionic liquid, *Appl. Phys. Lett.* **100**, 122402 (2012).

99. M. Kawaguchi, K. Shimamura, S. Ono et al., Electric field effect on magnetization of an Fe ultrathin film, *Appl. Phys. Express* **5**, 063007 (2012).

100. M. Oba, K. Nakamura, T. Akiyama, T. Ito, M. Weinert, and A. J. Freeman, Electric field modification of the magnon energy, exchange interaction, and Curie temperature of transition-metal thin films, *Phys. Rev. Lett.* **114**, 107202 (2015).

101. A.-M. Pradipto, T. Akiyama, T. Ito, and K. Nakamura, Mechanism and electric field induced modification of magnetic exchange stiffness in transition metal thin films on MgO (001), *Phys. Rev. B* **96**, 014425 (2017).

102. L. Gerhard, T. K. Yamada, T. Balashov et al., Magnetoelectric coupling at metal surfaces, *Nat. Nanotech.* **5**, 792–797 (2010).

103. A. Okada, S. Kanai, M. Yamanouchi, S. Ikeda, F. Matsukura, and H. Ohno, Electric field effects on magnetic anisotropy and damping constant in Ta/CoFeB/MgO investigated by ferromagnetic resonance, *Appl. Phys. Lett.* **105**, 052415 (2014).

104. A. Okada, S. He, B. Gu et al., Magnetization dynamics and its scattering mechanism in thin CoFeB films with interfacial anisotropy, *Proc. Natl. Acad. Sci.* **114**, 3815–3820 (2017).

105. M. Yamanouchi, A. Jander, P. Dhagat, S. Ikeda, F. Matsukura, and H. Ohno, Domain structure in CoFeB thin films with perpendicular magnetic anisotropy, *IEEE Magn. Lett.* **2**, 3000304 (2011).

106. T. Dohi, S. Kanai, A. Okada, F. Matsukura, and H. Ohno, Effect of electric field modulation of magnetic parameters on domain structure in MgO/CoFeB, *AIP Adv.* **6**, 075207 (2016).

107. A. L. Sukstanskii and K. I. Primak, Domain structure in an ultrathin ferromagnetic film, *J. Magn. Magn. Mater.* **169**, 31–38 (1997).

108. F. Ando, H. Kakizakai, T. Koyama et al., Modulation of the magnetic domain size induced by an electric field, *Appl. Phys. Lett.* **109**, 022401 (2016).

109. H. Kakizakai, F. Ando, T. Koyama et al., Switching local magnetization by electric field-induced domain wall motion, *Appl. Phys. Express* **9**, 963004 (2016).

110. T. Dohi, S. Kanai, F. Matsukura, and H. Ohno, Electric field effect on spin-wave resonance in a nanoscale CoFeB/MgO magnetic tunnel junction, *Appl. Phys. Lett.* **111**, 072403 (2017).

111. D. Chiba, M. Kawaguchi, S. Fukami et al., Electric field control of magnetic domain-wall velocity in ultrathin cobalt with perpendicular magnetization, *Nat. Commun.* **3**, 888 (2012).

112. U. Bauer, S. Emori, and S. D. Beach, Voltage-controlled domain wall traps in ferromagnetic nanowires, *Nat. Nanotech.* **8**, 411–416 (2013).

113. H.-B. Chen and Y.-Q. Li, Electric field-controlled suppression of Walker breakdown and chirality switching in magnetic domain wall, *Appl. Phys. Express* **9**, 073004 (2016).

114. K. J. A. Franke, B. Van de Wiele, Y. Shirahata, S J. Hämäläinen, T. Taniyama, and S. van Djiken, Reversible electric field-driven magnetic domain wall motion, *Phys. Rev. X* **5**, 011010 (2015).

115. M. Goto, K. Nawaoka, S. Miwa, S. Hatanaka, N. Mizuochi, and Y. Suzuki, Electric field modulation of tunneling anisotropic magnetoresistance in tunnel junctions with antiferromagnetic electrodes, *Jpn. J. Appl. Phys.* **55**, 080304 (2016).

116. P. X. Zhang, G.F. Yin, Y.Y. Wang, B. Cui, F. Pan, and C. Song, Electrical control of antiferromagnetic metal up to 15 nm, *Sci. China-Phys. Mech. Astron.* **59**, 687511 (2016).

117. S. Shimizu, K. S. Takahashi, T. Hatano, M. Kawasaki, Y. Tokura, and Y. Iwasa, Electrically tunable anomalous Hall effect in Pt thin films, *Phys. Rev. Lett.* **111**, 216803 (2013).

118. K. C. Nowak, F. H. Koopens, Y. V. Nazarov, and L. M. K. Vandersypen, Coherent control of a single electron spin with electric fields, *Science* **318**, 1430-1433 (2007).

119. M. Ono, J. Ishihara, G. Sato, Y. Ohno, and H. Ohno, Coherent manipulation of nuclear spins in semiconductors with an electric field, *Appl. Phys. Express* **6**, 033002 (2013).

120. T. Nozaki, Y. Shiota et al., Electric field-induced ferromagnetic resonance excitation in an ultrathin ferromagnetic metal layer, *Nat. Phys.* **8**, 491–496 (2012).

121. J. Zhu, J. A. Katine, G. E. Rowlands et al., Voltage-induced ferromagnetic resonance in magnetic tunnel junctions, *Phys. Rev. Lett.* **108**, 197203 (2012).

122. A. A. Tulapurkar, Y. Suzuki, A. Fukushima et al., Spin-torque diode effect in magnetic tunnel junctions, *Nature* **438**, 339–342 (2005).

123. K. Mizunuma, M. Yamanouchi, H. Sato et al., Size dependence of magnetic properties of nanoscale CoFeB-MgO magnetic tunnel junctions with perpendicular magnetic easy axis observed by ferromagnetic resonance, *Appl. Phys. Express* **6**, 063002 (2013).

124. S. Kanai, M. Gajek, D. C. Worledge, F. Matsukura, and H. Ohno, Electric field-induced ferromagnetic resonance in a CoFeB/MgO magnetic tunnel junction under dc bias voltages, *Appl. Phys. Lett.* **105**, 242409 (2014).

125. E. Hirayama, S. Kanai, H. Sato, F. Matsukura, and H. Ohno, Ferromagnetic resonance in nanoscale CoFeB/MgO magnetic tunnel junctions, *J. Appl. Phys.* **117**, 17B708 (2015).

126. M. Shinozaki, E. Hirayama, S. Kanai, H. Sato, F. Matsukura, and H. Ohno, Damping constant in a free layer in nanoscale CoFeB/MgO magnetic tunnel junctions investigated by homodyne-detected ferromagnetic resonance, *Appl. Phys. Express* **10**, 013001 (2017).

127. E. Hirayama, S. Kanai, J. Ohe, H. Sato, F. Matsukura, and H. Ohno, Electric field induced nonlinear ferromagnetic resonance in a CoFeB/MgO magnetic tunnel junction, *Appl. Phys. Lett.* **107**, 132404 (2015).

128. M. Harder, Y. Gui, and C.-M. Hu, Electrical detection of magnetization dynamics via spin rectification effect, *Phys. Rep.* **661**, 1–59 (2016).

129. H. Mazraati, T. Q. Le, A. A. Awad et al., Free- and reference-layer magnetization modes versus in-plane magnetic field in a magnetic tunnel junction with perpendicular magnetic easy axis, *Phys. Rev. B* **94**, 104428 (2016).

130. Y. Shiota, S. Miwa, S. Tamaru et al., High-output microwave detector using voltage-induced ferromagnetic resonance, *Appl. Phys. Lett.* **105**, 192408 (2014).

131. X. Zhang, T. Liu, M. E. Flatté, and H. X. Tang, Electric field coupling to spin waves in a centrosymmetric ferrite, *Phys. Rev. Lett.* **113**, 037202 (2014).

132. K. Nawaoka, Y. Shiota, S. Miwa et al., Voltage modulation of propagation of spin waves in Fe, *J. Appl. Phys.* **117**, 17A905 (2015).

133. K. Nawaoka, S. Miwa, Y. Shiota, N. Mizuochi, and Y. Suzuki, Voltage induction of interfacial Dzyaloshinskii-Moriya interaction in Au/Fe/MgO artificial multilayer, *Appl. Phys. Express* **8**, 063004 (2015).

134. S. Mühlbauer, B. Binz, F. Jonietz et al., Skyrmion lattice in a chiral magnet, *Science,* **323**, 915–919 (2009).

135. Y. Nakatani, M. Hayashi, S. Kanai, S. Fukami, and H. Ohno, Electric field control of Skyrmions in magnetic nanodisks, *Appl. Phys. Lett.* **108**, 152403 (2016).

136. W.-G. Wang, M. Li, S. Hageman, and C. L. Chien, Electric field assisted switching in magnetic tunnel junctions, *Nat. Mater.* **11**, 64–68 (2012).

137. Y. Shiota, T. Maruyama, T. Nozaki, T. Shinjo, M. Shiraishi, and Y. Suzuki, Voltage-assisted magnetization switching in ultrathin $Fe_{80}Co_{20}$, *Appl. Phys. Express* **2**, 063001 (2009).

138. S. Kanai, Y. Nakatani, M. Yamanouchi, S. Ikeda, F. Matsukura, and H. Ohno, In-plane magnetic field dependence of electric field-induced magnetization switching, *Appl. Phys. Lett.* **103**, 074208 (2013).

139. S. Kanai, F. Matsukura, and H. Ohno, Electric field-induced magnetization switching in CoFeB/MgO magnetic tunnel junctions, *Jpn. J. Appl. Phys* **58**, 0802A3 (2017).

140. S. Kanai, Y. Nakatani, M. Yamanouchi et al., Magnetization switching in a CoFeB/MgO magnetic tunnel junction by combining spin-transfer torque and electric field-effect, *Appl. Phys. Lett.* **104**, 212406 (2014).

141. C. Grezes, F. Ebrahimi, J. G. Alzate et al., Ultra-low switching and scaling in electric field-controlled nanoscale magnetic tunnel junctions with high resistance-area product, *Appl. Phys. Lett.* **108**, 012403 (2014).

142. S. Kanai, F. Matsukura, and H. Ohno, Electric field-induced magnetization switching in a CoFeB/MgO magnetic tunnel junctions with high junction resistance, *Appl. Phys. Lett.* **108**, 192406 (2014).

# 14

# Topological Insulators
## *From Fundamentals to Applications*

**Matthew J. Gilbert and Ewelina M. Hankiewicz**

## 14.1  INTRODUCTION

One of the biggest challenges of physics is the classification of different states of matter. Until recently, the phases of matter could be understood using Landau–Lifshitz theory [1], which characterizes states in terms of their underlying symmetries that are spontaneously broken. For example, a magnet spontaneously breaks rotation symmetry, although the fundamental interactions within the magnet itself are isotropic in nature. Starting from 1980 with the discovery of the quantum Hall effect (QHE) [2, 3], new phases have emerged that are not characterized by broken symmetry, but rather by the presence of a global, or topological, invariant that is contributed to by all of the states in the system. The QHE appears in large magnetic fields when the two-dimensional density of states becomes broken into successive, highly degenerate Landau levels. As is the case in the traditional Hall effect, the presence of an applied in-plane electric field drives a current from one side of the system to the other. In the case of the QHE, the position of the Fermi level relative to the Landau levels determines the type of transport that is observed. When the Fermi level is inside a Landau level, there are many states available to carry current across the system, both within the bulk and at the edge. However, when the Fermi level is between two successive Landau levels, then the kinetic energy of the bulk is quenched and becomes insulating. The resulting conduction becomes quantized, and appears through chiral (unidirectional) edge states observed at the boundary of the sample that are immune against backscattering. The connection between the QHE and topology is made via the TKKN (after Thouless, Kohomoto, Nightingale, and den Nijs) invariant, in which the first Chern number describes the winding of the corresponding Bloch wave functions over the magnetic Brillouin zone [4]. This quantized number corresponds exactly to the number of propagating edge states in the QHE, and is invariant to the details of the underlying system, so long as the two edge states remain spatially segregated. The TKNN invariant thus provides the means of calculating the topological invariant for the QHE.

While it is clear from the relationship between the TKNN invariant and the number of edge states in the system that the QHE is topological in nature, there is no explicit symmetry that protects the edge states. Therefore, the question then becomes whether topological phases in nature exist which have a non-trivial, or non-zero, topological invariant, but are characterized by particular symmetries that are preserved. In an early effort to answer this question, Haldane constructed an artificial model on the hexagonal lattice, characterized by a non-zero Chern number, but without any net magnetization [5]. This development of a topological phase that has broken time-reversal symmetry, yet preserved inversion symmetry, inspired Kane and Mele to define quantum spin Hall (QSH) systems, commonly referred to as two-dimensional (2D) topological insulators (TIs), that preserve time-reversal symmetry [6, 7]. Indeed, in a more general sense, one can define TIs as materials that possess an insulating gap in the bulk, while on the boundary, there are gapless metallic spin-polarized edge states whose gapless nature is stabilized by the presence of an underlying symmetry.

Specifically, within 2D TIs [6–11], the observed metallic electric conduction is associated with propagating states that occur only near the sample edges, while the conduction in the interior is suppressed by a band gap, like in ordinary band insulators. These edge states originate from intrinsic spin-orbit (SO) coupling, and are profoundly different from those appearing in quantum Hall systems in a strong perpendicular magnetic field [12, 13]. The key distinction between the edge states observed in the QHE, and the edge states of 2D TIs lies in the role of time-reversal symmetry. In 2D TIs, the SO coupling preserves time-reversal symmetry, resulting in a pair of counter-propagating channels on the same edge, as opposed to the chiral edge states in integer quantum Hall systems. The strong SO interactions in the 2D TI force the spin and momentum components of the two independent TI edge channels to be locked in opposite directions, so that one spin-polarized edge state propagates in one direction, while the other oppositely spin-polarized edge state flows in the opposite direction. Such spin-momentum locked edge states are referred to as helical. The helical edge state consisting of a single massless Dirac fermion is "holographic", in the sense that it cannot exist in a purely 1D system, but it can only exist as the boundary of a two-dimensional system [14]. Moreover, as a result of the presence of the time-reversal symmetry, the helical edge states have a nodal band dispersion (see also Figure 14.1) that is topologically protected against any structural or sample imperfections that do not break time-reversal symmetry [15, 16].

Nonetheless, TIs are not strictly limited to 2D systems, and we now introduce the three-dimensional (3D) analogues of 2D TIs. In 3D TIs, the topologically protected electronic states appear on the surface of a bulk material. These surface states have a nodal band dispersion in the form of a 2D Dirac-like cone reflecting a continuum of momentum directions on the surface, as shown in Figure 14.2. The family of materials and heterostructures which can host 2D surface states is large, and continues to increase. 2D surface states were first predicted in inverted semiconductor contacts, but without realizing their topological protection [17, 18]. The coexistence of metallic surface states and a bulk gapped band structure, normally referred to as the 3D TI phase, has been established theoretically for many materials including: the semiconducting alloy $Bi_{1-x}Sb_x$ [19], strained 3D layers of $\alpha$-Sn and HgTe [19], the tetradymite semiconductors $Bi_2Se_3$, $Bi_2Te_3$, and $Sb_2Te_3$ [20], thallium-based ternary chalcogenides $TlBiTe_2$ and $TlBiSe_2$ [21–23], as well as Pb-based layered chalcogenides [24, 25]. Experimentally, topological surface states in 3D materials have been observed by means of angle-resolved photoemission spectroscopy (ARPES) in many of the theoretically predicted

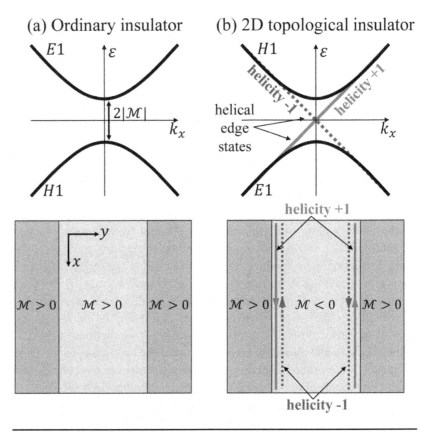

**FIGURE 14.1**    Band structure of (a) an ordinary insulating material, and (b) a 2D topological insulator with inverted band order based on HgTe QWs. In (b), the helical edge states of 2D TIs and their dispersion (for a single edge) are shown (depicted as solid and dashed lines, respectively).

materials, such as: $Bi_{1-x}Sb_x$ [26, 27], $Bi_2Se_3$ [28], $Bi_2Te_3$ [29], $TlBiSe_2$ [30–32], $TlBiTe_2$ [32], strained HgTe [33, 34], $Pb(Bi_{1-x}Sb_x)_2Te_4$ [35], and $PbBi_2Te_4$ [36].

Subsequent to the work on TIs in both 2D and 3D, the notion of topological protection was then extended to encompass topological phases of matter that are protected by symmetries beyond time-reversal. An example of topological protection beyond that provided by the presence of time-reversal symmetry, are symmetries such as crystalline symmetries. These crystalline symmetries include, for example: rotations [37], reflections [38], and glide planes [39]. The most common topological phases stabilized by the presence of crystalline symmetry are mirror-symmetric TIs [40–42]. As in the case of a TI that has preserved time-reversal symmetry, a topological crystalline insulator is generally defined as a bulk insulator with gapless edge or surface states that cannot be removed so long as the preserving crystal symmetry is intact.

Recently, further examples of topological phases have emerged in materials that are not insulating in their bulk, but rather semimetallic. Much like their 2D cousin graphene, Dirac semimetals contain degenerate 3D gapless Dirac points that are centered in the bulk of the material, rather than just on the surface. In contrast to graphene, however, Dirac semimetals have

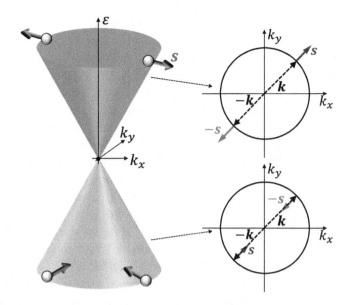

**FIGURE 14.2**  Two-dimensional gapless Dirac cone which is the dispersion of the surface state of 3D TIs. The helicity of the upper and lower parts of the cone is opposite.

Hamiltonians that are comprised of two counterpropagating Weyl fermions. Nevertheless, as two counterpropagating Weyl fermions may annihilate one another if they come into contact within momentum space, resulting in an avoided band crossing, there must be an additional symmetry present in the system that stabilizes the band crossing. Early theoretical predictions stated that $A_3Bi$ ($A$ = Na, K, Rb) compounds would be candidate materials to host the requisite 3D Dirac dispersion in the bulk [43]. These materials crystalize in the simple honeycomb lattice in-plane stacked in the $c$-axis direction orthogonal to the plane. The heavy atoms associated with the $A_3$ compound were predicted to invert the respective conduction and valence bands. However, the mere presence of band inversion is not sufficient to be able to predict topological behavior, as the band crossing will simply form an avoided level crossing without the presence of an additional stabilizing symmetry. In the case of $Na_3Bi$, the stabilization that allows for the preservation of the bulk band crossing comes from the time-reversal and inversion symmetries. Following the theoretical predictions, the first Dirac semimetal that has been experimentally observed is, naturally, $Na_3Bi$ [44]. Dirac semimetals with additional symmetries beyond those present in $Na_3Bi$ have also been theoretically predicted and subsequently experimentally confirmed. Examples of additional Dirac semimetals include: $Cd_3As_2$, that has preserved time-reversal symmetry and $C_4$ rotational symmetry [45], and CuMnAs [46, 47]. CuMnAs is a particularly interesting example of a Dirac semimetal, in that it is an antiferromagnetic semimetal that breaks both time-reversal symmetry and inversion symmetry. At first glance, it would seem that CuMnAs would then possess a gap due to the broken symmetries, yet CuMnAs preserves the combination of time-reversal and inversion symmetry, thereby allowing a protected bulk 3D band crossing point to persist.

As we have mentioned, Dirac semimetals are characterized by a low-energy effective Hamiltonian that consists of two copies of counterpropagating Weyl fermions. Weyl fermions are massless fermionic quasiparticle excitations, as in the case of a Dirac electron. However, Weyl fermions have a definite chirality in real space associated with them, essentially rendering them similar to that of the edge state encountered within the integer QHE. Another manner within which one can understand the relationship between Weyl fermions and Dirac fermions is that the Weyl fermion two component spinor is exactly half of the normal four component Dirac spinor. For many years, these excitations have been sought as another example of the intimate connection between high-energy physics and condensed matter physics. To be more specific, neutrinos had been assumed to be Weyl fermions; however, it was later realized that they do, in fact, possess a small mass that invalidates them as candidate particles that possess Weyl fermion characteristics. While finding high-energy realizations of Weyl fermions has not yet resulted in confirmed examples, the search has been fruitful within the context of condensed matter realizations of quasiparticle excitations that have characteristics consistent with Weyl fermions in semimetallic materials. Weyl semimetals are characterized by exotic Fermi open surface projections, or Fermi arcs, that connect sources and sinks of Berry curvature, known as Weyl nodes, that are characterized by an integer Chern number. To date there have been several different mechanisms through which Weyl fermions have been realized in semimetals by breaking different symmetries that allow the typical Dirac spinor to be broken into a Weyl spinor. Specifically, one can break: time-reversal symmetry, as in the case of topological heterostructures [48] or $HgCr_2Se_4$ [49], inversion symmetry as in the case of TaAs and related compounds [50, 51], or within compounds that break Lorentz invariance such as $W_{1-x}Mo_xTe_2$ and LaAlGe that are also referred to as Type-II Weyl semimetals [52].

The remainder of the chapter is constructed as follows: In Section 14.2, we describe in detail the construction of the bulk Hamiltonian for 2D TIs, as this forms the foundation for the understanding of the 3D TI and allows one to extrapolate to the construction of more complicated Hamiltonians containing additional symmetries. In our construction, we focus on the salient features associated with 2D TIs, including the formation of the metallic spin-polarized edge states and their topological protection. Using this formulation as a base, in Section 14.3, we move on to discuss the properties of 3D TIs, building on the knowledge that we gained in Section 14.2. The unique physics of TIs naturally leads to the development of many new proposals for applications. From the spintronics point of view [53], the spin-polarized edge channels in 2D TIs could be used to inject spin currents into doped semiconductors or metals [54]. Further, ferromagnet/3D TI junctions have shown a significant promise for potential spintronics applications due to: the large spin-orbit torque that the Dirac surface can generate [55, 56], electrical manipulation of spin-polarized currents in ferromagnet/3D TI junctions [57–59], as well as an unusual in-plane tunneling Hall effect which could be possibly used for spin-valves [60]. Nevertheless, while there are many proposals for applications of TIs, one of the most intriguing is their possible application in the field of topological quantum information processing. In Section 14.4, we address this interesting application through an introduction to the formation of the requisite states. In Section 14.5, we present a mathematical formulation of the process through which the combination of normal $s$-wave superconductivity and time-reversal invariant TIs can be utilized to form the excitations, known as Majorana-bound states, that form the basis of popular implementations of the topological quantum computing introduced in Section 14.4. We conclude with Section 14.6, where we discuss some of the recent progress that has been achieved in the field of topological superconductivity within TIs, so as to realize the necessary components for topological quantum information processing. Some related discussion on topological insulators can be found in Chapter 15, Volume 2, and on triplet superconducting pairing needed for Majorana-bound states in Chapter 16, Volume 1.

## 14.2 TOPOLOGICAL INSULATORS IN 2D: QUANTUM SPIN HALL SYSTEMS

### 14.2.1 EFFECTIVE BULK HAMILTONIAN FOR 2D TIs

We begin by describing the construction of the model Hamiltonian for 2D TIs. In this endeavor, there are two main models describing 2D TIs: the Bernevig/Hughes/Zhang (BHZ) model [8], which describes HgTe/CdTe and InAs/GaSb quantum wells (QWs), and the extensions of the Kane–Mele model which describes 2D TIs on hexagonal lattices [6, 7]. From a historical perspective, the Kane–Mele model of graphene is the first material that was proposed to be a 2D TI [7]. However, as the main orbitals at the Fermi energy are $p_z$-like, the atomic-SO interaction is between next-nearest neighbors, the induced gap is small, only around 20 μeV [61, 62], and the predicted effect is not experimentally measurable. Recently, in the context of new types of purely 2D materials on hexagonal lattices similar to that of graphene, there are proposals that may be understood via extensions of the Kane–Mele model that purport to show larger SO interaction, and, therefore, may yield topological phases that are more experimentally accessible. These materials include: silicene [63, 64] (with a band gap of around a few meV), germanene [63] (with a band gap of around 20 meV), the heterostructure of germanene on $MoS_2$ [65], functionalized stanene [66] (with a band gap of around 0.3 eV), and single layers of $WTe_2$ [67, 68, 69] (with a band gap of around 0.1 eV). Additionally, bismuthene on SiC has recently gained a lot of attention as a potential candidate for room-temperature topological behavior [70–73] where the on-site SO interactions give a band gap of the order of 0.7 eV.

As most of the physics surrounding the Kane–Mele model has yet to be experimentally verified, we will focus our attention on the BHZ model, and discuss 2D TIs in the context of their first experimental realization in HgTe/CdTe QW [9–11, 54]. We start our examination by considering the structure of CdTe. CdTe is a zinc-blende-type semiconductor whose band structure can be described by the eight-band $\mathbf{k} \cdot \mathbf{p}$ model. CdTe has normal band structure, i.e. its conduction band ($\Gamma_6$) consists of s-wave orbitals, while the valence band ($\Gamma_8$) consists of $p$-like orbitals. Therefore, a thin layer of CdTe produces the well-known semiconducting band structure that is similar to the one shown in Figure 14.1a with the $E1$ band (conduction band) having mainly electron-like properties, and the $H1$ band (valence band) having heavy hole-like character and positive band gap of value $2\mathcal{M}$. However, this is not the situation that one finds in bulk HgTe. Unlike conventional zinc-blende semiconductors, due to large relativistic corrections, including relativistic velocity corrections and the SO interaction, HgTe has an inverted band structure, by which we mean that the $\Gamma_6$ band, that originates from metallic s-orbitals and usually forms the conduction band in normal semiconductors, has a lower energy than the $\Gamma_8$ band which derives from chalcogenide $p$-orbitals [75]. Consequently, in the HgTe layer, the energetic subbands are also inverted and the band gap $2\mathcal{M}$ at $k = 0$ is negative, as shown in Figure 14.1b. In this case, the conduction band has heavy hole-like ($H1$) character, while the valence band has mainly electron-like character.

Now, let us build heterostructures where HgTe is sandwiched between CdTe layers to form a HgTe/CdTe QW. Intuitively, one can say that as long as the layer of HgTe is very thin, the full heterostructure has normal band ordering, like in Figure 14.1a, while for thick enough HgTe layers, the band structure can be inverted, like in Figure 14.1b. Indeed, this intuitive picture is confirmed in Figure 14.3, where the band structure of the QW as a function of the thickness, $d$, of the HgTe layer is shown. Here, $\Gamma_8$-derived subbands are denoted by $H1, H2,...$ as they correspond to heavy-hole subbands in non-inverted semiconductors, whereas the $E1, E2,...$ subbands are mainly derived from $\Gamma_6$ subbands and, therefore, are mainly electron-like subbands. Indeed, with decreasing HgTe thickness $d$, the energies of the $E$ subbands shift to positive energies as a result of quantum confinement, whereas those of the $H$ subbands shift to negative energies, as shown in Figure 14.1a. This results in the normal band ordering for thin QWs. The different $d$-dependence of the $E1$ and $H1$ subbands, shown in Figure 14.3, demonstrates this principle that there exists a critical thickness, $d_c$, at which point the subbands change their energetic order, leading to a positive band gap in CdTe and a negative band gap in HgTe (see Figure 14.1b). The effect of band ordering that we have been describing here implies something interesting should happen as one moves from one material that has normal band ordering (band gap $\mathcal{M}$ larger than zero), and HgTe that has inverted band ordering (band gap $\mathcal{M}$ smaller than zero).

To better understand the situation where we have a heterostructure that contains a transition between non-inverted and inverted materials, as we have in the case of CdTe/HgTe QW, we now begin to construct a simple model. Assuming that $d \approx d_c$ and near the $\Gamma$ point, and integrating growth direction (z-direction), we can derive an effective four-band model for the HgTe/CdTe QWs involving double degenerate $E1$ and $H1$ subbands from the eight-band $\mathbf{k} \cdot \mathbf{p}$ model [8, 76]. We now introduce the relevant basis states $|E1, j_z = 1/2\rangle$, $|H1, j_z = 3/2\rangle$, $|E1, j_z = -1/2\rangle$ and $|H1, j_z = -3/2\rangle$, where

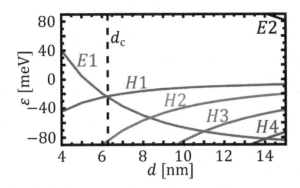

**FIGURE 14.3** Ordering of electron $E1, E2, \ldots$ and heavy-hole $H1, H2, \ldots$ subband energies versus well thickness $d$ from $\mathbf{k} \cdot \mathbf{p}$ calculations for a HgTe/Hg$_{0.3}$Cd$_{0.7}$Te quantum well. One can see the transition from the topologically trivial to the topologically non-trivial regime for a critical thickness of $d_c = 6.3$ nm, which corresponds to the inversion of the bands between $E1$ and $H1$ [74]. (After Büttner, B. et al., *Nat. Phys.* 7, 418, 2011. With permission.)

$j_z$ denotes z-component of the total angular momentum. Now using these basis states, one can write the effective four-band Hamiltonian for the QW system as follows [8, 76]:

$$H_{\mathrm{HgTe}} = \begin{bmatrix} h(\mathbf{k}) & 0 \\ 0 & h^*(-\mathbf{k}) \end{bmatrix}. \qquad (14.1)$$

In Equation 14.1,

$$h(\mathbf{k}) = \mathcal{A}(\sigma_x k_x - \sigma_y k_y) + \mathcal{M}_{\mathbf{k}} \sigma_z + \mathcal{D} \mathbf{k}^2 \sigma_0, \qquad (14.2)$$

and

$$\mathcal{M}_{\mathbf{k}} = \mathcal{M} + \mathcal{B} \mathbf{k}^2. \qquad (14.3)$$

We must now define the individual contributions to the Hamiltonian. We note that the two diagonal blocks of $H_{\mathrm{HgTe}}$ (14.1) describe pairs of states related to one another by time-reversal symmetry, in other words, Kramers partners. Indeed, for a half-integer spin, as in our case, every energy level is at least double degenerate (Kramers theory) and the two states corresponding to this degeneracy are related by time-reversal symmetry and called Kramers partners. Each of the blocks has a $2 \times 2$ matrix structure, with Pauli matrices $\sigma_{x,y,z}$ and unit matrix $\sigma_0$ representing the two lowest-energy subbands $E1$ and $H1$. $\mathcal{M}_{\mathbf{k}}$ is an Einstein mass term whose value can be manipulated to yield the band gap $2\mathcal{M}$ at the $\Gamma$ ($\mathbf{k} = 0$) point of the Brillouin zone (corresponding to the gap between electron and positron bands $2mc^2$ for a relativistic electron in vacuum). The linear terms in Equation 14.2, proportional to the constant $\mathcal{A}$ and in-plane wave-vectors $k_{x,y}$, describe the $E1$–$H1$ hybridization. The positive quadratic terms $\mathcal{B}\mathbf{k}^2$ and $\mathcal{D}\mathbf{k}^2$ are related to the effective (Newtonian) masses (band curvature) of the $E1$ and $H1$ bands in HgTe QWs [8]. The Hamiltonian (14.1) can be extended to include the SO coupling between the Kramers partners [10, 76].

Using the unitary transformation $H \to UHU^\dagger$ with $U = \begin{pmatrix} 0 & \sigma_z \\ -i\sigma_y & 0 \end{pmatrix}$, we can recast the Hamiltonian (14.1) into a Dirac-like form

$$H = \tau_z \boldsymbol{\sigma} \cdot (\mathcal{A}\mathbf{k} + \mathcal{M}_\mathbf{k}\mathbf{z}) + \mathcal{D}\mathbf{k}^2 \tau_0 \sigma_0, \tag{14.4}$$

where $k = (k_x, k_y, 0)$, $z = (0, 0, 1)$ and the Pauli matrix $\tau_z$ and the unit matrix $\tau_0$ act on the Kramers partners. We note that despite the Einstein mass term $\mathcal{M}_\mathbf{k}\tau_z\sigma_z$, the Hamiltonian (14.4) is invariant under time reversal, i.e. $T^\dagger H T = H$, where $T = i\tau_y \sigma_x \mathcal{C}$ is the time-reversal operator, with $\mathcal{C}$ denoting complex conjugation. The band dispersion for both normal and topologically non-trivial insulators is described by

$$E = \mathcal{D}k^2 \pm \sqrt{\mathcal{A}^2 k^2 + (\mathcal{M} + \mathcal{B}k^2)^2}, \tag{14.5}$$

where the energy $E$ is double degenerate. Therefore, to identify the topologically non-trivial regime, one needs to calculate a topological invariant of the system, or introduce a boundary in the problem to check the formation of edge states. Since time-reversal symmetry is preserved in 2D TIs, the Chern number, $C$, (winding number of the Bloch wave functions over the magnetic Brillouin zero) is zero, due to the cancellation of the Chern numbers for two time-reversal related blocks ($C_\uparrow + C_\downarrow = 0$) for Equation 14.1, where $C_\uparrow$ is the Chern number for a block with the positive $j_z$ and $C_\downarrow$ is the Chern number for the block with the negative $j_z$. However, the spin Chern number describing the difference between the winding number of two corresponding Bloch wave functions for the Kramers partners over the whole Brillouin zone is non-zero i.e. $C_s = (C_\uparrow - C_\downarrow)/2$. This spin Chern number defines the topological invariant of the BHZ model.

## 14.2.2 HELICAL EDGE STATES IN 2D TIS AND THEIR PROTECTION

Edge states can be realized when a QW with an inverted gap $\mathcal{M} < 0$ borders with a normal insulator with $\mathcal{M} > 0$ (see Figure 14.1b). They originate from the fact that one cannot deform the insulator with the negative band gap (topological insulator) with spin Chern number $C_s = 1$ to the one with the positive band gap (trivial insulator) with spin Chern number $C_s = 0$, without going through gapless states at the boundary (edge states). Through the bulk boundary correspondence [15, 16], the number of pairs of edge states (called also $Z_2$ invariant [6, 15, 16]) is exactly equal to $\nu = C_s \bmod 2$. Therefore for the BHZ model, we expect one pair of edge states at the interface between a topologically trivial and non-trivial insulator, as shown in Figure 14.1b.

In this subsection, we explicitly show the existence of such edge states and discuss the following properties:

✦ the edge-state spectrum is gapless and merges into the bulk spectrum above the band gap;

✦ the edge states are the orthogonal eigenstates of the helicity operator $\Sigma = \tau_z\sigma\cdot\hat{\mathbf{k}}$, where $\hat{\mathbf{k}}$ is the unit vector in the direction of the edge-state momentum. For this reason, the QSH edge states are called helical;

✦ a local static perturbation $H'$ preserving time-reversal symmetry does not couple the QSH edge states.

In order to illustrate these properties, we will make further simplifications. To get a simple analytical model, we will omit all terms $\propto \mathbf{k}^2$ in Hamiltonian (14.4). This is justified since the edge states occur in the vicinity of the $\Gamma(\mathbf{k}=0)$ point. Although without quadratic terms, one cannot define a topological invariant uniquely, for the particle-hole symmetric case, the edge states only exist in the topologically non-trivial regime [77]. Then Hamiltonian (14.4) takes the form $H = \tau_z\sigma\cdot(A\mathbf{k} + \mathcal{M}\mathbf{z})$, which in position representation corresponds to the following equation for the four-component wave function $\Psi(\mathbf{r})$:

$$[\varepsilon\sigma_0 - \tau_z\sigma\cdot(-iA\nabla + \mathcal{M}\mathbf{z})]\Psi(\mathbf{r}) = 0. \tag{14.6}$$

Second, one cannot confine the Dirac spectrum with hard wall boundary conditions when the quadratic terms (lattice normalization) are missing. For a linear differential equation, like the Dirac equation, only one boundary condition is needed, and one requires that particles do not leak through the boundary, i.e. the normal component of the particle current at the boundary vanishes [78]. In our case, we assume vanishing of the current for $y=0$. This yields the following condition:

$$J_y(x,y=0) = (e/\hbar)A\Psi^\dagger(x,y=0)\tau_z\sigma_y\Psi(x,y=0) = 0 \tag{14.7}$$

and correspondingly for the wave function:

$$\Psi(x,y=0) = \tau_0\sigma_x\Psi(x,y=0). \tag{14.8}$$

We seek solutions to Equation 14.6 in the form of the two eigenstates, $\Psi_{k,\pm}(r)$, of the diagonal matrix $\tau_z\sigma_0$ propagating along the edge (in $x$-direction) of the TI and decaying exponentially away from it in the $y$-direction. The decaying solution is enforced since there is a finite band gap on both sides of an interface in $y$-direction (see Figure 14.1).

$$\Psi_{k,+}(\mathbf{r}) = \begin{pmatrix} 1 \\ 0 \end{pmatrix} \otimes \begin{pmatrix} \Psi_{1k+} \\ \Psi_{2k+} \end{pmatrix} e^{ikx - y/\lambda}, \tag{14.9}$$

$$\Psi_{k,-}(\mathbf{r}) = \begin{pmatrix} 0 \\ 1 \end{pmatrix} \otimes \begin{pmatrix} \Psi_{1k-} \\ \Psi_{2k-} \end{pmatrix} e^{ikx - y/\lambda}, \tag{14.10}$$

with a real positive decay length $\lambda > 0$. The symbol $\otimes$ denotes a tensor product of an eigenstate of $\tau_z$ (first column) and the wave function in $\sigma$ space (second column). The conditions for the non-trivial solutions for the coefficients $\Psi_{1k\pm}$ and $\Psi_{2k\pm}$ follow from Equations 14.6 and 14.8, yielding two equations for $\lambda$ and $\varepsilon$:

$$1/\lambda^2 - M^2/\mathcal{A}^2 = k^2 - \varepsilon^2/\mathcal{A}^2, \tag{14.11}$$

$$1/\lambda + M/\mathcal{A} = k - \varepsilon/\mathcal{A}\tau, \quad \tau = \pm 1. \tag{14.12}$$

We notice that the left-hand side of Equation 14.12 does not contain the index $\tau$, whereas the right-hand side does. This can only be true if both sides of Equation 14.12 (and those Equation 14.11) vanish independently, which yields a solution with a gapless linear dispersion and a real decay length:

$$\varepsilon_{k\tau} = \mathcal{A}k\tau, \quad \lambda = -\mathcal{A}/M, \quad M < 0. \tag{14.13}$$

Since $\lambda$ must be positive, the edge states exist only in a system with an inverted negative gap, disappearing when $M$ turns positive. Their propagation velocity $v = \mathcal{A}/\hbar$ coincides with that of the bulk states above the gap (see also Figure 14.1).

The edge-state wave functions normalized to the half-space $0 \leq y < \infty$ are given by

$$\Psi_{k,+}(\mathbf{r}) = \begin{pmatrix} 1 \\ 0 \end{pmatrix} \otimes \begin{pmatrix} 1 \\ 1 \end{pmatrix} \sqrt{\frac{|M|}{\mathcal{A}}} e^{ikx - |M|y/\mathcal{A}}, \tag{14.14}$$

$$\Psi_{k,-}(\mathbf{r}) = \begin{pmatrix} 0 \\ 1 \end{pmatrix} \otimes \begin{pmatrix} 1 \\ 1 \end{pmatrix} \sqrt{\frac{|M|}{\mathcal{A}}} e^{ikx - |M|y/\mathcal{A}}. \tag{14.15}$$

The key feature of the edge states (14.14) and (14.15) is that they are orthogonal eigenstates of the helicity operator $\Sigma = \tau_z \sigma_x$:

$$\Sigma \Psi_{k,\tau}(\mathbf{r}) = \tau \Psi_{k,\tau}(\mathbf{r}), \quad \tau = \pm 1. \tag{14.16}$$

The helicity $\Sigma$ is defined as the projection of the vector $\boldsymbol{\Sigma} = \tau_z \boldsymbol{\sigma}$ on the direction of the edge-state momentum $\hat{\mathbf{k}} \parallel x$. Since the matrix structure of $\Sigma$ derives from the SO-split energy bands, it is also called the spin helicity. Equation 14.16 is a manifestation of time-reversal symmetry, and the fact that the QSH state is generally characterized by a $Z_2$ topological invariant, as mentioned at the beginning of this subsection [6]. Therefore, indeed the $Z_2$ topological invariant can only have two values: $v = 1$ if the number of pairs of Kramers partners is odd, or $\nu = 0$ if the number of Kramers pairs is even. Further, using properties of the time-reversal operator, one can easily show why the topological protection appears for an odd number of pairs of Kramers partners.

It is easy to see that one helical channel can be obtained from the other by simply applying the time-reversal operator $\mathcal{T} = i\tau_y \sigma_x C$:

$$\Psi_{k,+} = \mathcal{T}\Psi_{-k,-}, \qquad \Psi_{-k,-} = -\mathcal{T}\Psi_{k,+}. \tag{14.17}$$

The helicity (see Equation 14.16) protects the edge states from local perturbations that do not break time-reversal symmetry. Concretely, let us consider a perturbation $H'$ which is invariant under time reversal:

$$[H', \mathcal{T}] = 0. \tag{14.18}$$

Using Equations 14.17 and 14.18, we can transform the matrix element $\langle \Psi_{-k,-} | H' | \Psi_{k,+} \rangle$ invoking the antiunitary properties of the time-reversal operator as follows:

$$\langle \Psi_{k,+} | H' | \Psi_{-k,-} \rangle = \langle \mathcal{T}(\Psi_{-k,-}) | H' | \Psi_{-k,-} \rangle = \langle \mathcal{T}\Psi_{-k,-} | H' | \mathcal{T}^2 \Psi_{-k,-} \rangle \tag{14.19}$$

For an odd number of pairs of Kramers partners, i.e for half-integer spin of electrons $\mathcal{T}^2 = -1$, and using the hermiticity of $H'$ one gets immediately:

$$\langle \Psi_{k,+} | H' | \Psi_{-k,-} \rangle = (-1)\langle \Psi_{k,+} | H' | \Psi_{-k,-} \rangle, \tag{14.20}$$

i.e. for a Hermitian perturbation $H'$, its matrix element is zero:

$$\langle \Psi_{-k,-} | H' | \Psi_{k,+} \rangle = 0. \tag{14.21}$$

Therefore, the helical edge states of 2D TIs are protected against elastic and time-reversal symmetry protecting single-particle perturbations. In particular, the spin-independent disorder potential cannot cause scattering between the helical edge states. However, two-particle scattering involving electron-electron interactions, or perturbations which break time-reversal symmetry, like magnetic fields or magnetic impurities, can cause backscattering between the helical edges. In the two-terminal geometry, this protection induces a quantized conductance $G = 2e^2 / h$ in the gap of 2D TIs. The quantized value in the band gap of 2D TIs originates from the spin-momentum locking enforcing, under an applied chemical potential difference, that one Kramers partner propagates at the upper edge, while the other one can only propagate at the lower edge. In contrast, in the normal insulator $G = 0$ in the band gap [9, 10]. Still, if the edge-state spin polarization is out-of-plane and one omits the terms which mix Kramers partners blocks, an out-of-plane magnetic field commutes with the Hamiltonian, and does not open a gap in the edge-state spectrum [79–81]. Only perturbations which can mix the Kramers blocks, like charge puddles where electrons lose the orientation of their spin, or inversion symmetry breaking terms can cause scalar disorder to mix the edge states for out-of-plane magnetic fields. Indeed, in experiments on HgTe QWs, the longitudinal two-terminal conductance in a perpendicular magnetic field and in the diffusive regime decays to zero when Rashba SO interactions and puddles are present [9, 10, 82].

Still, experiments in magnetic fields do not directly confirm the helicity of the edge states. The helical edge transport in a QSH system can be seen in four-terminal devices shown in Figure 14.4 and distinguished from the usual QH effect. Using the Landauer–Bütikker approach, we express the current, $I_i$, injected through contact $i$ in terms of voltages $V_j$ induced on all contacts as follows [11, 83]:

$$I_i = \frac{e^2}{h} \sum_{j=1}^{N} (T_{ji} V_i - T_{ij} V_j), \tag{14.22}$$

where $T_{ji}$ is the transmission probability from contact $i$ to contact $j$. For a chiral QH edge channel, $T_{ji}$ connects the neighboring contacts only in one propagation direction (see Figure 14.4a), such that

$$T(QH)_{i+1,i} = 1, \quad i = 1,...,N, \tag{14.23}$$

where $N$ is the number of terminals, for example, $N = 4$ in Figure 14.4a, with the convention that $T_{N+1,N} = T_{1,N}$ describes the transmission from terminal $N$ to terminal 1. Assuming, for concreteness, that the current flows from terminal 1 to terminal 4, while leads 2 and 3 are used as voltage probes,

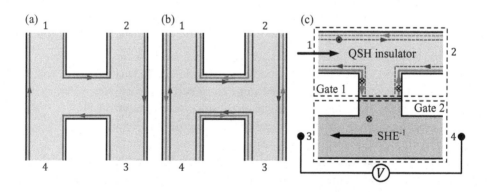

**FIGURE 14.4** Schematic of (a) chiral and (b) helical edge transport in a four-terminal device based on 2D TIs. (c) Verifying the spin polarization of edge channels in double gate H-bar structures. In (c) the upper part of H-bar structure is in the quantum spin Hall state, while the lower part is in the metallic phase characterized by the inverse spin Hall effect (SHE⁻¹). The current is driven between contacts 1 and 2 leading to the injection of a spin current into the middle part of the H-bar structure. This spin current is further converted into a voltage difference due to SHE⁻¹. (After Brüne, C. et al., *Nat. Phys.* 8, 486, 2012. With permission.)

yields a finite two-terminal resistance $R_{14,14} = \dfrac{V_1 - V_4}{I_{14}} = h/e^2$ and *zero* four-terminal (non-local) resistances $R_{14,12} = R_{14,23} = R_{14,13} = 0$. In contrast, in a QSH system, the helical edge channels connect the neighboring contacts in both propagation directions (see Figure 14.4b), such that

$$T(QSH)_{i+1,i} = T(QSH)_{i,i+1} = 1, \qquad i = 1,...,N, \tag{14.24}$$

with the conventions $T_{N+1,N} = T_{1,N}$ and $T_{N,N+1} = T_{N,1}$. Consequently, for a current flowing from 1 to 4, we find the two-terminal resistance [11] $R_{14,14} = \dfrac{V_1 - V_4}{I_{14}} = \dfrac{3}{4}\dfrac{h}{e^2}$, and the four-terminal resistances [11] $R_{14,12} = R_{14,23} = \dfrac{1}{4}\dfrac{h}{e^2}$ and $R_{14,13} = \dfrac{1}{2}\dfrac{h}{e^2}$

The *non-zero* non-local resistances are unique to the 2D TI state, allowing its unambiguous experimental detection [11]. The universality of the non-local resistances is just a consequence of time-reversal symmetry, and, therefore, is also expected for other proposed realizations of 2D TIs, for example, inverted InAs/GaSb QWs [84].

Yet other experiments studied the spin polarization of the edge states in QSH systems [54, 85]. The first experiment confirming the spin polarization of edge states involved an H-bar structure with two gates [54]. The upper part of the structure was in the QSH regime, while the lower one was in the metallic regime, which, due to strong SO interaction, exhibits an inverse spin Hall effect (SHE$^{-1}$) (see Figure 14.4c). The current was injected (contacts 1 and 2) into the QSH regime, while a non-local voltage drop was measured across the metallic leg (contacts 3 and 4). In this configuration, the spin-polarized helical edge channels injected a spin-polarized current into the metallic leg, causing a local imbalance in the chemical potential of spin-up and spin-down polarized carriers. Due to the inverse spin Hall effect (generation of a transverse voltage difference due to a spin current) [86–90], the spin current in the metallic region induced a voltage between contacts 3 and 4. This voltage can only develop provided the helical edge channels are spin polarized, and the metallic leg exhibits the inverse spin-Hall effect. To interpret the experiments, semiclassical Monte Carlo calculations were provided for the geometry shown in Figure 14.4c, demonstrating that indeed the experimental non-local resistance signal originated from the all-electrical detection of the spin polarization of the edge states [54].

## 14.3 THREE-DIMENSIONAL TOPOLOGICAL INSULATORS

Having discussed the case of 2D TIs in detail, we will now also provide a brief overview of their 3D counterparts. Similarly to 2D TIs, 3D TIs are characterized in the bulk by a gapped 3D Dirac Hamiltonian [20, 91]. When this 3D Hamiltonian is then projected into the (001) plane, the following Hamiltonian for the 2D gapless Dirac state emerges (see Figure 14.2):

$$H_{\text{surf}} = \int d^2 r \, \psi^\dagger v_F \left( p_x \sigma^x + p_y \sigma^y \right) \psi, \tag{14.25}$$

where $\psi^\dagger$ and $\psi$ are electron field operators with $\psi = \left( \psi_\uparrow, \psi_\downarrow \right)^T$ including two spin indices, $\sigma = \left( \sigma^x, \sigma^y \right)$, while $p_x$ and $p_y$ are electron momenta, and $v_F$ is the Fermi velocity. By examining the commutator $\left[ T, H_{\text{surf}} \right]$, we can observe that the Hamiltonian does indeed obey time-reversal symmetry. Similarly, as in 2D TIs, the spin and momentum of 2D surface state are locked, i.e. at every point of the Fermi surface spin is parallel to the momentum (see Figure 14.2), defining the helicity of the conduction band opposite to the helicity of the valence band. Further, any perturbation which is elastic and does not mix the spin cannot connect states $|s,p\rangle$ with states $|-s,-p\rangle$ (see Figure 14.2), and therefore backscattering (180-degree scattering), for example, by scalar disorder is prohibited. This lack of backscattering has many physical consequences, i.e. transport scattering in this system is two times larger than elastic scattering, giving rise to a non-monotonic mobility dependence as a function of the density of electrons [92], as well as weak antilocalization [93]. Further, the spin-momentum locking for surface states of 3D TIs gives rise to a revised Hikami–Larkin–Nagaoka formula when SO impurities and the Dirac dispersion are combined [94].

To further characterize 3D TIs, we will now introduce the concept of topological invariant for these materials. As we explained in Section 14.2, the $Z_2$ invariant for 2D TIs can have only two values, i.e. $\nu = 0$ or $\nu = 1$. Correspondingly, the extension of this definition for 3D TIs with inversion symmetry includes four $Z_2$ invariants $(\nu_0; \nu_1 \nu_2 \nu_3)$ [95, 96], especially $\nu_0$ which defines whether the TI is weak (consisting of many 2D TIs weakly coupled), or strong can be determined by calculating the product of $\delta_i = \pm 1$ in eight time-reversal symmetric points of the Brillouin zone $\Gamma_i$, i.e.

$$(-1)^{\nu_0} = \prod_i \delta_i \tag{14.26}$$

with $\delta_i = \prod_{m=1}^{N} \zeta_{2m}(\Gamma_i)$, where $\zeta_{2m}(\Gamma_i)$ is the parity eigenvalue of $\pm 1$ of the $2m$-th occupied band. Translating more explicitly Equation 14.26, if the system possesses inverted bands at an odd number of high symmetry points (every inversion is related to the change in the parity eigenvalue) in their bulk 3D Brillouin zone, then it should also exhibit an odd number of surface state crossings. This gives $\nu_0 = 1$, and prohibits adiabatic continuation of the band structure to the ordinary insulator. Since these crossings can be observed in the photoemission experiment, this gives direct confirmation of the non-trivial band structure of 3D TIs. Indeed, angle-resolved photoemission directly confirmed that $Bi_{1-x}Sb_x$ [26, 27], $Bi_2Se_3$[28], $Bi_2Te_3$ [29], $TlBiSe_2$, [30–32], $TlBiTe_2$ [32], $Pb(Bi_{1-x}Sb_x)_2Te_4$ [35], $PbBi_2Te_4$ [36], and strained HgTe [33] are 3D TIs. Further, recent experiments on the spin-resolved photoemission confirm the spin-momentum locking for the surface states of 3D TIs.

Another way to characterize the unusual properties of 3D TIs is using transport experiments. Indeed, applying an out-of-plane magnetic field causes an opening of the gap in the Hamiltonian (Equation 14.25) inducing the first Chern number corresponding to the Hall conductance $\sigma_{xy} = 2(n+1/2)$ due to the Dirac dispersion. Therefore, one should expect an odd Hall quantization for surface states of 3D TIs in magnetic fields. This has indeed been observed in transport measurements in strained HgTe [33, 34] and in BiSbTeSe$_2$ [97]. Further consequences of this non-trivial quantization are, for example, the Faraday and magnetoelectric effects [98–102].

# 14.4 TOWARDS TOPOLOGICAL QUANTUM COMPUTATION USING 3D TIME-REVERSAL INVARIANT TOPOLOGICAL INSULATORS

Armed with an understanding of the basic properties of time-reversal invariant TIs, it is now interesting to begin to explore the various possibilities associated with these fascinating new materials, in particular as a result of their strong SO interactions. In this chapter, we do so by examining the possibility of finding Majorana fermions on the surfaces of 3D TIs when they are coupled through the proximity effect with a conventional s-wave superconductor. As mentioned previously, Majorana fermions have been predicted to be the fundamental basis upon which topological quantum computing has been built. Therefore to be able to find and manipulate these quasiparticles within 3D TIs would indeed be quite interesting. Prior to exploring how these excitations may be manifested in TIs when paired with s-wave superconductors, it is important to mention some of the basic properties of Majorana fermions. First and foremost, these quasiparticles are comprised of an equal superposition of particle and anti-particle and, as such, may be described by a real wave equation. In terms of operators, we may write this equivalence as $\hat{\gamma} = \hat{\gamma}^{\dagger}$, where $\hat{\gamma}^{\dagger}$ is the Majorana creation operator and $\hat{\gamma}$ is the Majorana annihilation operator. The presence of superconducting materials is important mathematically, as it doubles the size of the Hilbert space, and, therefore, quasiparticles in the superconductor naturally come with their anti-particles. This particle-hole redundancy is referred to as an imposed, rather than inherent, symmetry that is present in the superconductor and is known as particle-hole symmetry. Thus, from a naive standpoint, it may be possible to find non-Abelian anyonic excitations in superconducting materials. This supposition has subsequently been theoretically proven for a variety of different systems, such as the fractional quantum Hall state at filling factor $v = \dfrac{5}{2}$ [103], semiconductor quantum wires coupled to a conventional s-wave superconductor when placed in a parallel magnetic field [104–107], one-dimensional strongly SO-coupled Fe atoms on an s-wave superconductor [108], an array of magnetic tunnel junctions coupled to a proximity induced superconductivity in a two-dimensional electron gas [109], and spinless $p_x + ip_y$ superconductors [110], where the Majorana fermions exist as zero energy bound states at the core of vortex excitations in these unconventional superconductors. Put another way, the existence of Majorana fermions as bound states at the vortex core implies that $[\hat{\gamma}, \hat{H}] = 0$, where $\hat{H}$ is the Hamiltonian for the system of interest. The preceding commutation relation is interesting, in that it further states that the action of Majorana fermions does not change the ground state energy of the system.

The presence of $N$ such vortices leads to a $2^{\frac{N}{2}}$ degenerate ground state, which may be non-trivially operated on through braiding processes in which the vortices are adiabatically rearranged. Majorana fermions have been proffered as the backbone of a potentially revolutionary type of computing, topological quantum computing, that offers significant potential technological advances, such as disruptive increases in computational efficiency at greatly reduced energy consumption through the manipulation and production of non-Abelian anyons [111]. Yet, beyond the possibility of technological relevance, the emergence of these new particles points to the fact that proximity-coupled topological systems provide a unique playground for exploring issues central to the foundation of quantum mechanics, such as supersymmetry and entanglement.

In the remainder of this chapter, we will explore the physics surrounding the initial theoretical predictions and current work surrounding the observation of Majorana fermions, and the underlying conditions that are necessary for their existence in proximity-coupled TIs. To do so, we will use the example system of a 3D time-reversal invariant topological insulator that is proximity-coupled to a conventional $s$-wave superconductor. In Section 14.5, we give a simple derivation following the pioneering work of Fu and Kane [112] to give a clear understanding of the equations and conditions that may lead to the observation of Majorana fermions, and the associated unconventional superconductivity. In Section 14.6, we discuss the current research into the pairing of TIs with superconductors to elucidate the progress achieved. In this final section of the chapter, we focus our attention on the experimental developments within 3D TIs in keeping with our discussion, though it should be noted that extensive theoretical and experimental progress has also been made within proximity-coupled 2D TIs [113–120].

## 14.5 MAJORANA FERMIONS IN 3D TIME-REVERSAL INVARIANT TOPOLOGICAL INSULATORS

We begin by considering the 2D surface state Hamiltonian for the 3D time-reversal invariant TI, for which the strength of the SO interaction has led to an inversion of the band gap at an odd number of time-reversed points within the Brillouin zone, written in real space as

$$H_{\text{surf}} = \int d^2 r \, \psi^\dagger \left( -i v_F \sigma \cdot \nabla - \mu \right) \psi. \tag{14.27}$$

where in Equation 14.27, $\psi = (\psi_\uparrow, \psi_\downarrow)^T$ are the electron field operators, in which we keep two spin indices, $\sigma = (\sigma^x, \sigma^y)$, and $\mu$ and $v_F$ are the chemical potential and Fermi velocity, respectively. We can see that Equation 14.27 is indeed time-reversal invariant by defining the time-reversal operator, here defined as $\mathcal{T} = i\sigma^y \mathcal{C}$, where $\mathcal{C}$ represents complex conjugation. By examining the commutator, $[\mathcal{T}, H_{\text{surf}}]$, we can observe that the Hamiltonian does indeed obey time-reversal symmetry. To proceed, we can rewrite Equation 14.27 in a more familiar form where we have replaced the partial derivatives with the momentum within the surface as in Equation 14.25 (see Section 14.3).

As we are explicitly concerned with a situation in which we are coupling the surface states of the 3D time-reversal invariant TI with an s-wave superconductor, then we must add the superconductive pairing that is associated with the presence of a superconductor. Formally, this is accomplished by adding a potential term of $V_{sc} = \Delta \psi_\uparrow^\dagger \psi_\downarrow^\dagger$ to the Hamiltonian of Equation 14.25. The Hamiltonian that results from this coupling must be rewritten in the expanded particle-hole basis using the Bogoliubov–de Gennes equation as

$$H_{\text{BdG}} =$$

$$\int d^2r \begin{pmatrix} \psi_\uparrow^\dagger & \psi_\downarrow^\dagger & \psi_\uparrow & \psi_\downarrow \end{pmatrix} \begin{bmatrix} v_F(p_x\sigma^x + p_y\sigma^y) - \mu & i\Delta^*\sigma^y \\ -i\Delta\sigma^y & v_F(p_x\sigma^x - p_y\sigma^y) + \mu \end{bmatrix} \begin{pmatrix} \psi_\uparrow \\ \psi_\downarrow \\ \psi_\uparrow^\dagger \\ \psi_\downarrow^\dagger \end{pmatrix}$$

$$(14.28)$$

The seemingly complicated equation above, which comprises the s-wave superconductor and the TI surface state Hamiltonian, can be simplified by changing the basis to $\tilde{\Psi} = (\psi_\uparrow, \psi_\downarrow, \psi_\downarrow^\dagger - \psi_\uparrow^\dagger)$. This allows one to eliminate the off-diagonal terms present in the pairing blocks of Equation 14.28 to obtain [112, 121],

$$H_{BdG} = \int d^2r \, \tilde{\psi}^\dagger \begin{bmatrix} v_F(p_x\sigma^x + p_y\sigma^y) - \mu & \Delta^*\sigma_0 \\ \Delta\sigma_0 & -v_F(p_x\sigma^x + p_y\sigma^y) + \mu \end{bmatrix} \tilde{\psi}, \quad (14.29)$$

where in Equation 14.29 $\sigma_0$ is the $2\times2$ identity matrix. Equation 14.29 can be expanded into a more useful form by representing the superconducting pairing as both a magnitude, $\Delta_0$, and a phase, $\phi$, in the following manner, $\Delta = \Delta_0 e^{i\phi}$, to obtain

$$H_{\text{BdG}} = v_F p_x \sigma_x \otimes \tau_z + v_F p_y \sigma_y \otimes \tau_z - \mu\sigma_0 \otimes \tau_z$$

$$+ \Delta_0 \sigma_0 \otimes \left( \cos\phi \, \tau_x + \sin\phi \, \tau_y \right),$$

$$(14.30)$$

where $\tau = (\tau_x, \tau_y, \tau_z)$ are Pauli matrices that mix the blocks of $\psi$ and $\psi^\dagger$ within the Hamiltonian. By defining the particle-hole symmetry that superconductors possess as an operator, defined as $\Xi = \sigma_y \otimes \tau_y C$, we can show that the Hamiltonian in Equation 14.30 obeys $\{\Xi, H_{\mathrm{BdG}}\} = 0$, thereby showing the system to be particle-hole symmetric, as one expects. If we assume that the superconducting order parameter is spatially homogeneous over the entire surface of the TI, then the spectrum of the Hamiltonian is

$$|E(k)| = \sqrt{\left(\pm v_F |k| - \mu\right)^2 + \Delta_0^2}. \tag{14.31}$$

Assuming that $\mu \gg \Delta_0$, then the low-energy spectrum of Equation 14.31 resembles that of a spinless $p_x + i p_y$ superconductor that obeys time-reversal symmetry. This can best be seen by rewriting Equation 14.28 in the single-particle eigenstates of Equation 14.27 and projecting away the lowest-energy band. The resulting Hamiltonian is then formally equivalent to a spinless $p_x + i p_y$ superconductor [112, 122]. As we are interested in determining the eigenvalues of this Hamiltonian in order to determine the zero energy modes associated with the presence of Majorana fermions, we first ask where the zero energy modes should exist. As Majorana fermions are topological objects, they cannot be defined by any local order parameter. Yet the surface of the proximity-coupled TI clearly has a superconducting order parameter on the surface that is well-defined everywhere. Thus, we look in the one place where the superconducting order is not well-defined: within the vortices proliferated on the surface of the TI. Furthermore, as we have made the case that the surface of the TI is proximity-coupled to a superconductor, we expect the presence of $h/2e$ vortex-bound states on the surface. Figure 14.5 shows a schematic view of the envisioned setup. To proceed, we assume that there is a vortex present at a real-space position of (0,0) with the other partner positioned off at $(\infty, \infty)$. It is easiest to solve for the zero modes in Equation 14.30 by rewriting the momentum $p = (p_x, p_y, p_z)$ as $-i\nabla = -i(\partial/\partial_x, \partial/\partial_y, \partial/\partial_z)$, and changing the coordinates

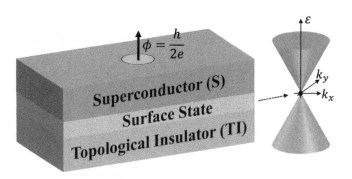

**FIGURE 14.5** Schematic picture of the proximity-induced superconductor on the top of the surface state of a 3D TI with an inserted $h/2e$ vortex.

from Euclidian to polar. With these transformations, the original surface state Hamiltonian becomes

$$H_{surf} = -iv_F \left( \cos \vartheta \, \sigma_x + \sin \vartheta \, \sigma_y \right) \frac{\partial}{\partial r}$$
$$-iv_F \left( -\sin \vartheta \, \sigma_x + \cos \vartheta \, \sigma_y \right) \frac{\partial}{r \partial \vartheta} - \mu \sigma_0. \tag{14.32}$$

In Equation 14.32, $\vartheta$ is the angle in position space that sweeps around the vortex that is located at position, $r$. Nonetheless, we are interested in solving for the eigenvalues of the combined TI and $s$-wave superconductor case, in which the Bogoliubov–de Gennes Hamiltonian is represented in polar coordinates as

$$H_{BdG} = -iv_F \left( \cos \vartheta \, \sigma_x + \sin \vartheta \, \sigma_y \right) \otimes \tau_z \frac{\partial}{\partial r}$$
$$-iv_F \left( -\sin \vartheta \, \sigma_x + \cos \vartheta \, \sigma_y \right) \otimes \tau_z \frac{\partial}{r \partial \vartheta} \tag{14.33}$$
$$+\Delta_0 \sigma_0 \otimes \left( \cos \vartheta \, \tau_x \pm \sin \vartheta \, \tau_y \right).$$

In Equation 14.33, it should be noted that we have explicitly set the chemical potential, $\mu$, to be zero and the superconducting pairing potential here is expressed in polar coordinates as $\Delta_0 e^{\pm i\vartheta}$. As we are particularly interested in the bound states of $H_{BdG}\xi = E\xi$ associated with the vortices, we postulate that the wave functions have the form

$$\Psi(r, \vartheta) = \alpha^{\pm}(\vartheta) e^{-\int_0^r \Delta_0(r')dr'/v_F}, \tag{14.34}$$

where $\alpha^{\pm}$ depends only on $\vartheta$ and the superconducting pairing magnitude only depends on the radial position, $r$. The four component vectors $\alpha^+$ and $\alpha^-$ denote solutions if the pairing potential is $\Delta_0 e^{i\vartheta}$ and $\Delta_0 e^{-i\vartheta}$, respectively. Due to the fact that $\Psi$ is an analytic function at $r = 0$ and $\Delta_0 = 0$, we expect $\alpha^{\pm}$ to be single valued and

$$H_{BdG}\Psi = 0$$
$$= \Delta_0 \left[ i \left( \cos \vartheta \, \sigma_x + \sin \vartheta \, \sigma_y \right) \otimes \tau_z + \sigma_0 \otimes \left( \cos \vartheta \, \tau_x \pm \sin \vartheta \, \tau_y \right) \right] \Psi. \tag{14.35}$$

In Equation 14.35, we have assumed that $\alpha^{\pm}$ is $\vartheta$-independent, as we are only interested in the lowest angular momentum state, or $p_\vartheta = 0$. By setting both the superconducting pairing amplitude, $\Delta_0$, and the position, $r$, to zero, we are now capable of solving the eigenvalue problem at hand. Rewriting Equation 14.35 slightly we have

$$\left[ i \left( \cos \vartheta \, \sigma_x + \sin \vartheta \, \sigma_y \right) \otimes \tau_z + \sigma_0 \otimes \left( \cos \vartheta \, \tau_x \pm \sin \vartheta \, \tau_y \right) \right] \alpha^{\pm} = 0. \tag{14.36}$$

In examining Equation 14.36, we can find that—up to a phase—the only possible $\vartheta$-independent eigenvectors for $\alpha^{\pm}$ are

$$\alpha^+ = \left(0, 1, -i, 0\right)^T \tag{14.37}$$

and

$$\alpha^- = \left(1, 0, 0, -i\right)^T \tag{14.38}$$

for $\Delta(r, \vartheta) = \Delta_0(r)e^{\pm i\vartheta}$. Therefore, we arrive at the final form of the zero energy Majorana mode for $\Delta(r, \vartheta) = \Delta_0(r)e^{+i\vartheta}$ as

$$\Psi = e^{i\pi/4}\left(0, 1, -i, 0\right)^T e^{-\int_0^r \Delta_0(r')/v_F dr'} \tag{14.39}$$

which is not spin-polarized, as the particle and hole possess the same spin. In Equation 14.39, we have added the phase $\pi/4$, such that $\Xi\Psi = \Psi$. Hence, the zero energy mode described by $\Psi$ is its own charge conjugate, and indeed obeys the properties associated with Majorana bound states. The field operator $\hat{\gamma}$ that is comprised of fermionic creation and annihilation operators and constructed from the wave function given by Equation 14.39 then satisfies $\hat{\gamma} = \hat{\gamma}^{\dagger}$.

## 14.6 EXPERIMENTAL PROGRESS IN PROXIMITY-COUPLED 3D TIME-REVERSAL INVARIANT TOPOLOGICAL INSULATORS

One may easily infer that when considering both topological materials and conventional $s$-wave superconductors, there are many available choices. The recent focus on the proximity effects, however, has naturally been on 3D time-reversal invariant topological materials, such as $Bi_2Se_3$. Prior to being able to actually observe Majorana fermions, one must be able to identify the underlying unconventional superconductivity that is necessary to produce vortex excitations capable of harboring Majorana bound states. Given the task of observing induced unconventional superconductivity in time-reversal invariant TIs, the initial experimental efforts have focused on depositing $s$-wave superconductors directly onto TI crystals, thin films, and nanostructures with the intent of observing the topology of the bands in the presence of superconductivity. This focus is due to the fact that it is difficult to produce insulating TIs i.e. where the Fermi energy lies within the bulk gap. Efforts quickly focused on thin films, nanowires, and exfoliated TIs where the intrinsic doping may be mitigated. In order to reach the topological regime, these devices were produced on substrates that allowed for backgating between superconducting contacts.

Nonetheless, an important first step towards the observation of Majorana fermions in proximity-coupled topological systems is to observe superconducting behavior in these systems. To this end, numerous groups have now demonstrated the presence of a proximity-induced superconducting state with transport behavior consistent with Cooper pair tunneling into the topological surface state of $Bi_2Se_3$ and $Bi_2Te_3$ [123–125], and strained HgTe [119, 126, 127]. These systems show clear evidence of superconductivity, such as a Fraunhofer pattern [123, 125], and the appearance of only even Shapiro steps under microwave radiation [117, 119], as well as universal skewness (non-sinusoidal current-phase relation) over a large range of parameters [127], which could indicate transport through the topological Andreev or Majorana bound states.

Often, however, in $Bi_2Se_3$ and $Bi_2Te_3$, the Fermi energy is well outside of the topological regime in the bulk bands. On the other hand, by combinations of electrical and chemical gating, the Fermi level may be pulled into the topological regime, as is evidenced by the observation of the Dirac point, as revealed through the ensuing peak in the resistance and change in the sign of the Hall voltage as the gate voltage is modulated in the presence of electrical transport [128]. With the ability to access the topological regime within the 3D TI, a supercurrent is observed whose critical current follows the gate-induced changes in the position of the chemical potential within the topological and non-topological bands. Thus, when the chemical potential is below the bottom of the conduction band, the supercurrent is carried by the surface states, until the chemical potential enters the bulk bands and the surface becomes sufficiently disordered that the surface orbitals hybridize with bulk orbitals, thereby removing the topological protection of the surface states. The topological and non-topological contributions to transport measurements are explained by 3D quantum transport calculations. These confirm that the supercurrents are carried by the topological surface states by demonstrating the robustness of the critical current within the topological regime in the presence of time-reversal preserving impurity disorder which disappears as the chemical potential enters the bulk band structure.

More recent advances in proximity-coupled systems have been enabled by the use of materials growth in which the TI is grown directly on top of a superconductor, as is the case with the TI $Bi_2Se_3$ and the $s$-wave superconductor $NbSe_2$ [129]. In this case, clear superconducting gaps appear on the surface of the heterostructure, indicating that superconducting Cooper pairs have indeed tunneled into the TI, while the band crossing in the surface states is still preserved, as the addition of $s$-wave superconductivity does not break time-reversal symmetry. Subsequent studies seeking signs of unconventional superconductivity examined the measured superconducting gap as a function of the angle at the Fermi surface. However, no signs of unconventional superconductivity were observed [130]. This is unusual, as one expects to see a clear angular dependence of the pairing magnitude as the Fermi surface is traversed. Therefore, in this case, it is proper to ask which order parameter develops in the TI when paired with an $s$-wave superconductor. To answer this question, we can make a very simple symmetry argument as to the correct form of the order parameter: within the TI, we have broken $SU(2)$ symmetry due to the presence of strong SO coupling, which allows both the singlet and triplet order to coexist within the TI. Furthermore, as we have broken inversion symmetry at both interfaces, inter-orbital and intra-orbital pairing is allowed by symmetry. Additionally, the symmetry constructed order parameter must obey time-reversal symmetry as this has not been broken. Thus, the order parameter that accumulates in proximity-coupled TI heterostructures is $s + p_x \pm ip_y$, which can essentially be understood as two copies of time-reversal breaking $p$-wave superconductivity with different chiralities, as discussed in the previous section, and one would expect to see isotropy in the Fermi surface, as the angle around the Fermi surface is changed as the $s$-wave order parameter is isotropic at the Fermi surface.

One potential way of eliminating the isotropic $s$-wave component so that the unconventional $p$-wave part may be more clearly observed is by breaking the underlying time-reversal symmetry. A simple method of breaking the underlying time-reversal symmetry is to add magnetic dopants to the TI. Yet this still does not produce purely $p$-wave superconductivity, as broken time-reversal symmetry solely guarantees that the two components of the superfluid are no longer equal to one another, and this is no guarantee that the isotropic $s$-wave component will disappear. Recent theoretical work has illustrated that when magnetic dopants are added to a thin-film proximity-coupled TI, the bulk states disappear as the magnetization increases, as the singlet state becomes energetically unfavorable with increasing magnetization. However, due to the strong SO coupling, this is not the case for the surface states. In the presence of the Zeeman field from the magnetic impurities, the spin of the surface electrons begins to move from being purely in-plane to out-of-plane, as one expects. However, there remains a finite projection of the spin onto the surface of the TI and this causes the $s$-wave component to persist up to larger Zeeman energies [131]. Yet, some of the most promising signatures of not only unconventional superconductivity, but also of Majorana fermions in TIs, arise when the proximity-coupled TI is in the quantum anomalous Hall insulator phase, which occurs when the system is both sufficiently magnetically doped and has broken translational invariance. When the chemical potential is in the gap in the energy spectrum opened by the magnetic impurities, the TI is in the quantum anomalous Hall phase (see also Chapter 15, Volume 2), and there exists a chiral state at the edge of the system. When this chiral edge state is proximity coupled with an $s$-wave superconductor, quantum transport measurements have shown conductance plateaus that appear at close to $0.5 \, e^2/h$ consistent with theoretical predictions and the existence of Majorana fermions [132].

## ACKNOWLEDGMENTS

E.M.H. thanks G. Tkachov for discussions and B. Scharf for a careful reading of the text. E.M.H. acknowledges financial support from the DFG via SFB 1170 ToCoTronics and the ENB Graduate School on Topological Insulators. M.J.G. acknowledges financial support from the NSF under CAREER Award ECCS-1351871 and DMR 17-20663 COOP.

# REFERENCES

1. L. D. Landau and E. M. Lifshitz, *Statistical Physics*, Pergamon Press, Oxford, 1980.
2. K. Von Klitzing, G. Dorda, and M. Pepper, New method for high-accuracy determination of the fine-structure constant based on quantized Hall resistance, *Phys. Rev. Lett.* **45**, 494 (1980).
3. R. B. Laughlin, Quantized Hall conductivity in two dimssensions, *Phys. Rev. B* **23**, 5632 (1981).
4. D. J. Thouless, M. Kohmoto, M. P. Nightingale, and M. den Nijs, Quantized Hall conductance in a two-dimensional periodic potential, *Phys. Rev. Lett.* **49**, 405 (1982).
5. F. D. M. Haldane, Model for a quantum Hall effect without Landau levels: Condensed-matter realization of the "Parity Anomaly", *Phys. Rev. Lett.* **61**, 2015 (1988).
6. C. L. Kane and E. J. Mele, $Z_2$ topological order and the quantum spin Hall effect, *Phys. Rev. Lett.* **95**, 146802 (2005).
7. C. L. Kane and E. J. Mele, Quantum spin Hall effect in graphene, *Phys. Rev. Lett.* **95**, 226801 (2005).
8. B. A. Bernevig, T. L. Hughes, and S.-C. Zhang, Quantum spin hall effect and topological phase transition in HgTe quantum wells, *Science* **314**, 1757 (2006).
9. M. König, S. Wiedmann, C. Brüne, A. Roth, H. Buhmann, L. W. Molenkamp, X.-L. Qi, and S.-C. Zhang, Quantum spin hall effect and topological phase transition in HgTe quantum wells, *Science* **318**, 766 (2007).
10. M. König, H. Buhmann, L. W. Molenkamp, T. Hughes, C.-X. Liu, X.-L. Qi, and S.-C. Zhang, The quantum spin Hall effect: Theory and experiment, *J. Phys. Soc. Jpn.* **77**, 031007 (2008).
11. A. Roth, C. Brüne, H. Buhmann, L. W. Molenkamp, J. Maciejko, X.-L. Qi, and S.-C. Zhang, Nonlocal transport in the quantum spin hall state, *Science* **325**, 294 (2009).
12. B. Halperin, Quantized Hall conductance, current-carrying edge states, and the existence of extended states in a two-dimensional disordered potential, *Phys. Rev. B* **25**, 2185 (1982).
13. A. MacDonald and P. Streda, Quantized Hall effect and edge currents, *Phys. Rev. B* **29**, 1616 (1984).
14. C. Wu, B. A. Bernevig, and S.-C. Zhang, Helical Liquid and the Edge of Quantum Spin Hall Systems, *Phys. Rev. Lett.* **96**, 106401 (2006).
15. M. Z. Hasan and C. L. Kane, Colloquium: Topological insulators, *Rev. Mod. Phys.* **82**, 3045 (2010).
16. X. L. Qi and S.-C. Zhang, Topological insulators and superconductors, *Rev. Mod. Phys.* **83**, 1057 (2011).
17. B. A. Volkov and O. A. Pankratov, Two-dimensional massless electrons in an inverted contact, *Pis'ma Zh. Eksp. Teor. Fiz.* **42**, 145 (1985).
18. O. Pankratov, S. V. Pakhomov, and B. A. Volkov, Supersymmetry in heterojunctions: Band-inverting contact on the basis of $Pb_{1-x} Sn_x$ Te and $Hg_{1-x} Cd_x$ Te *Sol. State Commun.* **61**, 93 (1987).
19. L. Fu and C. L. Kane, Topological insulators with inversion symmetry, *Phys. Rev. B* **76**, 045302 (2007).
20. H. Zhang, C.-X. Liu, X.-L. Qi, X. Dai, Z. Fang, and S.-C. Zhang, Topological insulators in $Bi_2Se_3$, $Bi_2Te_3$ and $Sb_2Te_3$ with a single Dirac cone on the surface, *Nat. Phys.* **5**, 438 (2009).
21. B. Yan, C.-X. Liu, H.-J. Zhang, C.-Y. Yam, X.-L. Qi, T. Frauenheim, and S.-C. Zhang, Theoretical prediction of topological insulators in thallium-based III-V-VI2 ternary chalcogenides, *EPL* **90**, 37002 (2010).
22. H. Lin, R. S. Markiewicz, L. A. Wray, L. Fu, M. Z. Hasan, and A. Bansil, Single-Dirac-Cone Topological Surface States in the $TlBiSe_2$ Class of Topological Semiconductors *Phys. Rev. Lett.* **105**, 036404 (2010).

23. S. Eremeev, Y. Koroteev, and E. Chulkov, Ternary thallium-based semimetal chalcogenides Tl-V-VI2 as a new class of three-dimensional topological insulators, *Piśma Zh. Eksp. Teor. Fiz.* **91**, 664 (2010).

24. S. Eremeev, Y. Koroteev, and E. Chulkov, On possible deep subsurface states in topological insulators: The $PbBi_4Te_7$ system, *Piśma Zh. Eksp. Teor. Fiz.* **92**, 183 (2010).

25. H. Jin, J. Song, A. Freeman, and M. Kanatzidis, Candidates for topological insulators: Pb-based chalcogenide series, *Phys. Rev. B* **83**, 041202 (2011).

26. Hsieh, D., Qian, D., Wray, L., Xia, Y., Hor, Y. S., Cava, R. J., and Hasan, M. Z., A topological Dirac insulator in a quantum spin Hall phase, *Nature* **452**, 970 (2008).

27. D. Hsieh, Y. Xia, L. Wray et al., Observation of Unconventional Quantum Spin Textures in Topological Insulators, *Science* **323**, 919 (2009).

28. Y. Xia, D. Qian, D. Hsieh et al., Observation of a large-gap topological-insulator class with a single Dirac cone on the surface, *Nat. Phys.* **5**, 398 (2009).

29. Y. L. Chen, J. G. Analytis, J.-H. Chu et al., Experimental Realization of a Three-Dimensional Topological Insulator, $Bi_2Te_3$, *Science* **325**, 178 (2009).

30. T. Sato, K. Segawa, H. Guo et al., Direct Evidence for the Dirac-Cone Topological Surface States in the Ternary Chalcogenide $TlBiSe_2$, *Phys. Rev. Lett.* **105**, 136802 (2010).

31. K. Kuroda, M. Ye, A. Kimura et al., Experimental Realization of a Three-Dimensional Topological Insulator Phase in Ternary Chalcogenide $TlBiSe_2$, *Phys. Rev. Lett.* **105**, 146801 (2010).

32. Y. L. Chen, Z. K. Liu, J. G. Analytis et al., Single Dirac Cone Topological Surface State and Unusual Thermoelectric Property of Compounds from a New Topological Insulator Family, *Phys. Rev. Lett.* **105**, 266401 (2010).

33. C. Brüne, C. X. Liu, E. G. Novik et al., Quantum Hall effect from the topological surface states of strained bulk HgTe, *Phys. Rev. Lett.* **106**, 126803 (2011).

34. C. Brüne, C. Thienel, M. Stuiber et al., Dirac-screening stabilized surface-state transport in a topological insulator, *Phys. Rev. X* **4**, 041045 (2014).

35. S. Souma, K. Eto, M. Nomura et al., Topological surface states in lead-based ternary telluride $Pb(Bi_{1-x}Sb_x)_2Te_4$, *Phys. Rev. Lett.* **108**, 116801 (2012).

36. K. Kuroda, H. Miyahara, M. Ye et al., Experimental verification of $PbBi_2Te_4$ as a 3D topological insulator, *Phys. Rev. Lett.* **108**, 206803 (2012).

37. C. Fang, M. J. Gilbert, and B. A. Bernevig, Bulk topological invariants in non-interacting point group symmetric insulators, *Phys. Rev. B* **86**, 115112 (2012).

38. T. L. Hughes, E. Prodan, and B. A. Bernevig, Inversion-symmetric topological insulators, *Phys. Rev. B* **83**, 245132 (2011).

39. C.-X. Liu, R. X. Zhang, and B. K. VanLeeuwen, Topological nonsymmorphic crystalline insulators, *Phys. Rev. B* **90**, 085304 (2014).

40. L. Fu, Topological crystalline insulators, *Phys. Rev. Lett.* **106**, 106802 (2011).

41. S.-Y. Xu, C. Liu, N. Alidoust et al., Observation of topological crystalline insulator phase in the lead tin chalcogenide $Pb_{1-x}Sn_xTe$ material class, *Nat. Commun.* **3**, 1192 (2012).

42. Y. Ando and L. Fu, Topological crystalline insulators and topological superconductors: From concepts to materials, *Ann. Rev. Conden. Mat. Phys.* **6**, 361 (2015).

43. Z. Wang, Y. Sun, X.-Q Chen et al., Dirac semimetal and topological phase transitions in $A_3Bi$ ($a$ = Na, k, Rb), *Phys. Rev. B* **85**, 195320 (2012).

44. Z. K. Liu, B. Zhou, Y. Zhang et al., Discovery of a three-dimensional topological Dirac semimetal, $Na_3Bi$, *Science* **343**, 864 (2014).

45. Z. K. Liu, J. Jiang, B. Zhou et al., A stable three-dimensional topological Dirac semimetal $Cd_3As_2$, *Nat. Mater.* **13**, 677 (2014).

46. P. Tang, Q. Zhou, G. Xu, and S.-C. Zhang, Dirac fermions in an antiferromagnetic semimetal, *Nat. Phys.* **12**, 1100 (2016).

47. P. Wadley, B. Howells, J. Zelezny et al., Electrical switching of an antiferromagnet, *Science* **351**, 587 (2016).

48. A. A. Burkov and L. Balents, Weyl semimetal in a topological insulator, *Phys. Rev. Lett.* **107**, 127205 (2011).

49. C. Fang, M. J. Gilbert, X. Dai, and B. A. Bernevig, Multi-Weyl topological semimetals stabilized by point group symmetry, *Phys. Rev. Lett.* **108**, 266802 (2012).

50. B. Q. Lv, H. M. Weng, B. B. Fu et al., Experimental discovery of Weyl semimetal TaAs, *Phys. Rev. X* **5**, 031013 (2015).

51. L. X. Yang, Z. K. Liu, H. Peng et al., Weyl semimetal phase in the non-centrosymmetric compound TaAs, *Nat. Phys.* **11**, 728 (2015).

52. A. A. Soluyanov, D. Gresch, Z. Wang et al., Type-II Weyl semimetals, *Nature* **527**, 495 (2015).

53. I. Žutić, J. Fabian, and S. Das Sarma, Spintronics: Fundamentals and applications, *Rev. Mod. Phys.* **76**, 323 (2004).

54. C. Brüne, A. Roth, H. Buhmann et al., Spin polarization of the quantum spin Hall edge states, *Nat. Phys.* **8**, 485 (2012).

55. A. R. Mellnik, J. S. Lee, A. Richardella et al., Spin-transfer torque generated by a topological insulator, *Nature* **511**, 449 (2014).

56. Y. Fan, P. Upadhyaya, X. Kou et al., Magnetization switching through giant spin-orbit torque in a magnetically doped topological insulator heterostructure, *Nat. Mater.* **13**, 699 (2014).

57. Z. Wu and J. Li, Spin-related tunneling through a nanostructured electric-magnetic barrier on the surface of a topological insulator, *Nanoscale Res. Lett.* **7**, 90 (2012).

58. C. H. Li, O. M. J. van't Erve, J. T. Robinson, Y. Liu, L. Li, and B. T. Jonker, Electrical detection of charge-current-induced spin polarization due to spin-momentum locking in $Bi_2Se_3$, *Nat. Nanotechnol.* **9**, 218 (2014).

59. J. Tian, I. Miotkowski, S. Hong, and Y. Chen, Electrical injection and detection of spin-polarized currents in topological insulator $Bi_2Te_2Se$, *Sci. Rep.* **5**, 14293 (2015).

60. B. Scharf, A. Matos-Abiague, J. E. Han, E. M. Hankiewicz, and I. Žutić, Tunneling Planar Hall Effect in Topological Insulators: Spin Valves and Amplifiers, *Phys. Rev. Lett.* **117**, 166806 (2016).

61. M. Gmitra, S. Konschuh, C. Ertler, C. Ambrosch-Draxl, and J. Fabian, Band-structure topologies of graphene: Spin-orbit coupling effects from first principles, *Phys. Rev. B* **80**, 235431 (2009).

62. S. Konschuh, M. Gmitra, and J. Fabian, Tight-binding theory of the spin-orbit coupling in graphene, *Phys. Rev. B* **82**, 245412 (2010).

63. C.-C. Liu, W. Feng, and Y. Yao, Quantum spin Hall effect in silicene and two-dimensional germanium, *Phys. Rev. Lett.* **107**, 076802 (2011).

64. P. Vogt, P. De Padova, C. Quaresima et al., Silicene: Compelling experimental evidence for graphenelike two-dimensional silicon, *Phys. Rev. Lett.* **108**, 155501 (2012).

65. L. Zhang, P. Bampoulis, A. N. Rudenko et al., Structural and electronic properties of germanene on $MoS_2$, *Phys. Rev. Lett.* **116**, 256804 (2016).

66. Y. Xu, B. Yan, H.-J. Zhang et al., Large-gap quantum spin Hall insulators in tin films, *Phys. Rev. Lett.* **111**, 136804 (2013).

67. Z. Fei, T. Palomaki, S. Wu et al., Edge conduction in monolayer $WTe_2$, *Nat. Phys.* **13**, 677 (2017).

68. Z.-Y. Jia, Y.-H. Song, X.-B. Li et al., Direct visualization of a two-dimensional topological insulator in the single-layer 1T'–$WTe_2$, *Phys. Rev. B* **96**, 041108 (2017).

69. S. Wu, V. Fatemi, Q. D. Gibson et al., Observation of the quantum spin Hall effect up to 100 kelvin in a monolayer crystal, *Science* **359**, 76 (2018).

70. F. Reis, G. Li, L. Dudy et al., Bismuthene on a sic substrate: A candidate for a high-temperature quantum spin Hall material, *Science* **357**, 287 (2017).

71. F. Dominguez, B. Scharf, G. Li, J. Schäfer, R. Claessen, W. Hanke, R. Thomale, and E. M. Hankiewicz, Testing topological protection of edge states in hexagonal quantum spin Hall candidate materials, *Phys. Rev. B* **98**, 161407(R) (2018).

72. G. Li, W. Hanke, E. M. Hankiewicz, F. Reis, J. Schäfer, R. Claessen, C. Wu, and R. Thomale, Theoretical paradigm for the quantum spin Hall effect at high temperatures, *Phys. Rev. B* **98**, 165146 (2018).

73. T. Zhou, J. Zhang, H. Jiang, I. Žutić, and Z. Yang, Giant spin-valley polarization and multiple Hall effect in functionalized bismuth monolayers, *npj Quantum Materials* **3**, 39 (2018).

74. B. Büttner, C. X. Liu, G. Tkachov et al., Single valley Dirac fermions in zero-gap HgTe quantum wells, *Nat. Phys.* **7**, 418 (2011).

75. J. Chu and A. Sher, *Physics and Properties of Narrow Gap Semiconductors*, Springer, New York, 2008.

76. D. G. Rothe, R. W. Reinthaler, C.-X. Liu, L. W. Molenkamp, S.-C. Zhang, and E. M. Hankiewicz, Fingerprint of different spin-orbit terms for spin transport in HgTe quantum wells. *New J. Phys.* **12**, 065012 (2010).

77. D. R. Candido, M. Kharitonov, C. Egues, and E. M. Hankiewicz, Paradoxical extension of the edge states across the topological phase transition due to emergent approximate chiral symmetry in a quantum anomalous Hall system, *Phys. Rev. B* **98**, 161111(R) (2018).

78. M. V. Berry and R. J. Mondragon, Neutrino billiards: time-reversal symmetry-breaking without magnetic fields, *Proc. R. Soc. Lond. A Math. Phys. Sci.* **412**, 53 (1987).

79. G. Tkachov and E. M. Hankiewicz, Ballistic Quantum Spin Hall State and Enhanced Edge Backscattering in Strong Magnetic Fields, *Phys. Rev. Lett.* **104**, 166803 (2010).

80. B. Scharf, A. Matos-Abiague, and J. Fabian, Magnetic properties of HgTe quantum wells, *Phys. Rev. B* **86**, 075418 (2012).

81. B. Scharf, A. Matos-Abiague, I. Žutić, and J. Fabian, Probing topological transitions in HgTe/CdTe quantum wells by magneto-optical measurements, *Phys. Rev. B* **91**, 235433 (2015).

82. J. Maciejko, X.-L. Qi, and S.-C. Zhang, Magnetoconductance of the quantum spin Hall state, *Phys. Rev. B* **82**, 155310 (2010).

83. S. Datta, *Electronic Transport in Mesoscopic Systems*, Cambridge University Press, Cambridge, UK, 1995.

84. I. Knez and R.-R. Du, Quantum spin Hall effect in inverted InAs/GaSb quantum wells, *Front. Phys.* **7**, 200 (2012).

85. K. C. Nowack, E. M. Spanton, M. Baenninger et al., Imaging currents in HgTe quantum wells in the quantum spin Hall regime, *Nat. Mater.* **12**, 787 (2013).

86. S. Murakami, N. Nagaosa, and S.-C. Zhang, Dissipationless quantum spin current at room temperature, *Science* **301**, 1348 (2003).

87. J. Sinova, D. Culcer, Q. Niu, N. A. Sinitsyn, T. Jungwirth, and A. H. MacDonald, Universal intrinsic spin Hall effect, *Phys. Rev. Lett.* **92**, 126603 (2004).

88. E. M. Hankiewicz, L. W. Molenkamp, T. Jungwirth, and J. Sinova, Manifestation of the spin Hall effect through charge-transport in the mesoscopic regime, *Phys. Rev. B* **70**, 241301(R) (2004).

89. E. M. Hankiewicz, J. Li, T. Jungwirth, Q. Niu, S.-Q. Shen, and J. Sinova, Charge Hall effect driven by spin-dependent chemical potential gradients and Onsager relations in mesoscopic systems, *Phys. Rev. B* **72**, 155305 (2005).

90. C. Brüne, A. Roth, E. G. Novik et al., Evidence for the ballistic intrinsic spin Hall effect in HgTe nanostructures, *Nat. Phys.* **6**, 448 (2010).

91. C.-X. Liu, X.-L. Qi, H. J. Zhang, X. Dai, Z. Fang, and S.-C. Zhang, Model Hamiltonian for topological insulators, *Phys. Rev. B* **82**, 045122 (2010).

92. G. Tkachov, C. Thienel, V. Pinneker et al., Backscattering of Dirac fermions in HgTe quantum wells with a finite gap, *Phys. Rev. Lett.* **106**, 076802 (2011).

93. G. Tkachov and E. M. Hankiewicz, Weak antilocalization in HgTe quantum wells and topological surface states: Massive versus massless Dirac fermions, *Phys. Rev. B* **84**, 035444 (2011).

94. P. Adroguer, W. E. Liu, D. Culcer, and E. M. Hankiewicz, Conductivity corrections for topological insulators with spin-orbit impurities: Hikami-larkin-nagaoka formula revisited, *Phys. Rev. B* **92**, 241402 (2015).

95. L. Fu, C. L. Kane, and E. J. Mele, Topological Insulators in Three Dimensions, *Phys. Rev. Lett.* **98**, 106803 (2007).

96. J. E. Moore and L. Balents, Topological invariants of time-reversal-invariant band structures, *Phys. Rev. B* **75**, 121306(R) (2007).

97. Y. Xu, I. Miotkowski, C. Liu et al., Observation of topological surface state quantum Hall effect in an intrinsic three-dimensional topological insulator, *Nat. Phys.* **10**, 956 (2014).

98. X.-L. Qi, T. L. Hughes, and S.-C. Zhang, Topological field theory of time-reversal invariant insulators. *Phys. Rev. B* **78**, 195424 (2008).

99. W.-K. Tse and A. H. MacDonald, Giant magneto-optical Kerr effect and universal faraday effect in thin-film topological insulators, *Phys. Rev. Lett.* **105**, 057401 (2010).

100. G. Tkachov and E. M. Hankiewicz, Anomalous galvanomagnetism, cyclotron resonance, and microwave spectroscopy of topological insulators, *Phys. Rev. B* **84**, 035405 (2011).

101. L. Wu, M. Salehi, N. Koirala, J. Moon, S. Oh, and N. P. Armitage, Quantized Faraday and Kerr rotation and axion electrodynamics of a 3D topological insulator, *Science* **354**, 1124 (2016).

102. V. Dziom, A. Shuvaev, A. Pimenov et al., Observation of the universal magnetoelectric effect in a 3D topological insulator, *Nat. Commun.* **8**, 15197 (2017).

103. G. Moore and N. Read, Nonabelions in the fractional quantum Hall effect, *Nuc. Phy. B* **360**, 362 (1991).

104. V. Mourik, K. Zuo, S. M. Frolov, S. R. Plissard, E. P. A. M. Bakkers, and L. P. Kouwenhoven, signatures of Majorana fermions in hybrid superconductor-semiconductor nanowire devices, *Science* **336**, 1003 (2012).

105. J. D. Sau, R. M. Lutchyn, S. Tewari, and S. Das Sarma, Generic new platform for topological quantum computation using semiconductor heterostructures, *Phys. Rev. Lett.* **104**, 040502 (2010).

106. Y. Oreg, G. Refael, and F. von Oppen, Helical liquids and Majorana bound states in quantum wires, *Phys. Rev. Lett.* **105**, 177002 (2010).

107. S. Albrecht, A. Higginbotham, M. Madsen et al., Exponential protection of zero modes in Majorana islands, *Nature* **531**, 206 (2016).

108. S. Nadj-Perge, I. K. Drozdov, J. Li, H. Chen, S. Jeon, J. Seo, A. H. MacDonald, B. A. Bernevig, and A. Yazdani, Observation of Majorana fermions in ferromagnetic atomic chains on a superconductor, *Science* **346**, 602 (2014).

109. G. L. Fatin, A. Matos-Abiague, B. Scharf, and I. Žutić, Wireless Majorana bound states: from magnetic tunability to braiding, *Phys. Rev. Lett.* **117**, 077002 (2016).

110. N. Read and D. Green, Paired states of fermions in two dimensions with breaking of parity and time-reversal symmetries and the fractional quantum Hall effect, *Phys. Rev. B* **61**, 10267 (2000).

111. C. Nayak, S. H. Simon, A. Stern, M. Freedman, and S. Das Sarma, Non-Abelian anyons and topological quantum computation, *Rev. Mod. Phys.* **80**, 1083 (2008).

112. L. Fu and C. L. Kane, Superconducting proximity effect and Majorana fermions at the surface of a topological insulator, *Phys. Rev. Lett.* **100**, 096407 (2008).

113. F. Crépin, B. Trauzettel, and F. Dolcini, Signatures of Majorana bound states in transport properties of hybrid structures based on helical liquids, *Phys. Rev. B* **89**, 205115 (2014).

114. S. Mi, D. I. Pikulin, M. Wimmer, and C. W. J. Beenakker, Proposal for the detection and braiding of Majorana fermions in a quantum spin Hall insulator, *Phys. Rev. B* **87**, 241405 (2013).

115. S. Hart, H. Ren, T. Wagner et al., Induced superconductivity in the quantum spin Hall edge, *Nat. Phys.* **10**, 638 (2014).

116. G. Tkachov, P. Burset, B. Trauzettel, and E. M. Hankiewicz, Quantum interference of edge supercurrents in a two-dimensional topological insulator, *Phys. Rev. B* **92**, 045408 (2015).

117. E. Bocquillon, R. S. Deacon, J. Wiedenmann et al., Gapless Andreev bound states in the quantum spin Hall insulator HgTe, *Nat. Nanotechnol.* **12**, 137 (2016).

118. S. Hart, H. Ren, M. Kosowsky et al., Controlled finite momentum pairing and spatially varying order parameter in proximitized HgTe quantum wells, *Nat. Phys.* **13**, 87 (2017).

119. R. S. Deacon, J. Wiedenmann, E. Bocquillon et al., Josephson radiation from gapless Andreev bound states in HgTe-based topological junctions, *Phys. Rev. X* **7**, 021011 (2017).

120. F. Dominguez, O. Kashuba, E. Bocquillon et al., Josephson junction dynamics in the presence of $2\pi$- and $4\pi$-periodic supercurrents, *Phys. Rev. B* **95**, 195430 (2017).

121. G. Tkachov and E. M. Hankiewicz, Spin-helical transport in normal and superconducting topological insulators, *Phys. Status Solidi* **250**, 215 (2013).

122. J. Alicea, New directions in the pursuit of Majorana fermions in solid state systems, *Rep. Prog. Phys.* **75**, 076501 (2012).

123. M. Veldhorst, M. Snelder, M. Hoek et al., Josephson supercurrent through a topological insulator surface state, *Nat. Mater.* **11**, 417 (2012).

124. B. Sacépé, J. B. Oostinga, J. Li et al., Gate-tuned normal and superconducting transport at the surface of a topological insulator. *Nat. Commun.* **2**, 575 (2011).

125. J. R. Williams, A. J. Bestwick, P. Gallagher et al., Unconventional Josephson effect in hybrid superconductor-topological insulator devices, *Phys. Rev. Lett.* **109**, 056803 (2012).

126. L. Maier, J. B. Oostinga, D. Knott et al., Induced superconductivity in the three-dimensional topological insulator HgTe, *Phys. Rev. Lett.* **109**, 186806 (2012).

127. I. Sochnikov, L. Maier, C. A. Watson et al., Nonsinusoidal current-phase relationship in Josephson junctions from the 3D topological insulator HgTe, *Phys. Rev. Lett.* **114**, 066801 (2015).

128. S. Cho, B. Dellabetta, A. Yang et al., Symmetry protected Josephson supercurrents in three-dimensional topological insulators, *Nat. Commun.* **4**, 1689 (2013).

129. M.-X. Wang, C. Liu, J.-P. Xu et al., The coexistence of superconductivity and topological order in the $Bi_2Se_3$ thin films, *Science* **336**, 52 (2012).

130. S.-Y. Xu, N. Alidoust, I. Belopolski et al., Momentum-space imaging of Cooper pairing in a half-Dirac-gas topological superconductor, *Nat. Phys.* **10**, 943 (2014).

131. Y. Kim, T. M. Philip, M. J. Park, and M. J. Gilbert, Topological superconductivity in an ultrathin, magnetically-doped topological insulator proximity coupled to a conventional superconductor, *Phys. Rev. B* **94**, 235434 (2016).

132. Q. L. He, L. Pan, A. L. Stern et al., Chiral Majorana fermion modes in a quantum anomalous Hall insulator–superconductor structure, *Science* **357**, 294 (2017).

# Quantum Anomalous Hall Effect in Topological Insulators

**Abhinav Kandala, Anthony Richardella, and Nitin Samarth**

## 15.1  BACKGROUND AND THEORY

The flow of an electrical current in a typical metal experiences dissipation due to scattering of electrons from defects in the crystal lattice. Systems where electricity can flow without resistance are rather unique, and require a mechanism that results in phase coherent transport. The best known examples are superconductors, where electrons form pairs and condense into a Bose–Einstein condensate (see Chapter 16, Volume 1). The quantum Hall effect (QHE) is another example of dissipationless electrical transport that can occur when a large magnetic field is applied to a high mobility two-dimensional (2D) electron gas at absolute zero. Here, the Landau levels caused by the magnetic field result in edge states with a Hall conductance quantized in integer multiples of $e^2/h$ and where the longitudinal 4-point resistance goes to zero, though the 2-point resistance does not. The integer multiples of the Hall conductance arise from the topology of the wave functions of the quantum Hall state, and are known as Chern numbers [1]. The realization that the QHE could be understood by using concepts of topology led to a profound reconceptualization of quantum phases of matter. It is now understood that distinct phases can be defined solely by differences in their topology, in the absence of any symmetry breaking.

As QHE depends on a large applied magnetic field, it was natural to ask if a material could possess similar quantized transport inherently, without Landau levels from external fields. This possibility was first raised in the context of a conventional semiconductor heterostructure [2]. Haldane provided another, better-recognized theoretical prediction of this possibility, showing that a zero-field quantized Hall conductivity could be achieved on a graphene-like lattice, if time reversal symmetry was broken by applying an opposite magnet flux on each sublattice, with the total flux summing to zero [3]. In graphene, there are two inequivalent Dirac cones at the K and K′ = -K points in the Brillouin zone. The opposite fields on each sublattice would open a gap at each Dirac point, adding an oppositely signed mass term to the Hamiltonian each cone. This in turn leads to each cone contributing a half integer quantum Hall conductance that sums to a total Hall conductance of $e^2/h$. The key to observing this in a realistic system is breaking time reversal symmetry in a system with a topologically non-trivial band structure.

Kane and Mele explored what happens in graphene when spin-orbit coupling is considered [4]. The spin-orbit term in the Hamiltonian was found to have a form similar to the staggered field in the Haldane model, except that it was spin-dependent. As in the spin Hall effect, spin-orbit coupling acts like an effective magnetic field that depends on the direction of the spin polarization. This takes the chiral edge state of the spinless Haldane model, and splits it into two oppositely rotating edge states with opposite spins. This was termed the quantum spin Hall (QSH) effect. Further, this state is topologically distinct from an ordinary band insulator, which guarantees that the edge states are robust against disorder (see Chapter 14, Volume 2). It can be characterized by a topological invariant called $Z_2$, which is similar to the

Chern number for the QHE state [5]. QSH was also predicted in strained semiconductors with large spin-orbit coupling, and was soon after experimentally observed in the HgTe/CdTe quantum wells [6, 7].

The QSH state can only be observed at low temperatures, because it is possible for spin non-conserving scattering, such as inelastic scattering, to scatter carriers from one edge channel into the other. The longitudinal resistance measured in a two-terminal device is determined by the quantum conductance of two channels, $2e^2/h$. It also differs from the Haldane model. Any non-zero Hall conductance breaks time reversal, so a system displaying it must break time reversal. Spin-orbit coupling preserves time reversal, so the Hall conductance is zero for QSH, in the absence of an applied external field.

The QSH state is also called a 2D topological insulator (TI), to distinguish it from 3D TIs which were identified soon afterwards [8]. In a 3D TI, there are surface states within the bulk bandgap that exist on all the surfaces of the crystal. These states have a Dirac dispersion, and a spin polarization that is locked perpendicular to their momentum. For a 3D TI, such as $Bi_2Se_3$, these surface states form a single Dirac cone centered in k-space at the $\Gamma$ point, where the spin rotates in a left-handed sense for energies above the Dirac point [9]. The degeneracy of the Dirac point in $Bi_2Se_3$ is protected by time-reversal and inversion symmetry. When time reversal is broken, however, a gap can be opened at the Dirac point which can be described by a massive Dirac Hamiltonian, where the sign of the mass is determined by whether the magnetization perpendicular to the surface is pointing outward, or inward into the surface [10].

Jackiw and Rebbi predicted long ago that the mass domain wall of a 1D Dirac system carries a bound state [11]. A simple generalization to a 2D Dirac system predicts the presence of a 1D chiral (one-way propagating) mode at the mass domain wall. The 2D Dirac surface states of 3D TIs are a natural test bed for these predictions. This is illustrated in Figure 15.1. Consider a 3D TI thin film in proximity with two oppositely oriented, perpendicular-to-plane magnetized domains of a ferromagnet. The surface states under the magnetic domains can be gapped by exchange coupling, and acquire a mass, whose sign is dependent on the direction of magnetization of the overlying domain. Therefore, the region underneath the magnetic domain wall is expected to create a mass domain wall that carries a chiral mode. This 1D mode exists within the magnetic gap, moves in one direction, protected from backscattering, and is therefore dissipationless. Practically, however, this requires that the chemical potential be placed inside the magnetic gap, and that there are no other states at the chemical potential to scatter into. Obviously, using a metallic ferromagnet on top of the TI is therefore problematic. One way to accomplish this is to make the TI itself ferromagnetic by magnetic doping, and to control the carrier density so that chemical potential lies inside the magnetic gap. If the magnetization points up, it will point outward from the top surface, and inward through the bottom surface, resulting in a 1D edge mode around the outside edges of the sample. This is the origin of the quantum anomalous Hall effect (QAHE), which was first

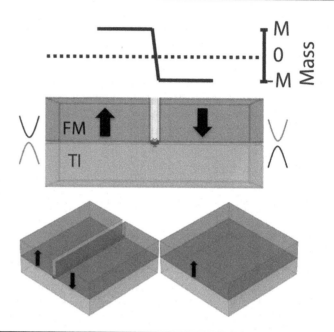

**FIGURE 15.1** A chiral state at the domain wall between two oppositely oriented ferromagnets on a TI surface. In the absence of a domain wall the whole surface is gapped and no state exists.

observed in Cr doped $(Bi_{1-x}Sb_x)_2Te_3$ [12, 13]. When two oppositely oriented magnetic domains are created in a ferromagnetic TI, as expected, quantized transport can be observed through a chiral mode localized at the domain wall [14, 15]. Another approach is using an insulating ferromagnet. As discussed later, much recent research has focused on interfacing 3D topological insulators with insulating ferromagnets, though to date such a chiral mode has not yet been observed using this method.

## 15.2 DEVELOPMENT OF QAHE MATERIALS

$Bi_2Se_3$, $Bi_2Te_3$, and $Sb_2Te_3$ are the most widely used 3D TI materials. All are narrow bandgap semiconductors, with a bulk bandgap up to ~0.3 eV for $Bi_2Se_3$, and have a single Dirac surface state cone centered at the $\Gamma$ point in k-space [16]. They are layered materials with a rhombohedral crystal structure and van der Waals bounds between layers. A major drawback of MBE grown TI thin films of $Bi_2Se_3$ ($Bi_2Te_3$) is unintentional doping arising from Se (Te) vacancies that results in the chemical potential lying in the bulk conduction band. These bulk carriers mask signatures of transport through the surface states. Furthermore, the Dirac point for $Bi_2Te_3$ lies submerged below the bulk valence band maximum. In contrast, thin films of $Sb_2Te_3$ have hole-type carriers due to Sb-Te antisite defects that place the chemical potential in the bulk valence band. Additionally, the Dirac point for $Sb_2Te_3$ lies exposed within its bulk band gap (Figure 15.2). Taking advantage of the similar lattice constants of $Bi_2Te_3$ and $Sb_2Te_3$, growth of thin films of the alloy $(Bi_{1-x}Sb_x)_2Te_3$ showed that the non-trivial topology of the band structure

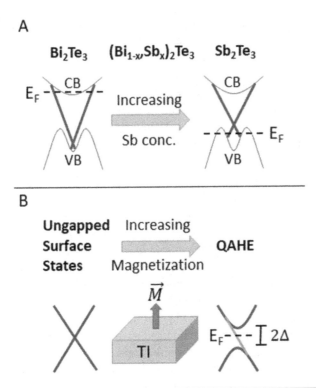

**FIGURE 15.2**    (a) In $Bi_2Te_3$ the Fermi energy of tends to be n-type, while the position of the Dirac point is below the valence band edge. $Sb_2Te_3$ is p-type and the Dirac point is in the bulk bandgap. By tuning the ratio of Bi to Sb, $E_F$ can be placed at the Dirac point, which can be above the valence band edge. (b) An out-of-plane magnetization, such as from magnetically doping a TI, can open a gap at the Dirac point. If the magnetization is sufficiently large it can create a chiral edge state around the outside edge of the sample that is the origin of the quantum anomalous Hall effect.

could be preserved over the entire composition range [17]. Importantly, it was shown that, by adjusting the Bi:Sb ratio, one could achieve ultra-low carrier density thin films with the chemical potential in the bulk band gap, with an exposed Dirac point. Further, tuning of the chemical potential could then be done using conventional gating, which is difficult to do effectively with materials like $Bi_2Se_3$. Gating a $(Bi,Sb)_2Te_3$ sample from the conduction band across the bulk bandgap and into the valence band reveals a gate dependence of the longitudinal and transverse resistivity that is reminiscent of the classic Dirac system—graphene. The longitudinal resistance reaches a maximum in the vicinity of the Dirac point and the carrier density changes sign [18]. This ability to control the carrier density was an important step towards separating surface state effects in electrical transport, and was critical to achieving the QAHE.

To measure a quantum of resistance associated with the QAHE 1D chiral mode requires freezing out any potential dissipative channels, which implies that the chemical potential should be tuned into the surface state magnetic gap. Growing these thin films on $SrTiO_3$ (STO) substrates is very helpful in this regard. The STO substrate presents itself as a natural backgate dielectric, due to STO's huge dielectric constant at cryogenic temperatures. Despite the large lattice mismatch between the film and STO substrate, the van der Waals nature of the bonds on the TI surface means that reasonable quality TI films can be grown. This is referred to as van der Waals epitaxy. Also, for the QAHE, the carrier density of the TI films needs to be low, despite the magnetic doping. Initial experiments in our group with Mn doping of $Bi_2Se_3$ and $Bi_2Te_3$ thin films revealed an unfavorable increase in the electron carrier density, and a propensity toward formation of secondary phases [19, 20]. This was in contrast to the p-type doping typically seen with Mn doping using bulk growth techniques. Another key requirement is the realization of a ferromagnetic phase with out-of-plane magnetic anisotropy. This has been shown with doping of $Bi_2Te_3$ and $Sb_2Te_3$ thin films with Mn, Cr and V. However, Mn doping of $Bi_2Se_3$ revealed a ferromagnetic phase with in-plane magnetic anisotropy [19]. The extent of magnetic doping is yet another important consideration. A large substitution of the "heavy" Bi has been shown to cause a transition to a topologically trivial band structure [21]. Finally, yet another important requirement is an exposed Dirac point. A Dirac point that is submerged in the bulk valence band would lead to back-scattering through these bulk states, and therefore a deviation from perfect quantization. In this context, band-engineering by controlling the Bi:Sb ratio is an important step. If the magnetic dopant introduces acceptors that requires too high of a Bi ratio to compensate for this, then the Dirac point could end up below the valence band edge.

The choice of magnetic dopant is thus crucial for the realization of the QAHE. Our group's initial attempts with Mn doping of $Bi_2Se_3$ thin films did not reveal an AHE in transport measurements, while Mn doped $Bi_2Te_3$ thin films demonstrated clear ferromagnetism with out-of-plane magnetic anisotropy [20]. Other attempts with Gd doping of $Bi_2Te_3$ or Cr doping of $Bi_2Se_3$ either did not show signatures of ferromagnetism or were in-plane easy axis [22–25]. Revisiting past work from U. Mich [26] that demonstrated the strong out-of-plane magnetic anisotropy of $Cr$-$Sb_2Te_3$ was a key step in the hunt for the QAH, creating the hope that Cr doping of $(Bi,Sb)_2Te_3$ could lead to low carrier density ferromagnetic thin films with an out-of-plane easy axis, and satisfying key requirements for the QAH. However, the extent of Cr doping is yet another important variable. Work from our group showed that in the high Cr doping limit, ultrathin films of $Cr$-$Sb_2Te_3$ grown on InP (111)A revealed $T_c$'s ~150 K, and square-shaped AHE curves, indicative of the strong out-of-plane magnetic anisotropy [27] (Figure 15.3). Interestingly, at $He^3$ temperatures, these films showed longitudinal sheet resistances in the $M\Omega$ range, raising the possibility of their use as low temperature ferromagnetic insulators. Similar results were seen more recently from the group at Tsinghua, in 5QL thick thin films of $Cr_y(Bi_xSb_{1-x})_{2-y}Te_3$ grown on STO. For a Cr concentration of $y = 0.44$, they observed a $T_c \sim 77$ K, with a gate tunable electrical resistivity taking maximum values over $10\ M\Omega$ [28]. In the limit of such large doping, the strength of the spin-orbit coupling is reduced, leading to a transition to a topologically trivial phase.

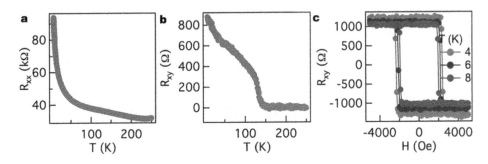

**FIGURE 15.3**  Transport of highly Cr doped Sb$_2$Te$_3$. (a) At low temperatures the resistance of the sample becomes strongly insulating. (b) A T$_c$ of ~150 K was observed in this sample from the anomalous Hall signal. (C) Square hystersis loops are observed, but with saturation values far from quantization. (After Kandala, A. Transport studies of mesoscopic and magnetic topological insulators, Ph.D. thesis, 2015. With permission.)

The requirements for realizing the QAHE were first met by the group from Tsinghua University, using a Cr doping concentration that was significantly lower than that discussed in the previous paragraph [13]. At 30 mK, 5 QL thick films of Cr$_{0.15}$(Bi$_{0.1}$Sb$_{0.9}$)1$_{.85}$Te$_3$ on STO revealed a zero-field plateau in the transverse resistance around h/e$^2$, when the chemical potential was tuned to charge neutrality. This was the first experimental realization of the QAHE. The temperature dependence of the AHE revealed a $T_c \sim 15$ K. A crucial point to note is that the regime of quantization is at temperatures almost two orders of magnitude lower than the onset of bulk ferromagnetism. We shall revisit this in later sections. When the chemical potential is tuned to the Dirac point, flipping the magnetization switches the Hall plateau between +h/e$^2$ and −h/e$^2$. The switching of the magnetization is accompanied by a sharp rise (2250%) in the longitudinal resistance, which, as in the QHE, is interpreted to arise from backscattering between adjacent edge modes via dissipative pathways in the bulk. Typical QAHE data, observed in magnetic field sweeps, is shown in (Figure 15.4). Similar quantization of the Hall resistance is seen in gate sweeps at zero field, with a plateau around h/e$^2$ near charge neutrality. The quantized Hall plateau is accompanied by a drop in the longitudinal resistance, which is another familiar signature of chiral edge transport from the QHE. This gate dependence is in stark contrast to the nonmagnetic TI thin films where the longitudinal resistance reaches a maximum at charge neutrality. In these samples there was a finite zero-field $\rho_{xx}$ of ~0.098 h/e$^2$, which is indicative of some dissipative current pathways. However, these dissipative channels were localized by the application of a strong external magnetic field (>10 T), that led to a vanishing longitudinal resistance. As the temperature is raised, the Hall plateau deviates from perfect quantization, and the drop in $\rho_{xx}$ is increasingly shallow. These are typical signatures of edge transport in the presence of increased dissipation. These signatures persisted up to 400 mK, indicating that the edge states persist to temperatures well beyond the regime of perfect quantization. Similar results were soon reproduced by several groups [29–31].

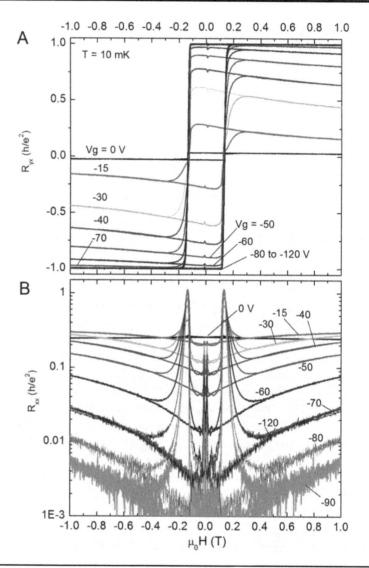

**FIGURE 15.4** QAHE in Cr doped $(Bi,Sb)_2Te_3$. Despite mobilities on the order of a few hundred $cm^2/Vs$, quantization to parts in $10^4$ can be achieved at dilution fridge temperatures. (After Liu, M. et al., *Sci. Adv.* 2, e1600167, 2016. With permission.)

First principles calculations predicted that doping of 3D TI thin films with Vanadium (V) would result in d-orbital impurity bands that lie in the band gap [12]. This would preclude the realization of a QAHE, due to the absence of an insulating ferromagnetic state. However, the experimental work of Chang et al. showed otherwise, demonstrating a robust QAHE at 25 mK, in 4 QL thick films of $(Bi_{0.29}Sb_{0.71})_{1.89}V_{0.11}Te_3$ grown on STO [32]. Key differences with Cr doping were a significantly higher $T_c$ and lower carrier densities for comparable doping concentrations, and a coercive field that is larger by an order of magnitude. Furthermore, key signatures of dominant edge state transport, such as a drop in $\rho_{xx}$, are seen at temperatures as high as 5 K, making it an important first step towards enhancing the temperature scale of the QAHE. Similar results with V doping were reproduced with growth on hydrogen passivated (111) Si substrates [33].

## 15.3 STRUCTURE, MAGNETISM AND TEMPERATURE SCALE

In both Cr doped and V doped films, precise quantization has been observed, with $\rho_{xy}$ deviating from 1 by a few 10 thousandths and $\rho_{xx}$ exceeding 0 also by a few ten thousandths, in units of $h/e^2$ [33, 34, 35, 36]. Deviations from these values arise from transport though parallel dissipative channels in the sample. As the system is tuned away from the ideal conditions, for instance, by changing the gate voltage or raising the temperature, dissipation increases rapidly, and can be understood as a quantum phase transition to an insulating state [37]. Tilted field measurements provide a convenient method to quantify ballistic edge vs. dissipative contributions in temperature and gate-voltage regimes that do not display full quantization [31]. When the external field rotates the sample magnetization from out-of-plane towards the in-plane direction, the magnetic gap that stabilizes the QAHE edge state closes, and the edge state is completely destroyed when the magnetization is forced in-plane. This field dependence is distinct from the typical anisotropic magnetoresistance (AMR) observed in conventional ferromagnets, and can be modeled using the Landauer–Büttiker formalism to quantify the proportion of ballistic and dissipative transport (Figure 15.5).

Despite the high degree of quantization that is achievable, remarkably, these materials are structurally far from perfect and contain a number of defects [38]. The QAHE has been observed in TI thin films grown on a variety of substrates, including InP, Si, and STO, but in all cases, structural defects such as dislocations and twin domains are common. Figure 15.6 shows the structural characterization of these thin films on STO by atomic force microscopy (AFM) and transmission electron microscopy (TEM). The AFM image of Figure 15.6 shows the presence of blobs that are almost ~20 nm in height, in addition to the well-studied layered structure of triangular QL's with twinned domains. The crystallinity of these features is seen in TEM; EDS reveals a chemical composition that is similar to the rest of the film, suggesting that the blobs are misoriented grains of $Cr-(Bi,Sb)_2Te_3$. Furthermore, TEM also reveals an amorphous layer at the substrate-film interface. Unsurprisingly, electrical transport measurements of these films show poor mobilities, in the range 100–200 $cm^2/Vs$ at cryogenic temperatures. Similar mobilities have been reported by other groups. This is in stark contrast to traditionally studied systems of 1-D transport in the QH regime, such as graphene and 2D electron gases that have required orders of magnitude larger mobilities. This is a remarkable feature of the QAHE, and highlights the difference in its origins from the QH effect, which requires high mobilities for the creation of Landau levels at modest magnetic fields.

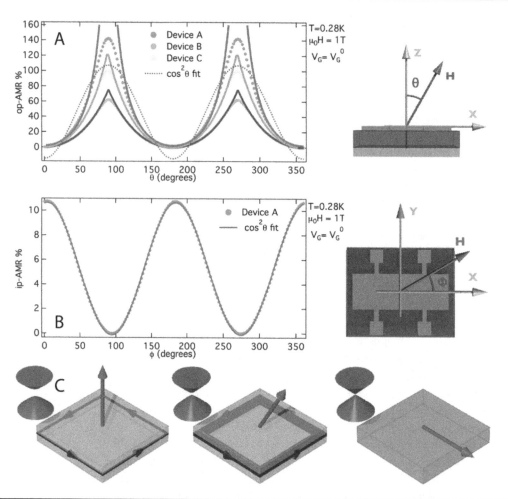

**FIGURE 15.5** Giant anisotropic magnetoresistance in a QAHE sample. (a) The magnetoresistance as the field is swept from out-of-plane to in-plane can be modeled by a Landauer–Büttiker formalism to extract edge and bulk contributions to the conduction. (b) In contrast, when the field is swept in-plane, the typical AMR dependence of a trivial ferromagnet is observed. (c) The spatial dependence of the edge state is shown schematically as the field is tilted into the plane of the sample where it is eventually destroyed as the magnetic gap closes. (After Kandala, A. et al., *Nat. Commun.* 6, 7434, 2015. With permission.)

Typically, ferromagnetism arises in dilute magnetic semiconductors via the carrier-mediated RKKY exchange interaction (see Chapters 4 and 9, Volume 2). For QAHE samples, which are purposely tuned and gated into the magnetic gap to eliminate bulk and surface state carriers, an interesting question arises: why are these samples ferromagnetic? Theoretically, it was found that the spin susceptibility of these TI materials is so large it can stabilize ferromagnetic order: this is known as van Vleck ferromagnetism [12]. When the carrier density is large, when the chemical potential is near the bulk band edges, RKKY also likely plays a role in mediating ferromagnetic order [39, 40].

Another interesting question to address is the reason for the relatively low temperatures required to access the QAH regime, despite a

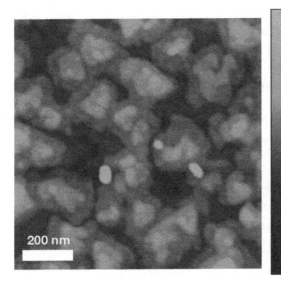
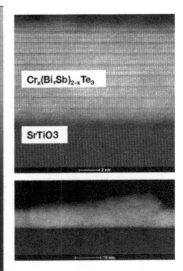

**FIGURE 15.6**   On the left, AFM image of the surface of a Cr-(Bi,Sb)$_2$Te$_3$ sample grown on STO and capped with Al. Despite roughness from step edges and defects, QAHE can be observed. On the right, corresponding high angle annular dark field scanning TEM images. Quintuple layer structure of the TI on STO is clearly seen, as is an amorphous region at the interface. Below, a defect in the film which appears to be a misoriented grain which may correspond with the type of tall defects seen in the AFM.

significantly higher $T_c$ for the onset of ferromagnetic order. Understanding this issue is critical for developing strategies for increasing the temperature scale. Initial samples required access to dilution fridge temperatures. The first results from Tshingua University were reported at 30 mK, from thin films with a $T_c \sim 13$ K, and similar temperature scales were observed in subsequent experiments [13]. This is suggestive of a significantly smaller energy gap for activation of dissipative channels. As in studies of QHE, an activation gap may be extracted from an Arrhenius fit to the metallic temperature dependence of the longitudinal conductivity. Measurements on our thin films with a $T_c \sim 15$ K reveal an activation gap corresponding to a temperature scale of ~190 mK, and similar numbers have been reported elsewhere [32, 35].

Some clues to the origin of this discrepancy are provided by nanoscale measurements. Scanning tunneling spectroscopy measurements of V-(Bi,Sb)$_2$Te$_3$ showed gaps around the Dirac point that varied widely in energy on the nanometer scale spatially [41]. These gaps were interpreted as magnetic gaps due to the V doping and implied that the distribution of the smallest gaps could allow percolative transport paths through the sample. Additionally, scanning nano-superconducting quantum interference device (nano-SQUID) measurements of Cr doped samples at 250 mK revealed that the switching behavior of the magnetization looks superparamagnetic in character [42]. On the 10s of nm scale, the switching is granular, with island-like domains switching independently of their neighbors, without any evidence of domain wall motion, but also displaying long timescale relaxation dynamics consistent with superparamagnetic or glassy behavior (Figure 15.7). At higher temperatures (5 K), magnetic force microscopy measurements of $V$-$Sb_2Te_3$ have reported disordered bubble-like domains but with more conventional domain wall propagation [43]. Note, though, that $V$-$Sb_2Te_3$ is heavily p-type and may be more representative of RKKY mediated magnetism than the situation when the chemical potential is near the Dirac point.

The picture that arises is one dominated by disorder, both in terms of local potentials due to random doping and defects, but also in magnetic inhomogeneity. One result that stands in contrast to this is the observation of large instantaneous jumps in the local and non-local transport resistivity when sweeping through the coercive field in the quantized regime [32]. At the lowest mK temperatures, the switching behavior appears smooth in both $\rho_{xx}$ and $\rho_{xy}$. However, in a temperature range of ~100 to 170 mK, jumps on the order of $h/e^2$ were seen in large macroscopic samples (Figure 15.7).

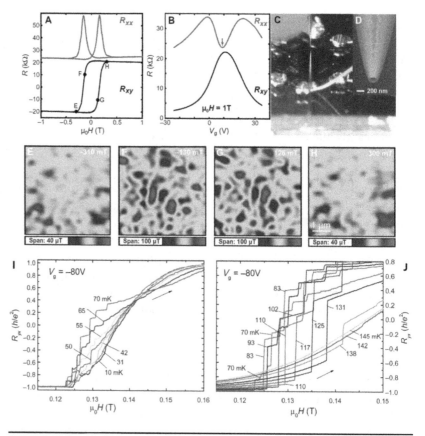

**FIGURE 15.7** (a)–(h) Scanning SQUID measurements showing the superparamagnetic-like nature of the local magnetization during the switching transition of $Cr$-$(Bi,Sb)_2Te_3$ at 250 mK [42]. (i)–(j) Instantaneous jumps in the Hall resistance when sweeping through the the coercive field observed between ~70 and 140 mK [34]. These jumps involve magnetization reversal in regions on the mm scale and contrast strongly with the local magnetization seen in the scanning SQUID. (After Liu, M. et al., *Sci. Adv.* 2, e16000167, 2015; Lachman, E.O. et al., *Sci. Adv.* 1, e1500740, 2015. With permission.)

From non-local measurements, it is clear the switching involved moving the magnetization domain wall over distances of hundreds of microns. This was interpreted as possible evidence of macroscopic quantum tunneling into the magnetic ground state as the field was swept through the transition. Such quantum coherence is difficult to reconcile with the superparamagnetic-like local magnetization seen ~100 mK higher in temperature, but could point to strongly temperature-dependent dynamics that again may be limiting the temperature scale of QAHE.

Another way to characterize the QAHE transition is looking at the scaling behavior of the flow trajectory diagram of the $\sigma_{xx}$ vs. $\sigma_{xy}$ [30]. In general, it is found that there is a very large increase in $\rho_{xx}$ during the transition when sweeping through the coercive field. This drives $\sigma_{xx}$ towards zero and, in some samples, results in a plateau at zero in $\sigma_{xy}$, which has been called the C=0 or zero Hall plateau [44]. In these samples, the flow diagram resembles that of the integer QHE following a semicircle running from (0,0) to (0, $e^2/h$) in $\sigma_{xx}$ and $\sigma_{xy}$. Tilted field measurements have also been used to show the same behavior [45]. We caution that the observation of $\sigma_{xy}=0$ can also result as an arithmetical artifact of converting measured resistivity into a deduced conductivity when a large value of $\rho_{xx}$ coincides with $\rho_{xy}$ approaching zero during the transition between the integer Chern states. Others observed that if $\rho_{xx}$ does not exceed $h/e^2$ during the transition, then a different flow diagram is obtained where the semicircle connects (0, $-e^2/h$) to (0, $e^2/h$) approaching ($e^2/h$, 0) during the transition [46]. Whether this implies a different phase of the QAHE is currently under debate.

This is relevant in the context of one fairly successful method to improve the temperature scale of QAHE, magnetic modulation doping near the surfaces of TI films. Tokura's group has shown that confining the magnetic dopants to the surfaces, or near surface regions, of the films can result in QAHE persisting to temperature as high ~2 K [47]. This appears to work by confining the magnetic doping to the surfaces, where it is needed to gap the surfaces' states, while reducing the disorder in the bulk that could lead to unwanted conduction pathways. Interestingly, since the magnetization of the top and bottom surfaces are now decoupled, it has been demonstrated that it is possible to switch only one surface, so that the magnetization is pointing outward on both surfaces, destroying the QAHE edge state while still leaving the surfaces gapped [48]. This has been called the axion insulator state and could be relevant for a host of exotic physics. The debate about the scaling flow diagrams is a question of whether this state has truly been achieved. Recently, using a Cr doped bottom layer and a V doped top layer with very different coercivities, it has been shown that the scaling behavior of the top and bottom surfaces can be plotted independently and that they each show the expected half integer Hall conductivity of a single gapped surface, which cancels to zero when each surface's magnetization is oppositely oriented [49, 50].

Another route towards increasing the temperature scale could deal with improved control of the chemical potential. TI materials are sensitive to exposure to air and water vapor, typically resulting in higher carrier densities as the films age and oxidize. To protect them from degradation, films are often capped *in situ* after growth to help preserve the properties of their surfaces. Al (which oxidizes on exposure to air), $Al_2O_3$, and Te are common capping materials. *Ex situ* atomic layer deposition (ALD) of $Al_2O_3$, to encapsulate the films prior to lithography, is also common if top gates are to be fabricated. To avoid possible deteriorating effects of exposure during conventional lithography, the Hall bars are often fashioned by mechanical scratching of the thin films. When STO is used as the substrate, a gated device can be made in this way with a minimum of steps. With most QAHE experiments to date, thin films with a single gate have been used. This effectively depletes the carriers near the closer interface, but can create a potential gradient across the film. The fabrication of dual-gated magnetic TI devices would enable independent control of the top and bottom surface states, and thereby improved control over the chemical potential, which would be useful given the small size of the magnetic gap. Our group's attempts in this direction led to the fabrication of thin film Hall devices of $Cr-(Bi,Sb)_2Te_3$ grown on STO, capped with a top gate dielectric of ALD deposited $HfO_2$ dielectric and evaporated Au metal. As shown in Figure 15.8, while the thin film could be tuned to its Dirac point with both gates, the films were likely degraded by the fabrication process, and the maximum gate-tuned AHE was ~3 KΩ. Others have seen similar reductions in the AHE signal after fabricating dual gates on samples on STO [51].

## 15.4 FERROMAGNETIC INSULATOR/TI HETEROSTRUCTURES

In our group, initial experiments exploring the magnetic heterostructure route were focused on the insulating ferromagnet GdN [52]. This is a low Curie temperature ferromagnet that was deposited on $Bi_2Se_3$ thin films by reactive ion sputtering at ambient temperatures. Controlling the nitrogen composition has been previously shown to be an excellent knob for manipulating its electrical conductivity. The ability to deposit GdN at ambient temperatures was of particular interest, since this prevents the diffusion of the magnetic species into the TI thin film, as confirmed by high-resolution electron energy loss spectroscopy measurements. This is an important consideration to differentiate proximity exchange effects with effects attributed to magnetic doping, which may have adverse effects on the topologically nontrivial band structure. The magnetic characterization of the heterostructures by SQUID magnetometry revealed a ferromagnetic phase with $T_c \sim 13$ K and an in-plane easy axis. We were also able to fabricate Hall bars for electrical characterization, with bare and GdN capped channels. This enabled a direct comparison of transport, and revealed the suppression of weak antilocalization in the magnetically capped layer. A key drawback of these initial attempts was the use of $Bi_2Se_3$ thin films which had their chemical potential up in the bulk conduction band, due to unintentional doping in the as-grown films. For transport experiments, this precludes the ability to observe clean signatures of time-reversal symmetry breaking effects in the surface states.

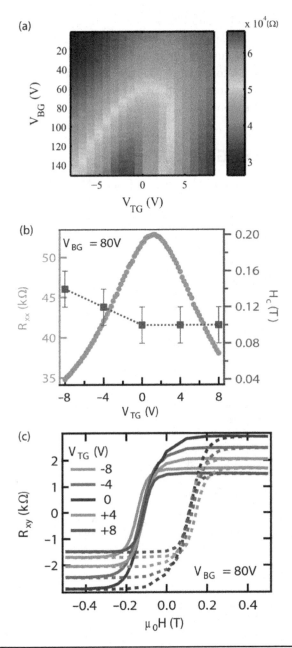

**FIGURE 15.8**  Dual gating of a Cr-(Bi, Sb)$_2$Te$_3$ TI film. (a) R$_{xx}$ vs. top and backgate voltages. (b) Maximum in R$_{xx}$ crossing the Dirac point. (c) R$_{xy}$ hysteresis vs. top gate voltage at a fixed backgate. (After Richardella, A. et al. *APL Mater.* 3, 83303, 2015. With permission.)

Our efforts since have focused on lowering the carrier density in TI thin films, and the use of alternate high resistivity magnets such as (Ga,Mn)As [53] and YIG. Spin pumping measurements on TI/YIG heterostructures have been useful to characterize the nature of the spin interaction at the interface [54]. Other groups have explored the use of insulating ferromagnets such as barium ferrite [55], EuS [56, 57], $Cr_2Ge_2Te_6$ [58], and thulium iron garnet (TIG). In particular, TI/EuS and TI/TIG heterostructures have recently shown high temperature, proximity induced ferromagnetism at the interface [59, 60]. However, a convincing signature of the dissipationless chiral modes in such heterostructures remains lacking, and in this context, magnetic doping has been a far more successful route.

## ACKNOWLEDGMENTS

This research is supported by grants from ONR (N00014-15-1-2370), NSF (DMR-1306510 and the Pennsylvania State University Two-Dimensional Crystal Consortium Materials Innovation Platform through NSF cooperative agreement DMR-1539916) and ARO MURI (W911NF-12-1-0461).

## REFERENCES

1. D. J. Thouless, M. Kohmoto, M. P. Nightingale, and M. Den Nijs, Quantized Hall conductance in a two-dimensional periodic potential, *Phys. Rev. Lett.* **49**, 405–408 (1982).

2. O. A. Pankratov, Supersymmetric inhomogeneous semiconductor structures and the nature of a parity anomaly in (2+1) electrodynamics, *Phys. Lett. A* **121**, 360–366 (1987).

3. F. D. M. Haldane, Model for a quantum Hall effect without Landau levels: Condensed-matter realization of the 'parity anomaly', *Phys. Rev. Lett.* **61**, 2015–2018 (1988).

4. C. L. Kane and E. J. Mele, Quantum spin Hall effect in graphene, *Phys. Rev. Lett.* **95**, 226801 (2005).

5. C. L. Kane and E. J. Mele, Z2 topological order and the quantum spin Hall effect, *Phys. Rev. Lett.* **95**, 146802 (2005).

6. B. A. Bernevig, S. Zhang, and C. Wu, Quantum spin Hall effect, *Phys. Rev. Lett.* **96**, 1–4 (2006).

7. M. König, S. Wiedmann, C. Brüne et al., Quantum spin Hall insulator state in HgTe quantum wells, *Science* **318**, 766–770 (2007).

8. L. Fu, C. L. Kane, and E. J. Mele, Topological insulators in three dimensions, *Phys. Rev. Lett.* **98**, 106803 (2007).

9. Y. Xia, D. Qian, D. Hsieh et al., Observation of a large-gap topological-insulator class with a single Dirac cone on the surface, *Nat. Phys.* **5**, 398–402 (2009).

10. M. Z. Hasan and C. L. Kane, Colloquium: Topological insulators, *Rev. Mod. Phys.* **82**, 3045–3067 (2010).

11. R. Jackiw and C. Rebbi, Solitons with fermion number ½, *Phys. Rev. D* **13**, 3398–3409 (1976).

12. R. Yu, W. Zhang, H.-J. Zhang, S.-C. Zhang, X. Dai, and Z. Fang, Quantized anomalous Hall effect in magnetic topological insulators, *Science* **329**, 61–64 (2010).

13. C. Z. Chang, X. Feng, J. Zhang et al., Experimental observation of the quantum anomalous Hall effect in a magnetic topological insulator, *Science* **340**, 167–170 (2013).

14. K. Yasuda, M. Mogi, R. Yoshimi et al., Quantized chiral edge conduction on reconfigurable domain walls of a magnetic topological insulator, *Science* **358**, 1311 (2017).

15. I. T. Rosen, E. J. Fox, X. Kou, L. Pan, K. L. Wang, and D. Goldhaber-Gordon, Chiral transport along magnetic domain walls in the quantum anomalous Hall effect, *NPJ Quantum Mater.* **2**, 69 (2017).

16. H. Zhang, C.-X. Liu, X.-L. Qi, X. Dai, Z. Fang, and S.-C. Zhang, Topological insulators in $Bi_2Se_3$, $Bi_2Te_3$ and $Sb_2Te_3$ with a single Dirac cone on the surface, *Nat. Phys.* **5**, 438–442 (2009).

17. J. Zhang, C.-Z. Chang, Z. Zhang et al., Band structure engineering in $(Bi_{1-x}Sb_x)_2$ $Te_3$ ternary topological insulators, *Nat. Commun.* **2**, 574 (2011).

18. D. Kong, Y. Chen, J. J. Cha et al., Ambipolar field effect in the ternary topological insulator $(Bi_xSb_{1-x})_2Te_3$ by composition tuning, *Nat. Nanotechnol.* **6**, 705–709 (2011).

19. D. Zhang, A. Richardella, D. W. Rench et al., Interplay between ferromagnetism, surface states, and quantum corrections in a magnetically doped topological insulator, *Phys. Rev. B* **86**, 205127 (2012).

20. J. S. Lee, A. Richardella, D. W. Rench et al., Ferromagnetism and spin-dependent transport in n-type Mn-doped bismuth telluride thin films, *Phys. Rev. B* **89**, 174425 (2014).

21. J. Zhang, C. Z. Chang, P. Tang et al., Topology-driven magnetic quantum phase transition in topological insulators, *Science* **339**, 1582–1586 (2013).

22. S. E. Harrison, L. J. Collins-McIntyre, S. Li et al., Study of Gd-doped $Bi_2Te_3$ thin films: Molecular beam epitaxy growth and magnetic properties, *J. Appl. Phys.* **115** (2014).

23. M. Liu, L. He, X. Kou et al., Crossover between weak antilocalization and weak localization in a magnetically doped topological insulator, *Phys. Rev. Lett.* **108**, 36805 (2012).

24. P. P. J. Haazen, J. B. Laloe, T. J. Nummy et al., Ferromagnetism in thin-film Cr-doped topological insulator $Bi_2Se_3$, *Appl. Phys. Lett.* **100**, 82404 (2012).

25. L. J. Collins-McIntyre, S. E. Harrison, P. Schönherr et al., Magnetic ordering in Cr-doped $Bi_2$ $Se_3$ thin films, *EPL*, **107**, 57009 (2014).

26. Z. Zhou, Y.-J. Chien, and C. Uher, Thin film dilute ferromagnetic semiconductors $Sb_{2-x}$ $Cr_x$ $Te_3$ with a Curie temperature up to 190 K, *Phys. Rev. B* **74**, 224418 (2006).

27. A. Kandala, Transport studies of mesoscopic and magnetic topological insulators, Ph.D. thesis, The Pennsylvania State University, 2015.

28. Y. Ou, C. Liu, L. Zhang et al., Heavily Cr-doped $(Bi,Sb)_2Te_3$ as a ferromagnetic insulator with electrically tunable conductivity, *APL Mater.* **4**, 086101 (2016).

29. X. Kou, S.-T. Guo, Y. Fan et al., Scale-invariant quantum anomalous Hall effect in magnetic topological insulators beyond the two-dimensional limit, *Phys. Rev. Lett.* **113**, 137201 (2014).

30. J. G. Checkelsky, R. Yoshimi, A. Tsukazaki et al., Trajectory of the anomalous Hall effect towards the quantized state in a ferromagnetic topological insulator, *Nat. Phys.* **10**, 731 (2014).

31. A. Kandala, A. Richardella, S. Kempinger, C. Liu, and N. Samarth, Giant anisotropic magnetoresistance in a quantum anomalous Hall insulator, *Nat. Commun.* **6**, 7434 (2015).

32. C.-Z. Chang, W. Zhao, D. Y. Kim et al., High-precision realization of robust quantum anomalous Hall state in a hard ferromagnetic topological insulator, *Nat. Mater.* **14**, 434. (2015).

33. S. Grauer, S. Schreyeck, M. Winnerlein, K. Brunner, C. Gould, and L. W. Molenkamp, Coincidence of superparamagnetism and perfect quantization in the quantum anomalous Hall state, *Phys. Rev. B* **92**, 201304 (2015).

34. M. Liu, W. Wang, A. R. Richardella et al., Large discrete jumps observed in the transition between Chern states in a ferromagnetic topological insulator, *Sci. Adv.* **2**, e1600167 (2016).

35. A. J. Bestwick, E. J. Fox, X. Kou, L. Pan, K. L. Wang, and D. Goldhaber-Gordon, Precise quantization of the anomalous Hall effect near zero magnetic field, *Phys. Rev. Lett.* **114**, 187201 (2015).

36. E. J. Fox, I. T. Rosen, Y. Yang et al., Part-per-million quantization and current-induced breakdown of the quantum anomalous Hall effect, *Phys. Rev. B* **98**, 075145 (2018).

37. C. Z. Chang, W. Zhao, J. Li et al., Observation of the quantum anomalous Hall insulator to Anderson insulator quantum phase transition and its scaling behavior, *Phys. Rev. Lett.* **117**, 126802 (2016).

38. A. Richardella, A. Kandala, J. S. Lee, and N. Samarth, Characterizing the structure of topological insulator thin films, *APL Mater.* **3**, 83303 (2015).

39. X. Kou, M. Lang, Y. Fan et al., Interplay between different magnetisms in Cr-doped topological insulators, *ACS Nano* **7**, 9205–9212 (2013).

40. Z. Zhang, X. Feng, M. Guo et al., Electrically tuned magnetic order and magnetoresistance in a topological insulator, *Nat. Commun.* **5**, 4915 (2014).

41. I. Lee, C. H. Kim, J. Lee et al., Imaging Dirac-mass disorder from magnetic dopant atoms in the ferromagnetic topological insulator $Cr_x (Bi_{0.1}Sb_{0.9})_{2-x}Te_3$, *Proc. Natl. Acad. Sci.* **112**, 1316–1321 (2015).

42. E. O. Lachman, A. F. Young, A. Richardella et al., Visualization of superparamagnetic dynamics in magnetic topological insulators, *Sci. Adv.* **1**, e1500740 (2015).

43. W. Wang, C.-Z. Chang, J. S. Moodera, and W. Wu, Visualizing ferromagnetic domain behavior of magnetic topological insulator thin films, *NPJ Quantum Mater.* **1**, 16023 (2016).

44. Y. Feng, X. Feng, Y. Ou et al., Observation of the zero Hall plateau in a quantum anomalous Hall insulator, *Phys. Rev. Lett.* **115**, 126801 (2015).

45. X. Kou, L. Pan, J. Wang et al., Metal-to-insulator switching in quantum anomalous Hall states, *Nat. Commun.* **6**, 8474 (2015).

46. S. Grauer, K. M. Fijalkowski, S. Schreyeck et al., Scaling of the quantum anomalous Hall effect as an indicator of axion electrodynamics, *Phys. Rev. Lett.* **118**, 246801 (2017).

47. M. Mogi, R. Yoshimi, A. Tsukazaki et al., Magnetic modulation doping in topological insulators toward higher-temperature quantum anomalous Hall effect, *Appl. Phys. Lett.* **107**, 182401 (2015).

48. M. Mogi, M. Kawamura, R. Yoshimi et al., A magnetic heterostructure of topological insulators as a candidate for an axion insulator, *Nat. Mater.* **16**, 516–521 (2017).

49. D. Xiao, J. Jiang, J.-H. Shin et al., Realization of the axion insulator state in quantum anomalous Hall sandwich heterostructures, *Phys. Rev. Lett.* **120**, 056801 (2018).

50. M. Mogi, M. Kawamura, A. Tsukazaki et al., Tailoring tricolor structure of magnetic topological insulator for robust axion insulator, *Sci. Adv.* **3**, eaao1669 (2017).

51. C.-Z. Chang, M. Kawamura, A. Tsukazaki et al., Simultaneous electrical-field-effect modulation of both top and bottom dirac surface states of epitaxial thin films of three-dimensional topological insulators, *Nano Lett.* **15**, 1090–1094 (2015).

52. A. Kandala, A. Richardella, D. W. Rench, D. M. Zhang, T. C. Flanagan, and N. Samarth, Growth and characterization of hybrid insulating ferromagnet-topological insulator heterostructure devices, *Appl. Phys. Lett.* **103**, 202409 (2013).

53. J. S. Lee, A. Richardella, R. D. Fraleigh, C. Liu, W. Zhao, and N. Samarth, Engineering the breaking of time-reversal symmetry in gate-tunable hybrid ferromagnet/topological insulator heterostructures, *Quantum Materials*, **3**, 51 (2018).

54. H. Wang, J. Kally, J. S. Lee et al., Surface state dominated spin-charge current conversion in topological insulator/ferromagnetic insulator heterostructures, *Phys. Rev. Lett.* **117**, 076601 (2016).

55. W. Yang, S. Yang, Q. Zhang et al., Proximity effect between a topological insulator and a magnetic insulator with large perpendicular anisotropy, *Appl. Phys. Lett.* **105**, 92411 (2014).

56. P. Wei, F. Katmis, B. A. Assaf et al., Exchange-coupling-induced symmetry breaking in topological insulators, *Phys. Rev. Lett.* **110**, 186807 (2013).

57. Q. I. Yang, M. Dolev, L. Zhang et al., Emerging weak localization effects on a topological insulator–insulating ferromagnet ($Bi_2Se_3$-EuS) interface, *Phys. Rev. B* **88**, 81407 (2013).

58. H. Ji, R. A. Stokes, L. D. Alegria et al., A ferromagnetic insulating substrate for the epitaxial growth of topological insulators, *J. Appl. Phys.* **114**, 114907 (2013).

59. F. Katmis, V. Lauter, F. S. Nogueira et al., A high-temperature ferromagnetic topological insulating phase by proximity coupling, *Nature* **533**, 513–516 (2016).

60. C. Tang, C.-Z. Chang, G. Zhao et al., Above 400-K robust perpendicular ferromagnetic phase in a topological insulator, *Sci. Adv.* **3**, e1700307 (2017).

# Index

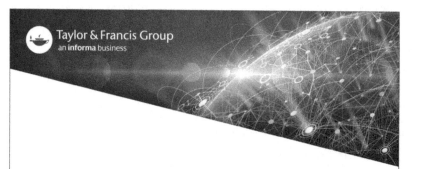